BIOLOGY
THE UNITY AND DIVERSITY OF LIFE

SECOND EDITION

CECIE STARR
Belmont, California

RALPH TAGGART
Michigan State University

with a chapter on plant physiology by
Cleon Ross
Colorado State University

Wadsworth Publishing Company
Belmont, California
A Division of Wadsworth, Inc.

Biology Editor: Jack C. Carey

Production Editor: Kathie Head

Art Director: Cynthia Bassett

Copy Editor: Don Yoder

Art Editor: Catherine Aydelott

Illustrators: John Dawson, Florence Fujimoto, Darwen Hennings, Vally Hennings, Julia Iltis, Marlene May, Virginia Mickelson, Nelva Richardson

Production Art: Lois Stanfield, Alan Noyes

Permissions: Peggy Meehan, Marion Hansen

Cover photograph: Plants, lichens, and an animal on a rock outcropping in Africa—symbolic of the diversity of life and the common environmental challenge. (George Holton/Ocelot)

A study guide has been designed to help students master the concepts presented in this textbook. Order from your bookstore.

Printed in the United States of America

3 4 5 6 7 8 9 10—85 84 83 82 81

Library of Congress Cataloging in Publication Data

Starr, Cecie.
  Biology.

  First ed. (1978) by D. Kirk, R. Taggart, C. Starr.
  Includes bibliographies and index.
  1. Biology.   I. Taggart, Ralph, joint author.
II. Kirk, David, 1934-   Biology. III. Title.
QH308.2.K57   1981   574   80-22273
ISBN 0-534-00930-1

PREFACE

Someone once said that textbook builders are madmen. Probably it's that they simply become driven, caught before they know it in the challenge of chasing after a story that unfolds faster than it can be told. General biology books can no longer be written during decent working hours, on weekends, on the side. All the diverse fields that the word *biology* encompasses have grown too broad and too deep for that. Researching the current thinking and investigations in each field, distilling the main concepts, perceiving connections between them are demanding tasks. There is no postponing them: within a few years, advances put much of what you have written out of date, and the research and distillations must begin all over again.

That is what happened to *Biology: The Unity and Diversity of Life*. The new edition is in many ways a new book. It has been updated and upgraded during three years of uninterrupted research, writing, and intensive reviewing by more than a hundred specialists and general biology teachers. The revision began shortly after the first edition came off the press; even now, bits of new information are being scribbled in as it goes once again to press. The new edition isn't harder; it *is* more current and more complete. For the moment.

Approach. In its general approach, the new edition remains *concept-oriented*, rather than a compendium of isolated facts. The most inclusive concepts are *energy flow* and *evolution*. The first unit shows how these two concepts explain the seeming paradox of life: its unity *and* diversity. The organization of remaining units parallels the levels of biological organization. With the insights provided by the initial treatment of energy flow and evolution, the significance of this hierarchy becomes evident. In addition, the new edition has a strong *systems integration* approach throughout. Instead of being asked simply to memorize names of the parts, students can sense how the whole multicelled individual—or cell, or community—adjusts to changing conditions. The concept of homeostasis, introduced in the first chapter, is used throughout the book.

On Content Changes. The following comments will give you some idea of the extent of changes in the second edition. A new introductory chapter distills *all* the main biological principles covered in the book. The cell biology unit is greatly improved in organization and coverage; energy transfers and transformations provide a strong conceptual thread throughout. Water chemistry, membranes, and transport mechanisms are integrated into a dynamic picture of cells surviving in ever-changing environments. Energy transformations, enzymes, and energy carriers are explained simply, as part of the same story of cell survival. With these topics already covered, the chapter on the main metabolic pathways is kept notably uncluttered. There is now a separate, comprehensive unit on genetics, organized in ways that will better serve the student. It begins with prokaryotic reproductive strategies, then proceeds through mitosis and meiosis. Coverage of the chromosomal basis of inheritance is greatly expanded and now precedes molecular genetics. A new chapter presents principles underlying the genetic basis of development, in anticipation of specialized developmental strategies of different groups of organisms discussed later on. Both the animal and the plant anatomy/physiology units are reorganized and expanded in text and illustrations; both now have a strong evolutionary/ecological approach. The treatment of evolutionary biology and ecology is almost entirely new, reflecting the recent conceptual developments and describing some of the exciting studies in these rapidly changing fields.

Illustrations. Students using the first edition appreciated the clear graphic presentation of such potentially confusing topics as mitosis, meiosis, and protein synthesis. The second edition improves on these and other basic illustrations. Also, optional boxed illustrations, with extended captions, describe *methods and experiments* crucial to discoveries and to the elucidation of principles. The number of illustrations has been expanded considerably. None was tacked on as decoration; all were developed and intensively reviewed with the manuscript.

Case Studies. Case studies encourage learning by involvement. Why just list infection-fighting cells and their products? Why not show students what happens when their bodies actually mount an immune response? Why take an encyclopedic approach to ecosystems (this is the tundra, lemmings live here)? Why not look closely at the delicate feedback relations between plants, animals, and microorganisms of specific ecosystems? Such is the rationale behind the book's unique Case Studies. You will find this material throughout the book.

Commentaries.   Special essays explore such thought-provoking topics as cancer, allergy, death, cardiovascular and lung disorders, human nutrition, starvation, human "races," in vitro fertilization, the energy cost of monocrop agriculture, recombinant DNA research, and phenotypic "cures." The Commentaries are not gimmicks for distracting a bored student reader. They are careful expansions of text discussions that should already be holding the student's interest.

Perspectives.   These end-of-chapter sections encourage students to take a moment for conceptual synthesis. The Perspectives provide bridges not only between chapters, but between chapter topics and the student's world. They invite reflection on past and possible futures for us and all of life.

Study Aids.   The new edition represents a fusion of two viewpoints—those of the student as well as the biologist. The English language has immense potential for complexity, and when its elaboration gets in the way of understanding, student readers get frustrated. No matter how exciting a subject may be, frustration can turn good minds away from it. The premise here was that students will tune into a subject with the same facility they display when tuning out—if the writing lets the subject itself shine through. So paragraph by manuscript paragraph, students were imagined to be present, asking "What is THAT supposed to mean?" Real students also reviewed the manuscript, sometimes, unbelievably, repeating that same question. The writing has been improved because of their real as well as perceived presence.

We never have assumed that students taking introductory biology courses already know enough about biology to spot all the key concepts in page after text page. In this edition, italic sentences within the text call attention to the main points that lead, step by step, to important concepts. These concepts are boxed within the text column and printed in boldface for emphasis. Taken together, the boldface key concepts provide an easy-to-identify, *in-context summary* for each chapter. *In-text questions* followed at once by answers (or best guesses) encourage alertness to the unfolding of ideas. *End-of-chapter questions*, new to this edi-

tion, test student understanding of the main points. The questions are keyed to italic and boldface sentences within the chapter, as a way of reinforcing recall. They also encourage understanding by asking the student to apply basic concepts to one or more specific problems. *Advance organizers* are built into the text. Chapters begin by stating what the chapter is about, how the topics will unfold, and how students may find the topics related to their interests. The new introductory chapter is an advance organizer in its entirety. It introduces the most basic of all biological concepts—ones that students will be using throughout the book. The *glossary* brings together the text's main definitions. The *index* is comprehensive, simply because students may find a door to the text more quickly among finer divisions of topics.

Finally, perhaps we are speaking now with more assurance. One never knows what one is going to get back when a first edition hits the desks. We received considerable encouragement from instructors who have been using the book. More telling, we received notes from students who were awakened to the idea that they don't have to work at "getting close to nature," or "going back" to it—that they are already one with the world of life. And that, in essence, is *the* message of biology.

ACKNOWLEDGMENTS

Our assurance comes not only from our evolution as authors. It comes from knowing that more than a hundred research specialists and biology instructors gave an amazing amount of their time and knowledge to improving the manuscript, through four rounds of reviews. It is strange to think about, but there *were* glimpses that the reviewers were as driven as we were. Although we alone are responsible for any inaccuracies that may have slipped through, we simply can't take all the credit for what is good about the book. Herman Wiebe, with his awful scribblings, would not let us off the hook. George Lefevre, with impressive calm, guided us back whenever we invented new genetics. Cleon Ross is a marvel of a friend and scholar; he not only reviewed, he went ahead and wrote Chapter Twenty himself. Dorothy Luciano gave an incredible amount of time, going line by line over the entire animal physiology unit. G. Tyler Miller, Jr., an outstanding author in chemistry and in environmental science, did the same for our chapters on basic biochemistry. Jane Taylor consistently came through with well-conceived test questions. How many times did Marian Reeve, Karen Van

Winkle-Swift, and Gary Wisehart bring the writing back down to earth with their insistence on keeping the words simple? Roger Burnard, Tom Hemmerly, Gary Grimes, Sally Faulkner—they and others whose names appear alongside the actual illustrations helped shape the book's graphics. We hope these limited anecdotes convey the sense of community enterprise that characterized this book's development. There never is enough space to acknowledge all those who helped make a book what it is; and the indebtedness increases with each edition. Again, the following list of reviewers is our way of thanking all those individuals who will know, in reading through the book, where they have left their imprint. To each one, thank you for your insights, your criticisms, your praise—whichever was most deserved on all those pages that crossed your desk.

Finally, we wish to acknowledge the Wadsworth individuals who turned the manuscript and sketches into a book. Kathie Head is a superior production editor. If only she could be cloned; few, if any, can match her kindness and competence. Cynthia Bassett, bless her, put up not only with the demands of a truly complex book but with truly demanding authors. Cathy Aydelott, as art editor, kept track of things when we did not. Marion Hansen is an outstanding photo researcher; whenever the illustration program showed signs of buckling, she did not. Neither did Peggy Meehan. And to you, Jack Carey, thank you for helping conceive of this book in the first place and shepherding it to completion. There is no magic in having a great many reviewers; on the surface, "100 reviewers" could even sound like overkill. But the magic resides in the way you can bring about a fusion of so many viewpoints; it resides in the remarkable system you have developed for analyzing trends in biology teaching; it resides in your deep commitment to developing useful, innovative books. You planned, you commiserated, you sometimes dragged us kicking and screaming to a conceptual point, your encouragement and enthusiasm never dwindled, you shaped this book more than anyone might believe. There can be no better editor, colleague, and friend.

Second Edition Reviewers

John N. Abelson, University of California, San Diego
Peter Armstrong, University of California, Davis
Bill Atkins, Jefferson State Junior College
Glenn D. Aumann, University of Houston
Allen A. Badgett, Southwestern Oklahoma State University
William Balamuth, University of California, Berkeley
William Barstow, University of Georgia
Penelope Hanchey Bauer, Colorado State University

Charles B. Beck, University of Michigan
Wayne Becker, University of Wisconsin, Madison
C. William Birky, Jr., Ohio State University
Brenda Blackwelder, Central Piedmont Community College
George Bleekman, American River College
James Bonner, California Institute of Technology
William Bowen, University of Arkansas
Robert N. Bowman, Colorado State University
Phyllis Bradbury, University of North Carolina
William L. Brown, Jr., Cornell University
Neal Buffaloe, University of Central Arkansas
Ruth Buskirk, University of Texas, Galveston
Allan M. Campbell, Stanford University
Thomas R. Campbell, Pierce College
John R. Capeheart, Downtown College
Arthur K. Champlin, Colby College
Bruce Christensen, Murray State University
Corbett D. Coburn, Jr., Tennessee Technical University
Marie Conklin, San Diego State University
Barbara Crandall-Stotler, Southern Illinois University
Ross H. Crozier, University of New South Wales
Mel Cundiff, University of Colorado
Winifred Dickinson, College of Steubenville
Gordon G. Evans, Tufts University
Wayland L. Ezell, St. Cloud State University
Richard Falk, University of California, Davis
H. S. Forrest, University of Texas, Austin
Jon Fortman, Mississippi University for Women
Douglas Fratianne, Ohio State University
Charles W. Gaddis, University of Arizona
Francis E. Gardner, Columbus College
Richard Garth, University of Tennessee
Wendell Gauger, University of Nebraska
Alan Gelperin, Princeton University
Michael T. Ghiselin, University of Utah
Ben Golden, Kennesaw College
Thomas M. Graham, University of Alabama
W. J. Graham, Monrow Community College
Charles R. Granger, University of Missouri
L. Greenwald, Ohio State University
Katharine B. Gregg, West Virginia Wesleyan College
James Grosklags, Northern Illinois University
William R. Hargreaves, Dupont Experimental Station
Jean Harrison, University of California, Los Angeles
W. K. Hartberg, Georgia Southern College
Daniel Hartl, Purdue University
Evan Hazard, Bemidji State University, Minnesota
Richard Heckmann, Brigham Young University
Thomas E. Hemmerly, Middle Tennessee State University
George Hennings, Kean College of New Jersey
Carol Jefferson, Minona State University
Burton Johnson, Ft. Steilacoom Community College
Patricia W. Johnson, University of North Carolina
Jane Kahle, Purdue University
Richard G. Kessel, University of Iowa
Robert T. Kirkwood, University of Central Arkansas
Charles J. Krebs, University of British Columbia
C. C. Lambert, California State University, Fullerton
Norma Lang, University of California, Davis

Andrew J. Lechner, University of Colorado, Denver
George Lefevre, Jr., California State University, Northridge
Herbert Levi, Harvard University
Dorothy Luciano, formerly with University of Michigan
Phil M. Mathis, Middle Tennessee State University
John H. McClendon, University of Nebraska
Jerry W. McClure, Miami University
Fred Meins, University of Illinois
James H. Menees, California State University, Long Beach
G. Tyler Miller, Jr., St. Andrews Presbyterian College
Helen Miller, Oklahoma State University
John Minnich, University of Wisconsin, Stevens Point
Margaret S. Misch, University of North Carolina
Barbara Murdock, University of North Carolina, Wilmington
Terrence M. Murphy, University of California, Davis
Steven Murray, California State University, Fullerton
David Nelson, University of South Alabama
B. Nicotri, University of Washington
William Odum, University of Virginia
James W. O'Leary, University of Arizona
Jane Overton, University of Chicago
Dennis Parnell, California State University, Hayward
Hugh J. Phillips, Mesa Community College
C. Ladd Prosser, University of Illinois
Marian Reeve, Merritt Community College
Rosemary Richardson, Bellevue Community College
Robert J. Robbins, Michigan State University
Robert C. Romans, Bowling Green University
Cleon Ross, Colorado State University
Frank B. Salisbury, Utah State University
Hosea Sanders, Jackson State University
Rudolf Schmid, University of California, Berkeley
William Schmid, University of Minnesota
Patricia Schulz, University of San Francisco
Judith Shea, North Shore Community College
Terry Shininger, University of Utah
William Sistrom, University of Oregon
David Smith, San Antonio District Junior College
E. E. Southwick, State University of New York,
    College at Brockport
Shirley R. Sparling, California Polytechnic State University
Irwin Spear, University of Texas
Pamela Sperry, California State Polytechnic University
Fleur L. Strand, New York University
Jane B. Taylor, North Virginia Community College
Donald Van Horn, University of Colorado
Karen Van Winkle-Swift, San Diego State University
C. Kenyon Wagner, Clemson University
Wendell E. Wall, University of Alabama
George Washington, Jackson State University
Terry M. Weidner, Eastern Illinois University
Ken Wells, University of California, Davis
Don Whitehead, Indiana University
Herman Wiebe, Utah State University
Mary Wise, Northern Virginia Community College
Gary Wisehart, San Diego City College
Stephen L. Wolfe, University of California, Davis
John Yopp, Southern Illinois University, Carbondale

First Edition Reviewers

We also wish to acknowledge the following reviewers of the
first edition: Donald Abbott, Stanford University; Kenneth Ar-
mitage, University of Kansas; George Ball, Gaston College;
Walter Becker, Washington State University; Marjorie Behringer,
University of North Dakota; Richard Boohar, University of Nebraska; Bel Dolbeare, Lincoln Land Community College; Frank
Einhelling, University of South Dakota; Harold Eversmeyer,
Murray State University; William Fennel, East Michigan University; Dorothy Frosch, Central State University; Berdell Funke,
North Dakota State University; Donald Garren, Lake Land College; Robert Grey, University of California, Davis; Harlo Hadow,
Coe College; Bruce Haggard, Hendrix College; Ted Hanes, Cali-
fornia State University, Fullerton; Laszlo Hanzely, Northern Illinois University; Vernon Hendricks, Brevard Community College;
James Henry, Illinois Central College; Joseph Hindman, Washington State University; J. Michael Jones, Emory and Henry College; Arnold Karpoff, University of Louisville; Norman Kerr,
University of Minnesota; Frederick Landa, Virginia Common-
wealth University; Frank Lang, Southern Oregon State College;
Daniel Lee, State University of New York, Plattsburgh; Robert
Macey, University of California, Berkeley; William Mason, Au-
burn University; Laura Mays, Occidental College; Francis Mc-
Carthy, California State College, Dominguez Hills; Margaret
McElhinney, Ball State University; Michele Morek, Brescia Col-
lege; Gerald Myers, South Dakota State University; Robert Sca-
gel, University of British Columbia; Ted Schwartz, University
of California, San Diego; Valerie Seeley, Queensborough Com-
munity College; Arthur Shapiro, University of California, Davis;
Frank Sivik, Broward Community College; Gordon Snyder,
Schoolcraft College; David Stronck, Washington State Univer-
sity; M. Camilla Suddreth, Gaston College; Jack Thomas, Los
Angeles Harbor College; Paul Thomson, Penn Valley Community
College; Frank Toman, Western Kentucky University; Norman
Tweed, Fort Steilacoom Community College; Warren Wagner,
University of Michigan; Jon Weil, University of California, San
Francisco; Jonathan Westfall, University of Georgia; and Richard
Woodruff, West Chester State College.

CONTENTS IN BRIEF

DETAILED CONTENTS

UNIT ONE

UNIFYING CONCEPTS IN BIOLOGY

1

REFLECTIONS ON LIFE:
SOME BASIC CONCEPTS

Buried somewhere in that mass of nerve tissue just above and behind your eyes are memories of first encounters with the living world. Still in residence are sensations of discovering your own two hands and feet, your family, the change of seasons, the smell of rain-drenched earth and grass. In that brain are traces of early introductions to a great disorganized parade of beetles, flowers, frogs and furred things, mostly living, sometimes dead. There, too, are memories of questions—*What is "life"?* and, inevitably, *What is "death"?* There are memories of answers, some satisfying, others less so.

Observing, asking questions, accumulating answers— in this manner you have acquired a store of approximate knowledge about the world of life. During the journey to maturity, experience and education have been honing your questions, and no doubt the answers have been more difficult to come by. What *is* life? What characterizes the living state? The answer may vary, depending, for example, on whether it comes from someone arguing for or against legalized abortion. When does life end? Again the answer may vary, depending on whether it comes from a physician, a clergyman, or a parent of a severely injured person who must be maintained by mechanical life support systems, because the brain no longer functions at all. Yet despite the changing character of the questions, the world of living things remains as it was before. Leaves still unfurl during the spring rains. Animals are born, they grow, reproduce, and die even as new individuals of their kind are being born. *The most important difference is in the degree of insight you now bring to your observations, questions, and answers about such events.*

It is scarcely appropriate, then, for any book to proclaim that it's your introduction to biology—"the study of life"—when you have been studying life ever since awareness of the world began penetrating your brain. The subject is the same familiar world that you have already thought about to no small extent. And there simply is no way, short of bad writing, that any book can make it strange and formidable, deserving of apprehension. That is why this book proclaims only to be biology *revisited,* in ways that may help carry your thoughts about life to deeper, more organized levels of understanding.

Let's return at the outset to the question, What is life? It happens that the answer has yet to be reduced to a simple definition. The word embodies a story that has been unfolding in myriad directions for several billion years! To biologists, "life" is what it is by virtue of its ancient molecular origins and its degree of organization. "Life" is a way of capturing and systematically using energy and materials. "Life" means individual adjustment to shifting conditions—it is *adaptive* to short-term environmental change. "Life" is also adaptive, through generations of individuals, to change over long spans of time. "Life" is a commitment to some specific program for growth and development; it is a capacity for reproduction. As you can see, a short list of definitions can do no more than hint at all that the word conveys. Only when you've completed this book will you have deeper insight into its meaning, *for life cannot be understood in isolation from its past history and its adaptive potential.*

Throughout this book, you will be coming across different examples of living things—how they are built, how they tick inside, where they live, what they do. You will also be coming across statements of concepts that explain what these examples might mean. The statements themselves are printed in darker type, and separated by lines from the text. *All* of these conceptual statements, taken together, will give you a pretty good idea of what "life" is.

With this in mind, let's now turn to a few examples that can illustrate the most general concepts of all. Although we will be taking closer looks at these concepts in later chapters, they are summarized here to give you perspective on things to come. You may also find it useful to return to them as a way of reinforcing your grasp of details later on.

Figure 1.1 A common egret on the wing, against a background of bald cypress and Spanish moss. These three kinds of organisms are diverse in appearance, yet they have much in common at deeper levels of organization. They speak of the unity and diversity inherent in *all* of life, which is the subject of this chapter.

ORIGINS AND ORGANIZATION

Suppose someone asks you to point out the difference between a frog and a rock. The frog, you might say, has a body of truly complex organization. Its hundreds of thousands of individual cells are organized into tissues. Its tissues are arranged into organs such as a heart and stomach. The frog can move about on its own. And sooner or later (given a receptive member of the opposite sex), it can reproduce. A rock shows no such complexity, it cannot move by itself, and it certainly cannot reproduce either on its own or in the company of another rock. If you deduce from this that "life" means complex organization, the capacity to move, and the capacity to reproduce, then the frog is alive and the rock is not.

Now suppose someone asks you to point out the difference between a bacterium and a rock. A bacterium is one of the' simplest kinds of organisms, no more than a single cell, really. Yet microscopic examination shows that bacterial bodies are patterned in ways that are far more organized than the insides of a rock. All bacteria have an outer wall and a plasma membrane (a saclike structure that helps control the kinds of substances moving into and out of the body). The membrane surrounds an inner, semifluid substance in which specific structures are embedded (Figure 1.2). Sooner or later, a bacterium can divide and reproduce itself. Some bacteria move on their own through their surroundings; others, however, cannot. The "movement" criterion is getting a little fuzzy now. But by the other two standards (complex organization, a capacity to reproduce), a bacterium is alive.

Suppose you are now asked to compare a virus with the rock. A virus is a peculiar particle in the shadowy world between the living and nonliving. Some viruses do have a distinct organization; some, for instance, have a "head" end, a sheathlike midsection, and a "tail" end. A virus has no means whatsoever of moving on its own. It cannot reproduce on its own. Yet all viruses contain instructions for producing more viruses just like themselves. The instructions have to be injected into a living host organism to be carried out. After that happens, the cellular machinery of the *host* starts churning out parts that will be used in building new viruses! By your initial criteria—complex organization, movement, reproduction—is a virus alive? In some respects yes, in others no.

Somewhere below that shadowed boundary to the living world are tiny spheres, called microspheres (Figure 1.2). Under the right conditions, they may form by themselves—they may assemble spontaneously from simple mol-

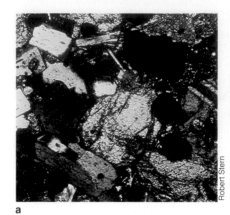
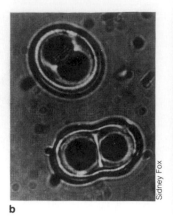

a   b

Figure 1.2 A hierarchy of structural organization between a frog and a rock. (**a**) Basalt rock, thin-section, from Marianas Trench. (**b**) Several microspheres, with their membranelike outer layer and their tendency to grow and fragment into new spheres. (**c**) Virus particles, with outer layers enclosing hereditary instructions. (**d**) Bacterial cell, sliced to show the inside. These single cells are unquestionably alive.
(**e**) Some of the complex structures within a single cell in a multicellular plant. (**f**) The many-celled frog body. (d from G. Cohen-Bazire)

ecules! Microspheres have an outer layer that acts like a simple membrane. Its structure passively controls the kinds of substances moving into and out of the sphere. This means that certain molecules can be isolated from random events in the surrounding environment. When concentrated inside, these molecules may become arranged in nonrandom ways relative to one another. Thus microspheres show at least some *potential* for organization. What about reproduction? When they accumulate some substances, microspheres can grow in size. They may even grow to the extent that parts break off and form new spheres, which grow in their turn. But the processes involved are not reproductive; they are only random chemical growth and fragmentation into (generally) nonidentical parts. So you are left to conclude that microspheres are not alive. Then again, the degree of organization they do show is far more intriguing than the organization we see in the rock.

And what about that rock? It does not seem any too remarkable when compared to a highly organized creature such as a bacterium. At the levels of viruses and microspheres, however, the difference begins to blur. At a still deeper level, the difference becomes nonexistent. Frog, bacterium, virus, microsphere, rock—all turn out to be

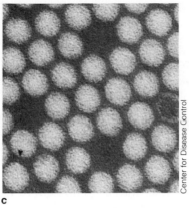

c

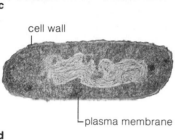

cell wall

plasma membrane

d

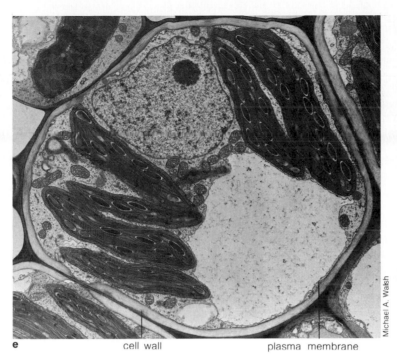

e

cell wall

plasma membrane

f

composed of the same raw materials. These materials are particles called protons, electrons, and neutrons. They become organized relative to one another according to the same rules of chemistry. At the heart of these rules is something called **energy**—a capacity for interaction between particles, a capacity to make things happen, to do work.

As you will see later, energetic interactions join these particles together, in predictable ways, and form atoms. They bind atom to atom in predictable patterns, thereby giving rise to the kinds of molecules that form (for example) all the frogs and rocks of our world. Energetic interactions hold a frog together and a rock together; the distribution of energy organizes and holds entire communities of organisms together. Thus we have a concept of profound importance:

The structure and organization of the living as well as the nonliving world arise from the fundamental nature of matter and energy.

This concept has implications for questions about the origin of life. It now appears that living things may have originated as increasingly organized constellations of matter and energy. A later chapter will tell you more about the evidence supporting this idea. For now, take a look at Figure 1.3, which outlines the levels of organization in today's world. Then consider the idea that these levels may echo successive stages in the history of life. According to this idea, interactions between atoms and molecules, under certain chemical conditions, led to the assembly of special large molecules that are found in all living things. Increasing molecular organization in some way led to structures possibly like microspheres. In microspheres, molecules became organized relative to one another in ways that allowed them to duplicate themselves—and lay the foundation for reproduction. This capacity led to the cell—the basic *living* unit. Reproduction of cells and their interactions led to populations of single cells and multicelled organisms, then to communities, ecosystems, and the biosphere. What we are unfolding here is a picture of increasingly complex, interrelated patterns of order in the use of materials and energy. It accounts, as any speculation about the origin of life must do, for this apparent fact:

The "difference" between the living state and the nonliving state is one of <u>degree</u>, not of kind.

Biosphere
Entire zone of earth, water, atmosphere
in which life can exist on our planet's surface

Ecosystem
All the energetic interactions and materials cycling
that link organisms in a community with one another
and with their environment

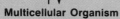

Community
Two or more populations of different organisms that occupy
and are adapted to a given environment

Population
Group of individuals of the same kind, occupying
a given area at a given time

Multicellular Organism
Individual composed of specialized, interdependent cells
arrayed in tissues, organs, and often organ systems

Organ System
Two or more organs whose separate functions
are integrated in the performance of a specific task

Organ
One or more types of tissues
interacting as a structural, functional unit

Tissue
One or more types of cells
interacting as a structural, functional unit

Cell
Smallest *living* unit; may live independently
or may be part of multicellular organism

Organelle
Structure inside a cell whose molecular
organization enhances specific cell activities

Molecule
Two or more identical or different kinds of atoms bonded together

Atom
Smallest unit of a pure substance that retains properties of that substance

Subatomic Particle
Unit of energy and/or mass; electron, proton, neutron

Figure 1.3 Levels of organization in nature.

UNITY IN BASIC LIFE PROCESSES

Metabolism

So far, we have touched on the nature of life's organization—the bricks, so to speak, that are assembled in orderly ways to form each living thing. Underlying the assembly of the bricks themselves, and underlying all activities of the organisms in which they are found, are *energy transformations.* As you will read later, energy may be stored in various substances. Afterward, that energy may be released to do work, such as building cell parts. There is only so much energy available to living things, however, and there is no way an organism can create "new" energy from nothing. To stay alive, it must "borrow" energy from someplace else. The organism must tap an *existing* energy source in its surroundings, then transform the energy into forms appropriate for its requirements.

For example, plant cells rely on an energy-trapping process called photosynthesis. They absorb sunlight energy, which is used in forming energy-rich molecules. Adenosine triphosphate, or ATP, is one such energy-rich molecule. When required, its stored energy may be released to power the cell's building and maintenance programs, as well as its reproduction. Two of the main energy-releasing pathways are called glycolysis and cellular respiration.

All single-celled and multicelled organisms have a capacity for acquiring and using energy in stockpiling, tearing down, building up, and eliminating materials in controlled ways. This capacity, called **metabolism**, is a unique feature of living things.

All forms of life show metabolic activity: they extract and transform energy from their environment, and use it for manipulating materials in ways that assure their own maintenance, growth, and reproduction.

Homeostasis

Any attempt to define the nature of life cannot focus exclusively on "the organism," for "the organism" cannot exist apart from its surroundings. And it most assuredly cannot exist without being able to respond to variations in those surroundings, whether random or self-induced. The living state happens to be maintained within rather narrow limits. Concentrations of such substances as carbon dioxide and oxygen must not rise above or fall below certain levels.

Toxic substances must be avoided or eliminated. Certain kinds of food must be available, and in certain amounts. Water with dissolved components must bathe each metabolically active cell. Water, oxygen, carbon dioxide, ions and foods of varied sorts, light, pressure, temperature—such environmental factors dictate the terms of survival. And such terms are subject to change.

How do living things respond to a changing environment? All organisms have built-in means of maintaining internal conditions within some tolerable range, even when external conditions fluctuate. This capacity is known as **homeostasis**. In cells, it may require participation of the plasma membrane, which selectively accumulates substances in short supply and expels excess amounts of others. Individual cells, in other words, have homeostatic controls. So do multicelled organisms. Birds, for instance, have sensors that signal the brain when outside temperatures drop. The brain may send signals to cells that control feather movements. Certain movements lead to feather fluffing—which helps retain heat and thereby helps maintain internal temperature.

We will be returning repeatedly to the concept of homeostatic control. Here, the concept may be summarized in this way:

All forms of life depend on homeostatic controls, which maintain internal conditions within some tolerable range even when external conditions change.

Growth, Development, and Reproduction

Through the precision of metabolic and homeostatic events, living things come into the world, they grow and develop, they reproduce. Most then move on through decline and death according to the timetable for their kind. Even as individual organisms die, reproduction assures that the form and function of the living state are perpetuated along the axis of time.

And yet, "an organism" is much more than a single organized form having a single set of functions during its lifetime. One example will make the point, even though actual details vary considerably from one kind of organism to the next. A tiny egg deposited on a branch by a female moth (Figure 1.4) is a transitional form. It's a compact package that contains all the necessary instructions for becoming an adult moth. But it does not become transformed at once into a miniature winged moth that need

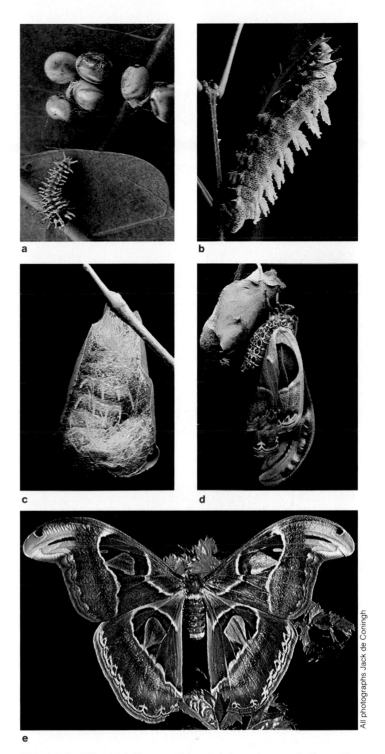

a   b

c   d

e

All photographs Jack de Coningh

Figure 1.4 "The insect"—a continuum of stages in organization, with new adaptive properties emerging at each stage. Shown here: the development of a giant moth, from egg (**a**) to larval stage (**b**), to pupal form (**c**), to emergence of the resplendent moth form (**d,e**).

only increase in size. Instead, developmental events going on inside the egg lead to an entirely different form: a wingless, many-legged larva called a caterpillar. The caterpillar hatches during a warm season when tender new leaves are unfolding. Not surprisingly, the caterpillar is an "eating machine" stage of this insect's life cycle. It comes equipped with mouthparts capable of tearing and chewing leaves, and with a metabolic capacity for extremely rapid growth. It eats and grows until some internal alarm clock goes off, setting in motion events that lead to profound changes in the living form. Some cells are disassembled, other cells multiply and are assembled in entirely different patterns. Tissues, too, are moved about during this stage of wholesale remodeling, the so-called pupal stage. From the pupa, the "reproductive machine"—the moth—emerges. No leaf-cutting mouthparts for the moth: instead it has a tubelike proboscis, which draws nectar from flowers. From the nectar comes energy that powers free-wheeling flights. For this insect, wings are emblazoned with colors and set to move at a frequency that can attract a potential mate. The moth is the form equipped with reproductive organs in which egg or sperm develop, and which enhance fertilization of an egg. With fertilization comes the beginning of a new life cycle.

None of these functional stages is "the insect." "The insect" is a progression of stages in organization, with new adaptive properties emerging at each stage. It's possible to talk about the units of energy required to power insect flight, the food required for caterpillar growth, the changes in chemical activity that trigger the transformation from caterpillar to pupa. All such metabolic and homeostatic events are important in describing the nature of life. But equally important is an understanding that they are no more, no less, than *parts* that are intermeshed in a whole cycle of events and circumstance:

Each living organism is a continuum of patterns that unfold during its life cycle. The patterns unfold in the same way for all of its kind, and they correspond to specific aspects of the environment.

DNA: Storehouse of Constancy and Change

Upon thinking about the preceding summary statement, you might wonder what could be responsible for **inheritance**—the transmission, from parents to offspring, of structural and functional patterns characteristic of each kind of organism. How is it that an ostrich hatching from an egg grows into a fairly exact replica of parent ostriches? How is it that a bacterium can divide and grow into two fairly exact copies of itself? Within each individual, there must be a molecular storehouse of hereditary information. It must be a remarkable storehouse, indeed, for it must contain all the details for a complete program of growth, maintenance, development, and reproduction.

There is another remarkable aspect of this storehouse of information. Although offspring generally resemble their parents in form and behavior, *variations* can exist on the basic plan. A newly produced bacterium may not be able to assemble (as it is supposed to) some molecule that is vital to its functioning. Sometimes humans are born with six fingers on each hand instead of five. Even though the hereditary molecule assures overall fidelity in its patterning, it somehow must be subject to change in some of its details!

As you probably have learned by now, the hereditary molecule has been identified. In almost all cases it is deoxyribonucleic acid, or DNA. We now know that changes occasionally occur in the kind, structure, sequence, or number of its component parts. These changes are **mutations.** Most mutations are harmful, for the DNA of each kind of organism is a package of information that is finely tuned to a given environment. In addition, its separate bits of information are part of a coordinated whole. When one crucial part changes, the whole living system may be thrown off balance. Such is the fate of the bacterium mentioned in the preceding paragraph, for the change probably means that it's doomed.

But sometimes a mutation may prove to be harmless, even beneficial, under prevailing conditions. For example, a mutant form of a light-colored moth *(Biston betularia)* is dark-colored. When it rests on soot-covered trees, bird predators simply don't see it. In places where there happen to be lots of soot-covered trees, its dark color can be advantageous: the mutant stands a better chance of reaching reproductive age than its light-colored kin.

In later chapters, more will be said about the twin features of constancy and change in the hereditary material. For present purposes, we can summarize these features:

DNA is a storehouse of patterns for all heritable traits. Mutations introduce variations in the patterns. The environment—internal and external—is the testing ground for the combination of patterns that comes to be expressed in each living thing.

a

Ron Nolan

DIVERSITY IN FORM AND FUNCTION

Until now, we have focused on the unity of life. We have suggested that all living things are linked together, in origins and organization, through the nature of matter and energy; that they all rely on metabolic and homeostatic processes; and that they have the same molecular basis of inheritance. These are fairly recent ideas. Before refinements in microscopy and the emergence of molecular biology, there was no reason to suspect that all living things hold these characteristics in common. What *was* apparent, and difficult to explain, was the tremendous sweep of life's *diversity*. Why is it that almost every environment on earth is host to an astonishing array of different organisms? What is the meaning of this diversity?

For example, imagine yourself exploring a tropical reef (Figures 1.5 and 1.6). Ages ago, tiny animals called corals began to grow and reproduce beneath the warm, clear waters near the land's edge. They left behind their skele-

b   c

Figure 1.5 (**a**) Underwater tropical reef. Two of the master builders: (**b**) green tube coral and (**c**) pillar coral—with individual animals sending out tentacles from their chambers. (Photographs by Douglas Faulkner)

Figure 1.6 Who eats whom on the reef. (**a**) Crown-of-thorns, a sea star that feasts on tiny corals. (**b**) Sea anemone, an animal with weapon-studded tentacles, which ensnare tiny animals floating past. (**c**) Sponges, with pores opened toward the oncoming food-laden currents. (**d**) Clownfish, curiously at home above the mouth of a sea anemone—a mouth through which other kinds of edible fish quickly disappear. (**e**) Green algae, plants that are food for various reef organisms. (**f**) Red algae, food for various animals (but not for this chambered nautilus, a shelled animal that swims expertly after shrimp and other prey). (**g**) A school of goatfish, which feed on small,

spineless animals on the sea floor. Goatfish are tasty to humans, also to large fish. (**h**) But some fish are not on the general menu. Here, a blue wrasse safely picks off and dines on parasites that prey on this large predatory fish. (**i**) Stone crab. Depending on the species, crabs eat plants, animals, and organic remains. The moray (**l**) prefers meat. (**j**) Lion fish, with its fanned, poison-tipped spines warning away intruders. (**k**) Find the scorpion fish—a dangerous animal that lies camouflaged and motionless on the sea bottom, the better to surprise unsuspecting prey.

tons as a foundation for more corals to build upon. As skeletons and residues accumulated, the reef grew. All the while, tides and currents carved ledges and caverns into it. Today, the reef's spine may contain any number of *750 different kinds* of corals: colorful, soft or stonelike animals shaped like staghorns, domes, brains, flowers, mushrooms, cabbages, folded draperies, and fans. Here also, plants called red algae may encrust the coral foundation. In shallow waters behind the reef, red algae give way to blue-green forms. Many small transparent animals feed on algae and other plants. These animals in turn are food for still larger animals, including some of the world's *20,000 different kinds* of fishes. All about are predatory sea anemones, each having a mouth fringed with weapon-studded tentacles that capture tiny fish. Yet, hovering above the tentacles *is* a certain kind of fish! It is as edible as most others, but somehow it is not recognized as prey. The fish moves out, captures food, and returns to the anemone's tentacles—which give it protection from predators. The anemone eats scraps of food that fall from the fish mouth. These two kinds of animals are, in effect, allies: one receives protection, the other receives food (Figure 1.6d).

The reef is also home for different kinds of sea stars. When feeding, the sea star extends its stomach *outside* its body, into coral chambers. Each chamber resident is digested in place before the stomach is pulled back out. When sea stars reproduce, millions of larvae emerge and feed on microscopic algae. Then, as the larvae grow, they become food for the meat-eating corals! Now the corals grow and reproduce. Eventually they repopulate the reef regions that the earlier generation of sea stars had stripped clean. The sea star larvae that do escape grow to become diner instead of dinner, and thereby initiate a new cycle of death and life.

Clearly, reef organisms are remarkably different in appearance and behavior. However, before speculating on what could account for the diversity, imagine yourself in another setting to see if the comparison yields any similarities or differences that might provide added insight. In the shadow of Kilimanjaro, a volcanic peak rising high above the edge of the East African Rift Valley, grasslands sweep out to the northeast. This is the African savanna, a region of warm grasslands punctuated with scattered stands of trees and shrubs (Figure 1.7). More large ungulates (hoofed, plant-eating mammals) live here than anywhere else. One form, the giraffe, browses on leaves some five meters aboveground, far beyond the reach of other ungulates. Another form is the Cape buffalo (Figure 1.8). An adult male can weigh a ton, it has formidable horns,

and its behavior is unpredictable. It is rarely troubled by predators. Other forms include zebra and impala—both smaller, more vulnerable, and far more abundant than Cape buffalo. They are constantly troubled by such predators as lions and cheetahs. Their remains (as well as the remains of lions and cheetahs) are picked over by scavengers—hyenas, jackals, vultures, and the marabou stork (see Figure 1.9).

Buffalo, zebra, impala, rhinoceros, giraffe—these and *eighty-five other kinds* of large, plant-eating animals live in the immense valley, as do predators and scavengers that feed on them. Somehow they exist side by side in time, moving westward, southward, and back again as dry

Figure 1.7 The East African Rift Valley, some 6,400 kilometers (4,000 miles) long. The sparsely wooded grasslands in this valley are home for a diverse array of animal life and, as you will read in later chapters, were the probable birthplace of the human species.

seasons follow rains, as scorched earth gives way to a resurgence of plant growth.

What tentative conclusions might be drawn from this brief comparison of diversity in the reef and the savanna? One thing their diverse organisms have in common is specialization in "who eats whom," beginning with plants and proceeding through specialized forms of animals that eat the plants and one another. In fact, you could spend years observing organisms in forests, deserts, the plains—and you would find that they speak eloquently of the same challenge. *All organisms must be equipped to obtain a share of available resources.* In large part, diversity in form, function, and behavior represents specialized ways of getting and

using resources—and ways to avoid being a "resource" for some other organism. In light of this observation, let's now address the question of how this diversity could have come about.

Imagine that since the time of origin, life of any form has required a constant supply of energy and materials. But we know today that the supplies have never been unlimited. In the past, then, *all the members of any group of organisms had to be demanding a share of limited resources.* Imagine, next, that *variant* individuals occasionally appeared (possibly through DNA mutations). Perhaps some variant forms were better equipped for getting a share of resources. Perhaps some were better equipped for re-

a

b

c

Figure 1.8 A sampling of the ninety kinds of large plant-eating animals that live in the savanna—a clear example of diversity in a single environment. (**a**) A herd of Cape buffalo. Imagine yourself a predator this close to the herd and you get an idea of one of the benefits of group living. (**b**) Zebra mother and offspring. (**c**) Male and female impala on the alert, ready to take cover in the nearby woods. (**d**) The rhinoceros, another formidably decked-out plant eater. (**e**) The giraffe, browsing on vegetation high up.

d

e

Norman Myers/ Bruce Coleman Inc.

a

sponding to predators, prey, or inadvertent allies around them. Being better adapted to prevailing conditions, they would tend to be the ones to survive and reproduce. Through reproduction, their successful variations would be perpetuated.

This line of thought amounts to one current explanation of the origin of diversity. The next chapter will describe how it is generally thought to be the best explanation at the moment, for all the available evidence points to its validity. Whatever the case, any idea about the nature of life must explain not only its unity but its diversity. Here, and in chapters to follow, the premise is this:

The diversity of life is the sum total of variations in form, function, and behavior that have accumulated over time—and that have proved adaptive in getting or using resources.

INTERDEPENDENCY AMONG ORGANISMS

The preceding summary statement is based on several assumptions about earth and life history. But the geologic record does indeed suggest that there was a time, early in life history, when groups of simple organisms floated about independently of one another in shallow lakes or seas. They must have fed on substances already present in the environment (such as carbon-containing compounds that had accumulated through geologic processes). Eventually, perhaps by chance when food supplies began to dwindle, they began relying more and more *on each other* as sources of materials and energy. Organisms, too, are stores of energy-rich molecules. Thus, by chance and by necessity, community interactions began and have continued in ever richer diversity. Today, through these interactions, few energy sources are unattended. One example

b

Figure 1.9 Predators and scavengers of the savanna. (**a**) An adult lioness standing over a fresh kill. These large cats stalk the herds at dusk or afterward, typically concealing themselves in dense or low-lying vegetation. (**b**) Vultures, together with marabou storks, feed on locusts, small birds, and small mammals—but they also clean up whatever carrion becomes available to them. In this dual predator-scavenger role, they are like other diverse animals of the savanna, including hyenas and jackals.

will make the point, even though the cast of characters seems of a most improbable sort.

First we have the adult male elephant of the African savanna (Figure 1.10). It stands almost two stories high at the shoulder and weighs more than eight tons. This grazing animal eats quantities of plants, the remains of which leave its body as droppings of considerable size. Appearances to the contrary, locked in the droppings are substantial stores of unused nutrients. With resource availability being what it is, even the waste products of one kind of animal are food for another. And so we next have little dung beetles rushing to the scene almost simultaneously with the uplifting of an elephant tail. With great precision they carve out fragments of the dung into round balls, which they roll off as a compact food supply and bury underground in burrows. In these balls they lay their eggs, which assures that their forthcoming offspring, too, will have a supply of nutrients. Also assured is an uncluttered environment. If the dung were to remain aboveground, it would dry out and pile up beneath the hot African sun. Instead, the surface of the land is tidied up, the beetle has its food source, and the remains of the dung are left to decay in burrows—there to enrich the soil that nourishes plants that sustain (among others) the elephants.

Such interactions of organisms with their environment and with one another are the focus of **ecology.** Today, everywhere you look you will find different organisms locked into patterns of ecological interdependency. Some patterns may be simple and some complex, and some may seem to border on the outlandish. But in all cases,

Almost all existing forms of life depend directly or indirectly on one another for materials and energy.

All photographs Roger K. Burnard

Figure 1.10 An interdependency of a most improbable sort, beginning with the plants that feed the elephants (**a**), the dung that leaves the elephants (**b**), the beetles that roll dung balls away and bury them (**c**), ending with the beetle larva (**d**) that hatches in the dung—and the remains of the dung itself, enriching the soil in which plants grow, eventually to feed the elephants.

ENERGY FLOW AND EVOLUTION

This chapter has touched on certain concepts about the living state. Taken together, these concepts suggest that *energy flow* is the foundation for the unity and diversity of life (Figure 1.11). At all levels of biological organization, energy is used in the organization of matter. Energy transformations and transfers occur within organisms, and energetic interactions occur between them. It is only recently that findings from investigations on many different levels have converged to afford us this insight. These findings are the legacy of work going on not only in biology but in fields ranging from chemistry to geology to astronomy

and physics. And it is only recently that a powerful, integrative principle has been shown to explain not only the basic fabric of energy use, but also the rich patterns of diversity that are embroidered into it. This is the principle of evolution by natural selection. It was first formulated by Charles Darwin and Alfred Wallace, then later refined (especially during the past few decades). It is the subject of the next chapter. A principle, in essence, is an idea whose validity holds up even when different observations and experiments are used to test it. In view of life's tremendous diversity, this is not an insignificant accomplishment. Consider that there may be many millions of different organisms, that only a fraction of these have been studied, and

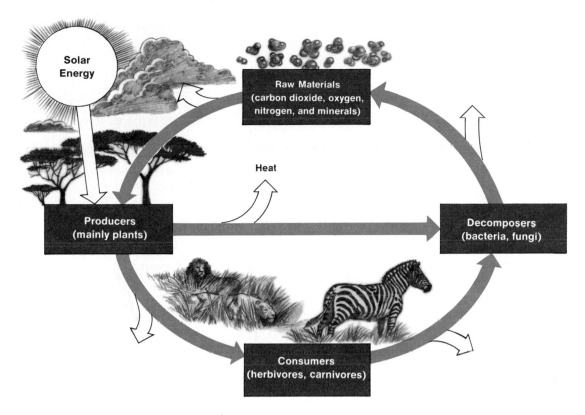

Figure 1.11 Energy flow and the cycling of materials through the biosphere. There is a one-way flow of energy from the sun, through producer organisms (mostly plants), and on through consumers (animals and some microorganisms) and decomposers (bacteria and fungi). The solid arrows represent links by which materials and energy are transferred among organisms. The open arrows depict energy lost to the atmosphere (usually in the form of heat).

that many millions more once existed and became extinct. *The strength of evolutionary theory is that, to date, it offers the most logical scientific explanation for the apparent contradiction inherent in life: its unity <u>and</u> diversity.*

Perhaps you plan on pursuing a career in some field other than biology, or in some specialized branch of biology such as medicine, forest ecology, microbial genetics, or plant pathology. No matter what your chosen specialization, the concepts of energy flow and evolution can be applied to yield insights of enduring value. No matter what the direction of inquiry, this knowledge can yield perspective on *life*—its characteristics in time and space, its past history, its adaptive potential. It will also remind you again and again of your own role in a story of magnificent dimensions.

Review Questions

1. Why is it difficult to give a simple definition of life?

2. What is meant by "adaptive"? Give some examples of environmental conditions to which plants and animals must be adapted in order to stay alive.

3. If the structure and organization of all things arise from the basic nature of matter and energy, then what is the essential difference between living and nonliving things?

4. Study Figure 1.3. Then, on your own, arrange and define the levels of organization in nature. What key concept ties this organization to the flow of life, from the time of origin to the present?

5. What is metabolic activity? List some sources of energy that keep your own metabolic machinery operating.

6. What is homeostatic control? What sorts of environmental changes call for adjustments inside your body?

7. What aspect of life is being overlooked when you talk about "the animal" called a frog? (Hint: What's a tadpole?)

8. What is DNA? What is a mutation? Why are most mutations likely to be harmful?

9. Why do you suppose organisms have become more and more interdependent since the origin of life?

10. Explain what is meant by "the unity and diversity of life." What theory is used in biology to reconcile the seeming paradox in this expression?

2

METHODS AND MIND SETS: A STUDY OF EVOLUTIONARY THOUGHT

Figure 2.1 Galápagos tortoises, one of many diverse species that Charles Darwin encountered on his first visit to the Galápagos Islands. (The Bettmann Archive Inc.)

Evolution, in essence, means change—progressive change through time. In biology, it is said to encompass an ongoing series of environmentally tested changes in the form and function of living things. The process is said to have begun with the origin of life, some 3.5 billion years ago. The idea that evolution has been occurring through this nearly incomprehensible span of time is a fairly recent one. No one saw it happen, obviously. For that reason alone, one might just as well say that life has always existed in the forms we see today; that humans and goatfish, Cape buffalo and giraffe have always been precisely the same since the time of creation. That is another idea, one that has been around for centuries. Yet the first is accepted as "scientific," the other as an article of faith, hence "nonscientific."

Why is this so? Both ideas have roots in human abilities basic to all of us. Every human has the potential for making observations about some aspect of the world, then summarizing those observations in a broad statement, or generalization. For instance, you might observe that the sun rose today, just as it did yesterday, and just as it did all yesterdays past. You might therefore conclude that "The sun always rises." This process of generalizing from specific observations is called **inductive reasoning**. Every human also has the potential for **deductive reasoning**—starting from a generalization assumed to be true, and reasoning from it to arrive at a specific conclusion. This is sometimes called the "if-then" process. For instance, you might deduce that "*if* the sun always rises, *then* it will rise tomorrow." The next morning, when the sight of the sun rising reinforces your conclusion, you may decide that you have hit upon a basic fact of nature. And you may let it go at that. But this is where ideas of science depart from ideas based on casual observation. For science goes on to *test* ideas. They are tested under as many related conditions as might help justify or disprove both the starting assumptions and the conclusions reached. *Will* the sun rise tomorrow? What of the Eskimo sledding across the frozen wastes during the time of near-perpetual darkness in the long arctic winter? What of the astronaut rapidly orbiting the earth or standing on the moon? Whether observations justify the idea that the sun will rise tomorrow depends on where one is standing—and on how one defines "tomorrow."

ON THE SCIENTIFIC METHOD

In the broadest sense, the **scientific method** of approaching questions such as "Does the sun always rise?" or "Does life evolve?" encompasses these steps:

1. Review all available observations. Be sure to note the range of conditions under which they have been made.

2. Use trained judgment in selecting and summarizing the relevant observations out of what could be nearly infinite observational trivia.

3. Work out a tentative explanation that seems in line with those observations. A tentative explanation is sometimes called a **hypothesis** (and sometimes an "educated guess").

4. Devise ways to test whether the explanation is valid. Think through how different but related conditions might affect the test outcome. Be sure the test you devise will address these so-called **variables**.

5. Carry out the tests. Repeat them as often as necessary to find out whether results consistently will be as predicted.

6. Report objectively on the tests and on the conclusions drawn from them.

The scientific method is a commitment not only to systematic observation but to systematic test. Of course, there is not just one scientific way of testing hypotheses. Even a brief look at scientific reports will yield descriptions of many ingenious and unique experiments. In all cases, though, the goal is to design the test so that effects of all variables (factors that might influence the results) can be accounted for.

For instance, perhaps you know that starlings, like other migratory birds, spend part of the year in one region and part in another some distance away. How do they know where to go? How do they find their way back? You might hypothesize that these birds use the sun's position in the sky as a reference point (a navigational cue) during their long journeys. If this is true, then their migratory flight should correspond not only to the sun's actual position but to what *appears* to be the sun's position. To test this prediction, you will have to "change" the sun's position under controlled experimental conditions. You also will have to eliminate variables, such as visible landmarks and prevailing wind direction, that might play a role in navigation. Say that you construct two large, screened cages. One cage is covered over, except for small openings through which mirrors deflect all incoming sunlight rays. You adjust the mirrors so that they all send rays ninety degrees from the normal migratory direction. Birds in this cage are an **experimental group**; test results of their behavior can be compared against your predictions.

The other cage blocks out winds and views of landmarks—but not views of the sun. Thus birds in the second cage *could* orient themselves by the sun's position, if that is how they do it. They are a **control group**, a basis of comparison against which differences that show up in the experimental group may be measured.

As you will read in a later chapter, the biologist Gustav Kramer devised such an experiment. As in most scientific work, the information gathered was expressed *quantitatively*: in precise numerical terms or measurements. It wouldn't be informative, for example, to report that "some" or "most" starlings oriented themselves in a "different" direction. It would be more informative to say something like this: "Of forty starlings in the experimental group, thirty oriented themselves in a direction displaced ninety degrees from normal. This orientation corresponds to the seeming ninety-degree displacement in the sun's position." Furthermore, it would not be very revealing to base the experiment on the behavior of only a few birds. In most scientific tests, enough organisms, specimens, or events are analyzed to avoid *sampling error*; in other words, to avoid basing conclusions on results that may be influenced by chance, in ways that would never show up under normal conditions.

It follows, from the above, that observations and tests must be made under well-defined conditions that may be reported to and repeated by any proficient person at any time. **Repeatability of experiments** is a cornerstone of science. The same test procedures and test conditions should yield the *same* results, regardless of whether people conducting the test live in Albuquerque or Kuala Lumpur or Anchorage; a valid scientific test transcends regional boundaries. That is one reason why science uses a shared "language"—a common set of terms for measuring and describing all aspects of experience. The metric system, for example, is a standard of measurement used by scientists in every country in the world. With such systems, different workers are able to use the same set of internally consistent terms for testing or challenging the accuracy, repeatability, and proper description of reported results.

Does such meticulous testing and reliance on precision *always* lead to a neat package of truths, complete and perfect and demanding to be believed? Not at all. There are no absolute truths in science. There are only high probabilities that an idea is correct *within the framework of the observations and tests from which it is derived*. Instead of absolutes, there is the **suspended judgment**. This means a hypothesis is tentatively said to be valid in that it is consistent with observations at hand. You won't (or shouldn't) hear a scientist say, "There is no other explanation!" More likely you will hear, "Based on present knowledge, this is our best judgment at the moment."

Now, sometimes the weight of evidence in favor of a hypothesis is so impressive that the explanation is elevated to the status of **theory**: a coherent set of ideas that form a general frame of reference for further studies in the field of inquiry. Sometimes the evidence seems so overwhelming that the explanation is considered a **principle**: a fundamental doctrine on which other concepts are based or from which they are drawn. Even so, ongoing observations and test results may not fit with the theory, even with the principle. New evidence may call for its replacement or modification. Far from being a disaster, such evidence stimulates the development of even more general, more adequate, yet always *revisable* explanations.

Obviously, individual scientists would rather come up with useful explanations than useless ones. But they must always ask: "Is my explanation consistent with all existing observations and tests of what I hope to explain?" *It is the external world, not internal conviction, that must form the testing ground for scientific beliefs.* Knowing this, scientists must keep reminding themselves to be objective. Of course, this doesn't mean all scientists are objective all of the time or even most of the time; no one can lay claim to that. It means only that scientists are expected as individuals to forsake pride and prejudice by testing their own beliefs, even in ways that might prove them wrong. Even if an individual scientist doesn't, or won't, *others will*—for science proceeds as a community that is both cooperative and competitive. There is a sharing of ideas, with the understanding that it is just as important to expose errors as it is to applaud insights.

Such is the nature of formulating and testing ideas in the community of science. But this is not to say that there is only one way to do scientific research. Science remains a *creative* process within this rigorous framework. Insights result from accident, from sudden intuition, or from methodical research. Some individuals adhere to standardized procedures, others may improvise as they go. Some tailor their work to reinforce an existing viewpoint, others deliberately take approaches that are likely to challenge prevailing views. It is simply that, no matter what the individual approach, the common element in all science is the process of testing existing knowledge, *with the understanding that knowledge is an open system.*

Systematic observations, hypotheses, predictions, relentless tests—in all these ways, scientific beliefs differ from systems of belief that are based on faith, force, authority, or simple consensus. It is not any "law" that forms the underpinnings on which science rests. Rather, it is the countless observations that the "law" attempts to explain. A "law" may be invalidated by new evidence, but the world remains, awaiting new and better tests of its meaning.

There are, in the history of science, a few individuals who challenged the long-standing beliefs held not only by the culture at large but by the scientific community within it. In biology, Charles Darwin and Alfred Wallace are among them. It will be useful to trace their story and its antecedents. Doing so will give insight into why the principle of evolution is considered to be one of the most powerful concepts of our time. Tracing their story will also show that the similarity between their separate journeys was not so much specific training as it was an underlying attitude. Both were willing to observe, to gather evidence, and to test their ideas—no matter how unsettling the outcome might be—with the reasoning that is the hallmark of the human species *and the discipline that is the hallmark of science.*

EMERGENCE OF EVOLUTIONARY THOUGHT

More than two thousand years ago, the seeds of biological inquiry were taking hold among the ancient Greeks. This was a time when popular belief held that supernatural beings intervened directly in human affairs. The gods, for example, were said to cause a common ailment known as the sacred disease. Yet from a physician of the school of Hippocrates, these thoughts come down to us:

It seems to me that the disease called sacred . . . has a natural cause, just as other diseases have. Men think it divine merely because they do not understand it. But if they called everything divine that they did not understand, there would be no end of divine things! . . . If you watch these fellows treating the disease, you see them use all kinds of incantations and magic—but they are also very careful in regulating diet. Now if food makes the disease better or worse, how can they say it is the gods who do this? . . . It does not really matter whether you call such things divine or not. In Nature, all things are alike in this, in that they can be traced to preceding causes. —On the Sacred Disease (400 B.C.)

Such was the spirit of the times; such was the commitment to finding natural explanations for observable events. And into this intellectual climate Aristotle was born.

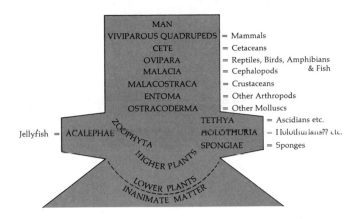

Figure 2.2 Scala Naturae—Aristotle's "ladder of life," the prototype of modern classification schemes. (From Singer, *A History of Biology,* © 1959, Harper & Row)

Aristotle was a naturalist who loved the world around him, and who described it in excellent detail. He had no reference books or instruments to guide him in formulating his descriptions, for the foundation of biological science *began* with the great thinkers of this age. Yet here was a man who was no mere collector of random bits of information. *In his descriptions we have evidence of a mind perceiving connections between observations, and constructing theories for explaining the order of things.* When Aristotle began his studies, he believed (as did others) that each kind of living thing was distinct from all others. Later he began to wonder about the bizarre forms that could not be readily classified. In structure or function, they so resembled other forms that their place in nature seemed blurred. Aristotle came to view nature as proceeding ever so gradually from lifeless matter through ever more complex forms of animal life. This view is reflected in his model of biological organization (Figure 2.2), the first such theoretical framework to appear in the history of biology.

By the fourteenth century, this line of thought had become transformed into a rigid view of life. A Great Chain of Being was seen to extend from the lowest forms, to humans, to spiritual beings. Each kind of being, or **species** as they were called, was seen to have a separate, fixed place in the divine order of things. Each was unchanged since the time of creation, a permanent link in the chain. Scholars believed they had only to discover, name, and describe all the links, and the meaning of life would be revealed to them. Contradictory views were not encouraged; scientific inquiry had become channeled into the encyclopedic assembly of facts.

As long as the world of living things meant mostly those forms existing in Europe, the task seemed manageable. With the global explorations of the sixteenth century, however, "the world" of life expanded by quantum leaps. Naturalists were soon overwhelmed by descriptions of thousands upon thousands of plants and animals discovered in Asia, Africa, the Pacific islands, and the New World. Some appeared to be remarkably similar to common European forms. But some were clearly unique to different lands. How could these organisms be classified? The naturalist Thomas Moufet, in attempting to sort through the bewildering array, simply gave up and recorded such gems as this description of grasshoppers and locusts: "Some are green, some black, some blue. Some fly with one pair of wings, others with more; those that have no wings they leap; those that cannot fly or leap they walk; some have longer shanks, some shorter. Some there are that sing, others are silent. . . ." It was not exactly a time of subtle distinctions.

Linnean System of Classification

The first widely accepted method of classification is attributed to Carl von Linné (also known by his latinized name, Linnaeus). This man was a naturalist to the extreme. He sent ill-prepared students around the world to gather specimens of plants and animals for him, and is said to have lost a third of them to the rigors of their expeditions. Although not very commendable as a student advisor, von Linné did go on to develop the **binomial system of nomenclature**. In this system, each organism could be classified by assigning it a Latin name consisting of two parts.

For instance, *Ursus maritimus* is the scientific name for the polar bear. The first name, which is capitalized, refers to the **genus** (plural, genera). Distinct but obviously similar species are grouped in the same genus. For example, other bears are *Ursus arctos,* the Alaskan brown bear, and *Ursus americanus,* the black bear. The second, uncapitalized name is the **species epithet**. The species epithet is never used without the full or abbreviated generic name preceding it, for it can also be the second name of a species found in an entirely different genus. The Atlantic lobster, for instance, is called *Homarus americanus.* (Hence one would not order *americanus* for dinner unless one is willing to take what one gets.)

The binomial system was the heart of a scheme that was thought to mirror the pattern of links in the Great Chain of Being. This classification scheme was based on

Table 2.1 Linnean System of Classification	
Category	Includes:
species	all organisms with distinct features that distinguish them from all other organisms
genus	collection of species that share some features but are distinct from one another in some other features
family	all closely related genera
order	all closely related families
class	all related orders
phylum (or division, in botanical schemes)	all related classes
kingdom	all related phyla; the most inclusive category of all

perceived similarities or differences in physical features (coloration, number of legs, and so forth). It was structured in the manner shown in Table 2.1.

In retrospect, we can say that the Linnean system of classification provided a concise, orderly method by which myriad organisms could be named and classified. It came at a time when ordering was desperately needed. Yet we must also say that the Linnean system reinforced the prevailing view—that species could be nothing but distinctly unique *and unchanging* kinds of organisms. To this day, it still works in subtle ways on our perceptions of the diversity of living things.

Challenges to the Theory of Unchanging Life

By the late eighteenth and early nineteenth centuries, the somewhat passive cataloging of life was giving way to intellectual ferment. Comparisons of the body structures of animals from different species were turning up all sorts of puzzles. For instance, most mammals have two forelimbs and two hindlimbs. When naturalists dissected and compared forelimbs ranging from whale flippers to human arms to bat wings, they found them to be **homologous structures**: they were constructed of the same kinds of materials, and they were constructed according to the same

basic plan (Figure 2.3). Even though such parts had a different function in different species, without doubt they had underlying similarities. What did the similarities mean?

Equally puzzling were **vestigial structures**: body parts that have no apparent role in the functioning of an organism (Figure 2.4). For instance, like humans, snakes have backbones. Unlike humans, they have no limbs. Now, if snakes were created as limbless creatures, why do some have what looks like remnants of a pelvic girdle—the set of bones to which hindlimbs are attached? Humans also have what looks like the remnants of a tail—and what could be the significance of *that?*

Even as these comparative studies were going on, geographers were turning up some curious variations in the world distribution of plants and animals. For instance, marsupials (pouched mammals such as kangaroos) are rare in most places, but they abound in Australia. Cactus plants thrive in North and South American deserts, yet they are nowhere to be seen in Australian and Asian deserts. Now, if all species had been created at the same time in the same place, as most scholars then believed, then how could so many be restricted to one part of the world or another? By the late eighteenth century, the zoologist Georges-Louis Leclerc de Buffon had an idea. Because species moving out from a single center would have been stopped, sooner or later, by mountain barriers or oceans, perhaps there were several "centers of creation." *Perhaps the origin of species was spread out in space.* At the same time, Buffon's work in zoology led him to suggest that organisms were not necessarily formed on an "original perfect plan." *Perhaps species have become modified through time.*

Was there any evidence in support of these two ideas? Buffon was keenly aware of layers of fossils in the earth. To him, the fossils suggested that there may have been a progression of epochs in earth history. Fossils had been known about from antiquity, when they had been considered no more than mysteriously marked stones. Eventually they began to be accepted as remains of once-living things. More than this, fossils of different types were found to be buried in different layers under surface rocks and soil. In older, underlying rock layers, fossils of marine organisms were relatively simple in structure. In layers above them, fossils of similar structure showed more complexity until, in the uppermost (most recently deposited) layers, they closely resembled living marine organisms. Some naturalists interpreted these patterns to be a record of a succession of modifications in life forms. The time required for such changes and the reasons for change were not known—*but the very concept of change was at variance with the concept of the fixity of species.*

Many naturalists now tried to reconcile these new observations of changing fossil patterns with a traditional conceptual framework that did not allow for change. The nineteenth-century anatomist Georges Cuvier had spent twenty-five years comparing fossils with living organisms. He did not deny that there were changes in the fossil record, and that the changes were somehow linked to earth history. But his explanation came out as a theory of **catastrophism**: There was only one time of creation, said Cuvier, which had populated the world with all species. Many had been destroyed in a global catastrophe. The few survivors repopulated the world. It was not that they were *new* species. Naturalists simply hadn't got around to discovering earlier fossils of them, fossils that *would* date to the time of creation. Another catastrophe wiped out more species and led to repopulation by the survivors, and so on until the most recent catastrophe—followed by the penultimate survivor, Man. However, investigations never turned up the kinds of fossils needed to support Cuvier's explanation. Rather they have turned up overwhelming fossil evidence against it (Chapter Twenty-One).

Lamarck's Theory of Evolution

One of Cuvier's contemporaries, Jean-Baptiste Lamarck, viewed the fossil record differently. Lamarck believed that life had been created in the past in a simple state. He believed further that it gradually improved and changed into the level of organization we see today. The force for change was a built-in drive for perfection, up the Great Chain of Being. For instance, Lamarck speculated that the ancestor of the modern giraffe was a short-necked animal. Pressed by the need to find food, this animal stretched its neck to browse on leaves that were beyond the reach of other animals. Stretching made the neck permanently longer. The slightly stretched neck was bestowed on offspring, which stretched their necks, also. Thus generations of animals desiring to reach higher leaves led to the modern giraffe. Conversely, a vestigial structure was an organ no longer being exercised enough. It was withering away from disuse, and the withered form was somehow passed on to offspring. Such was the Lamarckian theory of **inheritance of acquired characteristics**—the notion that changes acquired during an organism's life are brought about by environmental pressures and internal "desires," and that the changes are transmitted to offspring.

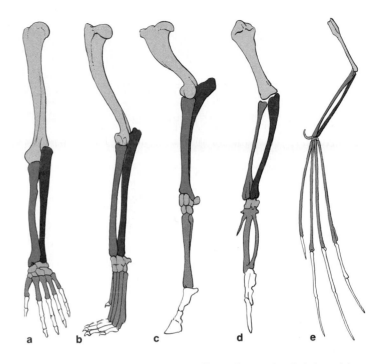

Figure 2.3 Homologous structures. Shown here, a forelimb from (**a**) a human, (**b**) a dog, (**c**) a horse, (**d**) a bird, and (**e**) a bat. The drawings are not to scale relative to one another. The homologous structures are shaded the same way from one animal to the next.

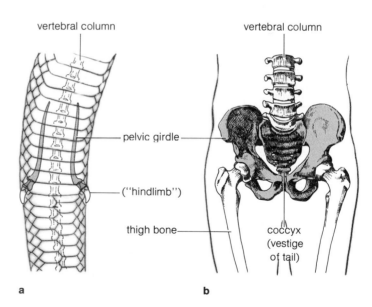

Figure 2.4 Vestigial structures. (**a**) Vestiges of a pelvic girdle in a python. The small hindlimbs protrude through the skin on the underside (ventral surface). (**b**) Pelvic girdle in humans. Also shown is the coccyx, a series of small bones at the end of the vertebral column that are considered to be the remnants of a tail.

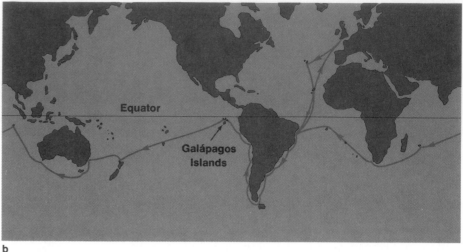

Figure 2.5 (**a**) The five-year voyage of the *Beagle*, shown as it appeared in the Straits of Magellan. (**b**) The Galápagos Islands, about 1,000 kilometers (600 miles) off the coast of Ecuador, support a number of unique plants and animals. The diversity of life on these isolated islands profoundly influenced Darwin's thinking.

Lamarck's contemporaries considered it a wretched piece of science, largely because Lamarck habitually made sweeping assertions but saw no need to support them with observations and tests. In retrospect, perhaps we can find kinder words for the man. His work in zoology was respected. And he did indeed put together a foundation for an evolutionary theory: *All species are interrelated, they gradually change through time, and the environment is a factor in that change.* It was his misfortune that he made some crucial observations but put them together in an explanation that was neither convincing nor, in times to follow, supportable by tests.

THE PRINCIPLE OF EVOLUTION BY NATURAL SELECTION

Naturalist Inclinations of the Young Darwin

Charles Darwin was destined to develop an evolutionary theory that would have repercussions through the whole of Western civilization. Surely his early environment influenced that destiny. His grandfather, a physician and naturalist, was one of the first to propose that all organisms are related by descent. Darwin's family was wealthy, which meant he had the means for indulging his interests. When Darwin was eight years old, he was an enthusiastic but haphazard shell collector. At ten, he was focusing on the habits of insects and birds. At fifteen his schoolwork seemed not nearly as important as the pursuit of solitude, hunting, fishing, and observing the natural world.

For a while Darwin followed his own inclinations toward natural history. Then his father suggested that a career as a clergyman might be more to his liking (and more "respectable" than natural history). So Darwin packed for Cambridge. His grades were good enough to earn him a degree. But most of his time he spent happily with companions of another sort: faculty members with leanings toward natural history. It was the botanist John Henslow who perceived and respected Darwin's real ambitions. It was Henslow who guided him in such ways as arranging for him to take part in a training expedition led by an eminent geologist. It was Henslow who, at the pivotal moment when Darwin had to decide once and for all on a career, arranged that he be offered the position of ship's naturalist aboard H.M.S. *Beagle* (Figure 2.5).

Voyage of the Beagle

The *Beagle* was about to sail to South America to complete earlier work on mapping the coastline. The prolonged stops at islands, mountainous regions, and along rivers would present many opportunities to study diverse forms of life. Almost from the start of the voyage, the young man who had hated work suddenly began working enthusiastically,

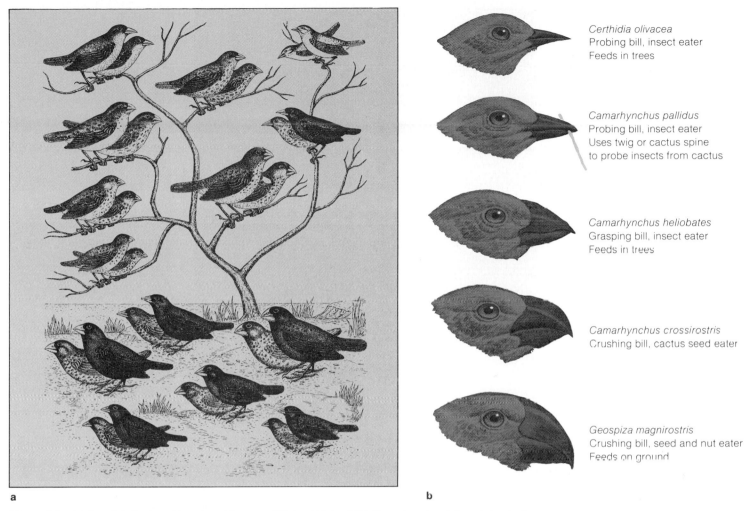

Figure 2.6 (**a**) Darwin's finches. (American Museum of Natural History) (**b**) Examples of variations in beak shape, as correlated with feeding habits.

Certhidia olivacea
Probing bill, insect eater
Feeds in trees

Camarhynchus pallidus
Probing bill, insect eater
Uses twig or cactus spine
to probe insects from cactus

Camarhynchus heliobates
Grasping bill, insect eater
Feeds in trees

Camarhynchus crossirostris
Crushing bill, cactus seed eater

Geospiza magnirostris
Crushing bill, seed and nut eater
Feeds on ground

despite lack of adequate training. Throughout the journey to South America, he collected and examined marine life. And he read—particularly Henslow's parting gift, the first volume of Charles Lyell's *Principles of Geology*.

Amplifying earlier ideas of the geologist James Hutton, Lyell argued that processes now molding the earth's surface—volcanic activity, the slow erosive action of wind and water, the gradual uplifting of mountain ranges—had also been at work in the past. This concept is called **uniformitarianism**. It convincingly extended earth history back in time, for known geologic processes would require not a few thousand years but millions upon millions of years to reshape the landscape—*time enough, then, for species to evolve.*

On the Galápagos Islands (Figure 2.5), Darwin began to see the implications of this concept. Astonishingly diverse organisms lived on that isolated cluster of volcanic islands. Darwin found the array of finches intriguing. The birds of these species had different kinds of beaks, which seemed related to different foods in the environment (Figure 2.6). In addition, even though each species was confined to a single island or cluster of islands, *all* were similar to a single finch species that Darwin had observed back on the South American mainland. Later it would occur to him that the finches might have descended from a single ancestral species, which winds might have carried over to the Galápagos. If there had been time enough for various

Figure 2.7 A few examples of the more than 300 varieties of domesticated pigeons. Such forms are thought to have been derived, by selective breeding, from the wild rock dove (**a**).

Cecie Starr / Fred Maenpa

descendants of the finch species to change—and Lyell's concept strongly suggested that there had been—then it would have profound implications for the whole question of species diversity!

Darwin and Wallace: The Theory Takes Form

The *Beagle* returned to England in 1836, after nearly five years at sea. In the years to follow, Darwin's writings established him as a respected figure in natural history. However, his consuming interest was the "species problem." *By what process do species evolve?* In searching for an answer, he patiently assembled his own data. Then he systematically analyzed the clues and difficulties they presented. *Whenever he developed an idea for explaining certain observations, he spent as much time trying to disprove it as he spent trying to support it.* Darwin desired only "to understand or explain whatever I observed, to group all facts under some general laws. . . . I have steadily endeavored to keep

my mind free so as to give up any hypothesis, however much beloved . . . as soon as facts are shown to be opposed to it." The outcome would be impressively objective evidence for his concept of evolution.

Given a problem so complex, Darwin turned to a simpler question. Among domesticated animals, how do various **breeds** (varieties within a species) originate? Domesticated pigeons proved illuminating. He first determined that none of these flamboyant varieties was found in the wild, except for some that recently had escaped from captivity. Then he noted that one pigeon species, the wild rock dove, shares certain features with all these breeds (Figure 2.7). Darwin concluded that the wild rock dove may have been their common ancestor.

But how did humans mold the rock dove into diverse breeds? The process, Darwin perceived, was one of *selection*. Rarely are any two animals of the same species exactly alike. In any generation, there is variation in such features as size, form, and coloration. Sometimes the variations are slight; now and then they are dramatic. In either case, plant or animal breeders select the features they consider most desirable. For instance, instead of randomly breeding dairy cattle, farmers choose bulls that have sired excellent milk-producing daughters. Then they breed the bulls with cows that also have a high milk production rate. In this manner, selective breeding encourages the perpetuation of some traits more than others.

Darwin reasoned that some sort of selection process must also be at work in the natural world. He gained insight into what that process might be when he read *Essay on the Principle of Population*, by Thomas Malthus. In his essay, Malthus had pointed out that populations tend to increase in ever greater numbers but food supplies do not. Thus, in the absence of controls, any population will outrun its resources. When that happens, individuals in the population must compete for available food. Gradually the meaning of Darwin's own observations came into focus. There was indeed a struggle for existence in the natural world, in which better adapted organisms had a competitive edge. It was this struggle that caused progressive adaptations to accumulate, hence that gradually brought about changes in species. Darwin's view of how such adaptations came into being is known as the **theory of natural selection**. The theory is expressed here in modern form:

Figure 2.8   Charles Darwin, at about the time he accepted the position of ship's naturalist aboard H.M.S. *Beagle*.

Royal College of Surgeons England

---

1. Members of a species vary in form and behavior. Some of this variation is inherited.

2. Each species produces more offspring than can survive long enough to reproduce.

3. Among offspring, some inherited traits improve chances for surviving and reproducing under prevailing environmental conditions. Hence, bearers of these traits tend to produce more descendants than other members of the species.

4. This tendency, called **differential reproduction**, means that the most adaptive traits will show up among more individuals of the next generation.

5. Thus, in **natural selection**, heritable traits that are most adaptive in a given environment appear increasingly among individuals of a species, for their bearers contribute proportionally more offspring to the next generation.

According to this view, natural selection doesn't *create* a new type of individual. Rather, selective agents in a given environment are at work on *existing* individuals, each with its own combination of traits. Let's go one step further

now, beyond the idea of how natural selection might bring about gradual change within a species. Let's consider one way in which it might bring about the **evolution of new species**—say, among sexually reproducing animals:

---

1. For one reason or another, a group of individuals sometimes gets separated from others of the same species. When that has happened, the group is a separate **population**.

2. In some large or small way, the environment for the separate population dictates unique physical or behavioral conditions for survival. Thus *different* physical or behavioral traits may be more adaptive here.

3. An isolated population gradually builds up its own pool of adaptive traits. An accumulation of traits that are adaptive to a different environment may lead to a separate line of descent.

4. Over time, the accumulated differences in form and behavior may be so great that members of the separated populations no longer can interbreed, even if they do get together again. They are recognized as separate **species**.

Figure 2.9 Alfred Wallace. Although Darwin and Wallace had worked independently, they both arrived at the same concept of natural selection. Darwin tried to insist that Wallace be credited as originator of the theory, being the first to circulate a report of his work. Wallace refused; he would not ignore the decades of work Darwin had invested in accumulating supporting evidence.

American Museum of Natural History

Identifying processes by which evolution of species may occur was a profound event in the history of biology. But now the theory had to be put to the test. For it is one thing to propose an idea; it is an entirely different thing to demonstrate its validity. *And how could convincing proof be given of the action of selective agents, without knowing exactly what they act upon—in other words, the physical nature of the heritable traits themselves?* The answer to that question would not be Darwin's to give. It would not be until the rise of genetics, decades later, that the physical basis for inheritance would begin to be understood.

Darwin continued his research, gathering notes and sifting his evidence for flaws in his reasoning. Then, in 1858, his careful search was interrupted. He received a paper from Wallace outlining the very theory that he had been developing for two decades! Like Darwin, Wallace was a respected naturalist. He had thirteen years of research

in South America and the Malay Archipelago to his credit. Like Darwin, he had been impressed with the writings of Malthus. But whereas Darwin had been working to document his ideas for more than twenty years, the concept of evolution by natural selection flashed into Wallace's mind and he wrote out his ideas in two days. This was the paper he sent to Darwin.

Despite the shock of seeing his theory presented by someone else, Darwin distributed Wallace's paper at once to his colleagues, suggesting that it be published (Figure 2.9). But Darwin's colleagues would not let him set aside the years that had gone into his own development of the theory. They prevailed on him to gather his notes into a paper that could be presented simultaneously with Wallace's. In 1858, both papers were presented to the Linnean Society, along with a letter Darwin had written several years earlier outlining the main points of this theory. The next year, Darwin published his book *On the Origin of Species by Means of Natural Selection.*

Many versions of the Darwin–Wallace story emphasize the controversy the book created in some quarters. However, Darwin's evidence was so overwhelming that the argument for evolving life was accepted almost at once by most naturalists and scholars from other disciplines. For as the naturalist Thomas Huxley commented, *the only rebuttal to the concept of evolution is a better explanation of the evidence—something which has yet to appear.*

Ironically, even though the idea of evolution had at last gained respectability, almost seventy years would pass before most of the scientific community would agree with Darwin and Wallace's remarkable insight—that natural selection is a means by which evolution occurs. Not before then would it become clear that their explanation holds up under many different tests, on many different levels of biological organization. Not before then would their explanation come to be regarded as one of the most basic principles of all. In the meantime, their names would be associated mostly with the concept that life evolves—something that others had proposed before them.

PHYLOGENY: CLASSIFICATION BASED ON EVOLUTIONARY RELATIONSHIPS

The theory of evolution by natural selection has profoundly changed our perception of the living world. All living things are now seen to be related by evolutionary threads extending back to the very dawn of life. Their many and diverse relationships are not random. They form a pattern of adaptive strategies that speaks of reasons why life has been able to inhabit almost every part of this planet.

Currently, classification schemes are based on observed similarities and differences among existing organisms which suggest similar ancestral lines of descent. These schemes attempt to reflect the flow of life from common beginnings, from unspecialized forms to the increased specialization in form, function, and behavior that we see today. A genus is now said to include only those species related by descent from a fairly recent, common ancestral form. A family includes all genera related by descent from a more remote common ancestor, and so on up to the highest (most inclusive) levels of classification: phylum and kingdom. A scheme that takes into account the evolution of major lines of descent is known as a natural system, or **phylogenetic system of classification**.

Constructing such systems is not easy. Throughout the past, environmental pressures have not been the same from group to group. Environments have changed. And the rates of evolution apparently have varied from one group to the next. In addition, the picture of interrelationships is not yet complete. Details must come from the fossil record, comparative anatomy, genetics, biochemistry, reproductive biology, behavior and ecology, even geology and geography. Some information, such as parts of the fossil record, is lost forever because of ancient geologic upheavals. Even so, now that we have a better idea of where to look, and of what we are looking for, the gaps are filling in fast.

Regardless of its strengths, no classification system should be viewed as the final word in our understanding. As long as there are observations still to be made, different people will interpret relationships among organisms differently. Some group all organisms into two kingdoms (plants and animals). Others group them into as many as twenty. In this book we use Robert Whittaker's five-kingdom model. (This scheme is sketched in Table 2.2 and in Figure 2.10.) Like the others, it helps summarize *current* knowledge about life's evolution. It, too, is subject to modification as new evidence turns up.

In this model, the earliest life forms have echoes in the kingdom **Monera**. The oldest known fossils suggest that the first forms of life were much like one-celled bacteria and blue-green algae in this category. Most existing members of the kingdom **Protista** are also single-celled. In some ways they are like monerans. But in other ways, some are like plants, others like animals, some like plants *and* animals, and still others like fungi! Protistans may be

Table 2.2	Classification Scheme Used in This Book*
Kingdom Monera	Single-celled. Includes autotrophs (depend on food self-assembled from simple raw materials such as carbon compounds and water) and heterotrophs (depend on tissues, remains, or by-products of other living things). Cells of monerans have no true nucleus or other membrane-enclosed organelles; they are prokaryotic. *Representatives: bacteria, blue-green algae*
Kingdom Protista	Single-celled. Heterotrophs and some autotrophs. Cells of protistans have a true nucleus and other membrane-enclosed organelles; they are eukaryotic. *Representatives include golden algae, diatoms, amoebas, sporozoans, ciliates*
Kingdom Fungi	Multicelled. Heterotrophs, most rely on digestion outside the fungal body, then absorption (they first secrete certain substances that break down food, then breakdown products are absorbed across fungal body wall). All eukaryotic. *Representatives include slime molds, true fungi*
Kingdom Plantae	Multicelled. Autotrophs; with few exceptions, plants build all of their own food through photosynthesis. All eukaryotic. *Representatives include red algae, brown algae, green algae, mosses, horsetails, lycopods, ferns, seed plants (such as cycads, ginkgoes, conifers, flowering plants)*
Kingdom Animalia	Multicelled. Heterotrophs of varied sorts. All eukaryotic. *Representatives include sponges, jellyfish, flatworms, roundworms, segmented worms, mollusks, arthropods (such as insects and lobsters), echinoderms, chordates (fishes, amphibians, reptiles, birds, mammals)*

*This scheme is expanded upon in Unit Six and Appendix I.

living examples of the kinds of organisms that must have existed at a major evolutionary crossroad in the past. At that point, one-celled forms began evolving in ways that gave rise to the multicelled forms in three separate kingdoms: **Plantae**, **Fungi**, and **Animalia**. Multicelled organisms are grouped into these three kingdoms on the basis of their energy-acquiring strategies. Plants assemble their own energy-rich food molecules. Fungi absorb particles that they have digested outside the fungal body. And animals ingest other organisms for food.

But the separate "pools" of life in Figure 2.10 are not to be viewed as one flowing out of the other. *Representatives of all five kingdoms are alive today, side by side in time.* The branching routes simply suggest how they might have arrived at where they are now. This is the evolutionary pattern of descent as many biologists now see it, as Darwin and Wallace might have anticipated it.

Figure 2.10 Simplified diagram of the five-kingdom model for classifying life forms. Only a few representative kinds of organisms are used to show the present scope of diversity. The general pattern suggests possible routes that may have led from the origin of life to this profusion of forms. Representatives of all kingdoms are still alive today.

flowering
plants

PERSPECTIVE

When Darwin disembarked from the *Beagle* with his observations and thoughts on life's diversity, he set in motion a chain of events that made the study of life simpler—and, at the same time, more complex. On the one hand, the concept of evolving life provides a clear intellectual path through the seeming maze of different kinds of organisms. On the other hand, even though species are no longer generally regarded as unchanging, they are in fact still widely regarded as distinct, nonoverlapping clusters of things. But the living world is not so tidy as the word "species" would suggest. Attempting to pigeonhole all organisms in some hierarchy of distinct clusters is a little like trying to build fences in order to divide up the expanse of nature. With observation and with educated guesses, we can make the fences follow the terrain, more or less. But it is important to keep in mind that *these boundaries are often imposed, artificial products of our incomplete knowledge.* In this sense, species names are more like convenient fenceposts scattered along a continuum of life down through the generations. They are useful, but only in the sense of providing perspective while the search goes on for more adequate explanations of the shifts in kinds and numbers of living things that we call evolution. We will be returning to some of the more recent investigations and explanations in a later unit of this book.

Readings

Darwin, C. 1957. *Voyage of the Beagle.* New York: Dutton. In his own words, what Darwin saw and thought about during his first global voyage.

Futuyma, D. 1979. *Evolutionary Biology.* Sunderland, Massachusetts: Sinhauer Associates. Well-written synthesis of modern evolutionary thought. Includes the clearest synopsis of evolutionary theory yet developed.

Mayr, E. 1976. *Evolution and the Diversity of Life.* Cambridge, Massachusetts: Belknap Press. Insights into a mind searching for ways to untangle the knot of biological diversity.

Mayr, E. 1978. "Evolution." *Scientific American* 293(3):46–55. Entire issue devoted to evolutionary theory, from chemical evolution to human behavioral evolution.

Moorehead, A. 1969. *Darwin and the Beagle.* New York: Harper and Row. Well-illustrated account of the places Darwin visited, what he observed, and the home to which he returned.

Singer, C. 1962. *A History of Biology to About the Year 1900.* New York: Abelard-Schuman (Harper and Row). Contains absorbing portrayals of the men and women who led the way in developing basic biological concepts.

Review Questions

1. Define inductive and deductive reasoning.

2. List the general features of the scientific method. How do beliefs derived from a scientific approach differ from beliefs based on faith, force, authority, or simple consensus?

3. These terms are important in scientific testing: control group, experimental group, variable, and sampling error. Can you define them?

4. How does a hypothesis differ from a theory? From a principle?

5. List and define the main categories in the Linnean system of classification. How do you suppose this system influences our perceptions of the diversity of living things?

6. What were the origins of the idea that species never change? How was this idea challenged by comparative anatomy studies and by studies of world distributions of plants and animals?

7. Uniformitarianism was a conceptual stepping stone to the idea that species can evolve. Can you say why?

8. State the key points of the theory of natural selection. What is meant by differential reproduction?

9. Define population. Explain how a population of one species may give rise to a new species.

10. What is a phylogenetic system of classification?

11. According to the Whittaker scheme, all organisms are grouped into five kingdoms. What is the main energy-acquiring strategy characteristic of each kingdom?

UNIT TWO

THE CELLULAR BASIS OF LIFE

3

ENERGY, ATOMS, AND CELL SUBSTANCES

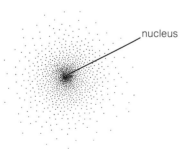

Figure 3.1 Model of a key building block for the substances of life—hydrogen, with a lone electron zipping about the nucleus. (The greater the density of dots, the more likely the electron would be there at any instant.)

In viewing the diversity of life, sooner or later we arrive at the question of how it came about. If diversity results largely from selection of traits that are adaptive to prevailing environments, what could have been the nature of those first environmental proddings? We can gain insight into the question by observing organisms alive today. Although they vary in nearly every way imaginable, they all have something in common. All depend on **energy** for staying alive. Energy isn't something you can see directly. And it isn't all that easy to define. A roundabout definition is that "energy is the capacity to do work." Food, for example, contains energy that can be put to use in driving life processes, such as breathing and running, growing and reproducing. As another example, sunlight contains a form of energy that plants harness and use in building their own food. At a more fundamental level, the energy of certain molecules found in every organism helps maintain the body and keep it alive. Some dormant seeds may even survive and then grow after hundreds of years, as long as the energy relationships between their molecules are not disturbed!

So energy is necessary for doing all the things that go into being alive. However, *energy is not available all the time to all organisms.* First, *sources* of energy change from time to time. A bacterium living in your gut may be able to use the energy of sugar molecules found in the milk that you drink—but what happens if you stop drinking milk? If the bacterium can't use other energy sources, that is the end of the bacterium. Second, *supplies* of energy normally are limited. The more a given population grows, the less food there is to go around. Eventually, members of the population have to compete for what remains. The ones better adapted for getting a share of it tend to survive and reproduce more than the others. Here, an adaptive trait might be broader leaves that capture more of the sunlight energy falling on plants crowded together in a meadow. Populations of these plants found in other places might retain narrow leaves. But here, in this crowded meadow, selection pressures could favor broader leaved variants. Another adaptive trait might be a stronger beak that crushes seeds, which skinny-beaked birds can't do. Skinny-beaked birds might continue specializing on, say, nectar or soft-bodied insects. But variant birds with stronger beaks might start relying on different foods (seeds) that are not being exploited by their kin. Gradually, diversity in leaf size or beak type would be fostered.

Examples of competition within and between populations can also be given for other sorts of things—living space, for instance, and other limited material resources.

But these examples vary considerably from group to group. It is the demand for *energy* that is basic to all of them. Hence we might say this: *Directly or indirectly, the major impetus behind diversification probably was competition for energy.*

Where does this idea take us? If we assume the idea is valid, then it becomes clear that insight into the meaning of life's diversity is going to require some digging into the relationships embodied in the word "energy." This chapter begins with the bits of matter called atoms and simple molecules. It is here that all energy transfers and transformations within and between living things originate. Then the chapter moves on to the character of carbohydrates, lipids, proteins, and nucleic acids—the most important large molecules of life. The discussions will give you an idea of why many of these molecules are the primary building blocks in cell architecture, why some can be readily broken apart so that their inherent energy becomes available for cellular work, and why still others can change shape or be zipped open with precision so that vital reactions can be played out on their surfaces. Later, you will read about cell structures assembled from these molecules, and about energy transformations essential for cell functioning. As you will see, there is unity in structure and function at both the molecular and cellular levels of life. And it is from such common molecular and cellular beginnings that the drama of diversity has surely been spun out.

ENERGY IN ATOMS

All substances in our everyday world are assembled from tiny particles called **atoms**. A pure substance made up of only one kind of atom is an **element**. Ninety-two different elements are known to occur in nature. Fourteen more kinds have been produced artificially under extreme laboratory conditions. All elements are composed of three major kinds of subatomic building blocks, yet each element displays unique properties. Why is this so? We can approach the question through this simple generalization:

Each atom in a given chemical element is unique because of its number of subatomic parts, and because of the arrangement of those parts.

The number and arrangement of subatomic parts dictate how two or more atoms combine to form **molecules**.

They also dictate what the properties of different molecules will be, and how (if at all) different molecules will interact. Thus a brief look at what goes on inside an atom will help explain why substances of life behave at they do.

How Parts of Atoms Are Arranged

Let's begin with the way that parts of an atom are held together. When we want to join two blocks of wood together, we can use some epoxy glue. One of the "glues" holding atoms together and joining them to other atoms is of an entirely different sort: it is called **electric charge**. When something is electrically charged, it has energy that allows it to push away or to attract something else, even without touching it. Electric charges are of two kinds: positive (+) and negative (–). Two identical charges (+ + or – –) tend to repel each other. Two opposite charges (+ –) tend to attract each other.

Every atom has one or more positively charged particles, called **protons**. Every atom (except a form of hydrogen) also has one or more electrically neutral particles, called **neutrons**. Protons and neutrons are both found in the atom's central region—the atomic nucleus. Whirling about this nucleus are one or more negatively charged particles known as **electrons**. No matter which atom you look at, the number of positively charged protons is exactly balanced by the *same* number of negatively charged electrons. In other words, an atom has no *net* charge.

How are these particles arranged in an atom? Being oppositely charged, an electron is attracted to protons in the nucleus. In addition, each electron is repelled by any other electrons present in the atom. Each one moves almost as fast as the speed of light in an **orbital** (a specific region of space around the nucleus, where you will find that electron most of the time). The simplest orbital is like a round cloud (Figure 3.1). It's easy to imagine how one or even two electrons might zip about a nucleus in a spherical orbital. But some atoms have over a hundred electrons! With so many negatively charged particles jockeying for position, their orbitals are shaped like balls, dumbbells, teardrops, even flabby inner tubes. *In these different orbitals, all the electrons get as close as possible to the protons and stay as far away as possible from each other.*

No two elements have the same number of protons (and electrons) in their atoms. That is why the 106 different elements may be assigned a unique **atomic number**, equal to the number of protons in the nucleus. (For instance, a carbon atom has six protons; it has atomic number 6.) Atoms may also be assigned a **mass number**: the total number of protons *and* neutrons in the nucleus. (The most common kind of carbon atom has six protons and six neutrons; its mass number is 12.) Table 3.1 lists the atomic number and mass number for elements found most often in living things, and gives the relative abundance of these elements in the human body.

All atoms of the same element must, by definition, have the same number of protons. But individual atoms of most elements may vary slightly in their number of *neutrons.* This means they can have different mass numbers. For example, a "carbon" atom may be carbon 12 (six protons, six neutrons), carbon 13 (six protons, seven neutrons), or carbon 14 (six protons, eight neutrons). The different forms of an element are called **isotopes**. As you will see, they have important uses in biological research.

One more simple concept may be introduced here. Even though the number of protons remains fixed in all atoms, sometimes the attraction between protons and electrons can be overcome. One or more electrons can be knocked out of the atom or pulled away from it. Sometimes, too, the nucleus may attract *additional* electrons. An atom that loses or gains one or more electrons is known as an **ion**. For example, a sodium (Na) atom has eleven protons and eleven electrons. It may lose an electron, thereby becoming a positively charged sodium ion:

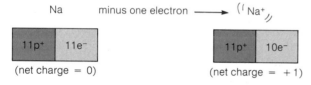

Similarly, a chlorine atom has seventeen protons and seventeen electrons. It may gain an electron, thereby becoming a negatively charged chloride ion:

Unlike the atom, then, which has no *net* charge, an ion has an overall positive or negative charge.

Interactions Between Atoms: Why They Occur

The hydrogen atom is found again and again in the molecules of living things. Its most common form has only one proton and one electron, so it's the simplest kind of atom

Table 3.1	Atomic Number and Mass Number of Elements Commonly Found in Living Things			
Element	Symbol	Atomic Number	Most Common Mass Number	Abundance in Human Body* (% Wet Weight)
Hydrogen	H	1	1	10.0
Carbon	C	6	12	18.0
Nitrogen	N	7	14	3.0
Oxygen	O	8	16	65.0
Sodium	Na	11	23	0.15
Magnesium	Mg	12	24	0.05
Phosphorus	P	15	31	1.1
Sulfur	S	16	32	0.25
Chlorine	Cl	17	35	0.15
Potassium	K	19	39	0.35
Calcium	Ca	20	40	2.0
Iron	Fe	26	56	0.004
Iodine	I	53	127	0.0004

Hydrogen, Carbon, Nitrogen, Oxygen = 96%

*Approximate values.

to think about (Figure 3.1). When left alone, its fast-moving electron is in the most stable, or lowest, energy level. But sometimes the atom is hit with enough incoming light or heat energy to excite the electron to a higher energy level. Another way of saying this is that *an electron can absorb certain amounts of incoming energy.* When it does, the electron spends more time farther away from the proton's pull. If the outside energy source is removed, the excited electron eventually gives off the extra energy it absorbed, and it returns to the lowest energy level. In many of the reactions of life, excited electrons give up some of their extra energy, which is harnessed to power life processes.

Only specific energy levels are available to an electron. An electron may be found in one level, or the next level, or the next, but never in between. It's something like standing on a ladder. The energy available to do work increases as you climb, and it decreases as you descend. You can stand on any of the rungs, and you can move your feet from rung to rung. But you certainly can't stand between rungs. It seems all electrons gain or lose energy like that: in steps of certain sizes. Living things respond to this property. For instance, electrons in special molecules found in some plant cells can be excited to higher energy levels by absorbing specific wavelengths (colors) of sunlight energy. These wavelength energies correspond exactly to specific "steps" on the energy ladder for the electrons.

As Figure 3.2 suggests, there is a limit to how many electrons can be squeezed into each energy level. The first level contains only one orbital, the second contains four, the third contains nine, and so on. *And each orbital can contain no more than two electrons.* When an orbital is filled with two electrons, more orbitals become filled at the next highest energy level.

It happens that electrons occupying the orbitals in the outermost (highest occupied) energy level of an atom are the ones that interact with other atoms to form a molecule. In addition, most molecules form through interactions between outer orbitals that contain only *one* electron. If the highest occupied orbitals are each filled with two electrons, an atom shows little tendency to combine with other atoms. Consider helium and neon (Figure 3.2). Under ordinary conditions, these elements do not react with other atoms at all. Helium has a filled orbital in its first energy level (which is the highest level occupied). For neon, all four orbitals in the second and highest occupied energy level are filled. In contrast, the other elements listed in Figure 3.2 have one or more partially filled orbitals in their highest occupied level. They all tend to interact with other atoms. From the points made so far, we may draw these conclusions:

Elements composed of atoms that have two electrons in each occupied orbital tend to be chemically <u>nonreactive</u> substances.

Elements composed of atoms that have one or more orbitals containing only one electron tend to be chemically <u>reactive</u> substances.

You can see why it's important to know about how electrons are arranged in orbitals. Their orbitals have different specific shapes, which means that atoms don't just combine any which way in forming a molecule. If they did, the results wouldn't be a dependable basis for the organization characteristic of life. Instead, atoms combine only in certain positions, and these positions give rise to specific shapes of molecules. In turn, the shapes of different molecules give rise to the shapes and structures of cells. Molecules and cell structures must be assembled according to specific patterns. If they are not, cells don't function properly—or they don't function at all.

CHEMICAL BONDS IN THE CELLULAR WORLD

In a **chemical bond**, the electron structure of one atom becomes linked with the electron structure of another atom (or atoms). Such links may depend on an atom giving up, gaining, or sharing one or more electrons. They may occur when an atom already bound in a molecule also comes under the influence of neighboring molecules. *Hence a chemical bond is not "an object," it is an energy relationship—and not necessarily a permanent one at that.*

Under the ranges of temperature, pressure, and moisture typically found inside and outside the cell, certain kinds of bonds predominate. Strong (covalent) bonds typically occur *within* molecules. Weaker bonds (such as ionic

| | | Number of Electrons in Each Orbital | | | | | | | | |
| | | First Energy Level (one orbital) | Second Energy Level (four possible orbitals) | | | | Third Energy Level (nine possible orbitals) | | | |
Element	Atomic Number									
Hydrogen	1	1								
Helium	2	2								
Carbon	6	2	2	1	1					
Nitrogen	7	2	2	1	1	1				
Oxygen	8	2	2	2	1	1				
Neon	10	2	2	2	2	2				
Sodium	11	2	2	2	2	2	1			
Chlorine	17	2	2	2	2	2	2	2	2	1

Figure 3.2 Distribution of electrons in atomic orbitals for a few elements.

and hydrogen bonds) often form *between* two or more different molecules. They also may form between different parts of the same molecule. Under cellular conditions, these weaker bonds are more readily broken. Even so, their effects are additive. Thousands of such bonds acting together help dictate the shape of many large biological molecules. They also influence the organization of these large molecules relative to one another in the cell.

In the cellular world, substances of life are held together by strong bonds that involve a sharing of electrons, and by weaker interactions between atoms or ions having opposite charge.

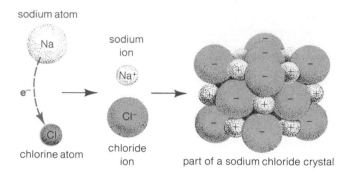

Figure 3.3 Ionic bonding in sodium chloride. Notice the orderly repeating pattern of the oppositely charged ions in the crystal lattice. (For simplicity, each atom and ion is drawn as a space-filling sphere.) .

Covalent Bonding

In a **covalent bond**, one or more electrons is *shared* between atoms or groups of atoms. Electron sharing is almost always done in pairs. For instance, hydrogen atoms (each with a lone electron) rarely exist by themselves. Most often, two are joined as a molecule, which may be written as H_2. In this molecule, each atom's orbital overlaps the other, and both electrons are found around both nuclei. When there is *equal sharing* of a pair of electrons, as there is here, atoms are said to be joined by a *nonpolar* covalent bond. The single covalent bond of the hydrogen molecule may be written as H—H. (This means the same thing as H_2. It just gives a better picture of the molecule's structure: the single line between them shows that the atoms share an electron pair. Such representations, in which lines signify bonds, are called structural formulas.)

Does this mean that *all* covalently bonded atoms share electrons equally? Not at all. Consider what happens in a water molecule, which has one oxygen and two hydrogen atoms joined in this way:

$$\underset{H \qquad H}{\overset{O}{|}}$$

The three atoms share electrons so that their orbitals are filled. But oxygen has more protons in its nucleus than hydrogen does. Hence it has a stronger pull on the electrons. The electrons average more time at the oxygen "end" than they do at the hydrogen "end" of the molecule. So the oxygen atom tends to carry more negative charge than it does positive charge. The hydrogen atoms, being deprived of a full share of electrons, tend to be positively charged. Because there is *unequal sharing* of electrons, the atoms are said to be joined by *polar* covalent bonds.

Ionic Bonding

In ion formation, an atom completely loses or gains one or more electrons. Such an event does not occur in isolation. Another atom of the right kind must be around to donate or accept electrons in the first place. For example, sodium tends readily to give up an electron to chlorine. As Figure 3.2 suggests, sodium has all orbitals filled in its first two energy levels—and one partially filled orbital in the third. Chlorine has seventeen protons—and requires only one more electron to fill the orbitals in its third energy level. What one kind of atom easily gives up, the other easily grabs. Both become ions in the process.

Depending on the environment in which the transfer occurs, an **ionic bond** may link the sodium and chlorine ions because of the positive–negative attraction between them. Thus Na^+ and Cl^- remain in association as NaCl: the basic repeating unit of the substance sodium chloride, otherwise known as table salt. Strictly speaking, there is no such thing as "a molecule" of NaCl, for the attraction is not restricted to a single pair of ions. As Figure 3.3 shows, the attraction can organize many NaCl units into a crystal lattice. In this lattice, oppositely charged ions are in orderly alignment with their immediate neighbors.

Hydrogen Bonding

In **hydrogen bonding**, a hydrogen atom already bonded covalently in a molecule is weakly attracted to an oxygen, nitrogen, or fluorine atom already bonded covalently in the same or another molecule. The attraction depends on the polar nature of unequally shared electrons at covalently

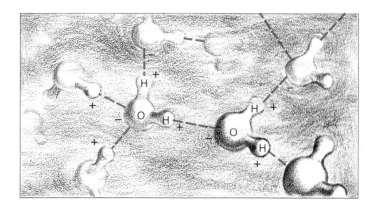

Figure 3.4 Hydrogen bonds between water molecules in liquid water.

bonded regions. For example, a hydrogen atom in one water molecule may form a weak hydrogen bond with the oxygen atom of another water molecule. In such interactions, oppositely charged parts between neighbors tend to orient toward each other, and parts of like charge tend to orient away from each other (Figure 3.4). This structuring of water has profound biological consequences, of the sort that will be described in Chapter Five. For now, the following simple example will help explain the kind of interactions on which water chemistry is based.

Sugar is a crystalline substance that dissolves in water. When dissolved, a substance is said to be in **solution**. For sugar, this means that its molecules have tended to become hydrogen-bonded with the surrounding water molecules rather than remaining bonded to each other. Why do they do this? Projecting from the surface of each molecule in a sugar crystal is a series of oxygen–hydrogen units, or —OH groups. Hydrogen bonds join neighboring sugar molecules at these groups. But the positive–negative interaction between sugar and water is *greater* than the attraction holding together the molecules of either substance. Hence the bonding pattern shifts, and the sugar dissolves.

THE ACCOMMODATING CARBON ATOM

Of all elements found in living things, carbon is the most versatile. The atoms of this element combine with one another and with many other kinds of atoms to form more kinds of substances than any other element. Its importance to life is so great that for more than a century, nearly all carbon-containing molecules have been called organic

("organism-related") to distinguish them from inorganic ("lifeless") molecules. Carbon's versatility can be traced to its electron arrangement and its strong tendency to form covalent bonds. When binding with other atoms, two of its six electrons reside in the lowest energy level. But each of the other four electrons zips about alone in its own teardrop-shaped orbital. In this configuration, other atoms may bond covalently to a carbon atom at all four of its "corners." Carbon bonded with four hydrogen atoms, for instance, forms the simple compound methane (CH_4). Methane is called a **hydrocarbon** because it's made only of hydrogen and carbon atoms:

Notice that even this simple compound has a specific three-dimensional shape. The more complex a molecule becomes, the more complex is its shape—which means it can't be depicted very well on flat paper. For convenience, structural formulas are generally drawn as if the molecules they represent are squashed flat. Thus methane is also depicted as:

$$H-\underset{\underset{H}{|}}{\overset{\overset{H}{|}}{C}}-H$$

The four-cornered pattern of carbon bonding typified by methane is the basis for an astonishing array of molecules, especially when carbon atoms bond covalently to other carbon atoms. Sometimes the basic carbon structure is a linear chain (or skeleton); sometimes it is branched:

$$-\overset{|}{\underset{|}{C}}-\overset{|}{\underset{|}{C}}-\overset{|}{\underset{|}{C}}-\overset{|}{\underset{|}{C}}-\overset{|}{\underset{|}{C}}-\overset{|}{\underset{|}{C}}-\overset{|}{\underset{|}{C}}-\overset{|}{\underset{|}{C}}- \quad \text{or} \quad \overset{\overset{C}{|}}{C}-\overset{|}{\underset{|}{C}}-\overset{|}{\underset{|}{C}}-\overset{|}{\underset{|}{C}}-$$

And sometimes carbon atoms are bonded to each other in a ring structure:

or ⬡ or ⬡

Notice that a ring structure may be shown flat on a page or as a tilted, stripped-down version of the same thing. When nothing else is shown on a ring, it's understood that a carbon atom occurs at every corner, with enough hydrogen atoms attached to give each carbon atom four covalent bonds.

Other possibilities include the formation of double and even triple covalent bonds between some of the carbon atoms (C=C or C≡C), as well as the attachment of other elements. In the molecules of life, carbon typically forms covalent bonds not only with itself but with atoms or atomic groups containing hydrogen, oxygen, nitrogen, sulfur, chlorine, and phosphorus (Figure 3.5).

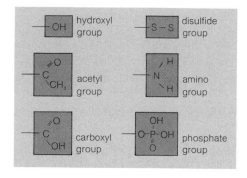

Figure 3.5 Some functional groups commonly found in the molecules of life. Such groups help give substances their distinctive properties. For instance, carboxyl groups help impart an acid taste to vinegar. Hydroxyl groups help give sugar and alcohol their properties.

BONDING AND UNBONDING IN THE MOLECULES OF LIFE

So far, the molecules described are fairly simple. At most they contain no more than a few dozen atoms. How do we get from such molecules to the larger molecules of life, some of which contain thousands or even millions of atoms? Many bonding processes are involved, but two can be outlined here to give you an idea of what goes on.

The first involves the covalent linkage of molecular subunits, such as short chains and rings of carbon-containing molecules. In this process, called **dehydration synthesis**, one subunit is stripped of an H atom and another is stripped of an —OH group; then the two subunits become covalently bonded. At the same time, the H and —OH released in the process may combine to form a water molecule (Figure 3.6).

Of course, life depends on more than building up large molecules. It also depends on tearing them down. How else would a cell dispose of worn-out parts? How else could it break down large food molecules for their stored energy? **Hydrolysis** is a cornerstone of such cellular activities. The process is something like dehydration synthesis in reverse. When bonds between certain parts of molecules are broken, an H atom and an —OH group derived from water become attached to the fragments (Figure 3.7). In cells, both dehydration synthesis and hydrolysis involve participation of a class of proteins called enzymes, to be described later. Both processes are important in the ways that cells constantly recycle the atoms and molecules of life. The two processes are representative of ways in which subunits are combined, pulled apart, and rearranged in the formation of large biological molecules, which will now be described.

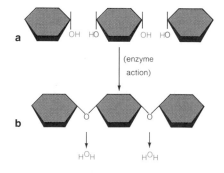

Figure 3.6 Dehydration synthesis of three molecular subunits (**a**) leads to a larger molecule (**b**). Water is formed during the process.

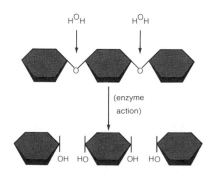

Figure 3.7 Hydrolysis of a molecule into three subunits.

CARBOHYDRATES

Trees, grasses, fruits and vegetables, cotton plants, weeds —all such things have carbohydrates as their main structural material. And almost all protistans, fungi, and animals—as well as the plants themselves—depend directly or indirectly on carbohydrates for their energy. A **carbohydrate** is made of only three elements: carbon, hydrogen, and oxygen. These elements are combined in about a 1:2:1 ratio. This means that for every carbon atom in a carbohydrate molecule, there typically are two hydrogen atoms and one oxygen atom, or $(CH_2O)_n$. The "n" represents whatever number of the subunits themselves are linked in a given molecule.

Monosaccharides

The building block of all carbohydrates is the **monosaccharide**: a simple sugar that typically has a skeleton of five, six, or seven carbon atoms. Monosaccharides that have five or more carbon atoms tend to form ring structures when they are dissolved in the watery environment of living cells. Among the most biologically important are the six-carbon sugars *glucose, fructose,* and *galactose;* and the five-carbon sugar *ribose.*

Notice, in Figure 3.8, that each glucose molecule has six carbon atoms, twelve hydrogen atoms, and six oxygen atoms—and that each fructose molecule does, too. Thus, these two simple sugars have the same molecular formula $(C_6H_{12}O_6)$. Yet, they are different molecules with different properties. For instance, fructose tastes much sweeter than glucose. These molecules are **isomers**: they have the same

number and kinds of atoms, but the atoms are grouped in different structural arrangements (hence the different properties). Several other six-carbon sugars have this same molecular formula; all are isomers of each other.

Sometimes there is more than one form of the *same* kind of molecule, simply because functional groups assume different positions in space around the carbon atom to which they are covalently bonded. Such molecules are **stereoisomers**. Thus we have two ring forms of glucose, which differ only in the position of an —OH group bonded to and projecting from a carbon atom:

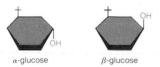

α-glucose   β-glucose

This small difference, you will see shortly, affects the architecture (hence the properties) of different substances in which these simple sugars are found.

Disaccharides and Polysaccharides

When two simple sugar molecules become covalently bonded through dehydration synthesis, the result is a **disaccharide**. For instance, a glucose and a fructose subunit may be combined into the disaccharide *sucrose.* Sucrose is the most abundant of all sugars, and the form that is transported about in the body of many leafy plants. Sucrose is also the substance called table sugar, which is extracted and crystallized from such plants as sugarcane. Another important disaccharide is *lactose,* or milk sugar. Lactose is composed of a glucose and a galactose subunit. Still another disaccharide is *maltose* (two glucose subunits). It's found in germinating seeds and is used in the brewing industry.

When more than two simple sugar molecules are bonded covalently by dehydration synthesis, the result is a **polysaccharide**. Many of these larger carbohydrates are formed from only one kind of simple sugar. Others are formed from two, three, or four different kinds.

Cellulose and amylose are two polysaccharides. Both are composed only of glucose subunits, yet they show markedly different properties. *Cellulose* is a fibrous substance, and it's insoluble in water. Cellulose is the main structural material in plant cell walls and is probably the most abundant organic substance in the living world. If you have ever tried to pull apart a tuft of cotton—which is almost pure cellulose—you know how tough its fibers

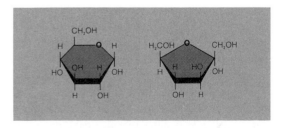

Figure 3.8 Ring form of glucose (left) and fructose (right). In fructose, two carbon atoms in the ring "share" an oxygen atom. Notice that both glucose and fructose have the same molecular formula $(C_6H_{12}O_6)$, but that their atoms differ in their structural arrangement.

a *Straight-chain bonding pattern of cellulose*

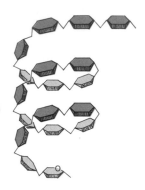

b *Coiled-chain bonding pattern of amylose (a starch)*

Jack Carey

Figure 3.9 (**a**) The β-form of glucose gives rise to the tough fibers of cellulose, a structural component of woody plants such as this weathered tree at Crater Lake. (**b**) The α-form of glucose gives rise to amylose, a substance having different properties.

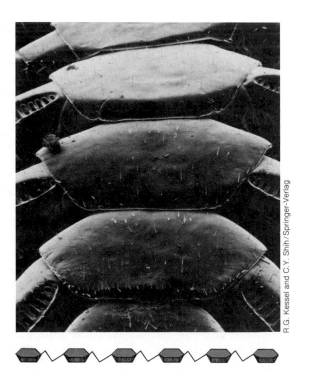

R.G. Kessel and C.Y. Shih/Springer-Verlag

Figure 3.10 Straight-chain bonding pattern for another polysaccharide, chitin. This substance is found in the tough external skeleton of insects, such as the armor plates of the centipede shown here.

can be. In contrast, *amylose* is a kind of starch—one of the most important of all foods! Amylose is readily broken apart, first into maltose and then into individual glucose molecules. Hence it is a reservoir of nearly instant energy. What gives rise to the differences between these two substances? Each is made of a different stereoisomer of glucose. Cellulose is made of long chains of about 3,000 β-glucose subunits, lying in parallel, and the bonding arrangement between these subunits resists hydrolysis. Amylose is made of coiled chains of about 250 α-glucose subunits, and the bonds between them are readily accessible to enzymes of hydrolysis (Figure 3.9).

Glycogen, another polysaccharide, has glucose subunits bonded together into a highly branched structure. Glycogen is the form in which carbohydrates are stored in such animal tissues as liver and muscle. Still another polysaccharide, *chitin,* is secreted by cells in the outer tissue layer of many animals. This carbohydrate is modified, in that it contains nitrogen atoms. In thin layers, chitin is flexible. When combined with calcium salts, it forms hard armor plates (Figure 3.10). For many insects and crustaceans (such as crabs), chitin is the main material in the external skeleton, biting mouthparts, and other specialized structures such as eye lenses and copulatory organs. For most fungi, chitin imparts a degree of firmness to cell walls.

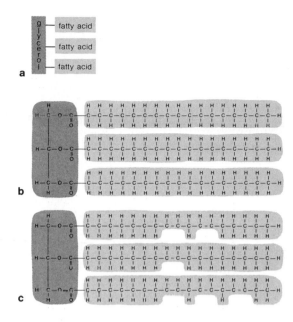

Figure 3.11 (a) General structure of triglycerides, fats formed by dehydration synthesis from glycerol and fatty acids. (b) Glyceryl tristearate, a fully saturated fat. (c) Linseed oil, an unsaturated fat.

LIPIDS

Lipids are a diverse assortment of oily or waxy substances, including true fats, waxes, steroids (such as cholesterol), terpenes, phospholipids, and glycolipids. Compared with a carbohydrate, a lipid molecule has fewer oxygen atoms relative to the number of carbon and hydrogen atoms. For the most part, its atoms are linked in nonpolar (C—C and C—H) bonds, which means that most regions of the molecule are insoluble in water.

True Fats and Waxes

The so-called **true fats** are lipids composed only of carbon, hydrogen, and oxygen atoms. These atoms are arranged in two kinds of subunits: glycerol and fatty acid. *Glycerol* is a kind of alcohol; it has a carbon skeleton to which three hydroxyl (—OH) groups as well as hydrogen atoms are attached. A molecule of glycerol may be represented in the following manner:

A *fatty acid* has an unbranched carbon skeleton to wh a carboxyl (—COOH) group as well as hydrogen atoms are attached. One kind looks like this:

Up to three fatty acid chains of this sort may be attached by dehydration synthesis to a glycerol molecule (Figure 3.11). The terms monoglyceride, diglyceride, and triglyceride refer to whether one, two, or three chains are attached.

Butter, bacon fat, and other animal fats are called *saturated fats*, because some maximum possible number of hydrogen atoms is covalently bonded to the carbon skeleton of their fatty acid "tails." The molecules of fully saturated fats are chains that pack tightly together. That's why these fats are usually solid or semisolid at room temperature. In contrast, vegetable oils (such as soybean, corn, safflower, and linseed) are called *unsaturated fats*. Their fatty acid tails contain a number of double covalent bonds between carbon atoms (typically one, two, or three in each fatty acid tail). At each double bond, the molecule bends in space. This bending prevents a snug fit between chains. That's why unsaturated fats are usually fluid (oils) at room temperature.

True fats are compact forms of stored energy. But the same fat molecules are not hoarded away indefinitely in cells. There is a constant turnover of fat reserves, with new molecules built up even as old ones are used. In studies of mice, for example, it was determined that half of the fat was replaced each week! Despite the turnover at the molecular level, an overfed body can stockpile fat molecules faster than it can use them up. This stockpiling contributes to obesity. What we call dieting means living off fat reserves faster than replacements are taken in.

In the lipids called **waxes**, long-chain alcohols are combined with long-chain fatty acids. In some organisms, waxes are secreted to the outer surface of cells or cell walls. The secretions help prevent water loss and impart some resistance to disease-causing or predatory organisms (Figure 3.12). Such waxes are important materials in *cutin*, which covers and waterproofs leaf, stem, and fruit surfaces. Some are materials in *suberin*, a water-resistant substance found in cell walls of cork (tree bark). Beeswax, a structural material secreted by honeybees, is used in the construction of honeycomb.

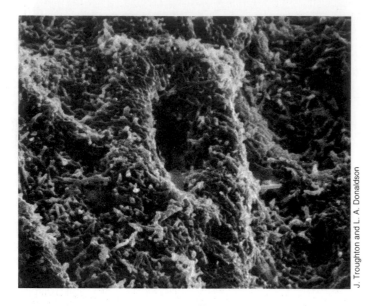

Figure 3.12 Extensive wax deposits on the leaf surface of a carnation (*Dianthus*).

Steroids and Related Substances

The lipids known as **steroids** have four carbon rings to which a variety of atoms may be attached. The steroid *cholesterol* has this structure:

Cholesterol itself is a component of most cell membranes. It may also undergo rearrangements that lead to formation of such substances as male and female sex hormones and bile acids (which function in digestion). Cholesterol molecules are insoluble in fluids such as blood. They must be bound to other substances before they can be transported to their destination. As you will read later, improper packaging of cholesterol may be a determining factor in the onset of such disorders as gallstones.

Terpenes are a diverse group of lipids assembled from five-carbon subunits. Often the subunits are strung together in a long, water-insoluble chain:

This kind of chain is part of light-trapping pigment molecules called *chlorophyll* and the *carotenoids*. The chain helps anchor these molecules to internal cell membranes in certain plants and microorganisms.

Phospholipids and Glycolipids

The **phospholipids** are among the most universally important substances in living things. These lipids are assembled from glycerol, fatty acids, a phosphorus-containing group, and usually a nitrogen-containing alcohol. Together with proteins, they are the basic fabric for all cell membranes. Most phospholipid regions are insoluble in water, which is why a cell membrane doesn't dissolve even though water is present on both sides. Also important are the **glycolipids**, which have water-soluble carbohydrate groups attached. These groups are one kind of site for chemical recognition between cells and substances present in the cellular environment (Chapter Five).

PROTEINS

Of all biological molecules, **proteins** are the most diverse in both structure and function. No matter what the process, proteins are sure to be involved. In this class of molecules are more than a thousand different known **enzymes**: agents that make some particular metabolic reaction proceed much faster than it otherwise would. Also in this class of molecules are the main substances underlying movements of subcellular structures and movements of cells. These cells range from sperm to amoebas to those making up every muscle in your body. Here, too, are storage molecules, and transport molecules such as hemoglobin (which carries oxygen in blood). Here are structural materials of the first rank, the stuff of bone and cartilage, hoof and claw. Here are molecules acting as chemical messengers, and others help protect the vertebrate body against disease agents.

Protein Structure

For all their diversity, proteins usually contain some combination of no more than twenty different subunits called amino acids. Generally, an **amino acid** has an amino group ($-NH_2$) and a carboxyl group ($-COOH$), both attached to the same carbon atom. Covalent bonds can form between the amino group of one amino acid and the carboxyl group of the other. This linkage is a **peptide bond** (Figure 3.13).

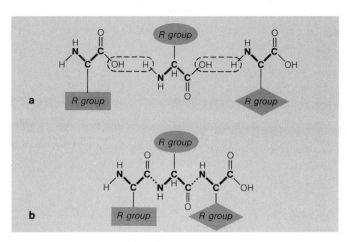

Figure 3.13 Assembling a protein from three amino acid subunits. (**a**) Through dehydration synthesis, H atoms and —OH groups are removed from the amino acids. The amino acids differ only in their side groups (designated R in the sketches). (**b**) Covalent bonds form between a carbon atom of one amino acid and a nitrogen atom of the next in line. These so-called peptide bonds are shown here as beads. The rest of the protein backbone is shown in boldface.

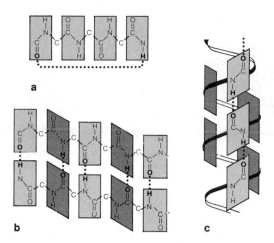

Figure 3.14 (**a**) Hydrogen bonding (dotted line) that forms between every fourth amino acid in a protein chain, which leads to an alpha pattern of protein structure. (**b**) The pleated, beta pattern in proteins. Dotted lines are hydrogen bonds that form between two chains. (**c**) The coiled alpha pattern of protein structure.

Any kind of amino acid can follow any other in a chain. Hence an extraordinary variety of proteins could, in theory, be assembled. Consider how many different dipeptides (molecules of two linked amino acids) may be built. Even with only twenty choices available for the first amino acid and twenty for the second, there are 20 × 20, or 400, possible dipeptide sequences. For a *small* protein having merely 100 amino acids, there are 20^{100} possible sequences. That number is billions of times larger than the number of stars in the universe!

What underlies this potential for variation? After all, the protein backbone itself (N—C—C—N—C—C—N···) seems rather monotonous in its structure. But each amino acid has a unique side group. Once amino acids are positioned in a protein chain, many of their side groups project into the environment. The chemical nature, size, shape, and ordering of side groups give rise to the three-dimensional arrangement of atoms in the protein. By extension, then, the following is true:

The chemical nature, size, shape, and ordering of side groups projecting from amino acids dictate how a protein will interact chemically with other substances—hence the role that the protein will play in a living system.

As a protein chain is being formed, the atoms taking part in each peptide bond tend to become rigidly positioned in space. Their rigid positioning imposes some limits on the protein structures that are possible. In some cases the backbone becomes coiled, with hydrogen bonds occurring between different groups along the molecule. This bonding pattern causes the chain to coil helically about its own axis, much like a spiral staircase. In other cases, hydrogen bonds between groups on *different* proteins result in fully extended chains being held side by side (Figure 3.14).

So far, we have been talking about two levels of protein structure. The specific sequence of amino acids, which sets the stage for everything else, is the **primary structure**. When a protein chain takes on a helical or an extended pattern because of repeated hydrogen bonding, it has assumed **secondary structure**. But most protein chains also assume **tertiary structure**: they become permanently folded in three-dimensional shapes when their side groups interact among themselves and with the backbone itself. For instance, wherever the amino acid proline occurs in a protein chain, the chain may bend sharply. Similarly, two amino acid molecules called cysteine may join covalently at their sulfur atoms. Hundreds of amino acids may separate the two molecules in the sequence. But twisting of the chain may allow them to approach each other, interact, and form

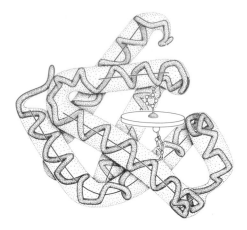

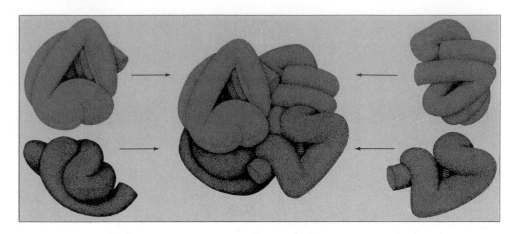

Figure 3.15 Three-dimensional structure of the protein component of hemoglobin. The disk-shaped structure represents an iron-containing component called a heme group, to which one oxygen molecule may bind. (From R. Dickerson and I. Geis, *The Structure and Action of Proteins.* Menlo Park, Calif. Benjamin)

Figure 3.16 Structure of the protein hemoglobin, a red pigment circulating in the blood of complex animals and carrying vital oxygen to and from tissues. In different animals, there are slight variations in the amino acid sequence in the hemoglobin molecule. Thus the molecule's backbone bends in a slightly different way in each case, which gives rise to variations in the way the molecule functions to pick up and release oxygen. Such variations are adaptations to different oxygen levels in different environments. (Redrawn from R. W. McGilvery, 1970. *Biochemistry: A Functional Approach.* W. B. Saunders Co.)

a strong disulfide bridge that locks protein structure at the site. Figure 3.15 is an example of the diverse shapes achieved through these and other interactions. Finally, two or more proteins may interact and form an aggregate structure, called **quaternary structure**. For example, four iron-containing proteins called globin interact and form the quaternary structure of hemoglobin (Figure 3.16).

Protein-Based Substances

Proteins may be linked with other substances. Iron-containing proteins include not only hemoglobin but also the **cytochromes**, which transport electrons in most kinds of cells. Sugar-containing proteins are called **glycoproteins**. They include hormones involved in female reproductive cycles. In human blood, gamma globulin is a glycoprotein that helps protect the body against invaders. Glycoproteins are also embedded in cell membranes. Cells interact at these sites.

Lipoproteins are compounds of lipids and proteins. Of recent interest are those to which cholesterol is bound as it is transported in blood. The proportion of lipid to protein in these molecules seems to vary among individuals. The variation is implicated in susceptibility to heart and circulatory disorders (Chapter Sixteen). Other protein-based substances include the **nucleoproteins**: com-

pounds of proteins and nucleic acids. For all organisms except monerans, nucleoproteins contain DNA destined to be shipped into new cells during cellular reproduction.

NUCLEIC ACIDS AND OTHER NUCLEOTIDES

Among the most important of all molecules of life are the nucleotides. Each **nucleotide** contains at least a five-carbon sugar (ribose), a nitrogen-containing base (either a single-ringed pyrimidine or a double-ringed purine), and a phosphate group. For example, one nucleotide has a sugar, base, and phosphate group hooked together in this manner:

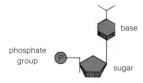

Three kinds of nucleotides or nucleotide-based molecules are the **adenosine phosphates**, the **nucleotide coenzymes**, and the **nucleic acids**. The first kind contains relatively small molecules that function as chemical messengers between cells, and as energy carriers. Cyclic adenosine monophosphate (cAMP) is one such messenger; adenosine triphosphate (ATP) is a nucleotide energy carrier. The second kind of nucleotide contains molecules that transport hydrogen ions and their associated electrons, which are so necessary in cell metabolism. Nicotinamide adenine dinucleotide (NAD+) and flavin adenine dinucleotide (FAD) are two of these coenzymes. The nucleotide-based molecules called nucleic acids are long, single- or double-stranded molecules. They consist of nucleotides strung into long chains, with a phosphate bridge connecting sugars and with bases sticking out to the side:

Deoxyribonucleic acid (DNA) and the ribonucleic acids (RNAs) are built according to this plan. DNA is usually a double-stranded molecule that twists about its own axis, with the bases of one strand connected by hydrogen bonds to the bases of the other strand. For a given protein to be built, these two strands must be temporarily unwound from each other to expose the protein-building instructions encoded in the bases sticking out from the sugar–phosphate backbone of one of these strands. Another group of nucleic acids "reads" the instructions, and guides their translation into actual protein structure. This group contains the RNA molecules. Another chapter will be devoted to the structure and function of nucleic acids. For now, it is enough to know that genetic instructions are encoded in the sequence of bases strung together in DNA, and that RNA is the means by which those instructions become translated into the proteins on which all forms of life are based.

Readings

Dickerson, R., and I. Geis. 1973. *The Structure and Action of Proteins.* Menlo Park, California: Benjamin. A classic in the literature.

Frieden, E. 1972. "The Chemical Elements of Life." *Scientific American* 227 (1): 52–64. Good summary of the biological roles of various elements.

Lehninger, A. 1975. *Biochemistry.* Second edition. New York: Worth. Advanced reading, but a standard reference for the serious student.

Miller, J. 1978. *Chemistry: A Basic Introduction.* Belmont, California: Wadsworth. If you are one of those who have absolutely no background in chemistry, Dr. Miller's book is about as accessible and interesting a treatment as you will find.

Watson, J. 1976. *Molecular Biology of the Gene.* Third edition. Menlo Park, California: Benjamin. More than genes; this classic includes a concise survey of chemical bonding and of the structure of large biological molecules. Up-to-date, accurate.

Review Questions

1. What is energy? What is the likely relationship between the universal requirement for energy and the rise of diversity among living things?

2. What is an atom? An element? Each atom in a given chemical element is unique because of its _____ and _____ of subatomic parts.

3. Define proton and electron. What happens when many negatively charged electrons jockey for position around positively charged protons in a nucleus? Why are the resulting, predictable arrangements important for life?

4. What is a positively charged ion? A negatively charged ion?

5. Electrons can absorb certain amounts of incoming energy that puts them into orbitals at higher energy levels. How many electrons can be squeezed into each orbital?

6. Atoms of chemically nonreactive elements have _____ electrons in each occupied orbital. Atoms of chemically reactive elements have _____ electrons in one or more orbitals.

7. Cell substances are held together largely by strong bonds that involve shared electrons, and by weaker interactions between atoms or ions of opposite charge. Name these kinds of bonds, and give examples of molecules in which they are found.

8. Describe the difference between dehydration synthesis and hydrolysis.

9. Why do carbon atoms tend to combine so readily with other atoms? Explain the difference between saturated and unsaturated fats in terms of the carbon skeletons of their fatty acid tails.

10. Is this statement true or false? Not all proteins are enzymes, but all enzymes are proteins.

11. How do its side groups dictate how a protein molecule will interact with other cell substances? (Hint: Describe the four levels of protein structure.) What consequences do these interactions have for cell functioning?

12. Define the general structure and function of the three kinds of nucleotides or nucleotide-based molecules so important to cell functioning and reproduction.

4

CELL STRUCTURE AND FUNCTION: AN OVERVIEW

Early in the seventeeth century, Galileo arranged two glass lenses in a cylinder. With this instrument he happened to look at an insect, and thereby came to describe the stunning geometric pattern of its tiny compound eyes. Thus Galileo, who was not a biologist, was the first to record a biological observation made through a microscope. The systematic study of the cellular basis of life was about to begin. First in Italy, then in France and England, biologists began to explore architectural details of a world whose existence had not even been suspected.

At mid-century Robert Hooke, "Curator of Instruments" for the Royal Society of England, was at the forefront of these studies. When Hooke first turned one of his microscopes to a thinly sliced piece of bark from a cork tree, he observed tiny, empty compartments. These he likened to the structure of honeycomb. He gave them the Latin name *cellulae* (meaning small rooms); hence the origin of the biological term, "cell." They were actually dead cells, which is what tree bark is made of, although Hooke did not think of them as being dead because he did not know that cells could be alive. He also noted that cells in other plant materials contained "juices." He did not speculate on what the juice-filled structures might represent.

Given the simplicity of their instruments, it is amazing that these pioneers saw all that they did. Antony van Leeuwenhoek, a shopkeeper, had the keenest vision of all and/or the greatest skill in glassworking. He devoted all of his spare time to constructing lenses and to observing everything he could get hold of, including sperm cells. He even observed a single bacterium—a type of organism so small it would not be seen again until the nineteenth century! But this was primarily an age of exploration, not of interpretation. Once the limits of those simple instruments had been reached, biologists gave up interest in cell structure without ever having thought to look for explanations of what it was they had seen.

Then, in the 1820s, improvements in lens design tempted biologists back into the cell. It now became clear that small structures were suspended within those cellular "juices." The botanist Robert Brown, for instance, reported the presence of a spherelike structure inside every plant cell that he examined. He called the structure a "nucleus." By 1839, the zoologist Theodor Schwann had confirmed the presence of cells in animal tissues. About this time, he began working with Matthias Schleiden, a botanist who had concluded that cells are present in all plant tissues, and that the nucleus is somehow paramount in the reproduction of cells. It was Schwann who distilled the meaning of these new observations into what came to be known as the first two principles of the **cell theory:**

All organisms are composed of one or more cells.

The cell is the basic living unit of organization for all organisms.

Another decade passed before the physiologist Rudolf Virchow finished exhaustive investigations into the reproduction of human cells, and completed the basic theory with this third principle:

All cells arise from preexisting cells.

Not only was a cell viewed as the smallest living unit, the continuity of life was now seen to be arising directly from the division and growth of single cells. *Within each tiny cell, events were going on that had implications of the most profound sort for all levels of biological organization!*

Identifying these events has been difficult, for most cells are very small. If you were to line up 2,000 or so of your red blood cells, you would have a string only about as long as your thumbnail is wide. Some cells are smaller still—less than 100 nanometers across. A nanometer is only a *billionth* of a meter (Appendix II). The finest detail that the modern light microscope can resolve is about 200 nanometers across (Figure 4.2). To be sure, some cells *are* large enough to be seen without a microscope. For example, some nerve cells running through a giraffe leg may be three meters long. And individual cells in a watermelon can be distinguished with the unaided eye. Yet even in these "giant" cells, details of key cellular activities and internal structures have been hidden from view. In fact, only in the past few decades have biochemical studies and electron microscopy really enlarged the picture of cell structure and function (Figures 4.2 and 4.3).

GENERALIZED PICTURE OF THE CELL

Today we know that cells are not the simple bags of juices that early investigators thought them to be. Each living cell shows astonishing degrees of organization. How can we keep the details of this organization in perspective? We may, at the outset, think of cellular activities as occurring in three interrelated zones: the plasma membrane, the cytoplasm, and the nucleus (or region of DNA). Pathways

of chemical communication weave through all three zones and link them together.

The **plasma membrane** is a differentially permeable boundary layer between the internal and external world of all cells. The term *differentially permeable* means that the plasma membrane is so contructed that it can bind and transport certain substances into and out of the cell, that it can keep other substances out at all times, and that it can admit still others only in required amounts. Enclosed by the plasma membrane is the **cytoplasm**—a semifluid substance in which various cell structures are embedded. (Early investigators dubbed these structures *organelles*, meaning "little organs.") These cytoplasmic structures do not float about randomly in the interior. As you will read later, some are even anchored by networks of filaments and moved about in controlled ways. Most reactions are linked in metabolic pathways, so cytoplasmic structures cannot be located haphazardly relative to one another in the cell. Finally, suspended in the cytoplasm is a zone of hereditary control, where most of the DNA molecules in the cell are formed. In most organisms, this zone is the **nucleus.**

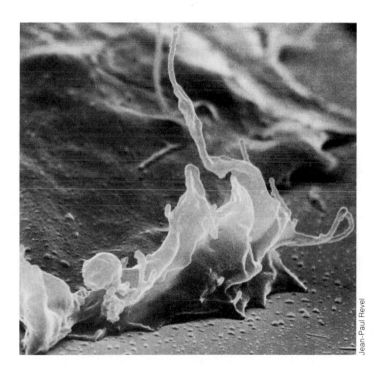

Figure 4.1 A "cell"—a vaguely dreary name more suggestive of carton boxes than of things seething with life. Shown here, the dynamic ruffling edge of a hamster cell on the move.

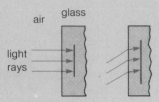

a *Refraction of light rays (The angle of entry and the molecular structure of the glass determine how much they will bend)*

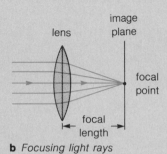

b *Focusing light rays*

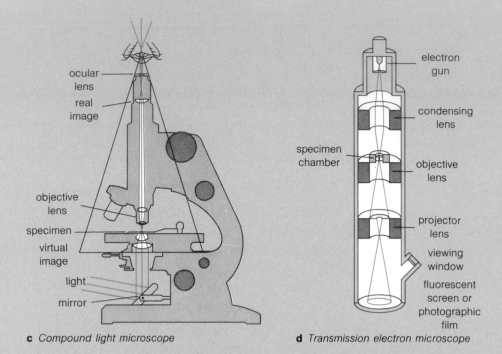

c *Compound light microscope*

d *Transmission electron microscope*

Figure 4.2 The microscope: gateway to the cell.

Light Microscopes Light microscopy is based on the bending (refraction) of light rays (**a**). Light rays pass straight through the center of a simple convex lens (**b**). The farther away they are from the center, the more they bend. A lens must be curved so all light rays coming from the object being viewed will be channeled to a single place behind the lens. A two-lens system, the *compound light microscope,* is shown in (**c**).

One problem with compound light microscopes is spherical aberration: when you bring tiny objects close to the objective lens, they can blur. Another problem is that light comes in different wavelengths (colors). For instance, red is a long wavelength, blue is a short one. Blue and red light rays are bent differently as they pass through the same piece of glass, so they don't end up at the same place. If you sharply focus the red part of an image, the blue part may be out of focus, and vice versa. The distortion (chromatic aberration) causes the color halos you sometimes see around images of small objects. Only in better microscopes does careful lens design overcome these problems.

If you wish to observe a *living* cell, it must be small or thin enough for light to pass through. (Some microorganisms and single cells are small enough; complex tissues are not. And when thin sections are made of cells in such tissues, the cells die.) Also, structures inside a cell can be seen only if they differ in color and density from their surroundings—but most are almost colorless and optically uniform in density. Specimens can be stained (exposed to dyes that react with some cell structures but not others), but staining usually alters the structures and kills the cells. Finally, dead cells begin to break down at once, so they must be pickled or preserved before staining. Most

observations have been made of dead, pickled, and stained cells. With the *phase contrast microscope* (in which small differences in the way different cell structures refract light are converted into larger variations in brightness), live cells *can* be observed as they actively move about. But the need for transparent specimens is even more critical.

No matter how good a lens system may be, when the magnification exceeds 2,000× (when the image diameter is 2,000 times as large as the object's diameter), cell structures will appear larger but not clearer. It's like what happens when you use a magnifying glass to enlarge a newspaper photograph. When you hold the magnifying glass too close, you see only black dots. There is no way to see a detail as small as or smaller than a dot; the dot would cover it up. In microscopy, something like dot size intervenes to limit *resolution* (the property dictating whether small objects close together can be seen as separate things). That limiting factor is the physical size of wavelengths of visible light. Red wavelengths are about 750 nanometers and violet wavelengths about 400 nanometers; all other colors fall in between. If an object is smaller than about one-half the wavelength, light rays passing by it will overlap so much that the object won't be visible. The best light microscope can resolve detail only to about 200 nanometers.

Transmission Electron Microscope The vibrations of electrons are much smaller than the smallest visible light wavelengths. How fast an electron vibrates depends on how much energy it has. The more energy, the shorter the wavelength. It takes very little energy to excite an electron to wavelengths of about 0.005 nanometer—about 100,000 times shorter than those of visible light!

Ordinary lenses can't be used to focus such accelerated streams of electrons, because glass scatters electrons. But each electron carries

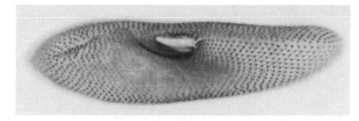

a

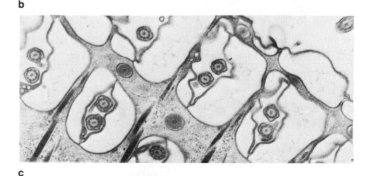

b

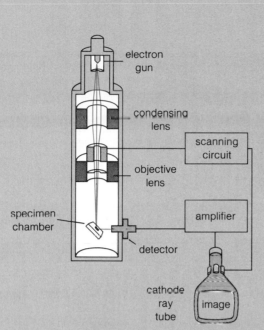

electron
gun

condensing
lens

scanning
circuit

objective
lens

specimen
chamber

amplifier

detector

cathode
ray
tube

image

e *Scanning electron microscope*

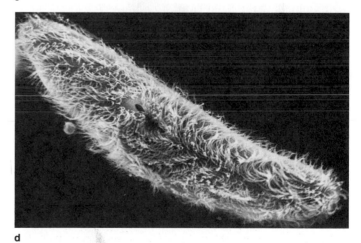

c

d

an electric charge, so a magnetic field can be used to divert it along certain paths. Accelerated electrons can be sent through an electromagnetic field designed in a way to channel them to a focal point. The *transmission electron microscope* (**d**) uses magnetic lenses of this sort.

The transmission electron microscope has greatly increased our knowledge of cell structure. But it, too, has limitations. Actual resolution of existing models ranges between 0.2 and 1.0 nanometer because there is an electrical equivalent to spherical aberration. Besides that, electrons must travel in a vacuum (otherwise they would be randomly scattered by molecules making up the air). Cells can't live in a vacuum, so living cells can't be observed at this higher magnification. In addition, specimens must be sliced extremely thin so that electron scattering will correspond to the density of different structures. (The more dense the structure, the greater the scattering and the darker the area in the final image formed.) Specimen fixation is crucial. Fine cell structures are the first to fall apart when the cell dies, and artifacts (structures that don't exist in a real cell) may result. Because most cell materials are relatively transparent to electrons, they must be stained with heavy metal ''dyes,'' which can create more artifacts.

Scanning Electron Microscope   With a *scanning electron microscope* (**e**), a narrow beam of electrons is played back and forth across a specimen's surface, which has been coated with a thin metal layer. This triggers the emission of secondary electrons from the metal. The emission pattern is detected by equipment similar to a television camera, and an image is formed. Scanning electron microscopy is restricted to surface views, and it does not approach the high resolution of the transmission instruments. But its images have fantastic depth (Figure 4.3).

Figure 4.3   Comparison of image-forming abilities of microscopes. The specimen in all cases is *Paramecium.* (**a**) Conventional light microscope (bright-field). 750×. (**b**) Phase contrast microscope. 400×. (**c**) Transmission electron microscope (glancing section through the ventral surface). 17,800×. (**d**) Scanning electron microscope. 550×. (All photographs Gary W. Grimes and Steven W. L'Hernault)

Figure 4.4 depicts how these three zones might be arranged in typical cells. It is important to keep in mind, however, that calling them "typical" is like calling a squid or a watermelon plant a "typical" animal or plant. Although Figure 4.4 typifies the basic plans, cells and their contents come in all manner of sizes, shapes, and elaboration. Variety among them is as remarkable as the variety that exists among large, multicellular organisms.

In looking at Figure 4.4, you can see at once that the plant and animal cells have far more complex internal organization than that of the bacterial cell shown. Specifically, they have many diverse, membrane-enclosed organelles in which special metabolic reactions or processes occur. This division of labor is so significant that every cell—hence every organism—has been classified as belonging in one of two major groups: the prokaryotes and the eukaryotes.

The word **prokaryote** means "before the nucleus." It refers to the fact that prokaryotic cells have no membrane-enclosed nucleus. Only bacteria and blue-green algae (which resemble bacteria) fall in this category. All of their life processes occur within a single, general chamber (the cell) bounded by a single membrane (the plasma membrane). In some photosynthetic prokaryotes, the plasma membrane is infolded extensively into the cytoplasm and studded with light-trapping units. Such infoldings increase the membrane surface. Hence they increase the number of reaction sites available for photosynthesis. But these internal infoldings are not the same thing as separate organelles. All other cells—those of protistans, fungi, plants, and animals—have a series of membrane-enclosed organelles in which specialized reactions are carried out. The most important of these organelles is the nucleus. Hence protistans, fungi, plants, and animals are called **eukaryotes,** a word that means "true nucleus."

Both the cytoplasm and the plasma membrane have truly intricate organization. For that reason, they will be described separately in the next chapter, after a description of the environment in which they must function. Here the concern will be other features typically found in cells: external walls, and the organelles which are, for the most part, characteristic of eukaryotes.

CELL WALLS AROUND THE PLASMA MEMBRANE

Among bacteria, blue-green algae, many protistans, fungi, and plants, the thin plasma membrane is surrounded by a more or less continuous layer called a **cell wall**. In wood,

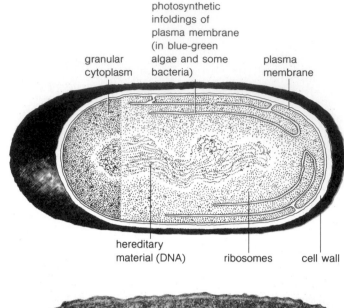

a *Generalized sketch of body plan for bacteria and blue-green algae. All components except photosynthetic membranes occur in the bacterium* Escherichia coli, *shown in long section. 28,500×. (G. Cohen-Bazire)*

Figure 4.4 Generalized plans for cellular organization. In looking at these sketches, keep in mind that they are extremely generalized. Immense variation exists among single cells, just as it does among multicellular organisms.

(**a**) Bacteria and blue-green algae, as typified by this sketch, do not have a separate, membrane-enclosed nucleus. In fact, they don't have any membrane-enclosed organelles in which specialized reactions may take place. Molecules are relatively free to move about in the entire enclosed space. And the region in which hereditary material is found is variable.

(**b**) Plant cells and (**c**) animal cells do have semi-isolated, membrane-enclosed organelles. In the evolutionary sense, the most significant of these organelles is the nucleus. It represents isolation of the cell's hereditary material (DNA) from the clutter of other cell systems. Isolation means independent control, and independent control means the amount and complexity of DNA—hence the size and complexity of the organism—can be increased.

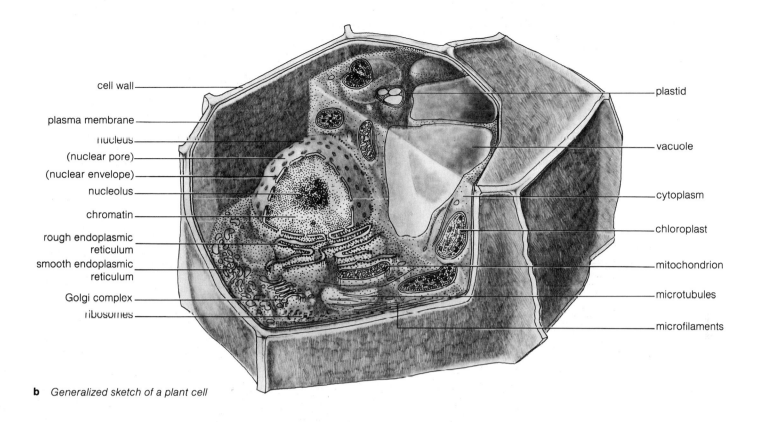

cell wall

plasma membrane

nucleus

(nuclear pore)

(nuclear envelope)

nucleolus

chromatin

rough endoplasmic reticulum

smooth endoplasmic reticulum

Golgi complex

ribosomes

plastid

vacuole

cytoplasm

chloroplast

mitochondrion

microtubules

microfilaments

b *Generalized sketch of a plant cell*

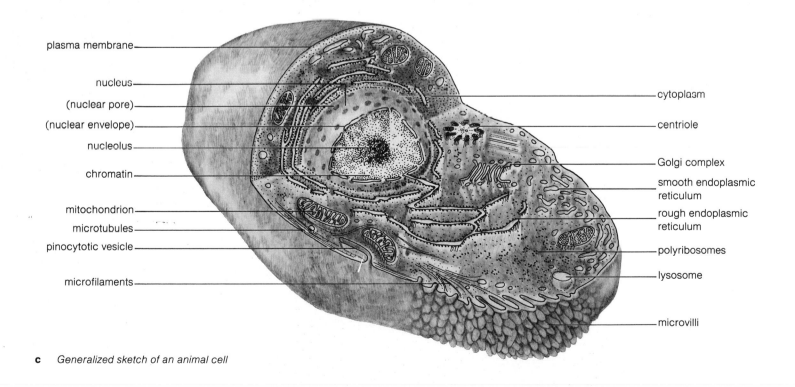

plasma membrane

nucleus

(nuclear pore)

(nuclear envelope)

nucleolus

chromatin

mitochondrion

microtubules

pinocytotic vesicle

microfilaments

cytoplasm

centriole

Golgi complex

smooth endoplasmic reticulum

rough endoplasmic reticulum

polyribosomes

lysosome

microvilli

c *Generalized sketch of an animal cell*

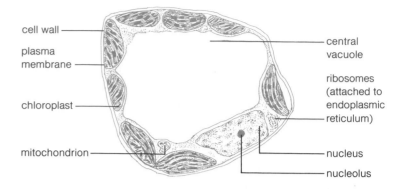

cell wall

plasma membrane

chloroplast

mitochondrion

central vacuole

ribosomes (attached to endoplasmic reticulum)

nucleus

nucleolus

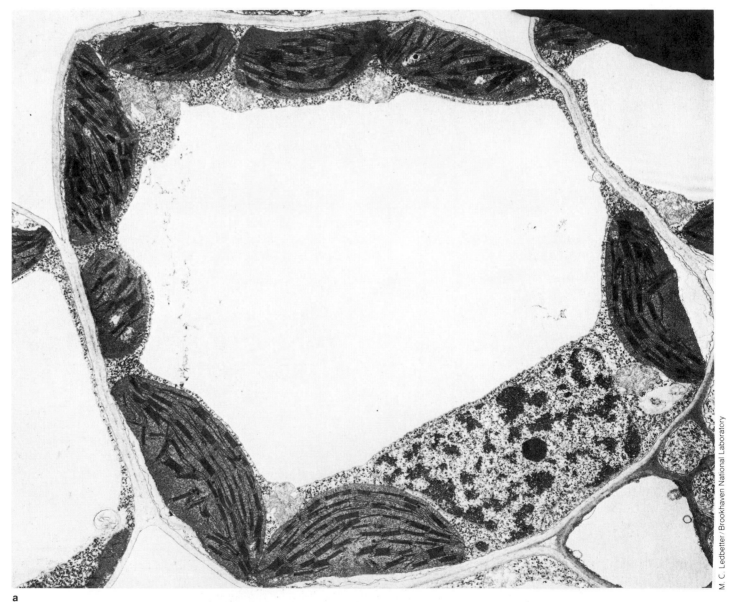

a

Figure 4.5 (**a**) Plant cell, from a blade of timothy grass. 11,300 ×. (**b**) Animal cell, from a rat liver. 15,000 ×.

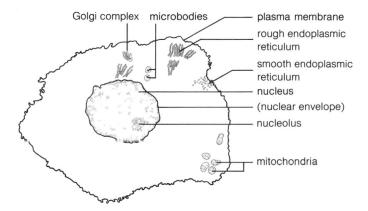

Golgi complex   microbodies   plasma membrane
rough endoplasmic reticulum
smooth endoplasmic reticulum
nucleus
(nuclear envelope)
nucleolus
mitochondria

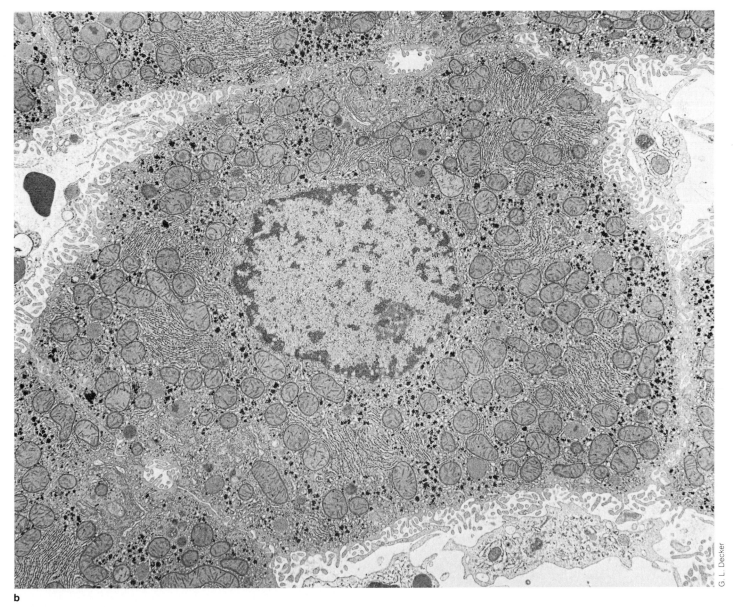

b

G. L. Decker

the cell wall layer confers rigidity and shape on the cell. In leaves, it confers high tensile strength when cells expand with incoming water. The cell wall also helps the plasma membrane hold in the cytoplasmic contents, without bursting under the pressure. Such walls are not impenetrable. They usually are porous, which means that water and other materials may be exchanged across them.

Most cell walls are rich in carbohydrates. Carbohydrate chains completely surround many bacterial cells, like a sack. Cellulose fibers are a component of cell walls in land plants, some fungi, and some protistans. In most land plants, cellulose is synthesized on the plasma membrane into thin strands. These strands are bundled together and added to a developing *primary cell wall* (Figure 4.6). After the main growth phase, many types of plant cells also deposit an inner, rigid *secondary cell wall* that includes cellulose and lignin. Cellulose walls are quite porous, but often they contain substances that alter permeability. Cutin, for instance, is a waxy substance that restricts water loss. It is deposited on outer walls of cells making up the surface layer of leaves.

Animal cells do not produce walls, although some secrete products to the surface layer of tissues in which they are found. Generally, the animal way of life demands great flexibility and movement, rather than rigid external support for its cells. Animal cells that do assume a well-defined shape (for instance, the *Paramecium* in Figure 4.3) usually have a framework of filaments just beneath the plasma membrane, not a cell wall.

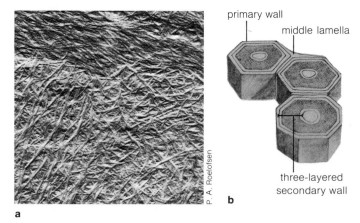

Figure 4.6 (**a**) Electron micrograph of the primary wall of a plant cell. The filaments are largely cellulose. (**b**) Primary and secondary cell walls, as they would appear in cross-section. The intercellular layer is mostly pectin.

ORGANELLES OF THE CYTOPLASM

Energy Metabolism: Chloroplasts

In photosynthetic prokaryotes, light-trapping reactions and ATP formation (Chapter Seven) occur on the plasma membrane. In photosynthetic eukaryotes (plants and some of the protistans), these reactions occur in an organelle called the **chloroplast**. Chloroplasts appear green in land plants. Among the algae they appear green, yellow-green, or golden brown. Their color depends on the kinds and relative numbers of light-absorbing pigment molecules embedded within their membranes. For instance, the pigment chlorophyll appears green because it absorbs all wavelengths of visible light *except* those corresponding to green light, which it transmits. Chlorophyll is found in all chloroplasts. Sometimes it is not immediately noticeable because there may be an abundance of pigments of other sorts.

Chloroplasts commonly are oval, round, or disk-shaped bodies about two to ten micrometers long. Some algae have only one chloroplast; other eukaryotic cells contain a hundred or more. Each chloroplast has a double-membrane system, which surrounds a relatively formless substance called the *stroma* (Figure 4.7). Suspended in the stroma of the most common types of chloroplasts are stacked, particle-studded membranes. Each stack is a *granum* (plural, grana). It looks like a pile of coins from the side. Its membranes apparently are sites where sunlight energy is trapped and converted to energy-rich molecules (ATP and $NADPH_2$). The newly formed molecules are then transported to the stroma, where they are used in building carbon compounds (such as sugar molecules) from carbon dioxide and water. Chapter Seven describes the nature and sequence of these reactions.

Energy Metabolism: Mitochondria

Energy stored in such food molecules as the sugar glucose must be released and converted to forms that the cell can use. In both prokaryotic and eukaryotic cells, some of the energy is released in the cytoplasm. But eukaryotic cells have the means for extracting a great deal more of the energy remaining. They do this in the membranous organelle called the **mitochondrion** (plural, mitochondria). In the case of glucose, more than ninety percent of its usable energy is released here.

Mitochondria commonly are about one to three micrometers long, or about the size of some bacterial cells.

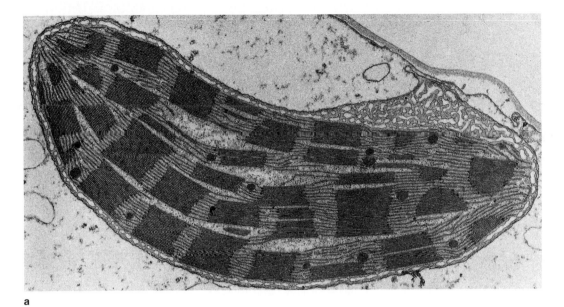

a

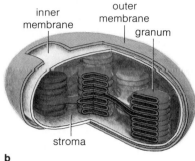

b

Figure 4.7 Micrograph (**a**) and generalized sketch (**b**) of a chloroplast membrane system. 20,200×. (L. K. Shumway)

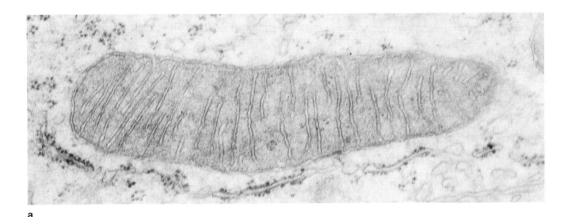

a

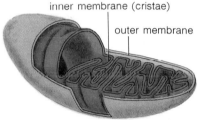

b

Figure 4.8 Micrograph (**a**) and generalized sketch (**b**) of a mitochondrial membrane system. 49,600×. (Richard Kolberg)

This places them near the limit of resolution in light microscopes. Some are round, some tubelike or threadlike, and some even have the shape of a potato. Their structures are not rigidly shaped, however. In less than a minute they may grow and branch out, even fuse with one another or divide in two.

Like chloroplasts, mitochondria have two distinct layers. The outer membrane forms a single boundary layer, which separates the inner membrane from the surrounding cytoplasm. The inner membrane folds in on itself. Most commonly it takes the form of *cristae,* or deep infoldings, of the sort shown in Figure 4.8. In the inner membrane

system are enzymes and molecules concerned with ATP formation.

Plant and animal cells typically contain anywhere from a dozen to a thousand mitochondria. Some cells, such as amphibian eggs, contain more than a hundred thousand. Then there is the protistan *Micromonas,* which gets along very well with only one mitochondrion. Muscle cells and others that demand high-energy output generally have many more mitochondria than less active cells. Their mitochondria have extremely convoluted cristae. These convolutions increase the membrane surface area, hence the number of reaction sites available.

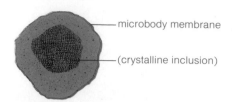
microbody membrane

(crystalline inclusion)

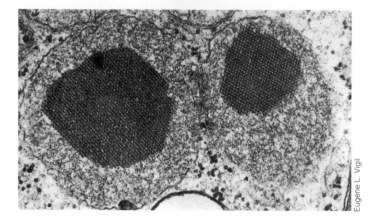

Eugene L. Vigil

Figure 4.9   Microbodies from a castor bean plant cell in thin section.

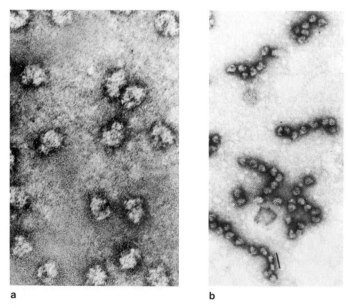

a

b

Figure 4.10   (**a**) Ribosomes and (**b**) polyribosomes. When the ribosomes are actively engaged in making protein, they are arranged in strings or whorls, called polyribosomes. (Nonomuri, Blobel, and Sabatini, *J. Molecular Biology*, 1971. 60:303–320)

Interconversion of Materials: Microbodies

Seldom in the real world is the supply of substances precisely matched to cellular needs. A cell must be able to convert excess amounts of some substances to different kinds that are required but not available. Cells of seeds such as castor beans have **microbodies** (Figure 4.9). In these organelles, fats can be converted to carbohydrates. In addition, excess amounts of some amino acids can be converted for use as cellular fuel, in place of sugar molecules. Some of these conversion reactions produce hydrogen peroxide, which can poison cells. But microbodies contain enzymes keyed to hydrogen peroxide breakdown as well as enzymes keyed to its production—all in the same organelle! Thus microbodies are sometimes called peroxisomes. As fast as hydrogen peroxide forms, some of the enzymes destroy it before it can destroy the cell.

Synthesis of Materials: Ribosomes

Proteins, again, are participants in nearly all cellular activities—either as enzymes or as structural materials. In both prokaryotic and eukaryotic cells, the site of protein synthesis is the **ribosome** (Figure 4.10a). A ribosome is not specialized for building one kind of protein only. It is more like a workbench on which all the different proteins can be built. An intact ribosome is composed of two subunits, each made of protein and ribonucleic acid. Unlike the other organelles described so far, a ribosome is not membrane bound.

Ribosomes are individually scattered through the cytoplasm of prokaryotic cells. In eukaryotic cells, they are either suspended in the cytoplasm or attached to a type of membrane which will be described next. When proteins are being assembled on ribosomal workbenches, these organelles are clustered in groups called polyribosomes (Figure 4.10b). Chapter Eleven will describe the assembly processes themselves.

Synthesis, Isolation, and Transport of Materials: Endoplasmic Reticulum

Eukaryotic cells have an interconnecting system of membranous tubes and flattened sacs. This is the **endoplasmic reticulum**. (The name sounds impressive, but the system actually was so named because it resembled "reticules," intricately netted ladies' purses that were popular during the eighteenth century.)

Rough endoplasmic reticulum has a grainy appearance because of ribosomes attached to the cytoplasmic

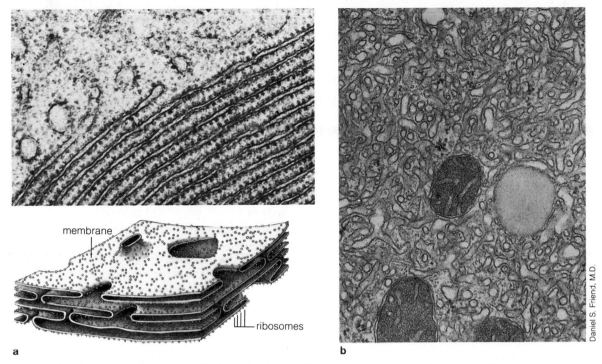

membrane

ribosomes

Daniel S. Friend, M.D.

a

b

Figure 4.11 (**a**) Rough endoplasmic reticulum, showing how the membrane surface facing the cytoplasm is studded with ribosomes. (From Bloom and Fawcett, *A Textbook of Histology*) (**b**) Smooth endoplasmic reticulum.

side of the membranes. Generally, proteins destined for export from the cell are assembled on these ribosomes. (In contrast, proteins used inside the cell generally are assembled on free ribosomes in the cytoplasm.) Even as they are being assembled, protein chains snake rapidly through some kind of opening into the tubes or flattened sacs of the membrane system. At the same time, the chains may have carbohydrate groups added to them. Once inside, the proteins become folded into their final shape; hence they cannot snake out again. Instead they are moved on through the channels to other membrane systems, then into vesicles that fuse with the plasma membrane and dump their protein cargo to the outside.

The membrane system called **smooth endoplasmic reticulum** has no ribosomes attached. Besides helping to isolate and transport materials, the smooth membranes contain enzymes that take part in the assembly of fats. Intestinal cells (which make triglycerides from lipid subunits) and adrenal cortex cells (which make steroid hormones) are rich in smooth endoplasmic reticulum. In liver cells the membranes have two additional functions. Some of their enzymes take part in glycogen breakdown. Others

inactivate toxic by-products of metabolism and foreign substances such as drugs. Figure 4.11 shows examples of both endoplasmic reticulum systems.

Synthesis and Secretion of Materials: The Golgi Complex

Besides the two kinds of endoplasmic reticulum, eukaryotes have another material-processing system. This membrane system functions in assembling substances, packaging them up, and transporting them. It is called the **Golgi complex** (after its discoverer). Figure 4.12 is an example of this system. The Golgi complex accepts membranous sacs pinched off the endoplasmic reticulum, modifies both the membranes and the substances inside, then releases them for distribution. The resulting products include lysosomes and secretory sacs, which will be described shortly. Most polysaccharides are synthesized in the Golgi complex. Glycoproteins partially assembled in the endoplasmic reticulum also undergo some final modification here. Figure 4.13 summarizes the interconnections between the three membrane systems.

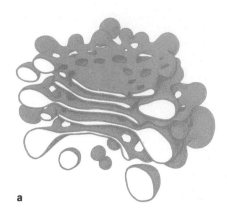

a

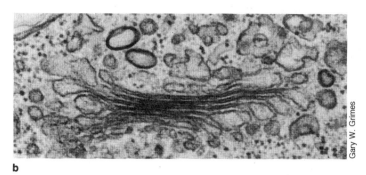

Gary W. Grimes

b

Figure 4.12   (**a**) Sketch of a Golgi complex. (**b**) Golgi complex, as it looks when sliced lengthwise. 53,300×. Notice the swollen sacs at the edges of the stack, and the free vesicles just beyond them.

Storage: Plastids and Vacuoles

Sometimes when materials assembled in the cell are not intended for export, storage organelles concentrate them in places out of the way of metabolic activity. In plant cells, **plastids** are such storehouses (Figure 4.14). *Leucoplasts* are colorless plastids that may accumulate starch grains. They are abundant in potato cells. *Chromoplasts* store plant pigments such as those giving carrots and fruits their characteristic color. The chloroplast, known mostly as the organelle of photosynthesis, is an example of this kind of plastid. In some cells, the chloroplast contains starch grains built up from sugar subunits. Other plastids contain oils and similar high-energy compounds as food reserves.

As you will read later, animal cells also contain storage organelles. These organelles are known as **vacuoles**. Typically they form through fusion of many smaller, membrane-bound chambers called *vesicles*. A vesicle is like a taxicab transporting particles through the cytoplasm, as well as to and from the plasma membrane.

Enhancing Environmental Contact: Plant Vacuoles

Although it also applies to membranous sacs in animal cells, the term "vacuole" is more commonly associated with a large, fluid-filled sac in many plant cells (Figure 4.14).

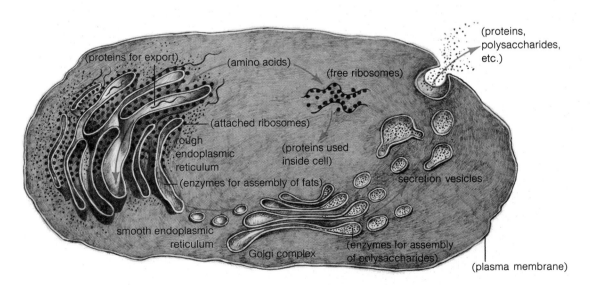

Figure 4.13   Overview of the interconnections between eukaryotic cell systems for assembling, isolating, transporting, and secreting substances.

Sometimes the plant vacuole takes up as much as ninety percent of the cell interior. Some plant vacuoles may be storage areas for metabolic by-products. But their main function appears to be one of increasing cell size and surface area. As fluid pressure builds up inside the vacuole, it presses against the rigid cell wall. Cells can become extended and grow by this force. The resulting increase in surface area enhances the absorption of essential ions and nutrients, which are present only in dilute amounts in watery environments or in the surrounding soil. Root hairs lengthen through vacuole enlargement, and enlargement does expose the root cell to more of the environment.

Dismantling and Disposal: Lysosomes

As part of normal maintenance activities, a cell routinely breaks down malfunctioning structures, worn-out organelles, and assorted foreign particles into their component molecules. In animals, the organelle of disassembly and disposal is the **lysosome**. It is a membranous container for a host of hydrolytic enzymes. These enzymes are able to break down virtually every large molecule used in cell architecture. Obviously, the enzymes could not be allowed to float about in the cytoplasm; they would destroy everything in sight. Instead they are isolated from it by the lysosomal membranes.

Figure 4.15 illustrates the disposal of worn-out organelles. A membrane is wrapped around the organelles and forms a sac. Several lysosomes fuse with the sac and their enzymes pour into it. Structures of the worn-out organelles are then hydrolyzed to ever smaller molecules, which become small enough for transport across the lysosomal membrane. The cell may reuse these small molecules in future building programs.

Perhaps you are wondering why the enzymes don't dismantle the lysosomal membranes along with everything else. Although the picture is far from clear, the cell apparently expends energy to maintain the membranes. This energy expenditure counters the constant enzyme attacks. A cell falls apart rapidly once it dies. Without energy outlays, lysosomal membranes cannot be maintained. Once the membranes disintegrate, the enzymes flood into the cytoplasm, there to destroy indiscriminately.

Lysosomes have been found only in animal cells. But some of the same hydrolytic enzymes are found in the central vacuoles of plant cells. The vacuoles may thus have a dual function: increasing cell volume, and disposing of wastes. Structures resembling lysosomes have also been found in fungi and protistans.

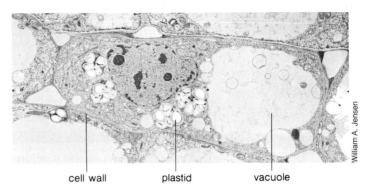

cell wall   plastid   vacuole

William A. Jensen

Figure 4.14 Plastids and a central vacuole in a cotton plant cell. Compare this micrograph with Figure 4.4.

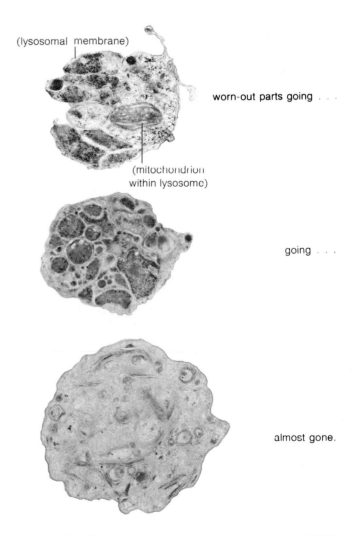

(lysosomal membrane)

worn-out parts going . . .

(mitochondrion within lysosome)

going . . .

almost gone.

Figure 4.15 Digestion of organelles as seen in a lysosome. 12,000×. (Gary W. Grimes)

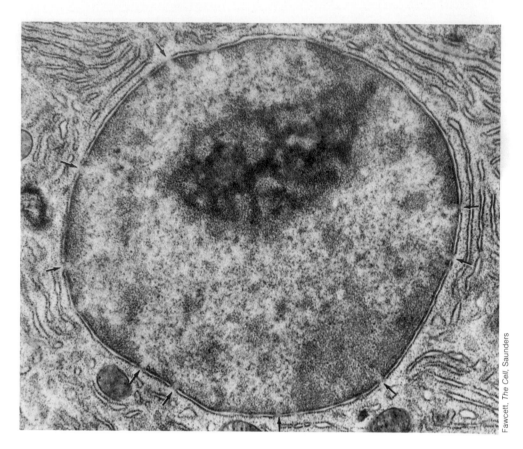

Figure 4.16 Electron micrograph of the nucleus from a bat's pancreatic cell. Arrows point to pores in the nuclear envelope. Dark region is the nucleolus. 12,600 × .

THE MEMBRANE-ENCLOSED NUCLEUS

All eukaryotic cells contain a **nucleus**: a large organelle isolated from the cytoplasm by a double membrane (Figure 4.16). It is the storehouse of information that assures survival and reproduction. How the nucleus controls reproduction as well as the everyday activity of the cell will be considered in later chapters. Here, we will simply consider some aspects of nuclear structure and function.

The Nuclear Envelope

The semifluid substance inside the nucleus is called the **nucleoplasm** to distinguish it from cytoplasm. The nucleoplasm is enclosed in a double membrane system known as the **nuclear envelope**. The outer surface of the membrane facing the cytoplasm is richly peppered with ribo-

somes. In some cases, this outer membrane is continuous with the endoplasmic reticulum.

At regular intervals, pores break through the nuclear envelope (Figure 4.16). A mass of material fills much of the pore and even extends beyond the pore diameter. In the center is a small opening. Little is known about the function of this organization. It seems to exert extremely selective control over the movement of molecules into and out of the nucleus. Intriguingly, the number of pores often increases with cell activity—as it does, for example, during cell division.

The Nucleolus

The most prominent structural feature of a nucleus is an internal mass that usually looks more dense than its surroundings in electron micrographs. This mass is the

nucleolus, the region where ribosomal subunits are synthesized (Figure 4.16). The mass is actually the raw material, including proteins and RNA, used in ribosome synthesis. During division, the nucleolus usually disperses throughout the nucleoplasm. After division is completed, a nucleolus forms once again in each of the two daughter cells.

Chromatin and Chromosomes

Anchored to the inner membrane of the nuclear envelope, on the side facing the nucleoplasm, are masses of DNA and associated proteins. These masses, which extend throughout the cytoplasm, are collectively called **chromatin**. They represent the uncoiled form of DNA as it is being used by the cell. Some of the proteins may lend structural support to the DNA molecules; others may regulate what information is read from the overall chemical message contained in DNA.

During division, chromatin becomes tightly coiled into distinct bodies called **chromosomes**. The name means "colored bodies," and refers to the fact that they are easily stained for viewing with the light microscope. Visible chromosomes are the form in which chromatin is condensed for distribution during the nuclear division of eukaryotic cells.

CELL SHAPE AND MOVEMENT

Microfilaments and Microtubules

Very few cells are spherical, even when freed from the physical confines of multicellularity. And few cells are internally rigid. If you were to look through a microscope at any living cell, you undoubtedly would be impressed with its contant motion. Cellular contents stream about endlessly in regular patterns. Chloroplasts move in response to changes in the sun's overhead position. Vesicles form at cell boundaries, pinch off from the plasma membrane, and move toward specific regions inside. Even animal cells from an organ that usually stays put (say, the liver) move actively when they are separated from the tissue and placed in a culture dish. (Figure 4.1, which shows the edge of a cell taken from a kidney of a small mammal, dramatically illustrates such movement.) Two kinds of structures have been implicated in example after example of cell shape and movements within the cell. They are microfilaments and microtubules.

A **microfilament** is an extremely fine structural element composed of *actin* (or actinlike material) and *myosin*. Both components are proteins that are known to play a major role in muscle contraction (Chapter Fifteen). If a cell having microfilaments attached to its plasma membrane is made to contract, the cell narrows at one end. If a cell having a ring of microfilaments around its middle is made to contract, the cell pinches in two (as it does during cell division). If a cell having microfilaments attached to the plasma membrane and to internal organelles is made to contract, the organelles are drawn toward the cell surface. In a cell crawling across a culture dish, microfilaments running from the lower front surface to the upper midsurface are involved in contractions that move the cell body forward. In addition, microfilaments are involved in *cytoplasmic streaming*, a constant motion of subcellular structures that is so evident in plant cells.

A **microtubule** is a hollow cylinder made primarily of proteins called *tubulins*. It is larger in diameter and more rigid than a microfilament. When tubulin subunits are added to or removed from one end, a microtubule gets longer or shorter; when microtubules are clustered together, they may interact and slide past one another. Both mechanisms apparently can cause an organelle attached to the microtubules to move. The list of events in which microtubules take part is impressive, and growing. Microtubules just beneath the plasma membrane dictate the shape of many cells and cell extensions. In growing plant cells, they may be involved in wall building. They help move chromosomes during nuclear division. Microtubules are also involved in the beating of flagella and cilia, which will be described shortly.

Recent evidence suggests that microfilaments and microtubules are highly organized in the ways in which they anchor and rearrange internal cell structures (Figure 4.17). In fact, the cell biologist Keith Porter sees the patterns of radial and parallel filaments, of fibrous mesh and bundles as no less than "cytomusculature." With such structures, organelles are constantly moved about during the dynamic life of a cell.

What mechanisms are responsible for these controlled movements? That is not known. In some manner, glycoproteins attached to the plasma membrane serve as receptors for chemical information about internal and external conditions. Pathways of communication exist between these glycoproteins and the filamentous network within the cytoplasm (Figure 4.17f). The signals transmitted to them somehow trigger changes in the distribution of microfilaments and microtubules—a distribution that under-

Figure 4.17 Glimpses of the cytomusculature of cells.

(**a**) Microfilaments. In many plants, cytoplasmic streaming of organelles such as the chloroplasts shown here seems to occur along tracks defined by microfilament bundles.

(**b**) Microtubule network that extends from the nucleus to the periphery of this animal cell. In *indirect immunofluorescence,* the location of a given substance is revealed by adding fluorescent-labeled antibodies (glycoproteins that combine preferentially with that substance) to the cell. Illumination of the cell with ultraviolet light shows the location of the antibodies, hence the substance. Here, the substance is tubulin.

(**c**) Microtubules, showing the regular packing array of their tubulin subunits.

(**d**) *Microvilli* (fingerlike bulges) and *blebs* (knoblike bulges) of cells from a hamster. These surface projections shoot out constantly, then retract, then shoot out elsewhere in a matter of seconds or minutes.

(**e**) Actin filaments visible in a cell taken from mouse tissue; the cell is moving from left to right.

(**f**) Components of cytomusculature. Shown here, the base and tip of a microvillus. Note the bridges between membrane glycoproteins and microfilaments.

(a from Y. M. Kersey and N. K. Wessels, *J. Cell Biology,* 1976. 68:264; b from M. Osborn and K. Weber, *Proc. Nat. Acad. Sci.,* 1976, 73: 867; c from R. Barton; d from D. Billen and A. Olson, *J. Molecular Biology,* 1976, 69:732. Copyright by Academic Press Inc. Ltd.; e from R. Pollack; f from Francis Loor; adapted from *Nature,* 264:272. Copyright ©️ 1976 Macmillan Journals l imited.)

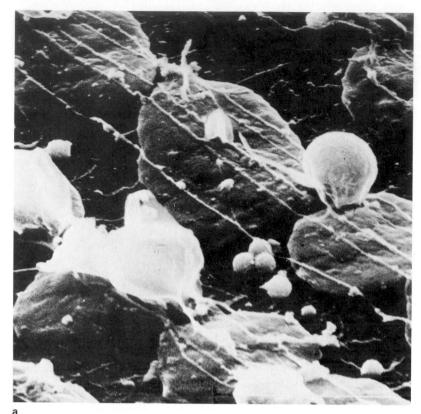

a

c

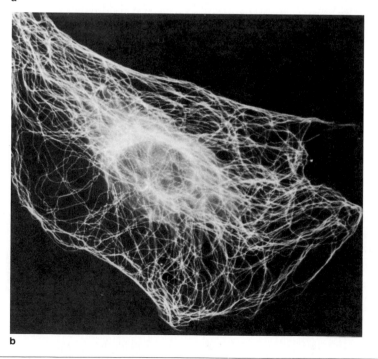

b

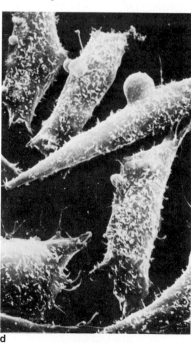

d

e

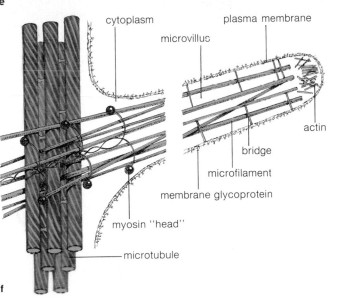

cytoplasm

plasma membrane

microvillus

actin

bridge

microfilament

membrane glycoprotein

myosin "head"

microtubule

f

lies cell shape, motion, and growth. Deciphering these signal mechanisms is sure to have profound impact not only on basic biology but on medical research. They are at the heart of cell functioning, cell division, the cellular response to injury, and the invasion of tissues by cells that have become cancerous.

Cilia and Flagella

So far, we have talked about frameworks and movements within the cell body. But many cells also move through their environment. How do they do it? Two types of cell structures provide the means for rapid propulsion. Among eukaryotic cells, these structures are the **cilium** and the **flagellum** (plural, cilia and flagella). Cilia are shorter than flagella, and more numerous where they do occur. Both are built from microtubules. Nine pairs of microtubules are arranged radially about two central microtubules. The pattern is called the *9 + 2 arrangement*. It extends outward from the cell body in a sheath that is continuous with the plasma membrane (Figures 4.18 and 4.19).

Motile structures also occur among prokaryotic cells. Usually they are long, ropelike threads composed of only one kind of protein filament (flagellin), and they do not contain a radial array of microtubules. Hence prokaryotic flagella have little in common with their eukaryotic counterparts.

Small, barrel-shaped structures called **centrioles** give rise to the microtubular array of cilia and flagella. As Figure 4.18 suggests, each centriole has nine radially arranged sets of microtubules. After the cilium or flagellum has formed, the centriole remains attached at its base and is henceforth called the **basal body** of the motile structure.

Centrioles occur only in the cytoplasm of eukaryotes having motile cells during some stage of the life cycle. You will find them among protistans, fungi, many plants, and most animals; all have flagellated reproductive cells. Just how important centrioles are to the continuation of life is clear from the fact that, like DNA molecules, they are duplicated and passed on to the next generation prior to cell division.

Propelling the cell body through watery worlds is one role carried out by cilia and flagella. Among many animal species, cilia carry out other roles. Cilia project in dense arrays from cells of organs such as nasal passages and tubelike bronchi of the respiratory system, where they help move fluids and dissolved particles. A great number of sensory organs depend on modified cilia, as you will read in Chapter Fifteen.

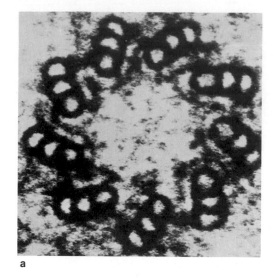

a

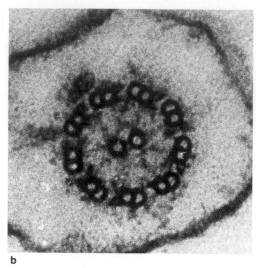

b

Figure 4.18 (**a**) Arrangement of microtubules in a basal body (centriole), shown in cross-section. 270,000×. (E. de Harven in A. J. Dalton and F. Hagenau, eds., 1968, *The Nucleus,* Academic Press, Inc.) (**b**) Arrangement of microtubules in a cilium, also in cross-section. 133,500×. (W. A. Jensen and F. B. Salisbury, 1972, *Botany*) (**c**) This drawing shows the relationship of these two structures relative to each other in a flagellum.

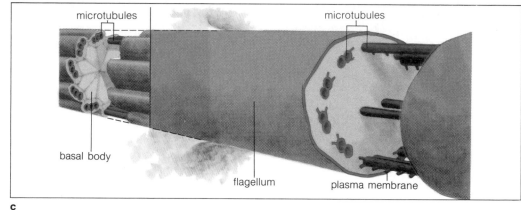

microtubules

microtubules

basal body

flagellum

plasma membrane

c

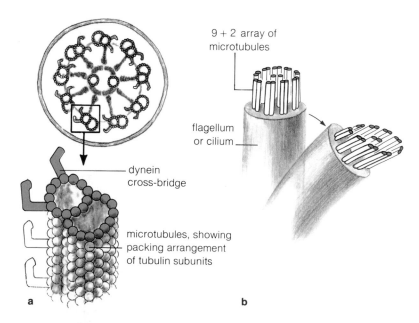

9 + 2 array of microtubules

flagellum or cilium

dynein cross-bridge

microtubules, showing packing arrangement of tubulin subunits

a

b

Figure 4.19 Movement of cilia and flagella. The mechanism underlying movement of complex flagella and cilia is not completely understood. According to one view, numerous clawlike ''arms,'' or cross-bridges (**a**), reach from one microtubule doublet in the outer ring to the next in line. Cross-bridges are made of *dynein,* a protein that takes part in converting the chemical energy of ATP to the mechanical energy of movement. When ATP energizes these molecules, they bend in a coordinated way. Cross-bridges between adjacent doublets are broken, and one doublet slides along the surface of the other. (**b**) Because all doublets in the ring are connected by radial spokes to the central doublet, the displacement caused by the sliding force is converted to a bending of the motile structure.

Table 4.1 Structures Typically Found in Prokaryotic and Eukaryotic Cells

Cell Structure	Primary Function	Prokaryotes	Eukaryotes			
		Monerans	Protistans	Fungi	Plants	Animals
Cell wall	Protection, support	✔	✔	✔	✔	
Plasma membrane	Selective boundary layer	✔	✔	✔	✔	✔
Photosynthetic pigments	Light-energy conversions	✔*	✔*		✔	
Membrane-bound chloroplasts	Light-energy conversions		✔*		✔	
Other plastids	Storage		✔		✔	
Mitochondria	Energy metabolism		✔	✔	✔	✔
Ribosomes	Assembling molecules	✔	✔	✔	✔	✔
Endoplasmic reticulum	Assembling molecules		✔	✔	✔	✔
Golgi complex	Assembling, secreting molecules		✔	✔	✔	✔
Vacuoles	Variable (e.g., storage, digestion)		✔	✔	✔	✔
Lysosomes	Breaking down molecules, structures		?	?	?	✔
Microbodies	Interconversions of molecules		✔	✔	✔	✔
Microtubules, microfilaments	Structural support, internal movement		✔	✔	✔	✔
Complex flagella, cilia	Movement through environment		✔*	✔*	✔*	✔
Centriole	Gives rise to microtubule systems		✔*	✔*	✔*	✔
DNA molecules	Information storage	✔	✔	✔	✔	✔
Membrane-bound nucleus	Genetic control		✔	✔	✔	✔
Chromosomes	Packaging DNA during division		✔	✔	✔	✔
Nucleolus	Assembling ribosomes		✔	✔	✔	✔

*Occurs in some groups but not in others.

PERSPECTIVE

All living things are made of one or more cells; each cell is a basic living unit of structure and function; and new cells can arise only from cells that already exist. Beginning with this cell theory, we have looked at the internal structures of this fundamental living unit. A later chapter describes the long history of chemical evolution that is thought to have preceded the appearance of the first cell. Given what we have covered so far, we can make the following points in anticipation of that story.

At the minimum, the basic cell form must have included all the parts necessary for keeping the simplest of living cells—the prokaryotes—alive. These parts are the plasma membrane, cytoplasm, ribosomes, DNA molecules, and enzymes. But suppose that mutations began giving rise to variant cell forms, and that environments began changing. Then agents of selection would have gone to work on the basic cell form. We can speculate that

eukaryotic cells almost certainly evolved from forms not much different from modern-day prokaryotes. The hallmark of eukaryotes is not only the membrane-enclosed nucleus, which isolates the hereditary material from the clutter of cytoplasmic activities. It is also an *abundance* of internal membranous compartments, put to use in diverse and highly specialized ways. In some primordial time, certain prokaryotic cell lines equipped with slight inward or outward dimplings in the plasma membrane may have evolved into cells having more pronounced infoldings and outfoldings. There could have been selective advantage in this, for cells having extensive membranes would have more surface area on which metabolic reactions could occur. Hence in some environments, they would have been more efficient at extracting and using the energy and materials available. In many environments, there would be still greater advantages in having internal membranes folded and fused into channels, tubes, and chambers. Such developments would have allowed such activities as material synthesis, packaging, transport, and secretion to be carried out more rapidly than could be done in the single, multipurpose "compartment" called the prokaryotic cell body. Whatever the pathways of cellular evolution, modifications to the basic plan have accrued on the same theme: *the acquisition and processing of energy and materials in highly controlled, specialized ways.* This theme will occupy our attention in chapters to follow.

Readings

Bloom, W. and D. Fawcett. 1975. *A Textbook of Histology.* Philadelphia: Saunders. Advanced reading, but an excellent reference book with beautiful electron micrographs.

Dyson, R. 1978. *Cell Biology: A Molecular Approach.* Second edition. Boston: Allyn and Bacon. Excellent, up-to-date text and graphics; packed with details yet well conceived in topical integration.

Fawcett, D. 1966. *The Cell: An Atlas of Fine Structure.* Philadelphia: Saunders. Outstanding collection of micrographs.

Kessel, R., and C. Shih. 1974. *Scanning Electron Microscopy in Biology: A Student's Atlas of Biological Organization.* New York: Springer-Verlag. Stunning micrographs of cell structures and products.

Lazarides, E., and J. Revel. 1979. "The Molecular Basis of Cell Movement." *Scientific American* 240(5): 100-113. Fine summary article on research into cell movement.

Ledbetter, M., and K. Porter. 1970. *Introduction to the Fine Structure of Plant Cells.* New York: Springer-Verlag.

Loewy, A., and P. Siekevitz. 1969. *Cell Structure and Function.* Second edition. New York: Holt, Rinehart and Winston. Good reference text. In paperback.

Novikoff, A., and E. Holtzmann. 1976. *Cells and Organelles.* Second edition. New York: Holt, Rinehart and Winston. Very nice introduction to cell structure and function. In paperback.

Satir, B. 1975. "The Final Steps in Secretion." *Scientific American* 233(12): 28-37.

Satir, P. 1974. "How Cilia Move." *Scientific American* 231(4):44-52. Correlates ciliary movement with microtubule activity.

Review Questions

1. What are the three main ideas of the cell theory?

2. The plasma membrane, cytoplasm, and nucleus are three functional zones of most cells. Can you briefly state what kinds of activities go on in each zone?

3. How do prokaryotes differ from eukaryotes? Give two examples of each kind of organism.

4. How do cell walls differ from plasma membranes in structure and function?

5. Describe the structure and function of a chloroplast and a mitochondrion. In what way are these two organelles similar?

6. Name the eukaryotic organelles involved in assembling, isolating, transporting, and secreting substances from the cell:

Which eukaryotic organelles are concerned with storage? What seems to be the main function of plant vacuoles?

7. Lysosomes dismantle and dispose of malfunctioning organelles and foreign particles. Can you describe how?

8. Describe the structure and main functions of the nucleus.

9. _____ and _____ are two structural elements that give rise to cell shape and movement. What mechanisms may control their assembly and disassembly?

10. How are microtubules arranged in cilia and in flagella? What structure apparently gives rise to this arrangement?

11. With a sheet of paper, cover the Table 4.1 column entitled Primary Function. Can you now name the primary functions of the cell structures listed in this table?

Describing cell structures as things unto themselves is like trying to explain the logic of the human body without even mentioning the environment in which it has come to exist. One might well be left to wonder about the selective advantages of our particular array of parts. Why, for instance, do we have forward-directed eyes instead of an eye on each side of the head, fishlike? Why a two-legged posture instead of a more stable four-legged stance? Why those bothersome wisdom teeth, which are four too many for such a small jaw? Such traits, for better or worse, bear the stamp of environmental pressures long since past. The human evolutionary story, with some of its environmental overtones, is told later in the book. But the same idea applies here, at the level of cells. Why does a cell have a selectively permeable membrane instead of a solid barrier to the outside? Why internal networks of microtubules and microfilaments in eukaryotic cells but not in prokaryotic? Why rigid external walls in some cells, and why microvilli in others? The logic of the *cell* body becomes evident only by taking into consideration the nature of the places where cells have come to exist.

THE CELLULAR ENVIRONMENT

To begin our story, cellular life must be maintained within a narrow range of environmental conditions. Life hinges not merely on a reliable food source. It hinges also on the availability and quality of such environmental factors as water, radiation, temperature, pressure, and various ions. For most cells in most places, these factors are not constant. Temperatures rise and fall. Sunlight intensity varies not only through the day but through the year. Water may be abundant in spring and scarce in autumn. In estuaries, where seawater mixes with currents from rivers and streams, ion concentrations change with the tides. Hence survival depends largely on tolerance to variable conditions:

Each type of cell has a range of structural and functional responses to some range of conditions typical of the environment in which its predecessors evolved.

In this chapter, we will be looking at some aspects of the ever-changing cellular environment. We will also be considering some examples of cellular responses to them.

5

THE CELL IN ITS ENVIRONMENT

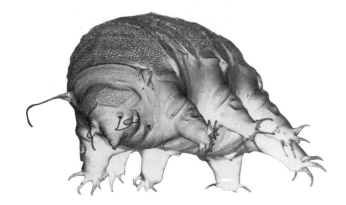

Figure 5.1 Without water there is no life—sort of. One of many organisms whose cells can adapt to extremely dried-out environments: a water bear on the prowl, magnified 300 times. (Robert D. Schuster)

WATER AND CELL FUNCTIONING

On the average, about seventy-five to eighty-five percent of an active cell is water. Considerable water is also present in the extracellular environment. Multicelled plants, fungi, and animals do have an outer layer of cells exposed to the surrounding air, but even these cells are in contact with water present inside the body.

It is not that life necessarily ends in the absence of water. Some plant spores can resume active growth after being dried out for centuries. Some worms called nematodes, when given enough time to prepare themselves for the occasion, can stay dried out for decades—only to resume activity upon being moistened again. Or consider the water bear (Figure 5.1). This tiny animal lives in ponds and damp soil. Its moist home may freeze over or bake beneath the summer sun, yet the water bear survives. It dries out very, very systematically and enters a state of suspended animation. Certain body parts, through which water normally escapes, are withdrawn from exposure to air. The body contracts so tightly it looks like a small barrel. As the body contracts, its cells produce a substance that replaces the water normally surrounding large biological molecules. The substance seems to hold these molecules in place and protect them from mechanical damage. Thus, through tight packing and controlled drying out, the structure of each cell is maintained through harsh times. With the return of favorably moist conditions, the water bear revives and actively goes about the business of living.

Water bears, nematodes, rotifers, lichens, spores, seeds, mosses—all are well matched to environments in which water periodically becomes scarce. Yet with their seeming independence of free-flowing water, the key word in all of this is *active* existence:

None of the <u>activities</u> associated with the term "living" can proceed without water.

What is it about water that makes it so central in cell functioning? Among other things, water has outstanding temperature-stabilizing properties and cohesive properties, and it happens to be the best solvent around.

Consider, first, water's *temperature-stabilizing properties.* The term **specific heat** refers to how much heat energy a single gram of a substance can absorb before its temperature increases by 1°C. Bodies of water absorb or give off quite a bit of heat energy before there is any drastic temperature change. What is the basis of water's high specific heat? All atoms and molecules are in constant motion, and their motion increases when they absorb extra energy. In fact, what we call temperature is a measure of the rate of molecular motion. But recall that hydrogen bonds organize water molecules relative to each other. Heat some water, and the hydrogen atoms locked within each molecule quickly absorb extra energy and begin moving faster. As they do, hydrogen bonds *between* neighboring water molecules are more readily broken and formed anew. With each making and breaking of bonds, some heat energy escapes and is absorbed by the hydrogen atoms of nearby molecules; it is not converted to the energy of motion of *whole molecules.* So the temperature does not increase as fast as it otherwise would. This temperature-stabilizing property is most fortunate for living things, which release considerable heat during metabolism. Without water's ability to conduct heat, cells would literally stew in their own juices.

Another temperature-stabilizing property of water is its high **heat of vaporization**: the amount of heat energy that must be absorbed before a single gram of liquid will be converted to gaseous form. For water, the amount is one of the highest known (539 calories are needed to bring a gram of water past its boiling point of 100°C). This property, too, can be traced to hydrogen bonds. Because of these bonds, individual water molecules resist moving faster under the prodding of incoming heat. Beyond the boiling point, though, the molecular vibrations *are* enough to break hydrogen bonds. Molecules at the water's surface begin moving fast enough to escape into the surroundings. This liberating process is called **evaporation**. It involves the release of considerable energy (the energy of motion of the escaping water molecules). Hence the surface of the water left behind drops in temperature. Oasis plants and many animals cool off through evaporative water loss. ("Sweating" in itself doesn't cool your body on hot days. Special glands actively secrete fluid to your skin surface, but the fluid must evaporate for surface cooling to occur. On extremely humid days, when atmospheric water content is high, molecular escape is hindered and sweat pools on your skin.)

Other examples could be given of water's notable temperature-stabilizing properties. But here it is enough to summarize their importance in this way:

Because of hydrogen bonds between its molecules, water can serve as a buffer against localized temperature extremes inside and outside the cell.

Figure 5.2 The water strider, an insect adapted to water's high surface tension (resulting from the tenacity with which water molecules cling to one another through hydrogen bonds).

In addition to resisting temperature changes, water shows remarkable *cohesive properties.* In pure water, each molecule is attracted to and hydrogen-bonded with other molecules. Such an attraction between like molecules is called **cohesion**. Because of the attraction, water molecules resist separating from each other when they are placed under tension. At air–water interfaces, hydrogen bonds exert a constant inward pull on molecules at the water's surface and impart a high surface tension to it (Figure 5.2). That is why beads of water form; that is why a pond surface resists penetration by small leaves drifting down on it. Cohesive forces and other factors also give water molecules the capacity to pull one another up through narrow passageways, to the top of even the tallest tree.

Perhaps the most remarkable feature of all, though, is water's capacity to dissolve other substances. This capacity underlies the organization of both cytoplasm and the plasma membrane—the cell's interface between its internal and external worlds. Water is an outstanding solvent. (A *solvent* is any fluid in which one or more substances is dissolved; *solutes* are what's dissolved.) Most substances that become dissolved in water remain unchanged by the association, because water is chemically inert. Thus molecules of varied sorts may be transported in blood or sap, or stored in solution, without danger of becoming chemically altered by the water. Even so, the *positioning* of solutes is influenced by the association. A water molecule, recall, shows polarity. Its oxygen "end" gets a bigger share of the electrons being shared with the two hydrogen atoms at the other "end." Sometimes the sharing gets even more unequal: the molecule separates, or *dissociates* into charged parts. The parts interact with solutes or with other water molecules, and orient them relative to one another. As you will see, water's solvency and polarity both help give rise to the complex organization of a cell—a largely fluid structure that exists in a largely fluid world.

ACIDS, BASES, AND SALTS

From time to time, with predictable frequency, a water molecule gets separated into these two charged parts:

becomes

The OH^- remnant is called a **hydroxyl ion**. The H^+ remnant is a **hydrogen ion**. Other substances, too, undergo dissociation when they are dissolved in water. They form a hydrogen ion and some kind of negatively charged ion. Carbonic acid is such a substance:

$$H_2CO_3 \rightleftharpoons HCO_3^- + H^+$$

carbonic acid   bicarbonate ion   hydrogen ion

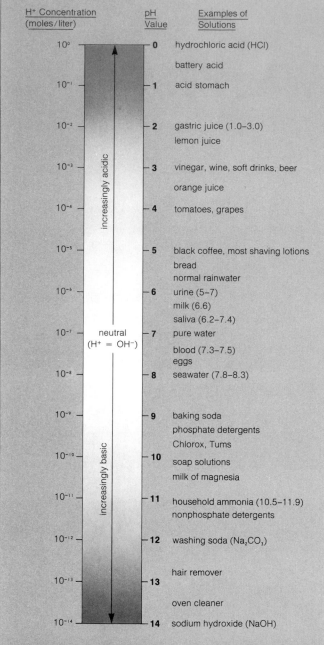

H+ Concentration (moles/liter)

H+ Concentration (moles/liter)	pH Value	Examples of Solutions
10^0	0	hydrochloric acid (HCl)
		battery acid
10^{-1}	1	acid stomach
10^{-2}	2	gastric juice (1.0–3.0)
		lemon juice
10^{-3}	3	vinegar, wine, soft drinks, beer
		orange juice
10^{-4}	4	tomatoes, grapes
10^{-5}	5	black coffee, most shaving lotions
		bread
		normal rainwater
10^{-6}	6	urine (5–7)
		milk (6.6)
		saliva (6.2–7.4)
10^{-7}	7	pure water
		blood (7.3–7.5)
		eggs
10^{-8}	8	seawater (7.8–8.3)
10^{-9}	9	baking soda
		phosphate detergents
		Chlorox, Tums
10^{-10}	10	soap solutions
		milk of magnesia
10^{-11}	11	household ammonia (10.5–11.9)
		nonphosphate detergents
10^{-12}	12	washing soda (Na_2CO_3)
		hair remover
10^{-13}	13	
		oven cleaner
10^{-14}	14	sodium hydroxide (NaOH)

increasingly acidic

neutral ($H^+ = OH^-$)

increasingly basic

Figure 5.3 The pH scale, in which a fluid is assigned a number according to the number of hydrogen ions present in a liter of that fluid. The scale ranges from 0 (most acidic) to 14 (most basic), with 7 representing the point of neutrality.

A change of only 1 on the pH scale means a tenfold change in hydrogen ion concentration. Thus, for example, the gastric juice in your stomach is ten times more acidic than vinegar, and vinegar is ten times more acidic than tomatoes.

A substance whose molecules *release* hydrogen ions in solution is known as an **acid**.

Notice, in the example just given, that the reaction is reversible: it can run in either direction (hence the double arrows). This means that bicarbonate can combine with hydrogen ions to form carbonic acid. Any substance that *combines* with hydrogen ions in solution is a **base**.

In a glass of pure water, the amounts of H^+ and OH^- ions are always the same. Hence pure water doesn't act like an acid or a base; it's said to be neutral. But suppose you pour some hydrochloric acid (HCl) into the water. This substance separates into Cl^- and H^+ ions. With more H^+ ions now in the glass, the OH^- ions are more likely to bump into and combine with them, thereby forming water molecules. The outcome is that the hydroxyl ion concentration decreases by as much as the hydrogen ion concentration increased!

Such changes in hydrogen ion concentration occur all the time in the cell environment. As you might imagine, it's often useful to be able to define the extent of change taking place. The trouble is, it would be cumbersome to work with numbers at the molecular level. (For example, that glass of pure water has 150,000,000,000,000,000 hydrogen ions in it!) That is why fluids are assigned a **pH value**: a whole number that is a shorthand way of referring to the number of hydrogen ions present in a liter of the fluid. Pure water, which is neutral, has a pH of 7. This is the midpoint in a scale ranging from 0 to 14 (Figure 5.3). A substance having a pH of less than 7 is *acidic*. If its pH is greater than 7, the substance is *basic*. All biochemical reactions are sensitive to shifts in pH.

Hydrogen and hydroxyl ions are so central to cell functioning that it is sometimes easy to overlook the fact that the cell also depends on many ionic substances which contain neither one. Such substances are called **salts**. For cells, essential salts include positively charged ions of potassium (K^+), sodium (Na^+), calcium (Ca^{++}) and magnesium (Mg^{++}), as well as negatively charged chloride (Cl^-). Together with water, these salts represent the main inorganic compounds present in cytoplasm. For instance, in most animals, nerve messages cannot travel without potassium and sodium ions, and muscle cell contraction is impossible without calcium ions. In the human body, about ninety percent of all ions dissolved in extracellular fluid are those of sodium and chlorine. In complex land plants, potassium ions are needed in activating many enzymes. And apparently they are involved in transporting such vital substances as nitrates and phosphates throughout the plant body.

CYTOPLASMIC ORGANIZATION OF WATER AND SOLUTES

In addition to ions and small molecules such as amino acids, the cytoplasm contains an array of proteins, lipids, polysaccharides, and nucleic acids. Most of these molecules have electrically charged groups of one sort or another. Hence they also interact with cytoplasmic water.

Consider how water becomes organized around ions and small molecules. Crystals of table salt (NaCl) get separated into Na^+ and Cl^- ions when dissolved in water. Water molecules tend to cluster around each positively charged ion, with their negative ends pointing toward it:

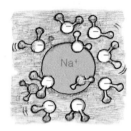

Similarly, water molecules tend to cluster around each negatively charged ion, with their positive ends pointing toward it:

These so-called "spheres of hydration" shield the charged ions and keep them from interacting. They also force ions to remain dispersed in the cytoplasm rather than settling in some part of the cell. *Thus water molecules help keep vital ions available for reactions throughout the cell.*

Interactions between water and large biological molecules are more intricate. Water molecules become oriented next to charged regions of these molecules, just as they do around Na^+ or Cl^-. In addition, some of these large molecules have polar regions, which form hydrogen bonds with (dissolve in) water. The regions are said to be *hydrophilic*, or water-loving. Nonpolar regions of these molecules, which have no net charge, show little tendency to form hydrogen bonds with water. They are said to be *hydrophobic*,

or water-dreading. For instance, phospholipids contain fatty-acid "tails" (which repel water) and a "head" with alcohol and phosphate groups (which dissolve in water). Phospholipid structure may be drawn in this generalized way:

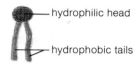

hydrophilic head

hydrophobic tails

In fluids, phospholipids sometimes become spontaneously arranged relative to one another in a two-layer film. In this **lipid bilayer**, hydrophobic tails point inward, tail-to-tail, and form a region that excludes water. The hydrophilic heads point toward the surrounding water molecules and are hydrogen-bonded with them:

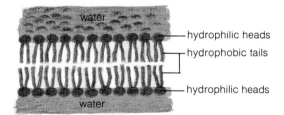

water

hydrophilic heads

hydrophobic tails

hydrophilic heads

water

As you will soon see, this arrangement is the framework for the plasma membrane, the thin barrier between the cytoplasm and the outside world. *Thus membrane organization arises largely in response to water molecules in the cell environment.*

Consider, finally, how water may become organized around the all-important proteins. For many proteins, negatively charged groups predominate on the surface. One result is that the protein surface attracts strongly positive ions present in the surroundings, which then attract negative ions. These ions in turn attract water molecules. Thus an electrically charged "cushion" of ions and water may be formed around the protein:

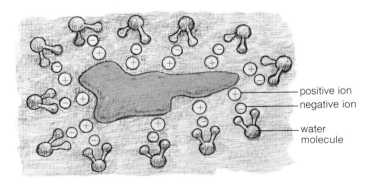

positive ion

negative ion

water molecule

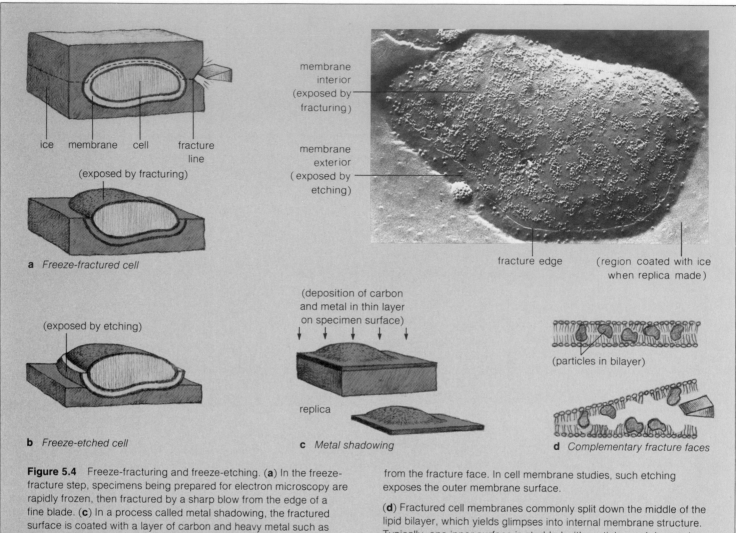

Figure 5.4 Freeze-fracturing and freeze-etching. (**a**) In the freeze-fracture step, specimens being prepared for electron microscopy are rapidly frozen, then fractured by a sharp blow from the edge of a fine blade. (**c**) In a process called metal shadowing, the fractured surface is coated with a layer of carbon and heavy metal such as platinum. This coating is thin enough to replicate details of the exposed specimen surface. The metal replica, not the specimen itself, is used for preparing electron micrographs.

(**b**) Sometimes before the metal replica is made, specimens are freeze-etched, which simply means that additional ice is evaporated from the fracture face. In cell membrane studies, such etching exposes the outer membrane surface.

(**d**) Fractured cell membranes commonly split down the middle of the lipid bilayer, which yields glimpses into internal membrane structure. Typically, one inner surface is studded with particles and depressions, and the other inner surface is a complementary pattern of depressions and particles. The particles are membrane proteins.

The micrograph shows a portion of a replica of a red blood cell, prepared by freeze-fracturing and freeze-etching. (P. Pinto da Silva and D. Branton, *J. Cell Biology*, 1970. 45:598)

When ions and water molecules surround a substance in this way, the substance remains in solution. It is said to be in the **sol state**.

Under some cellular conditions, enough ions become attracted to an oppositely charged protein to neutralize its surface charge. Then, proteins of overall like charge don't repel each other. Instead they begin adhering in a continuous, spongy network known as a **gel state**. The process of blood clotting is an example of this kind of transformation (Chapter Seventeen). In this case, an enzyme strips certain water-soluble proteins of several negatively charged groups, and the proteins stick together in a spongelike net.

The gel state is best typified by gelatin desserts, which are largely protein, sugar, and water. Although gelatin is somewhat dense, solutes readily move through it. (Put some drops of food coloring on a cube of Jell-O and watch what happens.) Indeed, gel structure has been likened to a pile of brush, within which various other solutes and water molecules are dispersed and tenuously arranged by chemical interactions.

The point of these examples is that the membrane-bound cytoplasm is no mere bag of free-floating water molecules and solutes. If we were to extrapolate from the sketch of the "cushioned" protein molecule, we could go on sketching more water molecules in predictable patterns of hydrogen bonding. We would see how these molecules would shift into new bonding patterns as we sketched in ions and different kinds of molecules, large and small, of different chemical reactivity. In fact, given all the interactions that surely must be going on constantly in cytoplasm, it seems probable that little, if any, "free" water is present in the entire cell:

Cytoplasm is not a formless substance. Regional systems of the cell are joined in a chemical continuum with cytoplasmic water—which has considerable structure of its own.

CELL MEMBRANES: LARGELY FLUID STRUCTURES IN A LARGELY FLUID WORLD

So far, we have looked at the chemical nature of both the internal and external environment of the cell. We are now ready to consider the remarkable barrier between them—the plasma membrane. Only in the past few decades have we begun to understand the structure of plasma membranes and to get an idea of how they function in constantly changing environments. Long before this, there were indications that membranes are fluid rather than solid structures. For instance, it was known that cells punctured with fine needles did not lose their cytoplasm when the needle was withdrawn. Instead, the cell surface actually seemed to flow over and seal the punctured region! Yet how could a *fluid* cell membrane remain functionally distinct from fluid surroundings?

For a time, researchers thought the membrane was composed only of lipids, possibly organized in the bilayer arrangement sketched earlier. But membrane properties could not be explained on the basis of lipids alone. Some researchers suspected that proteins were spread out on both sides of a bilayer, with lipids sandwiched in between. But

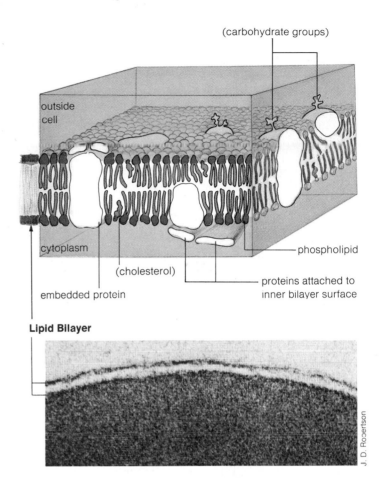

Figure 5.5 Fluid mosaic model of membrane structure, as described in the text. The micrograph shows a red blood cell membrane in thin section, at high magnification.

it was not until the late 1960s that the true nature of cell membranes started becoming apparent, largely through electron micrograph studies. For example, freeze-fracturing and freeze-etching, two new methods of preparing specimens for study, yielded finer details of cell structure (Figure 5.4). The surprising discovery was that each lipid layer was *not* uniformly smooth, as a "sandwich" model would have it. It was studded with membrane-associated particles!

By 1972, S. Singer and G. Nicolson put together the first summarization of what has come to be known as the **fluid mosaic model** of membrane structure. In this model, the bilayer is still the framework for cell membranes. But proteins do not cover both sides of it. Instead, *the lipid bilayer forms a sort of fluid sea, in which diverse proteins are suspended like icebergs.* Figure 5.5 is a sketch of the model, keyed to a micrograph of the sort that aided in its formulation.

In the fluid mosaic model, the lipid bilayer is seen to have a consistency like light machine oil. Some of the lipids have short or kinky tails (Figure 5.6). Both types disrupt the rigid hydrocarbon-chain packing that is typical of long-chain fatty acids clustered together (as they are in the lipid bilayer). Hence both types impart fluidity to the lipid matrix. Proteins are embedded within this fluid foundation or weakly bonded to either surface. Together, the lipids and proteins form the "mosaic."

Some proteins that penetrate the bilayer may act like transport channels across the membrane. Other proteins have water-repelling parts buried within the lipid bilayer and hydrophilic parts protruding from it. The protrusions may be sites where substances attach to the membrane surface. Certainly glycoproteins that span the bilayer are known to be linked with the cell's network of microtubules and microfilaments (Chapter Four, Figure 4.17). Perhaps they are sensory receptors, triggering cell movement in response to chemical changes in the environment.

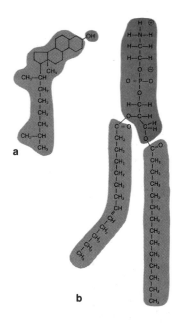

Figure 5.6 (**a**) Structural formula for cholesterol. (**b**) Structural formula for one of four main phospholipids in cell membranes. In both sketches, the dark brown area is the hydrophilic head, and the light brown areas are water-repelling tails. Notice the kink in the phospholipids's fatty-acid tail at the double bond, and the relatively short cholesterol tail. Both disrupt the tight packing otherwise possible in clusters of straight hydrocarbon chains. The spaces created by these disruptions impart some fluidity to the cell membrane.

MOVEMENT OF WATER AND SOLUTES INTO AND OUT OF THE CELL

The fluid mosaic model of membrane structure gives us several clues about how the flow of substances between the cell and its environment may be governed. As you will see, the plasma membrane is *differentially permeable*: some molecules travel rapidly across the membrane, others cross it more slowly, and some are kept from crossing it at all. For instance, water is one of the few molecules that can move freely into and out of the cell. Because of their size and chemical nature, many ions and molecules (such as large, negatively charged proteins) are prevented from doing so on their own. But the plasma membrane is a *regulatory* structure, which helps maintain appropriate amounts of ions and molecules inside the cell. It does this even when certain ions and molecules are scarce or too abundant. In fact, we can say this:

Built into the plasma membrane are mechanisms that help assure homeostasis at the cellular level.

Maintaining internal concentrations of water and solutes depends on mechanisms of two general sorts. "Passive" transport mechanisms (diffusion, bulk flow) do not alter in any way the direction in which some substance happens to be moving on its own. *Hence they operate without any direct energy outlay by the cell.* "Active" transport mechanisms (active transport, exocytosis, endocytosis) work to move some substance in a direction in which it shows little or no inclination to move on its own. *They cannot operate without direct energy outlays by the cell.* Let's look first at the ways individual molecules are transported across the membrane, then at the bulk movement of materials.

Diffusion

When you first drop a sugar cube into a cup of water, the sugar molecules stay stuck together in one region of the cup, and the water molecules are everywhere except in the spot that the sugar cube occupies. For each substance, you have created a **concentration gradient**: its molecules are concentrated more in one region than in another. But soon the molecules of each kind of substance undergo **diffusion**: a random movement of like molecules from their region of greater concentration to the region where they are less concentrated. Gradually, sugar molecules tend to become dispersed through the cup, just as water molecules tend to become dispersed through the space that the sugar cube once occupied.

Diffusion of individual molecules occurs not only in fluids (either gases or liquids) but also in solids to some extent. In either a fluid or solid, diffusion is driven by the unceasing motion inherent in all individual molecules. In a substance such as water, molecules are constantly moving and colliding with one another millions of times each second. The collisions send them off in new, random directions. Eventually these random movements carry some of the molecules into a different region. As more molecules join them, collisions send molecules back and forth until at some point, there is no more *net* change in concentrations between the two regions. The molecules are uniformly dispersed.

The *rate* of diffusion depends on several things. Diffusion speeds up with increases in temperature, for heat energy causes molecules to move faster. The size of molecules also affects diffusion rates, because small particles move faster than large ones at the same temperature. In addition, the greater the difference in concentrations between two regions, the faster the net movement.

Water, oxygen, carbon dioxide, and a few other simple molecules and ions diffuse readily across plasma membranes. In fact, diffusion alone accounts for the greatest volume of substances being moved into and out of the cytoplasm. Diffusion is also an important transport process *within* cytoplasm. Enclosed within the membrane boundary of the cell are millions of molecules of different substances. All jostle their neighbors and proceed along whatever concentration gradients exist for them. Over short distances (for instance, from one cytoplasmic region to another some fifty nanometers away), diffusion alone can distribute many substances within a fraction of a second. It only works in cells having a diameter of less than a centimeter or so. In fact, it is because of the constraint that diffusion imposes on cell volume that you might see very long, very *thin* nerve cells in, say, your legs—but never a fat, round cell having the volume of a watermelon. Through such an immense cell, diffusion of some vital substances would take months or years, and the cell would long since be dead because of the wait.

Diffusion accounts for the greatest volume of substances being moved into and out of cells. It is also an important transport process within cells.

Aside from free diffusion, there is another molecular transport process driven by concentration gradients. It

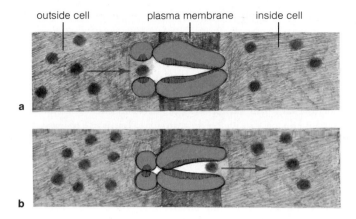

Figure 5.7 How a channel protein may be arranged in the plasma membrane, shown as if it were sliced down through its midsection. On the membrane surface, another protein may act like a gate. When a small molecule becomes bound to the surface protein (**a**), the channel protein spreads open and passively allows the molecule to pass through (**b**). When the channel is clear, the gate closes and the channel protein returns to its original shape. Here, cross-hatched areas represent hydrophilic regions of each protein's interior.

depends on proteins that are partly buried in the lipid bilayer of the plasma membrane. The proteins (and possibly other molecules) seem to assist the passage of small molecules across the membrane, in the same direction that gradients would take them. The process is called **facilitated diffusion**.

According to one view, facilitated diffusion occurs when a protein binds with some molecule on one side of a membrane, then flipflops to deliver it to the other side. However, not only do proteins seem reluctant to engage in flipflopping, the action itself would require considerable energy outlay. In another view, some membrane proteins act as fixed **channels** across the membrane (Figure 5.7). These proteins have hydrophilic groups lining their interior in a way that forms a channel. Channel opening and closing could be accomplished by changes in protein shape. The channels themselves aren't very large. Typically they are only wide enough to allow passage of one molecule at a time.

Active Transport

Many ions and molecules that a cell requires—potassium ions, for instance—simply are not all that abundant in the environment. The cell must employ some mechanism for *accumulating* potassium ions in concentrations that are greater on the inside than they are on the outside. Many other ions and molecules—sodium ions, for instance—are overly abundant in such environments as saltwater and the body fluids of animals. These ions can diffuse across a plasma membrane. But they would kill any cell if they were to reach the same concentration inside as exists outside! So the cell must have some mechanism for *preventing* the accumulation of sodium.

Moving individual ions and molecules into or out of a cell against a concentration gradient is called **active transport**. This transport process also enables cells to move materials across the plasma membrane against gradients of pressure and temperature. Like facilitated diffusion, active transport depends on proteins that serve either as carriers or as fixed channels across the plasma membrane. Unlike facilitated diffusion, active transport depends on cellular energy. Usually this energy becomes available through hydrolysis of energy-rich molecules, notably ATP (Chapter Six). As much as seventy percent of the readily available cell energy may be devoted to active transport.

Two active transport mechanisms have been studied intensively. The first, the sodium-potassium pump, primes nerve cells and makes them stand ready to receive messages from other cells. The second, the calcium pump, is vital in muscle contraction. Because both are important aspects of animal functioning, we will postpone looking at details of how they operate until later chapters. For now, it is enough to say that active transport is highly selective in conducting certain molecules but not others across the membrane. The selectivity can be traced to the transport proteins embedded in the membrane. Each binds only those molecules or ions that it is supposed to help transport. There is great variation in how the proteins function, but in each case an energy-yielding reaction is coupled with the transport process.

Whether it is stockpiling some ions or molecules, or getting rid of excess amounts of others, a cell must expend energy to get those substances moving <u>against</u> their concentration gradient. This energy-requiring process is called active transport.

Bulk Flow

When you talk about diffusion and active transport, it's possible to talk about the movement of one kind of molecule at a time. But sometimes all sorts of molecules and ions are present in some fluid and they all move together across a cell membrane. Perhaps they do so because of a *pressure gradient* (here, a difference in pressure between two regions of the fluid). In response to the gradient, different kinds of molecules present in the fluid may move together in the same direction; they may undergo **bulk flow**. For example, bulk flow is involved in the transfer of some ions and small molecules (water, salts, gases, glucose) between blood capillaries and surrounding body tissues. Blood capillary walls are only a single layer of very thin, flat cells. The cells themselves are permeable to some of these particles; gaps between cells let others pass through. Here the pressure gradient is created by heart muscle contractions. The contractions cause blood to flow from a region of high pressure (the heart chambers) to the region of lower pressure (arteries, then capillaries). Bulk flow does not alter the blood concentrations of such substances as glucose, sodium ions, and chloride ions, for they are all being moved together by the process.

In bulk flow, different ions and molecules present in a fluid move together in the same direction, often in response to a pressure gradient.

Osmosis

Diffusion and bulk flow because of a pressure gradient are two general types of passive transport mechanisms. A special case of passive transport, called osmosis, may involve both of them. **Osmosis** is the movement of water across any differentially permeable membrane in response to solute concentration gradients, and/or to a pressure gradient. The term applies *only* to the movement of water molecules.

What are the effects of osmosis? Imagine you have just made a plastic bag permeable, in a way that permits the passage of water molecules but blocks the passage of larger molecules. Suppose you fill the bag with water containing a two-percent concentration of table sugar. Now put the bag in a container of distilled water. Distilled water is the most concentrated kind, because all solutes have been removed from it. Hence water molecules can get much closer to one another. Water, however, is less concentrated inside the bag (where sugar molecules take up space) than it is on the outside. In this case, diffusion does *not* lead to uniform distribution of substances throughout the system. The net movement of water molecules is inward—

but sugar molecules can't move out (they are too large to pass through the holes you made in the "membrane"). Soon the bag swells with water, as Figure 5.8 suggests. Because the bag can hold only so much, fluid pressure builds up and the bag bursts or springs a leak.

What would happen if the water were more concentrated inside the bag than outside? Obviously the net water movement would be out of the bag, which would shrivel up (Figure 5.8). Only if the water concentration were the same on both sides of the bag would there be no net movement of water in either direction.

The same kind of thing happens to living cells. Imagine a red blood cell being immersed in distilled water. Because the cell contains many large organic molecules that can't move across its plasma membrane, internal pressure builds as water moves in. This kind of cell simply cannot function under such conditions. It is doomed to burst, because it has no mechanism for disposing of excess water. When such a cell is immersed in a solution having a sugar concentration that is greater than the one inside, the cell is still doomed: it will lose water and shrivel up. Only when red blood cells are immersed in an environment where solute and water concentrations are the same as on the inside will there be no change in cell volume (Figure 5.8).

In response to solute concentration gradients and/or pressure gradients, water moves across plasma membranes. Such water movement across differentially permeable membranes is called osmosis.

Endocytosis and Exocytosis

Neither passive nor active transport systems can carry more than individual ions and molecules across the plasma membrane. But some cells have the means to move whole bits of substances about, even to engulf other cells. Such **bulk transport** occurs in membranous compartments that form from the plasma membrane.

Two examples may be given of **endocytosis**, whereby vesicles form around particles at or near the cell surface and then discharge their contents in the cell interior. The amoeba, a single-celled organism, relies on *phagocytosis*. The word refers to its penchant for "cell-eating." When an amoeba encounters an appetizing cell or particle in its environment, one or two extensions of its membrane-enclosed cytoplasm moves outward in the manner shown in Figure

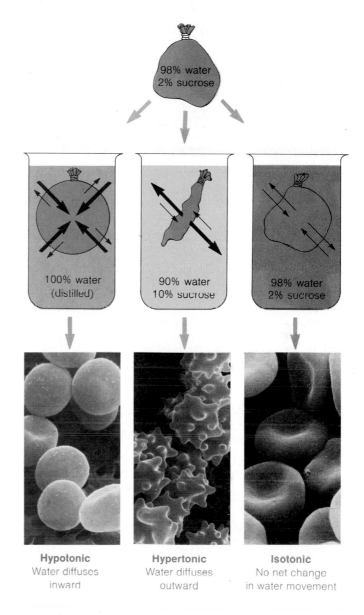

Hypotonic
Water diffuses inward

Hypertonic
Water diffuses outward

Isotonic
No net change in water movement

Figure 5.8 Effects of osmosis in different environments. The sketches show why it is important for cells to be matched to solute levels in their environment. (In each sketched container, arrow width represents the relative amount of water movement.)

The micrographs correspond to the sketches. They show the kinds of shapes that might be seen in red blood cells placed in *hypotonic* solutions (influx of water into the cell), *hypertonic* solutions (outward flow of water from the cell), and *isotonic* solutions (internal and external solute concentrations are matched).

Red blood cells have no special mechanisms for actively taking in or expelling water molecules. Hence they swell up and burst, or shrivel up, if solute levels in their environment change. (Micrographs from M. Sheetz, R. Painter, and S. Singer, *J. Cell Biology*, 1976. 70:193)

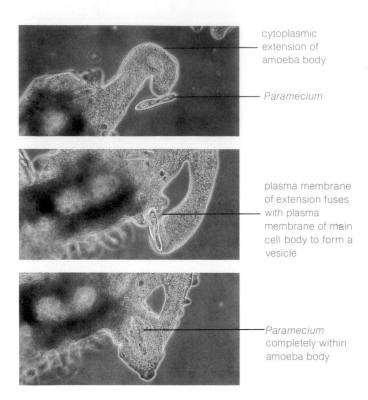

cytoplasmic extension of amoeba body

Paramecium

plasma membrane of extension fuses with plasma membrane of main cell body to form a vesicle

Paramecium completely within amoeba body

Figure 5.9 An example of endocytosis. *Amoeba* engulfing *Paramecium* (Eric. J. Gravé. National Audubon Society/PR).

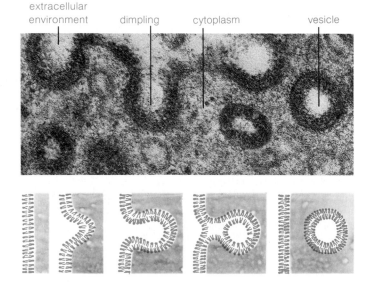

extracellular environment

dimpling

cytoplasm

vesicle

Figure 5.10 Pinocytosis in a red blood cell (Everett Anderson). The diagram (left to right) depicts vesicle formation. Exocytosis essentially occurs in reverse (in a sequence from right to left).

5.9. The lobelike extensions curve back, toward the amoeba's body, and thereby surround the particle. The membrane "pocket" created in this way becomes a vesicle that is moved into the amoeba's interior. There it fuses with lysosomes. Lysosomes, recall, release enzymes that break down particles into manageable bits. These bits readily diffuse across the membranous vesicle. Once in the cytoplasm, the bits become available for reuse. Phagocytosis is also the means by which certain white blood cells devour disease agents invading your body.

Another process for moving materials into the cell is *pinocytosis*, or "cell-drinking." In pinocytosis, a depression forms where small particles in the surrounding environment adhere to the plasma membrane. Vesicles seem to form as a response to this adhesion (Figure 5.10). As the membrane dimples inward, it surrounds some of the extracellular fluid. The vesicle pinches off and moves inside the cytoplasm. Once inside, the vesicle may dump its contents into the cytoplasm or fuse with others like itself into large storage vacuoles.

Substances also are moved out of the cell by **exocytosis**. This process requires fusion of vesicles with the plasma membrane. Many cells secrete substances by exocytosis. Special intestinal cells, for instance, secrete droplets of mucus in this manner. Figure 5.10 depicts this bulk transport process.

ADAPTING TO CHANGES IN SOLUTE LEVELS: A FEW EXAMPLES

At any time, a cell may be relying on some combination of the active and passive membrane transport mechanisms just described. At any time, it may be regulating the inward and outward movement of dozens of different substances. The mechanisms it employs are adapted to water and solute levels in the surroundings. Consider the cells of land plants. In most cases, their environments offer fresh water, which contains very little dissolved materials compared to the cytoplasm. Water follows its concentration gradient and moves into these cells by osmosis. Incoming water causes pressure to increase within the cell. This osmotically induced internal pressure, called **turgor pressure**, builds up against the inner surface of the cell membrane and forces water molecules back out. The strong but somewhat pliable cell wall around the plasma membrane keep the cell from bursting. When the outward movement of water equals the rate of inward movement, the internal pressure is constant. The plant cell is in equilibrium with its surroundings.

When environmental water becomes scarce, turgor pressure drops. Leaves wilt. So do some stems and roots, as you may have noticed at parties, where carrot and celery stick snacks that had been crisped earlier in water get progressively limp as the evening wears on.

Some single-celled protistans are adapted in a special way to fresh-water environments. They have their own water-pumping system, a **contractile vacuole**. As water accumulates in the cell body as a result of osmosis, these cells expend energy to pump the excess across a special region of membrane that forms a permanent vacuole in the body. When the vacuole is made to contract, water squirts back outside.

Even in environments where water and solute concentrations predictably undergo dramatic change, we find some cells getting along quite well. An estuary may be almost entirely fresh water during low tide. Under these conditions, the single-celled protistans living there use contractile vacuoles to rid themselves of excess water flooding into their systems. During other periods, the estuary may contain water with solute concentrations equal to those inside the cells, so water regulation is not necessary. During high tide, the estuary becomes much more salty. The protistans must again expend energy, this time in ways that lead to an inward movement of water. If conditions really get dry, some will even dry up and go dormant—remember the water bear?

PERSPECTIVE

This chapter has focused on some intriguing aspects of the plasma membrane—a largely fluid structure that maintains its integrity in a largely fluid environment. With a consistency no greater than machine oil, this thin boundary layer around the cytoplasm interacts with a variable environment from which the cell must derive substance and sustenance. Through interplays between passive and active transport mechanisms, it helps govern cell volume and cytoplasmic pH, vital ion concentrations and the stockpiling of nutrients, as well as the efficient release of toxic compounds. In chapters to follow, you will be reading about many other events played out across not only the plasma membrane but membranes in general. For the membrane is also the physical stage for such diverse events as photosynthesis, cellular respiration, nerve message propagation, and communication between cells in the multicellular body. The membrane, in short, is central to our understanding of life processes.

Readings

Dyson, R. 1978. *Cell Biology: A Molecular Approach.* Second edition. Boston: Allyn and Bacon. Excellent introductions to topics covered in this chapter.

Giese, A. 1973. *Cell Physiology.* Fourth edition. Philadelphia: Saunders. Advanced reading, but excellent descriptions of the cellular environment.

Salisbury, F., and C. Ross. 1978. "The Water Milieu." In *Plant Physiology.* Second edition. Belmont, California: Wadsworth. One of the few clear introductions to the properties of water as they relate to plant cell functioning.

Singer, S. J., and G. Nicolson. 1972. "The Fluid Mosaic Model of the Structure of Cell Membranes." *Science* 175:720-731.

Staehelin, L., and B. Hull. 1978. "Junctions Between Living Cells." *Scientific American* 238(5):140-152.

Review Questions

1. What are three characteristics of water that make it so central in cell functioning?

2. What kind of bond between its molecules enables water to serve as a buffer against localized temperature extremes inside and outside the cell?

3. Define and give an example of an acid and a base. Is the following statement true or false? Only a fraction of biochemical reactions are sensitive to pH levels.

4. What is a salt? Give examples of salts important in plant and in animal cell functioning.

5. How do water molecules help keep vital ions available for reactions throughout the cell? How do they influence the regional organization of biological molecules in the cell? (Hint: Describe hydrophilic and hydrophobic reactions that organize cell membranes.)

6. Describe the fluid mosaic model of the plasma membrane. What makes the membrane fluid? What parts constitute the mosaic?

7. What is meant by differentially permeable? What two general kinds of mechanisms associated with the plasma membrane assure homeostatic control at the cellular level?

8. Diffusion accounts for the greatest volume of substances moving into and out of cells. How does diffusion work?

9. Under what circumstances would active transport mechanisms come into play? What *is* active transport?

10. What is bulk flow? What does bulk flow accomplish in a cell?

11. How does osmosis differ from bulk flow in general? What is the only kind of molecule that moves by osmosis? What causes osmosis?

6

ENERGY TRANSFORMATIONS
IN THE CELL

When you look at a single living cell beneath a microscope, you are watching a form seething with activity. In its movements it is seeking out and taking in small molecules—food, raw materials—suspended in the water droplet on the slide. To power those tiny movements, the cell is tapping into internal reservoirs of molecules it had stored away earlier. It is dismantling those molecules now for their chemical potential energy. Even as you observe it, the cell is using the molecules in building and maintaining its membranes and extracellular structures and organelles, its pools of enzymes, its concentrations of chemical compounds, its information-storage system. It is alive; it is growing; it may reproduce itself. Multiply all this activity by *60,000,000 billion cells* and you have an inkling of the activity going on in your own body even as you are sitting quietly, doing nothing more than observing that single cell! Thus, how cells trap and use energy is not merely a story of interest only to biologists. It is also the story of how *you* trap and use energy—how you have the means to move, to think, to do all the things that go into being alive.

Soon enough, you will be following the main metabolic trails through cells. Before setting out, you may find it useful to get a general idea of how energy is transformed and transferred within cells, and of how enzymes and special energy-carrier molecules take part in what goes on. Often these topics are intertangled with presentations of the main metabolic pathways themselves. But the premise here is that the pathways are a lot easier to follow when they are kept as uncluttered as possible. If you study this chapter first, as you would study a map before traveling through some unfamiliar place, you'll probably have a much easier time on the cellular roads that lie ahead.

ON THE AVAILABILITY
OF USEFUL ENERGY

All events large and small, from the birth of stars to the death of a microorganism, are governed by two laws of energy. Both laws are concerned with the nature of energy in a given system (a "system" being the matter in some defined region) and its surroundings ("surroundings" being, if you carry it far enough, the rest of the universe). The **first law of thermodynamics** deals with the *quantity* of energy available. It may be expressed in this way:

There is some total amount of energy in the universe, and that amount never changes.

To be sure, energy exists in many forms. And it's possible to convert one form to another. Your body, for instance, can convert food energy to a form of energy that can power all your movements. But energy can't be created from nothing. The first law tells us that we can add to our own energy stores only if we take it away from someplace else. We can't, in short, get something for nothing.

The **second law of thermodynamics** deals with the *quality* of energy available. Some forms of energy happen to be more concentrated than others. The energy stored in a gallon of gasoline is more concentrated than, say, heat energy that has spread through a large room away from a fireplace. One is readily accessible for conversion; the other is so dispersed that it is not. The second law has this to say about the availability of useful energy:

Left to itself, any system (along with its surroundings) spontaneously undergoes energy conversions to a less organized form. Each time that happens, some energy gets randomly dispersed in a form that is not readily available to do work.

If the first law tells us that we can't get something for nothing, this second law tells us that we can't expect to break even. For although the total amount of energy in the universe stays the same, the amount available in forms that can be used to do work—to make things happen—is dwindling. The main reason it's dwindling is that, sooner or later, all energy tends to distribute itself *so uniformly* throughout a system that it can't be converted back very easily to another form.

For instance, some energy streaming away from the sun is still highly concentrated when it reaches the earth. Plants intercept and convert some of this sunlight energy to the chemical energy of sugar molecules. Some heat escapes to the surroundings during the conversion process, but enough energy gets captured so that the sugar molecules represent stores of high-quality energy. When you eat plants, some of this energy is converted to high-quality mechanical energy (such as muscle movements). But again, some is also given off as heat to the surroundings. (In fact, your body steadily gives off heat equal to that from a hundred-watt light bulb.) In other words, as concentrated

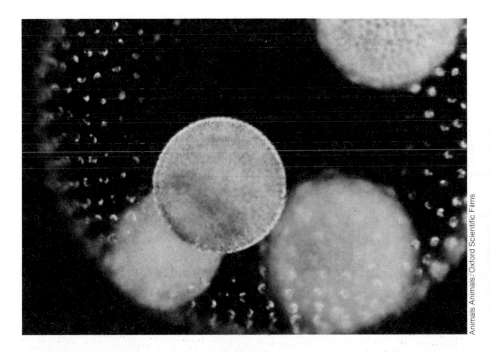
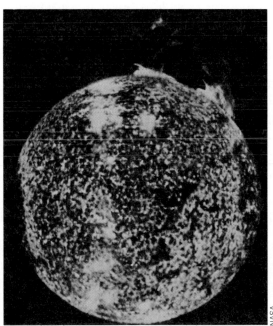

Figure 6.1 All events large and small, from the birth of stars to the life and death of a microorganism, are governed by the laws of energy. Shown here, eruptions on the sun's surface and, to the left, *Volvox*—a colony of single cells that capture sunlight energy necessary to drive their life processes.

energy is tapped, some of it inevitably flows into the environment as low-quality heat. The term **entropy** refers to how much energy in a system has become so dispersed (usually as evenly distributed heat) that it's no longer available to do work.

Entropy is constantly on the increase. As far as anybody can foresee, the universal rise in entropy will reach a maximum, some billions of years hence, and nothing will ever change again. It happens, however, that life is one glorious pocket of resistance to this somewhat depressing flow toward oblivion. For the entropy of any local region can be lowered—*as long as that region is being resupplied with usable energy being lost from someplace else.* Plants harness energy lost from the sun; plants lose energy to other organisms that feed, directly or indirectly, on plants. There is a steady flow of energy *into* the interconnected web of life, which compensates for the steady flow of energy *leaving* it. Thus, through energy transfusions from the sun, the universal trend toward entropy can be postponed on earth.

Energy at Rest and On the Move

When we look more closely at the nature of energy conversions, we find that energy exists in two general forms: potential and kinetic. **Potential energy** is any form of energy "at rest." It is energy in a potentially usable form that is not, for the moment, being used. Each substance has a certain amount of potential energy. Because of its position relative to the earth's center, water behind a dam has gravitational potential energy. Because of its composition, food has chemical potential energy. But with any transfer or transformation, there is a drop in potential energy and a rise in **kinetic energy**—the energy of motion. Kinetic energy can be transferred from one region (or system) to another. It can also be changed from one form to another.

For instance, a rabbit about to be pounced on by a coyote is a "system" containing chemical potential energy. A twig snaps under the coyote's feet and the rabbit becomes startled. It bounds away. Some of the rabbit's chemical potential energy thus becomes converted to kinetic energy. If the coyote is quicker than the rabbit, that's the end of the rabbit. But that is *not* the end of all the rabbit's energy. Part gets transferred into the coyote's system, where it is gradually stored away as chemical potential energy of muscles and fat.

We know, from the second law of energy, that a rise in entropy occurred with each transformation along the route. For instance, when the coyote ran down the rabbit,

some of its mechanical energy (as well as the rabbit's) was lost to the surrounding air as body heat. A stealthier coyote would not have had to expend as much energy securing dinner; and the dinner would have contained more potential energy upon reaching its gut. This example illustrates an important principle of energy use:

How efficiently energy is used in any system depends on the precise route by which one energy form is transferred or converted to another form.

Now, this does not mean that the most direct route is *always* the best route! Consider how small amounts of gasoline are exploded in a car engine. When gasoline is ignited, some of its chemical potential energy is converted to kinetic energy. Some kinetic energy is converted to mechanical energy of the pistons, the drive shaft, then the wheels. Some energy is lost as heat with each conversion. More is lost through friction, as moving parts rub together and become heated. The same amount of potential energy could be converted directly to kinetic energy by dropping a match right in front of the gas tank and exploding all the gasoline at once. Of course, little would get converted to *useful* mechanical energy. There would be a steep rise in entropy, and a disastrously disorganized car. As you will read soon, living cells also take a less direct (but safer) approach to using available energy.

Coupled Reactions: A Way of Conserving Cellular Energy

For one reason or another, molecules move about. They collide with one another. When they do, they often undergo chemical reaction: bonds holding each molecule together are broken, and parts may become rearranged into new molecules. When we talk about conversions of the sort going on in the coyote and rabbit, we are talking about two kinds of chemical reactions. One kind releases usable energy, and the other demands extra energy in order to occur. In **exergonic reactions**, some energy is lost during molecular rearrangements, so the new molecules (the products) end up with less energy than the old ones (the reactants):

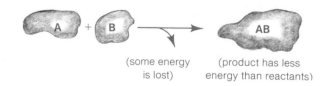

(some energy   (product has less
is lost)   energy than reactants)

In **endergonic reactions**, extra energy is fed into the reaction from the surroundings, so the products end up with more energy than the reactants:

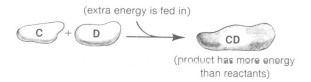

(extra energy is fed in)

(product has more energy than reactants)

In living things, exergonic and endergonic reactions commonly occur along sequences of coupled reactions, called **metabolic pathways**. In these pathways, products from one reaction serve as starting substances for the next reaction in line; and some energy "lost" as a result of a reaction can be harnessed by others. Thus,

---

Through reliance on coupled reactions, a cell temporarily conserves some energy that might otherwise be lost during its ongoing chemical conversions.

Equilibrium and the Cell

You may be thinking that reactions proceed only one way in cells. Under certain conditions, though, they may also proceed in what seems to be an improbable direction. Product molecules of lower energy may *revert* to reactant molecules having higher energy:

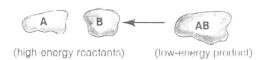

(high-energy reactants)   (low-energy product)

An increase in concentration, temperature, or pressure may trigger such reversible reactions. How does the increase do this? Because cells normally aren't bombarded with high heat or flattened under high pressure, let's focus on the first case. When a lot of reactant molecules are concentrated together, they are more likely to move about and collide with enough energy to react, which they do. As time goes on, more and more product molecules form, so not as many reactant molecules are left. The reaction rate drops accordingly. But all the while, *product* concentration has been increasing—which means that product molecules, too, start colliding more frequently. And some fraction will have enough collision energy to be driven in the reverse direction.

As long as both reactant and product molecules aren't funneled somewhere else, almost any reaction proceeds in both directions. Eventually, if left to itself, a reaction approaches **equilibrium**: it runs about as fast in reverse as it does in the forward direction. Unless conditions change, there is no further change in net concentrations. It's important to understand, though, that "equilibrium" doesn't necessarily imply *equal* concentrations. At equilibrium, product concentrations may be greater or lower than that of the reactants—*depending on how much energy is fed into or released from the reaction.* For instance, reactions that give off considerable energy are common in metabolic pathways. The point at which equilibrium is reached here favors the products so much that the reverse (energy-requiring) reactions hardly occur at all.

The word equilibrium implies a static rather than a dynamic state—yet the hallmark of the cellular world is constant, dynamic change. Availability of raw materials shifts. Requirements for different products may vary from minute to minute, from day to day, even during different stages of the life cycle. Depending on cellular conditions, sometimes products don't accumulate: they may be fed into new reactions as they become available. At other times, built-in controls shut down reactions when products do accumulate. At still other times, stockpiled products are broken back down into their component parts. In short, as long as a cell is alive and growing, there is no such thing as *total* equilibrium. The point is summarized here:

In a changing environment, cellular homeostasis is maintained only through constant adjustments in the reactions of life —through never-ending approaches toward and retreats from equilibrium along thousands of metabolic pathways.

ENZYME FUNCTION

Activation Energy and Enzymes

For any reaction to occur, molecules not only must collide, they must collide with a certain minimum energy. This minimum energy is the **activation energy** for the reaction. In spontaneous reactions, it is reached sooner or later as a result of the kinetic energy inherent in all molecules. Also, some reactions that never would occur on their own will indeed occur when they get a boost of extra energy. Consider what happens when you strike a match to start wood burning. The heat energy makes wood molecules

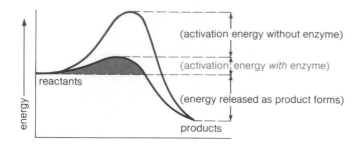

Figure 6.2 Enzymes and activation energy. Activation energy is like a small hill over which reactants must be pushed before a reaction will proceed. Enzyme action lowers the required activation energy in either direction.

jostle faster. They start colliding with oxygen molecules in the surrounding air. When they collide with enough energy to react, bonds are broken, new bonds form, and light and heat energy are released. The released energy increases the collision rate between molecules, so the reaction continues. Without that initial boost from the match, though, the average kinetic energy of the molecules is not enough for wood to begin reacting with oxygen in the atmosphere.

As it happens, the starting substances for reactions that normally take place in a cell tend to react slowly on their own. For life to continue, spontaneous reactions that would take years or decades to reach completion must be made to occur within a fraction of a second. Of course, if you've ever burned your fingers on a match, you know that boosting the temperature is not a good way of speeding up reactions in living things. Metabolic reactions must occur within a certain range of low temperatures. Enzymes make this possible. They are *catalysts*—substances that speed up a chemical reaction. They do so not by giving the reaction an energy "boost," but by making conditions so favorable that the reaction proceeds faster than it otherwise would (Figure 6.2).

The effect of enzymes is staggering. A single molecule of the enzyme **carbonic anhydrase** can combine water and carbon dioxide to form 100,000 molecules of carbonic acid (H_2CO_3) in *one second*. Yet throughout this activity, the enzyme itself is not permanently altered in any way. Once product molecules have formed and are released, enzymes are ready to unite a new set of reactant molecules.

What is it about enzymes that make them such efficient catalysts? All enzymes are proteins, with or without

other chemical groups attached. Like all proteins, they are folded into three-dimensional shapes. The folding creates a groove, or pocket, called the **active site** (Figure 6.3). At this site, a given set of reactants (or *substrates*) becomes temporarily bound to the enzyme. There are several explanations of why the association enhances the reaction rate.

First, atomic groups projecting from an enzyme's active site can match up and temporarily bond with groups projecting from a specific set of reactants. During the brief encounter between these mutually attractive groups, certain bonds *within* the substrate molecules are weakened. Not as much energy is needed to break these weakened bonds. Thus enzymes help *lower* the activation energy required.

Second, although reactant molecules collide on their own, they do so at random. This means that their mutually attractive groups often don't make contact, hence reaction doesn't always occur. In contrast, an enzyme's active site *orients* its substrates, thereby promoting interaction of their reactive parts. And an increased concentration of molecules primed to interact effectively also lowers the activation energy required.

Finally, enzyme shape may change during catalysis. One idea is that when an enzyme first binds to substrates, the active site's chemical groups are closely but not perfectly matched to their counterparts. The near-fit induces changes in enzyme shape. The changes allow the enzyme to bind tightly with the substrates—so tightly that the substrates briefly assume a highly unstable form. The changes may so distort or strain some of the bonds within substrate molecules that they are forced to break apart, giving way to the molecular form of the product. Activation energy for the reaction is thus lowered.

Whatever the precise nature of the association between an enzyme and its substrates, the outcome is this:

Enzymes speed up the rate of a metabolic reaction by lowering the activation energy that must be reached before the reaction will occur.

Enzymes and Equilibrium

To illustrate some of the main points made so far, we can give an example of conditions under which equilibrium is approached, how equilibrium can be disturbed, and the effect of enzymes on achieving equilibrium. We'll first describe how carbonic acid forms and breaks down in a bottle of cola. Then we'll look at how it forms and breaks down in your blood.

When soft drinks are manufactured in a bottling plant, carbon dioxide is pumped under high pressure into each bottle before it is sealed. Some carbon dioxide molecules dissolve at once in the liquid. There they combine with water to form carbonic acid, which then dissociates into H^+ ions and bicarbonate:

$$CO_2 + H_2O \longrightarrow \underset{\substack{\text{carbonic} \\ \text{acid}}}{H_2CO_3} \longrightarrow \underset{\text{bicarbonate}}{HCO_3^-} + H^+$$

After the bottle is sealed, this reaction continues until the increased product concentrations reverse the reaction:

$$CO_2 + H_2O \longleftarrow H_2CO_3 \longleftarrow HCO_3^- + H^+$$

Eventually equilibrium is reached, and there is no further net change in concentrations of reactants or products. When you pry off the bottle cap, though, pressure in the bottle is reduced and carbon dioxide gas remaining above the cola rushes outside. Some dissolved carbon dioxide fizzes out, too, so that the carbonic acid breakdown rate exceeds its formation rate. Thus the reaction runs faster in reverse until a new equilibrium is reached. That happens when the rate of carbon dioxide leaving the cola is the same as the rate at which carbon dioxide wanders in from the surrounding air. It takes many, many hours, and by that time the drink is unappetizingly flat.

Similar reactions go on in your body. When your cells use food molecules, carbon dioxide is given off as a by-product. As carbon dioxide accumulates, a pressure gradient develops between the cell's insides and surrounding tissues (Chapter Sixteen). Hence the carbon dioxide moves out of the cell, and eventually moves into your bloodstream—where the carbon dioxide concentration is lower still. The carbonic acid reacts at once with buffer molecules in the cells. (*Buffers* act like sponges in soaking up or giving up H^+ ions when cellular pH changes, so that the original pH is restored.) The outcome is the formation of bicarbonate and H^+ ions. When blood reaches the lungs, these concentrations are pulled in reverse, because the carbon dioxide concentration in air held in the lungs is lower still. When you breathe, the excess carbon dioxide is expelled from your body.

The rate at which carbonic acid forms and breaks down on its own is so low that cells would be poisoned

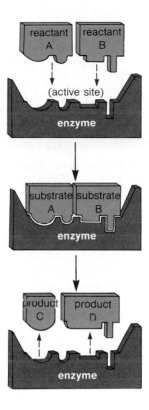

Figure 6.3 Active site of enzymes. A precise pattern of chemical groups projects from the active site. This pattern is an exact, three-dimensional complement of groups present on the specific substrates that the enzyme is meant to bind.

long before the reactions were completed. Your blood is pumped so fast through your body that it has only an instant to pick up carbon dioxide at any given place. Seconds later it is in the lungs; and there it has only an instant to unload its cargo. As it happens, carbonic anhydrase causes the forward and reverse reactions to proceed 250 times faster than they would on their own. In its action, this enzyme resembles the thousands of different enzymes that catalyze other reactions:

Enzymes speed up the rate at which a reaction approaches equilibrium. But they do not change the <u>proportions</u> of reactants and products that will be present once equilibrium is reached.

Thus, if a bottle of cola were to contain carbonic anhydrase, it would go flat with amazing speed once you opened it—but it would not get any flatter than it otherwise would.

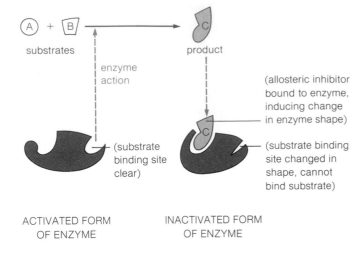

substrates

product

enzyme action

(allosteric inhibitor bound to enzyme, inducing change in enzyme shape)

(substrate binding site clear)

(substrate binding site changed in shape, cannot bind substrate)

ACTIVATED FORM OF ENZYME

INACTIVATED FORM OF ENZYME

Figure 6.4 An example of control over enzyme activity: feedback inhibition based on allosteric enzymes. Such enzymes have an active site where substrates are bound. They also have another binding site for intermediate or end-product molecules formed from the substrates. When such inhibitor molecules are bound in place, the enzyme's shape changes. The structural changes prevent the substrate molecules from matching up with and binding to their active site.

Factors Influencing Enzyme Activity

As you might suspect from the earlier description of protein structure and function (Chapter Three), enzymes are adapted to specific environments. For instance, all enzymes function only within a certain pH range. Most are effective at or near pH 7, which is generally optimum for most cell types. Pepsin, which is found in the stomach, functions only in an extremely acid setting. Trypsin, found in the small intestine, functions best at about pH 8.5. As another example, enzymes generally can't tolerate high temperatures. The rate of enzyme activity does increase as the environment heats up. But past the optimum temperature for a given enzyme, the reaction rate plummets. **Denaturation** occurs: hydrogen bonds holding the protein in its coiled secondary structure break, hydrophobic bonds shift, and the protein chains unwind. With these structural changes, the enzyme stops working. Even brief exposure to temperatures above optimum will destroy enzymes. Without enzymes, metabolism grinds to a halt and cells die. (This happens to cells during the extremely high fevers that accompany severe viral infections.) We can make this generalization about enzyme activity:

Enzymes work only on certain sets of substrates, and they work only under certain environmental conditions.

Because enzymes govern the flow of substances through the cell, control over enzyme activity is vital for all forms of life. Think about what enzymes do in energy metabolism alone. Food-energy molecules such as sugar are seldom present in exactly the right amounts at all times. The cell must do a juggling act with its enzymes and with substances that *are* available. For instance, when food molecules are scarce, structural proteins in the cell may be disassembled, so that its component parts can be fed into pathways concerned with energy breakdown. Carbohydrates, fats, proteins, nucleic acids, various inorganic molecules—all may be pulled out of one pathway and diverted to others. Thus we have another generalization:

Controls over enzyme activity help regulate which sets of substances will be formed in a cell, and in what amounts.

We will look at some of these control mechanisms in later chapters. Here, let's consider just one important control to get an idea of what goes on. **Allosteric enzymes** are control agents. The name allosteric implies that these enzymes can take on "other shapes." Such enzymes have an active site for substrate molecules—and another site that binds with intermediate or end-product molecules. When a molecule is bound at the "other" site, it so distorts the active site that the enzyme no longer matches up with its substrates. Thus intermediate or end-product molecules act as regulators over the very enzyme that helps lead to their assembly! Only when such molecules are released from the "other" site does the enzyme resume normal activity. This kind of control is called **feedback inhibition**. Figure 6.4 depicts how it works.

ENERGY CARRIERS

Actively transporting ions and molecules across membranes, building and rearranging molecules, moving body parts such as muscles and microtubules—these are the main energy-requiring activities of living things. Because of their chemical nature, enzymes speed up the reactions involved in these activities. But how does the potential energy stored in molecules such as glucose actually get transferred to places where energy is needed? Here we must turn to *energy carriers:* compounds whose structures enhance energy exchanges in the cell.

A major energy carrier in all living things is **adenosine triphosphate**, or **ATP**. As Figure 6.5 shows, this nucleotide consists of adenine (a nitrogen-containing compound), the five-carbon sugar ribose, and three phosphate groups. The key parts of the molecule are two P—O—P linkages, otherwise known as *pyrophosphate bonds.* Often you will hear them called high-energy bonds. The expression refers to their highly *accessible* energy, which is easily transferred to other molecules in the cell.

What's the reason for this accessibility? In pyrophosphate, charge repulsions exist within the array of phosphorus and oxygen atoms. This means that ATP is highly charged and somewhat unstable. It readily undergoes hydrolysis, which splits away one or two phosphate groups. We may depict a phosphate group in this manner: (P) The products of this hydrolysis reaction contain less chemical energy than the initial reactant. And the greater the difference in energy between reactants and products, the greater the potential for energy transfer to *other* molecules when the reaction occurs.

In most biological reactions, only one of the two pyrophosphate bonds in ATP is split. The result is **adenosine diphosphate** (ADP) and one phosphate group:

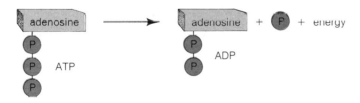

Under certain conditions a second hydrolysis occurs, from ADP to **adenosine monophosphate** (AMP) and one phosphate group:

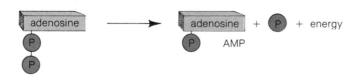

Other kinds of nucleotides take part in energy transfers. Some accept electrons from donor molecules,

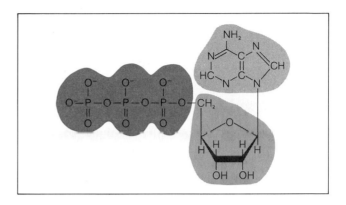

Figure 6.5 Molecular formula for ATP, composed of the sugar ribose (gray), adenine (tan), and a triphosphate (gold).

then release those electrons as part of a reaction series. **Nicotinamide adenine dinucleotide** (NAD^+) is such an electron transporter. When it has two electrons attached, it's abbreviated NADH. The H refers to a hydrogen ion that comes along for the ride. A related nucleotide is $NADP^+$, which has one phosphate group attached to an NAD^+ molecule. When two electrons (and two hydrogen ions) are attached, it becomes a high-energy carrier known as $NADPH_2$. **Flavin adenine dinucleotide** (FAD) also accepts two electrons at a time. Here again, two hydrogen ions may come along for the ride; hence the abbreviation $FADH_2$.

Aside from the energy-carrying nucleotides, some protein compounds also transport electrons. **Cytochromes**, found largely in chloroplasts and mitochondria, are such molecules. They have an iron-containing part that is the site of electron acceptance and transfer.

SOME ENERGY TRANSFER STRATEGIES

Coupled Reactions Revisited

When ATP is hydrolyzed into ADP and phosphate, the energy released would, by itself, do little more than heat the surroundings. It would be like turning on an electric motor without connecting it to anything. The motor would whir, the surrounding air would warm up, and electricity

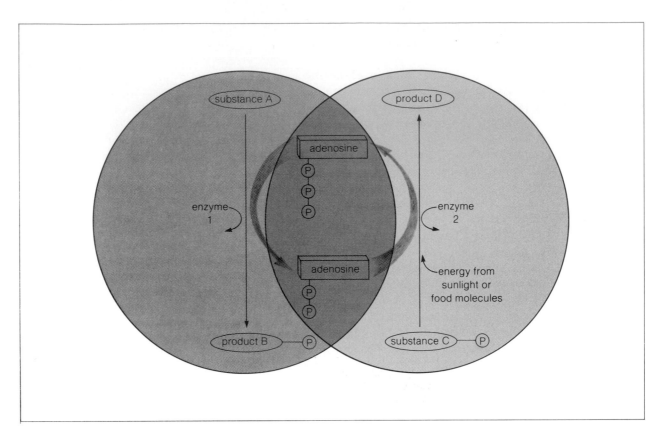

Figure 6.6 Coupled reactions in which ATP plays a role. Some of the energy captured in ATP when substance C is converted to D is used to drive the reaction in which substance A is converted to B. Here, the brown circle represents the energy-releasing reaction; and the gray circle represents the energy-requiring reaction.

would be dissipated—yet nothing would get done. But if you take a pulley belt and couple that motor to, say, a sewing machine, that same energy can be harnessed for use. In cells, enzymes make such connections. In the process, they help conserve much of the energy that would otherwise be lost:

Enzymes couple energy-yielding reactions such as ATP hydrolysis to the energy-requiring reactions involved in performing useful cellular work.

For example, a phosphate group from ATP can be transferred, through enzyme action, to another molecule in the cell. A molecule that gets a phosphate group attached

is said to have undergone **phosphorylation**. Thus phosphorylated, the molecule acquires some of the energy released from ATP. It thereby becomes more reactive:

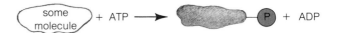

Perhaps, in this activated state, the molecule will take part in some reaction that might not otherwise proceed on its own (Figure 6.6). As another example, the energy released may bring about changes in the shapes of enzymes associated with muscle contraction. When your arm bends, this happens to billions of enzymes taking part in the movement of many billions of tiny protein filaments in muscle

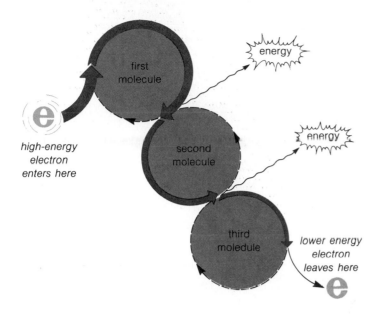

Figure 6.7 Electron transport chain. When the first molecule in the chain accepts an excited electron, it becomes reduced (solid line). When it gives up the electron to the next molecule in line, it becomes oxidized (dashed line) and thereby becomes ready to pick up another electron entering the chain. Each time the excited electron is transferred, it gives up some of its extra energy. At the end of the chain, a neighboring electron acceptor molecule carts the electron away.

cells. Active transport mechanisms, too, depend on enzymes that couple ATP breakdown to specific tasks. Here, changes in enzyme shape are geared to thrusting some unwanted molecule out of the cell or bringing in some vital substance. In all such cases, the following is true:

An energetically unfavorable reaction (one that doesn't readily proceed on its own) can be driven <u>indirectly</u> by an energetically favorable reaction, such as the splitting of ATP.

Figure 6.6 shows how enzymes and energy-carrier molecules interact in driving a reaction. Although each kind of enzyme works only for a specific reaction, ATP is used

in many different reactions as an intermediate energy source. That's why ATP is sometimes called the "universal energy currency" of the cell.

You may be wondering where all the ATP comes from. If ATP were used up like nonreturnable bottles, cells would be in trouble. For instance, the mass of ATP that *your* cells break down each day adds up to more than your total body weight! Obviously the leftover molecules are not discarded. Instead they are recharged with energy from some outside source and prepared for reuse by phosphorylation. Thus a single molecule may be phosphorylated, broken down, and rephosphorylated thousands or even millions of times in a single day.

Oxidation–Reduction Reactions

Among some of the carrier molecules described earlier, energy is embodied not in phosphate bonds but in some of their electrons, which are readily transferred about. Many reactions of life are based on these electron transfers. In an *oxidation* reaction, one or more electrons is stripped from an atom or molecule. In a *reduction* reaction, one or more electrons is gained by an atom or molecule. The molecule undergoing reduction is usually one that has accepted an electron from a hydrogen atom present in the cell. In fact, sometimes the whole hydrogen atom becomes attached, as you saw earlier. Similarly, a molecule becomes oxidized when it gives up an electron (or hydrogen atom). Of course, electrons don't float about randomly in cells. At the same time that a molecule acts as an electron donor, another must be around to act as an electron acceptor. That is why these two events are said to be coupled in an **oxidation–reduction reaction**.

Often, oxidation–reduction reactions occur one right after another, in organized sequence. Such organization is vital if the cell is to produce certain substances in dependable amounts. It is based on the arrangement of electron carrier molecules in series, otherwise known as an **electron transport chain**. Major electron transport chains occur in the pathways of photosynthesis and cellular respiration. In these pathways, the first molecule in line accepts an excited electron from a donor molecule outside the chain. The electron gets transferred to a cytochrome molecule. The cytochrome transfers the same electron to another cytochrome molecule in the line. The last molecule in the chain gives up the electron to an oxidized acceptor molecule. Removing the electron in this way keeps the transport chain clear for operation (Figure 6.7).

The point of all such oxidation–reduction reactions is the generation of usable forms of energy. Because no energy transformation is perfect, some of an excited electron's extra energy is released each time the electron is transferred from molecule to molecule. At certain transfer points, the amount of energy given off is enough to do useful work—such as attaching a phosphate group to ADP. Thus we have another generalization about energy use:

By moving electrons down an electron transport chain, a cell is able to build up a store of energy in carrier molecules such as ATP.

Later on, you will be coming across different examples of energy transfer strategies. Here the main concern has been to introduce you to the connections between energy, enzymes, and energy-carrier molecules. These connections are vital in the energy-trapping and energy-utilization pathways of the cell, two topics that will occupy our attention in the chapter that follows.

Readings

Dyson, R. 1978. *Cell Biology: A Molecular Approach.* Second edition. Boston: Allyn and Bacon. See especially Chapter 4: "Cellular Homeostasis." Good descriptions of cell energetics.

Ferdinand, W. 1976. *The Enzyme Molecule.* New York: Wiley.

Koshland, D., Jr. 1973. "Protein Shape and Biological Control." *Scientific American* 229(4): 52–64.

Miller, G. 1978. *Chemistry: A Basic Introduction.* Belmont, California: Wadsworth. Outstanding introductory book, with clear graphics and interesting analogies. Good for students with little or no background in chemistry.

Stryer, L. 1975. *Biochemistry.* San Francisco: Freeman. Chapter 11 treats the material covered in this chapter at a more comprehensive level.

Wolfe, S. 1981. *Biology of the Cell.* Second edition. Belmont, California: Wadsworth. Excellent discussion of enzyme structure and function.

Review Questions

1. What does the first law of thermodynamics state? The second law? How do these two laws affect living things?

2. Distinguish potential energy from kinetic energy. Give a few examples of potential and kinetic energy in terms of a living system. What determines how efficiently energy is used in any system?

3. What is meant by exergonic? Endergonic? How does teaming up an exergonic reaction with an endergonic reaction conserve energy that might otherwise be lost to the cell?

4. How is cellular homeostasis maintained if the environment changes constantly?

5. What is an enzyme, and what role does it play in cells? How are enzymes thought to carry out this role?

6. What is the relationship between enzymes and chemical equilibrium?

7. Name two factors that influence enzyme activity. What is a substrate? What is meant by denaturation?

8. What is a high-energy bond? Name a molecule in which such bonds exist. What is an energy carrier? Give an example of an energy-carrier molecule.

9. What is ATP hydrolysis? How does it relate to endergonic reactions? How can ATP hydrolysis be used to drive an energetically unfavorable reaction?

10. What is reduction? What is oxidation? If a substance gains electrons or hydrogen atoms, does it gain or lose energy?

11. How does a cell build up a store of energy in carrier molecules?

Just before dawn in the Midwest the air is dry and motionless; the heat that has scorched the land for weeks still rises from the earth and hangs in the air of the new day. There are no clouds in sight. There is no promise of rain. For hundreds of miles in any direction you care to look, crops stretch out, withered or dead. All the sophisticated agricultural methods in the world can't save them now, for in the absence of one vital resource water life in each cell of those many thousands of plants has ceased.

In Los Angeles, a student reading the morning newspaper complains to no one in particular about the hike in food prices that the Midwest drought will mean. In Washington, D.C., economists busily calculate the crop failures in terms of decreased tonnage available for domestic consumption and for export, and what it means to the nation's balance of payments. In Africa, a child with bloated belly and spindly legs waits passively for death. Even if food from the vast agricultural plains of America were to reach her now, it would be too late. Deprived of vital food resources, cells of her body will never grow normally again.

This chapter is about the acquisition and use of materials and energy. It gets into cellular pathways that might at first seem to be far removed from the world of your interests. But knowledge of these pathways is a key to understanding what it means to acquire food, why energy must be expended to get it, why food supplies are not keeping up with the increasing demands, and what might be done to replenish the supplies. Directly or indirectly, these concerns will touch your life in the decades to follow.

7

THE MAIN METABOLIC PATHWAYS

FROM SUNLIGHT TO CELLULAR WORK: A MAJOR STRATEGY

In this chapter, we will be tracing the events of three major pathways, which are linked by a flow of energy through them. These are the pathways of photosynthesis, glycolysis, and cellular respiration.

In **photosynthesis**, sunlight energy is trapped by pigment molecules (such as chlorophyll), and is used in the formation of intermediate energy carriers (ATP and $NADPH_2$). These carriers transfer some of their energy to reactions in which carbon dioxide and water from the environment are converted into food molecules. The reactions may be summarized in this way:

$$6CO_2 + 6H_2O \xrightarrow{\text{sunlight energy}} C_6H_{12}O_6 + 6O_2$$

(carbon dioxide)   (water)   (typical food molecule)   (oxygen as by-product)

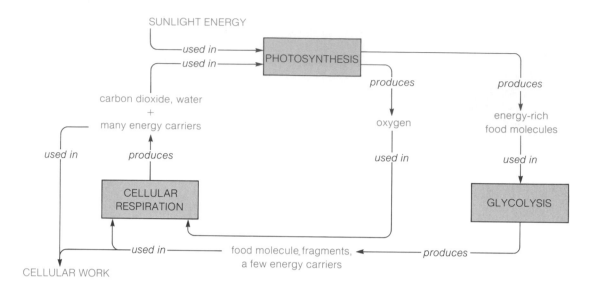

Figure 7.1 Links between three major energy-trapping and energy-releasing pathways.

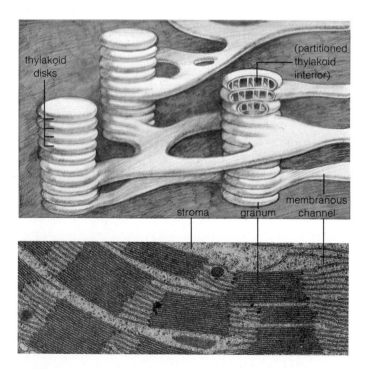

Figure 7.2 Functionally distinct zones of a chloroplast. The grana and the membranous channels connecting them are sites where light absorption and the formation of energy carrier molecules occur. The stroma is the region in which food molecules are actually assembled. (L. K. Shumway)

In **glycolysis**, food molecules are partially broken down. Some of their stored energy is released during the breakdown reactions and is used in forming a few energy carriers (ATP and NADH). These reactions take place in cytoplasm.

In **cellular respiration**, the food molecule fragments are broken down into carbon dioxide bits. During the reactions, hydrogen atoms (with their electrons) are stripped from the fragments and are transferred to energy carriers (FAD and NAD$^+$). The electrons are then sent down a transport chain, and energy they give off is used in forming ATP. Electrons leaving the chain combine with hydrogen ions and oxygen to form water. These reactions take place only in mitochondria.

When taken together, glycolysis and cellular respiration are something like photosynthesis in reverse:

$$C_6H_{12}O_6 \xrightarrow{\text{energy from carriers}} 6CO_2 + 6H_2O + \text{many energy carriers}$$

(food molecule)   (waste product)   (waste product)

As Figure 7.1 suggests, products of one pathway are starting materials for others; and energy—harnessed from the sun—flows through all three on its way to becoming available for cellular work.

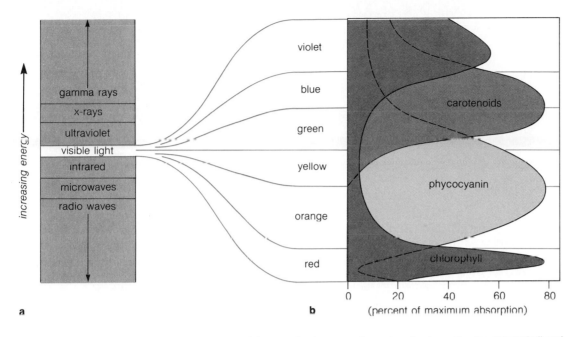

Figure 7.3 (**a**) Electromagnetic spectrum and (**b**) generalized ranges of wavelength absorption for chlorophyll and for two accessory pigments.

PART I. ENERGY-TRAPPING PATHWAYS

Photosynthesis: Where the Flow of Energy Begins

Sunlight-trapping activities are limited to a few kinds of autotrophs. An *autotroph* is an organism that can synthesize its own food, given an energy source and some simple inorganic compounds. Plants, blue-green algae, and some bacteria fall in this category. With few exceptions, a non-photosynthetic organism is a *heterotroph*. Being unable to build their own food, heterotrophs eat autotrophs, each other, or organic waste products. They include animals, fungi, and many microorganisms. Here, we'll be looking at photosynthesis as it occurs in the chloroplasts of autotrophic plants.

Figure 4.7 showed the overall structure of a chloroplast. Its semifluid matrix (stroma) is a construction zone: it contains enzymes and other molecules needed in assembling food molecules. In most land plant cells, chloroplasts have another functionally distinct zone. This zone consists of different stacks of membranous disks connected to one another by membranous channels. Each stack is called a granum; each disk, a thylakoid (Figure 7.2). Within this continuous membrane system are pigment molecules, enzymes, and reaction centers. All function in absorbing and transforming sunlight energy into the energy of carrier molecules.

Sunlight reaches the chloroplast as packets of energy called **photons**. The sizes of photons correspond to different wavelengths (colors) of light. The more energetic the photon, the shorter its wavelength. Different pigments inside the chloroplast absorb photons of different wavelengths. For instance, Figure 7.3 shows the absorption range of chlorophyll, the main chloroplast pigment. Chlorophyll is most efficient at using light from the red end of the visible spectrum, and it's also capable of using blue light from the other end. **Accessory pigments** absorb wavelengths in between red and blue light, and pass on some of the energy they absorb to chlorophyll.

These pigments are not sprinkled randomly through chloroplast membranes. Together with enzymes, they are organized into tiny light-trapping units of two sorts: **photosystems I and II**. It is here that the first stage of photosynthesis—the light reactions—begins.

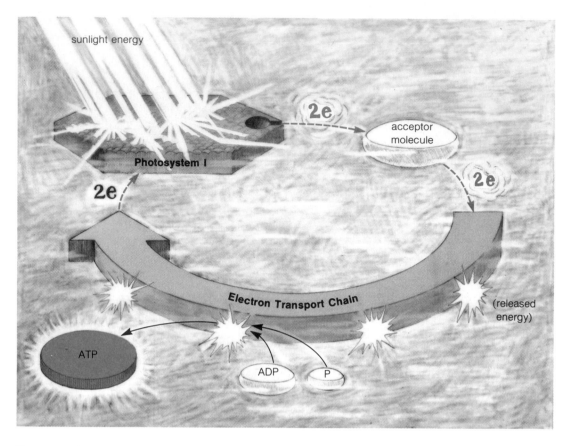

sunlight energy

acceptor molecule

2e

2e

Photosystem I

2e

Electron Transport Chain

(released energy)

ATP

ADP   P

Figure 7.4 Cyclic photophosphorylation, which yields one energy-rich ATP molecule.

The Light Reactions

In the **light reactions**, photon energy is harnessed and converted to usable forms of chemical energy. Initially, light-harvesting assemblies of pigment molecules transfer photon energy to a special chlorophyll–protein molecule. When enough energy is transferred to one of these molecules, some of its electrons become so excited that they pop out of the system in pairs. The energy of these electrons is used in the formation of energy carriers.

Before considering the way that electrons move through the whole set of light reactions, let's first look at a simpler, more ancient pathway. It is based only on photosystem I and a single electron transport chain (Figure 7.4). Electrons excited by light energy in photosystem I are passed to an electron acceptor molecule, then through a transport chain. Some of the energy released as electrons

is transferred down through the chain and used in forming ATP from ADP and inorganic phosphate (P_i). Eventually, the electrons lose their excess energy and return to the place they came from—photosystem I.

Because the electrons travel full circle each time, the process is said to be cyclic. Because they first gain energy from photons and then contribute that energy to phosphorylating ADP, the entire pathway is known as **cyclic photophosphorylation**. The important thing about it is this:

For every two electrons entering the cyclic photophosphorylation pathway, the energy yield is one ATP molecule.

Cyclic photophosphorylation probably represents the way that ancient bacteria and blue-green algae first harnessed light energy for producing usable cellular energy.

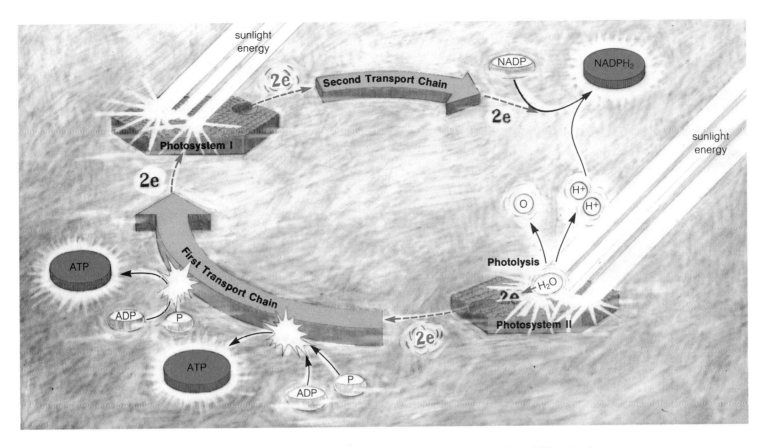

Figure 7.5 Noncyclic photophosphorylation, with its greater energy yield of one NADPH$_2$ molecule and two ATP molecules.

In existing plants, it is now used mainly in times of stress. Apparently natural selection long ago went to work on the cyclic pathway, expanding it to include a second photosystem and a second transport chain. When the *two* photosystems function together, electrons don't flow in a cycle. Instead, *new* electrons obtained from water molecules enter the pathway in photosystem II and ultimately end up in a molecule of NADPH$_2$. This expanded pathway is known as **noncyclic photophosphorylation**. It is the heart of photosynthesis.

The first step in the noncyclic pathway begins when water is split into oxygen, hydrogen ions, and their associated electrons. The process is often called *photolysis*, because photon energy indirectly drives the reaction (Figure 7.5). Two electrons from each water molecule flow into photosystem II. There they are excited to a higher energy level, and passed to the first electron transport chain. In the chain they give up energy in several oxidation–reduction reactions, and two ATP molecules are formed. But when electrons reach the end of this chain, they don't return to photosystem II. They now enter photosystem I. They are excited once again and passed to a second transport chain. The flow of electrons through this chain doesn't yield more ATP. Instead, the two electrons apparently combine with an NADP$^+$ molecule and with two H$^+$ ions (released from some water molecule at the start of the noncyclic route). The importance of this energy-harnessing strategy is outlined in the *Commentary*. Its outcome is this:

For every two electrons entering the noncyclic photophosphorylation pathway, there is an energy yield of one NADPH$_2$ and two ATP molecules.

COMMENTARY

Harnessing More Energy—An Evolutionary Breakthrough

Herman Wiebe, Utah State University

Early in the history of life, when bacteria were first tapping into the abundant light energy streaming down from the sun, cyclic photophosphorylation was probably the energy-trapping strategy of the day. Energy from photons could excite electrons in whatever pigment molecules existed in the membranes of those early photosynthetic prokaryotes. The excited electrons could then give up extra energy in amounts that could be used in phosphorylating ADP. Now, although ATP formation meant that energy was available for cellular work, the amount wouldn't have been enough to build sugar molecules except by roundabout and rather inefficient pathways. It happens that the energy-carrier $NADPH_2$ has about seven times as much useful energy as ATP. In addition, $NADPH_2$ has enough reducing power to rapidly break apart and re-form covalent bonds. Thus $NADPH_2$ funneled into oxidation–reduction reactions that convert carbon dioxide to sugar would make the conversions proceed much faster and more efficiently than ATP alone. Where could more hydrogens come from to make more $NADPH_2$?

The early atmosphere probably contained hydrogen gas (H_2) and methane (CH_4), and NADPH production would have depended at first on their somewhat limited atmospheric concentrations. But imagine how much hydrogen was present in H_2O—in the waters of oceans, streams, and seas! Why wasn't it being tapped?

Those ancient photosynthetic bacteria probably had pigment molecules that were different from chlorophyll. And the photons those pigments could trap as an energy source would not have been energetic enough to drive the synthesis of $NADPH_2$ from $NADP^+$ present in the cell and H_2O. Why couldn't these early organisms harness photons of higher energy? After all, light rays are streams of many different kinds of vibrating photons. The greater the frequency of their vibrations, the shorter the wavelength (and more energetic) the light rays. What

we sense as "visible" light are wavelengths ranging from about 400 to 750 nanometers—which is about the range of sensitivity shown by photosynthetic organisms, too. Why is photosensitivity limited mostly to this range? It happens that shorter wavelengths (strong ultraviolet rays, x-rays, cosmic rays) scarcely warrant a welcome mat in a cell. They are so energetic that they can break bonds holding biological molecules together—hence they can destroy living cells. Longer wavelengths (infrared and below) are not energetic enough to power the chemical changes in molecules necessary for $NADPH_2$ formation.

With their energy source so limited, these bacteria would have been candidates for natural selection. We might well imagine that mutations affecting the light-trapping photosystems themselves would have been fostered. The fact is, *what one photosystem alone couldn't harness, two linked together might harness very well.*

Getting the extra energy would have been like trying to carry a bucket of water up a ladder to wash some high windows. One person could run up and down the ladder. Or two people—one carrying in the water and one already standing on the ladder—could do the same thing faster and more easily. Even though one photon alone was not enough to "lift" $NADP^+$ to the high-energy carrier $NADPH_2$, two photons harnessed simultaneously would be quite enough. That is what happens in noncyclic photophosphorylation, with its yield of ATP *and* $NADPH_2$.

At some point, two photosystems did become linked in some ancient bacterium, which thereby may have become the forerunner to the photosynthetic blue-green algae—perhaps even to the chloroplasts of eukaryotic plants (Chapter Twenty-Two). The modification was a milestone in evolution. It is doubtful that more advanced plants, with their greater energy demands, could have emerged without it.

The Dark Reactions

Once the light reactions produce energy carriers, a photosynthetic cell can build food molecules. The pathways now followed are called the **dark reactions** because they don't depend directly on sunlight. (They *can* proceed without it, as long as ATP and NADPH$_2$ are available. But these molecules normally are produced only during daylight, so the "dark" reactions don't usually proceed for very long in the dark.)

To see how these reactions work, let's follow the events whereby six carbon dioxide molecules ($6CO_2$) are used to form one glucose molecule ($C_6H_{12}O_6$).

The reactions begin when an enzyme hooks up carbon dioxide to ribulose diphosphate (RuDP), a five-carbon compound. The result is a highly unstable intermediate that breaks apart at once into two molecules of a three-carbon compound, phosphoglyceric acid (PGA). This reaction sequence, in which carbon dioxide is removed from the air and combined with organic molecules, is called **carbon dioxide fixation**. For every six carbon dioxide molecules fixed, twelve PGA molecules are produced (Figure 7.6).

The twelve PGA molecules now enter a reaction series by which they are first converted to phosphoglyceraldehyde (PGAL). This conversion alone uses up no less than twelve ATP and twelve NADPH$_2$ molecules from the light reactions! The PGAL is then broken apart and rearranged into different intermediates. Two eventually are rearranged to form a glucose molecule. In reactions that use up six more ATP molecules, the remainder are reassembled into RuDP. This entire reaction series is called the **Calvin–Benson cycle** in honor of its discoverers, Melvin Calvin and Andrew Benson. It yields enough RuDP molecules to replace the six used up in carbon dioxide fixation, as well as one glucose molecule. The ADP, phosphate, and NADP$^+$ leftovers from the cycle are returned to the sites of the light reactions—where they can be converted once again to NADPH$_2$ and ATP. Figure 7.7 relates the dark reactions to the overall process of photosynthesis. The two stages of the dark reactions themselves may be summarized in this way:

In the first stage of the dark reactions, carbon dioxide from the air is combined with RuDP, then the carbon atoms are locked into stable intermediate compounds.

In the second stage, chemical energy in NADPH$_2$ and ATP is used in the conversion of the carbon-containing intermediates to food molecules, and RuDP used up at the start of the reactions is replaced.

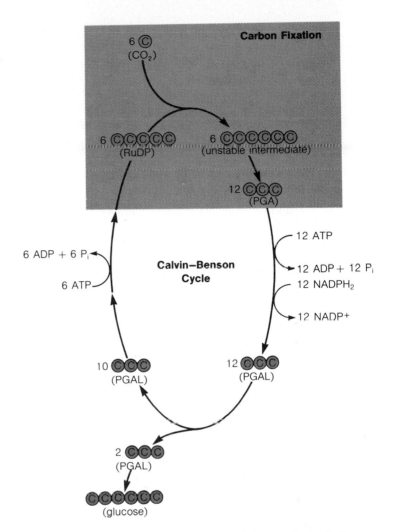

Figure 7.6 Summary of the dark reactions of photosynthesis. (Only the carbon atoms of the different molecules are depicted.)

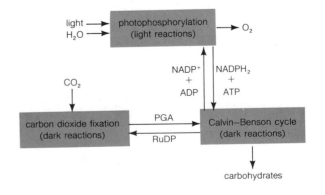

Figure 7.7 Summary of the reaction systems in photosynthesis.

What does an autotroph do with glucose produced in this way? It uses some as building blocks for cellulose walls. As soon as night falls, its cells use sugars (assembled from glucose and other subunits) as an energy source until the following day. But usually an autotrophic cell produces more sugar during the day than it can possibly use or even hold. The excess is converted into a different chemical form that can accumulate and that may be used in the future. Some plants (sugar beets and sugar cane, for example) store excess sugars as sucrose. By far the most common storage form is starch. In simple plants such as algae, starch is assembled and stored within the chloroplast that produces the glucose. In certain land plants such as potatoes, the sugars produced in leaf cells move down to special stem or root regions, where they accumulate as starch.

Chloroplasts are also able to assemble lipids and amino acids, in addition to carbohydrates such as glucose. The PGA produced in carbon dioxide fixation can be used as a carbon source in amino acid synthesis. Amino acids are used in building cell proteins and all the lipids required for cell membranes. Indeed, more than ninety percent of the carbon fixed by some green algae is used in constructing proteins and lipids. Having a brief life cycle and benefiting from plenty of water and sunlight, these algae put most of their photosynthetic activity into growth and reproduction, and very little into storage molecules.

Chemosynthesis

Photosynthesis so dominates the energy-trapping pathways that it is sometimes easy to overlook other, less pervasive routes that nevertheless hold importance for the world of life. Like photosynthesizers, some bacteria are autotrophs ("self-feeders"), for they can live completely on an inorganic diet. Unlike photosynthesizers, they harness energy released during the oxidation of such inorganic substances as ammonium ions (NH_4^+) and sulfur compounds.

An influential group of chemosynthesizers are the nitrifying bacteria. Some use ammonia (NH_3) molecules as their energy source, stripping them of hydrogen ions and their associated electrons. Nitrite (NO_2^-) ions are the remnants of this energy-securing route. Nitrite in turn is the energy source for still other nitrifying bacteria. All of these bacteria may be found in soil, where their activities enhance its fertility. The cycling of nitrogen through natural communities of life depends in large part on their activities. The chemosynthesizers convert nitrogen-containing compounds to a form that can be taken up by plant cells, hence by animals (Chapter Twenty-Eight).

PART II. ENERGY-RELEASING PATHWAYS

So far, we have energy-rich carbohydrates and other food molecules being stockpiled in autotrophic organisms. These are the molecules that will be dismantled by autotrophic cells and by the cells of heterotrophs that dine, one way or another, on the autotrophs. These are the molecules from which stored energy is released to drive life processes.

Overview of Glucose Metabolism

Depending on the organism, there is more than one kind of resource a cell can use, and there is more than one way to use them. To keep things simple, we'll focus on the way that energy extracted from a glucose molecule is converted to the energy of ATP, for it can be used as an example of energy-extraction pathways in general. The steps are outlined in Figure 7.8. You may find it useful to look at this illustration before we begin, and to keep this important point in mind:

Glucose breakdown and the associated energy transfers leading to ATP formation are central to the functioning of almost all prokaryotic and eukaryotic cells.

The word glycolysis refers to the *initial* breakdown of glucose (glyco-, "sugar"; -lysis, "break apart"). Briefly, a glucose molecule (which has six carbon atoms) is split in half (into two three-carbon fragments). The fragments are rearranged into two pyruvate molecules, which together have a little less energy than the glucose. During this juggling act, some energy is transferred to ADP, and two ATP molecules form.

Some of the simpler prokaryotes that depend on glycolysis live in places where there is no oxygen. They must rely solely on *anaerobic* pathways (which can proceed in the absence of oxygen). In one such anaerobic route, a carbon atom is split away from each pyruvate fragment. The carbon atoms end up in carbon dioxide molecules. The compound remaining is converted to ethyl alcohol (ethanol). And some energy released during the reactions gets transferred to two ATP molecules. This pathway, called **alcoholic fermentation**, is summarized below:

$$C_6H_{12}O_6 \longrightarrow 2C_2H_6O + 2CO_2$$
(ethanol)

energy yield = 2 ATP

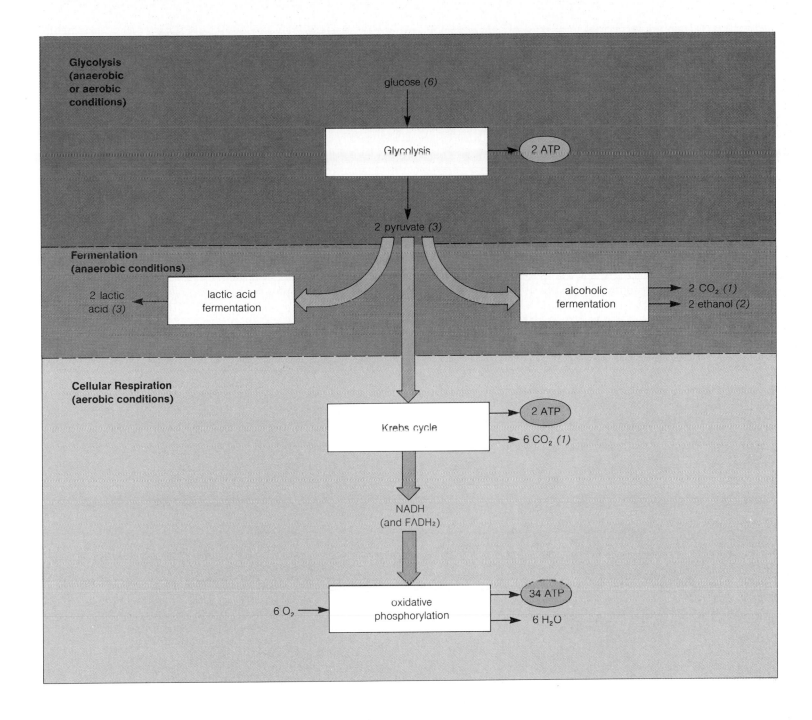

Figure 7.8 Overview of glucose metabolism. The gray arrows indicate three routes by which a glucose molecule may be further dismantled following its initial breakup in glycolysis. The route taken depends on the organism and on environmental conditions. (Italic numbers in parentheses refer to the number of carbon atoms in each molecule of the compound listed.)

A simple way to remember the *net* energy yield of the three routes is to add up the ATP molecules from start (in all cases, glycolysis) to finish. Only two are produced on the route ending in lactic acid fermentation; the same is true for alcoholic fermentation. Thirty-eight ATP molecules are produced on the route ending with the pathways of cellular respiration.

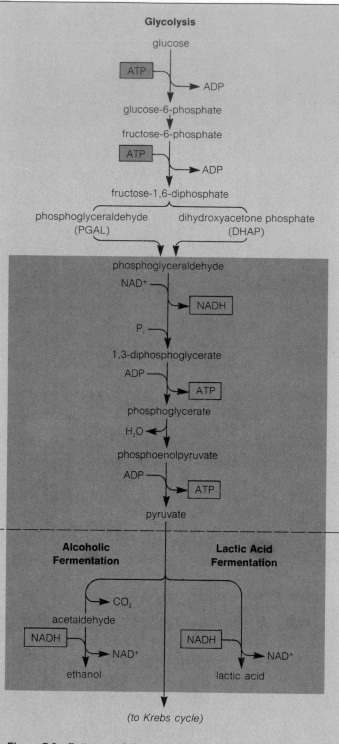

Glycolysis

glucose

A stable 6-carbon sugar

ATP → ADP

picks up a phosphate group from an ATP molecule,

glucose-6-phosphate

thereby becoming a reactive 6-carbon sugar phosphate

fructose-6-phosphate

that is promptly modified into another 6-carbon sugar phosphate.

ATP → ADP

The modified form picks up a phosphate group from another ATP molecule

fructose-1,6-diphosphate

to become an even more reactive 6-carbon sugar diphosphate,

phosphoglyceraldehyde (PGAL)   dihydroxyacetone phosphate (DHAP)

which is split into two 3-carbon sugar phosphates: PGAL and DHAP.

phosphoglyceraldehyde

NAD^+

NADH

DHAP can be converted to PGAL, so beyond this point, each step occurs <u>twice</u> for each glucose molecule being torn down.

P_i

PGAL gives up hydrogen to an NAD^+ molecule, which produces the energy carrier molecule NADH.

1,3-diphosphoglycerate

ADP

ATP

In the process, PGAL is converted to a 3-carbon acid and picks up another phosphate group,

which is promptly used to produce ATP from ADP. At this point the original energy investment of ATP has been repaid.

phosphoglycerate

H_2O

The resulting molecule undergoes rearrangements

phosphoenolpyruvate

which produce phosphoenolpyruvate (PEP).

ADP

ATP

PEP is capable of donating its phosphate group to ADP, so more ATP is formed

pyruvate

and this leaves only pyruvate (the ionized form of pyruvic acid).

Alcoholic Fermentation   **Lactic Acid Fermentation**

CO_2

acetaldehyde

NADH → NAD^+

NADH → NAD^+

ethanol   lactic acid

If the pyruvate now enters the alcoholic fermentation sequence, the whole chain of reactions leading from glucose to ethanol ends up with a net energy yield of two ATP molecules.

If the pyruvate enters the lactic acid fermentation sequence, the net energy yield is only two ATP molecules.

In both cases the NADH is restored to NAD^+, which can take part again in the reaction with PGAL.

(to Krebs cycle)

Figure 7.9 Pathways of glycolysis. All the steps in the shaded area must occur twice for every glucose molecule being dismantled along the glycolysis and fermentation pathways.

Yeasts (small, single-celled fungi) use this pathway. The gaseous by-product of the reactions (carbon dioxide) helps yeast dough to rise. The alcohol produced is the reason that beer, wine, and distilled spirits have the kick that they do.

In another anaerobic route, the pyruvate coming from glycolysis is rearranged into lactic acid, a three-carbon compound. This pathway is called **lactic acid fermentation**:

$$C_6H_{12}O_6 \longrightarrow 2C_3H_6O_3$$
$$\text{(lactic acid)}$$

energy yield = 2 ATP

Milk or cream turned sour is a sign that microorganisms are present and using pyruvate in lactic acid fermentation. The sourness depends on how much lactic acid has accumulated. Like the other fermentation route, this pathway has a net energy yield of two ATP molecules.

It happens that considerable energy is still locked up in the disassembled parts of glucose at the end of glycolysis. Neither of the fermentation pathways just outlined extracts very much of that energy. It is only when we get to cellular respiration that we cross a truly major threshold in energy-extraction processes. Cellular respiration consists of two closely linked pathways: the Krebs cycle and oxidative phosphorylation. Many prokaryotic cells and almost all eukaryotic cells can use these pathways, which begin with pyruvate from glycolysis.

In the **Krebs cycle**, pyruvate is completely broken down; all that remains is carbon dioxide. In itself, the cycle yields only two more ATP molecules for each glucose molecule. But during the reactions, hydrogen atoms with their high-energy electrons are transferred to energy-carrying nucleotides, such as NAD¹. The nucleotides cart them off to a transport chain similar to the one used in photosynthesis. *And as the high-energy electrons move through an electron transport chain, they give off energy in increments, some of which are enough to phosphorylate ADP.* This is the pathway of **oxidative phosphorylation**. It yields thirty-four ATP molecules for every glucose molecule being dismantled.

To keep the electron transport chain cleared for operation, something must take away all the hydrogen ions and electrons that were fed into it. Oxygen serves this role as hydrogen and electron acceptor. Hence oxidative phosphorylation is said to be an *aerobic* process. It doesn't proceed unless oxygen atoms are present.

Glycolysis, the Krebs cycle, and oxidative phosphorylation together constitute the major route by which usable energy becomes available in most living things. This route is summarized by the following equation:

$$C_6H_{12}O_6 + 6O_2 \longrightarrow 6CO_2 + 6H_2O$$

energy yield = 38 ATP

By now, you might be wondering why a cell bothers to go through all this trouble. Wouldn't it be far easier to break down the whole glucose molecule all at once? If you think back on what happens when a lit match is dropped in a gasoline tank (Chapter Six), you already have an inkling of why the most direct energy conversion route is *not* the best route for a cell. When glucose is converted to carbon dioxide and water, a great deal of energy is released—about 686 kilocalories per mole. (A "mole" is a certain number of atoms or molecules of any substance at all, just as a "dozen" can refer to any twelve doughnuts, roses, or any other object.) If a cell were simply to pop off the molecule in a single step (instead of the 140 steps taken along the aerobic route), most of the energy would be released as a sudden burst of heat. The burst of heat would literally cook the cell to death. It is this potentially destructive release of heat energy that calls for a moderate approach to energy metabolism.

And now, if you would like to (or are asked to) take a closer look at the energy-releasing pathways just outlined, the next sections are included to round out the picture.

A Closer Look at Glycolysis and Fermentation Routes

Why does a cell operating under anaerobic conditions even bother to convert pyruvate to lactic acid (or ethanol) and carbon dioxide? It gets no more energy once pyruvate has formed—so why doesn't it quit and leave well enough alone? For the answer, let's go back to where glycolysis begins.

Figure 7.9 lists the reaction steps in glycolysis. Glucose happens to be a relatively stable (unreactive) molecule. It is made reactive when a phosphate group is attached to it—not once, but twice. The resulting sugar diphosphate is split into a phosphoglyceraldehyde (PGAL) molecule and a dihydroxyacetone (DHAP) molecule. These molecules are structural isomers, and the DHAP is readily converted to PGAL. So now we effectively have two PGAL molecules.

A series of rearrangements now begins that will lead to the formation of new ATP. Another phosphate group

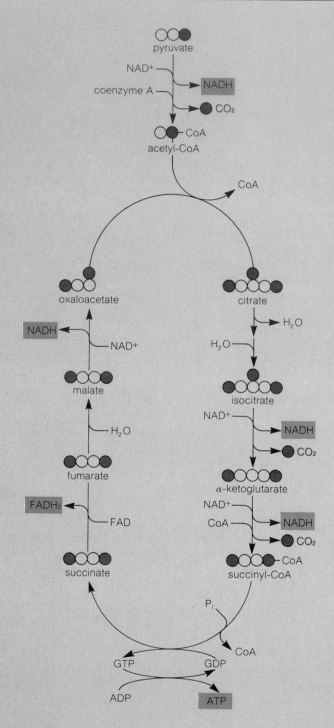

Step 1. As pyruvate enters the mitochondrion, a carbon atom is removed as CO_2. At the same time, two electrons and an H^+ ion are removed and transferred to NAD^+, producing NADH. The result is the conversion of one of the two remaining carbon atoms to an acid group. The two-carbon compound is linked to a molecule called coenzyme A, or CoA.

Step 2. The Krebs cycle proper begins when the two-carbon acid is coupled with a four-carbon compound having two acid groups (oxaloacetate). This produces citrate, a six-carbon compound having three acid groups. CoA is released for reuse.

Step 3. After two intermediate reactions, an acid group is split off the six-carbon compound and is released as CO_2 At the same time, two electrons and an H^+ ion are removed, forming a second molecule of NADH.

Step 4. Now the third and last carbon atom is split off as another CO_2 molecule. Once again electrons and an H^+ ion are given off to produce NADH. In the process, a new acid group is linked to coenzyme A. The rest of the cycle works to convert this new four-carbon molecule with two acid groups to the related compound oxaloacetate, with which the cycle begins once again.

Step 5. During the next conversions, the energy released when the CoA is split off is used to produce an ATP molecule. (Actually the related compound GTP is formed by addition of a phosphate group to GDP, but GDP then transfers its phosphate to ADP.)

Step 6. Two more hydrogen atoms are removed, but this time they are passed on to FAD, producing $FADH_2$. (There isn't enough energy by now to make more NADH.)

Step 7. And another NADH molecule is produced by the transfer of two more hydrogen atoms. The product of this last reaction is the same four-carbon compound that the cycle starts with. Thus, we are ready for another turn of the Krebs cycle.

Figure 7.10 The Krebs cycle, the first stage of cellular respiration. In this sketch, each ◯ signifies the presence of a carbon atom in the molecule, and each ● signifies an acid group ($-COO^-$) or a carbon dioxide molecule produced when an acid group is split off. Remember that for each glucose molecule being metabolized, two pyruvate molecules have been formed. So two turns of the Krebs cycle must occur for each molecule of glucose oxidized.

becomes attached to each PGAL. At the same time, two electrons and a hydrogen ion are released and picked up by NAD$^+$ present in the cytoplasm. The result is NADH (Chapter Six). The problem is that there is only so much NAD$^+$ around. Without NAD$^+$, there wouldn't be anything to accept the hydrogen atoms released during PGAL breakdown. Thus glycolysis would grind to a halt, and the cell would be doomed.

How does the cell maintain enough NAD$^+$? When pyruvate from glycolysis travels the alcoholic fermentation route, acetaldehyde formed during the conversion acts as a hydrogen *acceptor*. Acetaldehyde strips the NADH of its hydrogen ion (acquired during PGAL breakdown), and becomes converted to ethanol in the process. In the lactic acid fermentation route, the pyruvate itself acts as a hydrogen acceptor. The pyruvate takes hydrogen ions and electrons from NADH on the way to becoming lactic acid. Thus, even though further conversion of pyruvate yields no more usable energy in the anaerobic routes, it has this advantage:

The final steps of the fermentation pathways serve to regenerate the essential carrier molecule NAD$^+$.

A Closer Look at Cellular Respiration

For some prokaryotes, and for eukaryotic cells deprived of molecular oxygen, glycolysis ends with some fermentation product. For eukaryotic cells supplied with plenty of molecular oxygen, the pyruvate from glycolysis is completely oxidized to carbon dioxide in the Krebs cycle (Figure 7.10). In these reactions, NAD$^+$ and FAD serve as temporary acceptors for hydrogen ions and electrons being released during the oxidations. In fact, the main usable products of the Krebs cycle *are* the carriers NADH and FADH$_2$. Each time energy is released during an oxidation step, some is conserved in these molecules for later use in ATP formation. As you can see from Figure 7.10, eight NADH and FADH$_2$ molecules are produced for every two pyruvate molecules entering the Krebs cycle, along with two molecules of ATP.

Let's now follow the twelve carrier molecules through the oxidative phosphorylation sequence. Each delivers its cargo of H$^+$ and electrons to an electron transport chain embedded in the inner mitochondrial membrane (Figure

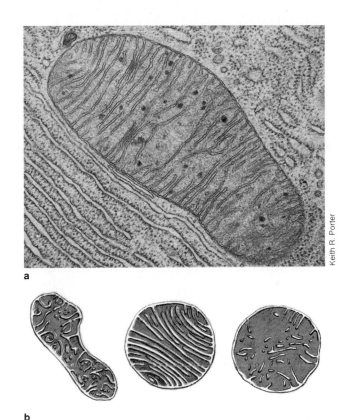

a

b

Figure 7.11 (**a**) Mitochondrion from a bat pancreatic cell thin section. (**b**) The inner membrane foldings, no matter how diverse they are in appearance, create two functional zones *within* the mitochondrion: an inner, isolated matrix (brown) and a zone of channels leading from the outer membrane into the mitochondrial body. (From Avers, *Basic Cell Biology*, Van Nostrand, 1978)

7.11). Regardless of how folded it may be, the inner membrane is like an envelope that creates a sealed compartment. According to the **chemiosmotic theory**, electron carriers are arranged in the membrane in such a way that H$^+$ ions are *expelled* from the compartment during the transfers down the transport chain. This sets up an H$^+$ concentration gradient across the membrane, as well as an electric gradient (a difference in charge between two regions). When H$^+$ ions flow down the gradient—*back* into the compartment—enzymes embedded in the membrane couple the energy of these ions to the phosphorylation of ADP. Experiments suggest that ATP formation in chloroplasts also depends on transport-chain activity and a hydrogen ion gradient across membranes.

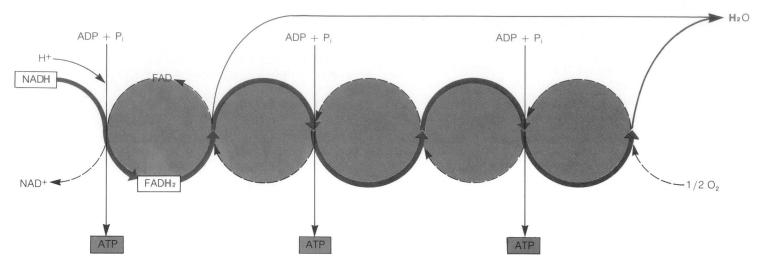

Figure 7.12 Simplified version of the energy flow through the oxidative phosphorylation sequence. NADH from the Krebs cycle enters the sequence, transferring its hydrogen atoms (with their electrons) to FAD. The energy released in the reaction couples P_i to ADP, forming an ATP molecule. (The sequence also can start directly with $FADH_2$ from the Krebs cycle, but the cell loses its chance to form an extra ATP molecule when it does.) The electrons derived from the hydrogen atoms are shuffled through the chain to an oxygen acceptor, which carts them off. Oxygen must be available to keep the chain clear for operation.

As Figure 7.12 shows, for every NADH molecule, the oxidative phosphorylation sequence yields three ATP molecules. For every $FADH_2$ molecule, the yield is two ATP. Hence thirty-four ATP molecules are formed with energy from the twelve carriers sent down the aerobic pathways of cellular respiration:

In glycolysis and the Krebs cycle, NAD^+ and FAD are reduced to NADH and $FADH_2$, which are used in the energy transfers of oxidative phosphorylation.

In oxidative phosphorylation, NADH and $FADH_2$ are stripped of hydrogen ions and the high-energy electrons associated with those ions. Energy released in the transfer of these electrons is used in forming thirty-four ATP molecules.

On Muscle Cramps and Energy-Extraction Efficiency

Under anaerobic conditions, again, cells use NADH to produce lactic acid (or alcohol). The reactions regenerate NAD^+, which is essential for cell functioning. During normal activities, your own lungs and circulatory system provide every cell in your body with enough oxygen to carry out aerobic metabolism. But what happens in muscle cells when you exercise so rapidly that oxygen uptake no longer corresponds to your pace? Then the transport chains of oxidative phosphorylation load up with electrons that oxygen normally would carry away. NADH is no longer stripped of hydrogen ions and electrons, so there is a shortage of NAD^+. Without NAD^+, the Krebs cycle stops. Pyruvate from glycolysis begins to pile up. At this point, muscle cells revert to the less efficient anaerobic pathway leading to lactic acid formation. Lactic acid builds up in cells, which are now forced to operate on the low ATP output from glycolysis. They can do this for a short time, assuming enough glucose is available. But lactic acid is toxic in high concentration and causes muscle cramps in your legs. When you stop to rest, the lactic acid is taken up in the bloodstream and is flushed away from your muscle cells. Then your muscle cells can revert to the more efficient aerobic pathways.

Just how efficient is aerobic metabolism compared to the anaerobic routes? As Table 7.1 shows, it's about *eighteen times* more efficient than anaerobic glycolysis in converting the energy in glucose to ATP.

Table 7.1 Summary of the ATP Yield From the Complete Oxidation of One Glucose Molecule		
Energy-Extraction Pathways		Energy Yield
Glycolysis:		2 ATP
Krebs Cycle:		2 ATP
Oxidative Phosphorylation		
2 NADH from glycolysis:	6 ATP	
8 NADH from Krebs cycle:	24 ATP	34 ATP
2 FADH₂ from Krebs cycle:	4 ATP	
		A total of 38 ATP

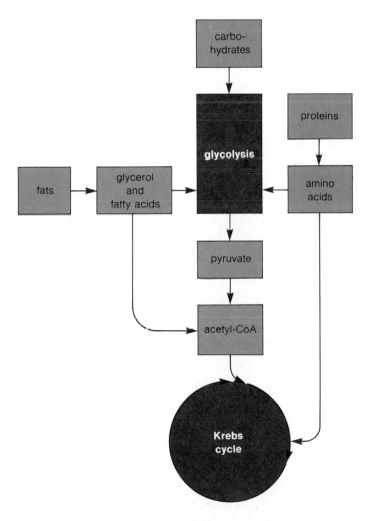

Figure 7.13 Where substances other than glucose flow into the energy-extraction pathways of a cell.

Some Conversion Possibilities

Cells can use energy sources other than glucose. For instance, complex carbohydrates such as starch can be converted to glucose or some other compound. These conversion products can flow directly into the glycolysis pathway (Figure 7.13). Similarly, fats can be metabolized into substances that flow into the glycolysis pathway or the Krebs cycle. Compared to sugar, fats are much more reduced (their ratio of hydrogen to oxygen is greater). This means that more NADH can be formed when a fat molecule is broken down. In fact, metabolizing a gram of fat yields about twice as much ATP as metabolizing a gram of carbohydrate. Fats are efficient storage molecules because they are such a concentrated energy source.

Proteins may be hydrolyzed and their amino acids used as another energy source. Amino acids can be converted to pyruvate or to one of the Krebs cycle intermediates. The energy yield from proteins is roughly the same as that from carbohydrates. Normally, a cell uses amino acids as raw materials in protein synthesis. But when protein intake exceeds cellular requirements, the excess is sent down energy-extracting pathways. Sometimes individuals do not take in enough carbohydrates and fats to meet their energy requirements, and their cells disassemble proteins that make up their own body. Among individuals suffering from starvation, such protein conversion is a last resort to secure energy for keeping the body alive.

Most heterotrophic cells take in a variety of carbohydrates, fats, amino acids, and nucleotides. But rarely are these molecules in precisely the right chemical form and proportions necessary to build new cell materials. Once again the basic metabolic pathways are called into action. For example, regardless of the form of carbohydrates eaten, intermediates of glycolysis and the Krebs cycle can be routed into synthesis pathways. These intermediates can serve as building blocks for all the kinds of sugar molecules that a cell might require for maintenance, growth, and reproduction. As another example, intermediates of the Calvin–Benson cycle, glycolysis, and the Krebs cycle can be diverted for the assembly of about twenty different amino acids. This is true for autotrophic cells, at least. Many heterotrophic cells have lost the ability to build all the different amino acids needed in protein synthesis. Our own cells are metabolically equipped to assemble only twelve of the twenty amino acids we require. The remainder are called *essential amino acids*, and they must be obtained from food. This aspect of human nutrition is discussed in the *Commentary* on page 110.

COMMENTARY

Resources and the Human Condition

From a biological perspective, the main limiting factor on our existence as a species is food. Said another way, where do we get the energy and materials needed to sustain 60,000 billion cells multiplied by 4 *billion* human bodies? It is beyond the scope of this book to address all the economic and social implications of this question. But we can outline the biological realities of what it takes to maintain a human individual. This information may help you in dealing with difficult decisions that we all must face in the near future.

dren, the growth rate slackens. Emaciation is one result; mental and physical deterioration are others.

When caloric intake remains below the minimum requirement, starvation sets in. The total number of deaths from starvation is impossible to estimate. Undoubtedly the annual figure reaches into the millions. Certainly it is greater than the toll of our greatest wars. Yet starvation is not a cyclic social aberration like war. It is a relentless commentary on the human condition and the environment that sustains it.

Calories: A Biological Energy Crisis

The first reality is that most humans are malnourished (see map). Most can't get enough energy (from fat and carbohydrates) or protein. Of 4 billion people, perhaps only 500 millions have adequate diets. A unique problem facing these well-fed few (mostly those living in North America, Western Europe, and the USSR) is obesity: energy intake exceeds body requirements and results in weight gain. Another 1 billion get enough energy but not enough protein. For the remaining 2½ billion, the situation is desperate.

As for all organisms, our energy needs are measured in kilocalories. (A kilocalorie is the same thing as a thousand calories—the amount of energy needed to heat 1,000 grams of water by 1°C. Because energy can be converted from one form to another, the energy of most reactions of life can be expressed in kilocalories even when the energy is used for some purpose other than simply heating water.) How much energy a human needs varies with size, age, degree of activity, and physiological state. An adult male of "average" size who engages in normal activities needs about 2,700 kilocalories/day. The "average" adult female needs about 2,000. The high metabolic activity of children in their midteens demands 2,400 to 2,800. When caloric intake falls below these energy requirements, weight drops. For growing chil-

Protein: A Matter of Quality and Quantity

Protein deficiency complicates the food resource picture. For us, essential amino acids are phenylalanine (and/or tyrosine), isoleucine, leucine, lysine, threonine, tryptophan, cysteine (and/or methionine), and valine. Our cells simply cannot make these materials; our diet must provide them. In addition, our cells can build the other "nonessential" amino acids only if the total amino acid intake is adequate. They can convert the "nonessential" ones, but they can't make them from nothing. It takes an amino acid of one sort to make another amino acid. To build one, a cell has to tear another down.

Protein deficiency is serious at any age. But it is most distressing among young children, for rapid brain growth and development occur during early life. Unless enough protein is taken in just before and just after birth, irreversible mental retardation occurs. Even mild protein starvation can retard growth, and affect physical and mental performance.

The minimum daily requirement for protein probably ranges between 0.214 and 0.227 gram for every 454 grams (1 pound) of body weight. To this, another 0.28 gram can be added as a safety factor against individual differences. This translates to about 43 grams (about 1½ ounces) of "pure" protein each day for an average adult male, and about 35 grams for an average female.

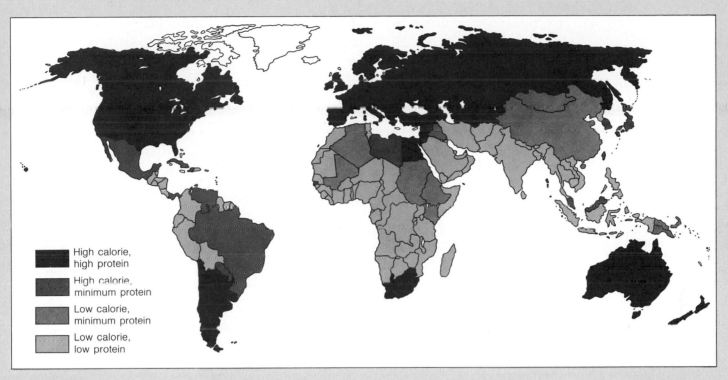

Caloric and protein utilization on a global scale. (From *Population, Resources, Environment: Issues in Human Ecology*, second edition, Paul Ehrlich and Anne Ehrlich. W. H. Freeman and Company. Copyright © 1972)

Legend:
- High calorie, high protein
- High calorie, minimum protein
- Low calorie, minimum protein
- Low calorie, low protein

But quantity alone is not enough. *Cells must receive essential amino acids at the same time, in just the right proportions, before they can assemble their own proteins.* Suppose the kind of protein you eat today has seven of the eight essential amino acids, in adequate amounts—but has only half the required lysine. Even if you eat 43 grams of that protein, you won't meet your daily protein needs. You must eat twice as much to get enough lysine.

To compare proteins from different sources, nutritionists use a measure called net protein utilization (NPU). NPU values range from 100 (all essential amino acids present, in ideal proportions) to 0 (one or more amino acid absent, which makes the protein useless when eaten alone). Balancing the diet with different proteins can make up for such deficiencies.

For much of the world, cereal grains are the main foods. As the Table suggests, cereal grains such as corn are low in protein content and NPU values. In contrast,

beans are high in protein. Although NPU values for beans are no higher than those for cereal grains, beans are deficient in *different* amino acids. When beans are eaten *with* grain, the one food enhances the other—and raises the overall NPU value.

The High Cost of Protein Conversion

It takes nearly 1,000 grams (about 2 pounds) of dried corn to meet the daily protein needs of an adult male (see Table). Multiply this amount by a large population and it's easy to see why even high-energy crops still can't satisfy protein needs unless there is enough variety in the diet. Of course, those of us in affluent countries don't think of corn when we hear the word protein. We think of meat. It's true that less than 257 grams (9 ounces) of meat are enough to satisfy daily needs. The problem with meat, though, is the large energy

Comparison of the Efficiency of Some Single Protein Sources in Meeting Minimum Daily Requirements

Source	Protein Content (%)	Net Protein Utilization (NPU)	Amount Needed to Satisfy Minimum Daily Requirement	
			(Grams)	(Ounces)
Eggs	11	97	403	14.1
Milk	4	82	1,311	45.9***
Fish*	22	80	244	8.5
Cheese*	27	70	227	7.2
Meat*	25	68	253	8.8
Soybean flour	45	60	158**	5.5**
Soybeans	34	60	210**	7.3**
Kidney beans	23	40	468**	16.4**
Corn	10	50	860**	30.0**

*Average values.
**Dry weight values.
***Equivalent of 6 cups. The figure is somewhat misleading, for most of the volume of milk is water. Milk is actually a rich source of high-quality protein.

investment needed to produce it. We use cattle as converters of "low-grade" plant protein into "high-grade" meat protein. That cattle can make the conversion at all is due largely to the activity of normally harmless microorganisms that live in their digestive tracts (Chapter Sixteen). The microorganisms break down tough cellulose fibers of plants and convert them to fatty acids. The fatty acids are used in building (among other things) the amino acids that cattle can't produce for themselves. Using plants as an amino acid source enables cattle to build proteins at a rapid rate.

Conversion costs depend somewhat on external factors. At one time, cattle were raised on the open range (semiarid land, at best only marginally suitable for raising crops). The raw materials (grasses) could not be used in other ways. Hence the conversion was profitable even though between 3,175 and 4,536 grams (7 and 10 pounds) of grasses were needed to produce only 454 grams (1 pound) of meat. In modern agriculture, cattle spend very little time grazing on the open range. They are confined to feedlots: pens in which they are fed high-quality

grains to bring them to market size. The conversion cost is about the same as before: about 10 to 1. But here we are talking about grains that can be used as a *direct* food source for humans. A tremendous part of the grain produced each year in the United States becomes feed for beef and dairy cattle, as well as for pigs, chickens, and turkeys. Much of our exported grain goes to developed countries, where it is used as livestock feed.

To some, the answer is simple. Humans can subsist most efficiently on plant proteins. *When domestic cattle are inserted between plant production and individual consumers, there is a 10-to-1 conversion loss both in calories and protein.* But can you honestly envision the entire nation giving up table meat? That simply will not happen, at least not in the foreseeable future. It will be more realistic to concentrate on working out alternatives to the ideal.

For example, consider that 454 grams of ground beef now cost about twice as much as the same amount of turkey. The cost difference stems from two things. Turkeys are more efficient than beef cattle at protein conversion (it takes less feed to produce the same amount of meat). And turkeys require less preparation to make them market-ready. More useful protein can be produced at lower cost; the market price reflects it.

Consider also the potential of fish harvesting. Fish harvests represent only five percent or so of the world protein supply for humans. Yet we are already harvesting more than half of what the world's oceans are apparently capable of supporting. Massive efforts to step up harvesting would seriously deplete fish populations —and it would not change the world protein picture very much as it did. Fish *farming* may be another matter. For centuries, small-scale fishponds have been part of Asian agriculture. Human and livestock wastes are emptied into the ponds, which contain controlled numbers of fish species. Nutrients in the wastes fan the growth of microorganisms that are food for some fish; plants thriving in the enriched water are food for other fish that can grow to about 1,575 grams (35 pounds). Such ponds often nurture disease-causing microorganisms, but development of proper waste treatment methods could eliminate the problem. Efforts are under way to develop a cycle in which (1) grains are used to feed livestock, (2) the runoff from feedlots is used to stimulate plant growth in nearby ponds that incorporate waste treatment facilities, (3) the plants feed selected kinds of fish, (4) pondwater is used in fertilizing grain fields, and (5) fish are harvested for food.

There may also be some shifts to more efficient use of plant proteins. Adults can meet their daily recommended allowances with only 210 grams of soybeans (see Table). If soybeans are processed into flour, the protein content jumps to 45 percent. This means a mere 158 grams of dried flour per day is enough to meet the body's protein needs. Soybean production is increasing. Efforts are also being made to breed plants for higher protein content or higher NPU values. For instance, corn and rice are low in lysine, which limits their NPU values. Varieties with higher lysine content are being developed by careful breeding and selection experiments. The problem is that these high-lysine varieties demand even more nitrogen-rich fertilizers than are used now. The fertilizer costs tend to be more than small subsistence farmers can bear. It also raises the agricultural energy deficit even more. Present efforts to incorporate nitrogen-fixing bacteria into the crop plant system (page 176) could improve matters.

Other possible (although radical) solutions include raising bacteria on petroleum or sewage sludge and harvesting edible by-products, or mass culturing protein-rich algae. Of course, such fundamentally different approaches—no matter how technologically or biologically feasible—will require monumental reeducation if well-fed people are to accept such proteins as alternatives to their T-bone steaks.

Other Variables

As valuable as they may be, research efforts in themselves are not enough to make the world food crisis diminish much. The problem is awesomely complex. Its variables must be unraveled on many levels besides that of improving NPU values and protein content. At each level, simplistic programs have already been proposed. But it's important to recognize that carrying them out would create new problems in their wake. To see why this is so, let's consider just four of these suggestions.

First suggestion: *Improve crop production on existing land.* In this view, we can (1) improve plant varieties for higher yields, and (2) export modern agricultural practices and equipment to developing countries. This thinking is the basis for the so-called green revolution. It has not been as easy to do as it sounds. High-yield crops require fertilizers, pesticides, and ample irrigation. The plain truth is that developing countries depend on subsistence farming for the agricultural base. And subsistence farmers can't make the investment needed to take widespread advantage of the new crop strains. Where improvements have become available, the farmers have become dependent on industrialized producers of fertilizers and machinery. Of necessity, the costs of these items are reflected in food market prices. Thus food becomes too expensive for the country's own population.

In the long term, even industrialized countries may have to reassess their own agricultural systems. One reason is that energy sources available to drive mechanized farming are dwindling and becoming more expensive for everyone. A realistic alternative is to build agricultural systems around harnessing solar energy (Chapter Thirty). Some devices are on the drawing boards now. (For example, low-cost, solar-powered microcircuits may be able to harness heat energy from the sun and convert it to direct currents of electrical energy.)

Second suggestion: *Open up new areas for agriculture.* Almost 3½ billion acres are now under cultivation. Some people have proposed that another 7 or 8 billion acres could be converted to agriculture. Aside from the environmental impact of such expansion (Chapter Twenty-Eight), the best land available for agriculture is already being used for agriculture. What remains as arable land (able to support cultivation) is less desirable. In some heavily populated regions such as Asia, food problems are already severe even though over eighty percent of the arable land is already intensively cultivated. Much of what remains is found in tropical regions of Africa and South America. However, land that supports a rich tropical forest simply will not support crops for more than a few years after it has been cleared. Desert areas are another possibility, assuming water (and fertilizers) can be brought in. But desert irrigation brings its own problems, such as salt buildup in the soil. (There isn't enough rain to wash away salts that accumulate because of the rapid evaporation of irrigation water in hot climates.) Besides, where will the water come from? Water, too, is not an unlimited resource (recall the serious draught in the central United States during 1980). Desalinizing ocean water (reducing its salt content) is not yet economically feasible for large-scale efforts. As long as desalinization processes are based on energy from petroleum, they won't help much in the short term.

The cost of opening up new lands must also be accounted for. These are costs of the land itself, clearing and building roads, developing transport and storage systems. Assuming each acre so cleared will support only one person, it might well cost about 30 billion dollars a year—every year—just to keep up with the current population growth rate.

Third suggestion: *Equalize food distribution*. This is the easy one—just make sure everybody gets an equal share. It is probably the most impractical suggestion of all, simply because we are dealing with human beings. In any nation with an agricultural surplus, farmers characteristically pay taxes on their land and buy seeds, fertilizers, and machinery. They expect a return for their labors in producing above and beyond what they personally need. They want to be paid for their crops. If a government wishes to give crops to other people, the government must directly or indirectly pay the farmers. In other words, it is the taxpayer who foots the bill, voluntarily or involuntarily. Each year sees a boost in energy costs and inflation, so it becomes more difficult to sell a program of international assistance to a hard-pressed general public. Lacking such programs, farmers will sell the surplus to whoever can pay the most—other industrialized nations, mainly.

One solution is to have developing countries with certain essential resources (such as oil) get a fair return for them on the world market. But not all countries have such resources. Many of those that do are as reluctant to use them to help their deprived neighbors as are industrialized nations. Oil-rich countries invest their profits in the developed world—even as they burn off natural gas in the Persian Gulf in amounts that could

produce all the fertilizer the developing countries could use.

Fourth suggestion: *Stabilize or reduce world population*. Ours is a finite world with finite resources. It can support a large number of individuals at a bare subsistence level. It can, as it does now, support a small number of individuals in first-class accommodations while confining the rest to steerage. Alternatively, it could support a smaller number in dignity and modest comfort. To achieve this alternative, we must do more than stop, we must back up in terms of population growth. We can dream of science pulling some new technological rabbit out of a hat. It is not a vain dream, for there probably are innumerable rabbits hidden among the folds and shadows. *But any solution must be in accord with the supply of material resources and the principles of energy flow*. So it's a numbers game we're playing, and the manner in which we choose to play will affect our own lives to some extent. More importantly, it will restrict the options of all those who follow. A question on the horizon is this: Who gets first-class or even second- and third-class tickets—and who gets left behind?

Our world contains vast resources—not resources that can be ripped from the ground and used for limited interests. It contains subtle sources of biological information on how to survive and lead productive lives on what is still available to us. Our species has dreamed of great accomplishments. But the luxury of dreams can be purchased only with a secure base for survival. Until now, our base has been a form of technological savagery—action without knowledge or regard for the consequence. With luck and with effort, future generations may look back on this as the first generation to seek the beginnings of a stable relationship with the total world of life.

PERSPECTIVE

In this unit, you have read about fundamental connections between matter, energy, and the living cell. These connections exist in the pathways by which organisms trap energy, store it in molecules, then break down those molecules to release energy when needed. They involve systems of producing and using food at the level of cells, organisms, species, and populations. Energy trapping through photosynthesis, energy extraction through glycolysis and cellular respiration—taken together, these are the central events of the living world.

Later chapters will turn to the greater picture of the origin and evolution of life. But the following overview of how these three metabolic pathways came to be linked together may give you a sense of how important they are for the continuation of life on this planet.

It now appears that when life originated, the environment was rich in molecules containing carbon, hydrogen, oxygen, and nitrogen. These carbon compounds would have been food for the first forms of life. Perhaps energy was extracted from these molecules by a glycolytic pathway similar to the one that living cells still use. In itself, however, a pathway like glycolysis would have been a one-course meal ticket. Breaking apart energy-rich compounds in this manner could result only in an accumulation of leftovers of lower potential energy. This pathway undoubtedly led to the world's first energy crisis. For as more and more organisms appeared, increasing demands must have been made on existing resources.

We can speculate that with intensified competition for dwindling energy, selective agents went to work on variations in existing cell structures. In some organisms, the machinery for breaking down carbon compounds was modified and extended. It was used for building up sugar molecules from carbon dioxide and water. How were the reactions made to run in this energetically unfavorable direction? Sunlight was harnessed as a source of extra energy. Such is the nature of photosynthesis, the trapping of photon energy and its conversion to forms that can be used to build large, energy-rich molecules from small, energy-poor ones.

But a combined reliance on glycolysis and photosynthesis was still a somewhat limited approach to extracting energy from available molecules. Following glycolysis, well over ninety percent of the potential energy locked in a sugar molecule was still locked in the disassembled parts—lactic acid parts, most probably. With the boost in growth and reproduction that surely accompanied the emergence of photosynthesis, lactic acid began piling up in the environment. In addition, oxygen by-products of photosynthesis were accumulating in the atmosphere. In high enough concentrations, lactic acid and oxygen poison the cells that produce them. Because of these accumulating waste products, early forms of photosynthesizers were polluting the very place in which they had to live! How would it be possible to utilize these by-products of life? The energy crisis was there—yet so was a solution. Oxygen, one of the pollutants, was a key.

In this changing environment, cells with new metabolic machinery appeared. With this machinery, the waste products could be combined with oxygen to yield carbon dioxide and water. And much more of the potential energy stored in sugar molecules could be extracted. Such is the nature of cellular respiration.

With cellular respiration, life became permanent and self-sustaining. For the final products of this metabolic pathway are precisely the materials needed in building sugar molecules in photosynthesis! Thus the cycling of carbon, hydrogen, and oxygen through the energy pathways of living organisms came full circle. Many similar cycles have come to exist for other essential elements, such as nitrogen and phosphorus. But it is through the so-called *carbon cycle* (Figure 7.14) that organisms are locked into the greatest interdependence in the search for materials and energy.

Perhaps one of the most difficult connections you are asked to perceive is the link between yourself—a living, intelligent being—and such remote-sounding things as energy, metabolic pathways, and the cycling of oxygen, hydrogen, nitrogen, and carbon. Is this really the stuff of humanity? Think back, for a moment, on the discussion of a single water molecule. A pair of hydrogen atoms competing with an oxygen atom for a fair share of the electrons joining them doesn't exactly seem close to our daily lives. But from this simple inequality, the polarity of the water molecule arises. As a result of the polarity, hydrogen bonds form between water molecules. And that is a beginning for the organization of lifeless matter which leads, ultimately, to the organization of matter in *all* living things.

For now you can imagine new kinds of molecules interspersed through the watery environment. Many will be nonpolar and will resist interaction with water. Others will be polar and will respond by dissolving in it—by forming hydrogen bonds with it. And certain molecules, such as phospholipids, contain both water-soluble and water-

insoluble regions. In swirling, agitated water (a primordial sea, perhaps, pounding against some ancient shore?), they form a two-layered film around a water droplet. Such phospholipid bilayers are the basis for all cell membranes, hence all cells. The cell has, from the beginning, been the fundamental *living* unit.

With membrane organization, the interior of the bilayer is separated from the environment. Through this isolation, chemical reactions can be contained and controlled. The essence of life *is* chemical control. This control is not brought about by some mysterious force. Instead, a class of protein molecules—enzymes—puts these molecules into action in precisely regulated ways. It is not some mysterious force that tells the enzymes when and what to build, and when and what to tear down. Instead it is a chemical responsiveness to the kinds of molecules present in the environment, and to changes in that environment. This responsiveness comes from the arrangements among thousands or millions of atoms making up the molecules of life. And it is not some mysterious force that creates the

enzymes themselves. DNA, the slender double strand of heredity, has the chemical structure—*the chemical message*—that allows molecule to faithfully reproduce molecule, one generation after the next. Those DNA strands "tell" the trillions of cells in your body how countless molecules must be built.

So yes, oxygen, hydrogen, nitrogen, and carbon represent the stuff of you, and us, and all of life. But it takes more than molecules to complete the picture. It is because of the way this stuff is organized and maintained by a constant flow of energy that you are alive. It takes outside energy from such sources as the sun to drive the formation of new energy-rich molecules. Once molecules are assembled into the cells of organisms, it takes outside energy derived from food, water, and air to sustain their organization. Individual plants, animals, and microorganisms are part of an interconnected web of energy use and materials cycling that threads through all levels of biological organization. Should energy fail to reach any part of any one of these levels, threads there will unravel; and life in that

region will join the trend toward ever-increasing disorder.

For energy flows through time in only one direction—from forms rich in potential energy to forms having progressively less usable stores of it. Only as long as sunlight flows into the web of life—and only as long as there are molecules to recombine, rearrange, and recycle with the aid of that energy—does life have the potential to continue in all its rich diversity.

Life is, in short, no more *and no less* than a marvelously complex system of prolonging order. Sustained by energy transfusions, it continues because of its capacity for self-reproduction—the handing down of hereditary instructions, and the means for organizing energy and materials to maintain life generation after generation. Even with the death of individual plants, animals, and microorganisms, life is prolonged. For their molecules are released to be recycled once more, providing raw materials for new generations. In this flow of matter and energy through time, each birth is affirmation of our ongoing capacity for organization, each death a renewal.

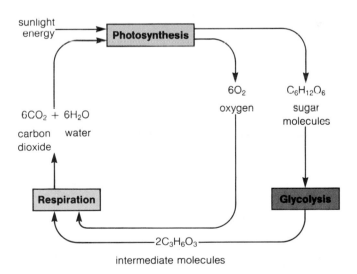

Figure 7.14 The carbon cycle, in which energy-rich molecules of carbon, hydrogen, and oxygen flow through all organisms on earth. In this recycling of matter through time, each birth is an affirmation of our ongoing capacity for organization, each death a renewal.

Readings

Armond, P., L. Staehlin, and C. Arntzen. 1977. "Spatial Relationships of Photosystem I, Photosystem II, and the Light-Harvesting Complex in Chloroplast Membranes." *Journal of Cell Biology* 73:400-418.

Avers, C. 1978. *Basic Cell Biology*. New York: Van Nostrand. See especially the chapter on mitochondrial form and structure.

Björkman, O., and J. Berry. 1973. "High-Efficiency Photosynthesis." *Scientific American* 229(4):80-93. Good explanation of the C4 pathway of photosynthesis.

Capaldi, R. 1977. "The Structure of Mitocondrial Membranes." In *Mammalian Cell Membranes* (G. Jamieson and D. Robinson, editors). Boston: Butterworth.

Govindjee, and R. Govindjee.1974. "The Primary Events of Photosynthesis." *Scientific American* 231(6): 68-82.

Hinkle, P., and R. McCarty. 1978."How Cells Make ATP." *Scientific American* 238(3):104-125.Good summary of the current chemiosmotic theory of ATP formation.

Kok, B. 1976. "Photosynthesis: The Path of Energy." In *Plant Biochemistry*, vol. 3 (J. Bonner and J. Varner, editors). New York: Academic Press.

Stanier, R., E. Adelberg, and J. Ingraham. 1976. *The Microbial World*. Fourth edition. New Jersey: Prentice-Hall.

Stryer, L. 1975. *Biochemistry*. San Francisco: Freeman.

Wolfe, S. 1981. *Biology of the Cell*. Second edition. Belmont, California: Wadsworth. Comprehensive coverage of the topics in this chapter.

Review Questions

1. Summarize the photosynthesis reactions in words, then as a chemical equation.

2. Is the following statement true? Glycolysis, the initial breakdown of glucose, can proceed *only* in the absence of oxygen.

3. Summarize the glucose breakdown route that starts with glycolysis and proceeds through cellular respiration. Do this in words, then as a chemical equation.

4. Distinguish cyclic from noncyclic photophosphorylation. How are electrons used in the noncyclic pathway? What is its energy yield?

5. The dark reactions of photosynthesis occur in two stages: carbon-dioxide fixation and the Calvin-Benson cycle. Describe what happens in each stage.

6. What is chemosynthesis? What sorts of organisms use this food-acquiring strategy?

7. Pyruvate molecules, the leftovers from glycolysis, can be sent down at least three different energy-extraction pathways, depending on the organism and on environmental conditions. What are these three pathways? If you include the two ATP molecules formed during glycolysis, what is the *net* energy yield of each pathway?

8. In anaerobic routes of glucose breakdown, further conversions of pyruvate don't yield any more usable energy. What, then, is the advantage of the conversions?

9. What happens to electron acceptors NAD^+ and FAD in glycolysis and in the Krebs cycle? How are the reaction products used?

10. What happens to NADH and $FADH_2$ in oxidative phosphorylation? Explain how electrons and hydrogen atoms stripped from them are used.

11. Why are oxygen atoms important in oxidative phosphorylation?

12. Explain what net protein utilization means. Do humans subsist most efficiently on a plant- or animal-based diet? List some policies that might be implemented if world starvation is to be alleviated.

13. Study Figure 7.14. Then, on your own, diagram the cycling of carbon, hydrogen, and oxygen through the biosphere.

UNIT THREE

THE ONGOING FLOW OF LIFE

8

GROWTH, DECLINE, RENEWAL:
THE LIFE CYCLES

At any given instant, a cell is drawing sustenance from its surroundings—through passive diffusion, perhaps, or through the dramatic engulfments of a predatory cell on the prowl. At any instant, it is engaged in a metabolic juggling act of great precision—dismantling some molecules, diverting the fragments from one pathway to another, and building new molecules into ever larger cell structures. We have, in the preceding unit, observed these and other aspects of the world of the cell. We have observed the cell frozen, as it were, at some random moment in time and space in order to chart the flow of energy and materials through it. With these observations behind us, we are ready to add still another dimension to our concept of this world: its continuity through time.

Whether single celled or multicellular, each living thing follows a **life cycle** characteristic of its species: it undergoes sequential changes in form and function, from the moment it emerges as a distinct entity to the moment of its reproduction. For many single-celled organisms, reproduction means renewal. One cell simply becomes two, then two cells become four, and so on. For many multicelled organisms, reproduction marks the onset of a period of decline that ends, sooner or later, in the death of the individual. No matter what the details of the life cycle, however, and no matter how complex an organism is destined to become, the turning point between old and new generations lies at the cellular level. Thus we begin our journey along the axis of time with the cell, and it is to the cell that we will be returning.

SOME MOLECULAR ASPECTS OF CELL REPRODUCTION

The point on which all life cycles pivot is **reproduction**: making a copy of oneself. When we extend the picture in time, we can say that reproduction means making a *series* of copies of oneself. These copies repeat the changes in form and function that must unfold during the prescribed life cycle for a species. The process begins at the molecular level. The molecular details, however, are a separate story. Here we will only look briefly at how all the parts of cells (hence of individuals) can be produced anew with each generation.

Instructions for producing each new cell reside in DNA. As you read in Chapter Three, DNA is assembled from four kinds of nucleotides, each with a different kind of nitrogen-containing base. Millions of nucleotides may be strung one after the other in a DNA strand, like beads of four different colors in a necklace. The order in which the "colors" are strung is different for each species of organism. (For instance, a DNA region for one species may be red-blue-red-green; for another species it may be red-red-red-yellow.) However, the order is the *same* in every cell of a given individual.

The sequence of nucleotide bases represents instructions for assembling certain RNA molecules, which in turn contain instructions for building proteins (Figure 8.1). Now, a cell doesn't need all possible proteins all of the time, so it certainly doesn't use all of its DNA instructions all of the time. Instead, **genes**—specific DNA regions that tell how to build different kinds of RNA (hence different kinds of proteins)—are activated from the whole collection. The ones activated function in specific cells at specific times.

Of course, it takes more than proteins to build a new cell. Carbohydrates are also needed, along with lipids and other organic molecules. But recall that enzymes, too, are proteins. With a DNA-derived capacity to make a given set of enzymes and other proteins, a cell can duplicate the same set of materials that went into building the previous cell generation. For enzymes give the cell a means to control the assembly of organic molecules from simple building blocks present in its surroundings. Thus,

---

Directly or indirectly, DNA governs which organic materials will occur in a cell and, in many cases, how those materials will become arranged as the cell grows and develops.

It follows, from the above, that the continuity of life depends on the *duplication* of a cell's set of DNA molecules before that cell reproduces. Duplication means the cell acquires two DNA sets. When the cell does divide, one DNA set ends up in each so-called daughter cell. Its presence assures that growth and development will proceed in the image of the parents.

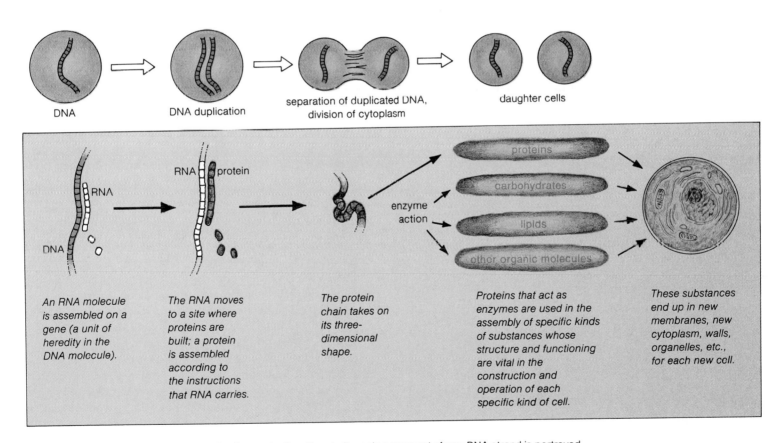

Figure 8.1 (Above) Simplified picture of cell reproduction. For clarity, only a segment of one DNA strand is portrayed. (Below) Hereditary material—DNA—must be duplicated prior to cell division if reproduction is to occur. In other words, each cell must have its own DNA, which contains the information for building and running just about everything else in the cell.

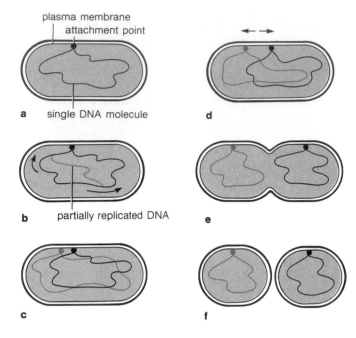

plasma membrane
attachment point

a   single DNA molecule

d

b   partially replicated DNA

e

c

f

Figure 8.2 The prokaryotic plan for replicating DNA and parceling it out to the next generation of cells. (**a**) Generalized drawing of a bacterium before replication of its single DNA molecule, which appears to be attached to the plasma membrane at a single point. (**b**) Replication begins at some point on the DNA; replication is thought to proceed in two directions away from the initiation site. (**c**) The replicated DNA is also attached closely nearby. (**d**) Membrane growth occurs between the two attachment points, which moves the two DNA molecules apart. (**e**) Once the DNA molecules are at opposite ends of the cell, the middle of the cell undergoes constriction. (**f**) With the growth of plasma membrane and cell wall material, the cytoplasm is divided in two.

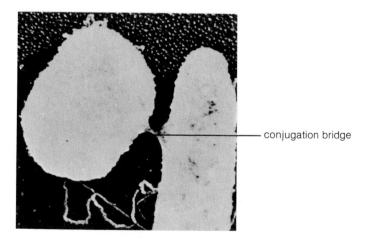

conjugation bridge

Figure 8.3 Conjugation between female strain (left) and male strain (right) of *Escherichia coli*, showing the conjugation bridge. (T. F. Anderson, E. L. Wollman, and F. Jacob)

When a cell reproduces, it gives up some of its existing cytoplasm to each daughter cell. And this is important, for the cytoplasm already contains metabolic machinery —especially the enzymes involved in reading the instructions contained in DNA and RNA. Thus, it is not enough for DNA duplication to precede division. Enzymes must also be ready and waiting for the hereditary molecules to help produce *more* enzymes for use in making a copy of the parent cell!

LIFE CYCLES OF THE SINGLE-CELLED PROKARYOTES

Binary Fission

Prokaryotic reproduction is about as simple as you might imagine, for reproduction of the cell and the organism are the same thing. Say that you put a single bacterium on a nutrient-rich medium in a culture dish. Over the next twenty or thirty minutes, the bacterium feeds and grows, adding plasma membrane here, cytoplasm and wall material there. And it duplicates its DNA. The DNA is not enclosed in a separate nuclear envelope, as it is in eukaryotes. Instead it is attached at a single point to the plasma membrane. The duplicated DNA also becomes attached nearby. Membrane growth occurs between the two attachment points. This growth moves the two DNA molecules apart. Membrane and wall material now start to grow across the cell's midsection, and the bacterial cell divides in two (Figure 8.2). In this **binary fission**, partitioning leads to two separate cells having basically similar parts.

Exchange of Hereditary Material Between Cells

Prokaryotic reproduction is not limited entirely to binary fission of a single cell. In the late 1940s, Edward Tatum and Joshua Lederberg showed that under certain conditions, some bacterial cells engage in *conjugation*. In this process, segments of genetic material are transferred from a "male" cell to a "female" cell by way of a bridge composed of cellular material (Figure 8.3). Under other conditions, some bacteria are capable of *genetic transformation*. They pick up intact DNA molecules from dead relatives and incorporate parts of that DNA into their own genetic material! Following viral attacks on some bacterial cells, *genetic transduction* takes place. When the cell is destroyed, new viruses escape from it and carry some of the dead cell's genes to the next cell that they infect. Conjugation, transformation, and transduction occur erratically, at best. Yet they all result in **genetic recombination**—the production of individuals having new combinations of genes that they acquired, one way or another, from more than one parent cell. Such primitive methods of exchanging genetic information probably helped bring about diversification among prokaryotes.

PROSPECTS AND PROBLEMS OF MULTICELLULARITY

Imagine what the world must have been like when it was inhabited by only the simplest prokaryotes. The fossil record suggests that those early single-celled forms were much like modern bacteria and blue-green algae. For about two billion years, the body plans and life cycles of these organisms remained much the same. The limited diversity perhaps speaks of relatively small amounts of genetic variation, and of a relatively stable environment. Pressures for change probably were limited to a few variable conditions in local settings—tidal movements, strong coastal currents, the flow of rivers and streams. In such settings, cells that by chance remained attached after binary fission could have enjoyed a selective advantage: strings of cells could wrap around a pebble or rock and thereby gain some resistance against the random battering of swirling waters. Sprinkled through the fossil record for that period is evidence of just such stringlike arrangements of cells.

We can assume that there has been persistent selection for this adaptation in certain environments. Even today, filamentous growth patterns occur among various blue-green algae and bacteria, which become anchored to rocks in rapidly flowing streams and hot springs. But the cells themselves remain simple, and the division of labor among them is rudimentary at best. Their stringlike association is not true **multicellularity**—a word that implies extensive division of labor and permanent interdependence among specialized cells. Instead, these single cells associate in the loosest fashion; each can reproduce independently.

About 1.2 billion years ago, however, the nature of the living community changed irrevocably. By that time, the earliest eukaryotes had appeared (Chapter Twenty-One). Among them were the world's first predatory cells. With their capacity for ingesting other organisms, these predators represented an evolutionary force of the first magnitude. No longer would most cells be able to "live and let live" in their cycles of growth and reproduction. No longer would it always be enough simply to become well adapted for tapping energy from the sun or from waste products of other cells. Survival for most would now depend largely on changes in life-styles that would allow

Gary W. Grimes and Steven W. L'Hernault

Figure 8.4 Mealtime for *Didinium*—a single-celled organism with a big mouth, and symbol of the sort of evolutionary pressures that may have brought about increased size and complexity among many forms of life. Dinner in this case is cucumber-shaped *Paramecium*, poised at the mouth (above) and swallowed (below).

organisms to avoid predators long enough to reproduce. There would have been selection not merely for forms that were adapted to the physical environment. There would have been selection also for the swift, the elusive, and the wary—and for the ones able to grow large enough to avoid being eaten by local predators (Figure 8.4).

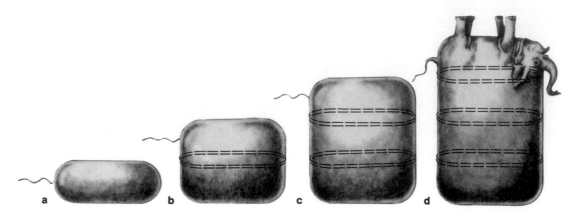

Figure 8.5 Why there never will be a single *Escherichia coli* as large as an elephant: a simple portrayal of the loss in surface area as cell volume increases. (**a**) In a small, one-celled organism, all of the cytoplasm is close to the active membrane surface, which is in contact with the outside environment on all sides. (**b**) If you double that cell in size, what would have been two sides of separate smaller cells (two dashed surfaces) are no longer at the surface. The cytoplasm doubled; the surface area did not. (**c**) And if you tripled the cell in size, what would have been four sides of separate smaller cells are no longer at the surface. (**d**) If you kept on enlarging it you'd have a dead cell, just as you'd have a dead cell in b and c: there would not be enough surface membrane area for efficient materials exchange.

On the Surface-to-Volume Ratio

Even if we conclude that increased size and complexity are traits with survival advantage, it could not have been a simple matter for those ancient cells to have increased in size. Consider, first, the simple fact that any cell can survive only as long as it can take in nutrients, transport them through the entire volume of cytoplasm, then get rid of the wastes. For a small single cell, the plasma membrane provides enough surface area to take up all required nutrients from the environment. Once nutrients have crossed the membrane, diffusion (through random motion of molecules) is enough to distribute them uniformly through the interior. Similarly, diffusion and active transport across the membrane are enough for waste disposal.

But when a single cell enlarges, it eventually encounters restraints imposed by the **surface-to-volume ratio**: as a cell's linear dimensions grow, its surface area does not increase at the same rate as its volume (Figure 8.5). Consider what would happen if a spherical cell were to enlarge four times in diameter, with no other changes. That would mean the cell volume would increase $4 \times 4 \times 4$, or sixty-four times. To keep the expanded cytoplasm functional, the amount of nutrients would also have to be increased sixty-four times. If you carry this line of reasoning further, you would see that sixty-four times as much waste products would also have to be removed.

The problem is that a fourfold increase in diameter (with no other changes) would increase the cell surface area by only sixteen times. This means that each unit area of the plasma membrane would now be called upon to serve four times as much cytoplasm! In addition, the cell center would now be four times as far from the active membrane surface. So it would take four times as long for wastes to diffuse back to the surface. Clearly, enlargement alone is not feasible; past a certain point, the cell simply would die.

You may already suspect, from Chapter Five, how the constraints of the surface-to-volume ratio were at least partly overcome during the evolution of larger organisms. For one thing, complex infolding of the plasma membrane increased the membrane surface area as cytoplasmic volume increased. The development of endocytosis was another way of bringing part of the environment into the cell. With exocytosis, part of the inside could be sent back out. Such developments work well enough in a single cell. But imagine the problems of accumulating and transporting nutrients and wastes in a multicelled body! *All living cells must be able to exchange materials with the environment.* And all can do this in multicellular organisms, whether growth takes place in one, two, or three dimensions (Figure 8.6).

Enlargement in one dimension leads to a long, threadlike body. This pathway apparently was taken by some ancient prokaryotes, certain protistans, some algae, and

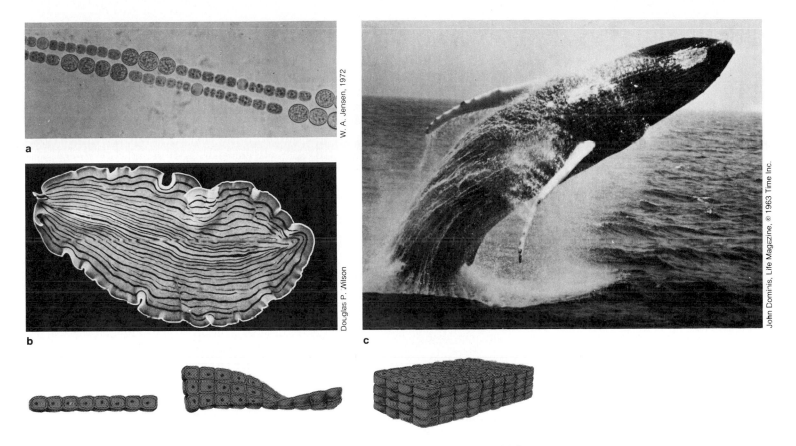

Figure 8.6 Multicellular strategies for getting around the constraints of the surface-to-volume ratio. (**a**) One-dimensional growth in *Anabaena*. (**b**) Two-dimensional growth in a flatworm, an animal about as big as your little toe. (Actually there is some three-dimensional growth in these animals, but most of the size increase comes from sheetlike growth.) (**c**) Three-dimensional growth in a whale. Here, internal transport systems assure even the internal cells of efficient materials exchange.

most fungi. Its advantage is that each cell is kept at the body surface, where it is capable of direct exchange with the environment. In some evolutionary lines, such as those leading to many existing algae, threadlike bodies became highly branched, creating rather massive bodies during one stage of the life cycle.

Enlargement in two dimensions leads to a flat, sheetlike body. This was the path taken by some protistans, a few algae, some fungi, and to some extent by animals called flatworms. As with one-dimensional growth, each cell remains at or near the surface. Even among organisms capable of extensive three-dimensional growth, there are times when enlargement into thin, sheetlike structures has advantages. (For instance, photosynthetic cells in the thin, flat leaves of a flowering plant are near the surface and therefore better exposed to sunlight.)

Even so, it is only through three-dimensional growth that multicellularity shows its greatest potential for extensive cell specialization and division of labor. At the same time, growth in three dimensions presents the greatest problem in terms of the surface-to-volume ratio. Massive bodies are possible only if there are also internal systems for transporting materials between the environment and cells deep within the body. Many multicelled organisms have such systems. Flowering plants have conducting tissues that transport fluids; many animals have blood and lymph vessels that do the same thing. Insects have internal air tubes that open to the body surface; we have tubes leading from the body surface to tiny air sacs in our lungs. Such are the responses of different groups adapting to the physical constraints of enlargement—of keeping every cell of the multicelled body nourished and alive.

Differentiated Cells: Integrating Parts into the Whole

Being more or less limited to growth in size and simple fission of the cell body, prokaryotic life cycles are relatively straightforward. By comparison, life cycles of multicelled eukaryotes are most intricate. Here, sequential changes in form and function depend not only on cell growth and division. They depend on a process called **differentiation**. In this process, cells having identical sets of DNA become structurally and functionally different from one another over time. How does this happen? Although these cells all have the *same* hereditary instructions, different *portions* of those instructions are used in different cells, at different times! Controls over this so-called gene expression (Chapter Eleven) lead to variations in the way that materials are patterned and arranged as the life cycle unfolds.

Among the most complex eukaryotes are plants and animals that live on land. Many become fairly large at some stage of the life cycle. And their cells become highly specialized for performing different tasks that help keep the whole body functioning. After growth and differentiation have first begun, cells of these organisms begin communicating and interacting. They do this largely through **hormones**: chemical substances produced by cells in one body region, and dispersed to affect the functioning of target cells in other regions. In complex animals, another communication system emerges that integrates cellular activities. This is the nervous system, a coordinative network that depends on electrical and chemical interactions between cells of many sorts (Chapter Fourteen).

On the Amount and Nature of Hereditary Instructions

How is it possible to build a larger, more complex body —and to maintain and control it through long life cycles? Among eukaryotes alive today, we find that an expanded reservoir of hereditary instructions is required. Not only do eukaryotes have more DNA than is found in prokaryotes, the DNA is more varied.

Even single-celled eukaryotes have a large amount of hereditary instructions, generally in the form of extra DNA copies. For instance, ciliates have two different kinds of nuclei. (The "micronucleus," concerned with reproduction, contains two sets of DNA. The "macronucleus" is an expanded store of genetic information; it contains DNA that has been duplicated many times even though the nucleus itself has not divided.) The increased hereditary information is sufficient for building, running, and reproducing the ciliate body. In some species, that body is one of the most complex single cells the world has ever seen. *Didinium* (Figure 8.4) is but one example.

Among multicelled eukaryotes, each cell typically relies on a set of DNA in its own individual nucleus. The same amount and kind of DNA are present in each cell of the organism's body. But the amount and kind vary from one species to the next. This variation doesn't correspond exactly to increased body complexity. (Some salamanders, for instance, have twenty-five times as much DNA as mammals.) In such complex eukaryotes, the DNA is concerned with far more than building a given set of proteins. It also contains large stores of instructions for *control* over the life cycle, including the timing, extent, and duration of its events. And considerable variation exists in the life cycles of different species.

More will be said about the nature of eukaryotic DNA in Chapter Eleven. *The point here is that eukaryotes depend on a larger amount of DNA per cell than prokaryotes.* And this dependence presents an entirely new challenge, for imagine what an expanded reservoir of hereditary material must mean in terms of the task of cell reproduction. The task cannot be entrusted to the somewhat simple mechanism of binary fission, with its solitary membrane attachment point for a solitary DNA molecule. More sophisticated mechanisms must be employed to parcel out precisely equal amounts of DNA to daughter cells—and to keep different strands from getting tangled up and thereby botching the process. Such mechanisms have indeed come to exist. They are called mitosis and meiosis, and they will now be described.

LIFE CYCLES OF SINGLE-CELLED AND MULTICELLED EUKARYOTES

Mechanisms for Handling an Expanded Reservoir of DNA

Parceling out DNA to a new generation of cells may not seem like such a remarkable feat. But stop to think about what is going on, and where it is taking place. Consider *Escherichia coli*, a relatively simple prokaryote. Although its body length typically is a mere 2 micrometers, its circular DNA molecule is 1,220 micrometers long! That molecule can't possibly be crammed haphazardly into the cell body if duplication and separation are to proceed in a reliable fashion. And indeed, it is twisted and folded in a specific way. A DNA molecule 1,220 micrometers long twists about 360,000 times on its long axis. One strand of *E. coli* DNA contains about 3,600,000 nucleotides. To make the duplicate strand, just as many new nucleotides must be assembled and joined together in precise order. The original and the duplicate strands must then be separated from each other in a way that prevents tangling. And all the unwinding, unfolding, untwisting, assembly of a duplicate strand, rewinding, refolding, and separation is finished in less than thirty minutes!

As you read earlier, prokaryotic DNA resides in cytoplasm and seems to remain attached at some point to the plasma membrane during duplication. Given this location, and given all the activity involved in replication, it's easy to see that existing prokaryotic cells probably can't accommodate much more hereditary material than they already have. Yet more than a billion years ago, the capacity to handle increased amounts of DNA—with all of its implied potential for increased size and complexity—apparently arose among the organisms leading to modern-day eukaryotes. This capacity was based on three developments: (1) the *nucleus*, with its double membrane providing physical isolation from unrelated metabolic activities in the cytoplasm; (2) the packaging of hereditary material and certain proteins into compact bodies called *chromosomes*; and (3) a system of *microtubules* that can handle the separation of more than one chromosome at a time. Together, these three developments underlie reproduction in single-celled eukaryotes, as well as physical growth in multicelled eukaryotes.

The physical isolation of DNA in the nucleus, chromosome packaging, and a microtubular system for moving many chromosomes at a time form the basis of reproduction in single-celled eukaryotes. They are also the basis of physical growth in multicelled eukaryotes.

Chromosomes: Eukaryotic Packaging of DNA

Let's first consider the events involved in duplicating and packaging eukaryotic DNA. Earlier, you saw a cross-section of the nucleus of a nondividing eukaryotic cell (Figure 4.16). From the appearance of the micrograph, you may have concluded that the nucleus is composed of tiny bits of grainy material. When early microscopists stained such nuclei for observation, the "granules" soaked up most of the stain and became quite distinct. At that time the nuclear material was named **chromatin** (the word means "colored material"). Sometimes, when the nuclei being observed were undergoing division, the dark-staining substance

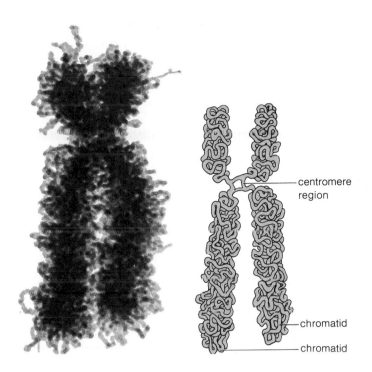

Figure 8.7 A human chromosome as it appears in the duplicated form just before nuclear division. The two chromatids remain attached at the centromere region. The corresponding sketch is drawn in this manner to suggest compactness and complexity. The actual organization of the nucleoproteins is undoubtedly of a highly regular, functional nature, but that organization is not yet identified. (Micrograph from E. J. DuPraw, *DNA and Chromosomes*, Holt, Rinehart, and Winston, 1970)

looked threadlike rather than grainy (see, for example, Figure 8.7). The distinct threadlike structures were called **chromosomes**, meaning "colored bodies." How were these observations explained? The tiny grains were thought to become joined together to form chromosomes when the nucleus was about to divide.

Refinements in microscopy have since changed the picture. Chromosomes are not assembled and disassembled each time the nucleus divides. Even in a nondividing nucleus, they are stretched out as long, almost invisibly thin threads. "Chromatin" and "chromosome," in short, are one and the same thing. The distinction has been retained simply to indicate the appearance of the dark-staining material of the nondividing and dividing nucleus.

Today we know that each chromosome contains a DNA molecule and different proteins associated *only* with eukaryotic DNA. (Prokaryotic DNA, in contrast, is a

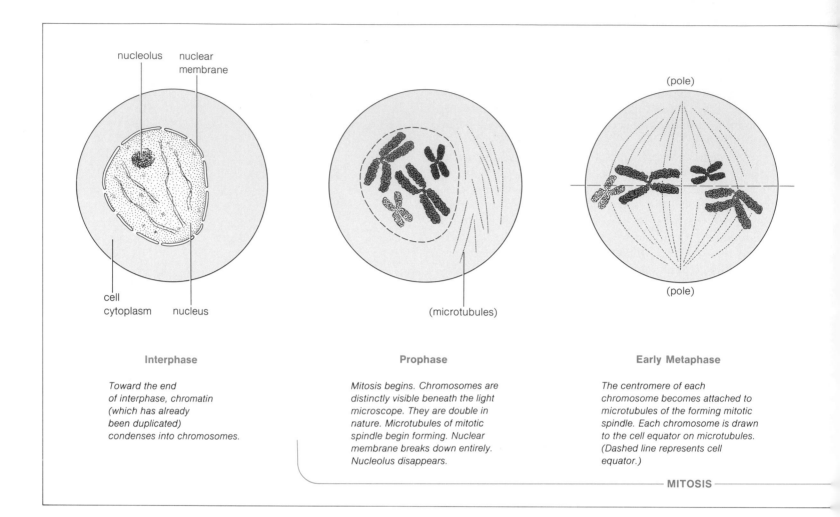

Interphase

Toward the end of interphase, chromatin (which has already been duplicated) condenses into chromosomes.

Prophase

Mitosis begins. Chromosomes are distinctly visible beneath the light microscope. They are double in nature. Microtubules of mitotic spindle begin forming. Nuclear membrane breaks down entirely. Nucleolus disappears.

Early Metaphase

The centromere of each chromosome becomes attached to microtubules of the forming mitotic spindle. Each chromosome is drawn to the cell equator on microtubules. (Dashed line represents cell equator.)

— **MITOSIS** —

Figure 8.8 Mitosis: the process by which the nuclear material of a eukaryotic cell becomes divided in two. Only two pairs of chromosomes are shown for this diploid cell. (The white and gold colored chromosomes were derived from one parent, the brown and gray colored chromosomes are their partners derived from another parent during sexual reproduction, as described on page 132.) Imagine how complicated the picture becomes for those cell types that contain more than a hundred different kinds of chromosomes!

"naked" molecule.) The DNA and proteins are thought to be complexed as a single, intact fiber—a nucleoprotein—that is continuous from one end of the chromosome to the other. During cell construction and maintenance activities, the nucleoprotein is extended. In this extended form, it is accessible to RNA molecules that can carry its messages to cell regions where proteins are assembled. Before and during nuclear division, the nucleoprotein is coiled into the condensed form called the chromosome.

The number of chromosomes differs from one species to the next. Each organism has a certain number of chromosomes characteristic of its species. For instance, each of your body cells normally has forty-six chromosomes (twenty-three pairs). The body cells of a fruit fly contain eight, and those of a frog contain twenty-six. The number ranges from 2 chromosomes for some sea-dwelling worms,

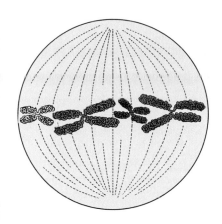
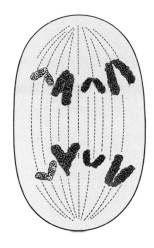
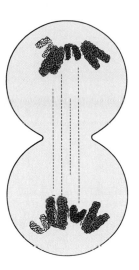
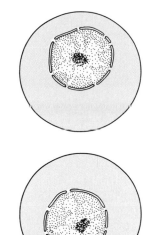

Late Metaphase	**Anaphase**	**Telophase**	**Interphase**
Now the fiber spindle apparatus is completed. The shorter microtubules attach at the centromere to connect each chromatid to one pole. Other microtubules run from pole to pole.	*Each centromere splits, which allows sister chromatids to separate. Microtubules of spindle apparatus guide the two chromatids of each chromosome to opposite poles.*	*Microfilaments begin to constrict at equatorial plane of this animal cell. (In plant cells, vesicles begin to condense into cell plate.) New nuclear membranes start forming; chromosomes unwind. Mitosis is now complete.*	*A distinct nucleus with a nucleolus has formed in each of two daughter cells, which now embarks on a cycle of growth.*

to 1,600 for one of their protistan neighbors. But the chromosome number isn't enough to identify a given species. For example, twenty chromosomes are characteristic of water fleas, asparagus, corn, and many other species.

As you will now see, the number of chromosomes *within* a eukaryotic cell whose nucleus is about to divide remains the same—although each nucleoprotein has a replica of itself attached to it, in the manner of Siamese twins, for part of the division period.

Mitosis: Dividing Up the Chromosomes

The entire process of **mitotic cell division** proceeds from DNA duplication and packaging into chromosomes, through nuclear division (mitosis proper), then through cytoplasmic division (cytokinesis) to form two cells. Thus

DNA duplication precedes mitosis, and actual cell division typically follows afterward at some point (Figure 8.8).

Mitosis, the separation of duplicated DNA into two equivalent parcels, assures each daughter nucleus of a complete set of hereditary instructions. It is the basis for reproduction in the single-celled eukaryotes, and the basis for physical growth in multicellular eukaryotes.

When a eukaryotic cell is about to divide, the DNA of each of its chromosomes becomes duplicated by mechanisms to be described in Chapter Ten. Suffice it to say here that each newly formed nucleoprotein is assembled as an exact replica of the original nucleoprotein. The two strands remain attached at one spot. When joined this way, the two strands are called **sister chromatids**. Together they

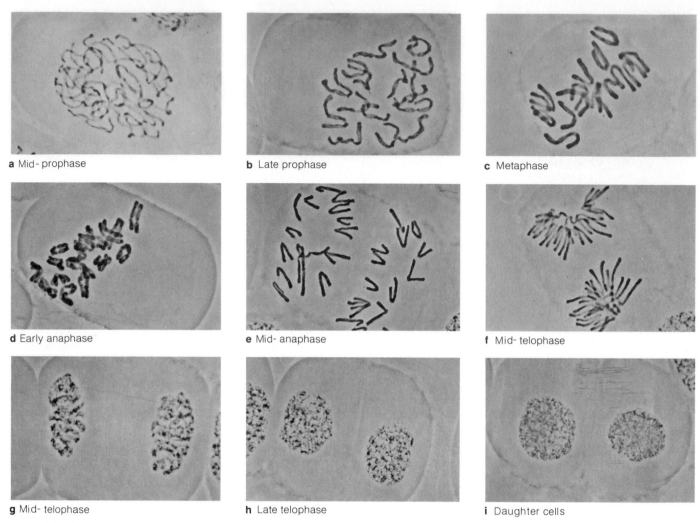

a Mid-prophase

b Late prophase

c Metaphase

d Early anaphase

e Mid-anaphase

f Mid-telophase

g Mid-telophase

h Late telophase

i Daughter cells

Figure 8.9 Glimpses of mitosis of a cell from an onion root tip, as viewed through the phase-contrast microscope. (William Tai, Michigan State University)

form one (duplicated) chromosome. At the spot where they remain joined is a differentiated region of chromosome called the **centromere**. (The location of this spot varies from one kind of chromosome to the next, but it's always the same for chromosomes of a given type.) With the onset of mitosis, the sister chromatids of each chromosome become progressively more coiled and compact.

During mitosis, all chromosomes are lined up at the midsection of the nuclear region. The centromeres of each chromosome move apart, and the two sister chromatids

separate from each other. Eventually, all of the separated chromatids form two equivalent parcels at opposite sides of the nuclear region. At the close of mitosis, new nuclear membrane forms around each parcel. Hence each new nucleus ends up with a complete set of identical hereditary instructions.

Figures 8.8 and 8.9 show the main events of this process. Although these events flow smoothly, one into the other, it is useful to consider them as four sequential stages: prophase, metaphase, anaphase, and telophase.

When the chromatin begins to coil into visible chromosomal bodies, the cell has arrived at **prophase**, the first stage of mitosis. During prophase, chromosomes become progressively shorter and thicker. The nucleolus disappears, and the nuclear envelope breaks down. Also during prophase, a series of microtubules begins to be assembled from proteins in the cytoplasm.

As **metaphase** begins, the microtubules increase in number and become organized into a **mitotic spindle**. This apparatus is composed of two groups of microtubules anchored at opposite poles of the cell. The microtubules extend toward the chromosomes, which by now have become lined up at the cell's equator. Some microtubules extend from pole to pole; others connect to the various centromeres. *The microtubules forming the mitotic spindle are responsible for separating sister chromatids from each other.*

In early metaphase, two sets of microtubules become attached to each chromosome at the centromere. One set extends from one pole to a chromatid, and one set extends from the opposite pole to the other chromatid. As more and more microtubules attach to each centromere, the chromosomes are subject to a tug-of-war, being pulled first toward one pole and then toward the other. By the end of metaphase, the tug-of-war ends in a draw. Each chromosome is now lined up halfway between the two poles at the mitotic spindle equator.

As **anaphase** begins, these events give way to what seems to be a pause in activity. But when the "pause" ends, it becomes clear that each centromere region has become divided. The once-joined sister chromatids have become separated. Now the sister chromatids of each chromosome move apart, and proceed toward opposite ends of the spindle. In most cells, the movement is the combined result of two different mechanisms. Microtubules attached to centromeres shorten, which *pulls* each chromatid toward its destination. Other microtubules that extend from pole to pole begin to elongate, which *pushes* the poles apart before cell division occurs.

Depending on the site of the centromere-microtubule attachment region, the chromatids may appear straight, U-shaped, or J-shaped as they move through cytoplasm:

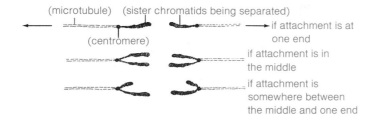

(microtubule) (sister chromatids being separated)

(centromere)

if attachment is at one end

if attachment is in the middle

if attachment is somewhere between the middle and one end

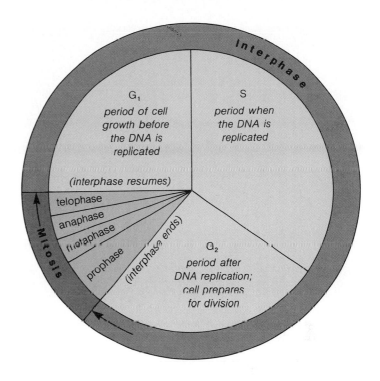

Figure 8.10 Where mitosis fits in the eukaryotic cell cycle. This drawing has been generalized; there is great variation in the length of different phases from one cell to the next. For instance, for some developing embryos, interphase may last for less than half an hour; yet human brain cells remain in interphase from birth to death.

Once the two parcels of separated chromatids arrive at opposite poles, anaphase is over. During the next stage, **telophase**, the chromosomes begin to uncoil. Newly forming nuclear membranes surround each parcel, and two nuclei eventually form. Mitosis is completed. The chromosome content of *each* daughter nucleus is identical with that of the original nucleus.

When does a eukaryotic cell actually undergo mitosis? The process takes up only a small slice of larger series of events known as the **cell cycle** (Figure 8.10). Depending on the cell type and on environmental conditions, mitosis may last anywhere from a few minutes to an hour or more. The cell spends most of the time in **interphase**. Unlike mitosis, interphase varies greatly in duration. It may last for only a few minutes, but in some cell types it can last more than a century! During interphase, most metabolic activities take place; it is the time of construction, growth, and maintenance of the cell body. Also, DNA is duplicated during interphase, prior to the onset of mitosis.

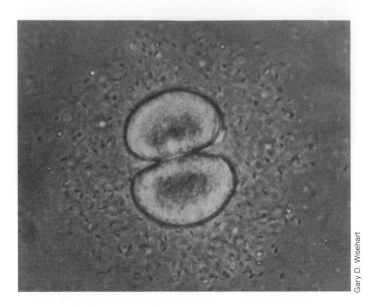

Figure 8.11 Cytokinesis in an animal cell.

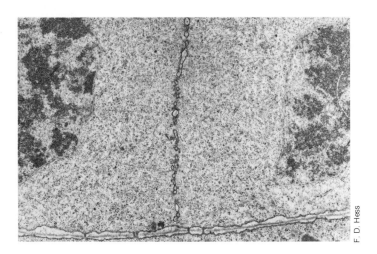

Figure 8.12 Cell plate formation in a plant cell.

Cytokinesis: Dividing Up the Cytoplasm

Cytoplasmic division, or **cytokinesis**, usually accompanies nuclear division and occurs during telophase (Figure 8.9). Exceptions do exist. For example, among single-celled protistans that rely on multiple nuclei, cytoplasmic division does not follow mitosis. Among your own skeletal muscle cells, mitosis without cytokinesis also leads to multinucleate cells, just as it does during certain stages of the life cycle for some insects, plants, and fungi.

When cytokinesis occurs in a typical animal cell, a **cleavage furrow** forms around the cell midsection. The furrow is created by a ring of microfilaments that forms just beneath the plasma membrane. When the ring contracts from the periphery to the interior, the cytoplasm is pinched in two (Figure 8.11). The daughter cells may or may not end up the same size, depending on where the microfilament ring actually forms.

When cytokinesis occurs in the cells of many plants, vesicles typically gather in the center of the region that formed the mitotic spindle equator. Upon fusing together, the vesicles form a **cell plate** that grows into a complete partition between the two newly forming cells (Figure 8.12). Plant cells divide in this way due to the inelasticity of plant cell walls, which don't lend themselves to the sort of pinching seen in the cytoplasmic division of animal cells.

On Sexual Reproduction

Some cells have only one of each type of chromosome characteristic of the species. They are **haploid** cells (the word means "single set"). Prokaryotes are mostly haploid cells. So are some special eukaryotic cells, as you will soon see. But most eukaryotic cells are **diploid**. This word means they have "two sets" of the chromosomes characteristic of the species. One chromosome of each pair has been inherited from one parent, and its equivalent has been inherited from another parent.

The diploid state of many eukaryotic life cycles arises through the process of **sexual reproduction.** At some stage of life cycle, diploid parent organisms produce haploid sex cells. Sex cells are known as **gametes**. The fusion of the hereditary material of two gametes, an event known as **fertilization**, leads to the formation of a new diploid cell. Sexual reproduction requires both gamete formation and fertilization. As you will see later, the precise timing of these separate events varies considerably during the life cycles of different species.

In most multicellular eukaryotes, the first diploid cell formed after fertilization is called a **zygote**. Through mitosis and cytokinesis, the zygote grows, and it develops into a new individual in which all cells are diploid. For most single-celled eukaryotes, the diploid cell *is* the new individual. Usually gametes are derived from two different parents, exceptions being such organisms as self-fertilizing plants. In contrast, both binary fission and mitotic cell division are considered to be forms of **asexual reproduction**, because gametes are not involved. In binary fission, a haploid cell simply gives rise to haploid copies of itself.

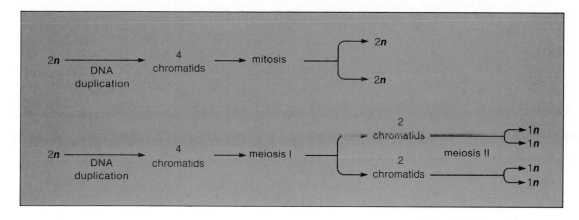

Figure 8.13 Mitosis *maintains* a parental diploid number of chromosomes (2n). Meiosis *reduces* the parental diploid number to the haploid state (1n).

Now, because haploid organisms have only a single set of chromosomes, they have only a single copy of each of the genes needed for cell construction, growth, and maintenance. As you might imagine, they are finely tuned for operating under a given range of environmental conditions. If a mutation occurs in one of their important genes (Chapter One), the outlook for survival is grim.

Prospects are better, under such circumstances, for diploid organisms—as, indeed, most eukaryotic organisms are. For them, even if a gene in one chromosome of a pair undergoes mutation, the equivalent gene on the other chromosome remains functional. For instance, if a mutation appears in a stretch of DNA that governs the construction of a key enzyme, the mutant DNA may give rise to an enzyme that functions less effectively under prevailing conditions. But the diploid cell may survive anyway because it has a corresponding nonmutated version in its second set of hereditary instructions. And that version may see it through.

Now suppose that the environment warms up for some reason. Suppose the mutant enzyme (defined by the mutant DNA) just happens to function better at higher temperature than the nonmutated version. Then the organism with the mutant DNA may have a definite survival advantage. We might well speculate that because it has a duplicate set of hereditary instructions, a diploid organism can accumulate a bank account of hereditary variability—which may be cashed in at a later date should environmental conditions change. As you will read later, the emergence of the diploid state more than a billion years ago corresponded to an explosive diversification of eukaryotic forms of life.

Meiosis: A Reduction–Division Process

Sexual reproduction presents a whole new set of problems in terms of how hereditary material is to be transmitted from one generation to the next. Quite simply, the number of chromosomes in a diploid reproductive cell must be *halved* prior to gamete formation. If there were no such reduction to a single (haploid) set of chromosomes in each gamete, then fertilization would result in a zygote with twice the amount of hereditary information required to build a new individual. Even if it could survive an overload of four chromosome sets, and even if it had the means to pass on four sets to the next generation, a zygote getting four chromosome sets from two parents would have eight sets. The next generation would have sixteen, and so on until there would be room for nothing else in the cell!

At some point in the distant past, a new method arose for handling chromosome sets in the reproductive cells of sexually reproducing organisms. It is called **meiosis**. Meiosis may have evolved through modifications to the already existing mitotic apparatus. All the chromosomes in the diploid nucleus are duplicated prior to division, just as they are before mitosis begins. Microtubules are used in moving chromosomes here, also. But meiosis differs from mitosis in some fundamental ways.

Essentially, mitosis *maintains* the parental diploid number of chromosomes (which can be represented as 2n). Meiosis serves to *reduce* the parental diploid number of chromosomes to the haploid state (1n). As Figure 8.13 illustrates, meiosis consists of *two* successive divisions of the nuclear material. The divisions commonly occur in rapid

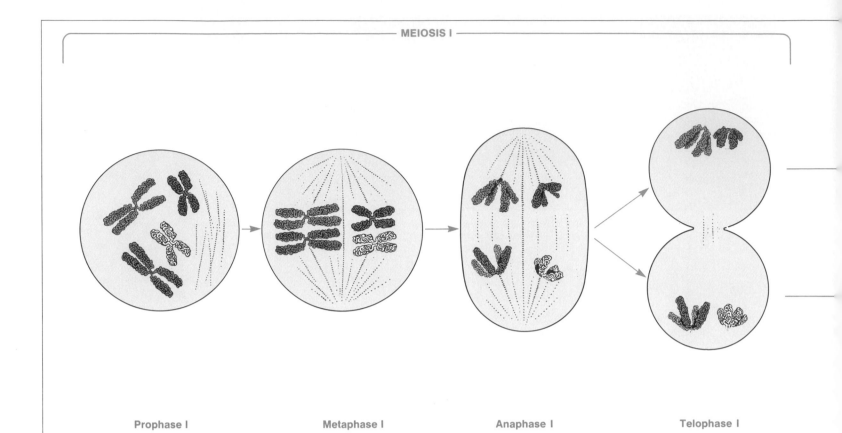

Prophase I

As in mitosis, chromosomes become clearly visible. We see each one has already duplicated itself, thus forming two chromatids that are attached at the centromere.

Metaphase I

As in mitosis, the chromosomes line up at the equator of the spindle apparatus. But <u>unlike</u> mitosis, homologous <u>pairs</u> of chromosomes match up with each other. Chromosomal microtubules connect one chromosome of a matched pair to one pole, and the other chromosome of the pair to the opposite pole.

Anaphase I

As the continuous microtubules push the poles apart, chromosomal fibers attached to the chromosomes shorten—the effect of which is to <u>separate the homologous pairs</u> and guide them to opposite poles. The centromere does <u>not</u> divide at this stage.

Telophase I

And now microfilaments begin constricting the cell at the equatorial plane. At this stage, meiosis I is complete.

Figure 8.14 Meiosis: halving the hereditary material in a reproductive cell of a sexually reproducing organism. To keep things simple, only two kinds of homologous chromosomes are shown here. The brown and gray chromosomes are derived from one parent; the white and gold ones are their equivalents from the other parent.

succession. Often the cell doesn't even divide in two at this point.

In *meiosis I*, all of the duplicated chromosomes in the reproductive cell are lined up at the spindle apparatus equator. They line up in **homologous pairs**: each chromosome derived from one parent lies side by side with the equivalent chromosome from the other parent. Then, as a result of microtubule activity, the two chromosomes of each homologous pair are separated from each other and moved to opposite poles of the cell (Figure 8.14). The centromere doesn't separate during meiosis I, so each chromosome is still composed of two joined sister chromatids.

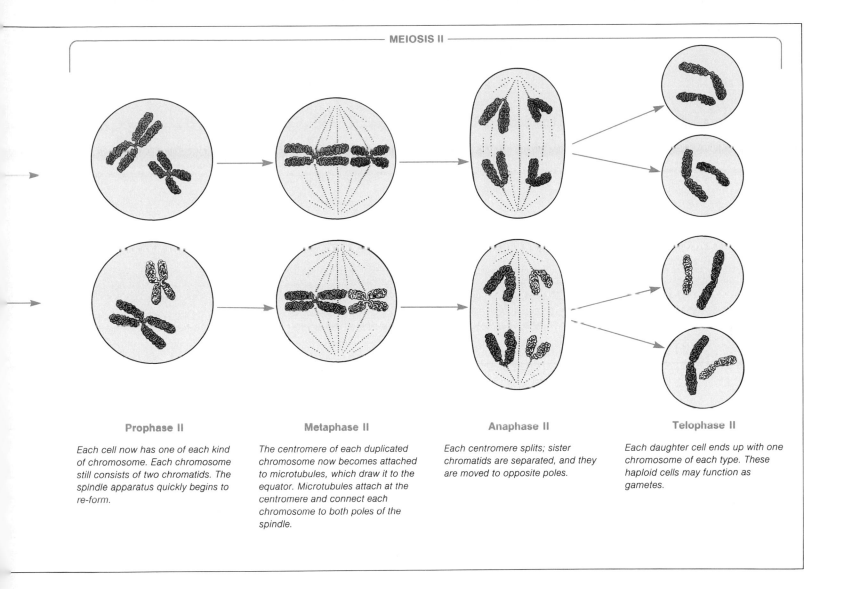

Prophase II

Each cell now has one of each kind of chromosome. Each chromosome still consists of two chromatids. The spindle apparatus quickly begins to re-form.

Metaphase II

The centromere of each duplicated chromosome now becomes attached to microtubules, which draw it to the equator. Microtubules attach at the centromere and connect each chromosome to both poles of the spindle.

Anaphase II

Each centromere splits; sister chromatids are separated, and they are moved to opposite poles.

Telophase II

Each daughter cell ends up with one chromosome of each type. These haploid cells may function as gametes.

In *meiosis II*, the two sister chromatids of each chromosome are separated from each other, as they are during mitosis. The result is that each of the four chromosome parcels produced by the two meiotic divisions is haploid in character.

In this manner, meiosis works to parcel out hereditary instructions from two diploid parents into haploid sets. Before leaving this topic, however, you should know that during meiosis, all manner of things happen that scramble up those instructions in ways that contribute to hereditary variation! Those scramblings are a key concern in the chapter to follow. When reading that chapter, you may find it helpful to refer to Figure 8.15 and to keep the following summary points in mind:

In the first division of meiosis, each duplicated (double-stranded) chromosome of a homologous pair is separated from its partner. In the second division, the two strands of each duplicated chromosome are pulled apart.

These two divisions occur <u>only</u> in the formation of reproductive cells of sexually reproducing organisms. They reduce the parental diploid number of chromosomes to the haploid state for each forthcoming gamete.

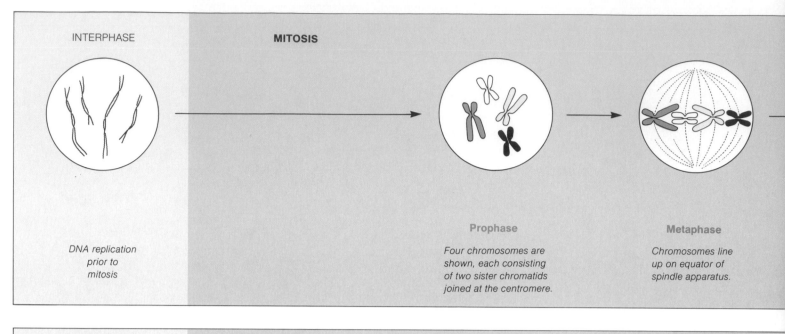

INTERPHASE

MITOSIS

DNA replication prior to mitosis

Prophase

Four chromosomes are shown, each consisting of two sister chromatids joined at the centromere.

Metaphase

Chromosomes line up on equator of spindle apparatus.

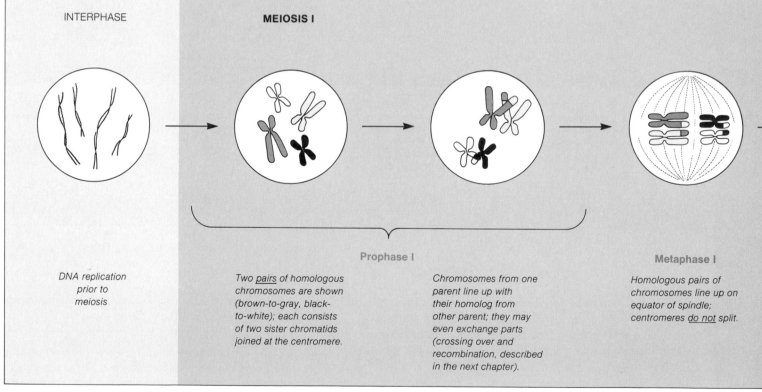

INTERPHASE

MEIOSIS I

DNA replication prior to meiosis

Prophase I

Two pairs of homologous chromosomes are shown (brown-to-gray, black-to-white); each consists of two sister chromatids joined at the centromere.

Chromosomes from one parent line up with their homolog from other parent; they may even exchange parts (crossing over and recombination, described in the next chapter).

Metaphase I

Homologous pairs of chromosomes line up on equator of spindle; centromeres do not split.

Figure 8.15 Summary of chromosome movements in mitosis and in meiosis. Remember that mitosis works to maintain the parental chromosome number; it is the basis for asexual reproduction in many single-celled eukaryotes and the basis for growth of multicellular eukaryotes. Meiosis works to reduce a parental diploid number of chromosomes to the haploid state; it is the basis for the formation of gametes, which are involved in the sexual reproduction of most eukaryotic organisms.

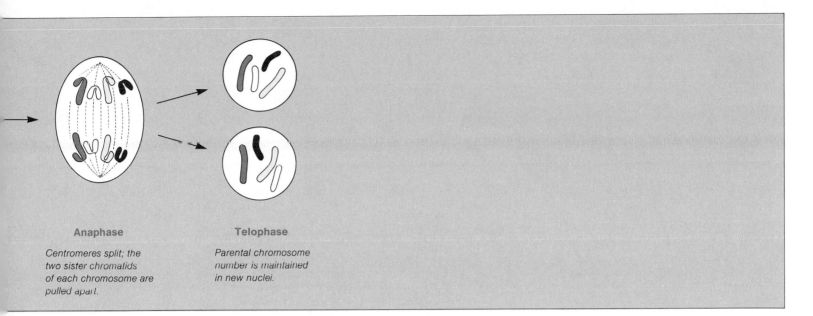

Anaphase

Centromeres split; the two sister chromatids of each chromosome are pulled apart.

Telophase

Parental chromosome number is maintained in new nuclei.

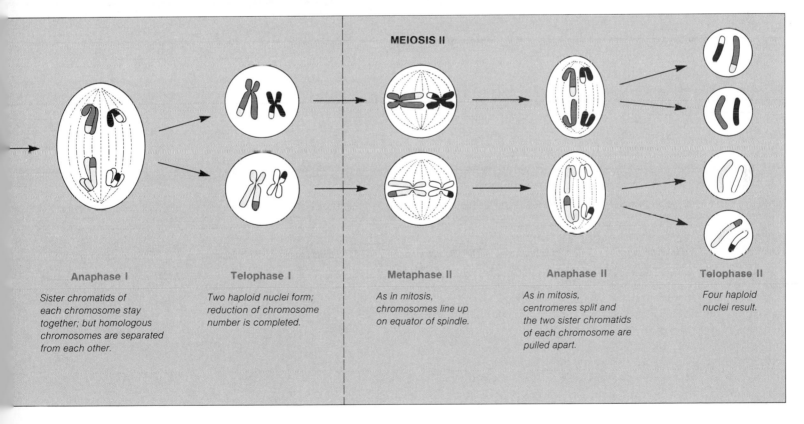

MEIOSIS II

Anaphase I

Sister chromatids of each chromosome stay together; but homologous chromosomes are separated from each other.

Telophase I

Two haploid nuclei form; reduction of chromosome number is completed.

Metaphase II

As in mitosis, chromosomes line up on equator of spindle.

Anaphase II

As in mitosis, centromeres split and the two sister chromatids of each chromosome are pulled apart.

Telophase II

Four haploid nuclei result.

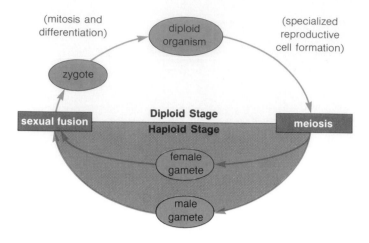

Figure 8.16 Generalized life cycle for animals.

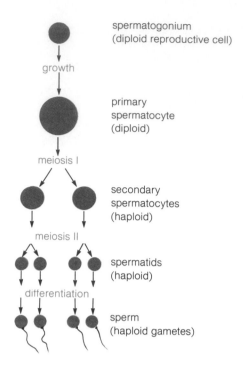

Figure 8.17 Generalized picture of spermatogenesis in male animals.

Strategies of Animals and Plants

We have come a long way from the simple life cycle of growth and fission for prokaryotic cells. Eukaryotes, we have seen, rely on both sexual and asexual reproductive processes that occur at various stages of the life cycle. Figure 8.16 shows a generalized life cycle for an animal. Typically, the cycle for most animals consists of the formation of haploid male and haploid female gametes, with their subsequent fusion. Gamete formation, or **gametogenesis**, occurs only in special sex cells in the male and female reproductive organs (Chapter Eighteen). These gametes take part in reproducing the whole diploid organism. Generally, a cell that undergoes meiosis produces four haploid cells. These cells then undergo some differentiation prior to sexual fusion.

In male animals, gamete formation is called **spermatogenesis** (Figure 8.17). A diploid reproductive cell (spermatogonium) grows in size and becomes a primary spermatocyte. This large, immature cell undergoes the first meiotic division, which leads to two secondary spermatocytes. Both of these cells undergo the second meiotic division, the result being four haploid spermatids. Each spermatid changes in form, develops a tail, and thereby becomes a **sperm**: the mature haploid male gamete.

Female gamete formation, or **oogenesis**, follows a similar sequence of events (Figure 8.18). However, a major difference is that considerable cytoplasmic substances (such

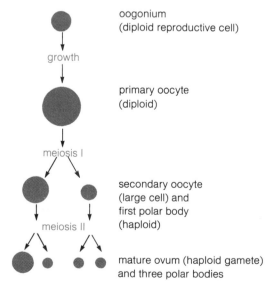

Figure 8.18 Generalized picture of oogenesis in female animals. This sketch is not drawn to the same scale as Figure 8.17. A primary oocyte is *much* larger than a primary spermatocyte.

as nutrients) accumulate in a female reproductive cell. In addition, the cells formed during the meiotic cell divisions are not equal in size or function. A primary oocyte undergoes meiosis I, but one of the resulting cells ends up with nearly all the cytoplasm. This cell is the secondary oocyte. Following meiosis II, it will become the mature **egg**, or **ovum** (plural, ova). The other cells produced during these meiotic divisions are called polar bodies. They are extremely small, compared to the ovum, and do not function in reproduction.

After the nuclei of male and female gametes fuse and a zygote has formed, a new diploid animal develops. No matter how large and complex that animal is destined to become, its body will be composed generally of diploid cells. At some point, though, the diploid organism will produce haploid gametes—and a life cycle may be repeated once more.

Like animal life cycles, plant life cycles proceed from meiosis to sexual fusion. This reliance on sexual reproduction probably originated with ancient protistanlike forms that gave rise to both plants and animals. Since then, however, plants and animals have gone their separate ways. Between the time of meiosis and sexual fusion in plant life cycles, a multicellular body develops that is composed exclusively of *haploid* cells!

At some point, meiosis is followed by the formation of haploid cells (Figure 8.19). In complex land plants, these cells then grow by mitosis and differentiate into a haploid plant body. This haploid body eventually will produce gametes; it's called the **gametophyte** ("gamete-producing plant"). Sooner or later, some or all of its cells simply divide mitotically and produce cells that function as gametes. Sexual fusion of gametes leads to a diploid zygote, which undergoes mitosis and differentiation into a diploid body. This diploid body is called the **sporophyte**. Some of its cells are destined to undergo meiosis, which is followed by the formation of haploid spores. A **spore** is a cell that develops directly into a new haploid body, which marks the onset of a new life cycle. Gametophytes and sporophytes are usually quite different in appearance. A pine tree, for example, is a sporophyte body. Pollen grains are immature male gametophytes; structures within pine-cones are female gametophytes.

Whenever a life cycle contains alternating haploid and diploid stages of a prolonged sort, it is known as an **alternation of generations**. Chapter Twenty-Three will outline the evolutionary trend toward increasingly prolonged sporophyte generations in ancient land plants, and speculate on why they may have come about.

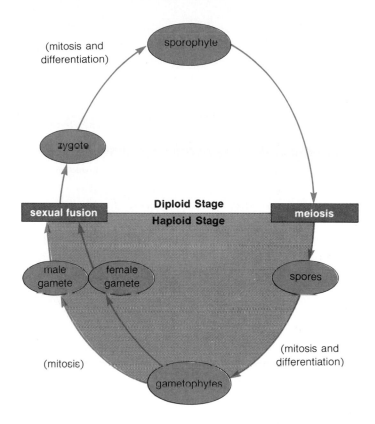

Figure 8.19   Generalized life cycle for complex land plants.

PERSPECTIVE

Each living thing is a continuum, in time and space, of patterns of growth, development, and reproduction. These patterns are characteristic of each kind of species, and they are adapted to some characteristic environment. Through mechanisms for transmitting complete sets of hereditary instructions to each new cell, each new generation grows and develops in the image of the species. In the DNA is information for assembling the necessary raw materials into like cells—into the architectural framework of the cell body and in the changing details of that framework through time.

Depending on the environment, depending on the reproductive strategies employed, tremendous variation has come to exist on this basic theme of producing copies of cells and organisms with each successive life cycle. Different life cycles have become molded in many ways to such environmental conditions as seasonal changes in the availability of food, water, and living sites. Among complex plants, sexual reproduction often depends on other organisms, such as insects, to carry gamete-producing bodies from one plant or plant structure to another. Among complex animals, predictable cycles of sexual reproduction are laced with more elusive patterns. These patterns change with age, with behavioral and physiological states, with the genetic makeup of the population to which the individual belongs. Thus, throughout this book, we will be looking at reproductive strategies not only in terms of individuals but in terms of populations and community interactions.

The focus, you will find, will be on the diverse structural and behavioral aspects of multicellular life cycles, especially among plants and animals. In reading through the details, however, keep in mind that underlying much of the diversity is a common set of problems associated with the increased size and complexity inherent in these groups: *(1) overcoming the constraints of the surface-to-volume ratio, (2) controlling and integrating the activities of differentiated body parts, and (3) handling enough hereditary material to assure reliable growth, development, and reproduction of large, differentiated bodies.* For most organisms, these common problems long ago shaped their life cycles into what they are today.

Readings

Berrill, N., and G. Karp. 1976. *Development.* New York: McGraw-Hill. Good coverage of gametogenesis and the events of fertilization.

Dyson, R. 1978. *Cell Biology: A Molecular Approach.* Second edition. Boston: Allyn and Bacon.

Grobstein, C. 1965. *The Strategy of Life.* San Francisco: Freeman. Provocative little book, with speculations on life's diversity. Paperback.

Wolfe, S. 1981. *Biology of the Cell.* Second edition. Belmont, California: Wadsworth. One of the most current overviews of what is known about mitosis and meiosis.

Review Questions

1. DNA directly or indirectly governs which organic substances occur in a cell. Refer to Figure 8.1, then explain in your own words how it does this.

2. What is binary fission, and what sorts of organisms rely on it?

3. Give three examples of genetic recombination that occur among prokaryotes.

4. Underlying much of life's diversity are three common challenges associated with increased size and complexity. What are these challenges?

5. What are some constraints on enlargement of single cells? Among multicelled organisms, what three growth patterns circumvent these constraints?

6. Compared with prokaryotes, eukaryotes generally have more complex and larger amounts of DNA. Handling all this DNA during cell growth and reproduction depends on three things: (1) _____, (2) _____, and (3) _____.

7. What is a nucleoprotein? During cell construction and maintenance, what form does a nucleoprotein take? Before and during nuclear division, what happens to it?

8. Define mitosis and its function. When does it take place in a cell cycle? Name the four main phases of mitosis, and characterize each phase.

9. Define meiosis and its function. When, and in which cells, does meiosis occur? Name the main phases of meiosis.

10. What is the difference between haploid and diploid cells? Between sexual and asexual reproduction?

11. Define cytokinesis. Describe what happens when it occurs in an animal cell, and in a plant cell.

12. Outline the steps involved in spermatogenesis; in oogenesis.

13. Distinguish between the sporophyte and gametophyte phases of some plant life cycles.

Biology toward the end of the nineteenth century was dominated by talk of Darwin and Wallace's theory of natural selection (Chapter Two). Many biologists were skeptical of whether natural selection of adaptive traits does indeed occur. After all, no one yet had presented a clear statement of what the physical basis for each trait might be, or how the hereditary material from two sexually reproducing parents could be combined in offspring. There were serious flaws even in the most "acceptable" hypothesis: that hereditary material from one parent was of a sort that could blend with the material from another at fertilization, much like cream blending with coffee. It was difficult, of course, to explain why offspring resembled one parent more than the other in specific traits—and even one grandparent more than another—rather than becoming the homogenized equivalent of *café au lait*. For example, if "blending" were a hard-and-fast rule, then a herd of white stallions and black mares would soon be replaced by uniformly gray horses. A village of tall people and short people would give way to people of middling height; and a field of some red-flowering and white-flowering plant would give way to pink-bloomed progeny. Blending scarcely explained the observable facts that in such populations, not all horses are gray, nor all people the same height, nor all flowers pink. It was considered a rule anyway. And Darwin had problems on this account, because uniform populations would present no variation whatsoever for selective agents to act upon. That being the case, "evolution" simply could not occur.

But even before the theory of natural selection was made public, evidence concerning the physical basis of inheritance was accumulating in a small monastery garden in Brünn, northeast of Vienna. Gregor Mendel, a scholarly, mathematically oriented monk, was beginning to identify rules governing inheritance.

The Monastery of St. Thomas was sufficiently removed from the European capitals to be out of the mainstream of scientific inquiry. But this is not to say that Mendel was a man of parochial interests, who simply stumbled by chance onto principles of great import. Having been raised on a farm, he was well acquainted with agricultural principles and applications. He kept abreast of new breeding experiments and developments described in the available journals. Mendel was a founder of the regional agricultural society, and won several awards for developing improved varieties of fruits and vegetables. After entering

THE GENETIC BASIS OF INHERITANCE

Figure 9.1 Gregor Mendel, founder of modern genetics. (The Granger Collection, New York)

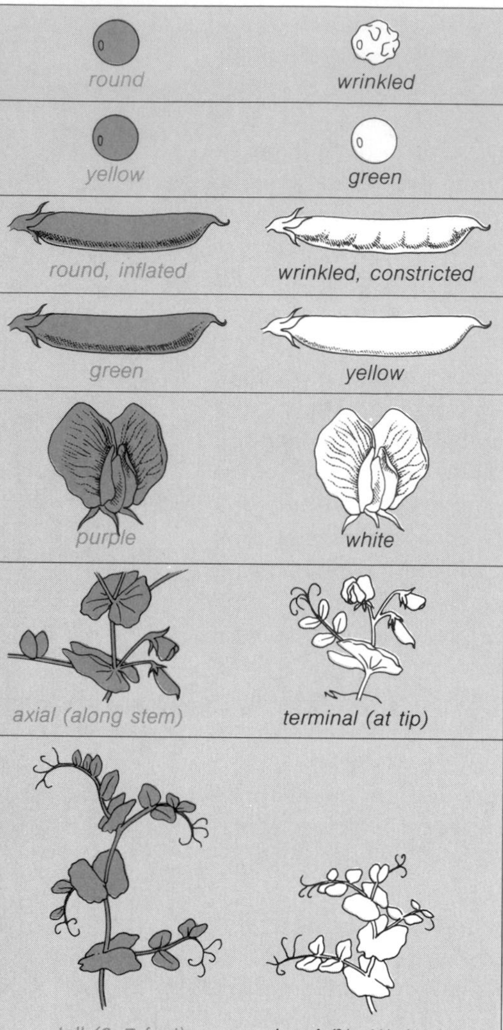

	Parental Dominant Trait	×	Parental Recessive Trait	=	Number of F_2 Dominant	Number of F_2 Recessive
seed shape	round		wrinkled		5,474	1,850
seed color	yellow		green		6,022	2,001
pod shape	round, inflated		wrinkled, constricted		882	299
pod color	green		yellow		428	152
flower color	purple		white		705	224
flower position	axial (along stem)		terminal (at tip)		651	207
stem length	tall (6–7 feet)		dwarf (¾–1½ feet)		787	277
Average ratio for all of the traits tested:					3:1	

Figure 9.2 Results from Mendel's series of seven monohybrid cross experiments.

the monastery, he spent two years at the University of Vienna honing his skills in mathematics. During those years he was influenced by a botany professor, Franz Unger, who was of a mind that plant diversity had not always existed but had come about gradually through natural processes. Shortly after his university training, Mendel began a series of experiments on the nature of plant diversity. It was through his combined talents in plant breeding and mathematics (as yet not considered even re-

motely relevant to plant breeding) that he perceived patterns in the expression of traits from one generation to the next.

For his experiments, Mendel chose different strains of the garden pea plant. This sexually reproducing plant is naturally self-fertilizing. But it also lends itself to **cross-fertilization** (combining of gametes from one individual with gametes from another). One suspects, from Mendel's choices, that he knew in advance what his experiments might prove. Recall, from Chapter Two, that the scientific method depends on trained judgment in selecting what appears to be relevant from a potentially immense range of observations on a given subject. Mendel didn't try to track the inheritance of all traits all at once. He initially concentrated on one trait at a time. First he obtained seeds from **true-breeding strains**, in which all self-fertilized offspring displayed the same form of a trait, generation after generation. For a given trait (flower color, for instance), he used two different strains (*white*-flowered and *purple*-flowered). He made sure that the two forms of the trait being studied were so different from each other that there would be no mistaking the offspring. Then he followed the course of inheritance through several generations. He did this in seven sets of experiments, following seven contrasting traits (Figure 9.2). It will be useful to retrace a few of the experiments, because the conclusions Mendel drew from them have turned out to apply, with some modification, to all sexually reproducing organisms.

The Concept of Segregation

When two true-breeding parent organisms displaying different forms of a trait are crossed, the offspring are called **hybrids**. Because Mendel first concentrated on the single trait by which parents differed, his initial experiments were **monohybrid crosses**. In one such cross, he transferred pollen from a purple-flowered plant to a white-flowered plant. The cross-fertilized plant produced seeds, which were planted the following season. All the seeds grew into plants bearing *only* purple flowers.

What had happened to the white-flower trait? Had it disappeared? Mendel allowed the purple-flowered plants of this first generation to undergo *self*-fertilization rather than crossing them with new plants. (Can you guess why?) Seeds from the self-fertilized plants were harvested, then planted the following spring. When the second-generation plants matured, some had white flowers! The white-flower trait had not been lost. For some reason, it simply didn't get expressed in the first-generation offspring.

(Here you might make note of a few symbols that have become entrenched in genetics. By convention, P stands for the parent generation. F_1 means first-generation offspring. F_2 means the offspring of two F_1 individuals, and so on. The F stands for the Latin *filius*, meaning "son.")

In all the monohybrid crosses that Mendel conducted, results were much the same. One of the contrasting traits "disappeared" in F_1 plants, only to show up again in some F_2 plants. He called the now-you-see-it-now-you-don't form of the trait **recessive**: in some way it was masked by the expression of the alternative form of the trait. In some way the fully expressed form was stronger, or **dominant**, in hybrid plants.

Mendel did more than make this general observation, though. For each trait, he *counted* how many F_2 plants showed the dominant form and how many showed the recessive. As Figure 9.2 illustrates, a regular pattern emerged. On the average, for every three plants expressing the dominant form of a trait, there was one plant expressing the recessive trait. How could such a 3:1 ratio arise?

Given his familiarity with mathematics, Mendel was able to arrive at a simple explanation for the results. He assumed that sexually reproducing organisms inherit from each parent one "unit" of hereditary material for each trait. Say that the unit for a dominant trait is designated A and that for the recessive form, a. If hybrid offspring inherit two units, one from each parent, then it follows that three combinations of units are possible in the F_2 generation: AA, Aa, and aa. Figure 9.3 shows why this is so. Mendel expected that on the average, for every four F_2 plants, one would show the dominant form of the trait (AA), two would show both forms ($2Aa$), and one would show the recessive form (aa). Because the dominant unit masks expression of the recessive one in hybrid F_1 plants, the ratio of dominant-to-recessive in the F_2 generation should be 3:1 *in outward appearance*; the underlying ratio of genetic combinations should be 1:2:1.

(The basis for this expectation, which is still valid for most cases, is that sperm and eggs combine at random. Any sperm is as likely to fertilize one egg as another. In hybrid F_1 plants, half the gametes formed following meiotic cell division will carry A; the remaining half will carry a. This means that about half of all eggs fertilized should contain the dominant hereditary unit, and about half the sperm fertilizing them should also contain the dominant unit. About one of every four zygotes, then, should be AA. The same reasoning can be used to show why about two of every four zygotes should be Aa, and about one of every four should be aa.)

It is important to realize that the actual results rarely will be in an exact 1:2:1 ratio. To understand why, flip a coin a few times. We all know that a coin is just as likely to end up "heads" as "tails." But often it ends up heads, or tails, several times in a row. Only if you flip the coin many times can you be assured that the head-to-tail ratio will be close to the predicted 1:1 ratio. Such random events illustrate three characteristics of **probability**. First, it's usually possible to predict the most probable ratio

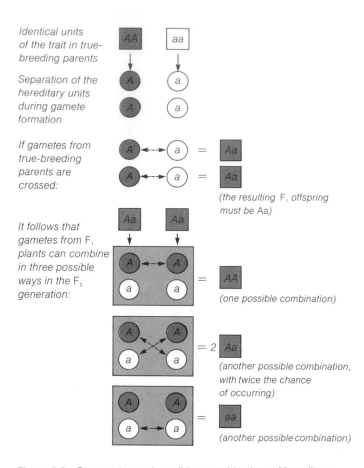

Identical units of the trait in true-breeding parents

Separation of the hereditary units during gamete formation

If gametes from true-breeding parents are crossed:

(the resulting F₁ offspring must be Aa)

It follows that gametes from F₁ plants can combine in three possible ways in the F₂ generation:

(one possible combination)

(another possible combination, with twice the chance of occurring)

(another possible combination)

Figure 9.3 Segregation and possible recombinations of hereditary units in gamete formation. Here, squares represent individual plants, and circles represent gametes. Notice the greater probability of the *Aa* combination in the *F₂* generation. Mendel was fully aware of the laws of probability, and came up with the empirical formula *AA + 2Aa + aa* for his monohybrid crosses of true-breeding diploid organisms.

of outcomes in any series of randomly ordered events. (See, for example, Figure 9.3.) Second, the actual ratio seldom will match the expected ratio exactly. Third, when only a few events are observed, the actual ratio may differ considerably from the predicted one. Mendel succeeded largely because he crossed hundreds of plants and kept track of thousands of offspring, rather than restricting his experiments to a few plants as others had done. Almost certainly his understanding of probability kept him from being confused by minor deviations from his predicted results.

On the basis of results from these well-conceived and carefully executed experiments, Mendel formulated a basic principle. It is still considered useful as a starting point for a discussion of the nature of inheritance:

Mendelian principle of segregation. In sexually reproducing organisms, two units of heredity control each trait. During gamete formation, the two units of each pair are separated (segregated) from each other and end up in different gametes.

Since Mendel's time, the process by which segregation occurs has become better understood. Studies at the molecular and cellular levels yielded insights into the mechanisms of meiosis (described in the preceding chapter), and into the behavior of chromosomes during meiosis (which will soon be described). As you might imagine, the terminology used to describe these events has also evolved, to the extent that Mendel's view of the physical units of inheritance would now be expressed in this way:

1. Distinct units of heredity—**genes**—are the physical basis for all traits of a new individual.

2. There is a haploid set of genes in a gamete. Following sexual fusion of two gametes, the new individual is diploid. It has *two* genes for *each* trait.

3. A gene for any given trait may occur in alternative states, known as **alleles**.

4. In a diploid individual, alleles may code for identical or different forms of a trait. For instance, both alleles for flower color may specify white. Then again, they may specify alternative forms. (One may specify purple, the other white, or orange, or yellow.)

5. If the two alleles of a given trait are identical, the individual is **homozygous** for that trait. If the two alleles are not identical, the individual is **heterozygous** for that trait.

6. In heterozygous individuals, expression of one allele may mask expression of the other (it's the dominant form of the trait). Even so, *both* alleles retain their physical identity throughout the individual's life cycle.

7. Each allelic pair represents the individual's **genotype**, or its genetic makeup, for a particular trait. The term **phenotype** refers to an individual's physical appearance.

Let's rephrase the description of Mendel's experiment with flower color in these current terms. Assume that both flower color alleles in a diploid plant specify purple (Figure 9.5). We can, as Mendel did, represent the plant as *AA*. The plant is homozygous dominant for the trait. Assume that both flower color alleles in another diploid plant spe-

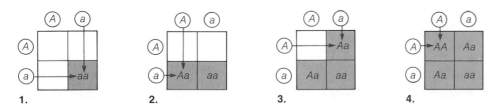

1. **2.** **3.** **4.**

Figure 9.4 Punnett square method of determining the probable ratio of traits that will show up in the offspring of self-fertilizing individuals known to be heterozygous (*Aa*) for a trait. The circles represent gametes; the letters inside them represent the dominant or recessive form of the trait being observed. Circles on the top of the grid represent female gametes; circles on the side represent male gametes. Each square depicts the genotype of one individual type of offspring.

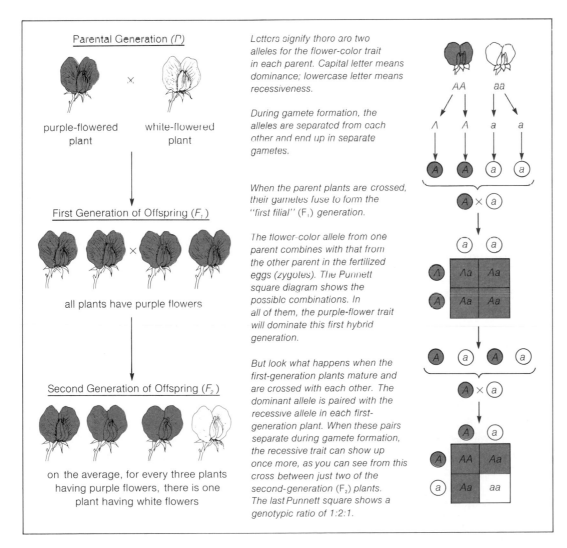

Parental Generation (P)

purple-flowered plant × white-flowered plant

First Generation of Offspring (F₁)

all plants have purple flowers

Second Generation of Offspring (F₂)

on the average, for every three plants having purple flowers, there is one plant having white flowers

Letters signify there are two alleles for the flower-color trait in each parent. Capital letter means dominance; lowercase letter means recessiveness.

During gamete formation, the alleles are separated from each other and end up in separate gametes.

When the parent plants are crossed, their gametes fuse to form the "first filial" (F₁) generation.

The flower-color allele from one parent combines with that from the other parent in the fertilized eggs (zygotes). The Punnett square diagram shows the possible combinations. In all of them, the purple-flower trait will dominate this first hybrid generation.

But look what happens when the first-generation plants mature and are crossed with each other. The dominant allele is paired with the recessive allele in each first-generation plant. When these pairs separate during gamete formation, the recessive trait can show up once more, as you can see from this cross between just two of the second-generation (F₂) plants. The last Punnett square shows a genotypic ratio of 1:2:1.

Figure 9.5 Results from one of Mendel's monohybrid cross experiments, showing the distribution of alleles through the F₁ to the F₂ generation.

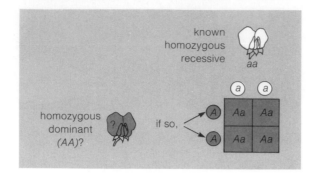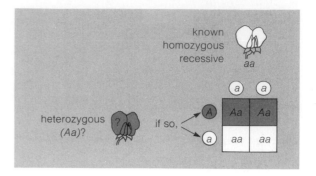

Figure 9.6 The two possible outcomes of a testcross between an individual known to be homozygous recessive for a trait (for example, white flower color), and an individual that shows the dominant form of the trait. If the dominant-appearing individual is homozygous dominant, all offspring will show the dominant form of the trait. If it is heterozygous, half the offspring will show the recessive form.

cify white. We may represent this plant as *aa*; it is homozygous recessive for the trait.

These diploid plants both produce haploid gametes prior to sexual reproduction. During meiosis, the two alleles for flower color are separated from each other. Because each haploid gamete formed has only one flower color allele, the zygotes formed upon sexual fusion will be *Aa*. The alleles are not identical in this kind of zygote; this kind is heterozygous for the trait.

All F_1 plants that mature from the zygotes will have purple flowers, because their dominant allele will mask expression of the recessive one. But now suppose these plants in turn form haploid gametes prior to sexual reproduction. Then half the gametes will, on the average, end up with the dominant allele, and half with the recessive. When sperm from one heterozygous plant fuse at random with eggs from another, the three possible combinations predicted by Mendel (*AA*, *Aa*, and *aa*) should occur in a 1:2:1 ratio (Figures 9.4 and 9.5).

Again, on the basis of physical appearance alone, we expect the ratio to be 3:1 (there are about three purple-flowered plants for every white-flowered plant). But homozygous dominant (*AA*) and heterozygous (*Aa*) pea plants *both* produce purple flowers. Thus the heterozygotes can't be distinguished from homozygous dominant individuals simply on the basis of phenotype. Their genotypes can be determined only by observing the kinds of offspring that they produce. The **testcross** often may be used for this purpose. The first step is to establish a true-breeding strain of individuals that are homozygous recessive for some trait (in this case, true-breeding white-flowered

plants). The next step is to cross the dominant-appearing individual of undetermined genetic makeup with a known homozygous recessive individual. There are only two possible outcomes for this testcross. If the unknown is homozygous dominant, all offspring will show the dominant trait (purple flowers). If the unknown is heterozygous, the recessive trait will show up in half the offspring. The Punnett square method (Figure 9.6) can be used to show the possible outcomes of a testcross.

The Concept of Independent Assortment

Through his monohybrid crosses, Mendel identified the course of a single trait through successive generations. He then turned to the problem of how multiple traits segregate and combine during sexual reproduction. To study this problem, he used **dihybrid crosses** (crosses between true-breeding plants that differ in *two* unrelated traits). When Mendel analyzed the results, he began to perceive how sexual reproduction might foster genetic diversity.

In one experiment, he crossed tall purple-flowered plants with dwarf white-flowered ones. Like the purple-flower allele, the "tall" allele was dominant. Thus, all first-generation plants were purple-flowered and tall. But as Figure 9.7 shows, when these plants were crossed in turn, every combination of traits that could occur in the second generation did occur. And they occurred in a distinct pattern for all of the dihybrid crosses. Mendel wrote that the phenotypic ratio in these dihybrid crosses was about

9:3:3:1. For every 9 tall purple-flowered plants, there were 3 dwarf purple-flowered, 3 tall white-flowered, and 1 dwarf white-flowered. What could the pattern mean? To be sure, each gamete from a heterozygous plant ends up with one allele for each trait. But in this case, a gamete containing the dominant allele for flower *color* could end up with either the dominant or recessive allele for *height!* On the basis of these results, Mendel proposed a second principle of inheritance, which is stated here in a simplified way:

Mendelian principle of independent assortment. When gametes are formed, distribution of segregated alleles for a given trait into one gamete or another is independent of, and doesn't interfere with, distribution of segregated alleles for other traits.

When independent assortment of alleles does occur, the resulting offspring possess some composite of parental traits. Its implications are staggering. Offspring from a cross between two parents differing from each other in twenty-three traits can have any one of 2^{23}, or about *8 million* possible combinations of those traits! Such possibility for variation is important in variable environments. There, selective agents may act to perpetuate those combinations of alleles that have made their bearers finely tuned to prevailing conditions (Chapter Eight).

It is important to keep in mind, though, that independent assortment does not *always* occur. Mendel's formulation of the "principle" of independent assortment was a major intellectual step in our understanding of inheritance. But, as you will soon see, later studies revealed that certain traits occasionally are inherited *as a group* rather than independently.

In 1865 Mendel outlined his studies and conclusions before the Brünn Society for the Study of Natural Science. His report made no impact whatsoever. The following year his paper was published, but apparently it was read by few and understood by no one. Remember that he was going up against the well-entrenched blending theory of inheritance. His mathematics and descriptions of statistical distributions of traits probably would not have made sense to anybody but mathematicians and statisticians—who probably would not have had the least bit of interest in pea plants. Worse yet, Mendel later turned to another plant, hawkweed, which repaid his interest by producing *only* the dominant form of certain traits in *both* the F_1 and F_2 generations! He did not know, as we do now, that hawkweed seeds can form without fertilization of gametes; diploid cells in the ovary can give rise directly to another

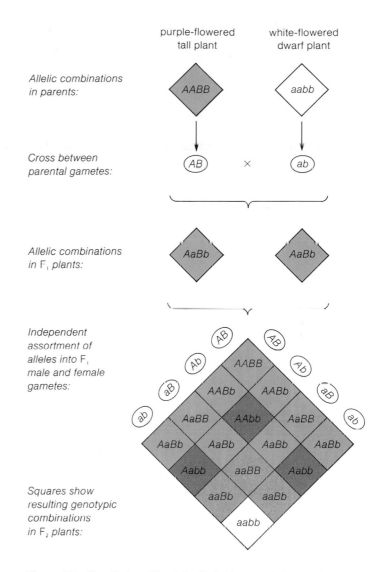

Figure 9.7 Results from Mendel's dihybrid cross of plants differing in flower color and in height. Here, *A* and *a* represent the dominant and recessive alleles for flower color; *B* and *b* represent the dominant and recessive trait for height. Notice that the phenotypic combinations in the F_2 generation occur in about a 9:3:3:1 ratio. (The nine gold squares are purple-flowered tall plants; three light-brown squares are purple-flowered dwarf plants; three gray squares are white-flowered tall plants; and the one white square is the white-flowered dwarf plant.)

plant having the exact genetic makeup as its parent. Mendel must have been bitterly disappointed by the results. By 1871 he became abbot of the monastery, and his experimental work gave way to administrative tasks. He died in 1884, the founder of but not a participant in modern genetics.

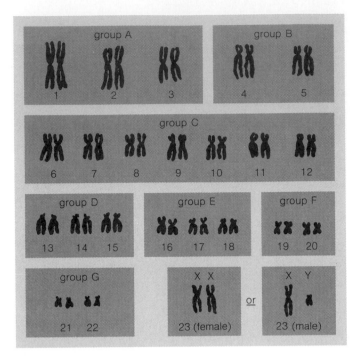

Figure 9.8 Karyotype for human chromosomes. (A *karyotype* is an arrangement and grouping of chromosomes on the basis of size, number, and shape.) Normally, human cells have a diploid chromosome number of 46, with 22 pairs being autosomes and one "pair" being the sex chromosomes. Each chromosome of the pairs shown here has already undergone replication, so that it consists of two chromatids held together at the centromere.

EMERGENCE OF THE CHROMOSOMAL THEORY OF INHERITANCE

In 1900 Eric Von Tschermak of Austria, Hugo de Vries of Holland, and Carl Correns of Germany published papers on their studies of inheritance in plants. It would be an understatement to say that the times must have been ripe for such studies, because all three had independently come up with much the same results. They also had searched the literature for related studies and came across Mendel's paper—with its news that his interpretations about the nature of inheritance preempted their own by more than three decades. What had happened in the interim to foster such interest, where before there had been almost none?

By about the mid-nineteenth century, interest in the nature of inheritance was fanned by cytology, the study of cells. Pioneers in Europe flocked to the fascinating fron-

tiers opened up by new developments in microscopy (Chapter Four). In 1882 Walther Flemming observed threadlike bodies in dividing salamander cells; he called the division process mitosis (after the Greek word for "thread"). Then Carl van Beneden discovered that daughter cells receive *identical* sets of the mitotic "threads" during division. He perceived further that gametes have one set of these threads, whereas fertilized cells have two. By 1887, August Weismann had proposed that a special division process must occur to reduce the number of threads by half during gamete formation. Sure enough, Flemming in that same year identified the process, which has become known as meiosis. Processes of inheritance were being observed in ever greater detail through the microscope; the significance of those processes had yet to be explained.

Then, in 1903, Walter Sutton suggested that the events observed during mitosis and meiosis dovetailed with Mendel's concepts of segregation and independent assortment. First, chromosome sets come in pairs, one from each parent. Second, the pairs separate during division. Third, sexual fusion of two haploid gametes restores the original diploid number of chromosomes. Flemming's "threads," in short, were likely candidates for being the carriers of Mendel's "hereditary units."

Enter the Fruit Fly

Sutton's hypothesis came to be confirmed by Thomas Hunt Morgan and his students at Columbia University. These workers chose, as the object of their experimental interest, the common fruit fly *Drosophila melanogaster*. These small flies can be grown in bottles, on nothing fancier than bits of rotting fruit and yeast. The life cycle is completed in less than two weeks. And the females lay hundreds of eggs in a few days. This means that hereditary traits can be tracked through nearly thirty generations of thousands of flies in the space of a year. Before long, Morgan's laboratory was glutted with bottles of fruit flies.

Normally, *D. melanogaster* is homozygous dominant for red eyes, and all the flies in all of Morgan's bottles had red eyes. One day there appeared a solitary male having *white* eyes. Morgan crossed the male with a red-eyed female. All of the first-generation offspring had red eyes. But inbreeding those offspring led to an F_2 generation in which both red-eyed and white-eyed individuals appeared! Through subsequent generations, the white-eye trait persisted. It had become established in the hereditary material the instant the mutation occurred.

Morgan's group began looking for other changes in single traits. A number of mutant flies were soon identified. Each had some distinct change in such traits as eye color, eye shape, body color, body size, and wing form. The explorations that Morgan now undertook with three of his students—Alfred Sturtevant, Calvin Bridges, and Hermann Muller—would lead to clear understanding and definitive proof of the chromosomal basis of inheritance.

Clues From the Inheritance of Sex

Most chromosomes are called **autosomes**: they are of the same number and kind in both males and females. By the time of Morgan's *Drosophila* experiments, it was known that most animal and some plant species also have chromosomes that differ in number or in kind between males and females. They were called **sex chromosomes**, for they undoubtedly were associated with sex determination.

For example, cytologists found that in cells from *male* grasshoppers, one chromosome doesn't have a partner. It was called the X *chromosome*. In cells from *female* grasshoppers, this chromosome does have a homologue. Hence the males came to be designated XO and the females, XX. In other species, the X chromosome in males does have a partner, usually of smaller size and different shape. Given the usage of "X," it was named the Y *chromosome*. Hence in these species, males are designated XY, and females, XX. This is by far the most common pattern among animals.

Humans have twenty-two pairs of autosomes and one pair of sex chromosomes (Figure 9.8). In females the twenty-third pair is XX; in males it's XY. Each normal female gamete carries one X chromosome. In males, the X and Y chromosomes act as homologues during meiosis, which means that half the sperm formed will carry an X and half will carry a Y chromosome. When an X-bearing sperm fuses with an X-bearing ovum, the zygote develops into a female. When a Y-bearing sperm fuses with the ovum, the zygote develops into a male (Figure 9.9).

It happens that the Y chromosome in humans doesn't carry many genes other than those determining maleness. But the X chromosome carries many genes besides the ones involved in sex determination. Any gene residing on an X chromosome is known as a **sex-linked gene**. This kind of gene was first understood in detail in Morgan's laboratory.

In *Drosophila*, XX specifies female and XY specifies male (Figure 9.10). Recall that when Morgan interbred red-eyed F_1 offspring from the mutant white-eyed male

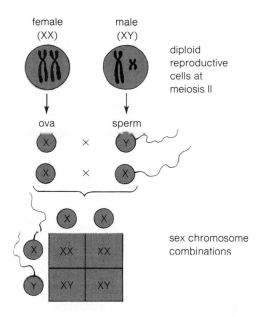

Figure 9.9 Sex determination in humans. Only the sex chromosomes are depicted here; each ovum and sperm also carries a haploid set of autosomes. Males transmit their Y chromosome to their sons, but to none of their daughters. Males receive their X chromosome only from their mothers.

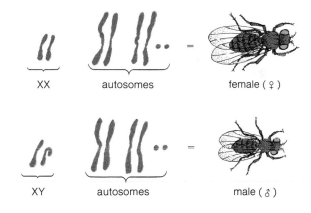

Figure 9.10 Sex chromosomes and autosomes of *Drosophila*, which together represent a diploid number of eight (four chromosome pairs).

fly, the recessive trait reappeared in some of the F_2 offspring. However, all the white-eyed F_2 individuals turned out to be *male*. Then, when a red-eyed F_1 female was backcrossed to the white-eyed parental male, white-eyed females appeared among the offspring (Figure 9.11). Now, *Drosophila* females (XX) obviously inherit one X chromo-

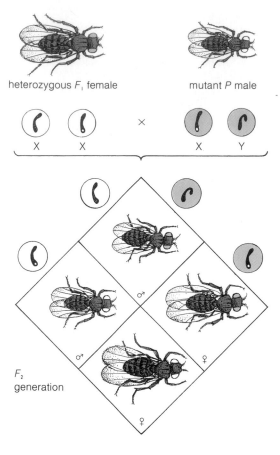

heterozygous F_1 female   mutant P male

×

X   X   X   Y

F_2
generation

♂   ♀

♂

♀

Figure 9.11 Morgan's backcross between an F_1 female *Drosophila* heterozygous for eye color (one dominant red allele and one recessive white allele) and the mutant white-eyed parental male. The recessive allele (depicted here by the white dot) must be carried on the X chromosome only, given the correlation between eye color and sex in the F_2 generation. Thus a male can carry only one recessive allele, whereas a female can carry two for this (or any other) sex-linked trait.

some from each parent, but males (XY) can inherit their single X chromosome *only* from their maternal parent. That being the case, the red-eyed first-generation females in Morgan's experiments had to be heterozygous for eye color, with the red-eye allele carried on one X chromosome and the recessive white-eye allele carried on the other. In addition, only homozygous recessive females would be white-eyed. Now, if the eye-color gene were carried on the X chromosome, it would be expressed in males *regardless of whether it is dominant or recessive*. Through further experiments, Morgan did indeed establish a clear link between this gene for eye color and the X chromosome.

In later studies, Bridges and others also found evidence that sex differences between males and females are not

the exclusive domain of sex chromosomes. **Sex-related genes**, which influence the expression of "maleness" *versus* "femaleness," can occur on autosomes as well. In most diploid organisms, major sex-related genes are concentrated on the sex chromosomes (the actual number in an individual is still not known). In humans and other mammals, the Y chromosome carries genes that are necessary to produce the male phenotype. Certain autosomal genes influence the growth and function of reproductive organs, body hair distribution, pitch of the voice, and breast size, to mention a few sex-related traits. These genes are not restricted to males *or* females; they occur in the autosomes of both. It's just that developmental events assure that these genes are expressed differently, largely through the stimulation or inhibition caused by different sex hormones. You will read more about this topic in chapters dealing with development and animal evolution. In this context, it is enough to say that early studies of sex-linked inheritance gave strong experimental support for Sutton's hypothesis, that genes are indeed carried on chromosomes.

Linkage: The Discovery of Some Not-So-Independent Assortments

During the first division of meiosis, recall, chromosomes are lined up at the equator of the spindle apparatus in homologous pairs: each chromosome derived from one parent lies side by side with the homologous chromosome from the other parent. Their alignment on one side of the equator has nothing to do with the alignment of other homologous pairs. In other words, there is nothing that says all the chromosomes derived from one parent must stay on the same side of the spindle equator and their homologues must stay on the other. Assume that we are talking about no more than three homologous pairs. Any one of the following combinations can occur through their alignment:

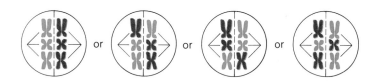

or   or   or

(Here, the brown chromosomes are derived from one parent, and the gold are derived from the other.) During meiosis I, these four arrangements occur with about equal likelihood. This means that when two chromosomes of each pair separate during meiosis, eight different combinations of whole chromosomes are possible in the forthcoming gametes. This random alignment of chromosome pairs is the basis for **independent assortment** as Mendel perceived it. However, the many genes on any one chromosome are physically linked and must travel as a group into gametes, just as their allelic partners on the homologous chromosome must travel into a different gamete. In addition, as you will now see, even the genes located on the same chromosome do not always *stay* together during gamete formation.

Morgan's group performed hundreds of dihybrid crosses involving mutant *Drosophila*. And they soon began to perceive that the traits being studied assorted into gametes as if they were somehow *linked* in four separate groups. They perceived further that these four groups probably corresponded to the four *Drosophila* chromosomes. For instance, one cross involved wing shape and body color. Normally, fruit flies have straight, flat wings (which can be designated C) and a gray body (B). One mutant form had curly wings (c) and a black body (b). According to Mendelian genetics, a cross between heterozygous flies (*CcBb*) should yield offspring in a 9:3:3:1 ratio, just as it did for the dihybrid cross depicted in Figure 9.7. Yet the actual ratio was closer to 3:1—*as if the experiments had been tracking only a single gene that produced two separate phenotypes.* The genes for wing shape and body color, then, were not assorting independently. It was as if they were physically linked by virtue of being located on the same chromosome. This occurrence came to be called **linkage**, with all genes located on a given chromosome comprising a **linkage group**. Through these and later studies with other organisms, it has become clear that the number of linkage groups corresponds to the number of chromosomes characteristic of the species—which provides further confirmation of Sutton's hypothesis that chromosomes are the vehicles by which genes are transmitted through generations.

Crossing Over

More surprises were still to come. In 1909, the cell biologist F. Janssens discovered that during meiosis, two nonsister chromatids of homologous chromosomes sometimes cross each other. Today we know that the following events are occurring when this happens.

During early prophase of meiosis I, a submicroscopic, thin band of RNA and protein is synthesized between the joined sister chromatids of each chromosome. Apparently there is chemical recognition between the band of one chromosome and the band of its homologous partner. Somehow, chemical attraction draws the two sister chromatids

of one chromosome very close to the two sister chromatids of its homologue. The attraction brings them into a precise, point-by-point alignment along their length. This alignment is known as **synapsis**:

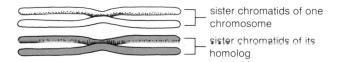

sister chromatids of one chromosome

sister chromatids of its homolog

In this arrangement, two *nonsister* chromatids often undergo breakage at exactly the same point along their length:

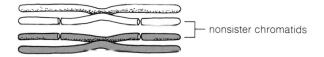

nonsister chromatids

These four broken ends do not simply stick back together. They stick back together crosswise:

crossing over at two sites

This event is called **crossing over**: two nonsister chromatids of homologous pairs of chromosomes exchange corresponding segments at breakage points. Its consequence is **genetic recombination**—the formation of new combinations of alleles in a chromosome, hence in gametes, and finally in zygotes.

During the last stage of prophase I, the two sister chromatids of each chromosome remain close to each other. But now the chromosomes of a homologous pair seem to repel each other. Even so, these homologues are held together at one or more regions as they are separated along most of their length. Such a region of apparent contact is called a **chiasma** (plural, chiasmata), which means a "cross." This sketch depicts two chiasmata:

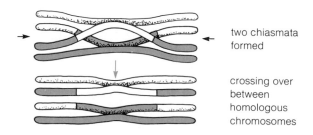

two chiasmata formed

crossing over between homologous chromosomes

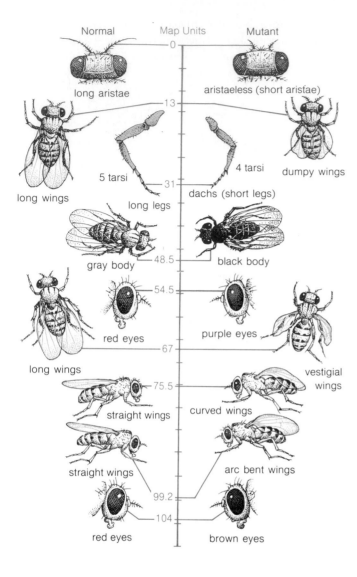

Normal   Map Units   Mutant

0 — long aristae / aristaeless (short aristae)

13 — 5 tarsi / 4 tarsi, dumpy wings

31 — long wings, long legs / dachs (short legs)

48.5 — gray body / black body

54.5 — red eyes / purple eyes

67 — long wings / vestigial wings

75.5 — straight wings / curved wings

99.2 — straight wings / arc bent wings

104 — red eyes / brown eyes

Figure 9.12 Genetic mapping of genes on a segment of chromosome 2 in *D. melanogaster*. Such maps don't show actual physical distances between genes. Rather they show relative distance between gene locations that undergo crossing over and other chromosomal rearrangements. Only if the probability of crossing over were equal along the chromosome's length (which it is not) would it be possible to calculate physical distance exactly.

Here, distances between genes are measured in map units, based on the frequency of recombination between the genes. Thus, if the frequency turns out to be 10 percent, the genes are said to be separated by 10 map units. The amount of recombination to be expected between "vestigial wings" and "curved wings," for instance, would be 8.5 percent (75.5 − 67). This works for genes close together on the map, but not for distant genes.

Each chiasma observed at this stage is the visible evidence that an exchange (crossover) occurred earlier between two nonsister chromatids. It does not necessarily indicate exactly *where* the actual breakage occurred, though, for chiasmata may slip down toward the ends of the chromosomes as the chromosomes are pulled apart.

Janssens's description of chiasma formation gave Sturtevant the clue he needed. He reasoned that the farther apart on the chromosome two linked genes may be, the more likely the chance that chiasma formation and crossing over can disrupt the original combination of linked alleles.

Mapping the Gene Sequence in Chromosomes

Through the work of Morgan and Sturtevant, the following picture has emerged. Crossing over apparently can occur at any point in the linear array of genes on each chromatid. *But the probability of crossing over and recombination occurring at a point somewhere between two genes located on the same chromatid is directly proportional to the distance that separates them.* If two genes are located very close together, in nearly all cases they will end up in the same gamete; they are tightly linked. If two genes are relatively far apart, chiasmata will form between them much more often than they do between tightly linked genes. Even so, in more than half the gametes formed, they will still be together on the chromatid. Such genes are loosely linked. Finally, if two genes are very far apart on a chromatid, crossing over and recombination can occur between them so often that they act as if they are assorted independently—even though they are located on the same chromosome.

The relationship between the organization of genes on chromosomes and their segregation patterns during meiosis is so regular that it can be used to determine the positions of genes relative to one another. Plotting their positions is called **linkage mapping** of genes.

For example, assume we find out that alleles of two different genes located on the same chromosome end up together in the same gamete about ninety-five percent of the time. We know, then, that they must be so close together that crossing over separates them only about five percent of the time. We could conclude that they are closer together than two genes showing ten percent crossing over —and farther apart than two other genes showing only one percent crossing over. When enough genes on a chromosome have been studied, two or three at a time, it is possible to draw a detailed map of their positions relative to one another. Figure 9.12 is an example of this.

Of the many thousands of genes contained in the four chromosomes of *Drosophila*, the positions of about 500 have been mapped—and *Drosophila* is one of the most intensively studied of all organisms. There are surely many more genes in a haploid set of human chromosomes, very few of which have been identified and their positions mapped. We have a long way to go in our mapping of the physical basis of heredity. But the work to date has yielded undeniable proof that genes are carried on chromosomes—*and that they are carried in linear array.*

Chromosomal Rearrangements

Crossing over normally occurs during gamete formation. On very rare occasions, chromosomal rearrangements known as deletion, duplication, translocation, and inversion also occur. A piece of chromosome lost entirely is a **deletion**. The deficiency means that during meiosis, the homologous partner of the affected chromosome must buckle, because it can't undergo proper synapsis:

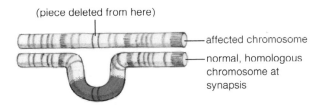

(piece deleted from here)
— affected chromosome
— normal, homologous chromosome at synapsis

The buckling may cause problems during crossing over, and during what would otherwise be an orderly separation of the partner chromosomes during meiosis. There may be problems even when the affected chromosome does manage to be transmitted to a zygote, if genes that control some vital trait happen to be the ones deleted. Then the zygote probably won't survive unless the alleles on the homologous chromosome are able to see it through.

What happens to segments so deleted? In some cases a deletion may become attached to the homologous chromosome. Because the recipient chromosome thereby contains a segment that corresponds to a segment it already had, the event is called a **duplication**. A duplication may lead to amplified expression of traits in offspring. Also on rare occasions, segments break and move from one chromosome to a nonhomologous chromosome. This event is known as a **translocation**. In another rare occurrence, the gene order in some segment of a chromosome may become reversed. A deleted piece of chromosome may get

Macmillan Science Company, Inc.

Figure 9.13 "Giant" chromosomes from a salivary gland cell of *Drosophila*. In these large cells, chromosomes replicate repeatedly but do not separate, so that thousands may remain packed, in parallel fashion, to form the "giant" chromosome structure in which density differences are magnified. (The dark bands so amplified are clearly visible in this light micrograph.)

turned around and may rejoin the chromosome at the same place—but backward. Such an event is called an **inversion**. Although an inversion doesn't change the total number of genes, the position and sequence of the genes do change. It's possible to note the occurrence of inversions in the giant chromosomes of *Drosophila* (Figure 9.13), because of the resulting flip-flop in banding patterns in the region corresponding to the change.

Besides these rare internal rearrangements, from time to time whole chromosomes or chromosome sets go astray. When the separation of a pair of homologous chromosomes is inhibited during meiosis, some gamete can end up with an **extra chromosome**. If we designate the normal haploid number as n, then the gamete having the extra chromosome is $n + 1$. Upon sexual fusion with a normal gamete

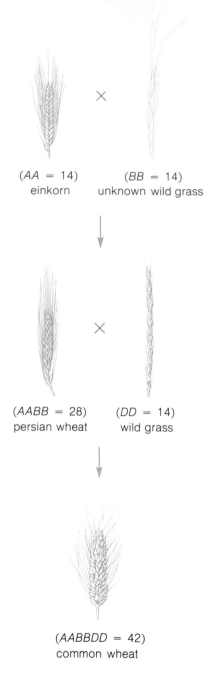

(*AA* = 14)
einkorn

(*BB* = 14)
unknown wild grass

(*AABB* = 28)
persian wheat

(*DD* = 14)
wild grass

(*AABBDD* = 42)
common wheat

Figure 9.14 Hybridization in wheat leading to chromosome doubling and polyploidy. Wheat grains dating from 7000 B.C. have been found in eastern Iraq. Several wild wheat species with 14 chromosomes (two sets of 7, designated *AA*) still grow there. Also growing in the region is a wild grass with 14 chromosomes (*BB*), which are different from *A* chromosomes, judging by their failure to pair with them during meiosis. One common wheat plant (persian wheat) has 28 chromosomes, and analysis during meiosis shows that they are *AABB*. Another common wheat has 42 chromosomes (six sets of 7) that can be designated *AABBDD*, the last set of chromosomes coming from *Aegilops squarrosa* (*D*), another wild grass. (After Jensen and Salisbury, *Botany: An Ecological Approach*, 1972)

to form a zygote, the number becomes $(n + 1) + (n)$, or $2n + 1$. Thus each cell arising through mitotic division during growth and development will be $2n + 1$.

Sometimes a whole extra set of chromosomes ends up in the same gamete, as a result of complete failure of meiosis. If this $2n$ gamete undergoes sexual fusion with another gamete, a **triploid** ($3n$) individual results. (The word means "three sets.") If two $2n$ gametes fuse, the result is a **tetraploid** ($4n$) individual with "four sets" of chromosomes.

The presence of more than two sets of chromosomes is called **polyploidy**. Polyploidy occurs far more commonly in plants than in animals. The condition may also be induced artificially by chemical treatment. Plant breeders use low concentrations of *colchicine*, a chemical known to inhibit microtubule formation. As one of its effects, colchicine inhibits spindle formation during nuclear division. Without a spindle apparatus, chromosomes can't separate from one another at anaphase. Many plants arising through induced polyploidy are larger and more vigorous, with showier blooms, than the diploid form. Compared with their relatives in the wild, they commonly are more vulnerable to stress (from drought, wind, storms).

Deletions, duplications, translocations, and inversions, along with the wayward distribution of whole chromosomes or chromosome sets, are **chromosomal mutations**. They invariably affect relatively large chromosomal segments of the total chromosome number characteristic of the species. Other kinds of changes also occur in the chromosome. **Point mutations** involve additions, deletions, or substitutions of individual nucleotide subunits of the DNA molecule itself.

In the next chapter you will be reading about the causes of point mutations. You will also read about some remarkable enzymes that monitor the DNA molecule and work to restore the original sequence when point mutations occur. Here it is enough to know that, like chromosomal mutations, the effect of their rearrangements is the introduction of some new combination of alleles in offspring. That combination may prove lethal, or beneficial, or merely innocuous under a given range of environmental conditions. But mutations are the source of all genetic variability, the grist for the evolutionary mill.

The Theory in Modern Form

So far, we have outlined the nature of observations, experiments, and hypotheses about hereditary mechanisms that unfolded in the decades after the rediscovery of Mendel's work. Taken together, they represent impressive evidence in support of what is known as the **chromosomal theory of inheritance:**

1. The chromosome is the vehicle by which hereditary information is physically transmitted from one generation to the next.

2. Each chromosome carries a linear sequence of genes, the units of hereditary information that govern the development of phenotype.

3. Diploid organisms have two sets of chromosomes, one from each parent. One gene for a given trait resides on a chromosome derived from one parent, and its allelic partner resides on the homologous chromosome from the other parent.

4. During meiosis, all homologous chromosome pairs become separated from one another, then are assorted into gametes. Because whole chromosomes are assorted independently of one another, there can be *different combinations of chromosomes* from both parents in different gametes.

5. The genes on one chromosome of a homologous pair tend to be inherited *as a group.* However, prior to gamete formation, *homologous chromosomes can exchange parts as a result of crossing over.* Thus the combination of alleles on any given chromosome that ends up in a gamete may not be exactly the same as that in either parent.

6. A chromosome may also undergo such internal rearrangements as deletion, duplication, translocation, and inversion of parts. All are considered to be chromosomal mutations.

7. Other chromosomal mutations involve the wayward movement of whole chromosomes or chromosome sets, which changes the chromosome number in the resulting gametes.

8. Point mutations (chemical changes in individual nucleotides of DNA) may occur to further change the nature of genes carried by a chromosome.

9. *Crossing over, independent assortment of whole chromosomes, chromosomal mutations, point mutations*—such events lead to new genotypic combinations. These combinations in turn give rise to new phenotypes upon which selective agents can act. In short, these events are wellsprings of diversity and evolutionary change.

CHROMOSOMAL ABNORMALITIES AND POINT MUTATIONS IN HUMANS

The chromosomal mutations we have been describing are not limited to fruit flies and polyploid plants. They may occur in all organisms. Consider the consequences of an improper separation of chromosomes during meiosis in a human reproductive cell. For example, on rare occasions, chromosome 21 wanders off course and a gamete ends up with two of these chromosomes instead of one. Thus fertilization with a normal gamete produces a zygote with three copies of the chromosome, a condition called **trisomy 21.** It leads to an affliction called **Down's syndrome.** This affliction is characterized by extreme mental retardation, as well as such distortions in outward physical appearance as a large, misshapen head and a sunken nose. Afflicted individuals frequently die at an early age. Those who do survive remain small in stature.

Some chromosome abnormalities involve the sex chromosomes. For instance, males suffering from **Klinefelter's syndrome** have testes that develop to only about half their normal size, and the deposition of body fat follows female patterns. There may be some breast enlargement. Afflicted individuals are always sterile. These individuals have one Y and two X chromosomes—a total of forty-seven chromosomes instead of the normal diploid complement of forty-six. In another sex chromosome abnormality, called **Turner's syndrome**, the secondary sex characteristics of adolescent females fail to develop at puberty, and the ovaries are nonfunctional. Such females received only one X chromosome. Hence they have only forty-five chromosomes instead of the normal diploid number.

Point mutations occurring in some key gene of a chromosome can be equally unfortunate. For instance, there is a recessive form of a gene governing fat metabolism. At birth, a child who is homozygous recessive for the gene appears normal. But gradually a certain lipid builds up in the myelin sheaths of the child's nerves (Chapter Fourteen). The accumulated fat blocks the transmission of nerve impulses, which leads to loss of muscle coordination. Brain cells cannot function properly and mental acuity begins to deteriorate. Within a few years, the afflicted child inevitably dies from this so-called **Tay–Sachs disease**.

A well-known example of a sex-linked abnormality in humans is the mutation leading to **hemophilia**. A gene on the X chromosome codes for one of the key proteins necessary for blood clotting. A mutation in this gene makes it code for a defective protein (or none at all). In a woman who is a "carrier" of (heterozygous for) the mutant gene,

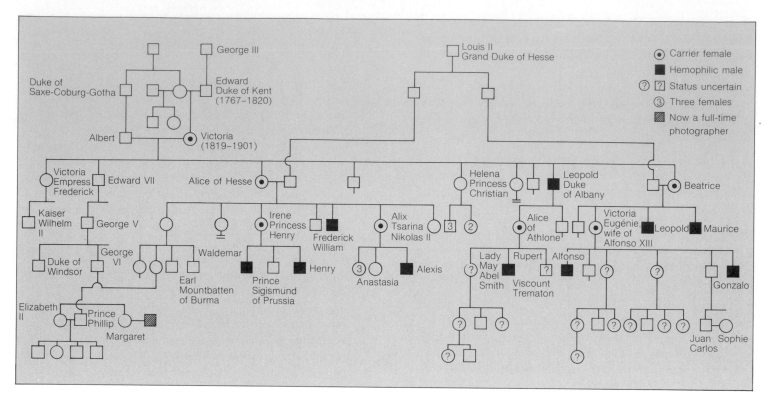

Figure 9.15 Descendants of Queen Victoria showing carriers and afflicted males that possessed the X-linked gene conferring the disease hemophilia. (After V. McKusick, *Human Genetics*, Second edition, © 1969. Reprinted by permission of Prentice-Hall, Inc. Englewood Cliffs, N.J.)

blood-clotting time is essentially normal, because the non-mutated allele on her normal X chromosome codes for enough of the normal protein. But if that woman gives birth to a son, he has a fifty percent chance of inheriting the X chromosome bearing the mutant allele. In human males, recall, the X chromosome is paired with a Y chromosome. Hence, if the mutant allele *is* inherited, it will be the only one that the male has for this trait. He will be unable to produce functional blood-clotting protein. Without medical attention, a cut, bruise, or internal bleeding could lead to death.

The mutant gene for hemophilia is rare in the human population. And many more males than females are afflicted by its phenotypic consequences. Only when a female who is a carrier of the defective gene marries a hemophilic male can daughters be born with the disease. Because the

gene occurs so rarely, such births are not likely to occur unless hemophilic males marry close relatives.

Hemophilia was a recurrent disease among the royal families of Europe during the nineteenth century. Queen Victoria of England was a carrier of the mutant gene. Unfortunately, of her sixty-nine descendants, eighteen were hemophilics or mutant-gene carriers (Figure 9.15). Particularly for Russia, the mutant gene and the cast of characters it indirectly drew together—Czar Nicholas II, Czarina Alexandra (a carrier who was a granddaughter of Victoria), their hemophilic son, and the power-hungry monk Rasputin who turned the disease to his political advantage—helped catalyze events that essentially brought an end to dynastic rule in the Western world. The historical novel *Nicholas and Alexandra* is a poignant account of these afflicted individuals and their children.

COMMENTARY

On Human Genetic Disease—Stopgap Measures Called Phenotypic Cures

Researchers have traced many of the diseases afflicting humans to genetic mutations, and we spend considerable time and money trying to find ways of getting around the consequences of these diseases. Until recently, our only approach to curing them has been at the phenotypic level. In a phenotypic cure, something intervenes to prevent the expression of a mutant allele. If the mutant allele cannot be expressed, then neither will the symptoms of the disease be expressed. In all phenotypic cures, of course, the mutant allele remains in the population.

Consider the disease phenylketonuria (PKU), in which a mutant allele codes for a defective enzyme. The normal enzyme converts the amino acid phenylalanine to the amino acid tyrosine. In individuals homozygous for the defective allele, phenylalanine accumulates and is diverted in large amounts to other pathways which gives rise to compounds in amounts large enough to interfere with normal body functioning. One of the compounds, phenylpyruvic acid, can be detected in the urine. In children afflicted with PKU, high concentrations of this compound can damage the central nervous system and lead to mental retardation. If diagnosed early enough, however, the disease symptoms can be alleviated simply by placing the individual on a special diet—one that provides only as much phenylalanine as required for protein synthesis, so that the body is not called upon to dispose of excess amounts. Aside from dietary restrictions, afflicted children can lead normal lives. Most hospitals in the United States routinely check all newborns for PKU, so the outward signs of the disease are rapidly disappearing.

Several other genetic diseases have yielded to phenotypic cures. Galactosemia leads to blindness and mental retardation. The disease is a result of a nonfunctional allele coding for an enzyme needed to break down the sugar galactose. If diseased children are detected early enough and placed on a synthetic (milk-free) formula, they grow up symptom-free. Some forms of diabetes are inherited disorders. In diabetes, the afflicted individuals cannot produce enough insulin, a hormone that is vital for cellular absorption of glucose. Their cells become starved for nutrients, and glucose levels rise in the blood. Symptoms may progress from weight loss to eventual brain damage, coma, and death. Yet children with diabetes can usually lead relatively normal lives through a combination of controlled diet and regular insulin injections.

For many genetic diseases, there is no phenotypic cure. Sometimes one or both parents come from families with a history of one of these diseases. They may worry about giving birth to an afflicted individual. Through genetic counseling, tests are made available to such individuals to help them determine whether they are heterozygous carriers, even though they themselves are disease-free. If both are found to be carriers, they may elect to employ birth control measures and adopt children. If the possibility that they will have a diseased child is established and if they still wish to proceed—or if the suspected disease is one for which the presence of the heterozygous state cannot be detected—another option remains. Through amniocentesis (sampling of the fluid in the mother's uterus), it is possible to obtain a few skin cells from the fetus at an early stage of pregnancy. The cells can be grown in sterile culture dishes and tested for the presence of any one of more than a hundred known genetic diseases. If the fetus is found to have a severe disorder, the parents may elect to request an abortion (an induced expulsion of the fetus from the uterus). Such decisions are bound by ethical considerations. The role of the medical community must be to provide information that the prospective parents need in order to make their own choice, which must be consistent with their own values within the broad constraints imposed by society.

In the Commentary section of the next chapter, more will be said about human genetic disease and the prospects of effecting cures through modifications to the genotype itself.

INTERACTIONS BETWEEN GENES IN THE EXPRESSION OF PHENOTYPE

Dominance Relations and Multiple Alleles

It was Mendel's genius to propose that the substance of heredity comes in distinct units that retain their physical identity during the life cycle of an individual. However, the average ratios obtained from his monohybrid crosses (3:1) and dihybrid crosses (9:3:3:1) were fairly clear-cut. The reason is that the dominance pattern for the particular traits studied arises through segregation of a fully dominant allele from a fully recessive allele in heterozygous offspring. Since Mendel's time, studies have shown that the possibilities are much more interesting than these cases of "all-or-nothing" dominance would suggest.

For instance, when homozygous red-flowering and white-flowering snapdragons are crossed, the first-generation plants all have *pink* flowers (Figure 9.16). (When the F_1 plants are subsequently crossed, their offspring have red, pink, or white flowers in about a 1:2:1 ratio.) This is an example of **incomplete dominance**, in which the activity of a so-called dominant allele is not completely able to mask the expression of a recessive partner. Apparently one "red" allele is not enough to form sufficient pigment to make the flowers appear red, as it does in homozygotes.

As another example, the blue Andalusian chicken is a phenotypic outcome of a cross between homozygous white-feathered and homozygous black-feathered parent birds. Neither the black nor the white allele is dominant. Both alleles are expressed in a mosaic color pattern that appears blue because of the way it refracts light (Figure 9.16). The pattern is an example of **codominance**, in which the characteristics of *both* phenotypes appear.

The preceding examples tell us that there can be a range of dominance between two alleles. Beyond this, there also can be a hierarchy of dominance among all the alternative forms that can exist for a given type of gene. Each gene has its own **locus**, a particular location on a chromosome. In diploid organisms, the allele at one locus specifies one of perhaps many possible forms of a trait. The allele at the matching locus on the homologous chromosome does the same. For instance, a gene locus for *Drosophila* eye color might bear a red or white allele—or an apricot, cherry, wine, or coral allele. Its partner, too, might bear any one of these possible alternatives. Red is the dominant phenotype, white is recessive, and the phenotypic expression of each of the other alleles listed is intermediate between the two. For any given population of organisms, all the possible alternatives for a given gene locus make up a **multiple allele system**.

Another multiple allele system exists for the **ABO blood group** locus on human chromosome 9 (Figure 9.8). This system has several alleles, with three—type A, type B, and type O—being the most common. Types A and B are codominant when paired with each other, and O is recessive when paired with A or B. This means that four phenotypes can occur based on these six possible genotypic combinations: A (either AA or AO), B (either BB or BO), AB, and O.

A and B alleles at the ABO locus code for enzymes that attach certain substances to human red blood cells. The substances act as **antigens** (they elicit a defense response when they enter the bloodstream of another individual of unlike blood type). Their invasion triggers the production or mobilization of molecules called **antibodies**, which chemically combine with specific antigens and thereby inactivate them (Chapter Seventeen). The body's ability to recognize and reject incompatible substances has been used to identify blood types in humans. Such iden-

Figure 9.16 Examples of the interactions between genes in the production of phenotype.

(**a**) Codominance at one gene locus. A white-feathered and a black-feathered parent chicken can give blue Andalusian offspring. Some feather areas are fine mosaics of black and white, which appear to be blue because of the way they refract light. Neither allele is dominant; both phenotypes are expressed.

(**b**) Incomplete dominance at one gene locus. Red-flowering and white-flowering parent snapdragons can give pink F_2 offspring. The red allele is only partially dominant in the heterozygous state.

(**c**) Random inactivation of chromosomes. In this female calico cat, one X chromosome carries a black allele for hair color and the other carries a yellow allele. In different body parts, random patches of these two colors occur, depending on which of the two chromosomes is active in the region. (The white patches result from interaction with *another* gene pair that determines whether any color appears at all.)

(**d**) Environmental effect on gene expression. In Siamese cats, fur on the paws, ears, and nose is darker than on the rest of the body. These regions normally are cooler than the main body. Some cats are homozygous for a recessive form of a key gene involved in the formation of the dark pigment melanin. The enzyme produced by this recessive allele is heat-sensitive: it is less active at warmer temperatures. Hence the lighter fur color on warmer body parts.

(**e**) Interactions between different genes. Comb shape in poultry can vary depending on interaction between two gene pairs. With complete dominance at the gene locus for pea comb and at the gene locus for rose comb, the interaction between dominants gives walnut combs. With complete recessiveness at both loci, the interaction gives single combs.

e (pea comb) (rose comb) (walnut comb) (single comb)

tification is vital for individuals requiring blood transfusions. Blood of the wrong type can cause clumping as well as bursting of red blood cells, which can lead to fever, jaundice, and tissue damage. It happens that type A persons always have antibodies that act against type B antigens. Hence a sample of their blood will cause clumping (agglutination) of type B and type AB blood cells. Similarly, a type B person has antibodies that react against type A antigens, which causes clumping of type A and type AB blood cells. A type AB person has neither antibody, which means that type AB individuals can receive blood from any donor. Because type O red blood cells have no antigen, type O blood can be given to anyone.

Interactions Between Different Genes

So far, most of our examples of heritable traits have been limited to ones that are largely controlled by a single gene locus. They have been fairly simple examples, for phenotype can be clearly related to interaction between two alleles. But even though a single gene may have a key effect on some trait, it does not have exclusive dominion over the expression of any trait. Part of the reason is that the environment influences phenotypic expression. In addition, genes simply do not exist by themselves. Their activity and inactivity depend on many genes, acting in concert to produce some effect on phenotype.

For example, many genes may have a small positive or negative effect on the same trait. Their interaction can produce a phenotype that falls somewhere in a continuous range of variation. Human skin color, for instance, is not limited to a few colors. There is a continuous and richly varied range from "whites" to "browns" and "blacks." Human skin color is known to be influenced by a number of similarly acting but different gene loci that control the formation of melanin pigment molecules. Their effect is roughly additive, with the intensity of skin pigmentation being determined by the total number of active alleles at all the different loci. In such **polygenic inheritance**, the small effects of different genes together produce observable, quantitative variations in some trait.

It is important to understand that polygenic inheritance is not a matter of "blending," a term that suggests loss of the original identity of hereditary traits. Rather, *all genes retain their physical identity, regardless of the phenotype produced by their combined positive and negative effects.*

Another kind of interaction between genes is called **epistasis**. Here, one gene *pair* masks the effect of one

or more other pairs. For example, many genes modify coat color in mammals. Among them is a gene pair that governs whether given colors show up at all. Calico cats (Figure 9.16) are heterozygous for black and yellow alleles at a gene locus for coat color. At a different gene locus, calico cats also have a "spotting" allele. Expression of the spotting allele at a particular coat region masks either color. Where it does, these cats have white patches on their coats.

Multiple Effects of Single Genes

The picture being sketched here shows that different genes have major or minor influences on the expression of a given trait. Another aspect of this picture is that a single gene often can have a major or minor effect on more than one trait. The multiple phenotypic effect of a single gene is known as **pleiotropism**. Examples are well known through studies of gene mutations. Hans Grüneberg, for instance, identified a mutation in the gene coding for a protein that is required for cartilage formation. Cartilage happens to be one of the most common structural materials in animals, especially during early development. A mutation in this gene leads to a host of deformed characters, ranging from blocked nostrils, narrowed tracheal passageways, and thickened ribs, to a loss of elasticity in lung tissue. The action of this single gene, we may conclude, is a key step in many pathways for the normal development of many traits—with its effects being far removed from the time and place of its initial activity.

Sickle-cell anemia provides an example of pleiotropic effects in humans. For most adult humans, the red oxygen-carrying pigment in blood is hemoglobin A, or HbA. Some individuals carry a variant form known as HbS. It is the result of a single substitution of one amino acid (valine) for another (glutamic acid) in the hemoglobin molecule. The variant molecule can still carry oxygen. But when HbS molecules give up their oxygen cargo to other cells in the body, they interlock chemically with one another and stack up like long, rigid rods. This causes the red blood cells to become distorted into "sickle" shapes (Figure 9.17). The deformed cells clump up in capillaries. There they block normal movement of oxygen into interstitial fluid and the removal of carbon dioxide from it. Severe damage to many internal tissues and organs soon follows. Because heterozygous (HbA/HbS) individuals also have a normal allele that is fully functional, they show few disease symptoms. It is the homozygote (HbS/HbS) who suffers serious consequences. More will be said about sickle-cell anemia in Chapter Twenty-Five.

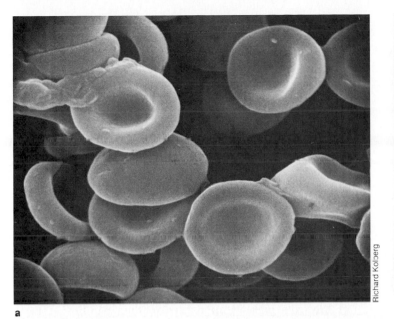
a

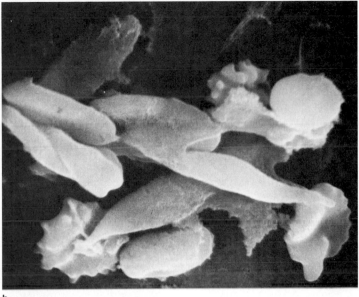
b

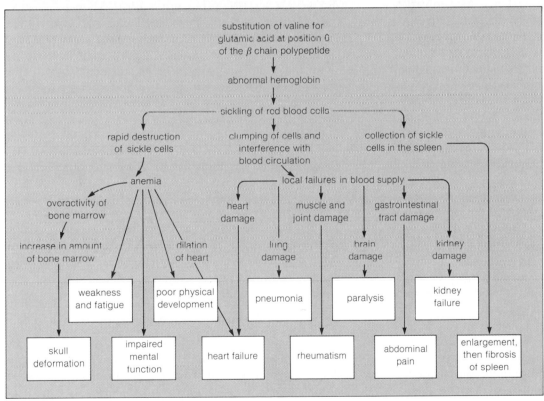

substitution of valine for glutamic acid at position 6 of the β chain polypeptide

↓

abnormal hemoglobin

↓

sickling of red blood cells

rapid destruction of sickle cells

clumping of cells and interference with blood circulation

collection of sickle cells in the spleen

anemia

local failures in blood supply

overactivity of bone marrow

heart damage

muscle and joint damage

gastrointestinal tract damage

increase in amount of bone marrow

dilation of heart

lung damage

brain damage

kidney damage

weakness and fatigue

poor physical development

pneumonia

paralysis

kidney failure

skull deformation

impaired mental function

heart failure

rheumatism

abdominal pain

enlargement, then fibrosis of spleen

c

Figure 9.17 (**a**) Normal red blood cells and (**b**) cells characteristic of the disease sickle-cell anemia. Because of their abnormal, asymmetrical shape, sickle cells do not flow smoothly through fine blood vessels. They pile up in clumps that block blood flow. The tissues served by the blood vessels become starved for oxygen and nutrients even as they become saturated with waste products. (**c**) Possible pleiotropic effects in sickle-cell homozygotes. (After Neel and Schull, in *Genetics,* Second edition. Copyright © 1976, Monroe W. Strickberger, Macmillan Publishing Co., Inc.)

ENVIRONMENTAL EFFECTS ON GENE EXPRESSION

We have looked at some of the ways in which phenotype can be altered through chromosome mutation, through point mutation, and through interaction between alleles and between genes at different loci. One more factor has profound effects on phenotype, and that is the environment. Throughout any individual's life cycle, genes must interact with the environment in the expression of phenotype. Genes provide the chemical messages for growth and development, but neither growth nor development can proceed without environmental contributions to the living form.

Earlier chapters pointed out the kinds of environmental factors—water, essential ions, food, suitable temperature—that affect the functioning of organisms. Here we can give examples of variations in the degree to which genotype is expressed in individuals confronted with environmental change. Simple changes in diet, for instance, lead to variable gene expression. The body fat in rabbits appears yellow in individuals homozygous for a certain recessive gene—but only if the rabbits also eat leafy plants containing the pigment xanthophyll. Otherwise body fat appears white. Or consider the effect of temperature on coat color in Siamese cats. In these animals, the main pigment molecule is a brown-black form of melanin. Melanin formation begins when the amino acid tyrosine undergoes successive modifications into a large molecule containing repeating units of the pigment. Each step along the way must be catalyzed by a specific enzyme, one of which is tyrosinase. One allele of the gene locus coding for tyrosinase produces a heat-sensitive form of the enzyme. In cats homozygous for this mutant allele, dark fur occurs in the relatively cool extremities (paws, ears, tail, nose). But light fur is present on warmer body parts, where the enzyme is less active.

Examples of this sort should not lead you to believe that interactions between genes and the environment are always clear-cut. More often the environmental factor is pervasive in ways that defy analysis. Even in the case of identical twins, effects of the intrauterine environment in which they develop cannot be reduced to simple terms. At the very least, twin embryos occupy different positions in space and their bodies are oriented differently. Also, their attachment sites to the mother and the nature of their connections to the maternal bloodstream cannot be identical. These and a multitude of individually small environmental effects undoubtedly influence their phenotypic potential in different ways even before birth.

The expression of any gene is influenced by more than the external environment. It is influenced also by the nature of changes within the organism itself. For example, you know that some genes are sex-linked, being carried on an X chromosome. All other genes are carried on autosomes and are present equally in both sexes. Yet, some autosomal genes are expressed in one sex but not the other! Such **sex-influenced genes** presumably are expressed beginning with the onset of hormonal activity. Male and female sex hormones produce different phenotypic expressions in males and females. Such genes influence the appearance of horns in some sheep species, and baldness in humans, to give two examples.

Or consider that each cell of an organism has the same set of genes, yet different cells proceed down different developmental pathways. For instance, during development of some kinds of animal embryos, paddlelike appendages are transformed into sets of fingers and sets of toes. Some of the cells making up tissues in the "paddle" seem to die on cue according to some internal scheme. What we call **aging** is really a final developmental stage for the specialized cells of multicellular organisms. Then, profound cellular changes occur through alterations in enzyme activity, hence in the activity of genes coding for the enzymes themselves. What transpires in the internal environment to bring about changes in gene activity? That we do not know, although Chapter Eleven looks at some intriguing possibilities.

What are the main points to remember from this section on factors that influence the expression of phenotype? They may be summarized in this way:

---

1. Gradations of dominance exist between two alleles for a given gene locus, so that the expression of one or both may be fully dominant, or one may be incompletely dominant over the other.

2. A hierarchy of dominance can exist among all the alternative forms that may exist for any given type of gene.

3. The activity of a gene in space and time may be influenced by other genes, with a sum total of their positive and negative actions producing some effect on phenotype.

4. The activity of a single gene may have major or minor effects on more than one trait.

5. The environment has profound influence over the expression of all genes and their contributions to phenotype.

PERSPECTIVE

We have, in this chapter, described observations and experiments that have spanned more than a century. We have mentioned a few important turning points in the road leading to our current understanding of the chromosomal basis of inheritance. Mendel received only belated recognition for his part in the journey; Morgan received a Nobel Prize. The names of other researchers are repeated almost arbitrarily and most are, in the long run, destined not to be repeated at all. Why is this so? The amount of biological information has grown to such monumental dimensions that there simply is too much for any one person to absorb.

Any introductory book on science is thus a lesson in compromise, a balancing between a description of some individual making a discovery, and the details of the discovery itself. Today, Mendel has his epithet in "Mendelian genetics." Morgan, Sturtevant, Bridges, Muller, and so many others are lumped together as "classical geneticists." Flemming and Van Beneden are, if mentioned at all, quickly dismissed as "nineteenth-century microscopists."

You might conclude, from all of this, that science is a dehumanizing endeavor with recognition bestowed only on the select few. A conclusion of that sort would be an artifact of having a limited encounter with the ways of science. The volume of biological information *is* monumental. But if by chance some part of this brief story caught your interest, then think about making further inquiries on your own into the literature. The names of individuals mentioned here are signposts along past journeys. And emerging through their own writings are the triumphs, the frustrations, the clear logic, and the stumblings of a very *human* enterprise. With commitment to exploring one line of research or another comes the time for probes of this sort, for then the monument of information is approached selectively. This is the way that specialists in science link minds with unseen predecessors; this is the way of continuity, and remembrance.

Readings

Grüneberg, H. 1963. *The Pathology of Development.* Oxford, England: Blackwell Publications.

Lerner, I., and W. Libby. 1976. *Heredity, Evolution, and Society.* San Francisco: Freeman. Intelligent introduction to genetics principles as they apply to society.

McKusick, V. 1978. *Mendelian Inheritance in Man.* Fifth edition. Baltimore: Johns Hopkins Press. A descriptive catalog of inherited human traits. Includes references under various entries.

Strickberger, M. 1976. *Genetics.* Second edition. New York: Macmillan. Probably the single most authoritative source of information on Mendelian genetics, from principles to applications.

Wolfe, S. 1981. *Biology of the Cell.* Second edition. Belmont, California: Wadsworth. Excellent discussions of chromosomes structure and function, as well as chromosomal malfunctions.

Review Questions and Problems

1. State the Mendelian principle of segregation. Does segregation occur during mitosis or meiosis?

2. Distinguish between the following terms:
   a. Genes and alleles
   b. Dominant and recessive
   c. Homozygous and heterozygous
   d. Genotype and phenotype

3. Define cross-fertilization, and distinguish between a monohybrid and a dihybrid cross. What is a testcross?

4. State the Mendelian principle of independent assortment. Does independent assortment occur during mitosis or meiosis?

5. State the difference between sex chromosomes and autosomes. In human females, the twenty-third pair of chromosomes is designated _____; in males, _____.

6. What is linkage? How is linkage related to independent assortment?

7. Define crossing over and its consequence. How does crossing over relate to the ability to map gene positions on chromosomes?

8. List the main ideas of the chromosomal theory of inheritance.

9. Give four examples of genetic diseases. How do genetic diseases usually arise?

10. How does polygenic inheritance differ from the notion of "blending" of heritable traits?

11. List the main factors influencing expression of phenotype.

12. *Albinism* is a heritable disorder that can occur among many animals and plants. It arises from a single recessive gene, and is characterized by an absence or near-absence of pigment molecules. Humans who are homozygous recessive for the trait have extremely light skin, white hair, and red or pink eyes. (With absorptive pigment absent from the retina, red light is reflected from blood vessels in the eye.) If a man whose mother was albino marries a woman whose father was albino, what is the probability that their first child will be albino? Their second? Their fourth?

13. In a cross of $AaBb \times Aabb$, what fraction of the offspring will be homozygous recessive for both pairs of genes? What fraction will be homozygous dominant for A? Homozygous dominant for B?

14. Children suffering from Down's syndrome possess forty-seven chromosomes rather than the normal forty-six (twenty-

three pairs). Suggest a mechanism that can produce an individual with an extra chromosome.

15. In zinnias, a cross of a red-flowered plant with a white-flowered plant produces all pink-flowered progeny. What types of offspring, and in what proportions, are expected from the following crosses:

    a. Red × red

    b. Pink × white

    c. Red × pink

    d. Pink × pink

16. Among the various species of wheat, some have a somatic chromosome number of fourteen. None of these species is economically important. Other species have twenty-eight chromosomes. Among these is a species (hard wheat) whose flour is used in making spaghetti. Finally, forty-two chromosome wheats are known. Common bread wheat, which is cultivated worldwide, is a forty-two chromosome species. Suggest how these related wheat species came into existence.

Answers to Problems

12. $Aa \times Aa = \frac{1}{4}\ aa$. (True for *each* child.)

13. Homozygous recessive for both pairs of genes ($aabb$) = $\frac{3}{8}$

    Homozygous dominant for $A = \frac{1}{4}$

    Homozygous dominant for B, none

14. Nondisjunction of a chromosome pair at meiosis, producing an $n + 1$ gamete

15.   a. $RR \times RR$ = all red

    b. $Rr \times rr = \frac{1}{2}$ pink, $\frac{1}{2}$ white

    c. $RR \times Rr = \frac{1}{2}$ red, $\frac{1}{2}$ pink

    d. $Rr \times Rr = \frac{1}{4}$ red, $\frac{1}{2}$ pink, $\frac{1}{4}$ white

16. Polyploidy. Two wild 14-chromosome species hybridize, and produce a 14-chromosome hybrid. Such hybrids are usually sterile. However, chromosome doubling (as with colchicine treatment) gives a 28-chromosome fertile wheat (chromosomes are now properly paired). Then another hybridization with 14-chromosome wheat gives a 21-chromosome sterile ($3n$) hybrid. Again, chromosome doubling gives the 42-chromosome ($6n$) bread wheat.

Each cell, in the time and space of its existence, deploys thousands of different enzymes in translating genotype into phenotype—in translating hereditary instructions into specific physical traits. Yet what could be the nature of the instructions for building the enzymes themselves, at the exact time and to the precise extent required? Cells certainly don't have an unlimited bank of instructions. If they did, then all cells—hence all multicellular organisms—would be able to survive just about anywhere, simply by churning out a new assortment of enzymes to meet each new challenge. But there is a direct and inescapable relationship between the living form and the environment in which it survives. For instance, many organisms abound in coastal waters. When accidentally washed onto the beach, where they are fully exposed to air and deprived of water, they perish. Still others crowd the water's edge, but they perish when they are accidentally swept into the water. None of them has any means whatsoever of instantly creating different kinds of enzymes to meet different conditions. *Somehow, each cell must come into the world with a specific set of instructions for survival in a given kind of place.* To the extent that the instructions help their bearer respond to both unchanging and unanticipated changes in the environment, they have the potential to be passed on. Thus,

---

Hereditary instructions must be of a sort that ensures constancy in structure and function, even while allowing room for subtle change.

From genetic research, we know that hereditary instructions reside in genes, which are located on chromosomes. *But in what molecular form do they exist, and how do they work?* As you will see, through the answers to these questions, we have come to appreciate the true meaning of life's basic unity and extraordinary diversity. Beyond this, retracing how the answers themselves were found may give you a glimpse of biologists in action.

THE SEARCH FOR THE HEREDITARY MOLECULE

In 1868, the German scientist Johann Miescher isolated an acidic substance from the nuclei of certain cells. He had discovered what came to be known as deoxyribonucleic acid, or DNA. His source of this substance was so unromantic as to belie the elegance of his discovery, for Miescher had deliberately chosen cells that were almost entirely nuclear in composition, with very little cytoplasm.

10

THE MOLECULAR BASIS OF INHERITANCE

live nonpathogenic strain (R)

mice live

live pathogenic strain (S)

mice die

heat-killed pathogenic strain (S)

mice live

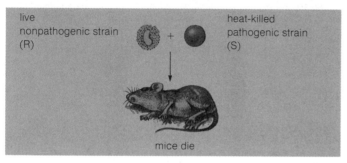
live nonpathogenic strain (R) + heat-killed pathogenic strain (S)

mice die

Figure 10.1 Griffith's experiments with harmless (R) strains and disease-causing (S) strains of *Diplococcus pneumonia*, as described in the text. (You may be wondering why the S form is deadly, and the R form harmless. The killer S form shrouds its cell wall with a smooth protective capsule, which resists attack by the host cell's normal defenses. The R form has no such capsule; hence the host's defense system destroys the R form before it can cause the disease.)

The cells he used came from the pus of open wounds and, later, the sperm of a fish.

At the time, few appreciated what this nucleic acid might mean for the flow of life, because only a few were beginning to suspect the nucleus as being the cell's hereditary control center. In fact, more than seventy-five years would pass before DNA would be recognized as a substance of profound biological importance.

The Puzzle of the "Transforming Principle"

One such research line involved studies of *Diplococcus pneumonia*, a bacterium that causes the lung disease pneumonia. In 1928 a British medical officer, Fred Griffith, isolated two distinct strains of the bacterium. He called one strain the "S" form, because it formed colonies having a smooth

appearance. The other strain was called the "R" form because of the rough surface appearance of its colonies on culture plates (Figure 10.1). Griffith began by determining the effects of these separate strains on laboratory mice. Living S cells formed one control group. When they were injected into mice, the mice promptly contracted pneumonia and died. Blood samples removed from the mice were teeming with bacteria; the S form had to be a disease-causing strain. Living R cells formed a second control group. When they were injected into mice, nothing happened; R cells were harmless.

Armed with these results, Griffith began experimenting. He heat-killed S cells, injected them into mice, and found that nothing happened. Then he mixed live R cells with *dead* S cells, and injected them into mice. Incredibly, the combination led to pneumonia and death—and blood samples from the dead mice were teeming with *live* S cells!

What was going on? Maybe heat-killed S cells weren't really dead. But if that were true, the mice injected with heat-killed S cells alone (the first experimental group) should have contracted the disease, too. Or maybe R cells had mutated into killer S forms. But if that were true, then why didn't the same thing happen in the control group of mice receiving only living R cells? It had to be that the dead S cells transferred their ability to cause infection to the harmless R cells. Through further experiments, it became clear that the "harmless" cells had become permanently transformed—for all their offspring also caused infections. *Thus the transformation had to involve a change in the bacterium's hereditary system itself.*

News of Griffith's results reached an American bacteriologist, Oswald Avery, and his colleagues. For a decade they worked on identifying the chemical nature of the "transforming principle" underlying the permanent change in the bacterial hereditary system. In 1944, they reported that DNA apparently was the unit by which this heritable feature was passed on from one bacterium to another. Their evidence, however, was almost uniformly ignored. The scientific community was adhering to a notion (by then a generation old) that only proteins could serve as hereditary molecules. Proteins, after all, are diverse and complex. In contrast, nucleic acids were thought to have a simple repeating-unit structure, like starch or cellulose; how, then, could nucleic acids encode all the instructions necessary to build and maintain any living organism? Despite this prevalent notion, Avery's message did not go entirely unnoticed. The American biochemist Erwin Chargaff, for one, came under its influence. As you will see later, his DNA studies would yield crucial information about the chemistry of heredity.

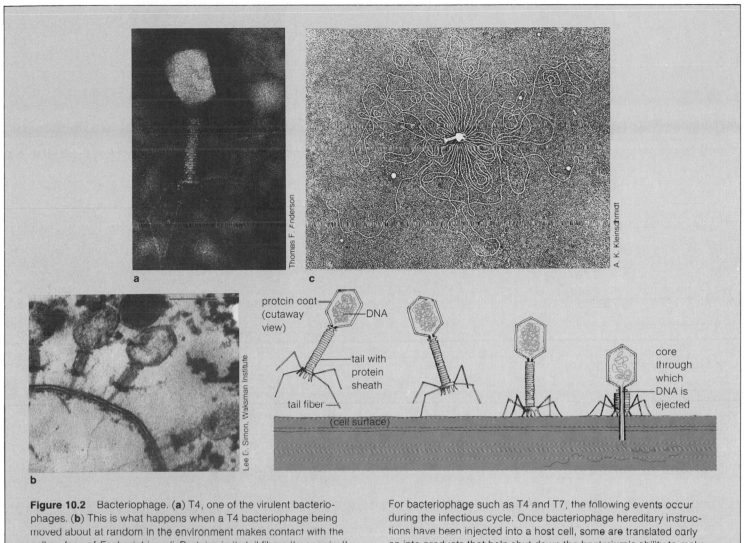

Figure 10.2 Bacteriophage. (**a**) T4, one of the virulent bacteriophages. (**b**) This is what happens when a T4 bacteriophage being moved about at random in the environment makes contact with the cell surface of *Escherichia coli*. Proteins in its tail fibers "recognize" proteins in the bacterial cell surface; the sheath contracts, and the contents of the bacteriophage head are injected into the victim. (**c**) T2 bacteriophage. In this now-famous electron micrograph, the contents of this bacteriophage's head appear to be a single tangled strand of seemingly incredible length, given the small size of the place from which it was released!

For bacteriophage such as T4 and T7, the following events occur during the infectious cycle. Once bacteriophage hereditary instructions have been injected into a host cell, some are translated early on into products that help shut down the bacterium's ability to make protein. Other parts of those instructions code for enzymes that digest the bacterium's DNA. Nucleotide subunits thus freed are used in building bacteriophage DNA. Other instructions are used for building new bacteriophage. Lastly, enzymes produced by the bacteriophage digest the host cell wall, which bursts and releases a new infectious generation. This sequence of events is a *lytic pathway*.

Enter Bacteriophage

Still more clues were about to converge. By the mid-nineteenth century, it had become clear that many infectious diseases (including influenza and the common cold) were not bacterial in origin. As tiny as bacteria are, some disease agents had to be smaller still. The structure of these agents, which came to be called viruses, was revealed through electron microscopy. A **virus** is a kind of infecting particle composed of nucleic acid and protein. Some viruses are spherical, others are rodlike, and still others have a distinct head, sheath, and tail fibers (Figure 10.2).

Figure 10.3 Hershey–Chase experiments to determine the destination of DNA and protein from T2 bacteriophage during infection of *E. coli.*

(**a**) One population of bacterial cells was grown on a medium containing a radioactive isotope of phosphorus (^{32}P), a building block for DNA but not for protein. Another population was grown on a medium containing a radioactive isotope of sulfur (^{35}S), a building block usually found in protein but not in DNA. The molecules in which ^{32}P and ^{35}S became incorporated were thereby tagged, or radioactively labeled. They could be distinguished, by certain methods, from unlabeled molecules in the cells being studied.

(**b**) Bacterial cells so labeled were exposed to and infected by T2 bacteriophages, which entered the lytic pathway (Figure 10.2).

(**c**) The bacteriophage progeny released upon lysis were radioactively labeled, for their protein and DNA had to have been assembled only from molecules available in the host cells.

(**d**) The labeled progeny were allowed to infect fresh, unlabeled bacteria. During infection, they remained attached to the host cell surface, as shown in Figure 10.2b.

(**e**) Suspensions of cells bearing their attackers were osmotically shocked by being churned rapidly in a kitchen blender. The shearing forces caused the bacteriophage bodies to break away from the cell surface. In the first experiment, analysis showed that the bacteriophage body, not the host cell, contained the ^{35}S. In the second experiment, the bacterial cells contained ^{32}P; the labeled DNA had been injected into the host cells.

(**f**) In other samples of the infected cells, the bacteriophage life cycle was allowed to proceed.

In the first experiment, analysis showed that bacteriophages released upon lysis contained very little ^{35}S. In the second experiment, however, bacteriophage contained radioactive DNA. The DNA, not protein, was being transmitted through generations of cells.

(These experiments did leave some room for doubt, for a small amount of protein is injected into bacterial cells along with DNA during the normal infection process. Later experiments used bacteria stripped of their cell walls, and pure labeled DNA stripped from bacteriophage. The results of these "clean" experiments did confirm Hershey and Chase's conclusion that DNA is the genetic material.)

first experiment

second experiment

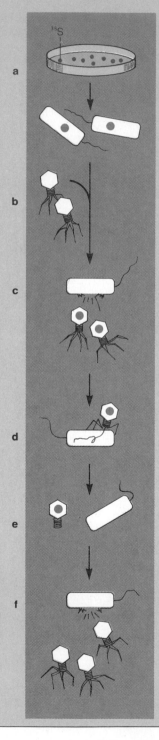
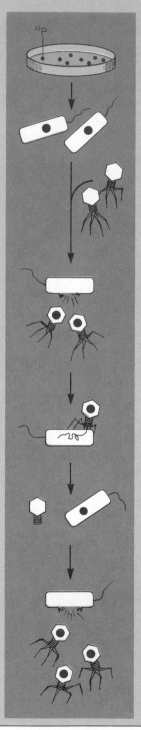

In the 1930s, Max Delbrück, Alfred Hershey, and Salvador Luria initiated research into the hereditary system of a kind of virus called **bacteriophages**. Bacteriophages infect certain types of bacterial cells, divert the metabolic machinery to other tasks, and eventually destroy their hosts. These awful guests go by such names as T2, T4, and T7. *Escherichia coli*, a normally harmless bacterium living in your gut, is one such target host.

Bacteriophages do not deliberately go about ambushing victims in a culture dish or in a plant or animal. They haven't any means whatsoever of moving about on their own. The only way that they can infect a potential host is to bump accidentally into it. But once contact does occur, a prescribed chemical behavior is set in motion. Bacterial surface receptors are no less than a chemical invitation to specific bacteriophages. If those bacteriophages have complementary molecules in their tail fibers, they can bind to the bacterial cell surface. Once binding occurs, the bacteriophage sheath contracts and the contents are injected into the victim. Within sixty seconds a remarkable thing happens. The bacterium stops making most of the things it normally would make, and produces an entirely foreign set of enzymes. Subsequently, all the bacterium's enzyme-mediated activities are devoted to building new bacteriophages! In less than a half hour, the infected cell undergoes **lysis**: its cell wall is degraded (by bacteriophage-produced enzymes) and the host cell bursts. By subverting the host's synthetic machinery, the bacteriophage has managed to reproduce itself.

For Alfred Hershey and his colleague Martha Chase, the intriguing question became this: What was the bacteriophage injecting into the host? Whatever it was, it had to be the chemical blueprint specifying "build bacteriophages." Knowing that a bacteriophage contains only DNA and protein, they narrowed their question to whether the injected substance was protein, or DNA, or a combination of the two. To find the answer, they relied on a chemical difference between the two kinds of molecules.

Protein contains sulfur but no phosphorus. DNA contains phosphorus but no sulfur. It happens that both chemical elements have radioactive isotopes (Chapter Three). Say that bacterial cells are grown on a culture medium containing radioactive sulfur. The cells must necessarily take up this isotope and use it as a building block in assembling amino acids, hence proteins. Suppose now that the cells become infected with bacteriophages. Following lysis, the new bacteriophage particles should contain labeled protein. Why? They can be built *only* of materials (in this case, including radioactive ones) available in their hosts. If these particles are used to infect unlabeled bacteria,

then it should be possible to determine whether the radioactive sulfur is left behind in the bacteriophage body or ends up in the bacterial cell as hereditary material. The same kind of experiment can be performed to determine the destination of radioactive phosphorus, hence of DNA.

Hershey and Chase performed these experiments. As Figure 10.3 shows, they found that the radioactive sulfur remained mostly outside bacterial cells, as part of bacteriophage bodies. They also found that the radioactive phosphorus ended up inside the host cells. More than this, they found that radioactive phosphorus was also present in the DNA of certain members of the next bacteriophage generation—but that the radioactive sulfur-labeled protein was not. Here was further proof that DNA, not protein, is transmitted as the hereditary molecule. Later work would show that some large viruses use RNA rather than DNA for their hereditary material. With this one exception, however, the following is true:

In almost all cases, DNA is the repository of chemical information governing inheritance.

DNA STRUCTURE: THE RIDDLE OF THE DOUBLE HELIX

Now the search was on to find out precisely how a DNA molecule is constructed. In that construction had to be the secret of life's capacity for self-reproduction. Recall, from Chapter Three, that only four different kinds of nucleotide bases are found in a DNA molecule. All four include the same five-carbon sugar, deoxyribose. Small numerals shown on its ring structure identify its five carbon atoms:

All four nucleotides also have the same phosphate group:

$$HO - \overset{\overset{\textstyle O}{\|}}{\underset{\underset{\textstyle O^-}{|}}{P}} - OH$$

But each nucleotide has one of four different nitrogen-containing bases. Two of these bases are single-ring *pyrimi-*

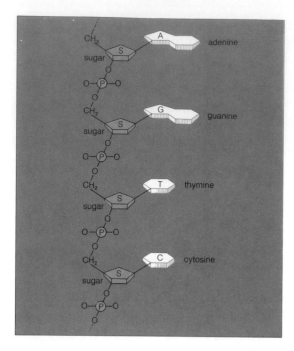

Figure 10.4 How nucleotides can become attached to one another in a chain structure. S designates the sugar ribose; P, the phosphate group.

dines, called cytosine and thymine. The other two are both double-ring *purines*, called adenine and guanine:

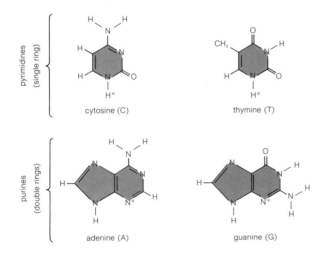

The asterisk indicates where each base bonds to the sugar ring structure. If you were to use a flat formula to show

how the phosphate group and one of these bases hook up with the sugar to form a nucleotide, you'd end up with something like this:

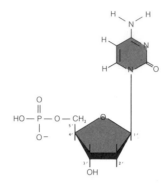

Notice how the phosphate group attaches to carbon 5' of the sugar ring structure. In a DNA molecule, nucleotides are strung together into a long chain, with each attached phosphate group hooking up with carbon 3' on the sugar ring (Figure 10.4). When nucleotides are strung together this way, the bases stick out to one side.

By the early 1950s, this much was known about the parts of DNA. In addition, through Chargaff's work, two portentous clues had been uncovered:

---

1. The four bases in DNA may vary greatly in relative amounts from one species to the next—yet the relative amounts are always the <u>same</u> among all members of a single species.

2. In every species, the amount of adenine present always equals the amount of thymine, and the amount of cytosine always equals the amount of guanine (A = T, and C = G).

At about the same time, Maurice Wilkins and his associates were using x-ray diffraction methods to determine DNA's structure. The atoms in a crystal of any substance can bend a narrow x-ray beam, and an atomic structure having a regular pattern bends the x-rays in a regular way. When a piece of film placed behind the crystal is exposed by the x-rays, a pattern of dots and streaks shows up on it. Each dot represents a beam that has been diverted by a given kind of repeating atomic unit. The distances and angles between these dots can be used to calculate the position of atomic groups relative to one another in the crystal.

One of Wilkin's coworkers, Rosalind Franklin, identified some intriguing aspects of DNA structure by analyzing its x-ray diffraction patterns. For one thing, the molecule had to be long and thin, with a constant 2-nanometer diameter along its entire length. For another, its structure had to be highly repetitive. Her data showed some structural element being repeated every 0.34 nanometer and another being repeated every 3.4 nanometers. Finally, the molecule had to be helical, with an overall shape like a circular stairway.

The promise of success was in the wind. It would be only a matter of time before someone would take the separate clues that had been accumulating and weave them together to find the answer. In 1953, at Cambridge University, James Watson and Francis Crick did just that. If the DNA molecule were indeed helical, it probably resembled the secondary structure of proteins. Linus Pauling had only recently identified protein secondary structure as being a single helical coil, held in place by hydrogen bonds. The important difference—and this was Watson and Crick's momentous insight—*was that the DNA helical coil had to be double, with two strands wound one around the other.*

The Pattern of Base Pairing

Watson and Crick knew that purines were double-ring structures, hence larger than pyrimidines. These nucleotides, they reasoned, could not be randomly arranged in DNA. If they were, then the molecule would bulge in purine-rich regions and narrow down in pyrimidine-rich regions—but Franklin's data suggested a uniform diameter for the whole molecule. Watson and Crick also knew that in terms of relative amounts, A=T and C=G. And no doubt purines must somehow be physically paired with pyrimidines in DNA. But *how* were they paired? The two researchers arranged and rearranged paper cut-outs of the nucleotides. They also checked with a chemist to determine the precise structure of each kind of nucleotide. Suddenly they came across the answer. In a certain orientation, adenine and thymine could form a pair of hydrogen bonds with each other—and in a very similar orientation, guanine and cytosine could form three hydrogen bonds with each other!

If *two* DNA chains were arranged in space so that their nucleotide bases faced each other, then hydrogen bonds might easily bridge the gap between them, like rungs on a ladder (Figure 10.5). Watson and Crick began constructing scale models of how this "ladder" might look. They found that the only arrangements possible in their model of a DNA double helix were purine-pyrimidine pairs: A—T and G—C! Chargaff's work was thus substantiated in what has become known as the **principle of base pairing**.

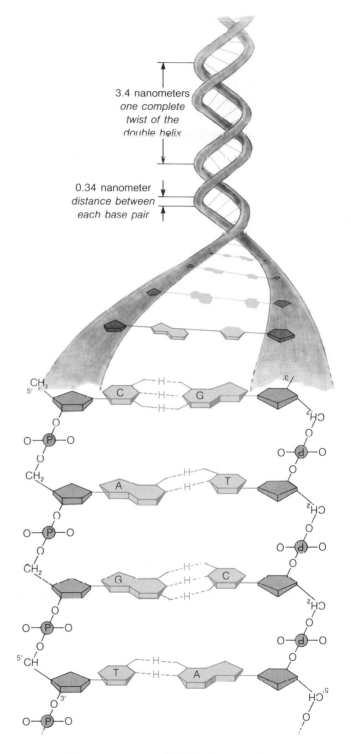

Figure 10.5 Representation of a DNA double helix. Notice how the two sugar–phosphate backbones run in opposite directions. This is the only arrangement in which one nucleotide base can become aligned with and bonded to its complementary base in the DNA molecule.

In the ladder model, the only way that purine-pyrimidine pairs could be aligned was to have two strands running in opposite directions and to twist each strand into a helix, as Figure 10.5 shows. Because the model was built to scale with known atomic sizes, it could be calculated that there was a base pair every 0.34 nanometer and a complete twist of the helix every 3.4 nanometers—and the helix diameter in the model came out to be exactly 2 nanometers. In all three measurements, then, the model fit the existing data.

With such base pairing, cytosine would always be present in proportions equal to guanine, and thymine in proportions equal to adenine, regardless of the species. Yet there still could be variation in the total amount of A—T relative to the total amount of G—C present in the DNA of different species. Why is this so? *Any pair could follow any other in the DNA chain, which means that the number of possible sequences for these two kinds of base pairs is staggering!* For instance, even in one small DNA region, the sequence might be:

```
T A T C T A C        G G G T G G C T
| | | | | | |   or   | | | | | | | |
A T A G A T G        C C C A C C G A
```

Thus the Watson–Crick model of DNA reflected at once the twin properties of unity and diversity required of the molecule of inheritance.

Unwinding the Double Helix: The Secret of Self-Duplication

Once DNA's double-helical structure had been deduced, Watson and Crick immediately saw how such a molecule might be duplicated prior to mitotic or meiotic division in a cell (Chapters Eight and Nine). If the two DNA strands were unwound from each other, their bases would become exposed to the nuclear environment. Such unwinding could occur readily enough, given that only relatively weak hydrogen bonds hold the two strands together. Assuming that appropriate enzymes were around, subunits (free nucleotides) could be gathered from the surroundings and attached to exposed bases on the two parent DNA strands. If the additions followed the pattern of base pairing, *then the only sequence that could be attached to one parent strand would have to be an <u>exact duplicate</u> of the base sequence occurring on the other parent strand.*

Figure 10.6 illustrates this view of DNA duplication. As you can see, when a region of the double helix is un-

wound, the two separated DNA strands can act as a pair of templates (or patterns), each being the complement of the other. *Each parent strand remains intact, and a new companion strand is assembled on each one.* One parent strand then winds up with a new strand, forming a double helix; and the other parent strand winds up with a new strand, forming another double helix. In each of the two molecules, one "old" DNA strand has been conserved as a partner for the newly formed strand.

Suppose there were some way that the "old" and newly synthesized DNA strands could be separately labeled. Then it would be possible to confirm this view of DNA duplication. In 1958, Matthew Meselson and Frank Stahl figured out a way to label DNA. Knowing that DNA contains nitrogen, they grew *E. coli* on a medium containing an isotope of nitrogen that is more dense than the isotope ordinarily found in the environment. They allowed the bacteria to grow and reproduce for many generations in order to obtain a population in which the "heavier" isotope would be the only nitrogen component in the DNA. They abruptly switched this population to a medium containing the more common isotope as the sole nitrogen source. At different intervals thereafter, they extracted DNA from the bacteria and subjected it to density-gradient centrifugation, a process described in Figure 10.7.

Analysis of the results showed that after one reproductive cycle, all the DNA molecules extracted from the first bacterial generation were hybrid: they contained equal amounts of the denser and the more common isotopes. After the second cycle, half the DNA molecules were hybrids—and half contained *only* the more common isotope. The experiment is illustrated in Figure 10.7. Its results were in agreement with the Watson–Crick hypothesis. Through many such experiments conducted with many different organisms, the semi-conservative replication of double-helical DNA is now known to be the nearly universal basis for duplicating the genetic material.

Enzymes Taking Part in DNA Duplication

Several classes of enzymes are known to play key roles in DNA duplication. Duplicative activities begin even before the whole DNA molecule unwinds. In circular prokaryotic DNA, some enzymes open up the double helix at a specific site (the initiation site). The two DNA strands are unwound in both directions from this point. In eukaryotes, some enzymes open up the double helix at many different points to expose many small regions

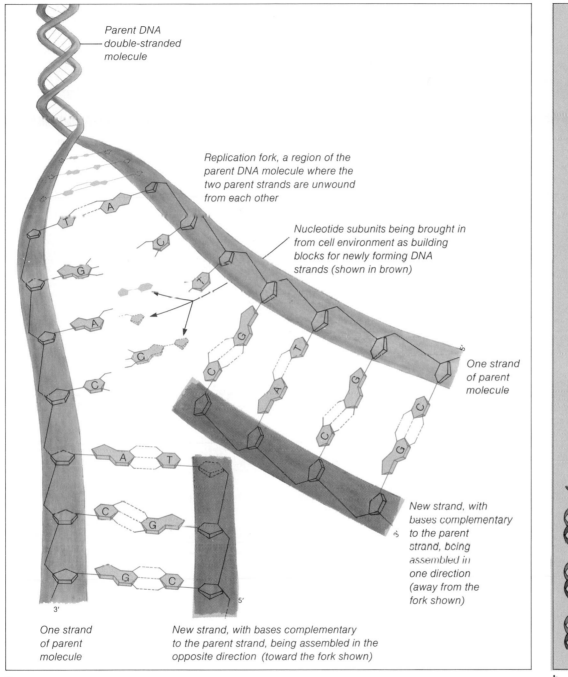

Parent DNA
double-stranded
molecule

Replication fork, a region of the
parent DNA molecule where the
two parent strands are unwound
from each other

Nucleotide subunits being brought in
from cell environment as building
blocks for newly forming DNA
strands (shown in brown)

One strand
of parent
molecule

New strand, with
bases complementary
to the parent
strand, being
assembled in
one direction
(away from the
fork shown)

One strand
of parent
molecule

New strand, with bases complementary
to the parent strand, being assembled in the
opposite direction (toward the fork shown)

a

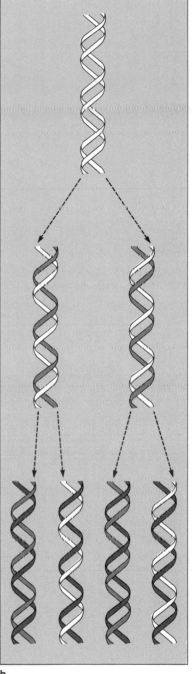

b

Figure 10.6 The semi-conservative nature of DNA duplication. (**a**) Here, the original two-stranded DNA molecule is shown in gray. The two new strands being assembled on them are shown in brown. Two ''half-old, half-new'' DNA molecules will be the result. (**b**) DNA duplication through two generations. White represents the parent molecule; gray represents the new strands in the first generation; and gold, the new strands in the second generation.

a *Bacteria are grown on ^{15}N-containing medium, then transferred to one containing ^{14}N:*

parent cells → F₁ cells → F₂ cells

b *DNA is extracted from cells at each stage, mixed with cesium chloride solution, and placed in centrifuge tube:*

c *Solution is centrifuged for two days. Centrifugal force drives cesium chloride toward the tube's base. But this is counteracted by diffusion. Thus a density gradient is created in the tube. DNA*

molecules band in regions where their density matches that of the cesium chloride solution:

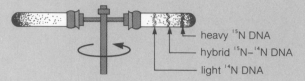

heavy ^{15}N DNA
hybrid ^{15}N–^{14}N DNA
light ^{14}N DNA

d *Results from tests on labeled DNA of cells taken from parent generation, F₁ and F₂ generations:*

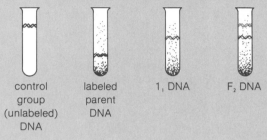

control group (unlabeled) DNA | labeled parent DNA | 1, DNA | F₂ DNA

Compare these results with the control group (unlabeled DNA) shown to the left in this series.

Figure 10.7 Meselson and Stahl's use of density-gradient centrifugation for testing whether DNA strands are separated and new ones formed upon each one during replication.

Density-gradient centrifugation is used to separate molecules of different densities. A centrifuge is a power-driven spinning device with tubes attached to its rotating arms. Typically these tubes are filled with concentrated solutions of heavy salts such as cesium chloride. High-speed spinning creates centrifugal forces that drive the heaviest molecules to the bottom of the tubes, intermediate-weight molecules toward the middle, and lighter molecules in some region closer to the top.

Meselson and Stahl grew bacteria on a medium containing a heavy isotope of nitrogen (a component of DNA). Some of these cells were transferred to a medium containing a lighter isotope

of nitrogen. DNA was extracted from cells of the parent generation and the F_1 generation; DNA was also extracted from F_2 generation cells.

By choosing a solution that has the same density as ordinary DNA, the researchers knew that centrifugation would cause DNA of lighter weight to band in one part of the tube, and DNA of heavier weight to band in another. If DNA replication did involve the separation of parent strands and the formation of new, complementary strands upon them, then the first-generation DNA would be composed of one heavy and one light strand. The density of these hybrid molecules would be midway between the heavy DNA and light DNA bands in the centrifuge tube. The results agreed exactly with these expectations.

---

simultaneously. (For some species, there may be thousands of these regions, each with its own initiation and termination point. The simultaneous activity allows a relatively long molecule to be duplicated within a reasonable time.) Once the two DNA strands are forced apart, construction enzymes move in. As you will see in the forthcoming chapter, they link free nucleotides together in a sequence

that is complementary to the exposed regions. Still other enzymes work at joining together short pieces of new DNA into an intact new strand. Others act like editors: they check the base pairs formed, clip out incorrect bases, and replace those parts with correct bases. In the laboratory, some of the enzymes associated with DNA assembly and repair are being used in remarkable ways (see *Commentary*).

COMMENTARY

More on Human Genetic Disease—Recombinant DNA and Prospects for Genotypic Cures

Most organisms bearing mutant DNA do not have much to say about the windfalls and pitfalls of the evolutionary process. As you probably suspect, humans are the only exceptions, especially when it comes to mutations leading to genetic diseases. As you read in the preceding chapter, phenotypic cures of human genetic diseases do alleviate individual suffering. They also circumvent the process of natural selection. Deleterious alleles not only are perpetuated, they may become represented with increasing frequency in the population as the afflicted persons marry and pass on their mutant DNA to the next generation. The "cure," in other words, is illusory. On the horizon is the possibility of modifying mutant DNA in gametes or in embryos, thereby effecting what is called a genotypic cure. Such modifications are not yet possible, but research is moving rapidly in this direction.

One avenue opening up has to do with breaking apart and recombining the DNA from different kinds of organisms. The resulting new genetic combination in the hybrid molecule is called recombinant DNA. In the natural environment, recombinant DNA produced from genes of two entirely different kinds of species occurs only rarely. In the laboratory, it is no longer a rare event.

In one laboratory technique, restriction enzymes are used to cut open the DNA double helix of a plasmid (a small DNA circle found in some bacteria in addition to the main DNA molecule). In living cells, these enzymes act as a line of defense against the injection of foreign DNA. (Thus, for example, the nucleic acid of bacteriophages that attack *E. coli* may be cut open and destroyed before having a chance to subvert the host's cellular machinery.) Restriction enzymes attack only those sites where a specific sequence of bases occurs (see figure to right). The broken ring takes the shape of a rod having two "sticky" ends—unpaired com-

plementary bases at both ends of the strand. In the laboratory, the nucleotide sequences at the sticky ends are used to form weak hydrogen bonds with complementary nucleotide sequences on the ends of fragments from foreign DNA, which has been cut with the same kind of enzyme. Through the action of a connecting enzyme (DNA ligase), the foreign DNA becomes incorporated in the plasmid, which is tied together again.

Under certain conditions, the hybrid DNA plasmid can be introduced into new *E. coli* cells. It will be reproduced every time the bacterium undergoes binary fis-

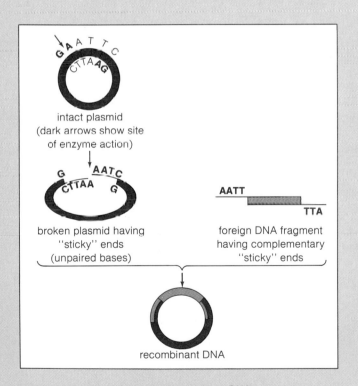

intact plasmid
(dark arrows show site
of enzyme action)

broken plasmid having
"sticky" ends
(unpaired bases)

foreign DNA fragment
having complementary
"sticky" ends

recombinant DNA

One method for assembling recombinant DNA.

sion. The bacterium's metabolic machinery can churn out many multiple copies, or <u>clones</u>, of the genes carried on the hybrid plasmid ring. With repeated doublings of a bacterial population, millions of these genes can be manufactured overnight.

Recombinant DNA techniques hold great promise for transferring genes from one organism to another. Consider, for example, that galactosemia results from a single nonfunctional enzyme that cannot break down galactose. It happens that *E. coli* carries a gene coding for this exact same enzyme. Already, investigators have introduced the gene from *E. coli* into plasmids from bacteriophages—which have been allowed to infect human cells growing in a culture dish. As another example, insulin genes from rats have already been cloned into bacteria.

Delivering a fully functional gene to humans is not yet possible. A difficult obstacle is isolating the exact sequence of DNA required. Restriction enzymes do not clip DNA precisely at the borders of any given gene. As a result of their activity, a vital regulatory signal for the gene might be lost, or the DNA fragment might include extra information which could have unforeseen effects on the host organism.

Recombinant DNA production raises questions of scientific ethics. Some molecular biologists feel that the release of new combinations of DNA may have serious effects on human health and on the environment. For example, it would not appear to be to our advantage if bacteria carrying recombinant DNA of a potentially hazardous sort were to escape from the laboratory. Such risks have not yet been accurately estimated. But neither have the benefits—and the potential benefits appear to be great. It is not merely that genotypic cures would halt the increase of a debilitating allele in the human population. By using recombinant DNA, plant geneticists hope to modify strains of corn and wheat so that these plants can harness nitrogen from the atmosphere. Such crops are the main food source for the world population, and presently they require expensive nitrogen-containing fertilizers. Atmospheric nitrogen is abundant, but it is in a chemical form that is converted to usable form only by a few kinds of soil microorganisms (Chapter Twenty-Eight). Through genetic engineering, this capability might be transferred to food crops. Beyond this, the regulation of such human genes as those implicated in cancer could alleviate terrible human suffering.

Most scientists agree that recombinant DNA research is safe as long as proper safeguards are followed. For instance, researchers are now using weakened strains of bacteria that are unable to survive at all in the absence of special laboratory conditions. Strict physical isolation is required, as it is now for research involving pathogenic microorganisms. Both scientists and nonscientists have a strong interest in the ethical issues raised by recombinant DNA research. Efforts to regulate this research must carefully balance between protecting public and environmental safety, and improving our ability to cure diseases permanently—even while allowing as much freedom as possible for scientific inquiry.

CONTINUITY AND CHANGE

For genetic information to be transmitted precisely from generation to generation, DNA would have to be copied precisely and distributed intact to offspring. But it isn't difficult to imagine that every so often, something might slip up during the fast-paced duplication process or during nuclear division (Chapter Nine). Also, agents such as cosmic or atomic radiation, ultraviolet radiation, and certain chemicals that enter a cell can cause point mutations in DNA strands.

Repair enzymes do monitor the DNA molecule for distortions. They work to restore the general form in accordance with the rules of base pairing. The recognition process, though, is not all that discriminating. Repair enzymes don't "memorize" the entire DNA molecule, and they can't always restore the sequence of bases to what it originally should have been.

For instance, sometimes a strand breaks. Sometimes a single base pair is left out at the break point, and usually the enzymes can take care of the omission. But if a whole series of nucleotides is left out at the break point, then the enzymes will have no way of recognizing and replacing what is missing. They will simply join the broken ends back together. Such defects, in which one or more base pairs are lost, are called **deletions**. Another kind of defect is an **insertion**: the incorporation of one or more extra base pairs into the original sequence. On rare occasions, too, the wrong base pair can be built into the DNA molecule. If, for example, A on the parent strand is accidentally paired with C on the new companion strand during duplication, repair enzymes will have no way of determining whether the correct base pair should be A—T or G—C! They simply will clip out a region of one DNA strand at the distortion and restore a sequence that adheres to the base-pairing rules. But there is a fifty-percent chance that they will perpetuate an error instead of duplicating the original sequence.

All such mutations are permanent, and they may mean the loss or major distortion of hereditary information that is normally carried by some gene.

The overall precision of DNA duplication and repair is the source of life's fundamental unity. The rare "mistakes"—the point mutations—account for much of its diversity. For although most mutations are probably harmful, some are not. In fact, some may confer on their bearers the ability to survive in different environments even as others perish—to leave more offspring even as others leave fewer when conditions change. Such mutations are perpetuated, and the variant phenotypes they foster become increasingly represented in the population. They clearly are central to the evolutionary process. We will be returning to the role of mutations in the flow of life through time (Chapter Twenty-Five).

PERSPECTIVE

DNA—deoxyribonucleic acid—is the hereditary molecule of every living cell. It is the helically twisted, double-stranded blueprint inside all prokaryotes and all eukaryotes. Every DNA molecule is composed of only three kinds of substances: a sugar, a phosphate group, and nitrogen-containing bases (adenine, thymine, guanine, and cytosine). Every DNA molecule is assembled according to the same rules. When the time comes for new bases from the cellular environment to be paired with old bases sticking out from DNA's sugar-phosphate backbones, adenine normally pairs only with thymine, and guanine only with cytosine.

What this means is that every living thing on earth shares the same fundamental chemical heritage with all others. Your DNA is made of the same kinds of substances and follows the same base-pairing rules as the DNA of earthworms in Missouri and grasses on the Mongolian steppes. Your DNA is duplicated in much the same way as theirs; occasional mistakes in its duplication are repaired in much the same way as theirs. In the evolutionary view, the reason you don't *look* like an earthworm or a flowering plant is largely a result of mistakes that appeared on rare occasions during the past 3½ billion years of life history—3½ billion years that led, in their unique divergent ways, to the three of you. Thus the *sequence* of base pairs along the DNA molecule has come to be different in the three of you.

Dinosaur DNA, too, was presumably assembled from the same chemical stuff as yours. But the mutations that gave rise to the unique sequence of base pairs that specified "build dinosaurs" made those creatures unsuitable, when environmental conditions changed, for continuing their journey.

In sum, DNA is the source of the unity of life, mutations in DNA structure are a fundamental source of life's diversity, and the changing environment is the testing ground for the success or failure of the mutations themselves.

Readings

Campbell, A. 1976. "How Viruses Insert Their DNA into the DNA of the Host Cell." *Scientific American* 235(6):102–113.

Chargaff, E. 1970. "One Hundred Years of Nucleic Acid Research." *Experientia* 26(7).

Jackson, D., and S. Stich (editors). 1979. *The Recombinant DNA Debate*. Englewood Cliffs, New Jersey: Prentice-Hall. Excellent source book for information on recombinant DNA research and the ethical questions it poses.

Lehninger, A. 1975. *Biochemistry*. Second edition. New York: Worth. Comprehensive but clearly written account of topics in this chapter.

Taylor, J. (editor). 1965. *Selected Papers on Molecular Genetics*. New York: Academic Press. Contains papers on major concepts forwarded by Avery, Hershey and Chase, Watson and Crick, Arthur Kornberg, Meselson and Stahl.

Vander, A., J. Sherman, and D. Luciano. 1980. *Human Physiology*. Third edition. New York: McGraw-Hill. Good treatment of heritable metabolic disorders.

Watson, J. 1976. *Molecular Biology of the Gene*. Third edition. Menlo Park, California: Benjamin. Outstanding source book; detailed and authoritative.

——— . 1978. *The Double Helix*. New York: Atheneum. A highly personal look at scientists and their methods, as well as an absorbing account of the discovery of DNA structure.

Review Questions

1. How did Griffith's use of control groups help him deduce that the transformation of harmless *Diplococcus* strains into deadly ones involved a change in the bacterium's hereditary system?

2. What is a bacteriophage? In the Hershey-Chase experiments, how did bacteriophages become labeled with radioactive sulfur and radioactive phosphorus? Why were these particular elements used instead of, say, carbon or nitrogen?

3. DNA is composed of only four different kinds of nucleotides. Name the three molecular parts of a nucleotide. Name the four different kinds of nitrogen-containing bases that may occur in a nucleotide. What kind of bond holds nucleotides together in a single DNA strand?

4. What kind of bond holds two DNA chains together in a double helix? Which nucleotide base pairs with adenine? Which pairs with guanine? Do the two DNA chains run in the same or opposite directions?

5. The four bases in DNA may *vary* greatly in relative amounts from one species to the next—yet the relative amounts are always the *same* among all members of a single species. How does the concept of base pairing explain these twin properties—the unity *and* diversity—of DNA molecules?

6. When regions of a double helix are unwound during DNA duplication, do the two unwound strands join back together again after a new DNA molecule has formed?

7. Name some of the molecules that monitor DNA duplication. What kind of tasks do they help perform? How do you suppose some of these molecules can be used as research tools in the study of DNA?

8. What is a recombinant DNA molecule? What are some of the potential hazards of recombinant DNA research? What are some of the potential benefits? Would you mind living next door to a laboratory where such research is going on? Would your answer be the same if you happen to be one of the millions of individuals afflicted with diabetes or some other genetic disease?

9. If genetic information were transmitted precisely from generation to generation, organisms would never change. What are some of the DNA mutations that give rise to phenotypic diversity?

10. How is your DNA like the DNA of earthworms, grasses, and (presumably) dinosaurs? How is your DNA different?

PART I. FROM DNA TO PROTEIN ASSEMBLY

DNA, in essence, is like a book of instructions that each cell carries inside itself. The alphabet used to create the book is simple enough: A, T, G, and C. But merely knowing what the letters are doesn't tell us *how* they are assembled into the language of life, with words evoking precise meaning, and with meaning controlled by punctuation. More than this, the letters alone can't tell us how a cell selectively reads the words, then puts different messages to work in assembling all the cell substances necessary for survival. To see how a cell puts its genetic heritage into action, let's turn first to the nature of genes and enzymes, and to a chemical link between them.

RNA: A Link Between Genes and the Proteins They Specify

In 1908 no one knew much about genes, other than that discrete units of inheritance probably exist and somehow give rise to an individual's physical traits. Probably no one even suspected that genetic information is built into the structure of DNA, a substance discovered just a few decades earlier. But in that year Archibald Garrod, an English physician, reported what he thought gene function must be. Garrod had studied many cases of childhood afflictions. He decided that in some families, there had to be a heritable basis for certain diseases that recur in particular patterns. What did these diseases have in common? Garrod inferred that within the cells of afflicted individuals, something was blocking one of the steps in a metabolic pathway. Normally, a cell converts a substance into several intermediate compounds before it ends up with the required product. When one of the intermediate steps is blocked, there sometimes is a buildup of the compound formed just before that step, and an absence of the compounds following it:

normal pathway: A → B → C → D

blocked pathway: A → B B C

Eventually, the compound being accumulated can be detected. In phenylketonuria, recall, excess amounts of an oxidation product of phenylalanine can be detected in the urine. In alkaptonuria, urine samples exposed to air turn a dark color because of concentrations of the intermediate

11

GENE EXPRESSION AND ITS CONTROL

alkapton. Garrod called such disorders "inborn errors of metabolism." He linked them to the absence or deficiency of an essential enzyme. He suggested that each enzyme must be specified by a single unit of inheritance—and that if the unit is defective, then the enzyme will be defective.

Thirty-three years later, research began that seemed to confirm Garrod's hypothesis. At that time, the geneticist George Beadle and the biochemist Edward Tatum were working together to discover how genes function. Earlier work with *Drosophila* seemed to suggest that genes function through the action of enzymes that they produce. But proof by way of *Drosophila*, a complex diploid organism, was slow in coming. Beadle and Tatum turned their attention to a much simpler organism, the bread mold *Neurospora crassa*. The effect of single gene mutations in *Neurospora* would be easy to identify. First, it is a haploid organism: it has only one gene for each trait. This meant that there would be no masking of a mutant gene by a functional allelic partner, as there can be in diploid organisms. Second, *Neurospora* can reproduce asexually, through spore formation. This meant that many genetically identical copies of a mutant organism could be grown for study. Third, *Neurospora* can grow on a simple chemically defined medium containing sugar and biotin (one of the B vitamins).

Every other substance *Neurospora* needs—other vitamins, amino acids—*it synthesizes for itself*. The biochemical steps in the synthesis of some of these substances were already known. If an enzyme catalyzing one of these steps were defective, then chemical analysis of mutant cell extracts would reveal a buildup of the intermediate compound known to be formed just before that step, and an absence of all compounds following. In other words, not only could a gene mutation itself be readily identified, any defective enzymes arising from mutation could also be identified.

Beadle and Tatum bombarded *Neurospora* spores with x-rays. They soon isolated two mutant strains, which were allowed to reproduce for several generations. One strain could grow only when vitamin B_1 was added to the culture medium. The other could grow only when vitamin B_6 was added. Some genes related to the production of these substances had to be mutant. Furthermore, analysis of cell extracts revealed a defective enzyme in one strain, and a *different* defective enzyme in the other. Because each separately inherited mutation corresponded to a different defective enzyme, Beadle and Tatum assumed that each gene controls the synthesis of one enzyme. Their assumption became known as the **one gene = one enzyme concept**.

Further research led to the extension and modification of this concept. For one thing, it soon became apparent that genes do not code directly for enzymes and other proteins. *Genes are carried on DNA, but the products they specify are not <u>assembled</u> on DNA.* Recall that DNA resides in the nucleus (or, in prokaryotes, in one cytoplasmic region). Gene products are built in the cytoplasm. In some way, then, the *messages* of genes had to be physically moved away from the genes themselves.

Today we know that certain kinds of ribonucleic acid (RNA) molecules serve as the message carriers. RNA is something like a single strand of DNA. It, too, is assembled from nucleotides (each nucleotide having a sugar, a phosphate group, and a nitrogen-containing base). However, as its name implies, the sugar in the RNA backbone is ribose, not deoxyribose. And in place of the base thymine, RNA has **uracil** (U). Uracil behaves chemically like thymine, for it can form hydrogen bonds with adenine. This means that RNA nucleotide bases can match up with those found in a DNA double helix. As you will see, with this complementarity, the cell has a means of leaving DNA molecules intact in one region even as it routes genetic messages to sites of protein synthesis.

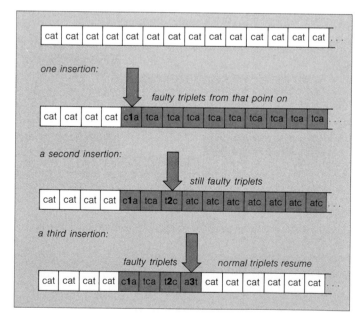

Figure 11.1 Crick's interpretation of Brenner's base insertion experiments. If translation of the genetic code depends on reading adjacent nucleotide bases three at a time, insertion of one or two extra nucleotides will change the sequence (hence the meaning) of all subsequent base triplets. But a third addition will restore much of the normal sequence.

The Genetic Code

Because RNA's nucleotide arrangement is complementary to that of DNA, we are still left with a central question. *How do we get from a linear sequence of nucleotides to a linear sequence of amino acids in a protein?* Recall, from Chapter Three, that a string of amino acids represents a protein's primary structure. The primary structure dictates the protein's final three-dimensional form. In turn, the three-dimensional form dictates which substances the protein will interact with in helping to build and maintain the cell. Thus, if we can figure out how nucleotide "letters" (bases) specify which amino acids should occur in a row, we have the key to the question of how a cell's hereditary molecule can give rise to proteins of specific structure and function.

Consider that there are only four different kinds of nucleotides in DNA or RNA, but that twenty different amino acids are commonly found in the proteins of living things. Obviously, each nucleotide letter doesn't call for a single amino acid, because that would give us only four different amino acids. Two nucleotide letters in a row couldn't call for an amino acid, because that would give us only sixteen different amino acids, not twenty. If three nucleotides in a row called for an amino acid, then sixty-four different combinations would be possible. That is far more than the twenty required to assemble amino acids. Can it be that different combinations specify the same amino acid?

For the answer let's consider, as Francis Crick and Sidney Brenner once did, the effects of deletions or insertions of one or more nucleotides into a gene. Recall that such point mutations give rise to partially or wholly non-functional enzymes, which can be detected because of the diminished amount or total absence of their product. Brenner discovered that one or two extra nucleotides inserted in the middle of a gene made the protein it specified completely defective. Yet, when a third nucleotide was inserted near the first two, gene function was partly restored! Why was the presence of three extra nucleotides less serious than the addition of one or two? Crick sensed that the results probably revealed the fundamental nature of the genetic code. *These results would be expected if the genetic code consists of nucleotide bases that are read linearly, three at a time, with the sequence of each triplet signifying an amino acid.* Figure 11.1 shows why this is so.

It is now known that the genetic code does indeed consist of sixty-four combinations of nucleotide triplets. In DNA and RNA molecules, these triplets are read in a line, one after the other. As Figure 11.2 shows, sixty-one triplets actually specify amino acids. (In nearly every case,

First Letter	Second Letter				Third Letter
	U	C	A	G	
U	phenylalanine	serine	tyrosine	cysteine	U
	phenylalanine	serine	tyrosine	cysteine	C
	leucine	serine	stop	stop	A
	leucine	serine	stop	tryptophan	G
C	leucine	proline	histidine	arginine	U
	leucine	proline	histidine	arginine	C
	leucine	proline	glutamine	arginine	A
	leucine	proline	glutamine	arginine	G
A	isoleucine	threonine	asparagine	serine	U
	isoleucine	threonine	asparagine	serine	C
	isoleucine	threonine	lysine	arginine	A
	(start) methionine	threonine	lysine	arginine	G
G	valine	alanine	asparagine	glycine	U
	valine	alanine	asparagine	glycine	C
	valine	alanine	glutamic acid	glycine	A
	valine	alanine	glutamic acid	glycine	G

Figure 11.2 The genetic code. Each triplet consists of three nucleotides. The first nucleotide of any triplet is given in the left column. The second is given in the middle columns; the third, in the right column. Thus we find (for instance) that tryptophan is coded for by the triplet U G G . Phenylalanine is coded for by both U U U and U U C . All such nucleotide triplets (codons) are strung one after another in messenger RNA molecules. Different sequences of triplets call for different protein chains.

two or more different combinations specify the *same* amino acid. Usually the ones coding for the same thing differ only in the last base in the triplet. For instance, CCU, CCC, CCA, and CCG *all* specify the amino acid proline.) The remaining three triplets act like punctuation points. Their occurrence in a genetic message signifies that the addition of amino acids to a growing protein chain must be terminated. Any nucleotide triplet in RNA that codes for an amino acid (or for chain termination) is now known as a **codon**.

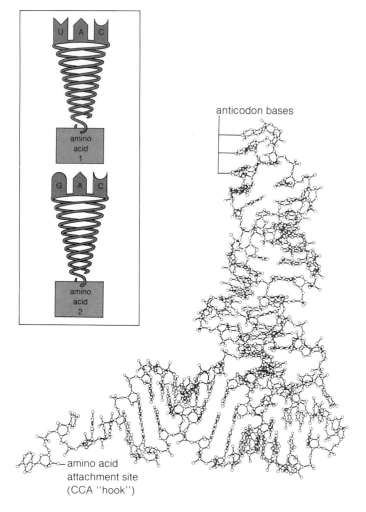

Figure 11.3 Messenger RNA. The diagram depicts some of the codons of an mRNA molecule. On the left, AUG specifies the amino acid methionine, a "start" codon that initiates the message.

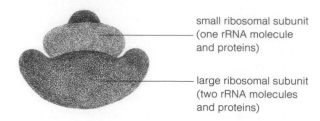

small ribosomal subunit (one rRNA molecule and proteins)

large ribosomal subunit (two rRNA molecules and proteins)

Figure 11.4 Ribosome structure, as it occurs during protein assembly.

anticodon bases

amino acid attachment site (CCA "hook")

Figure 11.5 Three-dimensional structure of a transfer RNA molecule, as determined by computer analysis. To keep things simple for later illustrations, we can portray tRNA molecules as stripped-down versions of this structure (see inset). (Three-dimensional structure courtesy of Sung Hou Kim)

The first evidence for the triplet nature of the genetic code came from experiments with *Escherichia coli* DNA. Further work with the DNA of yeasts and human mitochondria revealed some substitutions in the code. However, the following is true:

In all living organisms studied to date, the same genetic code is the basic language of protein synthesis.

This is compelling evidence for the fundamental unity of life at the molecular level.

A Closer Look at the Protein Builders: Three Kinds of RNA

So far, you have read about an RNA molecule that is assembled on DNA and is complementary to it. This molecule is called **messenger RNA** (mRNA). It alone carries protein assembly instructions away from the nuclear region to construction sites in the cytoplasm. Two other general kinds of RNA molecules are also required for protein synthesis. Both are also assembled on DNA, but neither becomes translated into protein structure. **Ribosomal RNA** (rRNA) becomes complexed with proteins to form ribosomes, which are the actual protein construction sites in the cytoplasm. **Transfer RNA** (tRNA) is a molecule that binds first with an amino acid present in the cytoplasm, then with an mRNA molecule that has arrived at some ribosome. Hence tRNA molecules are like trucks on which parts are loaded—parts that will be used in the construction zones.

Let's first look at the structure of mRNA. As Figure 11.3 suggests, mRNA molecules are linear, unbranched chains. They typically contain at least 300 nucleotides. (Few proteins are made of less than 100 amino acids, and each amino acid is specified by three nucleotides.) At the beginning of mRNA strands are more nucleotide sequences of variable length that do not code for specific proteins. Some of these "leader" sequences are known to function in control over gene expression.

Let's now look at the structure of the rRNA "workbench" on which proteins are assembled. Figure 11.4 shows the large and small ribosomal subunits. Both are rRNA–protein complexes. When protein synthesis begins, these subunits combine into an intact two-part structure. When protein chain assembly stops, they fall apart. Within or on the surface folds between the ribosomal subunits are binding sites for both mRNA and tRNA. At these sites, in some as yet unexplained way, the movements of mRNA and tRNA molecules as well as their alignment with each other are chemically controlled.

Finally, let's look at the structure of a tRNA molecule. As Figure 11.5 shows, one end of a tRNA molecule has an **anticodon**: three adjacent nucleotide bases that readily hydrogen-bond to the complementary three-base codon of mRNA. At the opposite end is a molecular hook for an amino acid. There are different types of tRNA molecules, and they all have the same general form. Yet between the anticodon and the amino acid hook, the nucleotide sequence can vary considerably. Some of the variation may promote interaction with the different ribosomal proteins, in ways that help maintain the overall precision of the protein assembly steps.

Summary of Protein Synthesis

What we have been describing so far are the elements of an information-processing system for putting DNA instructions to work. These elements may be summarized in this way:

---

1. A source of systematically arranged information (*the linear sequence of nucleotide triplets in DNA*).

2. An encoding process, or some means of assembling bits of that information into a coherent, transmittable message. (*Here we are talking about specific enzymes and their interactions with DNA regions.*)

3. The encoded message itself (*messenger RNA*).

4. Decoding mechanisms (*ribosomal RNA, transfer RNA, and enzymes*).

5. An information receiver (*cytoplasmic substances and cell organelles*).

Let's turn now to the ways in which these elements are used in assembling a given protein. First, enzymes open

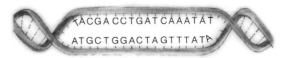

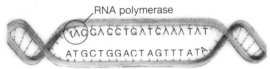

a *Strands of the DNA double helix separate in the region of a gene*

b *RNA polymerase attaches to the beginning of the gene on one strand only*

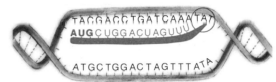

c *A strand of RNA complementary to one strand of DNA is formed on the exposed bases*

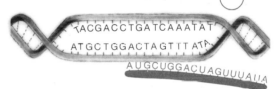

d *The newly made RNA and the RNA polymerase are released into the cellular environment*

e *The unzipped section of DNA resumes the shape of a double helix*

Figure 11.6 Transcription of an mRNA molecule from a DNA region, as described in the text. Only one DNA strand of the double helix is transcribed in this way.

up the two strands of the DNA double helix, much as they do during DNA duplication. But in this case the whole DNA molecule doesn't get unwound. The strands are separated only in the region corresponding to the beginning of the gene being read (Figure 11.6). Then an enzyme, RNA polymerase, attaches to one of the separated strands. It moves along the strand, all the while speeding up the assembly of RNA nucleotide subunits onto complementary regions of the exposed DNA. Then the subunits are linked together to form an mRNA strand. The DNA region and

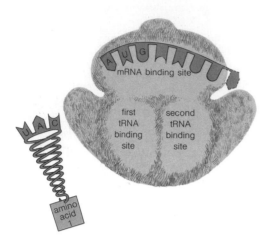

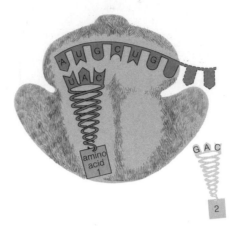

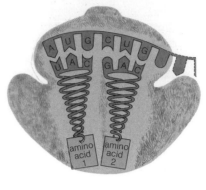

a *Three binding sites inside an intact, two-part ribosome. Here the start of an mRNA molecule is bound in place. A tRNA able to base-pair with the mRNA "start" codon (AUG) is moving in from the environment.*

b *The tRNA has become positioned in the first (acceptor) site of the large ribosome subunit. Another tRNA is approaching the ribosome.*

c *The second tRNA moves in and base-pairs to the second mRNA codon. As it does, its attached amino acid aligns with the amino acid that is attached to the first tRNA.*

Figure 11.7   Protein synthesis: the translation of a genetic message into protein structure.

the mRNA being assembled on it stick together for a time, but the union is only temporary. The mRNA strand falls away and moves into the cytoplasm. (The DNA strands wind up again and resume their helical shape.) Each mRNA molecule so formed contains a nucleotide sequence that is complementary to the DNA region on which it was assembled. This entire process of mRNA synthesis is called **transcription**. The word means "to make a copy"—although it might help to keep in mind that the "copy" is *complementary* to, not identical with, the DNA region on which it was formed.

Transcription is the first step in putting a gene into action. In the next step, the transcribed copy is translated into protein. Although the message remains the same, the language of proteins is entirely different: it is written out in amino acids instead of nucleotides! (It's like writing the same sentence first in Morse code and then in standard English. The meaning doesn't change, but "words" of different structure are used.) That is why protein synthesis, which is directed by a sequence of information in an mRNA

molecule, is called **translation**. The term simply means "to change from one language into another."

Translation begins with the convergence of the two ribosomal subunits, some proteins (which assist in initiating the process), and a tRNA molecule to which a specific amino acid is attached. This kind of tRNA is able to recognize a "start" codon (AUG or GUG) on an mRNA molecule. One of these becomes the first tRNA to become hydrogen-bonded to the mRNA molecule. As Figure 11.7 shows, this bonding event occurs when the mRNA occupies the small ribosomal subunit binding site, and when the tRNA occupies the first binding site in the large ribosomal subunit. Then another tRNA, with its amino acid attached, becomes positioned in the second binding site of the large subunit. It hydrogen-bonds with the second codon of the mRNA molecule. Once the two tRNA molecules are aligned with each other in the ribosome, then the two amino acids they carry are linked together (as a dipeptide). At the same time, the bond between the first tRNA and its amino acid is broken, and this decoupled tRNA is ejected from the

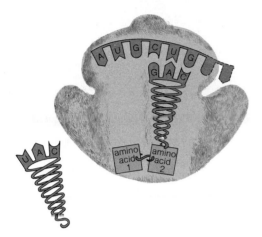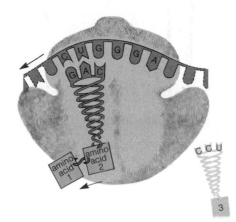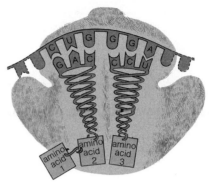

d *The bond between the first tRNA and its amino acid cargo is broken. The tRNA falls away. A peptide bond forms between the two amino acids.*

e *The vacated binding site in the large ribosome subunit is now filled because the mRNA and the second tRNA are moved to the left.*

f *A third tRNA able to base-pair with the third codon is bound in place. A peptide bond can form between amino acids 2 and 3. In this way, a polypeptide chain grows until the end of the mRNA message.*

first binding site. The tRNA remaining behind now has two amino acids attached. It moves over, into the first binding site. At the same time, the mRNA molecule bound to the small ribosomal subunit moves over, so that a third codon becomes aligned above the vacated second binding site. A new tRNA with its amino acid cargo moves in and the same steps are repeated.

As more and more amino acids become linked together, a polypeptide grows in length. In this way, protein chains are assembled from a set of twenty common amino acids. Each chain has its own unique sequence of anywhere from 20 to 3,000 amino acid subunits. The following sketch summarizes the steps involved in the transformation of genetic messages into proteins:

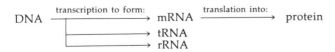

Figure 11.7 shows protein synthesis in action.

PART II. CONTROLS OVER PROTEIN SYNTHESIS

If all of its genes were being transcribed and translated at once, a cell would waste appalling amounts of energy, for it would be making far more enzymes and other proteins than it could possibly use. Besides, some enzymes would be working against one another, with one simultaneously helping to tear down what another was helping to build up.

Cells depend on controls over which proteins are assembled at any given moment of the life cycle. These controls govern transcription, translation, and enzyme activity.

The idea of control over enzyme activity was introduced earlier (Chapter Six). Here, we can give another example of how this works in regulating the construction of the amino acid isoleucine. The first enzyme involved in the biochemical pathway for isoleucine is allosteric (page 90): it has a second active site to which isoleucine, the

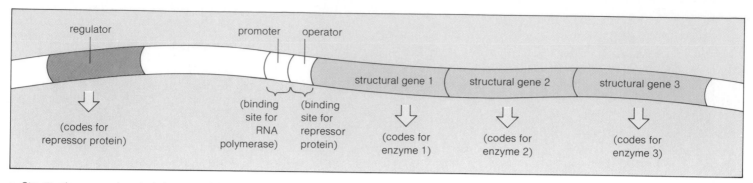

a *Structural genes and control elements of the lactose operon*

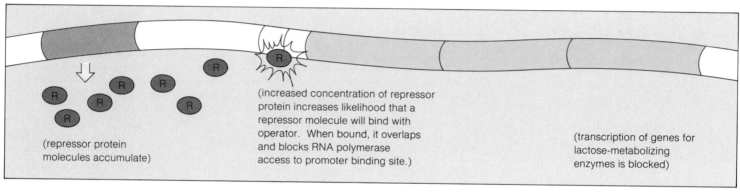

b *When lactose is absent from environment (repression occurs)*

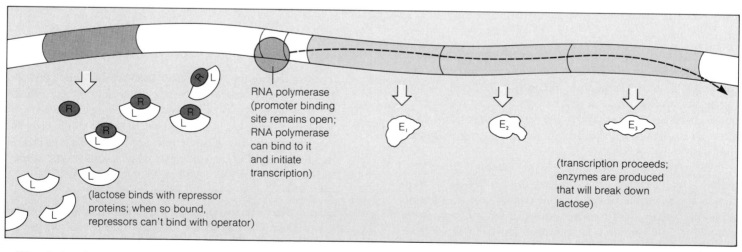

c *When lactose is present in the environment (induction occurs)*

Figure 11.8 Control of gene transcription with the lactose operon, as described in the text. (Illustrations based on conversations with Robert Robbins, Michigan State University)

end-product molecule, can bind. When isoleucine levels rise, excess molecules combine with the enzymes. When they do, the enzymes can't perform their first function, and isoleucine assembly slows down. It doesn't shut down completely, for feedback inhibition of these enzymes is not an on-again, off-again kind of control. For instance, in a bacterial cell, many thousands of enzyme molecules may have been synthesized for use in isoleucine assembly. As more isoleucine molecules accumulate, more of the enzymes will be shut down. Control in this case is proportional to the amount of end-product present.

Both prokaryotes and eukaryotes depend on such mechanisms as allosteric enzyme inhibition in controlling protein synthesis. But when we turn to controls over gene transcription and mRNA translation into the enzymes themselves, we find a significant difference between these two major kinds of organisms:

In prokaryotes, controls over gene expression are basically short-range responses to changing environmental conditions.

In eukaryotes, not only short-range but long-range controls over gene expression govern the events of development and differentiation.

Let's look briefly at the nature of the control mechanisms for these two kinds of organisms.

Gene Regulation in Prokaryotes

Living in the gut of a newborn mammal are populations of *E. coli*. This type of bacterium is destined to encounter an abrupt change in diet. For a few weeks or months, the host in which *E. coli* reside takes in nothing but milk, which contains large amounts of the sugar lactose. Once the weaning period is over, the mammalian host never takes in milk again. (The only exceptions are ourselves and a few of our pets.) During the first part of the mammalian life cycle, then, *E. coli* present in the gut must be able to metabolize a sugar that subsequent *E. coli* generations never will encounter. Those later bacterial generations will have no need to make thousands of copies of lactose-degrading enzymes.

Three French biologists, Andre Lwoff, Francois Jacob, and Jacques Monod, looked into this intriguing aspect of *E. coli* nutrition. Jacob and Monod grew bacteria on a culture medium in which lactose was absent. Then they abruptly switched the bacteria to a medium in which lactose was

the only energy source. They found that within minutes, three lactose-metabolizing enzymes were being produced! Such enzymes obviously were not built until the cells had need of them. Their assembly was somehow controlled by the presence of an "inducer" substance—in this case, one of the very substrates that the enzyme was destined to act upon!

Jacob and Monod identified the genes coding for these three enzymes, and found that they are adjacent to one another in the bacterial DNA molecule. More importantly, they concluded that these **structural genes**, which code for specific products, are controlled in a coordinated way. This coordination depends on three **control elements**: a *regulator* (a gene coding for a repressor protein that can block transcription), a *promoter* (a DNA site to which RNA polymerase may bind), and an *operator* (a DNA site that interacts with the repressor protein to dictate whether transcription proceeds). The term **operon** was coined to signify any set of structural genes that operate as a coordinated unit under the direction of DNA control elements.

How does the lactose operon work? Figure 11.8 outlines the two control mechanisms that regulate it: repression and induction. When lactose is absent from the cell environment, the repressor protein binds with the operator. Being a rather bulky molecule, it also overlaps the RNA polymerase binding site. Transcription is *repressed*; lactose-metabolizing enzymes are not produced.

When lactose is present in the cell environment, a molecule of lactose (or of an intermediate derived from it) binds with and *induces* change in the repressor protein's shape. In its changed shape, the repressor can't bind to the operator, it no longer overlaps the RNA polymerase binding site, and transcription proceeds apace. (Once all the lactose is degraded by these enzymes, repressor protein is free to bind to the operator again—and thereby repress the production of enzymes that are no longer needed.)

In *E. coli*, another operon codes for the enzymes necessary to build the amino acid tryptophan. When tryptophan is present in the cell environment, a tryptophan molecule can bind with and change the shape of the repressor protein for this operon. But it is the *changed* shape that is able to bind with the operator and thereby block transcription! When tryptophan is absent, the protein shape goes back to what it was, and the repressor protein sooner or later falls off the RNA polymerase binding site. Transcription then proceeds at a constant rate. Thus the end product is a *corepressor:* it helps shut down its own synthesis when it is present in enough quantity to do so, because its presence *activates* the repressor protein.

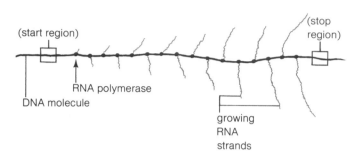

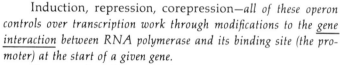

(start region)

(stop region)

RNA polymerase

DNA molecule

growing RNA strands

Figure 11.9 One moment in the transfer of genetic information. This historic electron micrograph shows genes isolated from a maturing egg of the spotted newt (*Triturus viridescens*). Each featherlike region spreads outward from the DNA molecule (the thin, long "shaft" of each feathered region). Each is an RNA strand. The short RNA strands are in early stages of synthesis; the longer are nearing completion. 27,000×. (Oscar Miller and Barbara Beatty, 1969, "Visualization of Nucleolar Genes," *Science*, 164:955–957. Copyright 1969 by the American Association for the Advancement of Science)

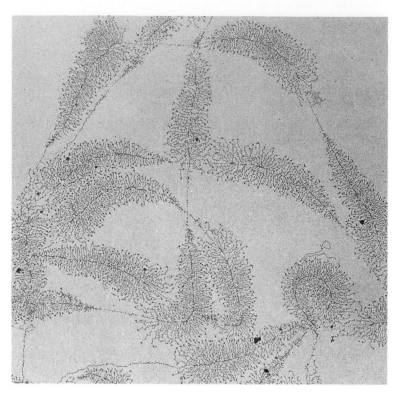

Induction, repression, corepression—*all of these operon controls over transcription work through modifications to the gene interaction between RNA polymerase and its binding site (the promoter) at the start of a given gene.*

Compared with straightforward transcription, operon control is an energetically expensive mechanism, and its use is limited to those metabolic events which, on balance, are worth the cost of building and maintaining DNA regions that do not directly yield useful cell products. Thus, being adaptive to changes in lactose availability means survival in a certain kind of environment and is worth the cost.

For many genes, more economical mechanisms exist that control the *rate* of transcription. Most of these mechanisms involve modifications to RNA polymerase or to its binding site, the promoter. For example, it happens that not all promoters are equally open-armed in their chemical invitation to RNA polymerase. Through variations in their nucleotide sequence, some promoters are better at binding this enzyme; we would expect that the genes they help control must be transcribed at a fairly steady rate. Other promoters show less binding affinity for the enzyme; we would expect the genes under their control to have a lower output, and their products to be required in lesser amounts.

Prokaryotes also appear to depend somewhat on *translational controls* over the rate at which protein products are assembled. For example, parts of mRNA molecules may be folded up tightly, which would make genes in those regions less accessible to translation enzymes. There is speculation, too, that some tRNA species are less abundant than others in the cell environment. If that is true, then ribosomal activity could be postponed where mRNA contains codons calling for those species; translation would idle until appropriate tRNAs could be brought in.

Organization of Eukaryotic DNA

The large, circular DNA molecule in *E. coli* contains about 5,000 genes. Most are structural genes, which code for specific proteins. Say you were to assume that most eukaryotic genes also code for specific proteins. If that were true, then given that there is about 2,000 times more DNA in your cells than in *E. coli* cells, you would have to find millions of different kinds of proteins being built in your body. Would you really discover *that* many structural genes, all coding for that many different proteins? Let's take a look.

For most eukaryotes, about half of the gene-sized segments are unique nucleotide sequences that occur only once in the total DNA. They represent **single-copy DNA**. Does transcription of all the different single-copy sequences lead to as many different proteins? It appears not, *for the RNA assembled on most single-copy DNA never leaves the nucleus*. For example, only about five percent of all single-copy *Drosophila* DNA gets transcribed into mRNA, which is then translated into proteins in the cytoplasm. In humans, more than ninety-five percent of all the RNA produced in a typical cell is destroyed in the nucleus within minutes of the time it is made. What this means is that most single-copy DNA is transcribed into an enormous amount of RNA of unknown function!

Other sequences occur abundantly in the DNA, in multiple copies; they are called **repetitive DNA**. The genes coding for rRNA and tRNA typically fall in this category. For instance, multiple copies of rRNA genes occur end to end, hundreds of times, in a region of the chromosome that forms the nucleolus (see, for example, the nucleolar genes in Figure 11.9). In addition, some short segments of repetitive DNA contain very similar nucleotide sequences. But they occur at many scattered locations along a chromosome. They are called *moderately* repetitive DNA. Although their function has not been established, these sequences are thought to contain control elements that govern the transcription of structural genes. Let's take a look at some ideas about the controls that seem to be built into eukaryotic DNA.

Gene Regulation in Eukaryotes

Recent evidence concerning the nature of nuclear RNA suggests that transcriptional control elements exist in the DNA sequence organization. The nucleus contains a large pool of long RNA strands during all stages of development. These strands vary greatly in length; hence their name, **heterogenous nuclear RNA** (hnRNA).

Hybridization experiments (Figure 11.10) show that hnRNA has a sequence organization similar to that of DNA, for most of it can match up to single DNA strands. The hnRNA does not hybridize with mRNA—which may mean that most hnRNA messages are not translated directly into proteins. Finally, the kinds of hnRNA molecules present in the nucleus are not constant from one developmental stage to the next. In samples ranging from maturing eggs to embryonic intestinal cells, the kinds of hnRNA present vary greatly—but there is no corresponding change of the same magnitude for most of the proteins being produced!

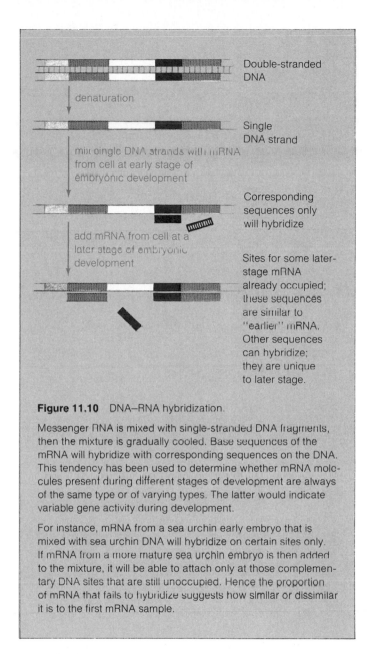

Figure 11.10 DNA–RNA hybridization.

Messenger RNA is mixed with single-stranded DNA fragments, then the mixture is gradually cooled. Base sequences of the mRNA will hybridize with corresponding sequences on the DNA. This tendency has been used to determine whether mRNA molecules present during different stages of development are always of the same type or of varying types. The latter would indicate variable gene activity during development.

For instance, mRNA from a sea urchin early embryo that is mixed with sea urchin DNA will hybridize on certain sites only. If mRNA from a more mature sea urchin embryo is then added to the mixture, it will be able to attach only at those complementary DNA sites that are still unoccupied. Hence the proportion of mRNA that fails to hybridize suggests how similar or dissimilar it is to the first mRNA sample.

Such evidence points to variable gene activity during development. It is being taken by many investigators to mean that the bulk of repetitive sequences in hnRNA—*hence in the DNA from which they are transcribed*—is necessary to drive control-element reactions.

Eukaryotic controls may also be inherent in the association of DNA with two classes of proteins: the histones, and other chromosomal proteins (collectively called "non-histones"). On the basis of weight, histones and DNA are

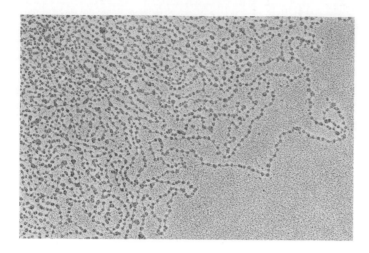

Figure 11.11 Nucleosomes of chromatin from *Drosophila melanogaster*. These beadlike structures are too small to be individual genes (which contain about 1,000 nucleotide pairs). Nucleosome formation may be a way in which long DNA molecules are packaged into more compact and organized—hence more manageable—form. 102,500×. (O. L. Miller, Jr., and Steve L. McKnight, University of Virginia)

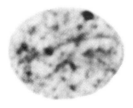

Figure 11.12 Barr body from a cell of a mammalian female. (Murray L. Barr)

always present in about equal proportions. The histones don't vary in overall composition from one chromosome region to the next, from one cell type to the next, or even from one species to the next! They also vary little in amino acid sequence relative to most other proteins. (Such uniformity suggests that histones have long been central to chromosome functioning, for mutations that would give rise to variations in their structure apparently have not survived during the evolution of eukaryotes.) Histones and DNA are tightly linked into a nucleoprotein fiber. This fiber looks something like a beaded chain, with each bead called a **nucleosome** (Figure 11.11). What does this orderly, beaded packing represent? It might be related to shutting down transcription, for the proteins would have to be sitting directly on both the control elements and the structural

genes. Might such complexes be loosened up, at prescribed times, so that transcription can proceed?

For a tentative answer, let's reconsider the behavior of chromatin just before nuclear division. About that time, it condenses into chromosomes, and most of the genes are not transcribed. Most remain silent until the compact chromatin once again becomes dispersed through the nucleus after division, whereupon transcription resumes. In addition, it happens that some chromatin stays clumped up as a chromosomal body between divisions—and it is known that genes on these special chromatin regions never are expressed.

For instance, in mammalian females, one entire X chromosome remains compact between divisions. It is called a **Barr body**, after its discoverer Murray Barr (Figure 11.12). The genes on this chromosome are not activated; only their counterparts on the other X chromosome of each diploid cell are expressed. Activation apparently occurs after the embryo has reached a multicelled stage of development. Which of the two chromosomes becomes turned on in a given cell seems to be a matter of chance. Once one of these embryonic cells has made its random chemical choice, though, the same choice is made by all of its descendants that come to form a given body tissue region. Mary Lyon was the first to figure out what was going on, hence the process has become known as **Lyonization**.

Consider what the process means for human males (XY) and human females (XX). Cells of *both* sexes have only one *active* X chromosome. Depending on which of the two is active in a given tissue region, this means also that female tissues are actually a "mosaic" of X-linked traits. Or consider the consequences for calico cats. Recall that these animals are heterozygous for black and yellow coat-color alleles, which reside on the X chromosome. Coat color in a given body region thus depends on *which* chromosome has been activated in a given region—in other words, on which of the two has been loosened up and is available to agents of transcription.

The transcription of a gene into an mRNA molecule doesn't have to mean the immediate and inevitable appearance of a specific protein. The nuclear membrane acts like a selective barrier between the hereditary instructions and the cytoplasm, where proteins are constructed. Through post-transcriptional controls, some regions of one RNA molecule may be completely preserved—but other regions may be completely degraded. Another RNA molecule may be extensively modified before it is passed into the cytoplasm. For instance, sometimes nucleotide sequences are clipped off while mRNA is still in the nucleus.

In addition, before mRNA can leave the nucleus, it apparently must be linked with a certain protein that assists it in moving across the nuclear envelope.

Finally, some controls are exerted after the mRNA molecules enter the cytoplasm. For example, levels of enzymes and of different tRNA species present may be varied at different times. Such variations would also influence the rate at which translation would proceed.

Beyond these possibilities, post-translational modifications can be made to the proteins themselves. Before a given protein can become fully functional, for instance, it may require the addition of a phosphate group. And such additions can be made at different rates, depending on controls over enzyme activity.

The study of gene regulation in eukaryotes is one of the frontiers in modern biology. Like all frontiers, it looks disconcertingly vast. The amount of DNA in, say, any mammalian cell is at least 800 times the amount in a prokaryotic cell! But more and more paths are connecting scattered outposts of research into molecular biology and into embryology, the study of early development and differentiation. Many prospects and problems await the convergence of these two lines of research. And that is the subject of the next chapter.

Readings

Benzer, S. 1962. "The Fine Structure of the Gene." *Scientific American* 206(1):70–84.

Dyson, R. 1978. *Cell Biology: A Molecular Approach.* Second edition. Boston: Allyn and Bacon.

Goodenough, U. 1978. *Genetics.* Second edition. New York: Holt. Good coverage of recent experiments in molecular genetics.

Miller, O., and B. Beatty. 1963. "Visualization of Nucleolar Genes." *Science* 164:955–957.

Nirenberg, M. 1963. "The Genetic Code: II." *Scientific American* 208(3):80–94.

Nomura, M. 1969. "Ribosomes." *Scientific American* 221(4):28–35.

Rich, A., and S. Kim. 1978. "The Three-Dimensional Structure of Transfer-RNA." *Scientific American* 238(1):52–62.

Spiegelman, S. 1964. "Hybrid Nucleic Acid Experiments." *Scientific American* 210(5):48–56. Describes experiments used to demonstrate base pairing between DNA and RNA.

Watson, J. 1976. *Molecular Biology of the Gene.* Third edition. Menlo Park, California: Benjamin.

Wolfe, S. 1981. *Biology of the Cell.* Second edition. Belmont, California: Wadsworth. One of the most up-to-date books available on material covered in this chapter.

Review Questions

1. Are the products specified by DNA assembled *on* the DNA molecule? If so, state how. If not, tell where they are assembled, and on which molecules.

2. How is a linear sequence of nucleotides in DNA translated into a linear sequence of amino acids in a protein?

3. Name the process by which RNA of three different types is assembled from the parent DNA code. Name the process by which the three different types of RNA cooperatively assemble a sequence of amino acids.

4. If sixty-one triplets actually specify amino acids, and if there are only twenty common amino acids, then more than one nucleotide triplet combination must specify some of the amino acids. How do triplets that code for the same thing usually differ?

5. Is the same basic genetic code used for protein synthesis in all living organisms? What significance is attached to that fact by most biologists?

6. Describe the general structure and function of the three types of RNA. What is a codon? An anticodon? Where is each physically located?

7. If you view protein synthesis as an information-processing system for putting DNA to work, then what are the elements of this system? (List each information-processing step, and name the molecules or structures that each represents.)

8. Cells depend on controls over which proteins are assembled at a given moment. Which three aspects of protein synthesis do these controls govern?

9. What is the basic difference between controls over gene expression in prokaryotes and in eukaryotes?

10. Define a structural gene. List and define three control elements that coordinate the operation of structural genes in prokaryotes.

11. Study Figure 11.8 which explains repression and induction of the genes involved in lactose breakdown in *E. coli*. In general, these controls work by modifying the interaction between _____ and _____ at the start of a gene.

12. In eukaryotes, most of the RNA in the nucleus seems to be concerned with driving control-element reactions. What does this suggest about the relative amounts of control elements and structural genes in the DNA itself?

13. What is it about nucleoprotein structure that suggests certain proteins may also be involved in controlling gene expression? As part of your answer, include an explanation of why calico cats have coat color variations.

12

THE GENETIC BASIS
OF DEVELOPMENT

If there is one thing that all the diverse prokaryotes have in common, it is their basic strategy for living: grow and divide, grow and divide, and grow again. Even this simple flow of living matter requires many controls at the genetic level, as you saw in the preceding chapter. By comparison, genetic controls in eukaryotes as complex as multicellular plants and animals must be awesome, indeed. After all, they must direct the unfolding of a single-celled zygote into all the specialized cells and structures of the adult form! The eukaryotic strategy, in essence, relies on controls through longer spans of time and over intricate cellular arrangements in space.

During **development** of all complex eukaryotes, cells first become different from one another in position, then in developmental potential, and finally in appearance, composition, and function. Cell division is of course necessary for these events. But what actually underlies **differentiation**—the processes by which cells of the same genetic makeup become structurally and functionally different from one another according to the prescribed program for the species? What controls are at work, coordinating the ways in which common predecessor cells give rise to the multicellular adult? The morphological aspects of development—the remarkable series of changes in body form—will be covered in later chapters, after you've had a chance to study different body parts and how they physically and functionally fit together in the whole organism. Here, the discussion will begin with a few general examples of what "differentiation" means before moving on to the kinds of genetic controls that are thought to govern it.

PATTERNS OF DIFFERENTIATION
IN EUKARYOTES

Dictyostelium discoideum, one of the cellular slime molds, is about as simple a eukaryote as you can get. At one stage of its life cycle, *D. discoideum* produces spores. Each spore gives rise to a single-celled amoeba, which divides into two amoebas, then four, and perhaps many more. The amoebas feed on bacteria. As long as plenty of food is around, they keep on growing and dividing. When food supplies dwindle, though, the amoebas begin streaming toward one another. Sometimes only a dozen or so are around to congregate; sometimes the number exceeds a hundred thousand. The mass of cells assumes the shape of a slug, which actually crawls about for a while! Then the slug differentiates in a remarkable way. Some of its cells are destined to form the upward stalk of a sporocarp, a mature fruiting body from which a new generation of spores will be discharged. Other cells are destined to become transformed into the spores themselves (Figure 12.1).

For this eukaryote, the change in environmental conditions obviously triggers differentiation. Some cells start producing quantities of a nucleotide, cyclic adenosine monophosphate (cyclic AMP). They secrete cyclic AMP into the environment. Then other cells some distance away move along the gradient, toward the region where the chemical is most concentrated. As cells contact one another during their movement, their outer membrane surfaces receive stimulatory signals. These signals must be transmitted through the cytoplasm to genes residing in the nucleus. At least, it's known that amoebas raised in isolation never receive such signals, their genetic program continues to specify only those enzymes and other gene products required for growth and division—and the cells never do differentiate. Thus, even for this "primitive" eukaryote, we can deduce the following:

Both environmental change and interactions between cells influence the activity of genes necessary for differentiation.

In corn, maple trees, cotton, and other flowering plants, environmental change often has this kind of drastic effect on gene activity. Here we're talking about more than responses to short-term shifts in external conditions such as nutrient availability. Seasonal changes in daylength, temperature, and moisture trigger processes leading to such major events as seed germination, flowering, leafing out in spring, and leaf drop in autumn. In flowering plants, we also find what appear to be *cytoplasmic effects* on differentiation. While they are forming, flowering plant zygotes are still attached to the parent plant. And even before a zygote begins to grow through mitotic divisions, it has already undergone **polarization**: cytoplasmic substances located at one end of this single cell are different from cytoplasmic substances located at the other. In a cotton zygote, for instance, most organelles (including the nucleus) reside in the top half; and a vacuole takes up most of the lower half. When the cell does divide, lower cells give rise only to a simple row of cells that transfer nutrients from the parent plant into the embryo. The upper cells give rise to the mature plant embryo. Thus, we might well speculate that *inherited differences in the cytoplasm contribute to later differences in structure and function.*

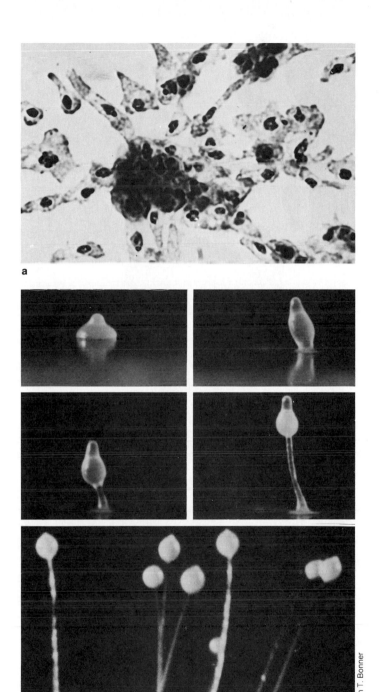

a

b

All photographs John T. Bonner

Figure 12.1 (a) Clumping of amoebas, preceding differentiation in *Dictyostelium discoideum*. (b) Formation of the mature fruiting body. The last photograph shows spores being discharged.

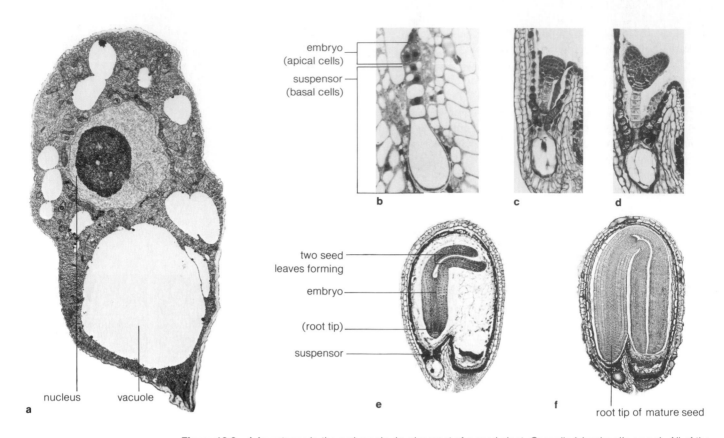

Figure 12.2   A few stages in the embryonic development of a seed plant, *Capsella* (shepherd's purse). All of the micrographs are in long-section. (**a**) Elongated, single-celled zygote, showing the polarization of cytoplasmic components. (**b**) Embryo after several mitotic divisions, with apical cells differentiated to form the embryo proper, and basal cells differentiated to form the suspensor. (**c**) Globular stage, containing differentiated cells that will make up the epidermal layer. (**d**) Embryo at heart stage, in which some cells are beginning to differentiate to form seed leaves (primary organs in these plants). (**e**) Continued divisions lead to an elongate embryo. (**f**) The embryo at maturity. (Micrographs a and b from Patricia Schulz; all others from Ray F. Evert)

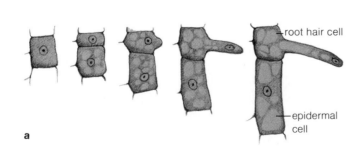

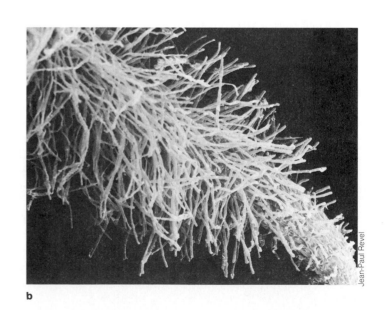

Figure 12.3   (**a**) Unequal cytoplasmic division in the differentiation of immature cells in a flowering plant. The unequal division helps dictate formation of a root hair cell (above) and an ordinary epidermal cell (below). (Sketch from Jensen and Salisbury, *Botany,* 1972)   (**b**) Scanning electron micrograph of root hair cells of a flowering plant. 46×.

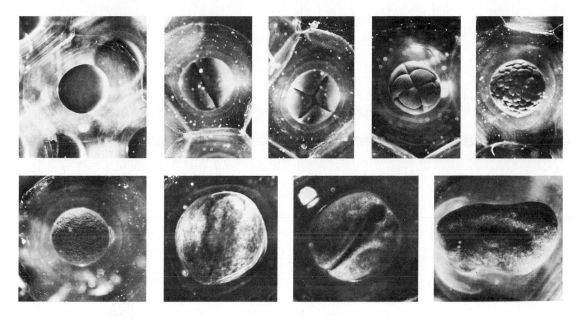

Figure 12.4 A single-celled frog zygote embarking on its genetically prescribed course of embryonic development and differentiation. The stages through which it must pass are outlined in the text; details of its morphological changes will be described in a later chapter. (All photographs © Carolina Biological Supply Company)

The plant embryo goes on to divide in ways that will form roots, stems, and leaves. Typically these mitotic cell divisions are unequal, as they are in immature cells that give rise to specialized root tissues (Figure 12.3). The unequal divisions lead to further variations in the way cytoplasmic substances are distributed in the growing cell mass. Beyond this, we now see cells dividing in different planes and expanding in different directions—and these differences lead to mature tissues and organs of diverse shapes. Can we identify any structural or chemical differences among these cells that might tell us about the nature of their variations? Yes we can.

First, by this time, individual cells vary in the number of microtubules present in their cytoplasm. Microtubules, recall, are assembled from protein subunits, with the aid of still other proteins (enzymes). *Hence transcription and translation of the genes coding for these proteins, as well as enzyme activity, may be variable in different parts of the developing plant.* Second, we can assume that by this time, the massed-together cells are interacting physically and chemically in ways that influence cell shape and function. It can be shown that substances released from cells in certain tissues induce changes in adjacent cells. Once target cells are stimulated, they become committed either to becoming just like the inducer cells, or to developing or functioning in entirely different ways. Some of the inducer substances are plant hormones. As you will read in Chapters Fourteen and Twenty, *hormones are known to have inhibitory or stimulatory effects on protein synthesis within target cell types.*

In complex animals, variable gene expression leads to the greatest cell specialization seen in the living world. And this specialization in turn affects gene expression. For instance, animal cells are able to concentrate on special metabolic tasks largely because of blood's stabilizing (homeostatic) effects. This circulating fluid flows past all body cells, helps nourish them, and thus helps keep them at their diverse tasks (Chapter Sixteen). At the same time, secretory organs such as the pituitary gland use the bloodstream as a highway for hormones that they release. When hormones reach target cells in one region, gene activity in those cells is coordinated with gene activity in related regions. In animal cells, then, sorting through the maze of gene interactions becomes a formidable task—*for organ systems interact in ways that alter gene activity.*

In animals, gene expression also varies through time. To get an idea of what may be going on at the genetic level, let's consider the kinds of changes that can occur during animal embryonic development. In the broadest sense, **gametogenesis** (gamete formation) is the first stage of animal development. During this time, recall, sperm and

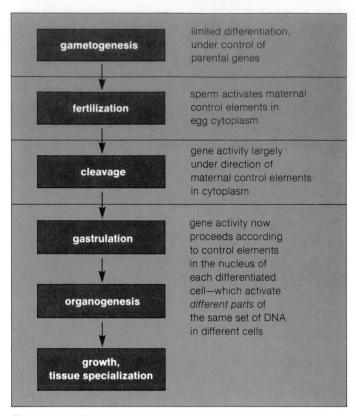

gametogenesis	limited differentiation, under control of parental genes
fertilization	sperm activates maternal control elements in egg cytoplasm
cleavage	gene activity largely under direction of maternal control elements in cytoplasm
gastrulation	gene activity now proceeds according to control elements in the nucleus of each differentiated cell—which activate *different parts* of the same set of DNA in different cells
organogenesis	
growth, tissue specialization	

Figure 12.5 Generalized picture of the stages of animal development. The same general pattern of gene activity applies to the development of complex, sexually reproducing plants. (As you will read later, plant cells don't undergo gastrulation; and organ formation, growth, and tissue specialization continue through the plant life cycle.)

eggs form and mature within the parent reproductive system. Sperm get decked out with a motile structure that will help move their DNA to an egg. In contrast, rich stores of substances become assembled in localized regions of the egg cytoplasm. The second developmental stage is triggered by sperm penetration into an egg. This stage, called **fertilization**, is completed when sperm and egg nuclei fuse. In **cleavage**, the third stage of animal development, the fertilized egg undergoes mitotic cell divisions to form the embryo. The embryo doesn't grow in size during cleavage. (Cells become smaller with each division, but together they occupy the same volume as the original zygote.) In some species, cleavage leads to a compact cell mass. More often, cells become arranged as a hollow ball (a blastula).

Gastrulation is the fourth stage of animal development. Through a series of cell movements, the embryo

becomes a gastrula (a structure having an inner cavity that opens to the exterior). Surrounding the cavity are two or more germinal layers of cells: differentiated forerunners for all the different organs that will make up the animal body. (For instance, three layers form in human and other vertebrate embryos. An outer ectoderm gives rise to skin epidermis and the nervous system. An underlying mesoderm is forerunner to such body parts as muscles, skeleton, and the circulatory system. An inner endoderm gives rise to digestive glands and the gut lining.) Once germinal layers have formed, cells divide into groups of cells, with each group destined to become a rudimentary organ. This is **organogenesis**, the fifth developmental stage. It is followed by **growth and tissue specialization**, a stage in which organs acquire the structure and chemical properties necessary for performing specialized tasks.

Figure 12.5 summarizes these stages of animal embryonic development. Let's now take a look at the main genetic events that help bring them about.

GAMETE FORMATION AND THE ONSET OF GENE CONTROL

In complex eukaryotes, the early stages of development are governed by control elements in the egg cytoplasm; they are not under the direct control of genes in the embryonic cell nucleus. What is the genetic nature of the messages harbored in the egg? Some information has been gathered from studies of different animal groups. During gamete formation in the female frog, the immature egg increases many thousands of times in volume. As it does, the nucleus becomes positioned at one end (the animal pole) and nutrients accumulate at the other (vegetal pole). This polarization determines where a forthcoming profusion of cellular components will be located. These components will include ribosomes, Golgi complexes, and endoplasmic reticulum required for protein synthesis, as well as mitochondria for rapidly producing the energy-rich molecules that will drive the synthesis reactions. Even while growth is occurring, tremendous amounts of RNA are being transcribed from the maternal DNA. Ribosomal, messenger, and transfer RNA molecules can be detected at this time, along with a large pool of heterogenous nuclear RNA (Chapter Eleven). Many of the mRNA molecules are translated at once into proteins (histones, for instance) that will be required for the first embryonic divisions. Other mRNAs become attached to ribosomes in different regions of the cytoplasm, and there they remain inactive. These are the **maternal messages** to be activated at fertilization. *Maternal messages direct the primary events of development, which in turn will set the stage for all that follows.*

Sperm penetration into the egg triggers fertilization. It sets in motion both morphological and chemical reactions within the egg cytoplasm. Soon, cytoplasmic substances in the fertilized egg undergo more rearrangement. These are the substances destined to take part in organ formation.

With the very first cleavages, different maternal messages may end up in different cells. Most genes in the nucleus are known to be inactive during cleavage. Hence the spatial and temporal differences in cell activity—and they are considerable differences—must be largely preprogrammed in the egg prior to fertilization! Some cells become flattened, some migrate about; communication is established between them. Microvilli and ridges appear, with their number and shape depending on the cell's location in the embryo. *All such changes in form and function of individual cells speak of changing patterns in protein synthesis from one cell to the next!*

Thus, with transcription of different maternal messages, different cells must rapidly acquire their own set of gene products—products for export or for incorporation into structures that give each differentiated cell type its special character. We can envision that chemical or structural differences of this sort may well induce changes in adjacent cells, activating gene after gene with cascading effects that spread through the developing embryo. Later chapters will describe the changes in form that accompany such early gene activity. Here we may say this:

The destiny of cell lineages is largely established according to which sector of the egg cytoplasm is inherited by the first embryonic cells formed during cleavage.

WHEN EMBRYONIC CONTROLS TAKE OVER

In animal cells, gastrulation marks the onset of major variations in gene activity. Now gene activity in the individual cell transcends the controls exerted by the maternal system. Transcription and translation proceed according to control elements in the nucleus of each differentiated cell, which activate different DNA regions. This pattern generally is maintained in each cell type.

During gastrulation, maternal controls over gene activity give way to control elements in each differentiated cell's nucleus.

One of the first examples of variable gene activity came from studies of *Drosophila*. In salivary glands of *Drosophila*, "giant" (polytene) chromosomes are formed when DNA is duplicated repeatedly and all the duplicated strands stick together in parallel array. A distinct banding pattern is characteristic of these chromosomes. The pattern is different for different types of chromosomes, but it's the same in all chromosomes of a given type. Each band is thought to be composed of homologous regions of tightly massed DNA on each strand, with each mass representing the location of one or perhaps several genes. In their highly condensed form, the DNA regions would be relatively inaccessible to enzymes of transcription.

However, some of these bands can be observed to uncoil into loops even as others remain condensed. Regions of polytene chromosomes apparently are like lampbrush chromosome segments (Figure 12.6), with open loops from homologous bands extending outward to form a **puff** (Figure 12.7). The transcription rate in cells containing giant chromosomes has been correlated with how large and diffuse these puffs become in appearance. In fact, it seems that puffing varies from one DNA region to the next during different developmental stages. This has been taken as further evidence that specific genes can be called into action at specific times in specific regions of differentiated cells.

CONTROLS MAINTAINING THE DIFFERENTIATED STATE

As the life cycle of a multicellular organism unfolds, genetic mechanisms that brought about the differentiated state remain active in maintaining the adult form. Cell interactions, too, are maintained by established communication networks, with electrochemical messages, hormones, and other signaling agents playing across intercellular junctions and the cell surface recognition factors characteristic of different cells. But in each differentiated cell, the DNA itself apparently remains the same, as you will now see.

Cloning Studies of Differentiated Cells

In 1958, F. Steward and his colleagues isolated small clumps of cells from a differentiated tissue in a carrot root. These cells they placed in a nutrient-rich liquid medium that was

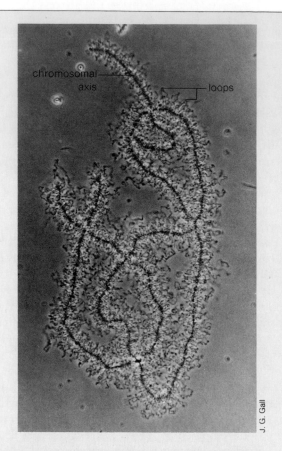

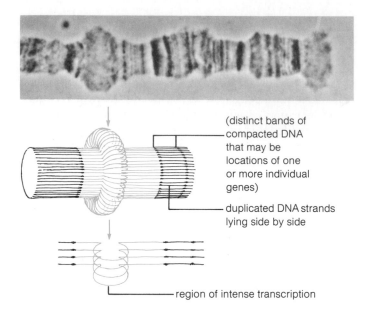

Figure 12.7 Chromosome puffing. The phase-contrast micrograph shows a prominent puff near the end of chromosome 3 from *Drosophila melanogaster*. (From J. G. Gall) The corresponding sketches represent the manner in which the parallel DNA strands become loosened and uncoiled in localized regions. (Sketches from A. Novikoff and E. Holtzmann, *Cells and Organelles*. Copyright © 1970, 1976 by Holt, Rinehart and Winston)

Figure 12.6 Micrograph of a pair of synapsed lampbrush chromosomes, taken from an immature egg of a female newt (*Triturus viridescens*). Notice how loops project laterally from the main chromosomal axis.

Much of what is known about the accumulation of maternal messages comes from studies of this kind of chromosome, which occurs in frogs, sea urchins, humans, and other animal groups. The chromosome takes on this form during oogenesis (Chapter 8). At that time, much of the DNA stays massed along the main chromosome axis, but some extends outward in long loops. In this state the chromosome resembles bristle brushes that were once used to clean lamps (hence the name). Its appearance heralds intense transcriptional activity. Dense clusters of RNA polymerases appear among the loops once they have formed. Soon, giant RNA transcripts peel off the loops.

The RNA assembled on lampbrush chromosomes apparently directs initial developmental events. Experiments have shown that early embryonic cells exposed to chemical inhibitors of gene transcription are still able to build most of the proteins characteristic of individual cell types. Such inhibitors have virtually no effect until late cleavage, when the embryo is composed of many thousands of cells that are beginning to transcribe their own genes in cell-specific patterns. During early cleavage, there simply is little RNA transcription for these inhibitors to suppress.

constantly rotated and aerated. The procedure soon led to a milky suspension of separate cells. The researchers discovered that any cell in the suspension could grow and develop normally into an embryo—even into an entire carrot plant, given the proper environment. In other words, these specialized cells were able to "de-differentiate" and repeat the developmental program! Any organism of this sort, which is reproduced asexually in such a way that it is genetically identical to its parent, is called a **clone**. Figure 12.8 shows the results from this kind of cloning experiment.

Equally spectacular were J. Gurdon's experiments. In the 1960s, Gurdon and his colleagues isolated intestinal cells from tadpoles of the South African clawed frog *(Xenopus laevis)*. They ruptured the plasma membrane of an individual cell in a way that left the nucleus and most of the cytoplasm intact. The broken cell was then introduced into an unfertilized *Xenopus* egg, whose own nucleus had been mechanically removed or destroyed by irradiation. In many cases, the transplanted nucleus was able to direct the developmental program leading to mature frogs!

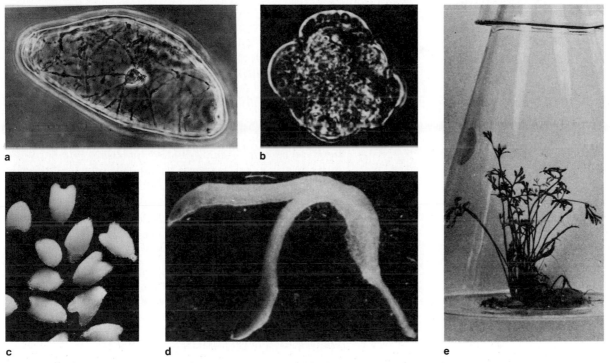

Figure 12.8 Development of a carrot plant from a single cell taken from differentiated tissue in a carrot plant. (a) Cell in suspension, (b) globular embryo growing in suspension, (c) beginnings of organ formation in developing cells, (d) embryo in which primary organs, the seed leaves, have developed, and (e) carrot plant grown from a "de-differentiated" cell. (a,b,d,e from Steward, Mapes, and Holsten, *Science*, 143:20–27. Copyright 1964 by the American Association for the Advancement of Science; c Walter Halperin, *American Journal of Botany*, 1966, 53:445)

The frogs appeared to be normal in every way—including their capacity to reproduce. More recently, Gurdon and R. Lasky performed similar experiments with nuclei derived from such differentiated *Xenopus* tissues as kidney, heart, lung, and testis—and normally differentiated, swimming tadpoles have been produced. Because the nucleus from a differentiated cell was embedded in a different kind of cytoplasm, its pattern of gene activity apparently was changed. But that nucleus must have had the same set of genes as the nucleus of the nondifferentiated egg—hence the development of mature frogs and tadpoles. *Such transplantation experiments lend strong support to the idea that all cells in a multicellular organism carry the same genes, with differentiation the result of different genes being expressed in different cells.*

Interactions Between Nucleus and Cytoplasm

That the cytoplasm helps control gene expression in differentiated cells became clear in the 1930s, through J. Hammerling's experiments with *Acetabularia*. These single-celled algae thrive in shallow tropical waters. Their nucleus re-

sides in the rhizoid, a tiny structure that anchors the cell to the seafloor. From this structure, a graceful stalk extends upward and then terminates in a caplike structure. In the species *A. crenulata*, the cap vaguely resembles a daisy. In the species *A. mediterranea*, the cap looks like an umbrella blown inside out. In one experiment, the stalk and cap were removed from the rhizoid of a mature daisylike alga about to undergo asexual reproduction. A stalk from *A. mediterranea* was grafted onto it. As Figure 12.9 shows, a daisylike cap was regenerated—a cap characteristic of the species from which the nucleus was derived. Similarly, when the stalk of *A. crenulata* was grafted onto the rhizoid of an alga of the blown-out umbrella sort, the cap regenerated looked like a blown-out umbrella.

How does the nucleus control these cytoplasmic differences? Perhaps hnRNA stored in the nucleus becomes activated by agents that signal the occurrence of cytoplasmic disruptions. Specific mRNA molecules could then be synthesized. They could leave the nucleus, diffuse through the cytoplasm, and become attached to those ribosomes upon which new proteins a given type must be built. At

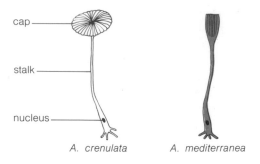

cap

stalk

nucleus

A. crenulata   A. mediterranea

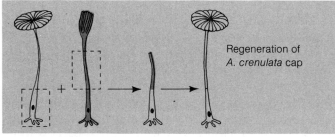

Regeneration of
A. crenulata cap

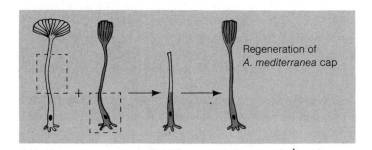

Regeneration of
A. mediterranea cap

Figure 12.9 *Acetabularia* grafting experiments implicating the nucleus as the control center for regeneration of cellular structures.

least, it's known that inhibitors of RNA synthesis injected into the rhizoid prevent cap regeneration. Furthermore, if a decapitated stalk and rhizoid are left intact for a few days—which would allow mRNA time to diffuse upward through the stalk—a new cap can be grown. It can be grown even when the stalk is severed from the rhizoid.

Such nuclear–cytoplasmic interactions are reciprocal. When a cap is severed from a mature alga about to undergo nuclear division, the nucleus never does divide. Once a new cap is regenerated, division proceeds. Similarly, an immature alga that has a mature cap grafted onto it thereby acquires the capacity to undergo nuclear divisions that lead to reproductive cell formation—as if it were developed to maturity (Figure 12.10). It is obvious, from these experiments, that cytoplasmic substances must regulate nuclear functioning; it is not obvious what those substances are.

CANCER: INSTABILITIES AMONG DIFFERENTIATED CELLS

Through nuclear–cytoplasmic interactions, variable gene expression is achieved and maintained in differentiated cells. Such interactions, however, are subject to breakdown. Although instabilities of this sort are extremely rare, when they do occur the results can be devastating.

For instance, at about the time of birth, brain cells cease dividing. From that point on, no new brain cells are formed. The stability of each brain cell depends on the stable state of its cytoplasm, and on the constancy of cytoplasmic signals being sent into its nucleus. Through these regulatory signals, all the genes governing cell growth and division are kept silent, and the genes governing synthesis of specific brain proteins are kept active. If something modifies the cytoplasm, then the cytoplasm will start sending abnormal signals to the nucleus. And the pattern of gene activity may change abruptly.

Consider what happens when nuclei from frog brain cells are transplanted into a fertilized frog egg. In a normal zygote, DNA duplication and cell division proceed rapidly; in a mature brain cell, they normally do not proceed at all. Almost from the moment of transplantation, though, brain cell nuclei respond to signals from the egg cytoplasm. Within five minutes, they begin duplicating their DNA and preparing for division!

Unfortunately, such gross modification of cell behavior is not merely a laboratory curiosity. Sometimes in human individuals, a single cell undergoes this kind of change. Instead of devoting itself to the normal program of gene expression, this differentiated cell begins duplicating its

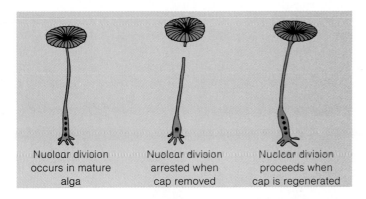

Nuclear division occurs in mature alga

Nuclear division arrested when cap removed

Nuclear division proceeds when cap is regenerated

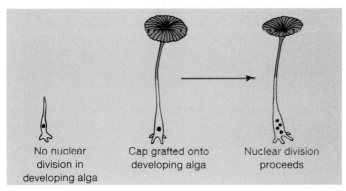

No nuclear division in developing alga

Cap grafted onto developing alga

Nuclear division proceeds

Figure 12.10 *Acetabularia* experiments implicating maternal control elements in the cytoplasm as having influence over nuclear activity.

DNA and growing. Then it divides. And it divides again and again. Soon the progeny of this one cell begin to crowd surrounding cells, exerting increasing pressure on them and interfering with vital functions. One single cell has gone out of control and has spawned a potentially lethal tumor.

Any **tumor** is an abnormal growth and massing of new tissue in any region of the body. It results from the breakdown of normal genetic control mechanisms. It is not that the affected cells grow and reproduce at some phenomenal rate. It is that they have lost the controls that tell them when to stop. If the tumorous growth is not removed, it may continue to destroy surrounding tissues and may bring about death of the individual.

If the only regulatory controls being modified are those governing cell growth and division, the situation is less serious than it might otherwise be. A tumor resulting from cells dividing more than they should is considered a **benign**

tumor. If the abnormal mass is surgically removed, its threat to the individual ceases.

If modifications in the control system lead to changes in the cell surface, the situation may be far more serious. Recall that membrane surface recognition factors identify a given cell as being of a certain type (Chapter Four). Such factors enable differentiated cells of like type to chemically recognize and interact with one another. Through these interactions, they remain bound together into tissues and organs. If something suppresses the genes coding for a given set of recognition factors, the cell will lose its identity. It will also show a disposition to adhere indiscriminately to any other cell type it encounters (Chapter Seventeen). Unchecked growth and division, and loss of surface properties necessary for proper intercellular communication—these are the characteristics of **malignant cells**, or **cancer**.

Malignant cells that have lost their normal surface markers can invade and destroy surrounding tissues. Not all malignant cells grow into one solid tumor. Some are capable of dispersal, or **metastasis**. In leukemia, white blood cells become malignant and divide, unchecked, in the bloodstream. Sometimes, too, cells slip away from a malignant tumor and enter the bloodstream; sometimes they wander through body tissues. They may become lodged in a variety of different tissues, there to begin multiplying and producing secondary tumors. Cancer cells exhibiting metastasis are more difficult to treat. Surgery in one region is not likely to remove all the malignant cells in the body. A tumor may be removed from one site, but another may appear elsewhere. Treatments other than surgery must then be used—specific chemical agents, radiation, or both—that selectively destroy uncontrollably dividing cells wherever they are in the body. As you will read in Chapter Seventeen, these treatments carry their own hazards. They must be administered with utmost caution in order to destroy cancer cells without destroying normally dividing cells (such as those which form blood).

What causes such traumatic changes in cellular controls? There are many known factors, and often they act in concert. Viruses cause cancer in experimental animals such as birds. Some viruses have been implicated in certain types of human cancer. Very rarely, common viruses normally associated with mild disease symptoms can initiate the loss of normal cellular controls. Malignancies may be triggered by other than infectious agents. Chemicals, environmental pollutants of many sorts, continuous exposure to sunlight, constant physical irritation of some tissue—all may increase susceptibility to, if not cause, some form of cancer. Chapter Seventeen will return to this topic.

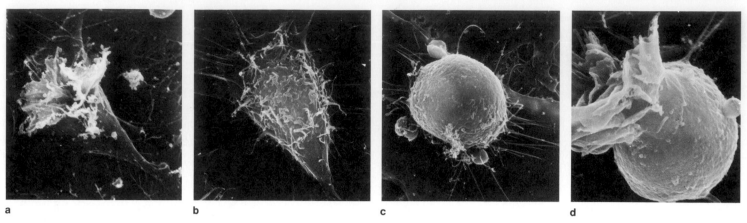

Figure 12.11 Transformation of a normal cell from a chick into a cancer cell. (**a**) After one hour, flowerlike ruffles appear on the surface of this normally spindle-shaped cell. (**b**) After two hours, microvilli and long, retractable fibers stud the surface. (**c**) After two days the cell has become spherical; notice the knoblike protrusions, or surface blebs. (**d**) New ruffles appear. In this case, the transformation was brought about by a temperature-sensitive mutant of a virus. (From E. Wang and A. Goldberg, *Proceedings of the National Academy of Sciences*, 73:4065)

Because we do not yet understand in detail the controls over the differentiated state in eukaryotic cells, neither do we understand in detail the changes occurring in cancerous transformations of those cells. But progress is now being made in identifying the processes involved, to the extent that many biologists are predicting that some presently unmanageable forms of cancer may be brought under control within the next few decades.

CELLULAR AGING

Curiously, cancerous cells have something in common with prokaryotic cells. As long as environmental conditions remain favorable, they simply go on dividing indefinitely. For instance, in 1951 a population of cancerous human cells was placed in culture—and today their descendants are still dividing, in cell culture laboratories all over the world. (These are the **HeLa cell** lines, an abbreviation of the name of the woman donor.) In contrast, normal cells of complex eukaryotes have limits on their life span—prescribed limits characteristic of their species. Following growth and differentiation, these cells begin to deteriorate. There is a gradual loss of efficiency in bodily functions, which culminates sooner or later in the death of the individual. The overall process of predictable cellular deterioration is called **aging**. It is built into the life cycle of all organisms in which differentiated cells show considerable specialization.

For instance, more than two decades ago, Paul Moorhead and Leonard Hayflick cultured normal embryonic cells from humans. They discovered that all the cell lines proceeded to divide about fifty times—and then the entire population died off. Furthermore, Hayflick took some of the cultured cells and froze them for a period of years. Afterward, he allowed them to thaw. These cells proceeded to complete the cycle of fifty doublings—whereupon they all died *on schedule.*

According to one hypothesis, aging is partly the result of an accumulation of environmental insults, which cause either structural changes or point mutations in chromosomes. Over time, such mutations would indeed hamper a cell's ability to produce functional enzymes and other proteins. Yet this assumption does not explain the fact that, regardless of induced damage, *different* cell types in the *same* organism have very different and entirely predictable life spans! Aging, in other words, cannot be explained entirely in terms of random environmental "hits."

Consider a process called **controlled cell death**, which is a built-in feature of many developmental events. In **morphogenesis**, groups of similar cells become coordinated in space so that they produce structures of predefined shapes. This development requires localized cell movements, localized controls over rates of cell division and differentiation, and localized contractions of some cells and extensions of others. The folding of parts, the hollowing-out of tubes, the shaping of bones, the separation of one part from another, the opening of eyes, mouth, nostrils, and ears—all involve plans of cell death. Perhaps you've noticed that kittens and puppies are born with their eyes sealed shut. From the time their eyes form until just after birth, the eyelids are an unbroken layer of skin. But then, certain cells stretching in a thin line across each eyelid respond to some internal clock and die on cue. As the dead cells degenerate, a slit forms in the skin, and the upper and lower lids part company.

Often, cell death is a response to a signal from a nearby developing structure. But there may be a long interval between receipt of a death warrant and cellular execution. In one experiment, John Saunders isolated two blocks of apparently similar cells from an embryo. One block came from a region where cell death normally occurs, the other came from a region where it does not. Both were cultured on a nutrient-rich medium. For days, both flourished. Then, at precisely the moment it was scheduled to die in the embryo, the predoomed block of cells suddenly died! The other block continued to thrive.

One more important example should be given here. Originally, hands and feet of developing vertebrates are shaped like paddles. In many species such as our own, skin cells between the lobes of the "paddle" die on cue, leaving separate fingers and toes. In other species such as ducks, cell death normally doesn't occur; that's why ducks have webbed feet instead of toes. Now, in some mice and some humans, a certain gene mutation blocks cell death; and hands and feet remain webbed. Experiments in which skin was grafted from normal mice to mutant mice showed that the mutation does not affect the ability of inner parts of hands and feet to *generate* the death signal. Rather, it changes the capacity of skin cells to *respond* to the signal. All other events of controlled cell death in all other parts of the developing body proceed on cue. Apparently the gene controls the response to one signal required for morphogenesis.

On the basis of such investigations, we can only speculate that aging and death may be coded in large part in DNA, like all other events of development and differentiation—and that signals from the cytoplasm of cells nearby or some distance away in the differentiated body activate those messages, telling one another that it is time to die. And we can only speculate that we may learn to come to terms with the inevitability of the prospect, accepting it as a natural process in the development and differentiation of an immense number of organisms on earth (see *Commentary*).

PERSPECTIVE

In this chapter, we have touched on some fundamental aspects of the genetic basis for eukaryotic development. We have looked at only a few examples, which have been limited largely to plant and animal developmental patterns. There are many variations on these basic patterns, as you will see in later units. Even so, the examples given are enough to convey the sheer enormity of the tasks confronting that set of genes residing in the microscopically small nucleus of the eukaryotic zygote. They have also been enough to convey the important fact that several principles are known to govern the genetic events underlying eukaryotic development:

1. Initially, embryonic cells differentiate largely because of localized differences that exist in the egg cytoplasm prior to fertilization.

2. Each differentiated cell contains a complete set of DNA, hence it contains *all* the genes necessary to produce a complete individual.

3. It follows that differentiated cells must be selectively expressing *different parts* of the same set of genes.

4. *Which* of these genes are expressed in a given cell depends on the nature of cytoplasmic changes induced by the environment and by interactions with other cells.

We have yet to discover how all the diverse cells in complex eukaryotes read the *same* genetic library in their own selective way, and thereby become unique. Somehow these cells respond not only to changes induced by the extracellular environment, but also to internal genetic cues. Somehow they exchange chemical signals that modify the cytoplasm. In turn, the cytoplasm of a given cell generates a new set of signals to the nucleus embedded within it, signals that cause some genes to be turned off and others to be turned on. That cell may then send out new signals, telling neighboring cells that it's time for them to change *their* patterns of structure and behavior. Thus eukaryotic cells embark on a most ambitious journey: they come into existence equipped with genetic controls specifying which species they are, where they are in relation to other cells, where they are headed, and what time it is as they move on a prescribed course from birth, through differentiation, to aging and death. And we now suspect that it is a journey that has its pathways—perhaps even its ultimate destination—encoded in the complex structure of eukaryotic DNA.

COMMENTARY

Death in the Open

Lewis Thomas
(Printed by permission from the New England Journal of Medicine,
January 11, 1973, 288:92–93)

Everything in the world dies, but we only know about it as a kind of abstraction. If you stand in a meadow, at the edge of a hillside, and look around carefully, almost everything you can catch sight of is in the process of dying, and most things will be dead long before you are. If it were not for the constant renewal and replacement going on before your eyes, the whole place would turn to stone and sand under your feet.

There are some creatures that do not seem to die at all; they simply vanish totally into their own progeny. Single cells do this. The cell becomes two, then four, and so on, and after a while the last trace is gone. It cannot be seen as death; barring mutation, the descendants are simply the first cell, living all over again. The cycles of the slime mold have episodes that seem as conclusive as death, but the withered slug, with its stalk and fruiting body, is plainly the transient tissue of a developing animal; the free-swimming amoebocytes use this organ collectively in order to produce more of themselves.

There are said to be a billion billion insects on the earth at any moment, most of them with very short life expectancies by our standards. Someone has estimated that there are 25 million assorted insects hanging in the air over every temperate square mile, in a column extending upward for thousands of feet, drifting through the layers of atmosphere like plankton. They are dying steadily, some by being eaten, some just dropping in their tracks, tons of them around the earth, disintegrating as they die, invisibly.

Who ever sees dead birds, in anything like the huge numbers stipulated by the certainty of the death of all birds? A dead bird is an incongruity, more startling than an unexpected live bird, sure evidence to the human mind that something has gone wrong. Birds do their dying off somewhere, behind things, under things, never on the wing.

Animals seem to have an instinct for performing death alone, hidden. Even the largest, most conspicuous ones find ways to conceal themselves in time. If an elephant missteps and dies in an open place, the herd will not leave him there; the others will pick him up and carry the body from place to place, finally putting it down in some inexplicably suitable location. When elephants encounter the skeleton of an elephant in the open, they methodically take up each of the bones and distribute them, in a ponderous ceremony, over neighboring acres.

It is a natural marvel. All of the life of the earth dies, all of the time, in the same volume as the new life that dazzles us each morning, each spring. All we see of this is the odd stump, the fly struggling on the porch floor of the summer house in October, the fragment on the highway. I have lived all my life with an embarrassment of squirrels in my backyard, they are all over the place, all year long, and I have never seen, anywhere, a dead squirrel.

I suppose it is just as well. If the earth were otherwise, and all the dying were done in the open, with the dead there to be looked at, we would never have it out of our minds. We can forget about it much of the time, or think of it as an accident to be avoided, somehow. But it does make the process of dying seem more exceptional than it really is, and harder to engage in at the times when we must ourselves engage.

In our way, we conform as best we can to the rest of nature. The obituary pages tell us of the news that we are dying away, while the birth announcements in finer print, off at the side of the page, inform us of our replacements, but we get no grasp from this of the enormity of the scale. There are 4 billion of us on the earth, and all 4 billion must be dead, on a schedule, within this lifetime. The vast mortality, involving something over 50 million each year, takes place in relative secrecy. We can only really know of the deaths in our households, among our friends. These, detached in our minds from all the rest, we take to be unnatural events, anomalies, outrages. We speak of our own dead in low voices; struck down, we say, as though visible death can occur only for cause, by disease or violence, avoidably. We send off for flowers, grieve, make ceremonies, scatter bones, unaware of the rest of the 4 billion on the same schedule. All of that immense mass of flesh and bone and consciousness will disappear by absorption into the earth, without recognition by the transient survivors.

Less than half a century from now, our replacements will have more than doubled in numbers. It is hard to see how we can continue to keep the secret, with such multitudes doing the dying. We will have to give up the notion that death is a catastrophe, or detestable, or avoidable, or even strange. We will need to learn more about the cycling of life in the rest of the system, and about our connection to the process. Everything that comes alive seems to be in trade for everything that dies, cell for cell. There might be some comfort in the recognition of synchrony, in the information that we all go down together, in the best of company.

Readings

Balinsky, B. 1975. *Introduction to Embryology*. Fourth edition. Philadelphia: W. B. Saunders. Outstanding introductory text; well written, and enough selective detail to enhance comprehension.

Behnke, A., C. Finch, and G. Moment. 1978. *The Biology of Aging*. New York: Plenum Press. Some of the current concepts and research pathways on the biology of aging.

Bonner, J. 1974. *On Development: The Biology of Form*. Cambridge, Massachusetts: Harvard University Press.

Croce, C., and H. Koprowski. 1978. "The Genetics of Human Cancer." *Scientific American* 238(2):117–125.

Davidson, E. 1977. *Gene Activity in Early Development*. Second edition. New York: Academic Press. Advanced text, but probably the only single source for quantitative information on molecular components of oocytes and early embryos.

Dyson, R. 1978. *Cell Biology: A Molecular Approach*. Second edition. Boston: Allyn and Bacon.

Goodenough, U. 1978. *Genetics*. Second edition. New York: Holt. A lot of current research is summarized here, but be prepared to face a forest of conceptual trees that will be thinned out, once many of the alternative theories are discarded.

Gurdon, J. 1974. *The Control of Gene Expression in Animal Development*. Cambridge, Massachusetts: Harvard University Press.

Hayflick, L. 1977. "The Biology of Aging." *Natural History* 86(4):22–116.

Lehninger, A. 1975. *Biochemistry*. Second edition. New York: Worth. Remarkably clear exposition on many of the topics covered in this chapter.

Salisbury, F., and C. Ross. 1978. *Plant Physiology*. Second edition. Belmont, California: Wadsworth.

Thomas, Lewis. 1974. *Lives of a Cell*. New York: The Viking Press. Thomas is both a scientist and poet; this collection of essays conveys optimism for the world of life in general and the human species in particular. A very special book.

Wollhouse, H. 1978. "Senescence Processes in the Life Cycle of Flowering Plants." *Bioscience* 28:25–31.

Review Questions

1. Define development, as it occurs in complex eukaryotes. Define differentiation.

2. What three major factors, besides the DNA instructions themselves, influence the activity of genes necessary for differentiation?

3. Can gene transcription and translation vary in different parts of a developing embryo?

4. Hormones affect protein synthesis within their target cells. If changes occur in one organ as a result of a hormone's activity, would you expect that the altered behavior of that organ might affect the development or activity of other organs?

5. List and describe what happens during the six main developmental stages for animal embryos. Do the embryos of all eukaryotes pass through all these stages?

6. Describe the maternal messages that accumulate in the egg cytoplasm. How do these messages direct the primary events of development?

7. As different cell types become specialized, what is happening to their particular assemblages of proteins?

8. During which developmental phase does maternal control over gene activity give way to control elements in each differentiated cell's nucleus?

9. What sorts of experiments suggest that all cells in a multicelled organism bear the same array of genes, but that differentiation results from *different* genes being expressed in different cells? Describe one such experiment.

10. Why do some researchers suspect that certain forms of cancer result from loss of control over gene activity?

11. What role does controlled cell death play in development? What do you suppose this process suggests about the process of aging?

12. Summarize the four main principles known to govern the genetic events underlying eukaryotic development.

ANIMAL SYSTEMS AND THEIR CONTROL

13

ANIMAL CELLS, TISSUES, AND SYSTEMS

SOME CHARACTERISTICS OF ANIMAL CELLS

Beneath the surface of the world's oceans and seas, attached to rocks and coral cliffs, tucked away in submarine canyons and proliferating in the sheltered waters behind tropical reefs are animals called sponges. It happens that the cells of these sponges display a tenacious capacity for self-recognition. The sponge body can be forced through a fine sieve or piece of cheesecloth, so that all that remains are isolated cells or tiny clumps of cells. These cells begin to move about, they bump into each other—and they begin adhering to each other. Their reunion is not random: channels and chambers lined with flagellated cells soon begin forming in a pattern characteristic of the intact sponge body! In the 1960s, the developmental biologists Tom Humphreys and Aron Moscona put two unrelated species of sponges through a fine sieve and mixed the isolated cells together in a medium containing calcium ions. The cells moved about, they bumped into each other—and they sorted themselves out according to their species!

Such experiments provided the first verification that something about *individual* cells in animal tissues encourages their chemical recognition of and adhesion to one another. The plasma membrane is now known to mediate these cell-cell interactions (Chapters Four and Five). As you read earlier, the external membrane surface contains glycoproteins and mucoproteins, which in many cases act as adhesion sites for like cells. In some cell types, the glycoproteins are so dense that the cell is said to have a sugar coating (or glycocalyx). The coating is soft and flexible, and has adhesive properties. In human red blood cells, the character of the coating varies from one cell type to the next, and this variation gives rise to the agglutination, or cell clumping, response between donors and acceptors of unlike blood types (Chapter Nine). Kidney cells, liver cells, nerve cells—all are known to have specific surface recognition sites for cells of like type. Mucoproteins, too, occur abundantly at the membrane surface and in the extracellular regions between many kinds of cells. Cells in the lungs, eyes, skin, skeletal joints—all are lubricated by or cemented to similar cells with the sticky, jellylike, or slippery mucoproteins.

Thus, chemical and structural bridges are assembled at the molecular level. These bridges weave through groups or layers of like cells, uniting them in form and function as a cohesive **tissue**. (The term, appropriately, is from the Latin word meaning "weave.") Groups of cells or tissues that interact as a unit in performing some specialized task represent what is called an **organ**. It is in the types of basic tissues present, and in the way they are joined, that organs take on their character. For instance, organs that must move more or less constantly (such as the stomach) are typically wrapped in layer after layer of muscle tissue, with different layers oriented at different angles for motion in more than one direction. Delicate organs such as capillaries are typically cushioned by connective tissue, which also helps hold them in place. It is not that there is tremendous diversity in animal tissues. Rather,

---

All the diverse body parts found in different animals may be assembled from a few tissue types, simply through variations in how those tissues are combined and arranged.

The animal cells making up these tissues are patterned according to the same basic plan described in Chapter Five. Most have a functional nucleus and other organelles embedded in the cytoplasm, which is surrounded by a plasma membrane. Somatic cells constitute the physical structure of the animal body. The other cells are germ cells, which develop into gametes. As you will now see, somatic cells become differentiated into the components of four main tissue types: epithelial, connective, muscular, and nervous tissues.

KINDS OF ANIMAL CELLS AND TISSUES

Epithelial Tissue

Covering the external body surface of all animals is one or more layers of cells. Such layers also line internal organs, ranging from simple gut cavities to vertebrate lungs. These are epithelial tissues: sheets of densely packed cells with little space or intercellular material at junctions between them. Epithelial cells are so well-stitched together at these junctions that they form a continuous barrier between the body parts they cover and the surrounding medium (water, air, or internal body fluids). Tight junctions in different tissues vary in number and extent. The more interconnected the cells are, the less permeable the tissue will be. And the more interconnected they are, the more the tissue can respond to stress associated with expansion and stretching movements.

In structural terms, cells in epithelial tissues look like flat floor tiles, cubes, or columns. Often these cells are ciliated on the side facing the surrounding medium, as they are in columnar epithelium that lines the earthworm gut. Epithelial tissue may be no more complex than one flat layer of cells; or it may be stratified into different layers (Figure 13.1).

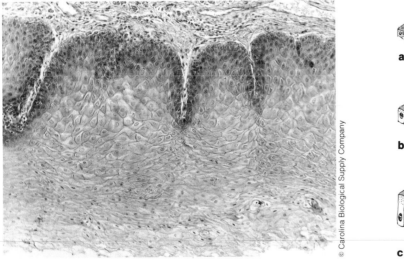
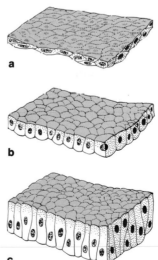

© Carolina Biological Supply Company

Figure 13.1 Types of epithelial tissue. Single-layer epithelium may be squamous (**a**), cuboidal (**b**), or columnar (**c**). Multilayered, or stratified, epithelium may be a composite of these types. The micrograph shows a section through the squamous (and, in this case, cornified) epithelium of mammalian skin.

Figure 13.2 Differentiation of a region of epidermal tissue into the organ we call a feather. The photograph shows the courtship display of a peacock, which relies on spectacularly specialized feathers to capture the attention of a peahen. (Sketches after T. Storer et al., *General Zoology*, sixth edition, 1979, McGraw-Hill)

Labels in figure: epidermis, dermis, sheath, follicle, cutaway of sheath to show shaft, barbs, quill, shaft, vane, barb, barbules

Roger K. Burnard

In functional terms, there are three kinds of epithelial tissues: protective, sensory, and glandular. **Protective epithelium** offers resistance to mechanical injury. It prevents loss of internal fluids. It also acts as a line of defense against microorganisms and other foreign particles. The outermost protective epithelium of all animals forms the **epidermis**. Compared to, say, a trout, the epidermis on amphibians, reptiles, birds, and mammals is stratified and somewhat harder (cornified). These animals spend all or some of their time on land, an environment that is more variable than marine or freshwater environments. Water supplies may not be constant, for example, and temperatures fluctuate. Hence the epidermis of land animals is tough enough to afford protection and to conserve internal body fluids. In addition, the nails, scales, claws, hairs, and feathers seen among these groups are all produced by differentiated dermal cell regions (Figure 13.2). Inside the body, protective epithelium acts as a boundary layer through which substances must pass in order to move into and out of the body or some body part. For instance, in most animals, food enters the body by moving across the epithelium that lines the digestive cavity. In many animals, oxygen enters and carbon dioxide leaves the body by moving across thin epithelium in the lungs.

Sensory epithelium contains cells that are highly specialized for receiving specific signals from the external environment, and sending them on to other parts of the body concerned with appropriate response to those signals. Part of your tongue consists of this kind of tissue, as do sensory cells of your eyes and your nose.

Glandular epithelium contains cells that are specialized for secreting extracellular products, such as mucus, sweat, and milk. Some, such as goblet cells, secrete their products on an individual basis. Others are clustered in an inwardly dimpled region of the epithelium, forming one kind of multicellular organ called a gland (Figure 13.3). If you have ever wondered why fish skin is moist and slippery, it is because of mucous secretions from glandular

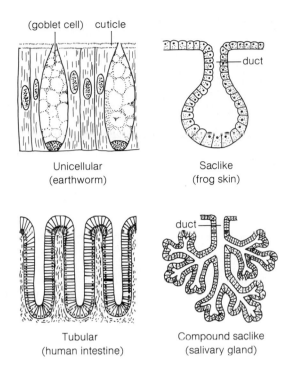

(goblet cell)   cuticle

Unicellular
(earthworm)

Saclike
(frog skin)

duct

Tubular
(human intestine)

Compound saclike
(salivary gland)

duct

Figure 13.3 Examples of glands found in glandular epithelium, as they would appear in long section. (After T. Storer et al., *General Zoology*, sixth edition, 1979, McGraw-Hill)

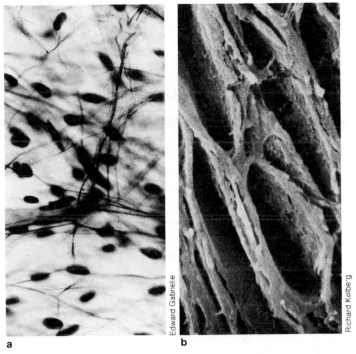

a   b

Edward Gabriele   Richard Kolberg

Figure 13.4 (**a**) Loose connective tissue. Cells and fibers are in a semifluid matrix. (**b**) Supportive connective tissue from ligaments. Parallel collagen strands give such tissues strength and flexibility.

epithelium. Secretions from such glands also form a cuticle, a thin, noncellular body covering of worms. They also form hard snail shells and the exoskeleton of the insect body.

Connective Tissue

Unlike epithelial cells, the cells of connective tissue are scattered through an extensive extracellular matrix. These tissues bind together and support other animal tissues.

Loose connective tissue is found throughout the body of most animals. It contains a weblike scattering of strong, flexible protein fibers (collagen) and a few highly elastic protein fibers (elastin). These collagen and elastin fibers are embedded in a semifluid matrix. Loose connective tissue is like packing material. It supports and holds in place blood vessels, nerves, and internal organs even while according them some freedom of movement. It also binds muscle cells together, and binds skin to underlying tissues.

Dense fibrous tissue attaches and holds movable body parts together. The structural organization of dense fibrous tissue is a key to its strength and function. It is typically made of parallel collagen fibers, tightly packed with very few cells and little matrix between them. Tendons, which attach muscle to bone, are made of these strong tissues. So are ligaments, which hold bones together at body joints. Sheets of this tissue also hold muscles in place.

Supportive connective tissue—cartilage and bone—is found in vertebrate skeletons. *Cartilage* (gristle) has small clusters of living cartilage cells suspended in a firm, rubbery matrix, which they themselves have secreted. *Bone* (osseous tissue) contains living bone cells, embedded in a dense, collagen-rich matrix that they have secreted. These cells are responsible for bone formation and repair. The collagen matrix is strengthened and made rigid by mineral deposits (mostly calcium salts). On the average, these deposits represent sixty-five percent of the total bone weight. Often, the deposits are thin layers (lamellae). Some of the bone cells reside in small spaces between layers, and become physically separated during growth and development. Even so, chemical communication between them is maintained through cytoplasmic processes and blood vessels (Figure 13.4).

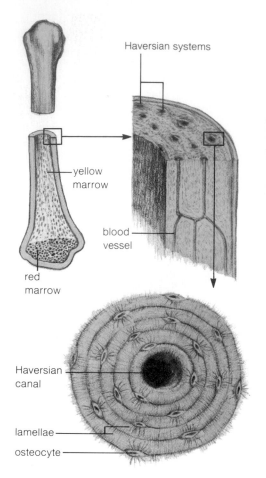

Haversian systems

yellow marrow

blood vessel

red marrow

Haversian canal

lamellae

osteocyte

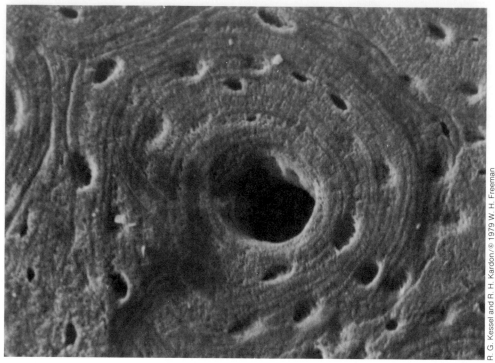

Figure 13.5 Scanning electron micrograph of a Haversian system in bone. The sketches show a long bone of a mammal with its distinct marrow regions and organized Haversian systems through which living bone cells receive nourishment (from blood vessels) and integrative signals (from hormones) by way of blood vessels.

In the long bones of mammals, lamellae are laid down around small channels called Haversian canals. These channels are interconnected and run more or less parallel to the bone (Figure 13.5). Haversian canals contain blood vessels, which wind inward from the bone covering and nourish bone cells embedded deep within the rigid matrix. Nerve cells also extend through Haversian canals, carrying chemical messages that control bone cell activities.

Other bone cells secrete hydrolytic enzymes that break down calcium deposits. The activities of these cells are necessary for changes in bone size and shape. Such changes occur during the life cycle of many animals, as they have during your own growth and development. Bone maintenance and remodeling both require that calcium removal be balanced with calcium deposition.

Wrapping each bone is a thin, fibrous covering to which muscles and tendons attach. The interior cavity of a long bone is filled with spongelike and fatty *yellow marrow*.

Bone ends are enriched with *red marrow*, a major site of blood cell formation. (Yellow marrow is like a reserve tissue. Whenever blood is lost, as it may be during bodily injury, this tissue also begins producing blood cells.)

Adipose tissue is another form of connective tissue. It consists of large cells, each having a single fat-filled vacuole. The fat in adipose tissue represents stored energy. It also cushions parts of the body against shocks and blows. In addition, adipose tissue insulates the body by helping to retard loss of heat generated during metabolism.

Blood, a circulating form of connective tissue, is found in almost all animals. Blood essentially consists of a fluid matrix (plasma) and free cells. Blood transports substances to and from body cells. Depending on the animal species, these substances include oxygen and carbon dioxide, nutrients and water, wastes, hormones, and infection-fighting cells. Blood also regulates pH by means of buffer molecules. It helps equalize body temperature by distributing metabolic heat. In many animal species, fluid filters through the walls of blood-transporting vessels and into the surrounding tissues that they service. This tissue fluid is called **extracellular fluid**. Some of it moves back through the blood vessel walls. But some is collected and returned to the blood by separate vessels, which are part of the **lymphoid system**. When it is in these vessels, the fluid is called **lymph**.

Vertebrate blood has four components: plasma, red blood cells, white blood cells, and platelets (Figure 13.6). The *plasma* part is more than ninety percent water, which acts as a solvent. Within this fluid are sixty or so different proteins. Each carries out a specific task, such as assisting in the body's defense mechanisms. *Red blood cells*, or erythrocytes, are the hemoglobin-rich oxygen carriers described earlier in the book. In an adult human, there are about 30 trillion red blood cells, or about 5 million per cubic millimeter of blood. Each has an average life span of about four months, but replacements are continuously formed in red bone marrow. *White blood cells*, or leukocytes, are not exactly permanent residents of blood. They use this circulating fluid as a highway for getting from their point of origin (bone marrow or lymphoid tissue) to their destination (some storage organ, for instance, or some battleground against infection). White blood cells function in the immune system, the physiological responses by which the body recognizes and eliminates foreign substances. They also help eliminate damaged, worn-out, and abnormal cells. *Platelets*, also produced in marrow, are cell fragments. They function in clotting (a reaction in which blood becomes coagulated, as it does around puncture wounds).

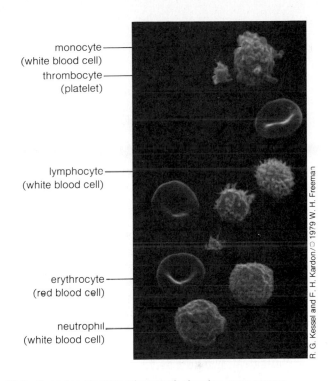

monocyte (white blood cell)
thrombocyte (platelet)
lymphocyte (white blood cell)
erythrocyte (red blood cell)
neutrophil (white blood cell)

R. G. Kessel and F. H. Kardon/© 1979 W. H. Freeman

Figure 13.6 Scanning electron micrograph showing some components of human blood: red blood cells (erythrocytes), platelets (thrombocytes), and white blood cells (leukocytes and lymphocytes). The leukocyte category contains cells known as monocytes and neutrophils. Although some white blood cells may look alike from their surface (depending on external conditions), they differ in nucleus shape and in the kinds and numbers of cytoplasmic particles. As you will read in Chapter Seventeen, they are important in the body's defense mechanisms.

Muscle Tissue

For most animals, movement through the environment depends on long, slender muscle cells. These cells contain *myofibrils*, threadlike structures that contract upon chemical stimulation. The contraction causes movements in body parts to which the muscle cells are attached. There are two kinds of muscle tissues: *striated* and *smooth*. The term "striated" refers to the striped appearance of myofibrils making up individual muscle cells (Figure 13.7). In the contracted state, the stripes look shorter and darker than they do when the cell is relaxed. Striated muscle forms both skeletal and cardiac muscle tissues.

Skeletal muscle tissue is composed of cylindrical cells that sometimes are longer than your thumbnail. The tissue

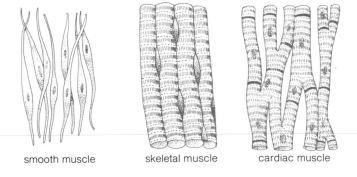

smooth muscle   skeletal muscle   cardiac muscle

Figure 13.7 Three kinds of muscle tissue.

itself is usually attached to vertebrate skeletons. It is responsible for voluntary movements that may be both rapid and intermittent. Typically, a number of skeletal muscle cells are enveloped in connective tissue to form a muscle bundle. Several of these bundles are usually enclosed in a tougher connective tissue sheath to form the functional units called individual muscles.

The cells of **cardiac muscle**, the contractile tissue of vertebrate hearts, are also striated (Figure 13.7). These cells are branched, and considerably shorter than skeletal muscle cells. In addition, membranes of adjacent cardiac muscle cells are fused, end to end, at regions called intercalated disks. Because of these fusion points, the cells do not function autonomously. Rather, when one receives a signal to contract, its neighbors are excited and contract in quick succession.

Smooth muscle tissue is composed of layers of spindle-shaped cells, held together by fibrous connective tissue (Figure 13.7). This tissue shows more prolonged and somewhat slower contractile behavior than striated muscle tissue. In vertebrates, smooth muscle is responsible for involuntary movements of such body parts as the digestive tract, blood vessels, the bladder, air passages to the lungs, and the genitals. (The term "involuntary" implies that the movements can occur on their own, without conscious intervention.)

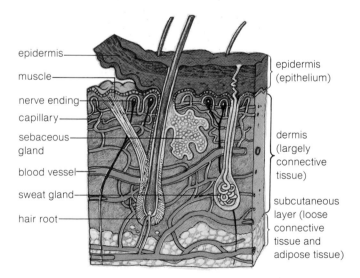

epidermis

muscle

nerve ending

capillary

sebaceous gland

blood vessel

sweat gland

hair root

epidermis (epithelium)

dermis (largely connective tissue)

subcutaneous layer (loose connective tissue and adipose tissue)

Figure 13.8 Some components of skin. (From *Vertebrate Body*, 5th ed., Romer and Parsons, © 1977 W. B. Saunders. Reprinted by permission of Holt, Rinehart and Winston.)

Nerve Tissue

With the exception of sponges, all animals depend on **nerve cells**, which have highly excitable plasma membranes. In response to certain chemical–electrical signals, nerve cells can stimulate neighboring cells (such as contractile cells) into activity. In cnidarians (jellyfish and their kin), nerve cells are dispersed through epithelium in a loosely connected way, so that messages spread diffusely from one body region to another without much orientation. In all other animals, nerve cells are functionally linked in pathways that have specific fields of origin, directional orientation, and specific destinations; they form a true nervous system. The **neuron**, an individual nerve cell, is the basis of such systems. A neuron's properties and functional connections with other cells enable it to receive and initiate signals in response to particular environmental changes, and to conduct messages specifically to another neuron, a muscle, or a gland. We will be taking a closer look at the neuron in the next chapter.

OVERVIEW OF ORGAN SYSTEMS AND THEIR FUNCTION

Having reviewed the basic kinds of animal cells and tissues, we may begin thinking about the ways that different tissues come together to form organs, then systems of organs. Human skin, for instance, is an organ system composed of all four basic kinds of tissues. As Figure 13.8 shows, the outer layer is the epidermis. Epithelial cells on the outermost layers are dead, but their dried-out toughness provides good protection in a land environment. The deeper epithelial layers are sites of active cell division. Here, new cells form through mitosis and push the overlying epithelial cells toward the surface. Dead cells are sloughed away from the surface even as their replacements are being formed. The renewal process goes on all the time, but we usually aren't aware of it unless confronted with the accelerated cellular activity associated with a case of dandruff or sunburned skin.

Beneath the epidermis but tightly connected to it is the dermis, a dense layer mostly of connective tissue. Scattered through the dermis are hair follicles (epithelial shafts in which hair grows), sweat glands, and oil glands. Smooth muscle cells attach to the follicles. When "goose pimples" form on your skin, these smooth muscle cells have contracted and pulled on the follicles. The follicles, which normally are positioned at shallow angles to the plane of

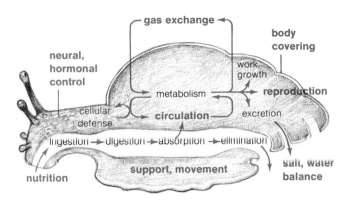

Figure 13.9 Simple portrayal of the basic functions in animals. In some animal groups, some of these functions (such as gas exchange) are accomplished only by individual cells. But as Table 13.1 suggests, there is in the animal kingdom a hierarchy of elaborations on this functional theme, involving organs and organ systems for the performance of the kinds of tasks listed.

the skin, stand up on end. The subcutaneous layer of human skin consists of loose connective tissue and adipose tissue. This layer binds the skin layers to the tissue below. Nerves wind through the dermis, reaching sweat glands, blood vessels, and hair follicles. Here, too, sensory neurons receive information about touch, temperature, and pain.

Like skin, all the specialized organs of the animal body are made of some combination of two or more basic tissue types. All of these organs have developed as adaptations to a particular environment. A dead animal could be laid out on the examination table, we could look at its gross structure, its component organs and tissues, its specialized cells—in other words, at its "anatomy." But that in itself wouldn't tell us why the animal is the anatomical composite that it is. Neither would it tell us much about how those composite parts work together. What kind of food is available for the animal? How does the animal capture it, extract nutrients from it, and get those nutrients to its individual cells to keep them functioning? How does the animal get rid of wastes? How does it keep warm or cool as temperatures change from night to day, or from one season to the next? How does it find shelter, or find a mate, or avoid predators? What adaptations allow it to survive on land, or in water, or both? How are its separate organs and tissues made to perform as a smoothly functioning whole?

Answers to such questions are the concern of "physiology"—the study of how organisms operate within the constraints of their environment. In this unit, we will be looking not only at the major organ systems of animals but also at how they work together to keep the animal alive in specific settings. These systems are more or less integrated in the representative animal sketched in Figure 13.9. Some assortment of these systems occurs in animals that belong to almost all major phyla, from cnidarians to chordates (Table 13.1). These are their main functions:

Body covering (protection from the external environment, protection from loss of internal fluids)

Nervous system (together with endocrine glands or system, integration of body functioning; detection of stimuli from the environment; control of responses to stimuli)

Endocrine glands or system (internal chemical control; together with nervous system, integration of body function)

Skeletal system (support; protection of some body parts; determination of some body shapes; in many animals, muscle attachment sites, blood cell production sites, calcium and phosphorus storage sites)

Muscular system (movement of internal body parts; movement of whole body through the environment)

Circulatory system (internal transport of materials to and from cells; pH and temperature stabilization)

Defense system (protection against foreign substances, such as those causing infection)

Respiratory system (gas exchange between surrounding medium and body cells, usually by way of circulating blood)

Digestive system (ingestion and preparation of food molecules for absorption; elimination of residues of digestion)

Excretory glands or system (disposal of certain metabolic wastes; regulation of salts and fluids in cellular environment)

Reproductive system (production of new individuals)

Not all of these systems are found in all animals, of course. And systems differ in appearance and complexity from one kind of animal to the next. Where they are found, however, their underlying function remains the same, as the categories of Table 13.1 suggest. This table gives an overview of the kinds of adaptations characteristic of the

Group (and some representatives)	Characteristic Environment	Characteristic Life-Style of Adult Form	Integration	Support and Movement
Porifera				
sponges	Most marine, some freshwater	Predatory; attached to substrate	Some contractile cells, no true nervous system	Support elements (needlelike spicules and/or protein fibers); contractile cells
Cnidaria				
hydras, sea anemones, jellyfishes, corals	Most marine, some freshwater	Predatory; some attached, some float or swim feebly; hydra somersault	Nerve nets; some have eyespots, statocysts (balance), sensory cells on tentacles	Muscle fibers in epithelium; some secrete hard external chamber material
Platyhelminthes				
flatworms, flukes, tapeworms	Marine, freshwater, body fluids of host, moist land	Predatory or parasitic; free-living or dependent on animal host tissue	Central and peripheral nervous systems; cephalization, dorsal-ventral differentiation, bilateral symmetry in systems	Well-developed muscle layers; no skeletal system
Nematoda				
nematodes, round-worms	Marine, freshwater, body fluids of host, land (deserts to ice)	Predatory or parasitic; free-living or dependent on plant or animal host tissue	Same as above	Longitudinal muscle fibers in body wall; tough cuticle; hydraulic skeleton
Annelida				
earthworms, leeches, polychaetes	Marine, freshwater, moist land	Scavengers, casual symbionts, or parasitic	Same as above	Longitudinal, circular muscles; some have parapodia (paddle-like appendages); hydraulic skeleton
Mollusca				
snails, slugs, clams; squid, octopus	Marine, freshwater, some on land	Predatory or scavengers; some crawling; squid and octopus, water-jet propulsion	Same as above	Muscular mantle; muscular foot in many; external shell in some
Arthropoda				
crustaceans, spiders, insects	Marine, freshwater, land	Predatory (carnivores, herbivores) or scavengers; some casual symbionts or parasitic; diverse life-styles from burrowing to flying	Same as above	Muscle bundles in jointed appendages; exoskeleton
Echinodermata				
sea stars, brittle stars, sea urchins, sea lilies, sea cucumbers	All marine	Predatory; some attached to substrate, others crawling	Central and peripheral nervous systems; nerve ring, dorsal-ventral differentiation, radial symmetry in systems	Endoskeleton of hard plates joined by muscles and connective tissue; tube feet for water-jet propulsion
Chordata				
tunicates, lampreys, true fishes, amphibians, reptiles, birds, mammals	Marine, freshwater, land	Predatory (carnivores, herbivores) or scavengers; few marine forms attached but most free-living; none strictly parasitic; highly diverse life-styles	Central and peripheral nervous systems; cephalization, dorsal-ventral differentiation, bilateral symmetry in systems	Antagonistic muscle system in many; axial, rodlike endoskeleton (notochord or vertebral column)

*Monoecious means having male and female reproductive organs, or gonads, in the same individual; hermaphroditic.
Dioecious means male and female reproductive organs in separate individuals.

Digestion	Gas Exchange	Circulation of Fluids	Water, Salt Regulation	Usual Reproduction
No digestive cavity; some cells transport food particles	Direct diffusion across cell membranes	Open system (inflow pores, central cavity, outflow channels for water)	Some cells regulate water flow	Asexual budding; sexual (monoecious or dioecious*)
Incomplete, saclike digestive tract (tentacles, mouth, gut)	Direct diffusion	No true circulatory system (gastrovascular cavity)	Direct diffusion	Asexual budding; monoecious or dioecious; simple gonads (reproductive organs)
Flatworms: incomplete straight digestive tract (mouth, gut); none in others	Direct diffusion	No true circulatory system (gastrovascular cavity)	Ciliated excretory cells (flame cells), excretory ducts	Monoecious; well-developed system of ducts, gonads, accessory organs
Complete straight digestive tract (mouth, gut, anus)	Direct diffusion (some internal parasites anaerobic)	No circulatory system	Bladderlike excretory organ	Dioecious; well-developed reproductive system
Complete digestive tract	Capillaries in parapodia; some have gills (thin filaments with capillary networks)	Closed blood-vascular system (heart, blood vessels)	Organ system with kidneylike structures (nephridia)	Dioecious; well-developed reproductive system
Complete digestive tract	Usually gills	Closed blood-vascular system in some	Organ system with nephridia	Dioecious; well-developed reproductive system
Complete digestive tract	System of gills, trachaea (finely branched tubes); some with book lungs (respiratory organs)	Open blood-vascular system (from heart, into tissues, to gills or book lungs, back to heart)	Malphigian tubules or secretory glands	Dioecious; well-developed reproductive system
Most have complete digestive tract; no anus in some	System of gills; or respiratory "trees"; also exchange with water moving through tube feet	Closed radial, blood-vascular system	Water-vascular system	Dioecious; some asexual by self-division; gonads, large, simple ducts
Complete digestive tract	Gills in some; respiratory system of trachaea, bronchi, bronchioles, lungs in others	Closed blood-vascular system; some with lymphatic system	Organ system that may include kidneys, ducts, tubules, bladder	Dioecious; well-developed reproductive system

main animal groups. As you continue with your reading, you may find it useful to refer now and again to this table. Otherwise, the necessarily limited examples in each chapter might give you a lopsided perspective on animal adaptations. For instance, much of the material in the next chapter has to do with the nervous systems of squids and mammals, simply because these systems have been studied more intensively and are better understood than others. But not all animals live in the complex, variable environments confronting squids and mammals, and not all have as complex a nervous system for response. This is not to say that the system they *do* have is any less adaptive, or even inherently less interesting; it is simply different. The table, moreover, can be a useful reminder of *diversity* in adaptations even though the forthcoming discussions use as their starting point the common challenges that all species of animals must face.

THE INTERNAL ENVIRONMENT

In scanning the list of functions carried out by separate organ systems, you may have recognized that these general activities are performed *within* individual cells. Animal cells have a biochemically protective covering (plasma membrane). They have their own regulatory control mechanisms, and their own cytomusculature. They have means for procuring energy-rich nutrients and other essential substances, for getting rid of wastes, and (in many cases) for reproducing. If each individual cell is able to perform all these basic activities, what is the contribution of organ systems to *cell* survival? The answer may be found by extending our earlier discussion of the increase in size associated with multicellularity (Chapter Eight). Most animal cells simply do not have direct access to the medium surrounding the body—in other words, to resources of the external environment. *What all organ systems do together is maintain a stable environment* within *the body that assures each cell of a steady supply of resources and favorable conditions for its activities.*

To give but one example, a heart muscle cell depends on cells in the bloodstream for its oxygen, even as blood cells depend on lung cells for access to oxygen from the atmosphere. At the same time, the heart muscle cell must perform its specialized role of contraction, thereby helping to assure that blood, with its oxygen cargo, will be pumped through the whole animal body. Thus we can readily envision each cell as operating on two levels at once:

Each animal cell must perform vital "housekeeping" tasks for itself, even as it performs some special activity that contributes to the stability of the internal environment for the entire multicellular body.

Let's briefly consider the nature of this internal environment. As first perceived by the nineteenth-century physiologist Claude Bernard, *each cell of each organ interacts with the extracellular fluid, a medium through which substances are continuously exchanged between cells.* For relatively simple animals such as the marine sponge, the sea itself bathes each cell. In more complex animals, the "sea" has been internalized. For instance, about sixty percent of a human's weight is water. About three-fourths of this water is inside individual cells, but the rest resides outside of them. In all vertebrates, the extracellular fluid is largely **interstitial fluid**, meaning that it fills the spaces between cells and tissues. The plasma portion of blood makes up the remainder of the extracellular fluid. Vertebrate blood constantly exchanges oxygen, nutrients, and metabolic by-products with the interstitial fluid, which then exchanges substances with the cells it bathes. It seems to be no coincidence that extracellular fluid closely resembles seawater, particularly in its high concentration of sodium ions. And in function, extracellular fluid is much the same as seawater that surrounds the cells of simple marine animals. Indeed, recognition of this similarity was one of the first clues that life may have evolved in the ancient seas.

Thus we have an extracellular medium containing water and essential ions—not only of sodium, but also ions of potassium, calcium, and hydrogen. Through this medium, nutrients and metabolic by-products move from cell to cell, and from one body region to another. As you read in Chapter Five, the levels of many extracellular substances must remain within a certain tolerance range if the cell is to survive. Each organ system contributes, in its way, to the preservation of optimal conditions. For instance, after more than a decade of research, Bernard discovered that the liver absorbs many of the nutrients carried to it by blood, and converts them to complex storage forms. When nutrient levels fall below a certain point in the blood—hence in the extracellular fluid—the liver chemically breaks down the stored substances and releases them for distribution throughout the body. Bernard also showed that the amount of blood being supplied to different body regions can be varied by the contraction and relaxation of small blood vessels—and that the variations are under local and neural control. This means that blood can be diverted to those regions where its cargo of oxygen and nutrients is in greatest demand.

HOMEOSTASIS AND SYSTEMS CONTROL

Through his studies, Bernard came to realize that ". . . all the vital mechanisms, varied as they are, have only one object, that of preserving constant the conditions of life in the internal environment." Much later, the American physiologist Walter Cannon carried this line of thinking further. Maintaining internal conditions, he stated, is possible only through systems of **homeostatic control:** coordinated systems whose operation works to keep some physical or chemical aspect of the body's interior within some tolerance range. These processes regulate the activities of all body parts, so that deviations from stable conditions trigger reactions that work to counteract or reverse the change.

By way of analogy, consider how a thermostatically controlled furnace works. A thermostat is a device that senses air temperature and that activates other devices for resisting temperature changes from a preset point. When the temperature drops below the preset level, the thermostat detects the change. Then the thermostat signals a modulator device, which turns on the heating unit. When air temperature has been raised to the proper level, the thermostat detects the change and signals the modulator, which shuts off the heating unit. The same kind of controls work inside an animal body. They are called **negative feedback mechanisms**, for an increase in some substance or activity will, at some point, inhibit the very processes leading to (or allowing) the increase.

Under some circumstances, **positive feedback mechanisms** may also be called into play. Here, some disturbance to the homeostatic state sets in motion a chain of events that throws things even further off balance! The blood-clotting mechanism is an example of this. (And it is an intricate enough example to warrant being postponed until later discussions.)

Homeostatic feedback mechanisms have three basic components: receptors, modulators, and effectors. Sensory cells or tissues act as *receptors* for some particular stimulus. A **stimulus** is any detected energy change in the environment. Some stimuli are changes in light or heat energy, sound wave energy, or chemical energy. Whatever the form, the energy of the stimulus is translated into electrochemical energy by the body's receptors. This energy

form is the signal sent to a *modulator*, a control point where responses to the stimulus may be selected. In some animals, the modulator is little more than some nerve cells clustered at the head end; in others, it is a complex spinal cord and brain. The modulator then sends signals to the body's *effectors*: muscles and glands. Effectors carry out the appropriate response to the original stimulus. In negative feedback mechanisms, the response works to counteract the stimulus. In positive feedback mechanisms, the response increases the stimulus.

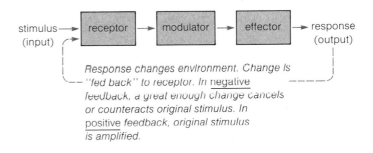

Response changes environment. Change is "fed back" to receptor. In negative feedback, a great enough change cancels or counteracts original stimulus. In positive feedback, original stimulus is amplified.

In addition to negative and positive feedback mechanisms at work within the internal environment, animals also have **feedforward mechanisms**, which amount to an early warning system. They depend on sensory receptors at the body's external surface. For instance, sensory cells in human skin may detect a drop in air temperature, and send messages to the brain. The brain, in turn, would send out signals to those metabolic and muscular systems by which internal body temperature is raised in response. With feedforward control, corrective measures can begin before a change in the external environment significantly alters the internal environment.

What we have been describing here is a general pattern of monitoring and responding to a constant flow of information about both the external and internal environment. All organ systems take part in the process. Thus, we can ask these things about the functioning of organs and systems: *What physical or chemical aspect of the internal environment are they working to maintain as conditions change? By what means are they kept informed of the change? By what means do they process incoming information? What mechanisms do they deploy in response?* With these questions in mind, we will now turn to the one system under whose dominion all others must fall—that of neural control.

Readings

Barnes, R. 1974. *Invertebrate Zoology.* Third edition. Philadelphia: W. B. Saunders. Good descriptions of organs and organ systems in invertebrate groups.

Bloom, W., and D. Fawcett. 1975. *A Textbook of Histology.* Tenth edition. Philadelphia: W. B. Saunders. Outstanding reference text.

Hickman, C., et al. 1979. *Integrated Principles of Zoology.* Sixth edition. St. Louis: Mosby.

Kessel, R., and R. Kardon. 1979. *Tissues and Organs: A Text-Atlas of Scanning Electron Microscopy.* San Francisco: Freeman. Outstanding; unique micrographs and well-written descriptions of major tissues and organs.

Romer, A., and T. Parsons. 1977. *The Vertebrate Body.* Fifth edition. Philadelphia: W. B. Saunders.

Staehelin, L., and B. Hull. 1978. "Junctions Between Living Cells." *Scientific American* 238(5):140–152. For the interested student, a good introduction to the kinds of cell–cell interactions that underlie tissue structure and functioning.

Storer, T., et al. 1979. *General Zoology.* Sixth edition. New York: McGraw-Hill.

Vander, A., J. Sherman, and D. Luciano. 1980. *Human Physiology: The Mechanisms of Body Function.* Third edition. New York: McGraw-Hill. Perhaps the clearest introduction to human organ systems and their functioning.

Review Questions

1. What is an animal tissue? An organ? Are there few tissue types, or a great diversity of different tissues? How can a bat wing and a human arm look and function so differently if they are constructed of the same basic tissues?

2. Name some functions of epithelial tissues. Describe protective, sensory, and glandular epithelia.

3. How do the five main types of connective tissues differ in structure and function? Name the four main components of blood, and their function.

4. Describe skeletal, cardiac, and smooth muscle tissue. How does nerve tissue differ from muscle tissue?

5. List the ten major organ systems that occur in animals. What is the main function of each system?

6. In the multicelled animal, cells operate on two levels at once. What are these two levels?

7. Together, all organ systems maintain a stable environment within the body. Why is their interaction important for individual cells?

8. What is the relationship between blood plasma and interstitial fluid?

9. What is homeostatic control? How is the internal environment kept more or less constant when conditions change?

10. Contrast negative and positive feedback controls.

11. Receptors, modulators, and effectors are the basic components of any homeostatic feedback mechanism. What kinds of body parts serve as receptors, modulators, and effectors in animals?

From time to time, the human body has been likened to a city, or state, or some other social unit composed of separate but interdependent parts. These analogies are wonderfully optimistic about our capacity for social organization. In truth, our cities and states don't begin to approach the degree of functional integration of *any* complex animal! In such organisms, lines dividing the operation of parts at each level—from cells, to tissues, organs, and organ systems—blur to insignificance. Whether the animal is asleep, relaxed, or alert to danger, some body parts are being called into action in a coordinated way even as activities of other parts are being suppressed. Throughout the life cycle, each body part is constantly monitored and evaluated—not for its sake alone but for how it is contributing to working patterns of the whole.

How is animal integration accomplished? For answers, we will have to consider how different environments present different sorts of challenges. Some animals live where conditions don't change much. For them, adjustments depend on relatively simple nerve nets (somewhat diffuse message-conducting pathways through the body). Other animals live in richly varied environments and must constantly adjust to changes of large and small sorts, outside and within the body. For the more complex forms, adjustment means the coordinative action of a nervous system and an endocrine system. As you will see, the foundations for this kind of control are *constellations of neurons* (nerve cells arranged relative to one another in precise message-conducting and information-processing pathways), and *endocrine cells* (which secrete chemical messages that affect target cells some distance away).

14

INTEGRATION AND CONTROL

THE NEURON: ITS STRUCTURE AND FUNCTIONING

Let's begin by looking at the **neuron**, for it is a nerve cell found in almost all animals. This kind of cell is able to receive and send messages about change. It is also able to weave together the meaning of *different* messages arriving from different communication pathways in the animal body.

In their structure, neurons are spectacularly diverse. Describing them all would be a hopeless task, if it were not for one thing. It turns out that most neurons have four message-handling zones. These zones occur on the cell body, and on its more or less permanent cytoplasmic extensions, collectively called "processes." All but the sim-

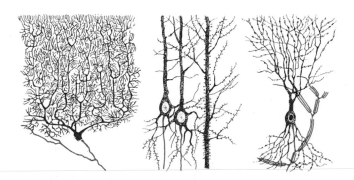

Figure 14.1 Examples of structural diversity in neurons. Those shown occur in the brain of mammals. Axons are shaded in gold. In some neurons, dendrites branch profusely; others form relatively thick trunks, tufts, or thin threads. Many more varieties exist, including the type of motor neuron shown in Figure 14.2.

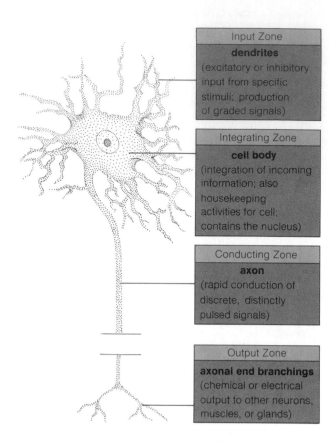

Figure 14.2 Main functional zones of many neurons.

plest neurons have an **input zone** consisting of one or more *dendrites:* relatively short, often branched, and sometimes spiny processes (Figures 14.1 and 14.2). Here the neuron receives most incoming messages. All neurons have an **integrating zone**, where information coming in simultaneously on different dendrites is summed. The summation may increase or decrease the neuron's activity. This integrating zone is the cell body. Most neurons have a **conducting zone**, which consists of a single, smooth-surfaced process called an *axon.* An axon may be as short as five microns or as long as five meters (as it is in a giraffe leg); when it is long, it sometimes has side branchings. The axon is a through-conducting pathway for messages that must travel rapidly, without alteration, from one region to another. At its terminal endings is an **output zone**, where chemical or electrical signals are released. These signals may affect another neuron, a muscle cell, or a gland.

Intimately associated with neurons are assorted cells, collectively called **neuroglia**. In vertebrates, neuroglial cells represent at least half the volume of the nervous system. Despite their abundance (and, by implication, their importance), the function of most neuroglial cells is not well understood. Some, though, are known to be metabolically associated with neurons, aiding in their nutrition. Others surround the axons of special neurons in ways that influence message propagation, as you will soon see.

What Is a Neural Message?

Like all cells, neurons have a plasma membrane, a lipid bilayer structure in which proteins and other molecules are embedded. You know from Chapter Five that a plasma membrane is differentially permeable. It passively permits some molecules and ions to move across the membrane, along their concentration gradient, and it passively prevents others from doing so. Sometimes it actively transports ions, so that they move against their concentration gradient. As you will see, this property is essential in message propagation from one region of the neuron to another.

A neural membrane is permeable in different ways to different ions.

Let's look at some ion distributions on both sides of the neural membrane. Among other things, a neuron "at rest" (one that isn't being stimulated) has far more positively charged potassium ions inside than out. It also has far more positively charged sodium ions outside than in

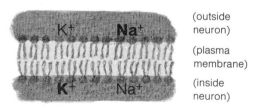

Figure 14.3 Simplified picture of ion distributions across the neural membrane, for potassium and sodium.

Figure 14.3. This means that concentration gradients are being maintained across the membrane:

Among other things, two steep concentration gradients—one for potassium ions, and one for sodium ions—exist across the neural membrane.

Steep potassium and sodium concentration gradients are only part of the picture, though. Another part is that an electrical gradient also exists across the neural membrane. Compared with the outside, the inside of a neuron at rest has an overall negative charge. For most neurons in most animals, this charge difference across the membrane is somewhere between sixty and ninety millivolts. The "millivolt" is simply a unit for measuring the potential of some separated electric charge to do work when it moves from one place to another—*as it can do across a membrane*. Thus the charge difference represents potential to do work; it is the **resting membrane potential**.

The electrical gradient as well as the concentration gradients influence the diffusion of ions across the neural membrane. Over time, potassium follows its concentration gradient, and shows a net outward diffusion from the neuron. Although sodium is simultaneously diffusing in, it is doing so at a much lower rate. So positively charged ions start accumulating outside. This shift in charge attracts potassium ions to the neuron's more negatively charged interior. Even so, they continue their net outward migration as long as their concentration gradient remains steep enough. Eventually, though, the charge difference across the membrane comes to exert an effect that's equal in magnitude and opposite in direction to the concentration force. When that happens, potassium ions undergo no more *net* movement across the membrane.

Now, if the concentration gradients for potassium and sodium are so steep, and if electrical attractions exist, then why doesn't net diffusion equalize concentrations across the membrane? The answer is a mechanism built into the neural membrane: the **sodium-potassium pump**. There is evidence that enzymes may be the actual "pumps."

For certain charged ions, concentration gradients across a neural membrane are maintained by enzyme pumps.

What effects do sodium-potassium pumps have on diffusion? Even though potassium ions tend to diffuse out of the neuron, enzymes actively pump them back in. Similarly, even though sodium ions tend to diffuse slowly into

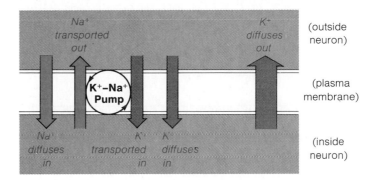

Figure 14.4 Sodium-potassium pump in the neural membrane. The relative widths of the arrows indicate the magnitude of the movements. As you can see, the combined effect of diffusion (in response to the electrochemical gradients) and of active transport by the pump is that the inward movement equals the outward movement for each kind of ion when the neuron is not being stimulated.

the neuron, enzymes actively pump them back out. These pumps help assure that the inward movements of each kind of ion are exactly balanced by its outward movements (Figure 14.4). Thus they help maintain the resting membrane potential. In summary, we can make the following points:

---

1. For certain charged ions such as sodium and potassium ions, concentration gradients exist across the neural membrane.

2. The inside of a neuron at rest has an overall negative charge with respect to the outside; an electrical gradient exists across the membrane.

3. The potential energy from these concentration and electrical gradients can be tapped to conduct a message.

So far, we have a neuron primed to conduct a message. What exactly is this "message," and how is it conducted along a neural membrane? In essence, it's a type of **current**—a movement of electrical energy. It arises largely through stimulus-induced movements of potassium and sodium, which change the resting membrane potential. A stimulus changes membrane permeability to these ions in the following way. The stimulus energy seems to cause a change in proteins that span the membrane width. Within

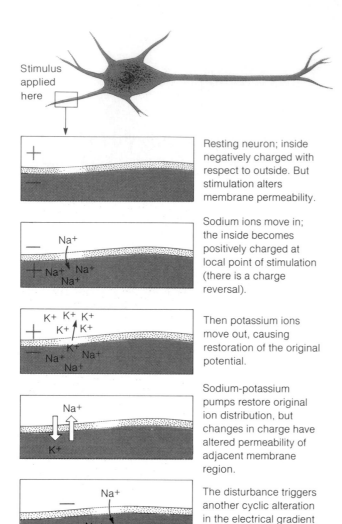

Resting neuron; inside negatively charged with respect to outside. But stimulation alters membrane permeability.

Sodium ions move in; the inside becomes positively charged at local point of stimulation (there is a charge reversal).

Then potassium ions move out, causing restoration of the original potential.

Sodium-potassium pumps restore original ion distribution, but changes in charge have altered permeability of adjacent membrane region.

The disturbance triggers another cyclic alteration in the electrical gradient across the membrane.

Figure 14.5 Simplified picture of how a message is propagated over the membrane surface of a neuron. Here the message is depicted as moving in only one direction; normally it radiates outward from the initial point of stimulation.

these proteins is a molecular "gate." The gates are normally closed, but they open under stimulation. Once the gates open, a given type of ion can rapidly move in or out of the neuron, in the direction of its chemical and electrical gradients. When sodium gates open, the membrane briefly becomes more permeable to sodium ions. With more sodium moving in, the neuron becomes more positive inside at the stimulated region. After a short delay, potas-

sium gates in the same region open, and potassium ions move out. When they do, the interior reverts to its original negative state. Sodium-potassium pumps then work to restore the original ion distribution. In the meantime, the disturbance has altered the permeability of adjacent membrane regions (Figure 14.5). There, sodium ions move in and potassium ions move out. This new disturbance affects the permeability of the next adjacent region, and so it goes, away from the original point of stimulation. What do all these temporary alterations in membrane potential mean?

Alterations in the electrical gradient across the membrane constitute the "message" that travels over the surface of a neural membrane.

Once they start, do messages travel without changing along the neural membrane? Some do, and some don't. To understand why this is so, it helps to think about how a nerve process compares with, say, a telephone cable. Electric current flows through conducting material (copper, for instance) in the cable's core. The cable is well insulated, which cuts down current loss to the outside. Also, its core diameter is large enough to reduce the longitudinal resistance that all conducting materials have to current flow. By comparison, whether a neuron is insulated or not, its plasma membrane would have to be called "leaky" to current flow. If the stimulus acting on the membrane is weak, then current flow isn't large enough to cause a significant disturbance in adjacent membrane regions; hence the message decays close to where it started out. (The reason is that sodium and potassium movements across the membrane quickly dissipate the small disturbance that a weak signal causes.) Also, many nerve processes are so small in diameter that they greatly resist current flow, so messages they carry decay quickly. But these properties are not flaws in neuron architecture! They are the reason why animals having complex nervous systems are able to show such tremendous diversity in response, as you will now see.

Message Conduction

Neurons have processes of varied length and diameter. Neurons concerned with fast responses usually have the largest and best-insulated axons. For example, giant axons innervate the muscular body wall (mantle) of predatory squids (Figure 14.6). Signals sent down these axons control rapid muscular movements required for the chase. As another example, the cockroach displays one of the fastest of all motor responses. It reacts within twenty-five milliseconds to a puff of air on receptors located at its tail end (such puffs normally being associated, we might suspect, with rapid-breathing cockroach-eating rodents). Cockroaches, earthworms, clams—such invertebrates as well as vertebrates are locked in predator–prey relationships. And all depend on fast, long-range transmission of unaltered messages.

Characteristic of neurons concerned with long-range message conduction is the **action potential**, also known as the *nerve impulse*. An action potential is an all-or-none, brief reversal in membrane potential (Figure 14.7). It doesn't decay with distance. Once an action potential has been triggered, it is self-propagating: its amplitude won't change even if the strength of the stimulus changes.

Messages that must travel unaltered and with great speed through the animal body are encoded as action potentials.

What is the chemical basis of the action potential? Recall that when a stimulus is applied to a resting neuron, the neuron's interior undergoes a brief change in membrane potential because of the inward sodium movement. Now, when a stimulus is strong enough, this initial change brings the membrane potential to a critical threshold value. A **threshold value** is the minimum change in membrane potential needed to trigger an action potential. The electrical disturbance associated with the change in ion permeabilities across the membrane triggers a chain reaction that leads to the opening of all sodium gates at the site. As Figure 14.7 suggests, the membrane potential reverses abruptly—which causes sodium gates to close and potassium gates to open. The cycle of events concludes with partial potassium gate closure when the membrane potential has returned to its resting value.

Nerve impulses always move in a direction away from the site of stimulation. There is no backflow, largely because each action potential is followed by a **refractory period**: a time of insensitivity to stimulation. It occurs while potassium is flowing out of the neuron and the resting membrane potential is being restored. After the refractory period ends, the electrical disturbance has moved far enough away that it no longer can cause sodium gates to open in the region of the original impulse.

An interesting variation on this theme is seen in sheathed neurons. Many vertebrate neurons don't have large-diameter processes, yet some of their thin axons prop-

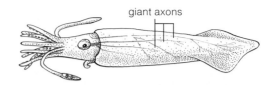

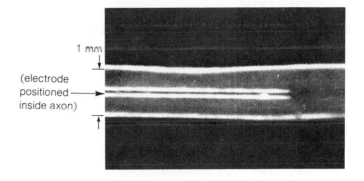

Figure 14.6 Giant axon that innervates the muscular body wall (mantle) of the squid *Loligo*. The sketch shows the approximate location of these axons in the squid body; they aren't drawn to scale. Being large enough to accommodate devices for measuring voltage changes (notice the electrode inside the axon in the photograph), this axon was the basis for experiments that revealed our first understanding of nerve functioning. (From A. L. Hodgkin, *J. Physiology* (London), 1956, vol. 131)

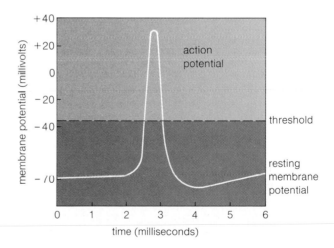

Figure 14.7 Recording of an action potential, or nerve impulse, as described in the text.

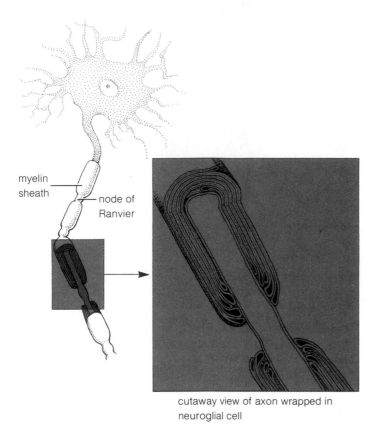

myelin sheath

node of Ranvier

cutaway view of axon wrapped in neuroglial cell

Figure 14.8 Myelinated axon of a neuron.

agate nerve impulses at a remarkable 120 meters per second! This high-speed conduction relates to an association between these axons and **Schwann cells** (a kind of neuroglial cell). Schwann cells wrap their surface membrane many times around a single axon. The membrane layers of Schwann cells form a lipid-rich *myelin sheath* (Figure 14.8). Each sheath cell is separated from the next by a *node of Ranvier:* a small gap where the axon is exposed to extracellular fluid. The lipid-rich wrappings between nodes are regions of high electrical resistance. This means that ions can't move readily across the membrane where wrappings occur, hence action potentials can't occur there. So current is forced down the axon until it reaches a node, where action potentials can occur. The current concentration at the node means that the threshold value (and the resting action potential) will be reached quickly. In this

manner, an action potential "jumps" from one node to the next in line. (This rapid node-to-node hopping is called *saltatory* conduction, after the Latin word meaning "to jump.")

Once started, action potentials can't change. Some neural messages do change, though, and they can do so within a short distance from the point of stimulation. These short-range messages are called **graded potentials**. They are conducted by sensory receptors, dendrites, and the branched endings of axons. A graded potential not only decays quickly in strength, it also can vary in size. The stronger the stimulus, the larger the current and the larger the associated change in membrane potential. The weaker the stimulus, the smaller the current and the smaller the change.

Why is the possibility for variation in the size of graded potentials so important? Consider that in your own brain, as many as 80,000 nerve processes can converge on a single neuron! At such convergence points, currents arriving from different processes may interact before transmission occurs. Without reinforcement, weak currents may simply subside. It also happens that some kinds of signals can cancel others. At each junction, *all* incoming messages are summed. And further action depends on which stimuli are stronger at a given moment. More will be said about this in the section to follow. In reading through this section, keep in mind the following point:

Graded potentials are short-range messages that can show variations in size. Hence graded potentials arriving at a neuron from different pathways can be <u>added up</u> to determine the course of further transmissions.

TYPES OF NEURONS AND FUNCTIONAL LINKS BETWEEN THEM

There is no such thing as one ant at a picnic. Neither is there any such thing as one neuron in an animal. In both cases, the functioning of one can't proceed for long without the company of others like itself. In some animals, *billions* of neurons are organized into an interconnected system. Some are **sensory neurons**, which are receptors for environmental stimuli. They serve as windows on changing conditions in the external and internal environments. Others are **interneurons**, which connect different neurons

Figure 14.9 Synapses between sending neurons and receiving cells. Messages typically move outward from each active synapse on the membrane surface of a receiving neuron. Hence the whole neuron surface, not just a process or two, can become excited. The *direction* in which a given message flows through the body depends on the way neighboring neurons and their processes are organized relative to one another, as the gold arrows suggest.

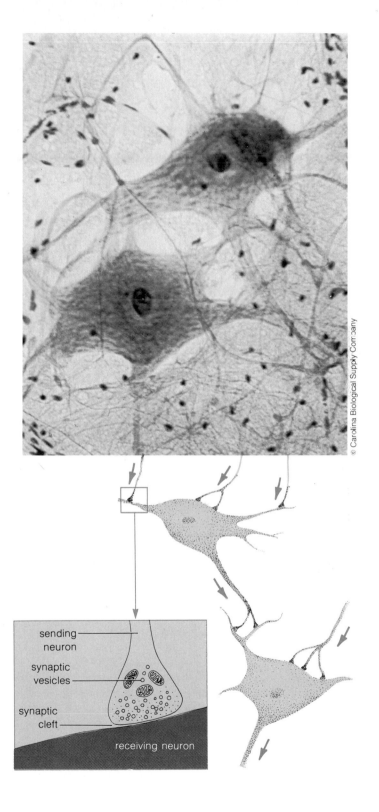

in the system. As you will see, your brain and spinal cord contain interneurons. Still others are **motor neurons**, which send action potentials down their axons to branched axonal endings. There, commands are transferred to effectors: to muscles or glands that increase or decrease their activity according to incoming signals.

Almost always, the functional connections between a neuron and another cell are near-contacts, with a small extracellular gap separating the two. Each such connection is specialized for transmission, and is called a **synapse**. The region between the two cells is a synaptic cleft. Here, a neuron transfers a signal to another cell (Figure 14.9).

Usually, the signal being sent to a receiving cell is carried by chemical messengers called **neurotransmitters**. The chemicals are produced near the tips of sending cell processes and stored there in small vesicles. Upon stimulation by an action potential, vesicles release neurotransmitter, which diffuses across the synaptic cleft to the receiving cell.

Signals are also produced and transmitted in muscle cell membranes. (In complex vertebrates, only nerve and muscle cells have "excitable" membranes.) Muscle contraction is described in the next chapter, but here we can consider the neural mechanism that triggers the event. For instance, activation of a vertebrate skeletal muscle cell begins when a signal reaches the end of an adjacent motor neuron (Figure 14.10). In these axonal endings, synaptic vesicles move toward and fuse with the neural membrane. Then their contents, molecules of neurotransmitter **acetylcholine**, are released into the synaptic cleft. Acetylcholine diffuses across the cleft and combines with protein receptor molecules on the muscle cell membrane. When that happens, ionic gates open and increase the membrane permeability to sodium and potassium. The resulting ion movements produce a current across the membrane. (The acetylcholine that triggered this change doesn't accumulate in the cleft; enzymes rapidly break it down.)

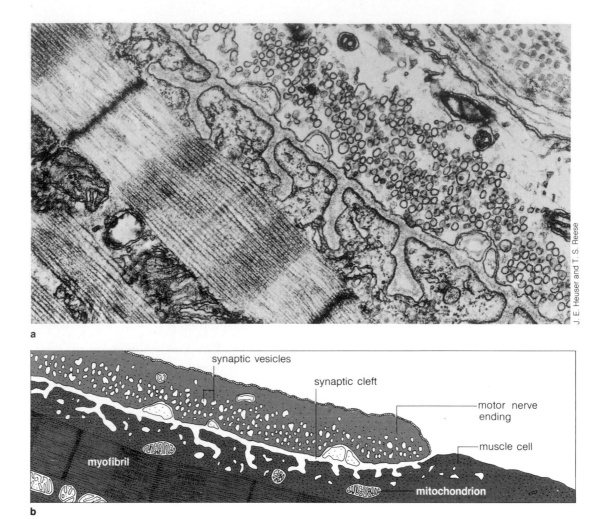

Figure 14.10 Transmission electron micrograph (**a**) and sketch (**b**) of a neuromuscular junction. The axon of this motor neuron terminates in a slight depression on the muscle cell surface. A narrow space (the synaptic cleft) separates the nerve ending from the muscle cell. Notice how the part of the nerve cell facing this space is not wrapped in a protective myelin sheath. Notice also how channels from the main synaptic cleft penetrate the region of the muscle cell membrane. An as-yet unidentified substance fills all these channels. And it is through this substance that chemical messages may travel from the nerve to bring about muscle contraction.

Labels in figure: synaptic vesicles, synaptic cleft, motor nerve ending, muscle cell, myofibril, mitochondrion

a

b

J. E. Heuser and T. S. Reese

FROM SIGNALING TO PERCEPTION

The Information Content of Neural Messages

So far, you have read that messages are carried along neurons as a kind of current, and that neurons receive and send messages to other cells. Let's now turn to the *content* of those messages, using your own body as an example. At all times, interneurons in your brain receive bits of information from receptors all over your body. These bits they piece together into a coherent mosaic—an understandable representation—of where you are in a changing environment. How do they do it?

The first step is encoding messages about the type of stimulus. There are many different types of receptor cells. And each kind normally carries information about only one form of stimulus. Even in the dark, you can "see stars"—experience visual sensations—when you stumble and bump your head hard enough. The blow triggers visual receptors in an unorthodox manner, but the receptors don't signal "pressure" or "pain." *They still announce what they are genetically equipped to announce.* Thus we can say this:

Neural messages contain information about <u>stimulus type</u>: about activation of particular receptors, which carry information about one form of stimulus only.

The second step is encoding messages about stimulus intensity. Sights, movements, sounds, temperature changes, taste—the perceived strength of such sensations depends largely on how many receptors in a given tissue region are being activated—and how often. For instance, how do action potentials convey to your brain the difference between a throaty whisper and a wild screech? Variations in stimulus strength are encoded as variations in the *frequency* of action potentials traveling one after another, in series. (Figure 14.11 gives examples of frequency variations corresponding to differences in sustained pressure applied to human skin.) In addition, a stronger stimulus usually acts on more nerve processes. The message of a very strong stimulus may even be carried by one or more **nerves**: bundles of separate axonal processes. Thus not only the frequency of signals but the number of receptors activated tells something about the stimulus.

Neural messages contain information about <u>stimulus intensity</u>, coded in the number of receptors activated and the frequency at which they are firing.

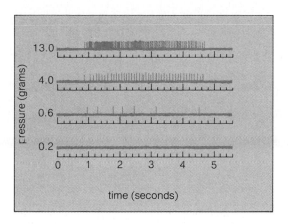

Figure 14.11 Impulses recorded from a single pressure receptor on the human hand, which correspond to variations in stimulus strength. A rod one millimeter in diameter was pressed against the skin with the force indicated. Vertical bars above each thick horizontal line correspond to the number of impulses. Notice the increase in impulse frequency which corresponds to increased stimulus strength. (From Hensel and Boman, *J. Neurophysiology*, 1960, 23:564–568)

The third step is identifying stimulus location. Your body's receptors are distributed through different tissue regions. One or more receptors of the same kind connect with a single sensory neuron. Together, the receptors and neuron form a sensory unit, and the tissue region they sample is called a **receptive field**. Stimulating a receptive field causes the sensory neuron to fire off. The neuron of each sensory unit is a private information channel leading into your spinal cord or brain. There, its message may branch out, through synapses with other interneurons. Its message may also converge with messages from other sensory units that synapse with the same interneuron. In either case, separate lines of interneurons may send messages to your brain. Just how accurately the brain pinpoints stimulus location depends on the physical track the messages have taken to reach it.

Pinpointing stimulus location also depends on the size of each receptive field, and on the overlapping of adjacent receptive fields. For instance, you are able to distinguish between the pressure of your thumb and that of your thumbnail even when you press them into the same tissue region of your fingertips. The variation is recorded through small, overlapping receptive fields that are abundant in fingertips. One reason why you have trouble localizing pain in your gut is that the walls of internal organs have fewer sensory processes and broader receptive fields, with less

overlap, than those found in your skin. From the points made so far, we can say this:

Neural messages contain information about <u>stimulus location</u>, coded in the number of signals on specific communication lines from a given body tissue region.

Integration: Summing Up Messages at Synapses

So far, we have information about stimulus type, intensity, and location. How is this information assigned relative importance? Not all stimuli are equally significant at any given moment. For example, some receptors in your eyes may be sending intense messages to your brain about the spectacular beauty of the Grand Canyon, which you are seeing for the first time. But if other receptors are sending messages that the edge of the canyon where you are standing is starting to crumble, then you really ought to have some means for giving the more urgent messages priority.

Different messages converging simultaneously on a neuron are summed to provide information of ever more precise content. A receiving neuron may have hundreds or thousands of synapses acting on it. Neurotransmitters secreted at these synapses can change the receiving

Figure 14.12 From signaling to visual perception.

Parts of the brain concerned with vision contain arrays of neurons, stacked in columns at right angles to the brain's surface. Connections run between neurons in each column and between different columns. Each column apparently analyzes only one kind of stimulus, received from only one location. The transformations from signaling to visual perception have actually been traced through eight levels of synapses.

What do eight synaptic levels tell us, given the *billions* of synaptic connections in the brain? Perhaps a great deal. In spite of their immense numbers, neurons in the brain's surface layer fall into a few basic categories. And those of each category seem to be tripped into action in the same way. For instance, excitatory signals from neurons climbing up through the brain's surface layer or running parallel to its surface activate particular neurons, which send out inhibitory commands to other neurons even as they themselves are being stimulated. Because of the precise vertical organization of these interactions, the excitatory and inhibitory feedback loops between neurons form narrow bands of activity—*the effect of which is a highly focused pattern of excitation through specific columns of neurons.*

Some experiments of David Hubel and Torsten Wiesel tell us something about this focusing. They implanted electrodes in individual neurons in the brain of an anesthetized cat. After the cat woke up, they positioned it in front of a small screen, then projected images of different shapes (such as a bar) onto the screen. Changes in electrical activity that accompanied different visual stimuli were recorded. As the above sketches suggest, the strongest signals were recorded for one kind of neuron when the cat observed an image of a vertical bar. When the bar image was tilted, electrical activity slowed down. When the image was tilted past a certain point, electrical activity stopped. Another neuron fired only when a block image was moved from left to right across the screen; still another fired when the image was moved from right to left. Such experiments suggest that the key to visual perception resides in the organization and synaptic connections between columns of neurons in the brain.
(Sketches from Kuffler and Nichols, *From Neuron to Brain*)

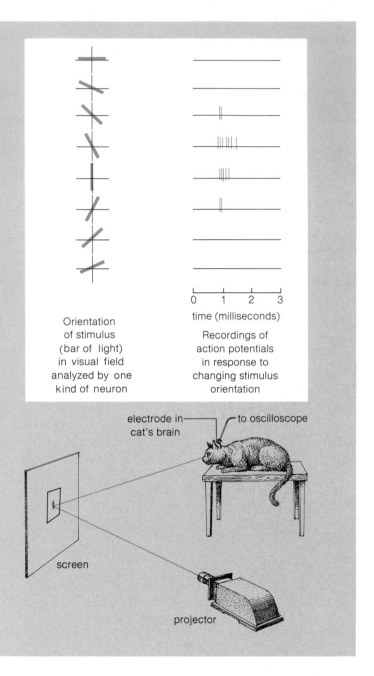

Orientation
of stimulus
(bar of light)
in visual field
analyzed by one
kind of neuron

time (milliseconds)

Recordings of
action potentials
in response to
changing stimulus
orientation

electrode in cat's brain — to oscilloscope

screen

projector

neuron's membrane permeability to some particular ion. The charge difference across the membrane is made to change in one of two ways: either the receiving neuron's activity is enhanced, or it is inhibited. The outcome depends largely on the kinds of ion involved, and on the state of the receiving neuron (for instance, on how close its membrane potential is to threshold).

At an **excitatory synapse**, a neurotransmitter such as acetylcholine increases membrane permeability to both sodium and potassium, which drives the membrane potential *toward* threshold. In other words, an excitatory synapse —alone or in combination with other excitatory synapses —goads a receiving neuron and increases the likelihood that it will fire off. At an **inhibitory synapse**, neurotrans-

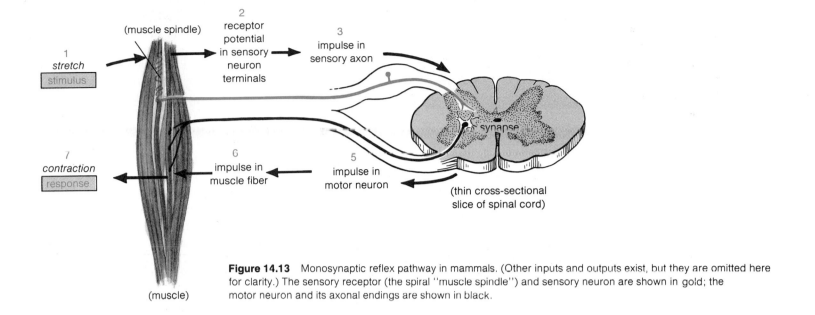

Figure 14.13 Monosynaptic reflex pathway in mammals. (Other inputs and outputs exist, but they are omitted here for clarity.) The sensory receptor (the spiral "muscle spindle") and sensory neuron are shown in gold; the motor neuron and its axonal endings are shown in black.

mitters increase membrane permeability to potassium and/or chloride, which drives the membrane potential *away* from threshold. In other words, inhibitory synapses decrease the likelihood that a receiving neuron will fire action potentials.

Imagine action potentials shooting toward your brain on separate communication lines from sensory receptor units. At each level of synapsing, some signals may be excitatory and others inhibitory. These signals, arriving as they do from branched endings of axons, are graded potentials. Graded potentials, recall, can be added up. Hence each convergence point for incoming messages represents a relay station where information may be sent on, reinforced, or suppressed.

What we call <u>integration</u> is, in its most fundamental sense, the moment-by-moment summation of all excitatory and inhibitory synapses acting on a neuron.

The brain happens to be the most complex integrative center in the living world. As Figure 14.12 shows, we are only beginning to understand the levels of synapsing that lie between the signaling of some environmental change and the brain's perception of that change. It is only recently that the excitatory–inhibitory balancing acts at each synaptic level have become recognized as a key to understanding what is going on.

THE REFLEX ARC: FROM STIMULUS TO RESPONSE

We have looked at short-range and long-range nerve messages, and at how their information is transferred from one cell to the next. Let's now look at one of the pathways in which messages are coordinated in complex animals. A **reflex** is a sequence of events elicited by a stimulus. During a **stretch reflex**, a muscle contracts involuntarily whenever conditions cause a stretch in length. For vertebrates, the stretch reflex assures a rapid and uncomplicated response. For instance, even when you aren't aware of the stretch reflex, it is at work helping you maintain an upright posture despite small shifts in your balance.

As Figure 14.13 shows, length-sensitive organs are located within skeletal muscles. These organs, called **muscle spindles**, are made of small muscle cells enclosed in a sheath and running parallel to the muscle itself. Sensory neurons connect with different muscle spindles. When a spindle is stretched strongly enough, action potentials are generated in the sensory axon. These potentials are conducted rapidly toward the spinal cord. There, axonal endings of the sensory neuron connect functionally with (among other things) motor neurons—which have axons leading right back to the muscle that was stretched. If the frequency of sensory impulses is high enough, action potentials will be generated in each motor neuron. They will travel to its axonal endings, there to activate the muscle membrane. This activation can lead to muscle contraction. Thus the stretch reflex prevents (within limits) the over-stretching of a muscle.

The picture just given is simplified, in that we have been looking at just one synapse in the stretch reflex arc (sensory neuron → motor neuron). However, other branches of the sensory neuron spread out in the spinal cord region and make other connections, leading (for instance) to the brain. It's true that stretch reflexes normally take precedence over other neural responses when you are standing upright. But even such one-to-one connections are subject to intervention by input from many inter-neurons. Say that you are standing up and someone stomps on your foot. Muscles in your leg cause your knee to bend rapidly in response. The injured foot is thereby lifted in the air—which is the opposite motor response of the stretch reflex.

NERVOUS SYSTEMS: INCREASING THE OPTIONS FOR RESPONSE

Evolution of Nervous Systems

A reflex arc is one of the clearest examples of message conduction between some point of excitation and the body's effectors. This is not to say that it's the simplest example. As you have seen, the activity of interneurons can alter the outcome. What *are* the simplest animal systems of message conduction? Some species of sponges contract slowly in response to a pinprick. But the response is diffuse, and never more than a few millimeters away from the point of irritation. It is not until we turn to the cnidarians that we find a system of neurons organized into reflex arcs.

Sea anemones, hydras, jellyfish—all such animals are cnidarians, and all live in water. Some forms attach themselves to the seafloor; others are free-floating. Regardless of whether they float or sit, all cnidarians face similar challenges and respond in similar ways. For them, food and danger are likely to appear not on the water's surface or on the bottom but anywhere in between (Figure 14.14). Their system for sensing and responding to the environment is arranged radially about a central axis, much like spokes of a bike wheel. Arrangements of this sort are said to show **radial symmetry**.

In the cnidarian system, many neurons are arranged in **nerve nets**: they are dispersed through epithelium, yet

are functionally linked to sensory cells, to each other, and to muscle tissue in ways that permit diffuse message con-duction and response. Most messages can travel either way along these neurons, so there is little orientation to infor-mation flow. That's why responses to most stimuli are somewhat diffuse.

Many cnidarian tissues have at least two nerve nets governing which muscles contract, and at what rate. One is a through-conducting reflex network controlling contrac-tile tissues (for example, between tentacles and mouth). Another network is localized, being able to spread graded potentials only a small distance. Cnidarians also have re-ceptors, such as statocysts that sense which way is up (Chapter Fifteen). At junctions with sensory receptors are **ganglia**: organized knots of neurons encased in connective tissue and forming integrative centers. Finally, many cni-darians show spontaneous neural activity in response to *internal* timing mechanisms! Some jellyfish ganglia, for in-stance, have internal "clocks," or **pacemakers**. These pace-makers are neurons. They become active during embryonic development and thereafter fire and reset each other in a rhythmic activity governing feeble swimming movements. Internal "clocks" also govern cyclic changes in the internal state. For instance, sea anemones "stand up" during active feeding phases, and "wilt" in between. Thus cnidarians show more than simple reflex responses to stimuli. They also show *patterns* of behavior.

We can use cnidarians as a model for the evolutionary beginnings of integration, but where do we go from here? A clue comes from the embryonic development of some modern-day cnidarians. Typically, offspring of these animals do not look at all like their parents. Each is a planula: a ciliated larval form. At first a planula uses its cilia as motile structures for swimming or crawling. Then it settles down on one end, and a mouth forms at the other. It becomes a polyp, a form typified by the lower left sketch in Figure 14.14.

If planulas settle down where food is plentiful, there is no problem. If they settle down where it is not, prospects are grim: they can't pick up and move on. Suppose that long ago, mutations gave rise to a planula that ignored the developmental signal to stop crawling, or in which the signal was never given. Suppose, further, that this planula's mouth opened up not on one end but on the underside. Such a mutant would resemble an existing animal: the flatworm. It would at once have selective advantages. At its mouth would be living microorganisms and the remains of others, in amounts more concentrated than food floating through the water. And with its "feet," it could now move to regions of organic feasts!

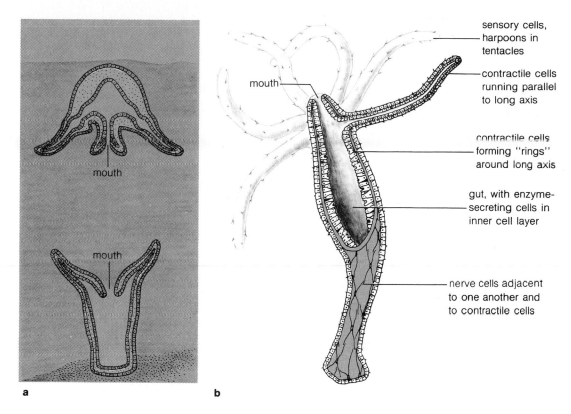

Figure 14.14 (**a**) Where cnidarians are in relation to their environment. Whether sedentary or free-floating, these animals have a radially symmetrical system of responding to food or danger that can come from any direction. (**b**) System of sensory reception and response in *Hydra*, arranged radially about mouth and gut—and suggestive of evolutionary pressures for efficient food-getting as a major impetus for the development of nervous systems.

The image labels read:

sensory cells, harpoons in tentacles

contractile cells running parallel to long axis

contractile cells forming "rings" around long axis

gut, with enzyme-secreting cells in inner cell layer

nerve cells adjacent to one another and to contractile cells

mouth

Now selective agents could work on individuals having variations in their diffuse, radially symmetrical nerve nets. Consider that, for forward-crawlers, food (and danger) would be encountered more frequently by the end going first. Variant individuals with sensory organs located up front could sense and turn toward food (and away from danger) more quickly than others, and would have selective advantages. Thus there would be **cephalization**: an increasing concentration of nervous structures and coordinative functions in the "front" end, or head.

We can speculate that shifts to a creeping life-style meant further changes in neural organization. Just as there would be selective advantages in having a head end different from the tail end, so would there be advantages in having the exposed top (*dorsal*) side different from the bottom (*ventral*) side. It would be more efficient to have

dorsal sensory pathways (the shortest distance between sensory structures and a brain), and to have ventral motor pathways (closer to motile structures). Of course, there would be little selection pressure for the body's right half to become much different from the left. Food and danger would most likely be encountered on both sides, not just one. So the forward-crawlers would have been under pressure also to become **bilaterally symmetrical**: the body's left half would become more or less equivalent to the right half. What does all this have to do with the evolution of nervous systems? Perhaps this:

A shift from radial to bilateral symmetry could have led, in some evolutionary lines of animals, to paired nerves and muscles, paired sensory structures such as eyes, and paired brain regions.

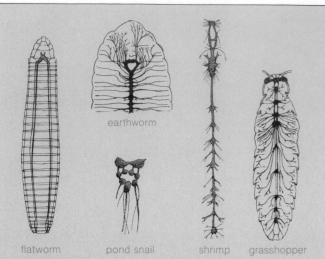

earthworm

flatworm   pond snail   shrimp   grasshopper

Figure 14.15 Examples of central and peripheral nervous systems from a few invertebrates. The sketches are not to scale relative to one another, the point being merely to show that even these supposedly simple animals have neurons organized into nerves as well as a brain and other ganglia. (Earthworm, pond snail, and shrimp from Bullock et al., *Introduction to Nervous Systems*. © 1977 W. H. Freeman and Co.)

In flatworms, a small brain tops a pair of longitudinal nerve cords. The cords are connected to each other by nerves that branch laterally in the body. The same branching pattern has echoes in the multiple body segments of annelids (such as earthworms) and arthropods (such as shrimps, crabs, and grasshoppers). Even the nerves branching from your own spinal cord show traces of this segmental patterning. A ganglion in each segment modulates reflex activities for that segment and for its immediate neighbors.

In mollusks such as abalone and snails, the central nervous system is a ring of paired ganglia to which peripheral nerves connect. In this group are the most advanced invertebrate nervous systems, those of the squid and octopus. Their central nervous system is dominated by a brain composed of a concentrated mass of ganglia—and there are regional differences in brain function. Some regions control internal organs such as the heart. Others control movements of eyes, head, and tentacles. One region governs the ability of these animals to retain information—they show both short-term and long-term memory—and to learn from it!

Cephalization, dorsal–ventral differentiation, bilateral symmetry—today these are features of almost all animals that are anatomically more complex than sponges and cnidarians. For them, the nervous system is organized in much the same way (Figure 14.15). Some parts are clustered in a central region, with the part exercising the most control concentrated at the head end. Other parts are in the body's

outlying regions, or periphery, but they have connections leading into or out from the neural center. The **central nervous system** is the brain and nerve cord (or paired cords). It contains specialized interneurons and cell bodies of motor neurons. The **peripheral nervous system** includes cell bodies of sensory neurons and all nerves (bundles of axons).

Differences that do exist among complex animals relate more to brain size and its degree of control over the rest of the nervous system. No matter how specialized a nervous system has become, though, the same basic principles apply:

1. Information flows by way of *graded potentials* (in receptors and at synapses) and *action potentials* (along axons).

2. There are pathways of *divergence*. A message may travel through branched endings of one neuron and activate many other neurons. Processes of those neurons may branch in different directions, and thus can disperse messages to different body regions.

3. There are pathways of *convergence*. Characteristically, each neuron receives both excitatory and inhibitory signals coming from many other neurons.

4. For all bilaterally symmetrical animals, each through-conducting pathway is oriented in one direction only: all sensory axons lead toward the central nervous system, and all motor axons lead away from it.

5. In both peripheral and central nervous systems, excitatory and inhibitory signals are summed at each synaptic transfer.

6. Signals about all events are sent to the brain. When modification in response is required, the brain sends out command signals. The brain's action represents the highest level of integration yet developed.

Vertebrate Nervous Systems

In all vertebrates, the central nervous system is enclosed in bony chambers and cushioned with fluid. Bony plates joined together form a skull, which protects the brain. A series of hard, bony segments joined into a vertebral column (backbone) protects the **spinal cord**: a region of local integration and reflex connections, along with nerve pathways leading to and from the brain. In cross-section, a spinal cord has two distinct regions (Figure 14.16). Its *white matter* contains myelinated sensory and motor axons;

this is the through-conducting zone. The *gray matter* includes nerve cell bodies, dendrites, and nonmyelinated end branchings of axons; this is the synaptic zone. Pair after pair of nerves lead into and out from different levels of the spinal cord.

In relatively simple vertebrates, the spinal cord is largely autonomous in handling reflex responses. For instance, a frog with its brain destroyed surgically can still maintain normal body position. It can even use its legs to kick away an irritant placed on its skin!

At the head end of the spinal cord are three distinct masses of neurons: the hindbrain, midbrain, and forebrain (Figure 14.17). The **hindbrain** is an enlarged extension of the upper spinal cord. The lower hindbrain region, the *medulla oblongata,* contains reflex control centers for breathing, heart rate, and blood pressure. Sensory nerves (except those dealing with sight and smell) pass through it, as do nerve processes that govern nearly all motor functions. The *cerebellum* is a hindbrain region that overlies the medulla. It integrates messages from receptors concerned with body position and motion, and from visual and auditory centers. The cerebellum helps coordinate motor responses associated with refined limb movements, with maintaining posture, and with spatial orientation. We get an idea of its importance by observing humans whose cerebellum is damaged. Uncontrollable tremors affect all of their movements—walking, speaking, even breathing.

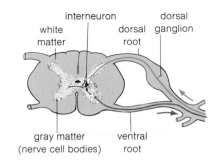

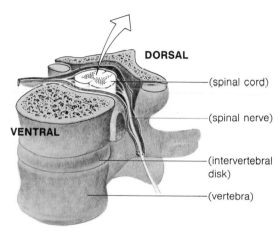

Figure 14.16 Organization of the vertebrate spinal cord. Compare this sketch with Figure 14.13 to get a better picture of the flow of information during reflex activity.

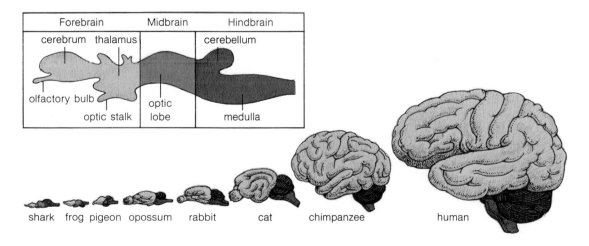

Figure 14.17 Brains of different vertebrates, drawn to the same scale. (From J. Eccles, *Understanding of the Brain*, 1977, McGraw-Hill) The inset shows the three divisions of the vertebrate brain, as well as a few of their component structures. (From *Vertebrate Body*, 5th ed., Romer and Parsons, © 1977 W. B. Saunders. Reprinted by permission of Holt, Rinehart and Winston.)

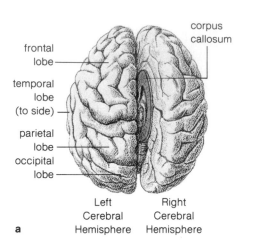

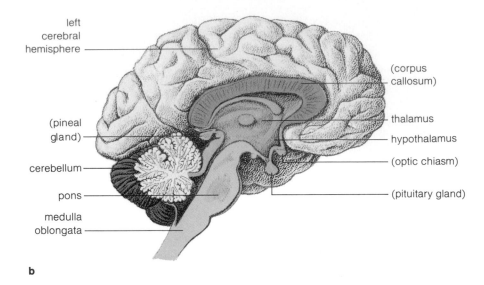

Figure 14.18 (a) Looking down on major regions of the cerebral hemispheres of the human brain, portrayed as if the main nerve tract connecting them (the corpus callosum) has been split so that the two hemispheres can be moved apart. (b) Main components of the brain, sliced through the midsection.

Early in vertebrate history, the **midbrain** developed as a major integrative center. Here, nerve pathways converge around a pair of *optic lobes*, structures concerned with visual reception. In the fish and amphibian midbrain, sensory information from different sources is integrated: it is the primary "association center." It is also the source of motor responses. In reptiles, birds, and mammals, the midbrain is reduced in importance. Messages still converge and are integrated in the midbrain, but they are sent on to the forebrain for further interpretation.

The **forebrain** contains thickened masses of gray matter divided into two halves, called the cerebral hemispheres. Underlying the cerebral hemispheres are forebrain regions called the thalamus, hypothalamus, and pituitary. Figure 14.18 shows these regions in the human brain. In the *hypothalamus*, hormones are produced that are secreted by the pituitary or that control the pituitary's secretion of its own hormones. (The pituitary is an endocrine gland that is considered later in the chapter.) The hypothalamus also contains centers concerned with body temperature regulation and with salt and water balance. Cells in the hypothalamus influence many forms of behavior, including those associated with hunger, thirst, aggression, sex, pleasure, and pain. The *thalamus* is the gateway to the cerebral

hemispheres. From here, impulses concerning sight, sound, taste, position, movement, and pressure are sent on to appropriate regions of the cerebral hemispheres, which will be described shortly. The thalamus also coordinates much of the outgoing motor activity.

Autonomic Pathways

We have seen how reflex responses can be channeled through ever more intricate pathways in the central nervous system. Taken together, all motor-to-skeletal muscle pathways and all sensory pathways form the **somatic nervous system**. Thus, skeletal muscles are its effectors. In humans, this system is usually under conscious control.

Another crucial aspect of integration is based on part of the peripheral and central nervous systems. It has to do with regulating cardiac cells, smooth muscle cells (such as those of the stomach), and glands. Because it generally isn't under conscious control, it is called the **autonomic nervous system**. Figure 14.19 diagrams the human autonomic nervous system. It is subdivided into two parts: *sympathetic* and *parasympathetic nerves*. Usually, sympathetic and parasympathetic nerves operate antagonistically (the activity of one system produces the opposite effect of the

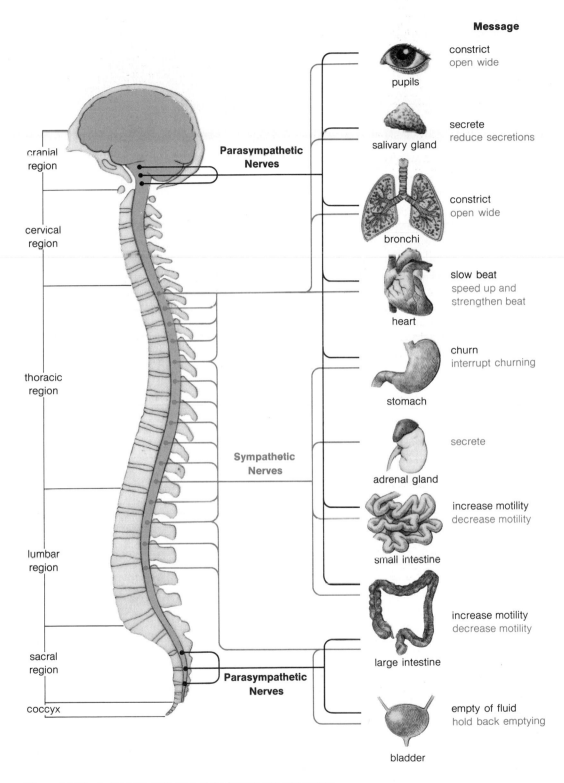

Message

pupils — constrict / open wide

salivary gland — secrete / reduce secretions

bronchi — constrict / open wide

heart — slow beat / speed up and strengthen beat

stomach — churn / interrupt churning

adrenal gland — secrete

small intestine — increase motility / decrease motility

large intestine — increase motility / decrease motility

bladder — empty of fluid / hold back emptying

Parasympathetic Nerves

Sympathetic Nerves

Parasympathetic Nerves

cranial region
cervical region
thoracic region
lumbar region
sacral region
coccyx

Figure 14.19 Autonomic pathways of the human nervous system.

other's activity). Through this antagonistic interaction, the body's organs and organ systems are adjusted to changing conditions.

Whether you are standing, sitting, sleeping, eating, digesting food, or eliminating wastes, your nervous system activates the organs required for the ongoing activity. Suppose you've just finished eating and are doing nothing more than sitting on the porch watching the sun go down. Messages flow out along parasympathetic nerves, stimulating into action all body parts that function in digestion. At this time, sympathetic nerves to the digestive system are relatively inactive. Blood vessels supplying your digestive tract are wide open, but those supplying limb muscles and kidney tissue are constricted. Suppose now that three fire engines with horns blaring race down the street. The strong stimuli trigger chemical messages that race through your body, inhibit parasympathetic nerves, and excite sympathetic nerves. Digestion stops; your eyes open wide and your heart starts pounding as you leap up to see what's going on. Only when things quiet down do the antagonistic pathways shift priorities, so that digestion resumes. We may summarize this kind of autonomic behavior in the following way:

Parasympathetic nerves generally dominate internal events when environmental conditions permit normal body functioning.

In times of stress, danger, excitement, or even heightened awareness, sympathetic nerves dominate internal events and mobilize the whole body for rapid response to change.

Of course, it's important to keep in mind that autonomic control generally isn't an "either/or" kind of thing. Just as your life is usually filled with complicating circumstances, so also does autonomic control over body function usually have shades of antagonistic dominance. Just because your autonomic nerves signal that it's time to empty your bladder doesn't mean that you are going to empty it on the front porch.

THE HUMAN BRAIN

Regions of the Cortex

In the human brain, as in the brain of other vertebrates, the **cerebral hemispheres** are the most advanced constellations of interneurons. Structurally, these paired masses appear to be mirror images of each other. As you will

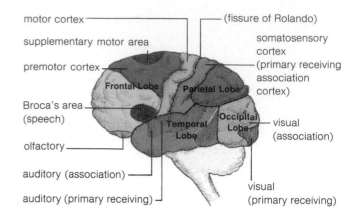

Figure 14.20 Primary receiving and association areas for the human cerebral cortex. Signals from receptors on the body's periphery enter primary cortical areas. Sensory input from different receptors is coordinated and processed in association areas. The text describes the main cortical regions. Also shown here are the *premotor area*, involved in intricate motor activity (as typified by a concert pianist performing); the *supplementary motor area*, which helps coordinate sequential voluntary movements; and *Broca's area*, which coordinates muscles required for speech. (After N. Lassen, D. Ingvar, and E. Skinhøj, "Brain Function and Blood Flow," *Scientific American*, October 1978)

see, functional differences do exist between them. Both have deeply folded, convoluted surfaces. At several turning points in vertebrate evolution, the brain underwent rapid enlargement. For humans, one such period corresponded to the emergence and immediate burgeoning of culture (Chapter Twenty-Four). Although the skull increased in volume, it did so much more slowly than the neural developments that occurred during this time span. The folds and convolutions of the brain's surface allowed for an increased number of neurons in the cerebral cortex—which could deal with the increasing demands for integration.

Inside each hemisphere is a core of white matter: axonal pathways connecting the rest of the central nervous system with the brain's surface layer. The surface layer, the **cerebral cortex**, is gray matter about a quarter of an inch thick. Certain cortical regions have specific functions (Figure 14.20). Some are primary receiving centers for signals from receptors on the body's periphery. Some are association centers, where sensory input is coordinated and processed. Others are motor centers, where instructions for motor responses are coordinated.

Some neurons in the *motor cortex* act as direct channels from the brain to motor neurons. Stimulation of different points on the motor cortex surface triggers contractions of muscle groups in different body parts. There isn't a one-to-one relationship between muscle size and the motor cortex area devoted to it. But there is a relationship between cortical area and function. For instance, muscles controlling thumb and tongue movements have a larger representation (Figure 14.21). This you might expect, given the control that must underlie our intricate hand movements and verbal skills.

Just behind the motor cortex is the *somatosensory cortex*, the synaptic zone for nerves concerned with signals from somatic receptors. These nerves relay news about mechanical and chemical stimulation, skeletal joint movements, and changing temperature. When your somatosensory cortex is stimulated, you experience sensations of touch, pressure, heat, cold, pain, and changes in body position. Visual sensations and sounds don't arise in this region. Communication pathways from your eyes terminate in a primary receiving center, the *visual cortex*, in the occipital lobe of each cerebral hemisphere. Pathways from your ears terminate in the *auditory cortex*, a small region of each temporal lobe.

Also lying outside the motor and somatosensory regions, but connected to them by neural pathways, are regions of *association cortex*. These largely unmapped regions may give rise to conscious perception of changing events in the surrounding world.

Conscious Experience

Our two cerebral hemispheres carry the stamp of our bilaterally symmetrical heritage. These two brain halves are strapped together deep inside the cleft between them by a thick tract of white matter, the **corpus callosum**. The corpus callosum, shown in Figure 14.18, consists of axons running from one hemisphere to the other. Thus you might assume that it functions in communication between them. Indeed, experiments such as those performed by Roger Sperry and his coworkers showed that this is the case. They also demonstrated some intriguing differences in perception between the two halves!

Pairs of sensory nerves enter the spinal cord from the body's right and left sides. These nerves run parallel to the brain. Similarly, paired sensory nerves from the eyes and ears run toward the brain from the body's right and left sides. But the information they carry is not all processed on the same brain side as the nerves. Instead, much of the information is projected onto the opposite hemisphere.

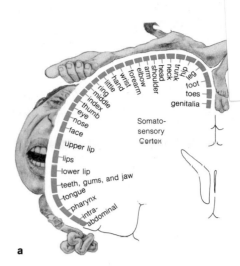

a

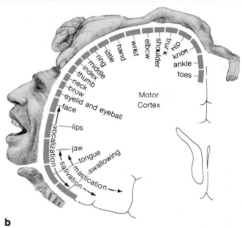

b

Figure 14.21 (**a**) Body regions represented in the somatosensory cortex. This region is a strip a little over an inch wide, running from the top of the head to just above the ear on the surface of each cerebral hemisphere. The sketch is a cross-section through the right hemisphere of someone facing you. (**b**) Body regions represented in the motor cortex, a region just in front of the somatosensory cortex. (After Penfield and Rasmussen, *The Cerebral Cortex of Man*. Copyright ©1950 Macmillan Publishing Co., Inc. Renewed 1978 by Theodore Rasmussen)

(The more ancient olfactory nerves don't cross over.) The fact is, *many of the lines leading into and out from one hemisphere deal with the opposite side of the body*.

Knowing this, Sperry's group set out to treat severe cases of epilepsy. Individuals afflicted with severe epilepsy are wracked with seizures—a kind of violent electrical storm in the brain—sometimes as often as every half hour of their lives. Clearly such individuals were desperate for relief.

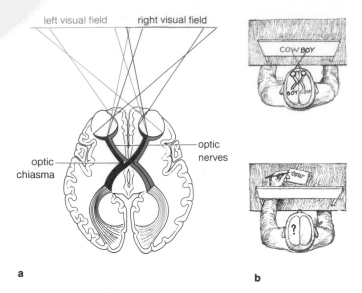

left visual field   right visual field

optic nerves

optic chiasma

a

b

Figure 14.22 (**a**) Optic nerve pathways into the left and right cerebral hemispheres. These nerve pathways cross, so that they inform each hemisphere of what is happening on the opposite side of the body. These pathways are not affected when the corpus callosum is cut, but the ability of the two hemispheres to coordinate the separate visual input is essentially lost. (**b**) The kind of experiment testing the perceptual abilities of split-brain individuals, as described in the text.

What would happen if their corpus callosum were cut? Would the electrical storms be confined to one hemisphere, leaving at least the other to function normally? Earlier studies of animals and of humans whose corpus callosum had been damaged suggested that this might be so. The surgery was performed. And the storms subsided, in both frequency and intensity. Apparently, cutting the neural bridge put an end to what must have been positive feedback loops of ever intensified electrical disturbances between the two hemispheres. Beyond this, the "split-brain" individuals were able to lead what seemed, on the surface, entirely normal lives.

But then Sperry devised some elegant experiments to determine whether the conscious experience of these individuals was indeed "normal." After all, the corpus callosum is a tract of no less than 200,000,000 through-conducting axons; surely *something* was different. Something was. "The surgery," Sperry would later report, "left these people with two separate minds, that is, two spheres of conscious-

ness. What is experienced in the right hemisphere seems to be entirely outside the realm of awareness of the left."

In Sperry's experiments, the left and right hemispheres of split-brain individuals were presented with different stimuli. Recall that visual connections to and from one hemisphere are mainly concerned with the opposite visual field. Sperry projected words—say, COWBOY—onto a screen. He did this in such a way that COW would fall only on the left visual field, and BOY would fall on the right (Figure 14.22). The subject reported seeing the word BOY. (The left hemisphere, which received the word, controls language.) The subject had no idea that another half of the word existed. However, if asked to write the perceived word with the left hand—a hand that was deliberately screened from view—the subject wrote COW. The right hemisphere, which "knew" the other half of the word, had directed the left hand's motor response. But it couldn't tell the left hemisphere what was going on because of the severed corpus callosum. The subject knew that a word was being written, but he could not say what it was!

Results from many experiments of this sort have revealed the following aspects of our conscious experience:

---

1. Each cerebral hemisphere can function separately, but it functions in response to signals mainly from one side of the body.

2. The main association regions responsible for language and analytical skills generally reside in the *left* hemisphere.

3. The main association regions responsible for nonverbal perception (sometimes called "intuition") generally reside in the *right* hemisphere.

Language, the verbal or written expression of our conscious experience, could not have emerged before 3 or 4 million years ago, the time of origin for the human lineage. Nonverbal perception must have been developing before then, in far more ancient animal lines. Animals living hundreds of millions of years ago were recognizing and responding to a world sensed by touch, by chemical clues of odor and taste, by patterns of sight and sound. The fossilized remains of skulls and sensory structures strongly imply that they had the neural fabric to do these things. Embroidered into this ancient fabric are increasing developments in the vertebrate forebrain, especially the thalamus. It seems probable that these developments reflect the arrival of nervous systems at the threshold of *conscious* experience. Indeed, clinical studies imply that the thalamus mediates our conscious awareness of painful or pleasant sensations. And now we have our most recent acquisition: an ability, generally residing in the left cerebral hemisphere, to verbalize what our "older" and "newer" brain structures piece together from incoming stimuli. It may be that the corpus callosum is a bridge between past and present ways of perceiving the world. Perhaps this bridge is crossed and recrossed in the neural interweavings that lead to art and science, music and technology, fantasies and mathematics—interweavings that are at once logical and intuitive, and uniquely human.

Memory

Conscious experience is far removed from simple reflex action. It entails *thinking* about things—recalling objects and events encountered in the past, comparing them with newly encountered ones, and making rational connections based on the comparison of perceptions. Thus conscious experience entails a capacity for **memory**: the storage of individual bits of information somewhere in the brain. The neural representation of such bits is known as a **memory trace**, although no one knows for sure what a memory trace is, or where it resides. So far, experiments strongly suggest there are at least two stages involved in its formation. One is a *short-term* formative period, lasting only a few minutes; then, information becomes spatially and temporally organized in neural pathways. The other is *long-term* storage; then, information is put in a different neural representation and permanently filed in the brain.

Observations of people suffering from **retrograde amnesia** tell us something about memory. These people can't remember anything that happened during the half hour or so before experiencing electroconvulsive shock, or before losing consciousness after a severe head blow. Yet memories of events before that time remain intact! Such disturbances suppress normal electrical activities in the brain. This may mean that whereas short-term memory is a fleeting stage of neural excitation, long-term memory depends on *structural* changes in the brain. In addition, information seemingly forgotten can be recalled after being stored for decades. This means that individual memory traces must be encoded in a form somewhat immune to degradation. Most molecules and cells in your body are used up, wear out, or age and are constantly being replaced—yet memories can be retrieved in exquisite detail after many years of such wholesale turnovers. Nerve cells, recall, are among the few kinds that are *not* replaced. Billions are formed, and as you grow older some 50,000 die off steadily each day. Only those nerve cells formed during embryonic development will be the ones present, whether damaged or otherwise modified, at the time of death.

The part about being "otherwise modified" is tantalizing. *There is evidence that neuron structure is not static, but rather can be modified in several ways—most likely depending on electrical and chemical interactions with neighboring neurons.* Electron micrographs show that some synapses regress as a result of disuse. Such regression weakens or breaks connections between neurons. The visual cortex of mice raised without visual stimulation showed such effects of disuse. Similarly, there is some evidence that intensively stimulated synapses may grow in size to form stronger connections, or may sprout buds or spines to form more connections!

Earlier studies had demonstrated that RNA and protein production rise during periods when memory traces are laid down. For a time, it was thought that one or both kinds of molecules might be the repository of memory. However, such new RNA molecules (or the proteins they specified) would have to contain novel sequences of information, corresponding to the memories that are *unique* to each individual. Yet how could this be? RNA must be synthesized on DNA—the hereditary molecule whose extraordinary stability assures a readout of essentially unvarying instructions! With newer evidence, it appears likely that the RNA and proteins are simply signs of increased metabolic activity and physical transformations that underlie changes in synaptic connections, with the structural changes themselves corresponding to memory storage.

Sleeping and Dreaming

Between the mindless drift of coma and total alertness are many *levels* of conscious experience, known by such names as sleeping, dozing, meditating, and daydreaming. Through this spectrum of consciousness, neurons in the brain are constantly chattering among themselves. This neural chatter shows up as wavelike patterns in an **electroencephalogram** (EEG). An EEG is an electrical recording of the frequency and strength of potentials from the brain's surface. Each recording contains contributions from thousands of neurons.

Figure 14.23 gives a few examples of EEG patterns. The prominent wave pattern for someone who is relaxed, with eyes closed, is an *alpha rhythm*. In this relaxed state of wakefulness, potentials are recorded in trains of about ten every second. Alpha waves predominate during the state of meditation. With a transition to sleep, wave trains gradually become longer, slower, and more erratic. This

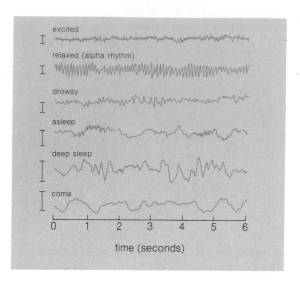

Figure 14.23 Examples of EEG wave patterns. The vertical bars indicate fifty millivolts each, with the irregular horizontal lines indicating the response with time. (After H. Jasper, 1941)

slow-wave sleep pattern shows up about eighty percent of the total sleeping time for adults. It occurs when sensory input is low and the mind is more or less idling. Subjects awakened from slow-wave sleep usually report that they were not dreaming but, if anything, seemed to be mulling over recent, ordinary events. With the transition from sleep (or deep relaxation) into wakefulness, EEG recordings show a shift to low-amplitude, higher frequency wave trains. Associated with this accelerated brain activity are increased blood flow and oxygen uptake in the cortex. The transition, called *EEG arousal*, occurs when individuals make a conscious effort to focus on external stimuli, or oven on their own thoughts.

Slow-wave sleep is punctuated by brief spells of *REM sleep.* The name refers to the *R*apid *E*ye *M*ovements accompanying it; the eyes jerk about beneath closed lids. Also accompanying REM sleep are irregular breathing patterns, faster heart rate, and twitching fingers. Waking someone from REM sleep is more difficult than waking someone from slow-wave periods. Most people who are awakened from REM sleep say that they were experiencing vivid dreams. In some experiments, individuals were made to wake up at the start of each REM period, so that most of their total sleeping time was spent in dreamless states.

Yet the next time they fell asleep, more time than normal was devoted to REM sleep. It has been suggested that the body has some as-yet unidentified need for dreaming.

What brain regions govern changing levels of consciousness? Deep in the brainstem, buried within ascending and descending nerve pathways, lies a mass of nerve cells and processes called the **reticular formation**. This neural mass forms connections within the spinal cord, cerebellum, and cerebrum, as well as back with itself. It constantly samples messages flowing through the central nervous system. The flow of inhibitory and excitatory signals along these circuits—and the chemical changes accompanying them—has a great deal to do with whether you stay awake or drop off to sleep. For example, when the reticular formation of sleeping animals is electrically stimulated, long, slow alpha rhythms are displaced by high-frequency potentials associated with arousal. Similarly, damage to the reticular formation leads to unconsciousness and coma.

Within this formation are neurons collectively called the **reticular activating system** (RAS). Excitatory pathways connect the RAS to the thalamus (the forebrain's switching station). Messages routed from the RAS arouse the brain and maintain wakefulness. Also in the reticular formation are *sleep centers.* One center, in the brainstem's core, contains neurons that release the neurotransmitter **serotonin**. This chemical has an inhibitory effect on RAS neurons: high serotonin levels trigger drowsiness and sleep. Another sleep center is connected to the pons, part of the brainstem that has been linked to REM sleep. Chemicals released from the second center counteract the effects of serotonin. Hence its action allows the RAS to maintain the waking state.

Yet why do these interactions occur? We spend as much as a third of our lives sleeping, and a good part of that time dreaming. Are these states of consciousness absolutely necessary? To be sure, lack of sleeping or dreaming has repercussions in our emotional state—but there is no evidence that we couldn't get along without them. To the contrary: some individuals who presumably suffered brain damage never fall asleep—yet they survive. Laboratory animals with a damaged midbrain never fall asleep, yet they survive. Are both states some unknown form of secondary information processing? Do they represent a sort of catalog time, during which the day's events are sifted through and committed to one or another memory corridor? Or are they times of neural interweavings and compromises between older and more recent layerings of the brain? No one knows, although speculations abound.

ENDOCRINE SYSTEMS: MORE RESPONSE OPTIONS

Animal Hormones and Their Action

As you read earlier, **neurotransmitters** are secreted from nerve cell endings and travel only a short distance, across a synaptic cleft, to an *adjacent* cell. Their chemical message to specific targets is this: "It's time to change your pattern of activity." Neurotransmitters include acetylcholine, epinephrine, norepinephrine, serotonin, and dopamine. All are released in minute amounts, they rapidly trigger change in the membrane potential of an adjacent cell, then they are degraded at once. They function, for a fleeting moment, as *information carriers*

Some nerve cells release different kinds of information carriers. Their carriers travel more slowly and travel farther, by way of the bloodstream, to many *nonadjacent* cells. These *neurosecretory cells* are concerned less with impulse conduction than they are with glandlike activities—with producing, storing, and secreting chemicals called **neurohormones**. In anatomically simple animals (cnidarians, annelids, probably flatworms), neurosecretory cells and their neurohormones are just about the only physiological controls over cells and tissues concerned with growth and reproduction.

In animals ranging from mollusks to humans, we find elaborations on the patterns of information-carrying secretions. These animals also have secretory *endocrine cells*. Some endocrine cells function individually. Others are organized into patches of secretory tissue and organs. Endocrine cell products, called true animal **hormones**, are transported by the bloodstream and regulate specific cellular reactions in tissues and organs some distance away. In general, hormones leave blood vessels and then diffuse through interstitial fluid; from there, they are picked up by target cells. Because of specific receptors on their surfaces, only "target" cells can respond to hormonal secretions. Some secretions trigger short-term body adjustments (for instance, by changing blood chemical composition or heart rate). Others trigger long-term adjustments, such as those involved in growth, differentiation, and reproduction.

(You may have heard about another kind of information-carrying secretion, the **pheromone**. Pheromones are products of specialized *exocrine glands*. Such glands have ducts that lead out to the body surface. The targets of their secretions lie *outside* the animal body. Pheromones may trigger behavioral changes in other animals in the same species. Some are released as sex attractants, others as trail markers, still others as alarm signals.)

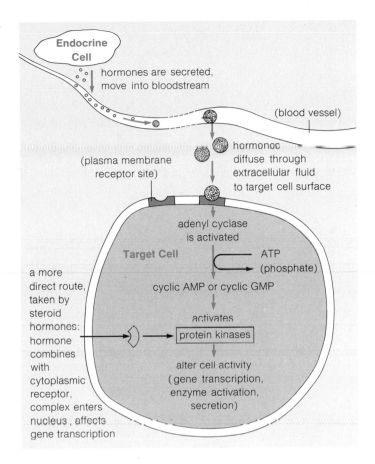

Figure 14.24 Proposed mechanisms of hormone action on target cell activities, as described in the text.

How do hormones induce changes in target cell activities? As Figure 14.24 suggests, many hormones can't penetrate the plasma membrane of their targets. They rely on functional partners—second messengers—inside a target cell, which mediate the endocrine instructions. Some of these hormones bind with a certain receptor site on the plasma membrane. The binding activates adenyl cyclase, an enzyme that transforms ATP present in the cytoplasm to **cyclic AMP.** (The full name is cyclic adenosine monophosphate; "cyclic" simply refers to the way this enzyme's phosphate group is arranged into a ring.) Cyclic AMP is one kind of second messenger; cyclic GMP (cyclic guanosine monophosphate) is another. Both act on protein kinases, which add or remove phosphate groups from proteins involved in diverse cellular responses. For instance, protein kinases must activate the proteins (enzymes) required for breaking down carbohydrates and lipids.

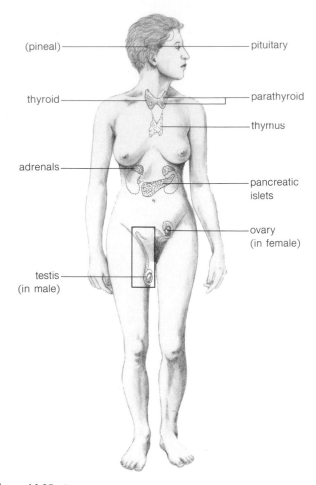

Figure 14.25 Location of endocrine elements in the human body.

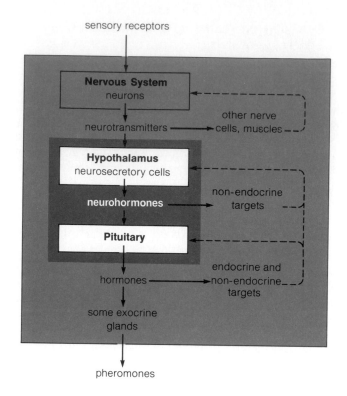

Figure 14.26 Main pathways leading to and away from the neuro-endocrine control center. Secretory pathways are shown in solid lines. Broken lines indicate negative feedback loops brought about by changing activities in the cells that are targets for the secretions indicated. Neurotransmitters, neurohormones, and hormones are secretions with internal targets. Pheromones are secretions with targets outside the animal body.

Some endocrine products, such as certain steroid hormones, *can* penetrate the plasma membrane of target cells. They don't require second messengers at all; they combine directly with cytoplasmic receptors. Gene transcription—hence production of a given protein in the cell—depends on these steroid–receptor complexes. The complexes seem to "loosen up" regions of eukaryotic chromosomes so that transcription can proceed (Chapter Eleven).

Aside from differences in their mode of action, hormones also differ in their degree of autonomy. Throughout this unit, you will come across examples of how the effect of one hormone may enhance or suppress the activity of another. Other examples will show how several different kinds of hormones typically must arrive at the cell, one after another in series, to produce the appropriate response by a given cell type. For now, the important point to remember is this:

Hormones help maintain the internal environment by regulating specific cellular reactions. Through their action, specific enzyme activities (or enzyme formation) are accelerated or slowed down.

On Neural and Endocrine Links

All vertebrates have much the same assortment of endocrine elements. Each comes equipped with a pituitary, adrenals, thyroid, parathyroids, pancreatic islets, endocrine elements in the gut, and gonads. The position of these endocrine elements in the body hasn't changed much during the course of vertebrate evolution. Hence a sketch of their approximate location in one kind of vertebrate—humans—applies as well for other kinds (Figure 14.25).

In looking at Figure 14.25, you might conclude (as researchers did for a long time) that the "endocrine system" wasn't much of a system at all. What could possibly be the functional link between cells, tissues, and organs that are physically separated, that secrete diverse hormones, and that often act on targets located nowhere near them? For possible answers to this question, consider that both the nervous system and the body's endocrine elements produce and secrete chemical messengers that influence the output of target cells. Consider also that neurohormones act on certain endocrine elements, and that some endocrine secretions influence the nervous system's activity. Most bodily states—and this includes mental and behavioral states—depend on their interaction. Could it be that the first endocrine elements arose as specialized regions of the nervous system?

Suppose that long ago, in animals not much more complex than flatworms, mutations led to structural or physiological changes in a sending or a receiving neuron. For example, what if chemicals diffusing across the synaptic cleft between them could not be degraded after the chemical message had been transferred? Then the secreted molecules might have started to diffuse through surrounding tissues. Such wayward molecules might even have begun slipping into and out of the bloodstream. What if the surface configuration of the molecules allowed them to stick to or penetrate the plasma membrane of cells in distant body parts? If the chemical recognition led to enhanced metabolic activity, then such mutant sending-cell types would have conferred a survival advantage on their bearer. Over time, they might have become modified in ways that allowed them to be more concerned with long-distance action; they would have been forerunners to endocrine cells and tissues. If endocrine elements did indeed evolve from neurosecretory beginnings, we might suspect that they must still retain close functional links with the nervous system. And today we know this:

The functioning of most existing endocrine elements is mediated either directly by a separate nerve supply or indirectly by secretions from particular centers of the nervous system.

In many animals, the hypothalamus and the pituitary form an all-important link between neural and endocrine activities. As Figure 14.26 suggests, this link provides a way of perceiving order in the seeming maze of integration and control pathways in the animal body. In fact, *the hypothalamus and pituitary are now being viewed as the neuroendocrine control center.*

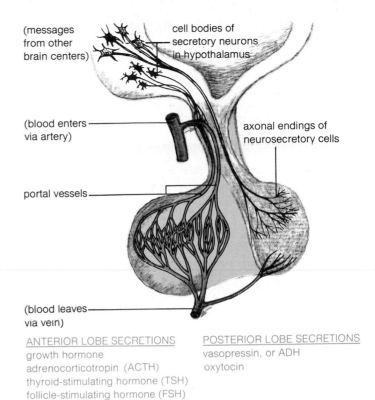

ANTERIOR LOBE SECRETIONS
growth hormone
adrenocorticotropin (ACTH)
thyroid-stimulating hormone (TSH)
follicle-stimulating hormone (FSH)
luteinizing hormone (LH)
prolactin

POSTERIOR LOBE SECRETIONS
vasopressin, or ADH
oxytocin

Figure 14.27 Pituitary gland in mammals. Shown here are the neural connections between the hypothalamus and the posterior lobe (shaded in gray), and the vascular connections between the hypothalamus and the anterior lobe (shaded in light brown).

Partners in Control: Hypothalamus and Pituitary

Information about changing conditions in the external and internal worlds flows constantly to the hypothalamus. Some information arrives along neural pathways from the cerebral cortex. Some arrives with the bloodstream—a highway for diverse secretions of the body's endocrine elements. In the hypothalamus, incoming neural messages are summed, shifts in hormonal concentrations are detected, and responses are sent out in the form of neurohormones (Figure 14.27). These neurohormones are routed to one of two pituitary regions.

The **pituitary** is a compound gland, about the size of a small jellybean in adult humans. A slender stalk attaches it to the floor of the hypothalamus (Figure 14.27). The connecting stalk and the pituitary's *posterior lobe* are nervous tissue. The *anterior lobe* is mostly glandular tissue.

Axons of some neurosecretory cells residing in the hypothalamus extend downward, into the posterior lobe. The axonal endings secrete two kinds of neurohormones, called oxytocin and vasopressin. (Vasopressin is also called ADH, in other contexts.) Unlike the posterior lobe, the pituitary's anterior lobe has no major neural connections with the brain. Instead, cells of the anterior lobe are functionally linked to the hypothalamus by an array of capillaries—called **portal vessels**. At least seven hypothalamic neurohormones are taken up by these portal vessels and conveyed to the anterior pituitary. Depending on the stimulus, the neurohormones may be *releasing* factors (which enhances the output of specific pituitary hormones), or they may be *inhibiting* factors (which suppress their output). Table 14.1 lists the secretions of the hypothalamus and pituitary. It also lists the secretions of other endocrine elements, which will now be described. The feedback loops connecting the nervous system and these endocrine elements will be traced in later chapters, so you may find it useful to scan this table as an overview of things to come.

Other Endocrine Elements

At the base of your neck, just below and in front of your windpipe, is a two-lobed **thyroid gland**. It stores and releases hormones that help govern growth, development, and metabolic rates throughout the body. Under the direction of the anterior pituitary, the thyroid gland secretes thyroxin (which affects oxygen uptake in carbohydrate and fat metabolism). When calcium levels in blood plasma rise, the thyroid also secretes calcitonin (which inhibits the release of calcium from bone storage sites). Embedded in the back surface tissues of the thyroid are four **parathyroid glands**. Their secretion (parathyroid hormone) counterbalances the effects of calcitonin by promoting the release of calcium in the bloodstream. Muscle action is one activity that depends on sufficient and carefully regulated amounts of calcium ions.

The pancreas is largely a digestive organ. However, about 2 million endocrine cell clusters, the **pancreatic islets**, are also scattered in the pancreas. Under the prodding of sympathetic and parasympathetic nerves, these cells synthesize hormones and secrete them directly into the bloodstream. The hormone insulin acts directly or indirectly on glucose, fat, and protein metabolism in cells throughout the body. The hormone glucagon counterbalances some aspects of insulin action by stimulating the breakdown of glycogen (a storage starch) into glucose subunits.

Above each of the two kidneys in the vertebrate body is a compound adrenal gland. Its outer layer, the **adrenal cortex**, develops embryonically in close association with reproductive organs. It secretes three types of hormones. The glucocorticoids help regulate carbohydrate, fat, and protein metabolism. The mineralocorticoids influence salt and water concentrations in the body. The adrenal cortex also secretes two classes of sex hormones (the androgens and estrogens).

The inner region of the adrenal gland, the **adrenal medulla**, is derived embryonically from nerve tissue; its cells are actually regarded as modified neurons. The hypothalamus, by way of sympathetic nerves, controls their secretion of epinephrine and norepinephrine. Normally, the adrenal medulla releases both epinephrine and norepinephrine in small amounts, which help regulate blood circulation and carbohydrate metabolism.

The **gonads** are the primary sex structures called testes (in males) and ovaries (in females). Gonads give rise to gametes. They also secrete hormones that prepare accessory reproductive structures (ducts, glands, and tissues) for reproduction. In the female, they help prepare the body for implantation and embryonic development. The gonads themselves will be described in Chapter Eighteen. Also in that chapter, you will read about the effects of their hormonal secretions.

The pineal gland and thymus have endocrine functions, but they are not usually listed among the major endocrine glands. The **pineal gland** takes part in reproductive physiology, although its precise role in the human body has not been identified. In other vertebrates, the pineal gland senses change in the length of daylight associated with changing seasons, and causes alterations in how fast certain hormones are produced and released. (This response to cyclic variations in light stimuli is a form of *photoperiodism*.) The **thymus** takes part in the immune response. It produces infection-fighting cells, and is a source of hormones that prepare other tissues for the production tasks.

Prostaglandins are a grab-bag of "local" hormones. They are secreted from tissues throughout the body—lungs, gut, prostate gland in males, and the liver. All prostaglandins may act as mediators between membrane receptors and the adenyl cyclase that activates cyclic AMP in diverse cell types. They have profound effects on reproduction. (They are being investigated as possible birth control

Table 14.1 Main Animal Hormones, Their Sources, and Their Actions

Source	Hormone or Neurohormone	Target	Primary Actions
Hypothalamus	Releasing and inhibiting factors (some identified as neurohormones)	Anterior pituitary	Regulate pituitary secretions
Anterior pituitary	Growth hormone	Bone, muscle	Induces bone and muscle growth
	Thyroid-stimulating hormone (TSH)	Thyroid gland	Stimulates thyroid secretions
	Adrenocorticotrophic hormone (ACTH)	Adrenal cortex	Stimulates adrenal secretions
	Prolactin	Mammary glands	Stimulates milk production; maintains corpus luteum; role in female reproductive behavior
	Gonadotropins:		
	Follicle-stimulating hormone (FSH)	Ovaries	Promotes early growth of ovarian follicles (in which ova mature); with LH, stimulates estrogen secretion and ovulation
			Maintains spermatogenic epithelium
	Luteinizing hormone (LH)	Ovaries	Stimulates corpus luteum
	Interstitial cell stimulating hormone (ICSH); same as LH in females	Testes	Stimulates secretion of male hormones
Hypothalamus (production) and posterior pituitary (storage, secretion)	Oxytocin (a neurohormone)	Uterus	Stimulates uterine contractions
		Mammary glands	Stimulates milk movement into secretory ducts
	Vasopressin (a neurohormone)*	Kidney	Promotes water reabsorption
Thyroid	Thyroxin	Most cells	Regulates carbohydrate and fat metabolism, growth, development
	Calcitonin	Bone	Lowers blood calcium level by inhibiting calcium reabsorption from bone
Parathyroids	Parathyroid hormone	Bone, kidney, digestive tract	Elevates blood calcium level by stimulating calcium reabsorption from bone and kidneys
Pancreatic islets	Insulin	General	Lowers blood sugar level by stimulating glucose utilization; fat storage, protein synthesis
	Glucagon	Liver, muscle, adipose tissue	Raises blood sugar level by stimulating glucose production
Adrenal medulla	Epinephrine (adrenalin)	Liver, cardiac muscle, adipose tissue	Raises blood sugar level, increases heart rate, promotes fat utilization
	Norepinephrine	Smooth muscle of blood vessels	Vasoconstriction
Adrenal cortex	Steroid hormones:		
	Glucocorticoids	General	Raise blood sugar level, regulate carbohydrate, fat, protein metabolism, counteract stress
	Mineralocorticoids (e.g., aldosterone)	Kidney	Promote sodium reabsorption, salt/water balance
	Sex hormones (androgens, estrogens)		Influence sexual characteristics
Testis (but hormone production is linked with adrenal cortex)	Testosterone	General	Stimulates development of genital tract, development and maintenance of accessory sex organs and secondary sex characteristics
Ovary	Estrogen	General	Same as above; also stimulates thickening of uterine lining for pregnancy
	Progesterone	Uterus, breasts	Prepares and maintains uterine lining for pregnancy, stimulates breast development

*Also called antidiuretic hormone (ADH).

agents.) Prostaglandins also influence nerve function, respiration, digestion, blood clotting, and the immune response.

CASE STUDY: FIGHT-OR-FLIGHT RESPONSE

Neurons, synapses, spinal cord, brain, sympathetic and parasympathetic pathways, hormones, neural and endocrine secretions—you have been exploring the domain of the most complex systems in the living world. Sometimes, to make the nervous and endocrine systems seem less challenging, their components are talked about in separate chapters—the brain here, the neuron there, and way over there something called chemical control (which is often limited to hormones, to the exclusion of neural secretions). But sometimes it is more rewarding to accept the challenge of focusing on the greater picture—to find coherence in seemingly separate parts of your own body. This is the way to catch glimpses of the awesome beauty of its functioning.

Consider, as an example, one of the most ancient, rapid, and thoroughly integrated responses of the nervous and endocrine systems—the fight-or-flight response. But this time don't think about it in the abstract. Imagine yourself scuba diving for the first time through the warm waters of a tropical reef, far below the water's surface and surrounded by spectacular coral formations. Beneath the shadow of a coral ledge, an extravagantly colored sea anemone catches your eye. You find a handhold and pull yourself toward it. Suddenly, out of the corner of your eye, you see glistening needlelike teeth—*MORAY!* Your arms and legs explode into action and your body shoots upward, out of danger. By violating a prime rule of diving—by allowing part of your body to come into contact with a place you had not yet examined carefully—you had come within a fraction of a second of losing your hand.

In that one instant, every part of your body had become mobilized for supernormal speed and strength. Your retinas registered a glimpse of flashing teeth. As soon as nerve impulses traveled from the retinas to your brain, they were processed and interpreted. Their meaning: *DANGER!*

From the cerebral cortex, signals shot out to the midbrain, hypothalamus, and other brain regions dealing with emotions and responses to them. From there, commands traveled to the pituitary, to motor neurons, to skeletal muscle, and along parasympathetic and sympathetic nerves.

Neural or hormonal signals quickly reached every vital organ. Epinephrine and norepinephrine from the adrenal gland and from sympathetic nerves signaled the heart to beat stronger and faster, thereby accelerating blood flow. They both signaled muscle cells in the tubing of your respiratory system to relax, and more oxygen flowed into your lungs. Your pupils dilated under their proddings. These messengers flowed to your liver and adipose tissue, stimulating an outpouring of glucose and fatty acids—which entered the bloodstream and thereby provided the body with urgently needed energy. At the same time, norepinephrine and epinephrine closed off blood vessels in your skin and in your gut wall, thereby diverting blood from those organs—which would play no part in the response—to those that would. Simultaneously, they caused the gut to relax (hence the gut would drain off less energy during the emergency). They caused blood vessels supplying skeletal muscles to open wide. Thus nourished, muscles could now respond with maximum speed and strength to whatever urgent commands descended from the brain.

With all systems alerted, your brain ran through its memory banks, searching for information stored away during your scuba diving lessons. "Don't panic!" "If it's a shark, freeze!" *"If it's a moray eel, get out of there!"*

And now the cerebral cortex fired off impulses by way of the motor cortex, where the correct detailed program was selected and initiated. From there, impulses flowed down the spinal cord and out the motor neurons in precisely the right sequence and frequency needed to activate muscles into producing a powerful swimming stroke. *And all of this happened in a fraction of a second!*

Your nervous system and endocrine system do more than work constantly to maintain your body in a homeostatic state. The also prepare your body to modify internal conditions instantly. The **fight-or-flight response** is an extremely rapid response to stress—to attack, trauma, hemorrhage, even to sudden and severe temperature shock. It is not unique to humans. All mammals display it whenever a sudden confrontation demands an immediate and unequivocal response.

DRUG ACTION ON INTEGRATION AND CONTROL

The fight-or-flight response switches the entire body into a state of heightened awareness. But the neural and endocrine secretions involved, and the systems that control them, are not deployed only during times of profound

Table 14.2 Examples of Drug- and Toxin-Induced Disruptions of Neural and Endocrine Functions

Messenger	Regions Where Identified	Known Effects	Some Known Drug-Induced Disruptions
Acetylcholine	Skeletal muscle (neuromuscular junctions)	Excitatory	*Botulinum toxin:* most toxic of all; disrupts release of acetylcholine; muscle contraction prevented; flaccid paralysis; death from respiratory paralysis
	Sympathetic ganglia	Inhibitory	*Curare toxin:* blocks receptor sites at muscles; muscle contraction prevented; flaccid paralysis; death from respiratory paralysis
	Parasympathetic nerves	Excitatory or inhibitory, depending on the target	*Tetanus toxin:* blocks inhibitory receptor sites in central nervous system; excitatory input unchecked; spasms, seizures, possibly death
Norepinephrine	Secreted by some neurons in brain, by sympathetic neurons, and by adrenal medulla cells (derived embryonically from neurons)	Excitatory or inhibitory, depending on the target	*Amphetamines:* increase brain activity; increase heart action, respiration; decrease REM sleep
Epinephrine	Secreted by some neurons in brain; forms 20 percent of adrenal medulla's secretions		*Antidepressants:* possibly work by blocking receptors for the neurotransmitter histamine
Dopamine	Mostly motor system		*Phenothiazines (major tranquilizers):* block dopamine receptor sites, prevent uptake of norepinephrine and serotonin; inhibit sensory input to reticular activating system; induce relaxation and calm
Serotonin	Central nervous system, especially hypothalamus and limbic system, parts of reticular activating system		*LSD:* may inhibit effects of serotonin and enhance activity of norepinephrine; extreme perceptual distortion and hallucination (ranging from "mind-expanding" to terrifying)

stress. Each day can bring some minor frustration or disappointment, some pleasure or small triumph—and the brain responds to the shadings of environmental stimuli with delicate interplays among the activities of norepinephrine, dopamine, and the like. These interplays translate into changing emotional and behavioral states. When stress leads to physical or emotional pain, the brain apparently deploys still other substances—**analgesics**, or pain relievers that it produces itself. Receptors for natural analgesics have been identified on neural membranes in many parts of the nervous system, including the spinal cord and limbic system. (The limbic system includes structures bordering the cerebral hemispheres, at the top of the brainstem.) When bound to their receptors, the pain relievers seem to inhibit neural activity. *Endorphins* and *enkephalins* are two brain analgesics that may have this inhibitory effect. High concentrations of endorphins ("internally produced mor-

phines") occur in brain regions concerned with our emotions and perception of pain.

Emotional states—joy, elation, anxiety, depression, fear, anger—are normal responses to changing conditions in the complex world around us. Sometimes, through neurotransmitter imbalances, one or another of these states becomes pronounced. For instance, schizophrenic individuals become miserable and despairing; they withdraw from the social world and focus obsessively on themselves. In an extreme form of the disorder, paranoid schizophrenia, afflicted individuals suffer delusions of persecution or grandeur. Yet by administering certain synthetic tranquilizers, the symptoms can be brought under control. It appears that these tranquilizers affect norepinephrine, dopamine, and serotonin levels in the brain, in ways that depress the activity of neurons utilizing these neurotransmitters (Table 14.2).

Tranquilizers, opiates, stimulants, hallucinogens—such drugs are known to inhibit, modify, or enhance the release or action of chemical messengers throughout the brain (Table 14.2). Yet research into the effects of drugs on integration and control is in its infancy. For the most part, we don't understand much about how any one drug works. Given the complexity of the brain, it could scarcely be otherwise at this early stage of inquiry. Despite our ignorance about these effects, one of the major problems in the modern world is drug abuse—the rampant, uncontrolled use of drugs for altering emotional and behavioral states. The consequences show up in unexpected places —among seven-year-old heroin addicts; among the highway wreckage left by individuals whose perceptions were skewed by alcohol or amphetamines; among victims of addicts who steal and sometimes kill to support the habit; among suicides on LSD trips who were deluded into believing that they could fly, and who flew off buildings and bridges.

Each of us possesses a body of great complexity. Its architecture, its functioning are legacies of millions of years of evolution. It is unique in the living world because of its nervous system—a system that is capable of processing far more than the experiences of the individual. One of its most astonishing products is language—the encoding of *shared* experiences of groups of individuals in time and space. Through the evolution of our nervous system, the sense of history was born, and the sense of destiny. Through this system we can ask how we have come to be what we are, and where we are headed from here. Perhaps the sorriest consequence of drug abuse is its implicit denial of this legacy—the denial of self when we cease to ask, and cease to care.

Readings

Axelrod, J. 1974. "Neurotransmitters." *Scientific American* 230(6): 58–71.

Bullock, T., R. Orkand, and A. Grinnell. 1977. *Introduction to Nervous Systems.* San Francisco: Freeman. Outstanding source book, and by far the best illustrated. Ranges from neuron fine structure, through message conduction, integration, the neural basis of behavior, development of nervous systems, to an evolutionary survey of nervous systems of major animal phyla. Highly recommended as a one-of-a-kind reference book for the serious student. Paperback.

Eccles, J. 1977. *The Understanding of the Brain.* New York: McGraw-Hill. Excellent, well-written account of neural activity in the brain; the descriptions of research into states of consciousness convey the excitement and frontier aspect of ongoing research. Paperback.

Gazzaniga, M. 1967. "The Split Brain in Man." *Scientific American* 217(2):24–29.

Hubel, D., and T. Wiesel. 1974. "Sequence Regularity and Geometry of Orientation Columns in the Monkey Striate Cortex." *Journal of Comparative Neurology* 158:267–294.

Kuffler, S., and J. Nicholls. 1977. *From Neuron to Brain.* Sunderland, Massachusetts: Sinauer. Unusual perspective; follows recent developments emerging from a few selected lines of research. Excellent discussions on neural signaling and the neural organization underlying visual perception.

Lassen, N., D. Ingvar, and E. Skinhój. 1978. "Brain Function and Blood Flow." *Scientific American* 239(4):62–72. New maps of the cerebral cortex, based largely on radioactive isotope tracings of changes of blood flow in different cortical regions during different states of activity.

Pastan, I. 1972. "Cyclic AMP." *Scientific American* 227(2):97–105.

Penfield, W., and T. Rasmussen. 1952. *The Cerebral Cortex of Man.* New York: Macmillan. Fascinating account of early mappings of cortical regions.

Romer, A., and T. Parsons. 1977. *The Vertebrate Body.* Fifth edition. Philadelphia: W. B. Saunders. Chapter 16 contains detailed pictures of vertebrate nervous systems. Excellent reference book.

Sagan, C. 1977. *The Dragons of Eden: Speculations on the Evolution of Human Intelligence.* New York: Random House. Marvelous musings about possible antecedents of our capacity for intelligence. Sometimes stunning in its insights, sometimes outrageous; nowhere is the book boring.

Shepherd, G. 1978. "Microcircuits in the Nervous System." *Scientific American* 238(2):93–103. For almost a hundred years message conduction has been described largely in terms of axonal pathways. This article describes recent discoveries of inhibitory feedback loops between dendrites only.

Sherrington, C. 1947. *Integrative Action of the Nervous System.* New Haven, Connecticut: Yale University Press. Sherrington was an outstanding neuroscientist, and a poet as well; his writing is near-lyrical, and his insights still rewarding.

——. 1933. *The Brain and Its Mechanisms.* New York: Cambridge University Press.

Sperry, R. 1970. "Perception in the Absence of the Neocortical Commissures." In *Perception and Its Disorders*, Research Publication of the Association for Research in Nervous and Mental Diseases, vol. 48.

Turner, C., and J. Bagnara. 1976. *General Endocrinology.* Sixth edition. Philadelphia: W. B. Saunders. Perhaps the best single reference on neuroendocrinology. Very well illustrated.

Vander, A., J. Sherman, and D. Luciano. 1980. *Human Physiology: The Mechanisms of Body Function.* Third edition. New York: McGraw-Hill. Contains very good introductions to what is known about information processing in the human brain.

Review Questions

1. Describe the functional zones of a neuron.

2. Two major concentration gradients exist across a neural membrane. What are they, and how are they maintained?

3. An electrical gradient also exists across a neural membrane. Explain what the electrical and concentration gradients together represent. What is a nerve message?

4. Distinguish between an action potential and a graded potential. What is meant by "all-or-none" and "self-propagating" messages?

5. Define sensory neuron, interneuron, and motor neuron.

6. What is a synapse? Explain the difference between an excitatory and an inhibitory synapse. Then define neural integration.

7. What is a reflex? Describe the sequence of events in a stretch reflex.

8. Contrast radially and bilaterally arranged nervous systems in terms of responsiveness to the environment.

9. Six basic principles govern the functioning of all nervous systems, from the simplest to the most complex. What are these principles?

10. Make a list of the main parts of the vertebrate nervous system. Can you define the main functions of each?

11. How do sympathetic and parasympathetic nerves regulate the autonomic pathways?

12. Define the source and function of neurotransmitters, neurohormones, hormones, and pheromones.

13. How do some hormones cause changes in the activities of their target cells?

14. How do the hypothalamus and pituitary link the activities of the endocrine system and nervous system?

15. Name five endocrine glands and a substance that each one secretes.

15

RECEPTION AND RESPONSE

Each living thing has its destiny linked to some small place on the earth's surface. No matter where that place happens to be, conditions are never constant, and survival means adjustment to change. The change may be as fleeting as minute shifts in ion concentrations, as stressful as confrontations with predators, or as gradual as the unfolding of seasons as the earth arcs through space. In complex animals, adjustments to change fall under neural and endocrine controls of the sort described in the preceding chapter. Yet how does information about change become channeled into the nervous system in the first place? In what ways are commands for response actually implemented? These questions bring us to sensory, motor, and behavioral adaptations for interacting with the external world. At first glance, some of these adaptations may seem bizarre, even frightful. There is, however, an underlying logic to the ways that animals sense and respond to their surroundings. To discover it, we must think about where each kind lives, where its probable forerunners evolved, and where its life cycle fits in the time frame of greater, planetary cycles.

For example, consider the bat. This is a winged mammal whose terrifying appearance, whose habits of hiding by day and emerging at dusk to swoop in the gathering darkness placed it, long ago, in the company of witches, demons, and other suspect figures of the human imagination. Where do bats live? What were conditions (and competition) like when their ancestors were evolving? How are their senses tuned to the rhythms of the planet? Only in the arctic or polar regions and on a few oceanic islands will you not find bats. One species or another lives in all manner of forests, grasslands, and deserts, although they are most abundant and diverse in the tropics. Those living in regions of pronounced seasons hibernate (their metabolism slows almost to a standstill) or migrate to warmer places during less hospitable times. Regardless of where they live, all bats are essentially immobile during the day. When the sun is up they roost upside-down in caves, barns, branches, or rocky cliffs. All become active at dusk, when they take to the air in search of fruit, pollen, nectar, or (depending on the species) insects, frogs, lizards, fish, even other bats. Thus they forage when the skies are cleared of the flights and sounds of almost all birds, which (not incidentally) eat the same kinds of food as bats.

What sensory adaptations have enabled bats to dominate the nighttime skies that birds leave vacant? How can bats detect food then, and why don't they fly smack into trees, cliffs, and other obstacles? For possible answers, consider that bats are generally social animals. They roost together, sometimes in colonies numbering in the millions. Males are equipped to make explosive, throaty calls that attract females during the breeding season; they use sound for communication. In addition, given that anywhere from dozens to millions of bats may roost together and fly together in the dark, you might expect them to have some means of keeping far enough apart to gain good, strong grips on their roosts, and of avoiding midair collisions—which they do. They make all manner of sounds by which they communicate their location to one another. Long ago, the use of sound waves for social communication may have been the basis for the evolution of sensory adaptations that allow most present-day bat species to forage in the dark. Bats **echolocate**: they emit high-frequency sound waves, and echoes from the waves bounce back to their ears from objects in the surroundings (Figure 15.1). In this way, bats orient themselves toward prey and away from obstacles. Bats and birds evolved concurrently many millions of years ago. Given that competition for the same food would have been fierce, they possibly came to coexist in the same places by dividing up the day-into-night cycle—with echolocation providing the competitive edge at night.

Thus some bats have big lips which they purse up during their flights, like a megaphone, to give off echolocating pulses—the echoes of which return to extremely sensitive ears. Others have a horseshoe-shaped structure around the nostrils that projects pulses straight ahead, much like a flashlight beam, for pinpointing objects. Bats that dine on small vertebrates have somewhat immense ears and (for a bat) large eyes, perhaps the better to hear faint rustlings and to observe surreptitious movements of their prey. Such exaggerated features of bat faces have nothing to do with intimidating other animals. They are no more and no less than windows through which this particular kind of animal has come to view the world.

Figure 15.1 Portrait of a moustached bat (*Pteronotus parnelli*) on the wing and, in the closer view, when it is listening to echoes of self-produced ultrasonic noises as they bounce back from objects in the environment. (Timothy Strickler and Terry Vaughan)

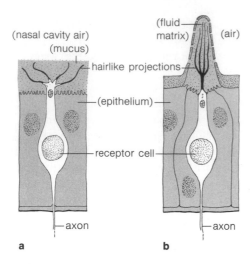

(nasal cavity air)
(mucus)
(fluid matrix)
(air)
hairlike projections
(epithelium)
receptor cell
axon
axon

a   **b**

Figure 15.2 Structure of an olfactory receptor cell in the epithelium of a vertebrate (**a**) and of an insect (**b**). (After Steinbrecht in C. Pfaffman, ed., *Olfaction and Taste*, vol. 3, Rockefeller University Press, 1969)

SENSING ENVIRONMENTAL CHANGE

Primary Receptors: Windows on External and Internal Worlds

The only windows between the nervous system and events going on within and around the animal body are **primary receptors**: cells or parts of cells that detect specific kinds of stimuli. A **stimulus** is any form of energy change in the environment that the body actually detects. If there were no such detection of change, the world would seem uniformly and perpetually devoid of detail. What you see as a blue flower in a golden wheat field is, in essence, a change in wavelengths of light energy—a difference in color between two regions of space. What you hear as sound are waves of change in air pressure—mechanical energy changes over time. What you feel as hunger is influenced by receptors inside your body that monitor shifting glucose levels—chemical energy changes—in your blood.

Receptor cells translate stimulus energy (any detected form of energy change in the environment) into electrochemical messages that can be dealt with by the nervous system.

Receptors reside in skin and body surfaces, in muscles and tendons, and in internal organ walls. Often they are arranged in epithelial and connective tissue to form **sensory organs**, such as the eye retina. Sensory organs amplify stimulus energy during its transformation into nerve signals. *Stimulus amplification* helps many animals detect weak but potentially important signals in the distance—for instance, to see or hear danger approaching. In addition, sensory organs provide information about *stimulus direction*. Thus, for example, your brain can estimate the source of some sound by comparing signals being sent to it from receptors in each of your two ears.

Receptors may be grouped according to the type of stimulus energy that they selectively detect:

Chemoreceptors. Detect impinging chemical energy (molecules or ions that have become dissolved in body fluids next to receptor). Include odor and taste receptors, and internal receptors such as those sensitive to blood oxygen levels.

Mechanoreceptors. Detect mechanical energy associated with changes in pressure, position, or acceleration. Include receptors for touch, stretch, equilibrium, and hearing.

Photoreceptors. Detect photon energy of visible and ultraviolet light.

Thermoreceptors. Detect radiant energy associated with temperature changes. Include infrared receptors.

Electroreceptors. Detect electrical energy movements (currents). Include receptors for electrical fields generated passively by external objects (such as prey), or for externally induced disturbances in a self-generated electrical field around the animal body (as in electric eels).

Nocireceptors. Detect energy changes that are injurious or painful to the body.

Let's focus on a few examples of these receptors to get an idea of how they channel environmental information into the nervous system.

Chemical Receptors

Insects and vertebrates have highly specialized chemoceptor cells. For instance, chemoreceptors on antennae of male silk moths *(Bombyx mori)* can sense one molecule of bombykol in 1,000,000,000,000,000,000 molecules of air! About forty chemoreceptors receiving only one molecule per second can trigger an action potential, through stimulus amplification. What's the adaptive value of this astonishing sensitivity? Bombykol is a pheremone that the female

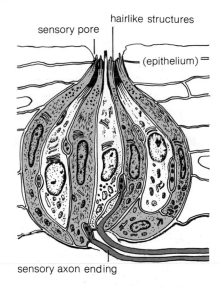

sensory pore

hairlike structures

(epithelium)

sensory axon ending

Figure 15.3 Structure and innervation of a taste bud, shown in long-section. Notice the different elongated cells with specialized hairlike structures at the tip. (After R. Murray and A. Murray, *Taste and Smell in Vertebrates*, 1970, Churchill)

moths secrete as a sex attractant. Chemoreception of bombykol permits the male to *find* a female, in the dark, as far away as a mile upwind.

Throughout vertebrate history, the nose has been most important in olfaction, the detection of odors. Olfactory receptors sampling odors from food in the mouth are important for our sense of taste. (That's why food seems tasteless when your nose is stopped up from a cold.) The only real distinction between odor and taste receptors seems to be in their sensitivity to different classes of molecules. Molecules reach both kinds of receptors by dissolving first in mucus or other liquids that coat the surface epithelium in which the receptors reside (Figure 15.2). Receptors for taste are more complex than olfactory receptors. Different cells having sensitive, hairlike projections at one end are clustered into tiny organs called **taste buds** (Figure 15.3). Your taste buds reside mostly in the mouth and pharnyx.

Tactile and Stretch Receptors

Threading through skin connective tissue are the nerve endings of **tactile receptors**. These sensory neurons detect mechanical pressure at or near the body surface. Tactile receptors occur in almost all body parts of birds and mammals; they are scarce in amphibians and nonexistent in fishes. This speaks of different environments. Animals living on land are more likely to interact with solid surfaces (the ground, for instance) than animals suspended in water. **Stretch receptors**, of the sort shown in Figure 14.13, occur in muscles of all four-legged animals. Impulses from these receptors help the brain determine the state of muscle contraction and the position of body parts in space. How does pressure or stretching trigger a potential in these receptor cells? One idea is that mechanical pressure deforms the plasma membrane enough to open additional sodium gates. The inward movement of extra sodium triggers a charge reversal across the membrane (Chapter Fourteen). Experiments with lobster axons show that stretching the plasma membrane does indeed increase membrane permeability to sodium ions.

Vibrations, Hearing, and Equilibrium

The mechanical energy detected by tactile and stretch receptors is fairly steady and intense in quality. Some forms of mechanical energy impinging on the body are not as constant or as intense. Consider the nature of a **vibration**, a wavelike change in stimulus strength. When you clap, you crowd many molecules together in the air between your hands. With the increased density they bump together more often and, because of the clapping's force, they bump together more rapidly. Molecules start flying outward and collide with molecules farther away, which bounce even farther away, and so on away from your hands. Meanwhile, back at your hands, the disturbed molecules keep jostling each other with less and less energy until they return to equilibrium density.

Now suppose you clap your hands repeatedly. Each time you do, you force air molecules together; you create a high-pressure region. At the same time, though, crowding a batch of molecules together means that there won't be as many left in the place they came from; you have also created a low-pressure region. These alterations in equilibrium density can be propagated, batch after batch, for long distances. They are a series of *longitudinal waves*—a form of mechanical energy transmitted outward from a stimulus source by molecules of air, water, even solids to some extent.

What we hear as **sound** is a form of vibration. It is our perception of periodic compressional waves, typically airborne, that are processed by specialized receptors. The perceived loudness, or *amplitude,* of sound depends on the

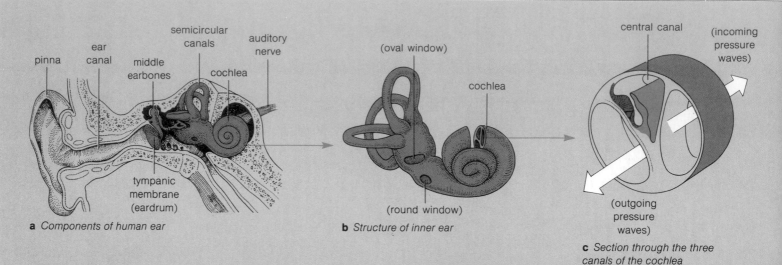

pinna ear canal middle earbones semicircular canals cochlea auditory nerve

tympanic membrane (eardrum)

a *Components of human ear*

(oval window) cochlea

(round window)

b *Structure of inner ear*

central canal (incoming pressure waves)

(outgoing pressure waves)

c *Section through the three canals of the cochlea*

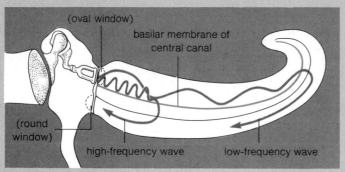

(oval window) basilar membrane of central canal

(round window) high-frequency wave low-frequency wave

d *Distribution of sound waves of different frequencies in the cochlea (shown here as if it were partially uncoiled)*

hair cell in organ of Corti tectorial membrane (auditory nerve)

basilar membrane

e *Close-up of hair cells in the organ of Corti, which is attached to the basilar membrane*

Figure 15.4 Case study: the human ear.

Your own perception of sound begins with a system of compressional wave amplifiers and receivers. Your external ear collects sound waves and channels them to the *ear canal* (**a**). The channeled waves arrive at a thin *eardrum,* the entrance to the middle-ear cavity. The eardrum bows inward slightly, under bombardment by a compressional wave, then springs back during the wave of rarefaction that follows. (With rapid changes in altitude, pressure differences can distort the membrane enough to cause pain. This can happen when you are in an ascending or descending airplane or elevator.) Attached to the back of the eardrum are three *middle earbones* (the hammer, anvil, and stirrup). The earbones amplify and transmit vibrations to the *oval window,* a membrane-covered gateway to the inner ear. The oval window is about twenty-five times smaller than the eardrum, hence mechanical energy of sound waves is greatly concentrated here. The energy so concentrated passes into the *cochlea,* a coiled tube vaguely reminiscent of a snail shell (**b**). Inside

this inner ear region are three fluid-filled canals (**c**). As shown in (**d**), there are two pathways for energy flow through these canals, depending on the pressure-wave frequency. Very low-frequency waves take the long way around: across the oval window, down an incoming canal, into an outgoing canal, then all the way up to the membrane-covered *round window.* Through this "window," any energy remaining is dissipated. The other pathway is a shortcut taken by higher frequency waves: from the incoming canal, then across a central canal and out the round window.

How are pressure waves sorted out this way in the inner ear? The answer lies in the way that one of the cochlear membranes, the *basilar membrane,* is constructed. At the cochlea's entrance it is narrow and somewhat rigid. But it gradually becomes broader and more flexible deep in the coil. High-frequency waves, which carry more energy, are able to displace the stiff region of the membrane. Most of their energy becomes transformed at once into membrane vibrations here, and these waves die out before traveling farther

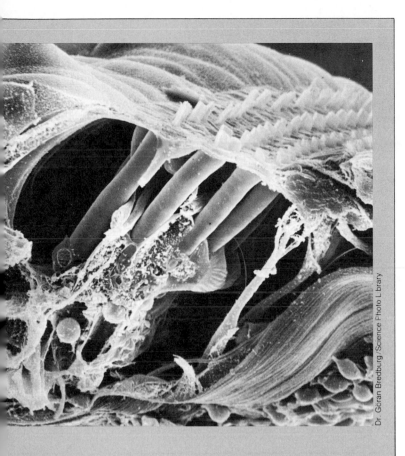

Dr. Göran Bredburg/Science Photo Library

down the coil. Low-frequency waves also set up vibrations in this region, but the vibrations are lower in amplitude. As a result, the low-frequency waves continue into the more elastic regions.

Precisely where along the basilar membrane these vibrations occur determine which mechanoreceptors in the ear will be stimulated. For it happens that perched on the basilar membrane, in the fluid-filled central canal, is the *organ of Corti.* And here we find *hair cells:* mechanoreceptors having hairlike vibration receivers at one end (**e,f**). As pressure waves travel inward, they displace the basilar membrane so that it moves. With this displacement, the hair cell processes are made to move in relation to an overhanging flap, the *tectorial membrane.* This movement is thought to change the permeability of the hair cell membrane, leading to excitation of the associated sensory neuron.

density difference between areas of high and low pressure in the wave trains. The more packed together the molecules are in each compressed region, the louder the sound. The perceived pitch, or *frequency,* of sound depends on how fast the wave changes occur. The faster the vibrations, the higher the sound.

Sound is a perception of periodic compressional waves—typically airborne, and typically varying in rate and in density differences between high- and low-pressure regions in the wave trains.

Which animals are attuned to sound as a stimulus? The ones with ears, you might say, thinking of your own flexible cartilage flaps on the sides of your head. But organs for sound reception are not always so readily apparent. Web-building spiders have no ears, yet they *are* attracted to the vibrations of a tuning fork. So are mosquitos. Crabs, shrimp, and other crustaceans produce growls, grunts, creaks, and rasping noises that evoke predictable behavioral responses in others of their species. So we may assume that they not only produce but perceive sounds—even though they have nothing that remotely resembles the sound receptors of vertebrates.

As a land vertebrate, you live in a setting where sound waves spread out through air, hence become weaker with distance. In your ears are membranes and bony structures that amplify these weak signals. Figure 15.4 explains how this amplification occurs. The human ear is capable of detecting some 400,000 different sounds reaching it. Some of these sounds are barely perceptible. Others are extremely intense and may cause structural damage to inner regions of the ear. Prolonged exposure to sounds of jet planes taking off, prolonged exposure to amplified live rock music—such recent developments have far outstripped the functional range of the evolutionarily ancient mechanoreceptor cells in the ear. These mechanoreceptors are **hair cells**: they have hairlike vibration receivers at one end. Figure 15.5 illustrates their inability to adapt to extremely loud sounds.

We tend to think of the ear as the hearing organ, but this is one of its more recently acquired tasks. In ancestral vertebrates, its primary function apparently was to detect body movements in different planes of space: it was an organ of equilibrium. An animal perceives itself to be in **equilibrium** when the forces acting on its body are so balanced that there is no change in position or acceleration relative to some direction.

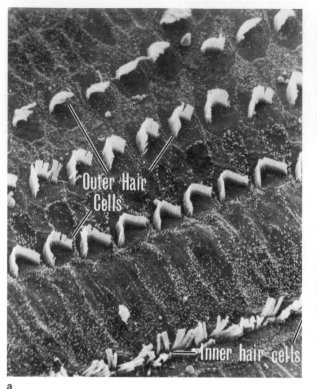 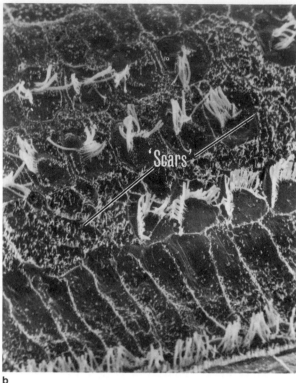

a

b

Figure 15.5 Effect of intense sound on inner ear. (**a**) Normal organ of Corti (guinea pig), showing three rows of outer hair cells and one row of inner hair cells. (**b**) Organ of Corti after twenty-four-hour exposure to noise levels approached by loud rock music (2,000 cycles per second at 120 decibels). (Robert E. Preston micrograph, courtesy Joseph E. Hawkins, Kresge Hearing Research Institute, University of Michigan Medical School)

How do organs of equilibrium work? Our internal ear contains a closed system of fluid-filled sacs and canals. In this system are mechanoreceptors—hair cells again, with their threadlike processes encased in a jellylike mass called a cupula. The **semicircular canals** (Figure 15.6) contain structures that detect rotational acceleration. As your body moves in a given plane in space, fluid in the canal corresponding to that plane is displaced and pushes against a cupula. The fluid pressure bends the hair cell processes encased in it. Because of the way these processes are oriented, turning the body in one direction triggers excitatory signals in an associated sensory neuron, and turning in the opposite direction decreases the rate of firing. Signals that do flow outward reach the cerebellum, where they are integrated with other information related to normal body position.

Other organs of equilibrium are present in the vertebrate ear and they, too, depend on specialized hair cells.

The cupula of these cells is thickened by deposits of calcium carbonate crystals. Such thickened structures are **ear stones** (otoliths). When the head is tilted or when the body accelerates in a straight line, the weighty ear stones press against hair cells, which signal the shift out of equilibrium. Similar organs occur in some invertebrates. (Examples are shown in Figure 15.7.)

Receptors for Light Energy

Light is a flow of discrete energy packets known as **photons**. Each photon is a bit of energy that has escaped from an excited atom—not only atoms of an original light source such as the sun, but atoms of objects in the environment. Trees, rocks, rabbits—the surfaces of all such objects absorb photons, the added energy raises electrons in atoms to higher energy levels, and these excited atoms give up new photons. Regardless of the source, photons always travel the same way. *Photons travel through air in a straight line.* Almost all animal groups have sensory adaptations for exploiting this property. Being able to sense the direction of light energy emanating from the sun, or another organism, or some other object enables an animal to move toward or away from it—not on a hit-or-miss basis, but with some precision.

All light-energy receptors depend on the absorptive properties of pigment molecules embedded in their cell membranes. In **photoreception**, the pigment molecules selectively absorb photons, which briefly alters their molecular structure. A consequence of this alteration is that photon energy becomes transformed into the electrochemical energy of a nerve signal. Here you may want to note the fact that photoreception is *not* the same thing as vision. All organisms, whether they see or not, are sensitive to light wavelengths. Shine a bright light on a single-celled amoeba that is moving about and it will stop abruptly in its pseudopodial tracks. Neither photoreceptors nor pigment molecules have been found in some small invertebrates, yet they display **phototaxis**: they orient toward or away from the direction of incoming light.

Only the most elaborate expressions of light sensitivity lead to vision. A **visual system** includes structures for focusing patterns of light energy onto photoreceptors, and a neural network that can deal with those patterns in the brain. As you read in the preceding chapter, different aspects of a visual stimulus (such as its position, brightness, shape, and distance) are detected by different receptors. A key part of most visual systems is a lens. A **lens** is typically a spherical or cone-shaped body of transparent protein fibers, which channels incoming light energy to photoreceptor cells located behind it. But a lens alone doesn't do the trick. Some invertebrates have eyes equipped with lenses, yet they can't see as we do. Their lenses channel light either in front of or behind their photoreceptors, the result being a very diffuse kind of stimulation. These invertebrates detect a general change in light intensity, as might occur when another animal passes overhead in the water. But they can't discern the size or shape of moving or motionless objects in their surroundings. We can summarize these points in the following way:

Vision requires precise light focusing onto a layer of photoreceptive cells that is dense enough to sample details concerning the light stimulus, followed by image formation in the brain.

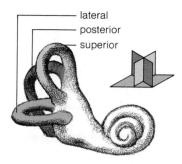

Figure 15.6 The three semicircular canals of the human ear, oriented in three planes of space. Mechanoreceptors that function in equilibrium reside in these canals.

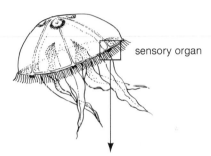

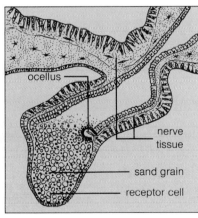

Figure 15.7 Invertebrate organs of equilibrium. Along the margin of the bell-like crown of the jellyfish *Aurelia* are a number of small sensory organs. Each contains a *statocyst:* a hollow organ containing sand grains that move when the body shifts in position relative to gravity. When the sand grains move, they stimulate receptor cells lining the organ.

Often, equilibrium receptors are ciliated, as they are in lobster statocysts. Resting on these hairlike projections is a *statolith,* a small mass of stuck-together sand grains or calcium crystals that weighs more than the surrounding water. When the lobster body is tilted, the statolith stimulates receptor cells on the downslope side. This causes a steady firing of their sensory processes, which synapse with muscle cells in a reflex arc that corrects body position.

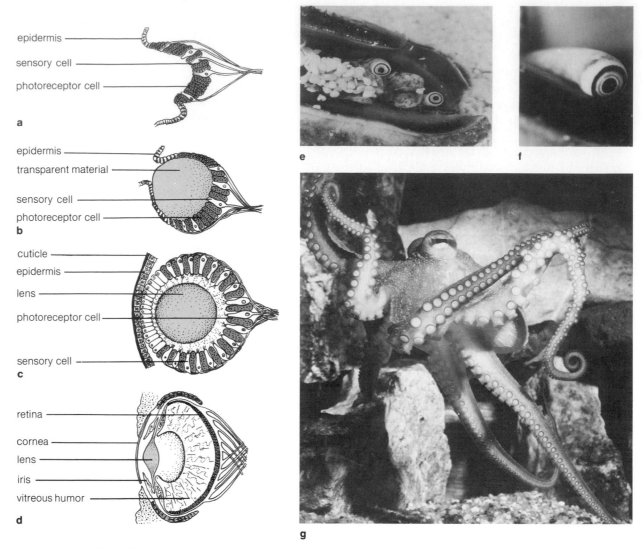

Figure 15.8 Sampling of molluscan photoreceptors, shown as if sliced lengthwise through the middle. (**a**) Limpet eyespot, no more than a shallow epidermal depression; (**b**) abalone eye, with secreted material that may serve as a lens; (**c**) land snail eye; and (**d,g**) octopus eye. (**e,f**) A red-mouthed stromb, peering about the Australian Barrier Reef. (a–d after M. Gardiner, *The Biology of Invertebrates,* 1972, McGraw-Hill; e,f Keith Gillett / Tom Stack & Associates; g, J. Grossauer / ZEFA)

Eyespots (ocelli) are found in many invertebrates. Figure 15.7 shows their location in the jellyfish *Aurelia.* Eyespots are simple clusters of photosensitive cells, usually arranged in a cuplike depression in the epidermis. In these cells, pigment-containing membrane is often folded into **microvilli** (tiny, fingerlike projections of surface membrane). This membrane pattern occurs in vertebrate photoreceptors as well.

Mollusks are the anatomically simplest animals with **eyes**: with well-developed photoreceptor organs that allow at least some degree of image formation. Some mollusks have closed fluid-filled vesicles, complete with transparent lens, **cornea** (a transparent cover), and **retina** (a tissue containing densely packed photoreceptors). Figure 15.8 shows a sampling of molluscan eyes. One group of mollusks, the cephalopods, includes the squids, cuttlefish, and octopuses.

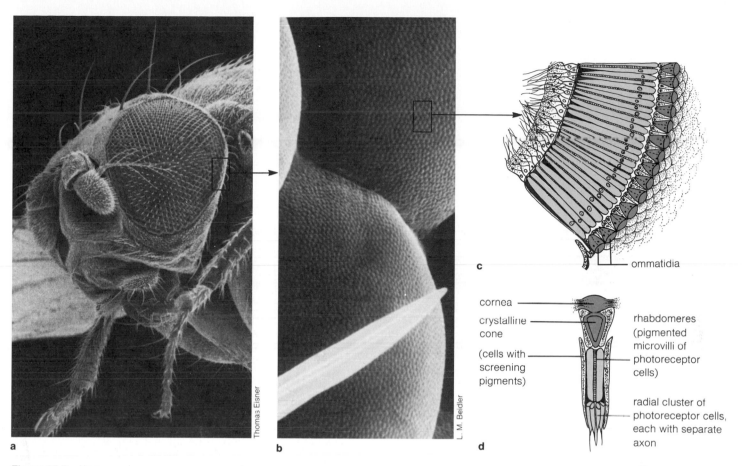

cornea

crystalline cone

(cells with screening pigments)

ommatidia

rhabdomeres (pigmented microvilli of photoreceptor cells)

radial cluster of photoreceptor cells, each with separate axon

a

b

c

d

Thomas Eisner

L. M. Beidler

Figure 15.9 Compound eyes, as occur among insects such as *Drosophila* (**a**). Each compound eye is composed of densely packed photoreceptor units (**b**) and (**c**), which are called ommatidia. The structure of one kind of ommatidium is shown in (**d**). The crystalline cone acts as a lens to focus light on the rhabdomeres, which contain light-absorbing pigments. It is in this central array of microvilli that light energy is transformed to nerve signals.

All are fast-moving predators that live in dimly lit underwater worlds—and all have large, paired eyes capable of effective image formation. Their eyes are positioned in the body region right behind prey-grabbing tentacles. Both eyes are used in aligning the tentacles at just the right striking distance from an edible morsel. Muscles control movements of both the eyeball and an **iris**, an adjustable ring of contractile and connective tissues. The open center of this contractile ring (the pupil) can be varied in size to admit more or less light. At the base of each eye is a retina of densely packed photoreceptors.

Bright light causes the cephalopod pupil to shrink in size—not into a tiny circle (as it does in your eyes) but into a narrow slit. In case you are ever snorkeling or diving and happen to stumble over a large octopus, you might like to know in advance that the pupils of its already large eyes can suddenly flare open, one or both at a time, in response to your unexpectedly close movements. The effect is startling, and no doubt adaptive; the octopus is known to use its enormous stare in securing the attention of a potential mate, and possibly to warn away a potential enemy. (You might also like to know, in case you are a little squeamish about the tentacles, that an octopus will attack only *moving* objects. Experiments have shown that an octopus will respond to changes in the size, shape, and distance of objects, but it is characteristically indifferent to stationary ones.)

The **compound eyes** of insects and crustaceans contain up to several thousand closely packed photosensitive units, of the sort shown in Figure 15.9. These units are known

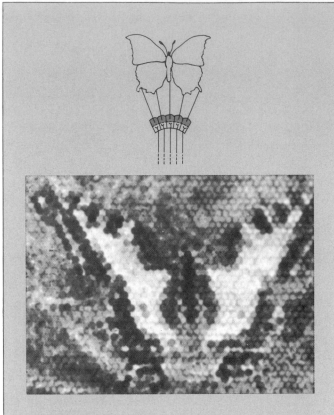

Figure 15.10 An approximation of light reception in the compound eye of insects, based on the mosaic theory of image formation. Each ommatidium receives light energy from only a small part of the visual field, which presumably is integrated into a larger visual pattern. This kind of image is actually formed when a photograph is taken through the outer surface of a compound eye that has been detached from an insect. However, it may not be what the insect actually "sees." Summation of potentials being sent to the brain from many ommatidia can strengthen some signals and inhibit others to produce a more crisply defined image. The representation shown here is useful insofar as it suggests how the overall visual field may be *sampled* by separate ommatidia. (After G. A. Mazokhin-Porshnykov (1958). Reprinted with permission from *Insect Vision*, © 1969 Plenum Press)

as *ommatidia* (singular, ommatidium). Usually, each has a hexagonal cornea and a cone-shaped, crystalline lens below that. Light entering through the lens is absorbed by pigment in anywhere from one to a dozen photoreceptor cells arranged radially below the cone (Figure 15.9d). Light ab-

sorption in microvilli of these cells triggers nerve potentials, which travel directly to the optic lobe from each photoreceptor cell. How these signals are processed to form visual images is not yet understood. According to the **mosaic theory** of image formation, each ommatidium detects information about only one small region of the visual field. An image is built up according to signals about different light intensities detected by all the ommatidia, with each contributing a separate bit to the whole visual mosaic (Figure 15.10).

Photoreception in the Vertebrate Eye

Almost all fishes, reptiles, birds, and mammals have eyes capable of effective image formation. For these groups, the eyeball has a **sclera** (a tough outer coat), then a **choroid** (a dark-pigmented tissue through which blood vessels course), and a densely packed retina. Part of the skin tissue overlying the eyeball meshes with the front of the sclera to form a transparent, light-focusing cornea. Toward the front of the eye, choroid tissue extends inward to form an iris. Here it is richly endowed with light-screening pigments, and with radial and circular muscle fibers used in controlling how much light enters the eye. A clear fluid (aqueous humor) fills the space between cornea and iris. A jellylike substance (vitreous humor) fills the chamber behind the transparent lens. Figure 15.11 shows how these parts are arranged in the human eye.

Among vertebrates, variations occur in the means by which light rays from both distant and close-up objects are made to converge precisely onto retinal photoreceptors. Convergence begins in the cornea. When light rays pass through its curved surface, they are bent so that they funnel toward some focal point. (Here you may want to review the description of light refraction in Figure 4.2.) But it is in the lens that *adjustments* are made. If the angle of bending is not great enough, the focal point will end up behind the retina. If it's too great, the focal point will end up in front. The lens helps bring the image into focus.

In the vertebrate eye, lens adjustments assure that the focal point for a given batch of light rays will land on the retina.

The term **accommodation** refers to lens adjustments that bring about precise focusing onto the retina. For example, fishes use eye muscles that move the entire lens forward or backward, thereby adjusting its distance from the retina. Increasing the distance moves the focal point forward; decreasing the distance moves it back. In other vertebrates, lens shape is adjusted under the coordinated stretching and relaxation of eye muscles and fibers attached to the lens (Figure 15.12). Birds that depend on rapid and intricate maneuvers during flight must have the means for rapid accommodation. Not surprisingly, the bird lens is highly elastic and ringed with muscles. The same is true for the lens in humans and other highly active mammals.

The retina is truly well developed in birds and mammals. Its basement layer, composed of pigmented epithelium, covers the choroid. These pigments prevent light scattering. Hence they prevent diffuse stimulation that could create blurred images. Resting on this basement layer is nerve tissue containing both photoreceptor cells and sensory neurons. Figure 15.13 shows how this arrangement works in photoreception in the human eye.

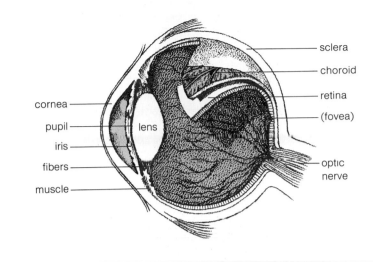

Figure 15.11 Main components of the human eye.

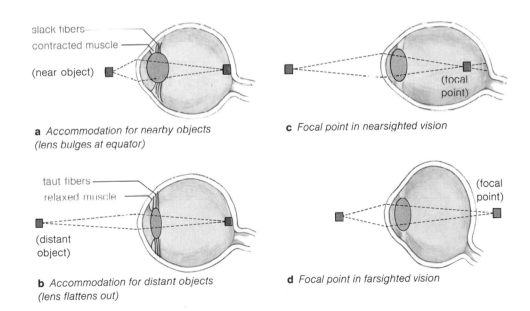

a *Accommodation for nearby objects (lens bulges at equator)*

c *Focal point in nearsighted vision*

b *Accommodation for distant objects (lens flattens out)*

d *Focal point in farsighted vision*

Figure 15.12 Visual accommodation, as it occurs in the human eye. (**a**) Close objects are brought into focus when eye muscles contract enough to slacken certain fibers interposed between them and the lens, which causes the lens to thicken in width at its equator. (**b**) Distant objects are brought into focus when these muscles relax, which puts tension on the fibers and stretches the lens into a flatter shape.

(**c**) People who are *nearsighted* have eyes in which the retina is too far behind the lens; light from distant objects is focused in front of the retina. (**d**) Those who are *farsighted* have light from nearby objects focused behind the retina.

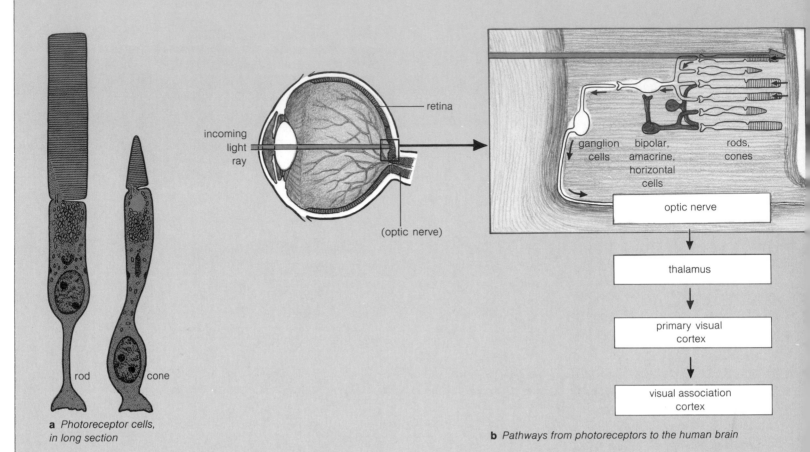

a *Photoreceptor cells, in long section*

b *Pathways from photoreceptors to the human brain*

Figure 15.13 Case study: photoreception in the human eye.

Two kinds of photoreceptor cells are intermingled in the retina's photoreceptive fields. They are called rods and cones because of their shape (**a**). A *rod cell* is concerned with perception of gray and black shading in very dim light, and for coarse perception of changes in light intensity that occur with movements across a visual field. Rods typically are packed in the retina's periphery. A *cone cell* is stimulated only by high-intensity light. It's concerned with sharp daytime vision and usually color perception.

Three types of cone cells have been identified. Each contains photopigments most sensitive to wavelengths corresponding to what we perceive as the colors red, green, and blue. (Colors in between are perceived when graded potentials from two or three different types of cones are summed at synapses. Thus excitation of both ''red'' and ''blue'' cones means we see the color purple.) In the human eye, cones are most densely packed in the fovea. A *fovea* is a funnel-shaped pit near the retina's center, where overlying nerve tissue is thinned away. This pit is only a millimeter across. Yet photoreceptors clustered here provide the greatest visual acuity (precise discrimination between adjacent points in space).

What happens to nerve signals generated in rods and cones? They move along at least two types of communication lines to the brain. In *through-conducting pathways,* only a few photoreceptor cells synapse with a few so-called bipolar cells, which relay signals to a ganglion cell (**b**). Axons of the ganglion cells lead out from the nerve tissue layer, and converge to form the optic nerve. From there, potentials travel to the thalamus, then on to visual processing centers in the cerebral cortex.

For the most part, signals from cones travel the direct lines. That's why cones provide precise detailing; their messages remain relatively unmixed. Signals from rods typically follow *lateral pathways:* side-to-side connections between many bipolar and ganglion cells. Summation of messages along these lateral pathways is one reason why rods can trigger an action potential even when light intensity is too low to excite cones. It's a reason why light-sensitive yet some-what fuzzily signaling rods let you sense where things are in a room too dark to activate their pointy-headed counterparts.

In the human retina, receptive fields for ganglion cells are circular. Some overlap, and they number in the hundreds. Stimulating different fields (even different parts of the same field) excites ganglion cells in in an ''on-center'' field triggers increased firing of the cells. Light

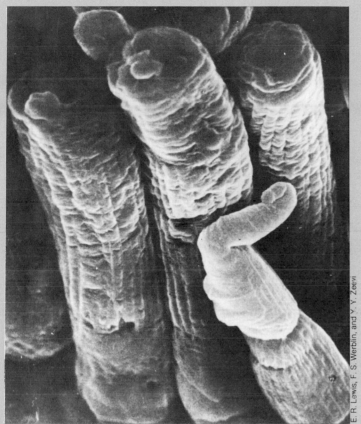

c Scanning electron micrograph of rods and cones (about 7,000×)

specific ways. For instance, light falling on an inner circle of receptors falling on the surrounding ring of receptors triggers a decrease in firing. Activity in an "off-center" field is reversed, and has the opposite effects. The organization of cells in such fields enhances our ability to see the boundaries between objects.

Also, lateral connections between retinal cells bring about graded, localized potentials. These messages, recall, decay unless amplified by other messages. Thus the *frequency* of firing in a given ganglion cell is modified by changes in size, location, or intensity of a light stimulus.

In this manner, the brain receives signals about different shapes and movements detected by receptors in a visual field—signals about lines of sight, orientations of those lines, their edges and contours, and their location over time. *Thus photoreception in the retina is more than an "on-off" activity. Preliminary integration also occurs here, even before messages about that activity course down the optic nerve and on to the brain.*

(a from "Visual Cells" by Richard W. Young, *Scientific American*, October 1970. Copyright © 1970 by Scientific American, Inc. All rights reserved.)

MOVEMENT IN RESPONSE TO CHANGE

Motor Systems

The sensory systems we have been considering range from simple arrays of photoreceptors to networks that sample the surroundings with great precision. Paralleling this sensory spectrum are motor systems adapted for response to a range of stimuli. All are based on the ability of certain cells to contract (shorten) and relax (extend in length). They are based also on the presence of some medium or structure—air, water, a skeletal framework—against which the force of contraction may be applied. Various motor systems will be described in Chapter Twenty-Four, which integrates these and other key animal developments into a sweeping evolutionary story. Here we will simply make acquaintance with mechanisms on which such systems are based.

All motor systems are based on effector cells able to contract and relax, and on the presence of a medium against which the contractile force may be applied.

In sea anemones, motor systems are little more than a network of T-shaped cells, each containing contractile proteins. Some of the cells run longitudinally through one body tissue layer. When they contract, the body shortens and looks fatter. Other contractile cells in another tissue layer run circularly about the central body axis. When they contract, the body lengthens and looks skinnier. (Figure 24.1 shows the results of such movements.) The longitudinal and circular muscle layers work as an **antagonistic muscle system**: action of one motor element opposes the action of another.

Annelids such as earthworms depend on a different arrangement. These worms are segmented: the body cavity is divided into a series of compartments, each with a flexible wall (cuticle) surrounding a fluid-filled chamber. Each segment has a set of longitudinal and circular muscles. First the contractile force of the circular muscles is applied against the fluid-filled interior which, because it resists compression, acts as a **hydrostatic skeleton**. As circular muscles contract, fluid is squeezed down the body, much like toothpaste being forced down a tube. When that happens, longitudinal muscles are made to stretch. Now sets of bristles (setae) projecting from the earthworm body grip the ground, acting like toeholds for the stretched-out worm. When the longitudinal muscles contract, their force is applied against the toeholds, and the body is pulled forward.

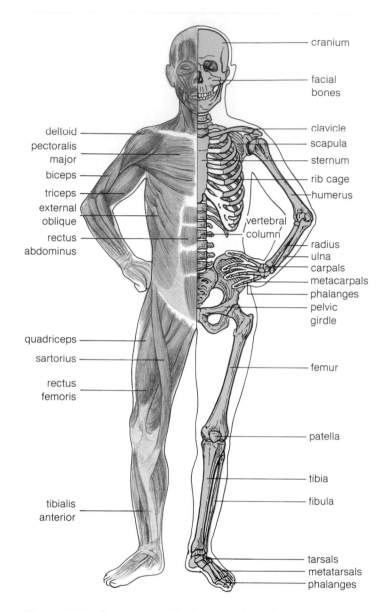

Figure 15.14 Components of the human motor system.

Insects, crabs, and other arthropods also have segmented body plans. A hardened cuticle covers each body segment. It forms an external skeleton, or **exoskeleton**, for a network of antagonistic muscles. The cuticle between body segments remains pliable and acts like a hinge for movement in different directions.

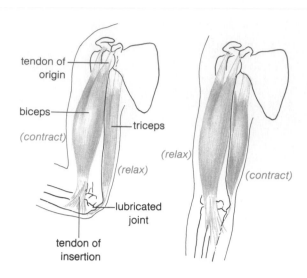

Figure 15.15 Antagonistic muscle movement, showing how two muscles of a pair can contract and relax in opposition to each other.

Vertebrates such as yourself have an internal skeleton, or **endoskeleton**. Your own internal framework consists of bone and cartilage (Figure 15.14). Bone and cartilage cells are arranged to form hollow tubes—structures that are lightweight yet afford strength. The endoskeleton functions in movement, in support, and in protection of vital organs. Even its ribs and the muscles between them help in maintaining an upright posture. The *axial* portion includes the vertebral column, skull, and rib cage. The *appendicular* portion includes the arms, legs, and associated appendages. Together with skeletal muscle, the appendicular portion acts as a system of levers for moving the body. Webs of muscles stretch over the framework. Tough connective tissue strands, called tendons, attach muscles firmly to skeletal bones. Connective tissue fibers, called ligaments, form flexible connections at joints.

Skeletal muscles contract and relax along one axis only. But the human body requires movement in multiple directions. Muscles in body limbs are arranged in antagonistic pairs, such as the biceps and triceps shown in Figure 15.15. Notice how they bridge the joint. When one member of this antagonistic pair contracts, the joint flexes (bends). As it relaxes and its partner contracts, the limb extends (straightens out again). Reciprocal reflex connections exist between the pair. When the biceps contracts, inhibitory

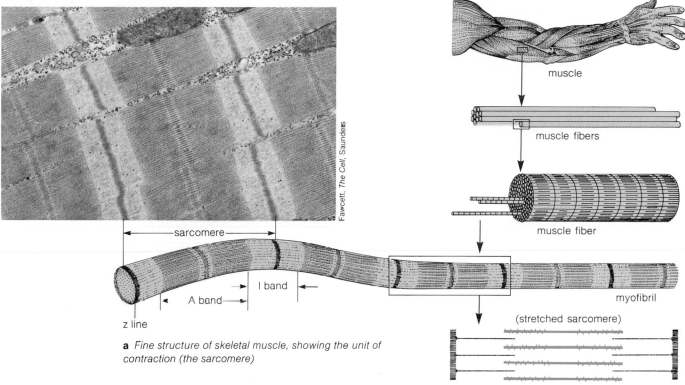

Fawcett, *The Cell*, Saunders

muscle

muscle fibers

muscle fiber

myofibril

(stretched sarcomere)

(contracted sarcomere)

myosin filament

actin filament

— sarcomere —

I band

A band

z line

a *Fine structure of skeletal muscle, showing the unit of contraction (the sarcomere)*

interneurons simultaneously act on the motor neurons of its partner, the triceps, which thereby relaxes. Conversely, when the triceps contracts, inhibitory signals simultaneously act on motor neurons of its partner. This **reciprocal inhibition** is the basis for coordinated movements in various skeletal–muscle systems.

In skeletal–muscle systems, reciprocal inhibition of reflexes between antagonistic muscle pairs is the basis for coordinated contractions.

Reciprocal inhibition can be overridden. You can contract biceps and triceps at the same time, when (for example) you hold your arm upright like a stiff pillar.

Skeletal Muscle Contraction

Skeletal muscles contain anywhere from a few hundred to thousands of **muscle fibers**: cylindrical cells, often several centimeters long. Connective tissue holds muscle fibers together in bundles, which are attached to bone. Figure 15.16a shows the fine structure of skeletal muscle. Each fiber consists of finer threadlike structures called **myofibrils**. Myofibrils contain two kinds of protein strands.

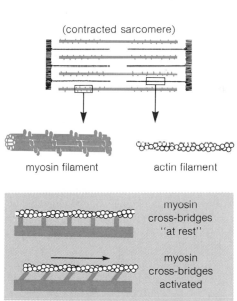

myosin cross-bridges "at rest"

myosin cross-bridges activated

b *Movement of myosin cross-bridges. Several actin filaments are arranged around each myosin filament. For clarity, only one is shown here.*

Figure 15.16 Proposed mechanism of skeletal muscle contraction, as described in the text.

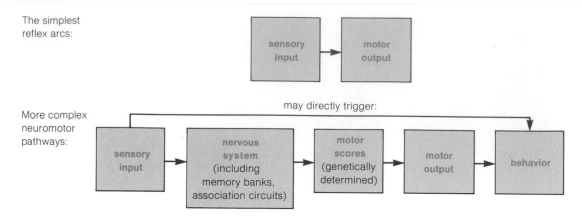

The simplest reflex arcs:

```
sensory          motor
 input    →     output
```

More complex neuromotor pathways:

may directly trigger:

```
sensory      nervous         motor           motor
 input   →   system    →    scores     →    output    →   behavior
             (including     (genetically
             memory banks,  determined)
             association
             circuits)
```

Figure 15.17 Pathways between sensory input and motor output for both simple and complex behavioral responses.

Myosin strands have side-projecting cross-bridges. They are bundled together to form relatively thick filaments. Two protein chains coiled together form each *actin* strand, which is thinner than myosin (Figure 15.16a). Repeating bands of actin and myosin translate into a light-dark-light repeating unit that gives skeletal muscle its striped appearance. Darker (Z) lines between each repeating unit define the **sarcomere**, the fundamental unit of muscle contraction.

Recall that motor neurons synapse with muscle cells (Figure 14.10). At a neuromuscular junction, the neurotransmitter acetylcholine interacts with receptors on the muscle cell membrane. It causes changes in ion permeabilities, which lead to charge reversal across the muscle membrane. This forms the basis of an action potential, which propagates along the muscle membrane into the fiber. The propagating muscle action potential leads to the release of calcium ions from a special membranous sac (the sarcoplasmic reticulum) surrounding the myofibrils. Calcium ions bind to the thin filaments in such a way that actin is free to interact with adjacent myosin cross-bridges. These interactions involve activation of enzymes contained in cross-bridges projecting from the myosin. The enzymes split ATP molecules, which releases usable energy. When activated, the cross-bridges attach to actin, bend, release their hold, and attach again (Figure 15.16b). In this way, actin strands are propelled past myosin strands, much like boats being propelled down a narrow stream by oarsmen alternately digging their oars into the bank and pushing on them.

Each cross-bridge movement uses up ATP. When you think of the tremendous number of cross-bridges involved, you can see why muscle activity consumes so much ATP. More ATP is used to run calcium pumps, which transport calcium ions back to the sarcoplasmic reticulum. With the removal of calcium ions from myofibrils, actin no longer interacts with myosin, and the sarcomere returns to the relaxed state.

BEHAVIOR: COORDINATED RESPONSE TO CHANGE

The Neuromotor Basis of Behavior

In the preceding chapter, you read about the stretch reflex, the simplest sort of reflex arc. Few neuromotor responses are as well understood. In most responses, integrative centers and memory banks in the brain intervene in many ways, as do the proddings of endocrine elements. Genetic programming has its effect; so do physiological states, diverse environmental changes, and learning. Thus there may be a range of variation within the same category of response. Its intensity, how rapidly it is carried out, and how long it lasts depend on external and internal cues—both of which vary. Integration of sensory, genetic, neural, and endocrine factors leads to the coordinated output called animal **behavior**. Figure 15.17 summarizes the pathways resulting in behavior.

Figure 15.18 A male finch fluffing his feathers not because it is cold outside but because a receptive female finch is on the perch.

Animal behavior is a coordinated neuromotor response to changes in the external and internal states. It is a product of the integration of diverse sensory, genetic, neural, and endocrine factors, and it may be modified through learning.

In many cases, coordinated movements are based largely on sequences of motor output that are determined by the central nervous system. Such a sequence is known as a **motor score**. For instance, in the sea slug *Tritonia*, individual interneurons control contractions in sets of muscle cells that give rise to predictable swimming patterns. Stimulation of one interneuron in the left cerebral hemisphere leads to contractions that bend the head and tail ends of the body to the left. Stimulation of another causes contractions that move the body up and down in a wavelike pattern.

Motor scores that predictably run their course once some stimulus sets them in motion are genetically determined. They represent **innate behavior** characteristic of all members of a given species. How do we know they are determined genetically, and not learned? Consider the feeding behavior of chickens. A newborn chick confronted with small particles on the ground will spontaneously peck at them, even if it has never observed another chicken performing this behavior. A newborn duckling raised in isolation and then shown mud for the first time in its life will spontaneously run through a motor sequence for sifting mud through its beak, which is a ducklike feeding behavior. In these examples, the particles and the mud act as stimuli. Any stimulus that triggers a motor score, or even some part of it, is called a **releaser**.

Perception of what constitutes the correct releaser for a given motor score also has a genetic basis. Consider the fruit fly *Drosophila*. Reproduction is not an inevitable consequence of encounters between a male and female. First the male must perform an intricate courtship dance—a highly stereotyped motor score. The dance acts as a releaser for a motor score in a receptive female, which causes her to assume a certain body position required for successful copulation.

As with all heritable features, motor scores may be modified through mutation. For example, a *Drosophila* male that is heterozygous for a particular mutation behaves the same as wild-type males. But a male homozygous for the mutation displays bizarre mating behavior. He ignores females and goes into a frenzied courtship dance for every male he meets. In this case, the motor score is normal, but a mutation has modified the fly's ability to perceive what constitutes the normal releasers! Because the mutant males are obviously not going to leave offspring, this behavior is quickly lost from the population.

Over time, selection pressures may restrict certain innate behaviors and other times extend the circumstances under which they are released. Feather fluffing in birds may be an example of a single behavior that has come to be released by more than one environmental stimulus. This innate behavior has a physiological advantage: feather fluffing helps keep body temperature constant when outside temperature drops. Somewhere along the line, feather fluffing has undergone divergence in function. Not only is it used to retain body heat, male birds of many species use it in both courtship and territorial behavior (Figure 15.18). By fluffing their feathers, the males create a larger silhouette. We might speculate that the first male birds to be genetically programmed to fluff feathers on sight of another member of their species were somewhat more effective in turning off a potential rival, or turning on a prospective mate. This example suggests it is not the animals themselves that consciously bring about innate behavioral diversity. Rather, such diversity probably arises through the greater reproductive success of individuals whose genetic programming caused the motor scores involved to be released by additional stimuli.

Even though they are genetically determined, motor scores often are modified in ways that are adaptive to changing times and circumstances. Even in supposedly simple animals, the same stimulus doesn't always trigger the same motor response. The response may be modified by secondary factors and by the recent past history of the individual. For instance, substances found in the flesh of all animals will, when squirted in the water near a hungry sea anemone, elicit a vigorous feeding reaction. But the same substances squirted near a well-fed sea anemone elicit an avoidance reaction. The internal state of the animal influences its responsiveness to external stimuli.

Feeding behavior has been studied in laboratory rats. If a region in the lower part of a rat's hypothalamus is destroyed surgically, the rat overeats so much that it becomes outrageously fat. This region must be a "satiety center" controlling the innate feeding response. Normally, a high glucose level in the blood is one of the signals that triggers the satiety response, because glucose molecules bind to receptors in that region. But when the satiety center is not functional, feeding is not turned off as it usually would be when enough food is eaten to meet normal body needs for glucose. These receptors are highly specific to glucose. In one experiment, glucose molecules that had been slightly modified were injected into a rat. The molecules could still bind to cells in the rat's satiety center, but the modification was enough to prevent the inhibitory effect, and the rat would not stop eating.

Other brain centers are known to take part in programs for sexual and aggressive behavior. Even when an animal observes a potential mate (or rival), it will not respond with sexual behavior (or with aggression) if activity of the corresponding brain center is artificially blocked. Yet artificial stimulation of the appropriate brain center has little effect unless the animal does observe the releaser—a potential mate (or rival). Analysis of such brain centers is far from complete. But work to date strongly suggests that "motivations" of animals, or the "drives" that compel them to eat, fight, and mate, will become understandable as interactions among the pathways outlined in Figure 15.17.

The activities of brain centers regulating innate behavior may themselves be products of natural selection. We may readily envision how different species may have come to display characteristically different motor patterns. For instance, how often have you spent a day at a zoo and noticed that large cats—lions, leopards, jaguars, pumas—are usually lethargic, but that most of the wild dogs—wolves, Cape dogs, coyotes—usually are pacing back and forth for hours on end? What ecological and evolutionary factors might have given rise to this difference in innate motor patterns between the two types of predatory animals (Figure 15.19)?

Learned Behavior

In many species, some forms of innate behavior show little capacity to be modified. As long as a motor score contributes to adaptability, selection would tend to favor individuals in which that behavior is rigidly locked into the genetic program. But say an individual has the genetic potential for modification in some innate behavior. If the modification makes it more successful than its relatives at finding food, escaping from predators, or producing offspring, then its plasticity of response will tend to be perpetuated. If such modifications are more than one-time motor responses to a new cluster of stimuli—if there is an enduring potential for adapting *future* responses as a result of *past experiences*—then the modification is called **learning**.

In constant or highly predictable environments, rigid genetically determined motor patterns are common.

In changing and unpredictable environments, the capacity for behavioral plasticity and learning tend to prevail.

As you will see, learned behavior ranges from simple modifications of motor patterns to the complex insight learning of some primates, especially humans. However, much of learned behavior is still a puzzle. One problem is that learning has generally been studied not in the natural environment but in controlled experimental situations. Because the natural setting contains many variables operating at once, it is not likely that the "pure" learning responses observed during experiments exist so crisply in the wild. We will survey some aspects of learning here, but it's useful to keep in mind that in the natural world, a behavior pattern may be an intricate mix of innate *and* learned components, and it is often impossible to make the distinctions we will be making here.

The classic work of Ivan Pavlov, a Russian physiologist who was interested in digestive juice secretion, grew into one of the first controlled studies of learning. Pavlov observed that his laboratory dogs salivated immediately after he placed a meat extract on their tongues. This he interpreted to be a simple reflex response. But then he found that if he rang a bell just before giving the dogs the extract, the dogs began salivating at the sound of the bell alone. Pavlov called this new response a *conditioned reflex*, for the dogs had come to associate the sound of the bell (a conditioned stimulus) with the taste of food (the reinforcing stimulus). This behavior, whereby a connection is made

Figure 15.19 Two kinds of large, predatory carnivores characterized by two kinds of activity patterns during the day. How might natural selection have shaped the difference in these innate behaviors?

between a new stimulus and a familiar one, is a form of **associative learning**.

Why is associative learning adaptive? If an animal learns to anticipate certain events, it can respond more rapidly to them when they do occur. If the sound of an airplane engine always precedes shotgun blasts from the hunter who picks off animals from the air for the fun of it, the arctic wolf that learns to respond quickly to the sound of an approaching airplane may have a better chance of surviving the pointless slaughter than its less adaptive relatives.

Another form of associative learning is *instrumental conditioning*. Here, a reinforcing stimulus (reward or punishment) appears after a behavioral response is given. The animal learns by trial and error. Earthworms show learned behavior in tests with a simple T-maze (a maze with a base and two arms shaped like a "T"). The earthworms enter the maze at the base, and if they turn down one of its arms, they encounter an irritating stimulus (say, an electric shock). If they turn down the other arm, they encounter a moist, darkened chamber. After many trials and enough shocks, they generally learn which motor pattern leads them to the more hospitable setting.

In the forms of learning just described, the behavior persists for as long as the reinforcement persists. But if the reinforcing stimulus is withdrawn (say, if the bell is rung again and again without being followed by the taste

of food), the learned behavior may soon become extinguished. That, too, is a learning process; it is called **extinction**.

Maze studies have indicated that some animals have a **latent learning** ability. A rat will learn its way through a complicated maze more quickly if it's first given a chance to poke about, even without any reward when it happens to find the way out. Such exploratory behavior is popularly called "curiosity." But what is actually causing the rat to explore? And without reward, why does it remember what it discovers? We can speculate that because curiosity leads to learning about the physical nature of the environment, it must be adaptive. Such learning might include knowing about a good hiding place, which might well mean the difference between life and death, given an unexpected encounter with a hungry predator. Thus there would be gradual selection for latent learning ability. (Of course, exploratory behavior may also leave an animal vulnerable. It may be perpetuated only if the rest of the behavioral potential of the individual allows it to cope with what it happens to come across while poking about.)

It is only among some primates that **insight learning** has been adequately demonstrated. With this behavior, alternative responses to a situation are first evaluated mentally. The evaluation may lead to a sudden connection of separate bits of stored information, which tells the animal what an appropriate response may be. Thus insight learn-

a   b

Figure 15.20 (**a**) Human imprinting objects. No one can tell these goslings that Konrad Lorenz is not Mother Goose. (**b**) A sexually imprinted rooster wading out to meet the objects of his affections. During a critical period of the rooster's life, he was exposed to a mallard duck. Although sexual behavior patterns were not yet developing during that period, the imprinting object became fixed in the rooster's mind for life. Then, with the maturation of sexual behavior, the rooster sought out ducks, forsaking birds of his own kind, and lending further support to the finding that imprinting may be one of the reasons why birds of a feather do flock together.

ing is a "trial-and-error" process that goes on in the brain. It is a synthesis of accumulated experiences that suggest what the appropriate responses may be in new situations.

For many animals, learning certain forms of behavior occurs only during limited time spans called **critical periods** of development. Consider the process by which newly hatched ducklings come to recognize members of their species. When they observe a large moving object shortly after hatching, they tend to show an enduring preference for associating with it. Normally, of course, the first object the hatchlings see is the mother duck. But experiments show that the birds will also attach themselves to artificial models of various sizes, shapes, and colors—even to human models (Figure 15.20). This preferential behavior toward a stimulus presented during a critical period of development is called **imprinting**.

Often, imprinting causes modifications in motor patterns that are expressed much later in the life cycle. Bird song is one example. The general structure of the chaffinch song, which has three parts, is imprinted on male chicks when they are a few weeks old as they hear adult male chaffinches singing. But they don't join the competitive chorus until the next spring, about eleven months later, when they set out to establish territories of their own.

During the critical period of their first spring, the young birds develop the song by imitating other males in the neighborhood. Although the general vocal movements underlying this song structure are similar from bird to bird, local "dialects" exist. And these are the versions that are learned and will be sung year after year, no matter how many versions are heard after that.

Humans appear to have an important learning period during their first two years of life. It had been known for some time that stimulus deprivation during early postnatal life leads to severe, irreversible learning deficiencies in laboratory animals. But it was not until Rick Heber's prolonged Milwaukee Project that stimulus enrichment during the first two years of human life was found to be crucial. Early enrichment programs have more beneficial effects than such extended programs as Operation Headstart, which normally have been initiated during later periods of development.

In Heber's study, a test population of socially underprivileged children was divided into two matched groups. The children were raised in their own homes. But Heber and his researchers visited the children in one group and provided them with intensive, individual stimulation from birth to age two. During the visits, they held the children, showed them illustrated books, talked to them, and sang to them. (The researchers had no contact with the other group.) When this sensory enrichment program was introduced early enough, the results were dramatic. The average IQ for children in the test group soared some twenty to thirty points above the control group by the end of preschool. The greatest difference did not even show up until the children had become teenagers. Around puberty, when the nervous system undergoes final maturation, the differences in learning ability and achievement between the two groups were indisputable. It is postulated that during early postnatal life, connections are made in the brain that do not become functional until much later. But in the absence of rich and varied stimuli during the early period, these connections simply are not made.

MIGRATIONS—BEHAVIOR ON A GRAND SCALE

During at least one stage of development, all animals move about—in search of food, protection, a mate, a place to give birth, a place to die. These movements may be individual interactions with the environment, with nothing more than short-term motor responses to changing conditions. They also may be movements of individuals or populations over great distances. Perhaps the most dramatic of all these movements through time is the **migration**, the cyclic surging of whole populations away from one region toward another that beckons with compelling force—and then a returning to the place from which the journey began.

Migration is a predictable behavior pattern of many species. Crustaceans of the open ocean heed a recurring call to move into water of different depths. Insect populations may show migratory patterns, but the life span of most insects is so short that they may not be able, as individuals, to complete the round trip. Most likely their offspring are the ones to journey home, thereby giving the illusion of cyclic migration. The monarch butterfly is an exception. A long-term study with thousands of tagged individuals proved that populations of these insects do indeed make the immense round trip from North America to their wintering grounds in Mexico each year.

Many fishes show migratory behavior. Each year, newly hatched salmon leave the headwaters of their inland home for the rich feeding grounds of the open seas, sometimes thousands of kilometers away. They live for several years in the open seas, feeding and developing to sexual maturity. Then, in the spring of their last year, they leave the seas and return to their birthplace, where they repro-

duce and die. Hormones trigger their quest, and production of those hormones has been correlated with the lengthening days of spring. Once migration is set in motion, chemoreceptors function in the salmon's unerring ability to return from its journey, following a trail of odors specific to its home stream and never forgotten. Perception of other environmental cues—water currents, temperature gradients, sounds of waterfalls and rapids—may also function in salmon migration. How these cues function remains largely unexplained.

Among bird populations, migration takes on impressive dimensions. Many bird species make an annual round trip between wintering feeding grounds and their breeding range. Birds have the advantage of flight, which means they have motor systems and physiological adaptations that allow them to traverse long distances quickly. Well-developed eyes allow them to sense details of the environment—distance, direction, visual depth, motion—that influence their flight path. The skeleton and musculature of migratory birds are studies in adaptive architecture (Figure 15.21). Bones are thin, hollow, and exceptionally light in weight. Breast and pelvic bones are so arranged that either wings or legs can support the body near its center of gravity. Strong, massive muscles attached to the wing bones are slung beneath the breastbone, with the main muscle mass low in the body—an arrangement that helps stabilize flight. Wings are decked out with lightweight, interlacing feathers that promote lift through the surrounding air. Long migratory journeys demand high energy expenditure. Just before the onset of migration, molting occurs: new, stronger feathers replace old ones. And energy reserves (fat) are accumulated.

Birds normally have a high rate of metabolic activity, and this might help explain why they undertake their immense migrations in the first place. Because they burn up energy quickly, they must always have a reliable food source. Few environments are so productive that they yield abundant food all year long. With the onset of winter, for example, food supplies dwindle in the north temperate regions, where many migratory birds reproduce. Even though birds have effective adaptations for withstanding temperature changes, the fact remains that the food must be there for birds to survive. That requirement alone would be enough of a selection pressure over time. To be sure, dangers are associated with their ambitious undertakings—flights over open oceans, through fog and storms, with predators along the way. But with the return of spring, they are rewarded for making the return trip by finding their breeding grounds rich in new plant growth and sustaining abundant insects.

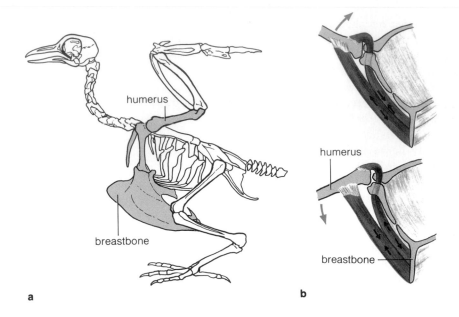

a

b

Figure 15.21 (**a**) Skeletal system of a bird, showing the thin bones, the flight-stabilizing heavy breastbone, and bone fusions in the breast, pelvic girdle, and backbone—all adaptations for flying. (**b**) The rope-and-pulley type of mechanism between powerful antagonistic muscles that help raise and lower the bird wing. The main muscle mass lies low in the body, which also helps stabilize flight. (a reprinted by permission from *Ornithology in Laboratory and Field* by O. S. Pettinghill, 1970, Burgess Publishing Company; b from T. Storer et al., *General Zoology*, sixth edition, 1979, McGraw-Hill)

Once migratory journeys begin, birds depend on external stimuli and internal mechanisms for arriving at their destination. It may be that latent learning plays a part in how birds find their way about. In their **homing behavior**, birds learn to recognize familiar landmarks in the area around the nest. Homing behavior is limited for nonmigratory birds; few can find their way back if they are taken very far away from their home range. For migratory birds, the range can be astonishing. The literature is filled with such examples as the attempt to remove albatrosses from their home on Midway Island in the South Pacific Ocean. Because the birds were interfering with military planes landing and taking off there, they were transported to potential new homes as far away as Washington and Oregon. One albatross released in Washington refused to be displaced by the United States Armed Forces. It returned to Midway in a little over ten days, having averaged 317 miles a day for the 3,200-mile journey—a journey never before

undertaken and most of it occurring away from any landmarks, over open ocean!

Underlying short-range and long-range homing and migratory behavior is a form of **piloting**. The term refers to an orientation toward home by topographic or meteorologic cues that a bird senses through either random or systematic explorations. Photoreceptors might provide visual cues—information about river systems, forests, or landforms such as mountains or coastlines. Mechanoreceptors might feed in information about prevailing winds, air turbulence patterns over specific landforms, even shifts in humidity of air masses through the year. Perhaps one of the clearest examples of environmental cues comes from Gustav Kramer's studies of caged starlings (Chapter Two). He proved that these day-flying birds use the sun's position as a compass. Throughout most of the year, his caged starlings fluttered randomly. But in the spring and autumn, their flutterings became oriented in the direction characteristic of migration for their species. Even when the sides of the cage were covered, so that landmarks could not be seen, the starlings still oriented in the normal direction—as long as the cage remained open to the sky. When the cage was entirely covered with material that let in only diffuse light, their flutterings became random once more. Kramer then built a cover for the cage. Shuttered windows were built into the sides of the cage cover. Next to each window was a mirror, which could be moved to shift the angle of incoming sunlight. When the mirrors were shifted so that light rays entering the cage were deflected ninety degrees, the starlings changed the orientation of their flutterings according to where the sun *appeared* to be, rather than where it really was.

These studies raise other questions. If starlings navigate by the sun's exact position, how do they take into account the variations in the sun's position in the sky—not only with the seasons but with the time of day? Because starlings migrate for at least six hours a day, they somehow must compensate for the sun's shifting position. They must have some internal timing mechanism for doing so—a "clock," so to speak. What could be the nature of biological clocks? For some vertebrates, at least, the answers appear to lie in the nervous and endocrine systems.

Behavior and Biological Clocks

Much of animal behavior is correlated with rhythms of the planet. Two such rhythms are particularly pervasive. Approximately every twenty-four hours, the earth completes one rotation about its long axis. This rhythm is called **circadian**, after the Latin *circa* (about) and *dies* (day). In addition, because the earth is tilted on its long axis relative to the sun's position, the length of day changes cyclically as the planet completes its annual journey around the sun. For example, much of North America experiences short days in winter and long days in summer. The ability to detect changes in daylength is called **photoperiodism**. A **biological clock** enables an animal to measure rhythmic changes of the sorts just defined, and to trigger physiological and behavioral adjustments to them.

In some vertebrates, the pineal gland has been shown to be one structure involved in communicating time information. This gland secretes the hormone melatonin, which is synthesized from the neurotransmitter serotonin. Serotonin, recall, appears to help regulate sleeping behavior. From studies of birds, it appears that melatonin output coordinates motor activity as well as metabolic activity. In some species, melatonin is also known to affect the increase in gonad size associated with reproductive cycles. When melatonin levels in the blood are high, motor and metabolic activities are low—as they are in animals that sleep by night. When melatonin levels are lowest, motor and metabolic activities are at their peak.

It appears that the enzyme N-acetyltransferase, which helps catalyze the synthesis of melatonin, is a key part of biological clocks that measure changes in light. The activity of this enzyme is regulated by the animal's perception of light—it is enhanced during the dark and inhibited during daylight. Light perception in rats depends on nerve pathways that course from photoreceptors in the eye to the pineal gland itself. In birds, both retinal photoreceptors and perhaps photoreceptorlike cells in the pineal gland may be involved. It is not known yet whether this time-keeping enzyme is reset directly by light or indirectly by neural or hormonal commands.

PERSPECTIVE

In this chapter, we touched on the links between the physiology of individual animals and the external environment—on the union of stimuli and the sensory systems filtering them, the motor systems that carry out responses to them. The coordinated motor responses to stimuli are called forms of behavior. The examples given here cover both the genetic and the learned aspects of behavior. Two major points have run through the discussion. First, all behavioral responses have their basis in the way that an animal is able to perceive its environment—in the limitations and the potential of its anatomy and physiology. Second, all the diverse ways of perceiving and responding to external cues have a common source. Each provides a selective advantage that somehow helps an animal escape predation, find food and mates, or survive changing environmental conditions. The external cues may be chemical, mechanical, visual, electrical, or thermal. There may also be internal signals, arising from neural and endocrine balancing acts. In all cases, cues are heeded in a selective way that enhances the individual's probability of surviving and reproducing.

In a later chapter, we will describe how these stimuli and responses come together in the rich tapestry of behavioral displays. These displays are elaborate and often ritualized signals that animals use to communicate information. They herald a new level of organization in nature—the extension of the individual "need to survive" to the *cooperation* of individuals in the environment.

Readings

Animal Behavior. 1972. Life Nature Library. New York: Time-Life Books. Entertaining, well-illustrated introduction to animal behavior.

Barnes, R. 1974. *Invertebrate Zoology.* Philadelphia: W. B. Saunders.

Binkley, S. 1979. "A Timekeeping Enzyme in the Pineal Gland." *Scientific American* 240(4):66–71. Well-written summary of research on biological clocks.

Bullock, T., and G. Horridge. 1965. *Structure and Function in the Nervous System of Invertebrates.* San Francisco: Freeman.

Daly, H., J. Doyen, and P. Erlich. 1978. *Introduction to Insect Biology and Diversity.* New York: McGraw-Hill.

Dorst, J. (translator C. Sherman). 1962. *The Migration of Birds.* Boston: Houghton Mifflin. Excellent descriptions of migrations that occur throughout the world.

Eckert, R., and D. Randall. 1978. *Animal Physiology.* San Francisco: Freeman. This is probably the best single source of information on comparative animal physiology. Outstanding text and illustrations.

Heber, R., and H. Garber. 1975. "The Milwaukee Project." In *The Exceptional Infant.* Volume 3: *Assessment and Intervention* (B. Friedlander, G. Sterritt, and G. Kirk, editors). New York: Brunner/Mazel. Study of the interplay between human intelligence and environmental deprivation.

Huxley, H. 1965. "The Mechanism of Muscle Contraction." *Scientific American* 213(6):18–27.

Romer, A., and T. Parsons. 1977. *The Vertebrate Body.* Fifth edition. Philadelphia: W. B. Saunders.

Vander, A., J. Sherman, and D. Luciano. 1980. *Human Physiology.* Third edition. New York: McGraw-Hill. Outstanding text on human functioning.

Vaughan, T. 1978. *Mammalogy.* Second edition. Philadelphia: W. B. Saunders. Everything you wanted to know about bats, and other mammals. Well written.

Welty, J. 1975. *The Life of Birds.* Philadelphia: W. B. Saunders.

Young, R. 1970. "Visual Cells." *Scientific American* 223(4):89–91.

Review Questions

1. What is a stimulus? Receptor cells detect specific kinds of stimuli. When they do, what happens to the stimulus energy?

2. Give some examples of chemoreceptors and mechanoreceptors. What kind of mechanoreceptor occurs repeatedly in sensory organs of different kinds of animals?

3. What is sound? How are amplitude and frequency related to sound? Give some examples of animals that apparently perceive sounds.

4. How does vision differ from photoreception? What sensory apparatus does vision require?

5. How does the vertebrate eye focus the light rays of an image? What is meant by nearsighted and farsighted?

6. What are the two basic components of any motor system? Distinguish between a hydroskeleton, exoskeleton, and endoskeleton.

7. What is antagonistic muscle action? Why is reciprocal inhibition of reflexes necessary in producing coordinated contractions?

8. Describe the fine structure of muscle fibers. Explain how the muscle fiber components interact in muscle contraction.

9. Name five factors that together shape the neuromotor responses called animal behavior.

10. Study Figure 15.17. Then, on your own, diagram the pathways between sensory input and motor output for both simple and complex behavior.

11. Define motor score and behavioral releaser. How do they give rise to innate behavior?

12. Contrast the motor patterns of animals in constant, highly predictable environments with those of animals in changing environments.

13. Define learning; then describe these learning processes: associative learning, extinction, latent learning, and insight learning. What is a critical learning period in the development of some animals? What is imprinting?

14. What are some of the sensory and motor adaptations involved in bird migration? What are some internal sensing mechanisms that may govern the directional orientation and timing of migration?

Nutrition—here is a word that has to do with all those processes by which the body ingests, digests, absorbs, and assimilates food. This word signals that you are about to begin one more educational trek through the animal gut. To follow is a march through the bloodstream, then through the lungs. This time around, however, you will be moving beyond the passive memorization of names for regionally specialized tissues and organs. This time your main concern will be with *systems integration*—with how whole systems function together in nourishing all the cells in animals that show increased size and complexity.

Consider the female bear in Figure 16.1, and the destination of the salmon in her mouth. Is it enough, really, to assume that the nutritional picture begins and ends with a description of her digestive tract? In many animals, the digestive tract is a more or less elaborated tube, extending from mouth to anus. The space inside the tube (the *lumen*) remains part of the external environment. To reach the internal environment, nutrients must cross the tube's membranous linings. Having crossed them, they must be distributed to all parts of the body. In the female bear, nutrients move into a bloodstream, which transports them to tissue regions. There, nutrients move out of the bloodstream and are taken up by individual cells. In those cells, the nutrients pass down one metabolic pathway or another. For the female bear, assimilation of nutrients depends on

16

NOURISHING CELLS OF THE ANIMAL BODY

Charles G. Summer, Jr./Tom Stack & Associates

Figure 16.1 Animal nutrition encompasses all those interrelated processes by which food is ingested, absorbed into the internal environment, then circulated to and assimilated by individual cells. These are the subjects of this chapter.

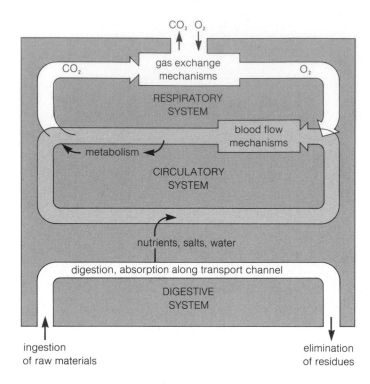

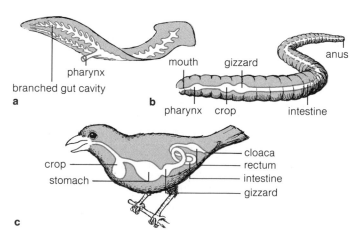

Figure 16.2 Interconnected systems for nourishing individual cells. The connections represented here are characteristic of the most complex animals, including mollusks, echinoderms, and chordates. Sensory receptors in each system channel information to the nervous system, which monitors and coordinates the interrelated activities.

Figure 16.3 (**a**) Incomplete digestive system of the planarian, a flatworm. Its pharynx is a muscular tube that protrudes from the body during feeding. (**b,c**) Complete digestive systems of animals ranging from earthworms to birds and humans have specialized food-processing regions, and a one-way movement of materials.

oxygen as an electron acceptor for keeping metabolic pathways open; it also means a possible stockpiling of carbon dioxide wastes that could hamper cell functioning (Chapter Seven). So nutrition may depend on more than digestion and circulation. It may also involve respiration: oxygen movements into the body, then into individual cells—and carbon dioxide movements out of them (Figure 16.2). That female bear has more than a salmon in her mouth. She has oxygen moving into her lungs and bloodstream, and carbon dioxide moving out of them—and this, too, is essential for nutrition.

In this chapter, you first will be reviewing the main components of digestive, circulatory, and respiratory systems, as well as their individual functions. Then you will see how activities of the three systems can be integrated, under neural and endocrine commands. The main examples will be from an organism with which you are already more or less acquainted: yourself. Later, in the evolutionary context of Chapter Twenty-Four, you will read about how these interrelated systems may have emerged and how they have become specialized in different animal groups.

DIGESTIVE SYSTEMS

On Feeding Strategies

In animals, food enters some form of body cavity or tube, where it is reduced (more or less) to particles small enough to be absorbed into and distributed throughout the internal environment. Flatworms have what is called an **incomplete digestive system**, for it has only one opening. What goes in but cannot be digested must go out the same way. A muscular organ (the pharynx) opens into a highly branched cavity that serves both digestive and circulatory functions. Here, food is partially digested and transported to individual cells even as residues are being sent back out. As long as two-way traffic is necessary, regions of this cavity cannot become specialized for any one task. These animals generally rely on *continuous feeding*. Consider the parasitic flatworms called tapeworms, which are little more than a tube-within-a-tube. They have no digestive system at all; they depend entirely on their hosts to digest their food for them. Unless their preprocessed food is always available, they die or go dormant. Nematodes, too, must feed constantly.

Annelids, mollusks, arthropods, echinoderms, and chordates have a **complete digestive system**—a tube with an opening at one end for taking in food, and an opening at the other for eliminating undigested residues. In between these two openings, food moves in one direction through regions specialized for transport and processing. For instance, earthworms and birds have a crop (a food storage organ). They also have a gizzard (a muscular organ in which food is ground into smaller pieces). They have an intestine (a region where chemical digestion and absorption occur). The specializations we see among animals with complete digestive systems depend on the animal's feeding habits. These animals show *discontinuous feeding* patterns. Food supplies may not be available all the time. Or predators or some other selection pressure may keep the animal from eating steadily. Thus, for example, grazing animals subject to predation have specialized storage regions for food that is eaten rapidly and digested later, in comparative safety. In some cases, too, the kind of food being eaten requires longer processing time. Cattle, sheep, goats—these so-called ruminants use plant cellulose as food. Ruminants have multiple stomachlike chambers. The first two chambers contain vast populations of symbiotic bacteria and other microorganisms. These symbionts have enzymes that can break down the tough cellulose fibers along with other nutrients. When the ruminants "chew their cud," they have regurgitated food from these two chambers and are grinding it up further in their mouth before swallowing it again. Gradually the food moves into two more stomachs, then into the intestine for absorption.

Regardless of the feeding pattern or the degree of specialization seen in different animal groups, these processes occur:

Digestion is the mechanical and chemical reduction of ingested food, into particles small enough to diffuse through epithelial cells and into the internal environment.

Absorption is the passage of digested nutrients from the gut lumen into the blood or lymph, which distribute them throughout the body.

Assimilation is the movement of digested nutrients from interstitial fluid, across the plasma membrane, and into the cytoplasm of cells.

Human Digestive System

The human digestive system is divided into a mouth, oral cavity, pharynx, esophagus, stomach, small and large intestines, rectum, and anus (Figure 16.4). All these regions have an inner lining (mucosa) backed by connective tissue.

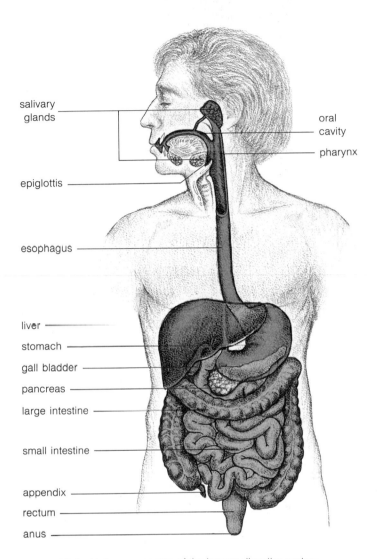

Figure 16.4 Main components of the human digestive system.

From the esophagus onward, the regions are encircled by tubes of circular, longitudinal, and sometimes oblique smooth muscle. Connective tissue forms the thin outer layer of the entire tube (Figure 16.5).

Like all mammals (the only vertebrates that chew food inside the mouth), humans have **salivary glands**. These glands secrete a nearly neutral or slightly alkaline fluid (saliva) into the oral cavity. Saliva contains an enzyme (amylase) that hydrolyzes starch. It also contains a glycoprotein that lubricates chewed and moistened food, which eases its passage to the stomach. When food is processed into a moistened ball, voluntary muscular contractions move the tongue upward. The movement forces the ball

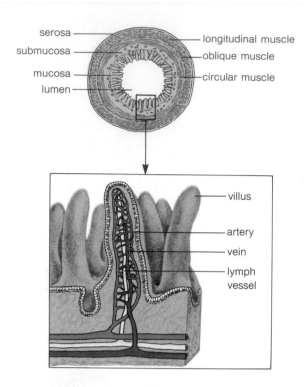

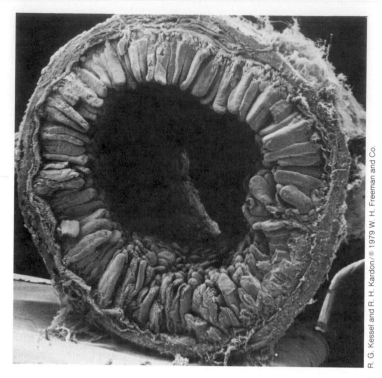

Figure 16.5 Scanning electron micrograph and sketch of a section through the small intestine. Like all parts of the digestive tube, it has an inner lining (mucosa). The relationship between each villus (a fingerlike projection of the intestinal mucosa) and the bloodstream is shown in the boxed inset.

into the pharynx. In humans, the **pharynx** is a muscular tube that opens not only into the esophagus but into the trachea—another tube, leading to the lungs. What keeps food from entering the trachea? During swallowing, pharynx muscles contract as a reflex response to incoming food. The contractions force the upper tracheal region (larynx) against a cartilaginous flap called the **epiglottis**, which closes off the respiratory tube. Normally, reflex action inhibits breathing during the brief time that food is in the pharynx.

From here on, voluntary controls of digestion generally give way to autonomic (involuntary) control. Once the food ball passes into the pharynx, waves of muscular contraction force it through the esophagus. The **esophagus** is a muscular tube specialized for propelling food to the stomach. Circular muscles contract behind each mass being moved, which pushes it toward the stomach. In coordination with the constriction, muscles ahead of the mass relax, which widens the tube region in front of it. Any alternating progression of contracting and relaxing muscle movements

along the length of a tube is known as **peristalsis**. As you will see, peristaltic movements occur in other regions of the digestive tract as well.

Food now passes into the **stomach**, a muscular, elastic sac. The stomach's most important function is in regulating the rate at which food reaches the intestine. Helping to control the directional movement of food are rings of muscles, or **sphincters**, that serve as gates. When a sphincter opens, food moves through it. When a sphincter is closed, food is prevented from moving back. Nerve impulses and hormones coordinate stomach and intestinal action. When food distends the stomach and small intestine, local reflex circuits are activated. The rate of stomach emptying is slowed, allowing digestive processes in the intestine to keep pace with food arriving from the stomach. Hormones coordinate peristaltic activity here, too. Recall that fatty acids are not readily broken down (Chapter Five). When fats enter the small intestine in large amounts, they stimulate some intestinal cells into secreting a hormone that travels, by way of the bloodstream, to the stomach.

There, the hormone inhibits stomach muscle cells involved in peristalsis—which decreases the rate at which the stomach empties. This allows more time for fat breakdown in the small intestine.

As Table 16.1 shows, the stomach is also involved in the initial mechanical digestion of proteins. Glands in the stomach mucosa secrete mucus (which forms a protective coat over the stomach lining), hydrochloric acid, and enzymes—three major components of the gastric juice. The hydrochloric acid destroys most of the bacteria inadvertently ingested with food. Hydrochloric acid also denatures incoming proteins (it breaks the hydrogen bonds that help hold proteins in their three-dimensional shapes). Thus denatured, proteins have more of their peptide bonds exposed for breakdown. In addition, hydrochloric acid converts the precursor molecule pepsinogen into **pepsin**, a major hydrolytic enzyme involved in splitting peptide bonds. Although proteins are initially broken down here, they are not absorbed across the stomach wall.

The stomach's main function is to regulate the rate of food movement into the small intestine. It is also responsible for the initial breakdown of proteins.

It is in the **small intestine** that digestion and absorption of most nutrients occur. Here, mechanical digestion involves peristaltic movements, as well as localized movements that mix food with digestive juices. Here, chemical digestion requires the action of enzymes released from cells in the intestinal mucosa, as well as from the liver and pancreas. Table 16.1 lists the main enzymes involved in breaking down food into molecules small enough to be absorbed into the internal environment. Absorption occurs in the dense outfoldings—villi—of intestinal mucosa (Figure 16.5). Villi are covered with millions of microvilli: threadlike projections that greatly increase the membrane surface area available for absorption.

Beneath its one-cell-thick epithelium, each villus houses blood and lymph vessels. As the small intestine moves, its microvilli sway slightly, thereby coming into contact with more small molecules that can be absorbed. Once small molecules move across the epithelium, they travel one of two routes. Amino acids, glucose, and water molecules enter the blood capillaries. Fats (re-formed from glycerol and fatty acids) enter lymph vessels, which drain eventually into the bloodstream.

Human digestion depends on secretions from several glandular organs, including the liver. This organ has many diverse functions, as you will read later. Its main digestive function is bile secretion. **Bile** is a solution containing bile

Table 16.1	Some Enzymes of Digestion		
Enzyme	**Source**	**Substrate**	**Main Breakdown Products***
Salivary amylase	Salivary glands	Polysaccharides (starch, glycogen)	Disaccharides (maltose)
Pepsin	Stomach	Proteins	Polypeptides
Trypsin, chymotrypsin	Pancreas	Proteins and polypeptides	Small polypeptides, dipeptides
Amylase	Pancreas	Polysaccharides	Disaccharides (maltose)
Lipase	Pancreas	Fats	Fatty acids, glycerol
Disaccharidases	Pancreas	Disaccharides (lactose, maltose, sucrose)	Monosaccharides (glucose, fructose, galactose)
Peptidases	Pancreas	Small polypeptides, dipeptides	Amino acids
Lipases	Pancreas	Fats	Fatty acids, glycerol

*White part of table identifies breakdown products small enough to be absorbed by cells of the intestinal mucosa.

salts, bile pigments, cholesterol, and lecithin. These substances are needed in fat digestion. Fats, recall, are insoluble in water. Once fat molecules leave the stomach, their hydrophobic "tails" clump together in large globules. Some of the bile substances separate the fat globules into an emulsion (small, suspended droplets) in the liquid contents of the small intestine. In this way, the surface area of fats is increased, and enzymes break them down more rapidly.

Bile enters the small intestine through a bile duct, which is ringed by a sphincter where it joins the digestive tube. When the sphincter is closed, bile is temporarily stored and concentrated in the **gallbladder**, a small sac branching off the bile duct. You probably have heard about **gallstones**. They form when something changes the concentrations of bile salts, lecithin, and cholesterol in the bile. Bile salts and lecithin normally keep cholesterol from forming fat globules. But if their concentrations rise or fall, or if cholesterol concentration rises, then cholesterol molecules aggregate into "stones." Large stones may become stuck in the bile duct, which effectively shuts down bile secretion—hence fat digestion. The presence of gallstones irritates the bile duct and can be extremely painful; surgery may be called for.

The other major glandular organ associated with digestion is the **pancreas**. Its enzymes degrade proteins, lipids, and carbohydrates. These digestive enzymes are produced in the main body of the gland. Small patches of pancreatic cells (the islets of Langerhans) also secrete the hormones insulin and glucagon into the bloodstream. As you will read later, these hormonal secretions are central in feedback relationships that govern the entire body's metabolism.

Within two to five hours after eating, the undigested and undigestible residues enter the large intestine, or **colon**. Like the small intestine, the colon functions in absorbing water and minerals. By the time residues leave the small intestine, they are in a highly liquid state, having become mixed not only with ingested water but with gastric and intestinal juices. Thus, some of the water being absorbed here is being reclaimed from earlier secretions. The colon secretes mucus, which helps lubricate the remaining residues as they are moved to the rectum. From there, residues move out through the anus (the digestive tube's terminal opening).

Digestion and absorption of carbohydrates, proteins, and lipids occur in the small intestine with the aid of secretions from the liver, pancreas, and gallbladder, in addition to intestinal secretions. Absorption of water and minerals occurs in both the small and the large intestine.

Symbiotic relationships exist between the colon and its resident bacteria. From the first few hours of your life until after your death, millions of bacteria (including *Escherichia coli*) are present in the colon. They are acquired during the birth process or immediately thereafter. Once they enter the relatively benign environment of the colon, they multiply in proportion to the amount of material traveling past. They form a necessary link in digestion, for they help convert the intestinal contents to a bulk form (feces) that stimulates the intestinal lining and increases peristalsis. Their attention to the intestinal slush assures the elimination of wastes in an efficient manner, assuming all else is well with the gastrointestinal tract (see *Commentary*). Intestinal bacteria also manufacture vitamin K, which is essential for blood-clotting.

COMMENTARY

Human Nutrition and Gastrointestinal Disorders

Good health means taking in carbohydrates, fats, and proteins in adequate amounts. It also means supplying the body with enough vitamins and minerals. The catch-all category vitamins refers to more than a dozen accessory substances that are required, in small amounts, for normal metabolic activity. Most plant cells are able to synthesize all of these substances. In general, animal cells have lost the ability to do so, hence animals must obtain vitamins from food. Human cells need at least thirteen different vitamins (Table 16.2).

In addition to vitamins, all cells require inorganic materials known as minerals. (Some minerals are called trace elements because they are needed only in extremely small amounts.) Most cells require both calcium and magnesium in a host of enzyme-mediated reactions. All cells need phosphorus for phosphorylation. They need sodium and potassium for maintaining osmotic balances, especially in muscle and nerve functioning. All cells require iron for building cytochromes. Red blood cells require still more iron to produce hemoglobin (Table 16.3).

The most reasonable way to supply your cells with essential vitamins and minerals is to eat a well-balanced assortment of foods that contain carbohydrates, fats, and proteins of the right sorts. A diet of about 32–42 grams of protein, 250–500 grams of carbohydrates, and 66–83 grams of fat should also assure you of getting enough of these accessory substances. In recent years, there have been claims that massive doses of certain vitamins and minerals are spectacularly beneficial. To date, there is no clear evidence that vitamin intake exceeding the recommended daily allowances leads to better health. To the contrary, excessive vitamin doses are often merely wasted. Individuals who take in large amounts of vitamin C don't realize that the body simply will not hold more vitamin C than it needs for normal functioning. It is not fat-soluble, and tends to be excreted. Direct chemical analysis shows that any amount above the recommended daily allowance ends up in the urine almost immediately after it is absorbed from the gut. Abnormal intake of at least two other vitamins—A and D—can cause serious disorders. The reason is that, like all fat-soluble vitamins, vitamins A and D can accumulate in the body. Consider the advice given in a recent edition of a popular nutrition book, *Let's Have Healthy Children*. The author recommended massive doses of vitamin A for infants. One mother followed the advice—and ended up with a stunted child. The same author advised that infants suffering from a gastrointestinal disturbance called colic be given massive doses of potassium—on the order of 3,000 milligrams. One mother did just that. Her infant son's heart immediately began to malfunction; four days later he was dead. Although potassium has indeed been used in treating colic, the author failed to add the vital warning that potassium should never be administered to individuals who are dehydrated, as they can be following bouts of diarrhea. *Both shortages and massive excess of foodstuffs can disturb the delicate metabolic balances that characterize physiological health.*

The United States harbors one of the best-fed populations in the world, yet digestive disorders among its individuals are on the increase. Aside from child deliveries and tonsillectomies, about a third of all surgeries performed in the United States have to do with correcting problems of the digestive tract.

Along with affluence, it appears we have picked up some bad eating habits. We skip meals, eat too much and too fast when we do sit down at the table, and generally give our gastrointestinal tracts erratic workouts. Worse yet, our diet tends to be rich in refined sugar, cholesterol, and salt—and low in bulk. (Here, bulk means the volume of fiber and other undigested food materials that can't be decreased by absorption.) Without a good percentage of fiber, food moving through the colon puts considerable internal pressure on the digestive tube walls. Such pressure may contribute indirectly to such diseases as appendicitis (inflammation of a blind tube off the large colon) and cancer of the colon. These diseases are practically nonexistent in rural Africa and India, where the inhabitants cannot afford to eat much more than whole grains—which are high in fiber content. When individuals from these rural areas move to urban centers of the more affluent nations, they tend to become more susceptible to these diseases. This suggests that diet, not genetic disposition, is a key factor here. In addition, what we eat is known to affect the distribution and diversity of bacterial populations living in the gut. Do these changes somehow contribute to gastrointestinal disorders? That is not known.

Certainly the emotional stress associated with living in complex societies seems to compound the nutritional problem. Urban populations seem to be more susceptible to the irritable colon syndrome (once called colitis)—abdominal pain, diarrhea, and constipation. Diarrhea (excessive, rapid movements of the colon's contents) can be brought on by emotional stress. There seems to be a genetic predisposition to some kinds of ulcers—inflammations of the stomach, the lower end of the esophagus, and the duodenum (the first region of the small intestine). But emotional stress apparently is a major factor in some, particularly duodenal ulcers.

Where does this leave us? Short of surgery, there may not be much we can do about many inherited structural disorders of the gastrointestinal tract. Learning to handle stress is one way that we can ease up on this tract, and certainly learning how to eat properly is another. Yet what is "eating properly"? In 1979 the United States Surgeon General released a report representing a medical consensus on how to promote health and avoid such afflictions as high blood pressure, heart disorders, cancer of the colon, and bad teeth. The report advised us to eat "less saturated fat and cholesterol; less salt; less sugar, relatively more complex carbohydrates such as whole grains, cereals, fruits, and vegetables; and relatively more fish, poultry, legumes (for example, peas, beans, and peanuts); and less red meat." As might be expected, the beef, dairy, and egg industries were quick to attack the report. So were some scientists, who point out that cholesterol has not been linked to, say, heart disease beyond a shadow of a doubt. Of course, in 1964, the Surgeon General released a report on possible links between smoking and lung cancer, which was promptly denounced by the tobacco industry—for the links had not been made beyond a shadow of a doubt. The links were indeed established, albeit many years later. The controversy over what constitutes proper nutrition rages on. In the meantime, it might not be a bad idea to think about your own eating habits and how moderation in some things might help you hedge your bets. Put the question to yourself: Do you look upon a bowl of bran cereal with the same passion as you look upon, say, french fries and ice cream, prime rib and chocolate mousse? Now put the same question to your colon.

Table 16.2   Vitamins Necessary for Normal Cell Functioning

Vitamin	RDA* (milligrams)	Dietary Sources	Major Body Functions	Possible Outcomes of Deficiency	Possible Outcomes of Excess
Water-Soluble					
Vitamin B₁ (thiamine)	1.5	Pork, organ meats, whole grains, legumes	Coenzyme (thiamine pyrophosphate) in the removal of carbon dioxide	Beriberi (peripheral nerve changes, edema, heart failure)	None reported
Vitamin B₂ (riboflavin)	1.8	Widely distributed in foods	Constituent of two flavin nucleotide coenzymes involved in energy metabolism (FAD and FMN)	Reddened lips, cracks at corner of mouth (cheilosis), lesions of eye	None reported
Niacin	20	Liver, lean meats, grains, legumes (can be formed from tryptophan)	Constituent of two coenzymes involved in oxidation–reduction reactions (NAD⁺ and NADP⁺)	Pellagra (skin and gastro-intestinal lesions, nervous, mental disorders)	Flushing, burning and tingling around neck, face, and hands
Vitamin B₆ (pyridoxine)	2	Meats, vegetables, whole grain cereals	Coenzyme (pyridoxal phosphate) Involved in amino acid metabolism	Irritability, convulsions, muscular twitching, kidney stones	None reported
Pantothenic acid	5–10	Widely distributed in foods	Constituent of coenzyme A, which plays a central role in energy metabolism	Fatigue, sleep disturbances, impaired coordination, nausea (rare in humans)	None reported
Folacin	0.4	Legumes, green vegetables, whole wheat products	Coenzyme (reduced form) in carbon transfer in nucleic acid and amino acid metabolism	Anemia, gastrointestinal disturbances, diarrhea, red tongue	None reported
Vitamin B₁₂	0.003	Muscle meats, eggs, dairy products	Coenzyme in carbon transfer in nucleic acid metabolism	Pernicious anemia, neurological disorders	None reported
Biotin	Not established. Usual diet provides 0.15–0.3	Legumes, vegetables, meats	Coenzyme in fat synthesis, amino acid metabolism, glycogen formation	Fatigue, depression, nausea, dermatitis, muscular pains	None reported
Choline	Not established. Usual diet provides 500–900	All foods containing phospholipids (egg yolk, liver, grains, legumes)	Constituent of phospholipids. Precursor of putative neurotransmitter acetylcholine	None reported for humans	None reported
Vitamin C (ascorbic acid)	45	Citrus fruits, tomatoes, green peppers, salad greens	Maintains intercellular matrix of cartilage, bone, and dentine. Important in collagen synthesis	Scurvy (degeneration of skin, teeth, blood vessels, epithelial hemorrhages)	Relatively nontoxic. Possibility of kidney stones
Fat-Soluble					
Vitamin A (retinol)	1	Provitamin A in green vegetables. Retinol in milk, butter, cheese, margarine	Constituent of rhodopsin (visual pigment). Maintenance of epithelial tissues.	Xerophthalmia (keratinization of ocular tissue), night blindness, permanent blindness	Headache, vomiting, peeling of skin, anorexia, swelling of long bones
Vitamin D	0.01	Cod liver oil, eggs, dairy products, margarine	Promotes bone growth, mineralization. Increases calcium absorption	Rickets (bone deformities) in children. Osteomalacia in adults	Vomiting, diarrhea, weight loss, kidney damage
Vitamin E (tocopherol)	15	Seeds, green leafy vegetables, margarines	Functions as an antioxidant to prevent cell membrane damage	Possibly anemia	Relatively nontoxic
Vitamin K (phylloquinone)	0.03	Green leafy vegetables. Small amount in cereals, fruits, and meats	Important in blood-clotting (involved in formation of active prothrombin)	Conditioned deficiencies associated with severe bleeding, internal hemorrhages	Synthetic forms at high doses may cause jaundice

*Recommended daily allowance, for an adult male in good health.
From ''The Requirements of Human Nutrition,'' by Nevin S. Scrimshaw and Vernon R. Young. Copyright © 1976 by Scientific American, Inc. All rights reserved.

Table 16.3 Minerals Necessary for Normal Cell Functioning

Mineral	Amount in Adult Body (grams)	RDA* (milligrams)	Dietary Sources	Major Body Functions	Possible Outcomes of Deficiency	Possible Outcomes of Excess
Calcium	1,500	800	Milk, cheese, dark-green vegetables, dried legumes	Bone and tooth formation Blood-clotting Nerve transmission	Stunted growth, Rickets, osteoporosis Convulsions	Not reported for humans
Phosphorus	860	800	Milk, cheese, meat, poultry, grains	Bone and tooth formation Acid–base balance	Weakness, demineralization of bone, loss of calcium	Erosion of jaw (fossy jaw)
Sulfur	300	(Provided by sulfur amino acids)	Sulfur amino acids (methionine and cystine) in dietary proteins	Constituent of active tissue compounds, cartilage and tendon	Related to intake and deficiency of sulfur amino acids	Excess sulfur amino acid intake leads to poor growth
Potassium	180	2,500	Meats, milk, many fruits	Acid–base balance Body water balance Nerve function	Muscular weakness Paralysis	Muscular weakness Death
Chlorine	74	2,000	Common salt	Formation of gastric juice Acid–base balance	Muscle cramps Mental apathy Reduced appetite	Vomiting
Sodium	64	2,500	Common salt	Acid–base balance Body water balance Nerve function	Muscle cramps Mental apathy Reduced appetite	High blood pressure
Magnesium	25	350	Whole grains, green leafy vegetables	Activates enzymes. Involved in protein synthesis	Growth failure Behavioral disturbances Weakness, spasms	Diarrhea
Iron	4.5	10	Eggs, lean meats, legumes, whole grains, green leafy vegetables	Constituent of hemoglobin and enzymes involved in energy metabolism	Iron-deficiency anemia (weakness, reduced resistance to infection)	Siderosis Cirrhosis of liver
Fluorine	2.6	2	Drinking water, tea, seafood	May be important in maintenance of bone structure	Higher frequency of tooth decay	Mottling of teeth Increased bone density Neurological disturbances
Zinc	2	15	Widely distributed in foods	Constituent of enzymes involved in digestion	Growth failure Small sex glands	Fever, nausea, vomiting, diarrhea
Copper	0.1	2	Meats, drinking water	Constituent of enzymes associated with iron metabolism	Anemia, bone changes (rare in humans)	Rare metabolic condition (Wilson's disease)
Iodine	0.011	0.14	Marine fish and shellfish, dairy products	Constituent of thyroid hormones	Goiter (enlarged thyroid)	Very high intakes depress thyroid activity
Cobalt	0.0015	(Required as vitamin B_{12})	Organ and muscle meats, milk	Constituent of vitamin B_{12}	None reported for humans	Industrial exposure: dermatitis and diseases of red blood cells

*Recommended daily allowance, for an adult male in good health.
From ''The Requirements of Human Nutrition,'' by Nevin S. Scrimshaw and Vernon R. Young. Copyright © 1976 by Scientific American, Inc. All rights reserved.

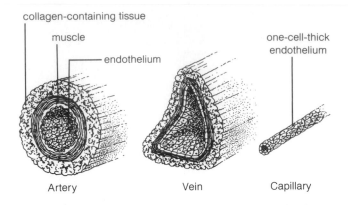

collagen-containing tissue

muscle

one-cell-thick endothelium

endothelium

Artery   Vein   Capillary

Figure 16.6 Examples of arteries, veins, and capillaries, emphasizing their relative degree of musculature (or, in the case of capillaries, the lack of it). These examples are from the human cardiovascular system. (From W. L. Clark, *Tissues of the Body*, Clarendon Press, 1952)

CIRCULATORY SYSTEMS

Internal Distribution Strategies

In the flatworm, recall, nutrients are partially digested in and distributed through the branchings of a central cavity that opens (by way of a pharynx) to the external world. Both digestion and circulation proceed inside the gut—which is "outside" the body. In animals more complex than flatworms, nutrients are circulated internally through a network of blood transport vessels. These vessels service the interstitial fluid bathing each cell (Chapter Thirteen). One or more hollow, muscular organs pump blood through these vessels. A muscular blood pump is called a **heart**. The blood itself, the heart, and blood vessels collectively are a **cardiovascular system**.

Many invertebrates, including insects, depend on **open circulation** systems. Blood pumped through a heart enters a large blood vessel. This vessel drains into a thin-walled cavity, forming a "pond" in which materials can be exchanged between blood and the surrounding tissue. From the pond, blood seeps somewhat sluggishly back to the heart. The distribution and circulation rate of blood are not controlled in any spectacular way in these animals. Either their life-style doesn't demand more efficient metabolic support systems, or other systems round out the transport process. Flying insects especially, which demand considerable oxygen for metabolism, have elaborate respiratory tubes that bypass the bloodstream and channel oxygen directly to body tissues.

Some invertebrates and all vertebrates depend on **closed circulation**: blood remains enclosed within the walls

of one or more hearts and blood vessels. For example, in the giant earthworm, peristaltic contractions of smooth muscles keep blood moving through this system. Also in the earthworm are five hearts that distend as they fill with blood, then contract and squeeze blood out.

Let's take a closer look at blood vessels and their function. These specialized tubes are called arteries, arterioles, blood capillaries, venules, and veins (Figure 16.6). An **artery** carries blood *away* from the heart. An artery has thick, impermeable walls containing smooth muscle and elastic connective tissue. Because of these components, it can expand under the surges of fluid pressure from blood leaving the heart. It also can recoil elastically, which forces blood onward. Arteries branch into smaller tubes, or **arterioles**. Within these tubes are rings of smooth muscle. Coordinated opening and closing of these muscle rings in arterioles throughout the body directs blood to regions of greatest metabolic activity. The more active the cells in a given region, the greater the volume of blood that must reach them. The role of arteries and arterioles in blood circulation may be summarized in this way:

Through controls over their musculature, arterioles can be used in varying the resistance to blood flow along different routes; hence they function in controlling blood flow distribution through the body.

A **capillary** is a blood vessel with such a small diameter that red blood cells must flow through it single file. Capillary walls consist of no more than a single layer of endothelial cells. Because these thin walls are permeable to many materials, a capillary serves as the exchange point between blood and surrounding tissues. An immense meshwork of capillaries is characteristic of vertebrate circulatory systems. Even in the largest vertebrates, most cells are no more than a few cells away from a capillary. Such extensive meshworks increase the surface area through which diffusion can occur.

In each capillary bed, small molecules move between the bloodstream and the interstitial fluid. These molecules include salts, ions, dissolved gases, glucose, amino acids, fatty acids, and hormones. The direction in which a substance moves depends on its concentration gradient. Water, too, is exchanged here. An additional force driving water out of capillaries is hydrostatic pressure generated by heart contractions. Water is forced back into capillaries by an osmotic pressure gradient. (Blood proteins are generally too large to leave capillaries, and the space they take up in the bloodstream means that less water is present in capillaries than in the surrounding tissues; hence the osmotic pressure gradient.) We may summarize the main function of capillaries in this way:

Because of their thin walls and their large numbers in vertebrate circulatory systems, capillaries represent an immense cross-sectional exchange area between blood and interstitial fluid.

Capillaries come together in **venules** and **veins**, which are the blood vessels leading *back* toward the heart. A vein has impermeable walls. Its walls are generally thinner than those of arteries and the lumen is larger (Figure 16.6). Being a distensible elastic tube, a vein enlarges when internal pressure is high. In vein walls are **valves**: tissue flaps that project into the lumen, in the direction of flow. Should fluid start flowing backward, it distends the flaps and pushes them into the lumen, which closes off the vessel and prevents backwashing.

Both venules and veins serve as a temporary reservoir for blood volume, precisely because they are so highly elastic. When the body's metabolic needs are low, blood volume is high in these vessels. As you will soon read, when metabolic needs increase, the heart draws on this reservoir at a faster rate. For now, keep in mind that this is the main function of venules and veins:

With their great distensibility, veins and venules function as blood volume reservoirs during low metabolic output.

Figure 16.7 diagrams the closed circulatory system of humans. As you can see, the heart is a pumping station for two major blood transport routes: *pulmonary circulation* (leading to and from the lungs) and *systemic circulation* (leading to and from the rest of the body). This figure also shows that the blood circulation system is supplemented by vessels called the **lymph vascular system**. Pressure generated by the heart forces slightly more water out of capillaries than osmotic pressure drives in. Given the tremendous number of capillaries in the human body, considerable fluid could accumulate in interstitial regions if it were not for the lymph vascular system. **Lymph vessels** are the drainage tubes that take care of the overflow. The fluid in these tubes is called lymph. The vessels themselves converge into larger vessels and ducts, which eventually return the fluid to the blood circulation. The lymph vascular system functions not only in reclaiming water lost from the bloodstream. It also transports fats from the small intestine to the bloodstream. As you will see in the next chapter, it also serves in the body's defense system.

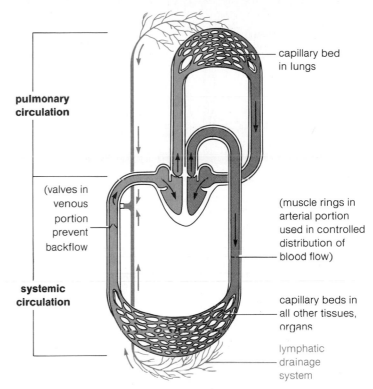

Figure 16.7 Generalized picture of systemic and pulmonary circulation. Also shown is the supplementary lymph vascular system, which drains into the bloodstream.

The Vertebrate Heart

In the broadest sense, the vertebrate heart has two functional zones: one for receiving blood that's being returned from the body, and one from which it is sent back out. As Figure 16.8 shows, a "receiving zone" is called an **atrium** (plural, atria); and a "departure zone" is called a **ventricle**. In fishes, the heart is no more than this: one atrium, one ventricle. In amphibians, the heart is three-chambered (two atria, one ventricle). One atrium receives oxygen-enriched blood directly from the lungs, and another receives blood from the rest of the body. Blood from both is forced into the ventricle, then is forced out into blood transport vessels. Birds and mammals have a four-chambered heart, with two thin-walled atria and two thick-walled ventricles (Figure 16.8c). The possible evolutionary relationship between these structurally varied pumps is explored in Chapter Twenty-Four. For now, let's take a look at how the four-chambered heart interacts with blood vessels in supplying cells of the human body with oxygen, nutrients, and other substances.

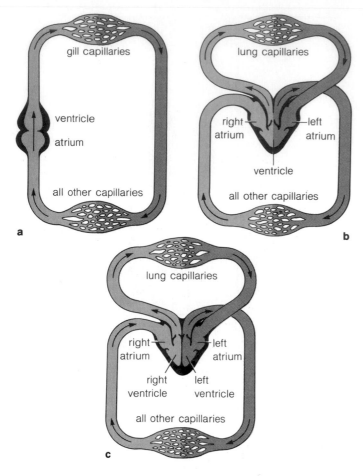

gill capillaries

ventricle

atrium

all other capillaries

a

lung capillaries

right
atrium — left
atrium

ventricle

all other capillaries

b

lung capillaries

right — left
atrium — atrium

right   left
ventricle ventricle

all other capillaries

c

Figure 16.8 Generalized sketches of the relationship between the heart and blood vessels for (**a**) fishes, (**b**) amphibians, and (**c**) birds and mammals.

During a seventy-year life span, the human heart (Figure 16.9) will beat some 2½ billion times to keep blood on the move. This remarkable pump is no larger than a clenched fist. As Figure 16.10 suggests, oxygen-depleted blood from the body collects in the right atrium and moves into the right ventricle. Then it is pumped to the pulmonary artery, which leads to the lungs. Oxygen-enriched blood from lungs then collects in the left atrium, moves into the left ventricle, and is pumped into the main artery (aorta) leading to the rest of the body. Valves within the heart prevent backflow as the blood is being circulated.

What is the basis of the pumping action? Each heartbeat is a coordinated contraction and relaxation of the whole muscle mass. Although the contraction rate can be altered by excitatory and inhibitory neurons acting on heart muscle cells, the heart itself contracts spontaneously. This has been demonstrated by heart muscle cells that have been cultured apart from the body: they continue to contract rhythmically on their own. The source of these contractions is a rhythmic firing of action potentials that began soon after heart muscle cells first appeared in the developing embryo. Where major veins enter the right atrium of the human heart, specialized muscle cells make up a region called the **pacemaker**. Action potentials in these cells show the fastest inherent rate of firing, hence they set the pace for other heart muscle cells. All heart muscle cells are electrically linked at intercellular junctions. Thus, when pacemaker cells are excited, the atria are stimulated into contracting as a unit.

Each wave of excitation from the pacemaker first spreads over the two atria, from right to left. So stimulated, the atrial tissue contracts and the atrial contents empty into the ventricles. Because of insulative (electrically neutral) connective tissue regions between the atria and ventricles, the wave of excitation is blocked *except* at a region called the atrioventricular node. The channeling of excitation to this region delays conduction just long enough to allow the atria to empty before the ventricles are stimulated to contract. From the atrioventricular node, bundles of muscle fibers carrying pacemaker excitation branch through tissues of both ventricles. The ventricles contract in response to the stimulation and force blood out from the heart. (Heart valves snapping shut generate the "lub-dup" sound that can be heard at the chest wall during this contraction period. Each "lub" marks the closure of atrioventricular valves; the "dup" marks the closure of semilunar valves behind blood leaving the heart.)

Each contraction period is called **systole**. It is followed by a rest period, **diastole**, in which heart muscle cells relax and the heart refills. The alternating periods of systole and diastole constitute the **cardiac cycle**.

Blood Pressure and Pulse Pressure

Blood flow through the pulmonary and systemic routes depends on pressure generated by heart contraction. Along these routes, however, fluid pressures drop. As blood courses first through arteries, then through arterioles, the decreasing diameters of these vascular tubes present more and more resistance to flow. At any point, blood flow relates to tube diameter. To complicate matters more, some of the tubing is elastic; and when tubing walls stretch,

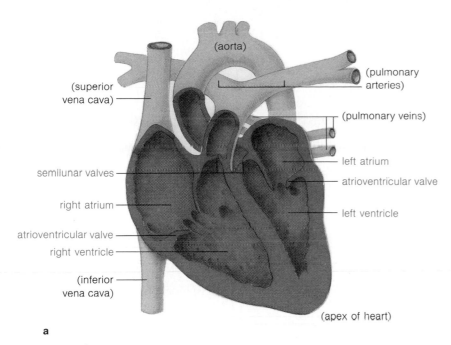

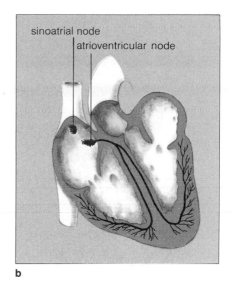

a

b

Figure 16.9 (**a**) Cutaway view of the human four-chambered heart. (**b**) Location of the heart's pacemaker, and the neural pathway by which its waves of excitation are channeled to the ventricles after first spreading through atrial tissue.

pressure drops. How, then, is blood flow equalized through a network of variable tubes?

Let's begin with the average pressure required to keep blood circulating. Blood pressure is generally measured at large arteries of the systemic circulation, such as the one in your upper arm. First a systolic reading is taken of the highest pressure generated by heart contractions. In young adults, systolic pressure is about 120 mmHg. (This is the distance, expressed in millimeters, that a column of mercury—Hg—will rise when subjected to that amount of pressure.) Next, a diastolic reading is taken of the lowest pressure in the artery, just before blood is pumped out of the heart again. Diastolic pressure is generally about 80 mmHg. The **pulse pressure** is the difference between systolic and diastolic readings (for example, 120 − 80 = 40 mmHg). This isn't the same thing as pulse *rate,* which refers to how many times the arterial wall distends and recoils in a minute (hence to the number of heartbeats per minute).

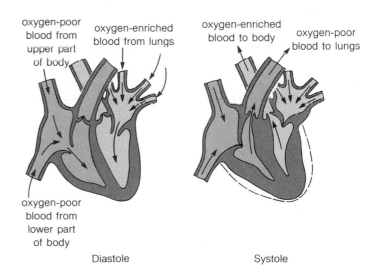

Diastole

Systole

Figure 16.10 Direction of blood flow through the four-chambered human heart.

Pulse rate averages 72 for most individuals at rest, although it varies considerably depending on such factors as physical condition and the degree of excitement.

Now, the volume of blood returning to the heart is the same as the volume being pumped out. *But blood pressure is not the same at the start and end of the circuit.* This is important, because if there were no pressure difference, blood would go nowhere. By the time blood has traveled back through the veins and has arrived at the right atrium, blood pressure is virtually zero. What happens is this. The arterial system, which channels blood toward capillaries, serves as a *pressure reservoir*. Thick-walled arteries are elastic, but only enough to smooth out pressure oscillations caused by heart contractions. Farther down the circuit, arterioles are muscular, but only enough to control blood flow to one body region or another. Under neural and hormonal commands, arteriole sphincters constrict and dilate, so that the blood flow to different capillary beds varies. Capillaries constantly open and close, but only five to ten percent of all capillaries are open at any one time. The venous system, in contrast, is a *volume reservoir*. Blood pressure here is low, and fully half the total blood volume is in the venous system at any moment.

What forces venous blood back to the heart? Heart contractions are the main force. Limb muscle movements, expansion of the rib cage during breathing—such activities also create some negative pressure that helps force blood through the venous system. We can summarize these forces in this way:

The pressure gradient that heart contractions establish is aided by pressure generated by muscle pumps and by respiration; these forces together work to return venous blood to the heart.

When you remain motionless—say, in a standing position for a long time—the return of venous blood to the heart is reduced, the volume of blood being pumped through the heart drops, hence arterial pressure drops. Mechanoreceptors in the heart chamber walls and arterial walls monitor blood pressure changes, and reflex responses can adjust blood circulation to a large extent. But if you stand still long enough, the prolonged reduction in blood flow to the brain can cause you to faint. Similarly, bedridden patients—even astronauts—whose limbs remain inactive for prolonged periods experience such temporary effects of blood pooling. Such problems are alleviated when normal activity resumes. The *Commentary* on this page describes a few other cardiovascular problems that are correctable, and a few that are not.

COMMENTARY

On Cardiovascular Disorders

During times of increased physical activity, systolic pressure rises because the heart rate accelerates and a greater volume of blood leaves the heart. Thus pulse pressure rises. Sometimes arterial walls undergo structural changes, and this, too, leads to increased pulse pressure. In arteriosclerosis, calcium salts and fibrous tissue gradually build up within arterial walls, leading to a "hardening of the arteries." As arteries become less and less elastic, they cannot recoil as they normally do and pressure builds up in the arterial system. If arterial blood pressure is high enough, small blood vessels supplying brain cells rupture, which is one cause of a brain stroke. If the brain cells deprived of oxygen and nutrients are in a region governing the coordination of body movements, paralysis may follow. If they are in regions controlling vital functions such as breathing, death will follow.

In Chapter Three, you read that lipids such as fats and cholesterol are insoluble in water. Such lipids are transported in the bloodstream, where they are bound to protein carriers that keep them suspended in the plasma. In atherosclerosis, abnormal smooth muscle cells multiply in arterial walls. Then lipids are deposited within them and in the surrounding extracellular space. Calcium salts are deposited on top of the lipids, and a fibrous tissue forms over the whole mass. This so-called plaque sticks out into the lumen. Sometimes platelets become caught on rough plaque edges, and are stimulated into secreting some of their chemicals. When they do, they initiate clot formation (Chapter Seventeen). As the clot and plaque grow, the artery can become blocked. This shuts off blood flow to tissues that the artery supplies. Heart muscles are supplied by coronary arteries, which branch off the aorta. A coronary occlusion may build up slowly, as it does from atherosclerosis. Or it may occur suddenly, as when clots dislodged from the coronary or other arteries travel to one of these heart supply lines and plug the passageway. Heart attacks may result.

Not all heart attacks are fatal. Survival depends on how much heart tissue is deprived of its blood support

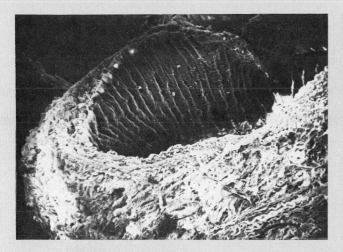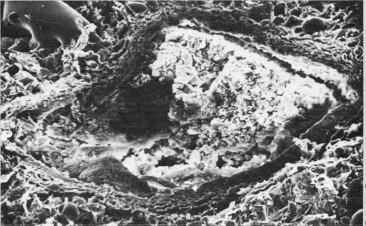

Robert La Porta

To the left, a normal artery; to the right, an artery clogged with plaques.

system, and on where the damage occurs. The right side of the heart is more critical, for it houses pacemaker cells. Survival also depends on immediate care. A pumping machine can be attached to a leg artery and synchronized with the victim's heartbeat, so that the heart has to work only to sixty percent of its normal capacity. If the victim is confined to bed rest, the deprived heart cells may recover as new blood supply channels develop and nourish them.

One of the most prevalent of all cardiovascular disorders is hypertension, a gradual increase in arterial blood pressure that can be brought on by atherosclerosis and other problems. Hypertension has been called the silent killer, because afflicted individuals may show no outward symptoms; in fact, they feel healthy. Thus, even when their high blood pressure has been detected, some hypertensive individuals tend to resist corrective changes in diet, exercising, and medication. Of 23 million Americans who are hypertensive, most are not undergoing treatment. About 180,000 will die each year as a result.

What causes cardiovascular disorders? Cholesterol, obviously, is under suspicion. When transported through the bloodstream, it is bound to one of two kinds of protein carriers: high-density lipoproteins (HDL) and low-density lipoproteins (LDL). Evidence is accumulating that high levels of LDL are related to a tendency toward heart trouble. It appears that LDLs, with their cholesterol cargo, show a greater penchant for infiltrating arterial walls. It may be that HDLs can attract cholesterol out of the walls and transport it to the liver, where it can be converted to fatty acids and glycerol. Atherosclerosis is uncommon in rats; rats have mostly HDL. In monkeys, pigs, and humans, which generally have mostly LDL, atherosclerosis is not uncommon.

Behavior patterns, too, may trigger conditions that lead to cardiovascular disorders. Some individuals are so-called Type A personalities: they tend to be competitive, aggressive, and impatient; they are least stressed when their lives are most organized. Type A individuals tend to become acutely stressed when some aspect of their life—personal or professional—swings out of control. And the stress may lead to behavioral and physiological precursors to heart and vascular disorders. Type B personalities are more relaxed about themselves and about life; they seem to be less prone to cardiovascular troubles. Intriguingly, Robert Nerem and Fred Cornhill of Ohio State University reported in 1979 on a most unusual side effect of their studies of drug action on high-cholesterol diets. Rabbits were the subjects. One group of rabbits was cuddled and played with; the other group received impersonal care. Arterial tissue samples from both groups showed that atherosclerosis was less prevalent by half in the cuddled rabbits. The experiment was repeated; the results were the same. Might we chalk one up for tender loving care?

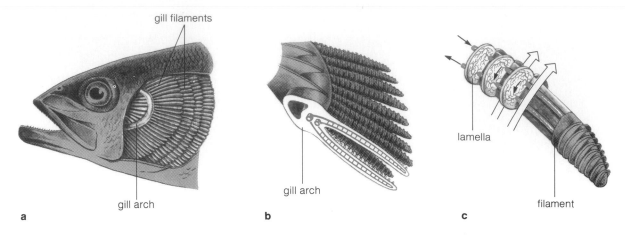

gill filaments

lamella

gill arch

gill arch

filament

a   **b**   **c**

Figure 16.11   Fish gills. (**a**) Location of gills beneath the bony, protective lid (operculum), which has been removed for this sketch to show the chamber below. (**b,c**) Each gill has an outgoing and an incoming blood vessel. Along their length, capillary beds are arranged in thin membrane folds (lamellae). In these capillary beds, direction of water flow (large arrows) is opposite the direction of blood flow (small arrows). (After C. P. Hickman, Jr. et al., *Integrated Principles of Zoology*, 1979)

RESPIRATORY SYSTEMS

Respiration Strategies

Almost all animals use oxygen for cellular metabolism, and they must dispose of the carbon dioxide wastes of metabolism. Almost all use the bloodstream as a highway to do this. The overall exchange of gases between cells, the blood, and the environment is known as **respiration**. Animal life apparently originated in water, and many animals live there still. Such animals depend on oxygen dissolved in the surrounding medium. Other animals have respiratory mechanisms adapted to life in moist land and dry land settings, where oxygen is available directly from the atmosphere.

In most animals, respiration is closely linked to a circulatory system. Even the "simple" earthworm has fine capillaries, right under the moist body wall, that carry inward-diffusing oxygen to deeper tissue regions. In segmented marine worms, sea stars, fish, and amphibian larvae, gas transfer occurs across gills. A **gill** is a thin, moist surface membrane, endowed with blood vessels and outfolded into a simple bump, flap, or filament.

For instance, each fish gill is a fragile filament, arranged with many other filaments in layers beneath a protective lid (Figure 16.11). Together, these filaments represent a large membrane surface area for gas exchange. In addition, gas transfer is enhanced enormously by a **countercurrent flow mechanism**, in which water flows past the bloodstream in the opposite direction. Oxygen-rich water passing over the gill first encounters the domain of a vessel carrying blood about to move back into the main part of the body (Figure 16.11c). Oxygen pressure in this blood is not as great as it is in the surrounding water, so oxygen leaves the water and diffuses into the blood. Just before it leaves the gill, this same water passes over the domain of a vessel carrying blood that is leaving the main part of the body—and which has even less oxygen than the oxygen-poor water! Hence oxygen still diffuses inward, down this second gradient. With countercurrent exchange, fish can extract as much as ninety percent of the dissolved oxygen from their surroundings.

In water-dwelling animals with gills, dissolved gases move readily across cellular membranes. But what happens in animals whose ancient ancestors made the transition to life on land? How can gases dissolve and move across

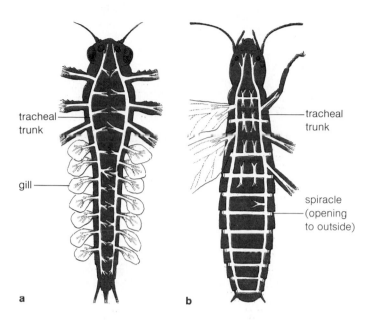

tracheal trunk

gill

tracheal trunk

spiracle (opening to outside)

a

b

Figure 16.12 Tissue adaptations allowing for gas exchange in arthropods. (**a**) Many aquatic insects have closed tracheal systems, which extend into lateral body wall extensions that serve as gills. (**b**) In land arthropods, tracheal systems open to the outside through tubes in each abdominal and thoracic segment. (After Gardiner, *Biology of Invertebrates*, McGraw-Hill, 1972)

a membrane between a *dry* external world and the internal environment? Consider the basic respiratory adaptation of insects, centipedes, and some spiders. These animals have **tracheae**: chitin-lined air tubes leading from the body surface to the interior (Figure 16.12). Each trachea branches into many finer tubes—which are filled at the end with fluid. Hence incoming oxygen can dissolve in this fluid and move down its pressure gradient into the extracellular fluid. (These animals don't rely on blood vessels for gas transport. The highly branched tracheal tubes extend like highways to all tissue regions.) In some insects, a lid tops the tracheal opening to the outside and prevents evaporative water loss.

In land vertebrates, too, respiratory membranes are kept moist on both surfaces, which facilitates gas exchange. These animals have what amounts to inward-directed gills. Air tubes leading in from the body surface branch elaborately in one or more chambers (air sacs or lungs). These chambers are lined with moist epithelium and intermeshed with blood capillaries. We can make these summary points about the requirements for gas exchange in animals:

Any respiratory membrane must be kept moist on <u>both</u> surfaces for gas exchange to occur.

When moisture retention is not a problem (which it is not for water-dwelling animals), gas exchange is enhanced by <u>evagination</u> (outfolded membrane, such as a gill).

When moisture retention is a problem (as it can be for land-dwelling animals), gas exchange is enhanced by <u>invagination</u> (infolded membrane, as occurs in air sacs and lungs).

Human Respiratory System

In humans, membranes of the respiratory system remain moist largely because the lungs are housed within an inner chamber, the chest (thoracic) cavity. In addition, air becomes moisturized, warmed, and filtered on its way through air tubes leading into the lung chamber. Normally, air enters the body through the nose and nasal cavity. Hairs at the entrance of this passageway, and cilia on its epithelial lining, filter out dust and other foreign particles. Numerous blood vessels embedded in the lining help warm incoming air, and mucous secretions moisten air before it flows into the lungs. From the nasal cavity, air moves through the pharynx. Then it moves down through cartilage-reinforced tubes called the **larynx** and **trachea** (Figure 16.13). Tracheal epithelium is lined with cilia, and mucus-secreting goblet cells are embedded within it (Figure 16.14). As inhaled particles become stuck against mucus of the tracheal membrane, the upward-beating cilia sweep debris-laden mucus back to the mouth or nasal cavity, where it can be expelled from the air passageway.

The trachea, or windpipe, branches into two **bronchi** (singular, bronchus). Bronchi lead to the paired lungs. The lungs themselves are elastic, spongelike sacs that can move freely in the chest cavity. They are attached only at the region where air tubes and major pulmonary blood vessels connect with them. One layer of a fragile membrane, the **pleura**, covers each lung's surface; the other pleural membrane adheres to the chest cavity wall. The very narrow interpleural space contains a fluid that escapes from capillaries and is constantly being reabsorbed through lymph networks.

In the lungs, bronchi branch into smaller and smaller tubes that divide the lung substance into even smaller compartments. The smallest are respiratory **bronchioles**, which end in thin-walled pouches. These pouches are the **alveoli** (singular, alveolus). The alveoli protrude between

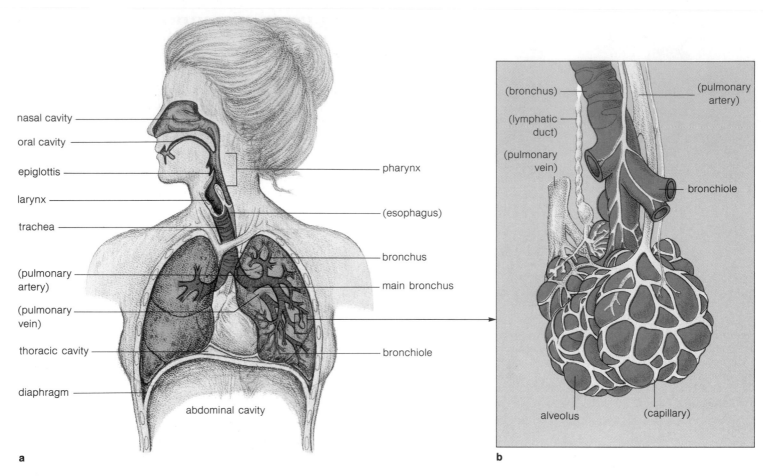

Figure 16.13 Human respiratory system (**a**). In this sketch, the paired lobes of the lung are shown in position, next to the heart. Lungs are located in the thoracic cavity, which is separated from the abdominal cavity by a muscular partition called a diaphragm. (**b**) Close-up of alveoli and lung capillaries. (After Avery, Wang, and Taeusch, Jr., "The Lung of the Newborn Infant," *Scientific American,* April 1973. Copyright © 1973 by Scientific American, Inc. All rights reserved.)

blood capillaries like bunches of grapes (Figure 16.13). More than 300 million alveoli are present in human lungs. Gas exchange occurs across the alveolar epithelium, through interstitial fluid in a thin layer of connective tissue, and on across blood capillary epithelium.

Gas Exchange Mechanisms

In the respiratory system just described, oxygen moves inward and carbon dioxide moves outward through the same branched tubes. Gas exchange in the *same* tubes is possible because of rhythmic changes in pressure gradients between the lungs and the atmosphere. The lungs themselves don't cause these changes; the outer wall of the chest cavity does. As the chest cavity volume increases and decreases, the lung sacs go along with the changes. Chest cavity volume becomes altered through the coordinated movements of a **diaphragm** (a thin muscle partition separating the chest and abdominal cavities), highly elastic intercostal muscles that move the ribs, even abdominal and neck muscles. During normal **inhalation**, the dia-

Figure 16.14 Ciliated epithelium of the human windpipe, the trachea. Of all the air-passage structures in the lung, only alveoli are not lined with this kind of membrane. The smooth-surfaced bumps, which protrude through the cilia like half-hidden Easter eggs, are mucus-secreting goblet cells. 1,310×.

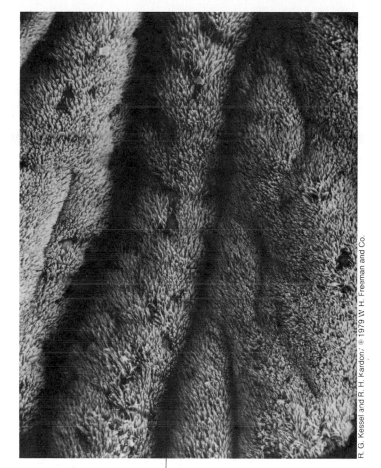

R. G. Kessel and R. H. Kardon; © 1979 W. H. Freeman and Co.

mucus-secreting cell

phragm moves downward and flattens; the rib cage moves outward and upward. When this happens, the chest cavity volume increases and its internal pressure drops. But atmospheric pressure remains the same, so air rushes in through the nose and trachea. During **exhalation**, the diaphragm and rib cage return to their resting position, and the chest cavity volume decreases sharply. The resulting rise in pressure on the lungs forces air outward (Figure 16.15).

The rate of inhalation and exhalation—breathing—is governed by a respiratory center in the brainstem. This center monitors signals coming in from lungs, blood vessels, and other brain regions. For instance, chemoreceptors in arterial walls detect changes in carbon dioxide levels in the blood. So do chemoreceptors located within the brainstem itself. When increases are sensed, nerve signals flow out from the brainstem to muscle cells in the diaphragm and intercostal muscles. The contraction rate and strength of both become altered. Breathing becomes faster and deeper, which enhances the movement of carbon dioxide out of the body.

Breathing itself normally occurs without thought, although the breathing rate can be consciously altered. It can't be consciously stopped for long. Children who attempt to intimidate their parents by holding their breath might pass out—but as soon as they do, basic breathing reflexes take over and breathing resumes.

During exhalation, lungs don't collapse completely; considerable air remains inside. In fact, each breath of "new" air becomes mixed with the equivalent of seven breaths of "old" air in the lungs. (If the volume of each breath increases, the fraction of "new"-to-"old" air would be greater than one-to-seven.) Thus, it is not only metabolic activity that determines the amount of carbon dioxide in the blood. It also the rate and depth of breathing—hence the quantity of new air entering the lungs.

With each new breath, air that is low in carbon dioxide and rich in oxygen enters the lungs and flows down into alveoli. Within adjacent lung capillaries is blood that is depleted in oxygen and rich in carbon dioxide. Immediately

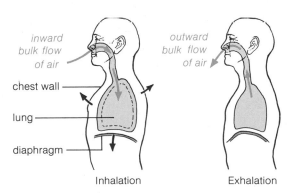

inward bulk flow of air

outward bulk flow of air

chest wall

lung

diaphragm

Inhalation

Exhalation

Figure 16.15 Lung expansion and collapse during normal breathing, correlated with changes in the volume of the thoracic cavity.

these two dissolved gases move down their concentration gradients. As oxygen moves into the blood plasma, and then into red blood cells, it forms a weak, reversible bond with hemoglobin. Four oxygen molecules bound to hemoglobin form the compound oxyhemoglobin. Oxygen transport is far more effective using oxyhemoglobin than it would be if plasma alone were used. Without hemoglobin, the plasma would be able to carry only about two percent of the oxygen that whole blood carries!

When oxygen-rich blood reaches a systemic tissue capillary bed, oxygen diffuses outward and carbon dioxide moves from tissues into the capillaries. Carbon dioxide is carried in blood in more than one way. Recall, from Chapter Six, that carbon dioxide combines with water to form carbonic acid, when then dissociates into bicarbonate and hydrogen ions:

$$CO_2 \;+\; H_2O \; \rightleftarrows \; \underset{\substack{\text{carbonic} \\ \text{acid}}}{H_2CO_3} \; \rightleftarrows \; \underset{\text{bicarbonate}}{HCO_3^-} \;+\; H^+$$

The reaction rate, again, is hardly spectacular. It is in red blood cells, which house the enzyme carbonic anhydrase, that the forward reaction proceeds 250 times faster than it does in blood plasma. In fact, almost all of the carbon dioxide moving into the bloodstream is moving into red blood cells. There, about fifty percent forms carbonic acid with intracellular water, which then splits into bicarbonate and hydrogen ions. The bicarbonate diffuses out and is transported by the plasma, and the hydrogen ions are buffered by hemoglobin. The rest of the carbon dioxide entering red blood cells combine with hemoglobin to form $HbCO_2$ (carbaminohemoglobin). Why does it matter that carbon dioxide is bound up rapidly in these ways? It happens that the carbon dioxide pressure gradient between blood and body tissues is not that steep. The difference amounts to a mere five mmHg. However, with enzyme action, the diffusion process is *facilitated* enough to keep carbon dioxide moving away rapidly from tissue regions that it might otherwise poison.

Once blood returns to lung capillaries, the lower carbon dioxide concentration in alveoli pulls the reactions in the reverse direction. Again with the assistance of carbonic anhydrase, carbonic acid dissociates to form water and carbon dioxide. The hemoglobin shows greater affinity for oxygen than it does for carbon dioxide, and promptly exchanges its cargo. The carbon dioxide so released is exhaled from the body, and the blood is ready for another round trip through the systemic highways.

COMMENTARY

When the Lungs Break Down

In urban environments, in certain occupations, even in the microenvironment surrounding a cigarette smoker, airborne particles and certain gases exist in abnormal amounts. They put an extra workload on the ciliated mucous membranes of respiratory passageways. These membranes are extremely sensitive to cigarette smoke, probably because of the chemical nature of the concentrated particles. Smoking and other forms of air pollution interfere with ciliary movement in the passageways. As a result, mucus—and the particles it traps, which include bacteria—begins accumulating in the trachea and bronchi. Coughing sets in as the body reflexly attempts to clear away the mucus. If irritation continues, the coughing reflex persists. Coughing aggravates the condition because it further irritates the bronchial walls. Bronchial walls become inflamed and infected. Tissue is destroyed by bacterial activity. Cilia diminish in numbers. And mucus-producing cells increase as the body works to fight against the accumulating debris. All of this aggravation leads to the formation of fibrous scar tissue. Such are the characteristics of bronchitis.

A person suffering from an acute attack of bronchitis who is otherwise in good health responds to medical treatment. But what happens if the irritation persists —if, for example, a chain-smoker continues to smoke? As fibrous scar tissue begins to obstruct the respiratory passageways, bronchi become progressively clogged with more and more mucus. Carbon-dioxide-rich gases then become trapped in alveoli, which tend to disintegrate under the pressure. The lungs fill with gases, which can't be expelled efficiently. The normally pink, elastic lung sacs become dry and perforated. The outcome is emphysema—the distension of lungs and loss of gas exchange efficiency to the extent that running, walking, even exhaling become painful experiences.

Why don't all cigarette smokers get emphysema? There is evidence that early environmental conditions—

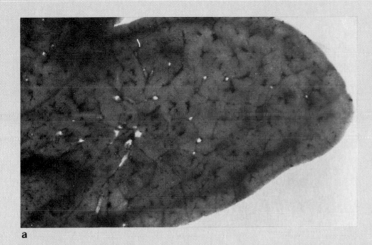
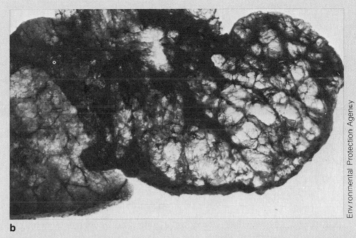

a

b

(**a**) Normal appearance of human lung tissue and (**b**) appearance of a lung taken from a person who suffered emphysema. Membranous walls between alveoli break down, which results in enlarged alveoli and less surface area for gas exchange. (Webb-Waring Institute for Medical Research, Denver, courtesy Environmental Protection Agency)

poor diet, chronic colds, other respiratory ailments—can create in some persons a predisposition to this disease later in life. In addition, many who suffer from emphysema have a hereditary deficiency in their ability to form antitrypsin, a substance that inhibits tissue-destroying enzymes produced by bacteria. These individuals may therefore be at a disadvantage in fighting off respiratory infections when they do strike. When such people have a smoking habit, their prospects unquestionably are grim. Part of the problem is that the potential threat seems exaggerated—what's so terrifying about coughing up mucus now and then? But emphysema is insidious. It can develop slowly, over twenty or thirty years; and few today seem to think much about what they will be doing twenty or thirty years hence, and about what shape their body will be in during their remaining years. By the time emphysema is detected, the damage to lung tissue is irreparable. The threat is not exaggerated: about 1,300,000 individuals in the United States alone now suffer from this disease, inflicting pain on themselves and stress on their families, who can do nothing to alleviate the condition.

Cigarette smoke is also known to contain compounds that can lead to lung cancer. These compounds, such as methylcholanthrene, are found in coal tar and cigarette smoke. It appears that they become chemically modified in the body, through the action of natural hormones, into highly reactive intermediates that are the real carcinogens. In these forms they act on cells in lung tissues. Either they cause irreversible alterations in the way that the cellular DNA is expressed, or they alter the DNA itself. Whatever the case, cell division goes out of control.

The more cigarettes smoked each day, the longer that individuals inhale, the more deeply they inhale—all of these behaviors increase the susceptibility to a disease that is agonizing in its terminal stages. At least ninety percent of all lung cancer deaths are the legacy of cigarette smoking. It is a disease that only ten out of a hundred afflicted individuals will survive, with varying degrees of tissue damage and physiological malfunctioning.

Table 16.4	Activities That Depend on Liver Functioning
1.	Carbohydrate metabolism
2.	Control over blood composition
3.	Protein assembly and disassembly
4.	Urea formation from nitrogen-containing wastes
5.	Assembly and storage of fats
6.	Fat digestion (bile is formed by the liver)
7.	Inactivation of chemicals such as hormones
8.	Detoxification of many poisons
9.	Immune response (liver is involved in antibody formation, removal of some foreign particles)
10.	Red blood cell formation (liver absorbs, stores factors needed for red blood cell maturation)

After F. Strand, *Physiology*, 1978.

ORGAN AT THE CROSSROADS: THE VERTEBRATE LIVER

So far, we have been looking at three major organ systems—those of digestion, circulation, and respiration. A central link between them is the liver. In vertebrates, the **liver** is the largest glandular organ. It is also one of the most important homeostatic organs in regulating the organic components of blood—hence in regulating metabolism itself. The liver is active in carbohydrate, fat, and protein metabolism; it is also a detoxifier for blood.

Lipids reach the liver by way of the general systemic circulation. Other nutrients are absorbed across the intestinal wall and are transported directly by capillaries to the **hepatic portal vein**, which leads to the liver. Here, in the liver, some lipids are deposited, stored, or broken down into compounds such as acetyl-CoA. Acetyl-CoA, recall, is used in such metabolic pathways as the all-important Krebs cycle. Glucose is transported to the liver, where the excess is stored as a polysaccharide (glycogen). Here, too, excess amino acids are converted to forms that can be sent through the Krebs cycle as an alternative energy source. It is in the liver that most hormones are finally turned

off—inactivated as the bloodstream carries them in, then sent to the kidneys for excretion. In addition, amino acid conversions in the liver form ammonia, which is potentially toxic to cells. The liver immediately converts the ammonia to urea—a much less toxic waste product that can be expelled, by way of the kidneys, from the body. Table 16.2 lists the liver's functions. The following case study will illustrate just how central the liver is in normal body functioning.

CASE STUDY: FEASTING, FASTING, AND SYSTEMS INTEGRATION

Suppose, this morning, that you are vacationing in the mountains and decide on impulse to follow a beckoning forested trail. You fail to notice the wooden trail marker that bears the intriguing name, "Fat Man's Misery." As you walk down the tree-lined corridor, you are enjoying one of the benefits of discontinuous feeding. Having stocked your digestive tract with a large breakfast, your body's cells are assured of ongoing nourishment; you don't have to forage constantly amongst the ferns as, say, a nematode must do. Food partly digested in your stomach has already entered the small intestine. There it is broken down further by enzymes secreted from glandular epithelium of both the pancreas and small intestine. Right now, amino acids, simple sugars, and fatty acids are moving across the intestinal wall, then into the bloodstream. Consider the fate of glucose alone. With the sudden surge of nutrients, glucose molecules are entering the bloodstream faster than your body can assimilate them. The level of blood glucose begins to rise. *Your body, however, has a homeostatic program for converting glucose into a storage form when it is flooding in, then releasing some of the stores when glucose is scarce.* Your liver plays a key role in this program. So do your hypothalamus, pituitary, pancreas, adrenal glands, and muscles.

Long before blood glucose approaches dangerously high levels, pancreatic cell clusters (islets of Langerhans) are called into action. These islets include hormone-secreting alpha and beta cell types, which function antagonistically. *Beta cells* secrete **insulin**, a hormone that enhances glucose uptake into your body's cells. *Alpha cells* secrete **glucagon**, a powerful hormone that prods liver cells into converting glycogen (a storage starch) into glucose subunits. When blood glucose levels are high, insulin secretions rise—which prods cells into quickly using or storing

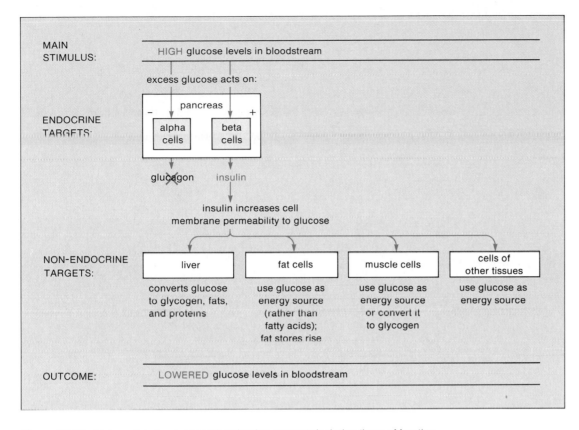

Figure 16.16 Main metabolic routes and endocrine commands during times of feasting.

away the incoming glucose. And glucagon secretions dwindle—which slows down the liver's conversions of starch reservoirs into glucose.

Right now, so much glucose is flooding into your body that not all of the molecules can be swept down metabolic pathways. Some of the glucose molecules bind to and *inactivate* alpha cells, so glucagon secretions dwindle. At the same time, glucose also binds to beta cells—and this union *activates* beta cells, which now increase their insulin secretions (Figure 16.16).

Insulin enters your bloodstream and travels to cells throughout your body. When insulin reaches target cells, it increases plasma membrane permeability to glucose. So more glucose enters cells. In liver cells, the glucose is converted to glycogen. In fat cells, the glucose is used in the formation of new fat. In muscle cells, some glucose is used at once and some is converted to glycogen. Increased glucose uptake, glucose metabolism, and storage of the excess—these cellular activities proceed even as you walk deeper into the woods. And higher in the mountains.

Even though you are no longer feeding your body, your brain cells have not lessened their demands for glucose. Neither have your muscle cells, which are getting a strenuous workout. Little by little, blood glucose levels drop lower and lower. Now endocrine activities shift in the pancreas. With less glucose binding to them, beta cells decrease their insulin secretion. With less glucose to inhibit them, alpha cells step up glucagon secretions. When glucagon reaches your liver, it causes the conversion of glycogen back to glucose—which is returned to the blood. This completes a feedback loop: blood glucose levels now stabilize.

But the best-laid balance of internal conditions can go astray when external conditions change. In your case, the "miserable" part of the trail has now begun. You find yourself scrambling up steep inclines, squeezing along narrow ledges, climbing higher and higher. Suddenly you stop, surprised, in great pain. You forgot to reckon with the thinner air in the mountains, and your leg muscles cramped. Your body has already detected the need for producing more oxygen-carrying red blood cells at this

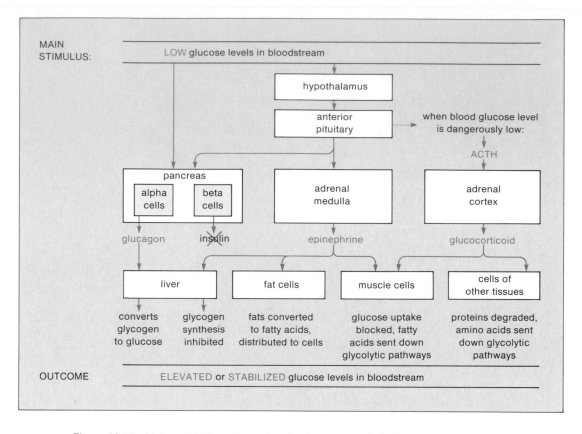

Figure 16.17 Main metabolic routes and endocrine commands during times of fasting (and starvation).

altitude. However, it will take days before enough additional red blood cells accumulate. In the meantime, your muscle cells aren't being supplied with enough oxygen for your strenuous climbs, and they have switched to the anaerobic pathway of glycolysis. An end product of glycolysis, recall, is lactic acid—which is toxic to your cells if allowed to accumulate. The sudden lactic acid buildup has seriously impaired muscle cell functioning.

Again, systems interact and work to return your body to a homeostatic state. Receptors in your arterial walls detect the reduction in oxygen pressure and an accompanying increase in hydrogen ion concentrations. Nerve impulses course toward the respiratory center in the medulla. The result: increased activity in the diaphragm and other muscles associated with inflating and deflating your lungs. You breathe faster now, and more deeply. Gradually muscle cells are purged of lactic acid and replenished with oxygen. Gradually lactic acid is carted off to the liver. In

the liver, lactic acid is converted to glucose, which is returned to the blood.

By now, blood glucose levels have dropped considerably, and the appetite center of your hypothalamus becomes activated. Nerve impulses shoot out to muscles that start a churning in your stomach—hunger pangs! On checking the sun's position, you see it's well past noon. And guess what: you forgot about lunch. When you start the long walk back, your liver's store of glycogen is nearly gone and blood glucose levels drop once more. In combination with your anxiety about getting back before dark, the decline calls on new homeostatic controls. Under hypothalamic commands, your adrenal medulla begins secreting norepinephrine and epinephrine. Its main targets: the liver, adipose tissue, and muscles. In liver cells, glycogen synthesis stops. In muscles, glucose uptake is blocked. In fat cells, fats are converted to fatty acids, which are routed to the liver, muscles, and other tissues as an alternative energy source (Figure 16.17). For every fatty acid molecule sent down metabolic pathways in those tissues, several glucose molecules are saved for the brain.

This new metabolic strategy is enough to get you back to the start of the trail by sundown. In fact, your body had enough energy stored away in fat to sustain you for many more days, so the situation was not really desperate. Even then, your energy supplies would not have run out. Another hypothalamic command would have prodded your anterior pituitary into secreting andrenocorticotrophic hormone (ACTH). In turn, ACTH would have prodded adrenal cortex cells into secreting glucocorticoid hormones, which have an extremely potent effect on the synthesis of carbohydrates from proteins. Slowly, in muscles and other tissues, your body's proteins would have been disassembled. Amino acids so released would have been used in the liver as an alternative energy source—and once more your brain would have been kept active. As extreme as this last pathway may be, it would be a small price to pay for keeping your brain functional enough to figure out how to take in more nutrients, and bring the body back to a homeostatic state.

Readings

Barnes, R. 1974. *Invertebrate Zoology*. Third edition. Philadelphia: W. B. Saunders. Good descriptions of organs and organ systems in invertebrate groups.

Hickman, C., et al. 1979. *Integrated Principles of Zoology*. Sixth edition. St. Louis: Mosby.

Kessel, R., and R. Kardon. 1979. *Tissues and Organs: A Text-Atlas of Scanning Electron Microscopy*. San Francisco: Freeman. Outstanding, unique micrographs, and well-written descriptions of major tissues and organs.

Romer, A., and T. Parsons. 1977. *The Vertebrate Body*. Fifth edition. Philadelphia: W. B. Saunders.

Storer, T., et al. 1979. *General Zoology*. Sixth edition. New York: McGraw-Hill.

Vander, A., J. Sherman, and D. Luciano. 1980. *Human Physiology: The Mechanisms of Body Function*. Third edition. New York: McGraw-Hill. Perhaps the clearest introduction to human organ systems and their functioning.

Review Questions

1. Study Figure 16.2. Then, on your own, diagram the connections between metabolism and the digestive, circulatory, and respiratory systems.

2. Explain the differences between digestion, absorption, and assimilation.

3. What are the main functions of the stomach? The small intestine? The large intestine? Define lumen, and peristaltic movement.

4. Which enzymes are involved in the breakdown of (a) polysaccharides, (b) proteins, and (c) fats? Name the four kinds of breakdown products that are actually small enough to be absorbed across the intestinal mucosa and into the internal environment.

5. What is a heart? A cardiovascular system? A lymph vascular system? Describe the difference between open and closed circulation.

6. Explain how arteries, arterioles, and capillaries help regulate blood flow to different body regions.

7. State the main function of blood capillaries. What forces drive substances out of and into capillaries in capillary beds?

8. State the main function of venules and veins. What forces work together in returning venous blood to the heart?

9. Describe the cardiac cycle in a four-chambered heart. Explain the difference between pulse pressure and pulse rate.

10. Define respiration. What is the main requirement for gas exchange in animals? What membrane adaptation enhances gas exchange in (a) water-dwelling animals and (b) land-dwelling animals?

11. Explain how a countercurrent flow mechanism works in a fish gill.

12. What are the main parts of the human respiratory system? By what mechanisms do carbon dioxide move out of the body and oxygen into it through the *same* system of branched tubes?

13. Within the lungs, gas exchange occurs across _____ epithelium, through _____ present in a thin layer of connective tissue, and on across _____ epithelium.

14. In what forms is carbon dioxide transported in the bloodstream?

15. What is the liver? Describe some of the functions that suggest its central role in metabolism.

16. What are the roles of insulin and glucagon in glucose metabolism? When blood glucose levels are high, glucagon secretions are (enhanced/inhibited). When blood glucose levels are low, insulin secretions are (enhanced/inhibited).

What keeps cells metabolically humming? In the animal body, their functioning is assured largely as a result of the way that they carry out specialized roles in keeping one another alive. Thus, as you have seen, cells of digestive, circulatory, and respiratory systems interact and keep one another supplied with essential substances. Keeping the animal cell well fed is no doubt a most vital aspect of homeostasis, but of course it is not the only one. What happens to that well-fed cell when the animal sits under the hot sun too long, or can't get warm on a cold winter night? What happens when the animal eats salty food, and salt loads increase in the bloodstream? What fate awaits that cell when the animal body comes under siege—by bacterial invaders, even by some of its own cellular kin gone cancerously berserk? Let's take a look now at a few more self-preservation mechanisms for coping with shifts and landslides in the internal state.

ADJUSTING TO TEMPERATURE CHANGE

Balancing the Insides With Temperatures Outside

For almost all animals, body temperature must be maintained within narrow limits. Above 50°C, protein denaturation occurs, enzyme activity is skewed, and nerve cells (for example) don't function at all. In humans, exposure to extremely high temperatures leads to convulsions, then death. Low temperatures are equally hazardous. Water freezes at 0°C. If water turns to ice in, say, the extremities, metabolism will grind to a halt. As the ice expands, tissues rupture and die. Hence the danger of fingers and toes frostbitten for too long. Most fish living in large lakes or the open seas are **ectotherms**: their body temperatures rise and fall with environmental changes, and there is not too much their insides can do about it. Fortunately for these fish, temperatures in large bodies of water generally don't vary by much more than ten degrees throughout the year. Reptiles, too, are ectotherms, but temperatures on land can change rapidly. That is why you see reptiles basking in the sun, changing body position with respect to incoming light rays, absorbing heat that was absorbed earlier in the day by a large rock, moving in and out of shade when internal temperatures go up and down. Before night falls and their metabolism drops enough to render them immobile, these animals tuck themselves into crevices and under rocks, where they are not as vulnerable to predators.

Birds, mammals, and a few large fish such as tuna are **endotherms**: through metabolic activity they generate

Figure 17.1 Keeping the warm-blooded body warmer on cold winter nights in a West German zoo. In complex animals, diverse behavioral adjustments supplement built-in adaptations in maintaining the homeostatic state.

17

MAINTAINING THE INTERNAL ENVIRONMENT

Everett C. Johnson

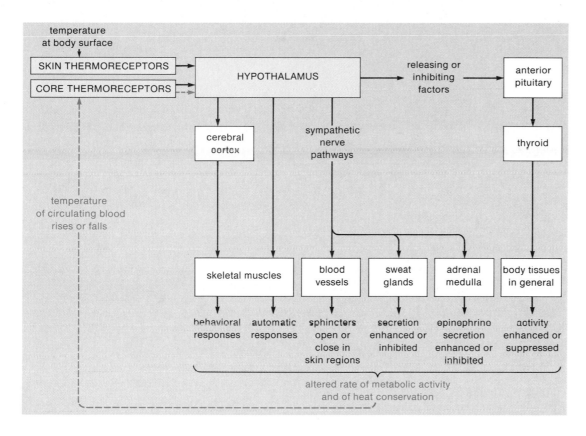

Figure 17.2 Summary of main homeostatic feedback relationships in controlling mammalian body temperature. The broken brown line completes the feedback loop. (After Vander, 1975, and Strand, 1978)

considerable internal heat. Hence they are not entirely at the mercy of variable surroundings. In these animals, neural and endocrine controls are at work, balancing the amount of heat being produced with the amount being lost from the body. Homeostatic control over internal temperature is a constant challenge. Not only do external temperatures vary, the production of internal heat varies with the degree of physical—even mental—activity. Consider that the normal temperature for humans averages 37°C (98.6°F) when recorded orally. Yet most of us show daily fluctuations around the norm. Our peak body temperature corresponds to the time when we are most active and alert. We are most lethargic when body temperature is at its lowest, whether it be in the morning or at night. People who are able to work midnight shifts have successfully reset their temperature cycles. In addition, females have monthly temperature cycles that correspond with hormonal secretions during the ovarian and menstrual cycles (Chapter Eighteen). In all such cases,

Heat loss must be adjusted to equal heat production if any given body temperature is to be maintained.

Let's now turn to some of the controls underlying thermal balancing acts, especially as they occur in land-dwelling mammals.

Controls Over Body Temperature

In mammals, the hypothalamus is the seat of temperature control. This means, of course, that sensory receptors scattered throughout the body must send their information about temperature into the central nervous system. Now, you might be thinking that such thermoreceptors are found only in skin—but you would be wrong. It isn't the body *surface* temperature that is so important, it is the *core* temperature that must be maintained (Figure 17.2).

COMMENTARY

Falling Overboard and the Odds for Survival

In 1912, the ocean liner *Titanic* set out from Europe on her maiden voyage to America. In that same year, a huge chunk of the leading edge of a Greenland glacier broke off and began floating out to sea. Late at night on April 14, off the coast of Newfoundland, the iceberg and the *Titanic* made their ill-fated rendezvous. The *Titanic* was considered unsinkable; four of the watertight compartments making up its hull could be ripped apart and the ship would remain afloat. The iceberg ripped through five. Lifeboats and survival drills had been neglected, and only about a fourth of the 2,000 people on board managed to scramble into the lifeboats that could be launched. What happened to the rest of the passengers? Within two hours, rescue ships were on the scene—yet 1,513 bodies were recovered from a calm sea. All were wearing life jackets. None had drowned. Every one of those individuals had died from hypothermia—

from a drop in body temperature below tolerance levels:

36°–34°C	Shivering response, increase in respiration. Increase in metabolic heat output. Construction of peripheral blood vessels, so that blood is routed to deeper regions.
33°–32°C	Shivering response stops. Metabolic heat output drops. Dizziness and nausea set in.
31°–30°C	Capacity for voluntary motion is lost. Eye and tendon reflexes inhibited. Consciousness is lost. Cardiac muscle action becomes irregular.
26°–24°C	Ventricular fibrillation sets in. Death follows.

Skin thermoreceptors provide early warning signals about shifts in environmental temperature. Thermoreceptors within the body—including the spinal cord and the hypothalamus itself—signal what is going on inside.

The main regulatory center is in the hypothalamus. When circulating blood temperature rises above or falls below normal, the hypothalamus causes alterations in metabolic rates and behavioral responses. It resets these rates and responses when receptors for blood temperature signal a return to normal body temperature. Figure 17.2 summarizes the feedback relationships involved.

What sort of tissue responses does the hypothalamus evoke? Here we can consider two kinds of responses to cold temperatures: *heat production* and *heat retention.* Under hypothalamic commands, the adrenal medulla steps up its epinephrine secretions—and epinephrine increases the metabolism of energy-rich lipids and carbohydrates. Nor-

epinephrine secretions act directly on fat cells, which increase the amount of fuel available to support the increased metabolism. In addition, muscles are made to contract. Because muscles represent almost half the total body weight in mammals, considerable heat is released when they contract. More heat is produced if they are made to contract very frequently, a response called the **shivering response**. Still more heat can be produced behaviorally, through foot stomping, hand clapping, and other sorts of conscious exercise. While these activities proceed, coordinated vasoconstriction of arterioles directs blood away from the body's surface and into deeper tissue regions. This response cuts down on the amount of heat radiating away from the body's surface. Mammals typically have a fat layer under the skin, and a hairy coat (fur)—two adaptations that also help hold heat close to the body. Also, muscles at the base of hairs cause fur to fluff, thereby forming a thermal blanket that holds heat near the body surface.

MAINTAINING THE INTERNAL SEA

Recall that activities associated with the term *living* cannot proceed without water. Water must be present inside a metabolically active cell and it must bathe the outside—if only as a thin film—if substances are to be exchanged across the plasma membrane. Thus, we might reasonably deduce that the first living cells must have originated in watery settings. We might further deduce that some kind of internal "sea" must bathe all the cells of multicellular animals, regardless of whether they live in water or on land.

It is not that the internal sea must be exactly the same in composition as water present in the environment. It is not even that body fluids must have the same composition in all animals. What is important is that animals must have mechanisms for *maintaining* whatever body fluid concentration and composition are necessary for cell functioning.

For all animals, internal water and solute concentrations must, on the average, remain constant within some fairly narrow range.

Such maintenance is simpler for some animals than for others. For example, water concentration inside the body of a sea star or oyster can conform, within limits, to changes in the surrounding seawater. If they encounter water that is a little more dilute than before, water moves into the body (down its concentration gradient). For sea stars and oysters, body fluid becomes more dilute (and the body swells up). Other animals, such as a sea-dwelling crab, must maintain high water concentrations inside even when the surrounding water becomes dilute; it must *regulate* osmotic concentrations. In addition, *all* animals must regulate individual ion concentrations in the body. It happens that water usually contains ions such as sodium, potassium, calcium, phosphorus, and magnesium. This is true of the salty seas, freshwater lakes, ponds and streams, the body fluids of plants, and the body fluids of animals (Table 3.1). Therefore any "water" moving into the animal body—for instance, by drinking, by eating plants or other animals—is going to have some effect on internal solute concentrations. Beyond this, metabolism produces such soluble by-products as carbon dioxide, nitrogen-containing ammonia, urea, and uric acid. All are potentially toxic. Thus, without the means of eliminating some solutes, an animal could poison itself. In Chapter Five, you read about active and passive transport mechanisms that govern water and solute concentrations in individual cells. Here we will look at some examples of controls over fluids in the multicellular environment.

On Worms, Fat Sand Rats, and the Human Kidney

Water and salts enter the body in different ways. Food and water enter by way of digestive systems. In water-dwelling animals, water and salts also diffuse across epithelial membranes, such as those found in fish gills. Water may be eliminated by way of the digestive system, and by evaporation from the body surface and lungs. Many animals also have tubular systems that are specialized for exchanging water and salts with the surroundings, in order to maintain internal conditions.

Earthworms, for instance, have this kind of tubular system. It is functionally linked with the earthworm's closed circulatory system. Except for the first three head units and last tail unit, each body segment contains a pair of water-regulating tubules called **nephridia** (singular, nephridium). A fluid-draining funnel at the beginning of each nephridium leads away from the segment just ahead (Figure 17.3). From the funnel end, the tubule loops up and around a capillary bed, enlarges into a storage bladder, then ends in its own pore to the outside. As body fluids pass through the tubule system, water and solute levels are controlled. Water is conserved in the storage bladder. When body salt levels drop, they can follow their concentration gradients and move out of the tubules, into the interstitial fluid, then back to blood capillaries. In other words, salts in the tubules can be *reabsorbed* by the body;

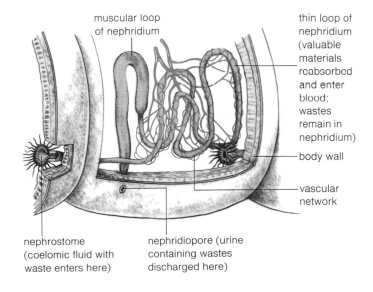

muscular loop of nephridium

thin loop of nephridium (valuable materials reabsorbed and enter blood; wastes remain in nephridium)

body wall

vascular network

nephrostome (coelomic fluid with waste enters here)

nephridiopore (urine containing wastes discharged here)

Figure 17.3 Earthworm system for salt and water regulation. The functional unit of this system is called a nephridium. (After C. P. Hickman, Jr. et al., *Integrated Principles of Zoology*, 1979)

they don't automatically leave with outgoing water. Earthworm nephridia may be taken as models for what must have been forerunners to the paired vertebrate kidneys. Let's examine the structure and function of these more complex organs, as they occur in mammals.

The mammalian **kidney** is an intricately structured organ for water and salt regulation. It is part of a tubular network called the **urinary system**. Mammalian kidneys occur in pairs in the lower back region, one to a side. In humans, each kidney is a dark red organ, about the size of a fist. Beneath its connective tissue coat is the kidney *cortex*, a rindlike layer in which initial functional links are made with the bloodstream. The cortex lies over an inner layer, the kidney *medulla*. This inner layer is filled with transport ducts that empty into a central cavity, the *renal pelvis*. (The word "renal" comes from the Latin *renes*, meaning kidneys.) The walls of this cavity converge into a *ureter*, a tube that empties into a storage organ called the *urinary bladder*. Two ureters (one from each kidney) empty into this bladder. Fluids then leave the bladder through the *urethra*, a tube leading out through the male penis or female vulva. Figure 17.4 shows how these parts are arranged in the human urinary system.

During every twenty-four-hour period, about 180 liters (190 quarts) of fluid filters from the blood through the paired human kidneys. On the average, only 1½ liters actually leaves the body through the urethra. The rest is taken back (reabsorbed) into the bloodstream. The fluid destined for discharge, called **urine**, consists of water, salts, nitrogen-containing wastes such as urea, and substances that the body can't metabolize. (These substances include excess vitamins and drugs such as penicillin.)

As you have probably noticed, the human way of life would be somewhat hampered by an incessant dribbling of urine, and built-in controls do exist over fluid discharge. As a urinary bladder fills, tension increases in its strong, smooth-muscled walls. The increased tension triggers a reflex relaxation of muscles located at the upper and lower ends of the urethra, and a coordinated contraction of the bladder walls. These contractions force fluid through the urethra. Although it is basically an involuntary response, the reflex discharge of fluid can be consciously inhibited by neural commands. This capacity for controls occurs once the cerebral cortex is developed enough to monitor signals from stretch receptors located in bladder walls. Despite the urgent promptings of parents who don't know better, this doesn't occur until infants are one or two years old.

Urine composition and volume are both adjusted as internal conditions change. These adjustments occur through exchanges between blood capillaries and kidney

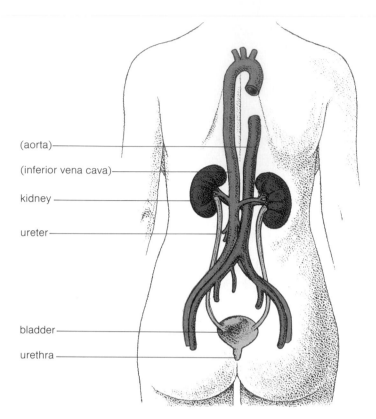

(aorta)

(inferior vena cava)

kidney

ureter

bladder

urethra

a *Human urinary system*

tubules called **nephrons**. In the human kidney, more than a million nephrons lie side by side, perpendicular to the surface. Blood from renal arteries makes its first functional link with nephrons in the kidney cortex. Here, the nephron walls balloon outward and form a cup around a set of blood capillaries (Figure 17.4c). This capillary cluster is called a *glomerulus*. The nephron cup is known as *Bowman's capsule*. The nephron isn't open at this end. Rather, the capsule is like a tennis ball that has been punched in on one side. From its cupped end in the cortex, the nephron plunges into the medulla, forms a hairpin turn (the *loop of Henle*), then shoots back up. A second set of blood capillaries laces around its tubular parts. The nephron terminates on a collecting duct, which services many neighboring nephrons.

Urine formation and its eventual composition are outcomes of three events in the nephron/capillary unit: filtration, reabsorption, and secretion. **Filtration** begins in Bowman's capsule. Here, all the components of blood

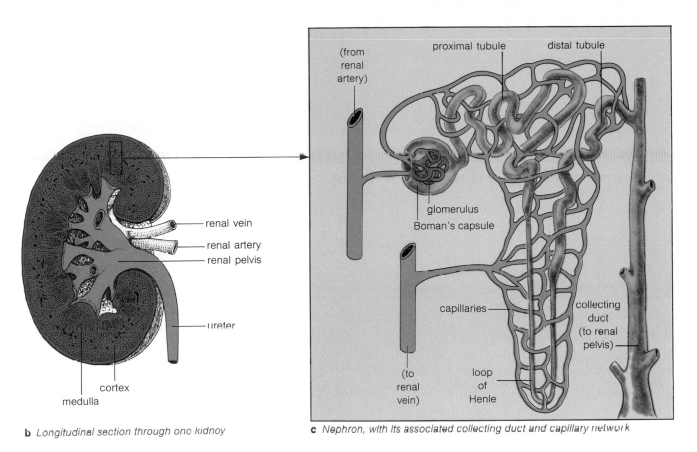

b *Longitudinal section through one kidney*

c *Nephron, with its associated collecting duct and capillary network*

Figure 17.4 Human urinary system (**a**); a closer look at one of its kidneys, the organ concerned with regulating salt and water exchange (**b**); and a closer look still at the functional unit within the kidney—the nephron/capillary unit (**c**), as described in the text.

except blood cells and other proteins can move across the one cell thick capillary epithelium, then across the one-cell-thick capsule epithelium. About one-fifth of the water and other noncellular components of blood do so. Active transport is not involved; filtration is an entirely passive event. What, then, causes the movement? Hydrostatic pressure (that is, blood pressure created by heart contractions) is greater in capillaries than it is in the capsule. This pressure drives the blood components out. The proteins being left behind in the bloodstream do create an osmotic pressure gradient in the opposite direction. Even so, water and solute flow across the membrane is about a hundred times greater than it is in the systemic capillaries. Thus, *membrane permeability and the pressure difference across it are the reasons why kidneys can filter an astonishing volume of fluid each day.*

During **reabsorption**, all but about one percent of the water and most of the solutes that entered the nephron

are returned to the bloodstream. As the filtrate flows along the nephron, selective reabsorption rapidly alters its composition. For instance, sodium ions (Na^+) are actively transported across the proximal tubule walls and into the interstitial fluid. Chloride ions (Cl^-), being oppositely charged, are "pulled" in the same direction. On the average, more than eighty percent of the sodium and chloride present in the filtrate leaves the proximal tubule. This outward solute movement sets up an osmotic gradient, and now water moves out, too. Fluid pressure in the extracellular region rises as a result. When it does, both the solutes and water flow into the lower pressure region—into adjacent capillaries. Thus, *although water reabsorption into capillaries is a bulk flow process, water is reabsorbed only because solutes are reabsorbed first.*

Some substances appear in urine in greater amounts than were carried into the nephron to begin with. They

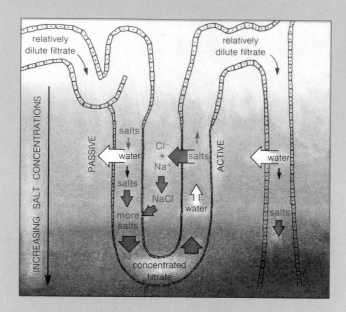

Figure 17.5 Urine concentration in the loop of Henle. What is this loop's function? It establishes a sodium concentration gradient in the kidney, and this gradient is necessary if water is to be conserved. The gradient starts at the cortex; it is steepest in the basement of the medulla. Exchanges with blood vessels surrounding the loop help maintain the gradient. The adjacent collecting duct depends on it in producing concentrated urine.

The gradient is established through a process called *countercurrent multiplication*. The word countercurrent, recall, simply refers to a flow in opposing directions through a pair of channels. The word multiplication refers to the way that sodium is moved not once, but twice through the loop for a given volume of filtrate: down through

the descending branch, up the ascending branch—*then through the surrounding fluid and back into the descending branch*. The longer the distance that sodium must travel, the more opportunity there is for its exchange with the surroundings. Hence the two-pass approach means that the concentrating power of the system can be multiplied, so to speak.

When glomerular filtrate moves into the loop's descending branch, it has about the same solute concentrations as the surroundings. But the descending branch is permeable to water and impermeable to sodium. Water moves out, and sodium stays in here. By the time the filtrate rounds the bend, its solute concentrations approach that of the surroundings.

The loop's ascending branch differs: it is impermeable to water—and permeable to sodium and chloride. Here, chloride ions can be actively transported *out* of the loop; sodium ions, being of opposite charge, can follow passively in their wake. With water staying in and salts moving out, the filtrate becomes more dilute again as it flows back up toward the cortex.

But it happens that the loop's descending branch is permeable to more than water. It's also permeable to *sodium chloride*—NaCl. When sodium and chloride combine to form NaCl, they can move back into the descending branch. Together with new sodium ions coming into the tubule from the nephron entrance, solute concentrations inside become steeper than on the outside—and salts move out again. Through this countercurrent process, a standing concentration gradient is established in the kidney.

When the body rids itself of excess water, urine is highly dilute (has a high ratio of water to salts). At this time, walls of the collecting duct are impermeable to water, which thus leaves the kidney.

When the body rids itself of excess salts, urine is highly concentrated. ADH action makes walls of the distal convoluted tubule and collecting duct permeable to water, which moves passively (down its osmotic gradient) into the salty interstitial fluid.

---

are actively transported from the second capillary set into the nephron tubules, an event known as **tubular secretion**. These substances include potassium and hydrogen ions, penicillin and other drugs, and some toxic substances. We can summarize the points made so far in this way:

Filtration, reabsorption, and secretion in the nephron/capillary unit of the kidney all influence the ultimate composition and volume of urine. Hence they influence how much water and solutes the body conserves.

Another factor influencing water and salt concentrations is the architecture of the nephron itself. Figure 17.5

provides a closer look at concentration mechanisms in the loop of Henle. Variations in the loop length are adaptive to different environments. For instance, a fat sand rat of Africa gets all of its food and water from saltbush leaves. A staggering 12.2 percent of the dry weight of these leaves is sodium, which makes the saltbush too salty for most plant-eating animals. Yet these rats dine with impunity on the leaves. The loop of Henle in these rats is one of the longest known, relative to body size. Recall that water is reabsorbed into the bloodstream only when solutes have been reabsorbed first. Because these loops are longer, a steeper concentration gradient is created by salt reabsorption—hence more water can follow and be reabsorbed. In this way, the rat can excrete small amounts of highly concentrated urine, and thereby avoid poisoning itself even as it conserves water—a scarce resource in desert regions. In general,

The longer the loop of Henle, the greater the animal's capacity to conserve water and to concentrate solutes for excretion in urine.

Controls Over Fluid Volume and Composition

Most of the exchanges between blood capillaries and nephrons are obligatory. They occur regardless of whether internal fluids need adjusting, simply because that is what the system is structurally equipped to do. Actually, though, the final urine composition is under complete homeostatic control. Through these controls, adjustments are possible when changes in diet, muscular activity, and metabolic activity alter body fluid volume and composition.

Any increase or decrease in water volume automatically affects solute concentrations. The more water there is, the less concentrated salts become. Thus homeostatic controls over water input and output can alter solute levels. Two interrelated mechanisms are at work here: hormonal action that influences the amount of water excreted in urine, and a thirst mechanism that regulates water intake. The hypothalamus governs both.

The main hormonal agent is **antidiuretic hormone**, or ADH. "Diuresis" is a Greek word meaning urination; hence "antidiuretic" signifies that the hormone opposes urination. As if that were not enough to remember, the hormone is also called vasopressin. The name refers to its other role of constricting blood vessels during the channeling of blood flow to different body regions (Chapter Sixteen). Neither name signifies the full extent of this hormone's action. It really ought to be assigned a name that better denotes its involvement with fluid movement through both vascular and urinary tubes.

In any case, the hypothalamus contains a nerve cell cluster that is sensitive to concentrations of sodium and some other solutes in blood. When solute concentrations rise (or when blood volume drops), these cells become excited. Hypothalamic commands then travel to the posterior pituitary and trigger ADH release. The bloodstream carries the ADH to the kidneys. There, ADH acts on cells of the distal convoluted tubule and the collecting duct, making them more permeable to water. With increased ADH secretions, more water is reabsorbed. As a result, urine volume decreases and its salt content is greater. When

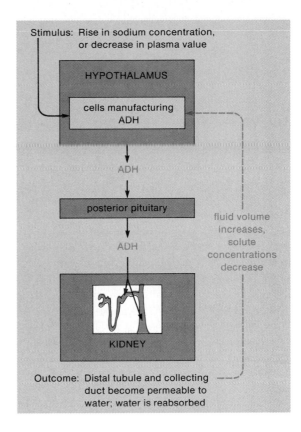

Figure 17.6 Homeostatic control over water and solute levels in body fluids. The hormone ADH acts on the nephron's distal convoluted tubule and the collecting duct walls, making them permeable to water. Water moves into the interstitial fluid and is reabsorbed into the bloodstream. Thus blood pressure (volume) rises. The change is detected through the associated decrease in solute concentrations. Here, the brown dashed line completes the feedback loop.

ADH secretions are decreased, urine volume increases and the fluid volume leaving the body is more dilute (Figure 17.6).

Also within the mammalian hypothalamus is a **thirst center**. This nerve cell cluster is stimulated at the same time that solute-concentration receptors call for ADH release. When these receptors detect a rise in salt levels (a drop in water volume), thirst center cells then signal that it is also time to take in more water.

Other hormones also influence solute concentrations. Aldosterone is one example. Among other things, it increases sodium reabsorption into distal tubules and collecting ducts by enhancing the activity of sodium-potassium

COMMENTARY

Kidney Failure, Bypass Measures, and Transplants

Sometimes, kidneys no longer can perform their filtration, reabsorption, and secretion tasks. For example, arteriosclerosis of blood vessels associated with the nephron can trigger formation of scar tissue that impairs kidney function. As another example, salts and cholesterol may aggregate into "stones." Such stones can become lodged in the kidney (and ureter, bladder, or urethra), where they can alter or block urine discharge. In the United States alone, an estimated 3 million individuals suffer from some kidney disorder. And these disorders occur in all age groups.

When kidneys malfunction, solute levels get out of balance. Substances such as potassium can accumulate in the bloodstream and become highly toxic. The buildup may lead to nausea, fatigue, loss of memory and, in advanced cases, death. A kidney dialysis machine can be employed to restore the proper solute balance. It is sometimes called an artificial kidney, not because it resembles the natural organ but because its end result is the same. Concentrations of substances can be regulated by their selective addition to and removal from the bloodstream.

The artificial kidney is based on dialysis: the separation of substances across a membrane between solutions of differing concentrations. In hemodialysis, a patient is plugged into the machine by tubes leading from an artery or vein. Blood is then pumped through narrow, coiled cellophane tubes. (Cellophane has pores of the same diameter as glomerular capillaries.) The cellophane tubes are located in a warm-water bath. The bath contains a precisely balanced mix of salts, glucose, bicarbonate, amino acids, and other substances that set up the proper gradients with the blood flowing past. Thus, substances at too high a concentration in the patient's blood will diffuse into the dialysis fluid. On the average, hemodialysis takes about five hours. The machine doesn't approach the natural kidney's efficiency, and blood must circulate over and over again through the tubes. Afflicted people must be treated with the machine three times a week.

For temporary disorders, the artificial kidney is used as a bypass measure until normal kidney function resumes. For chronic kidney disease, it must be used for the remainder of the person's life, or until a functional kidney is transplanted. With treatment and with controlled diet, many are able to resume normal activity. Kidney dialysis, however, is controversial. Even after it had appeared in hospitals, hundreds died because they could not afford the treatment costs, which were then about 30,000 dollars a year. The costs are still a staggering 20,000 dollars annually, although Medicare will now pick up the tab, at taxpayers' expense. The catch is that the machines themselves are expensive to build, to maintain, and (with soaring hospital costs) to operate. Hospitals are equipped to handle only 1,000 of the 10,000 or so individuals who become stricken by kidney failure each year. Who selects the chosen few? Hospital committees—a mix of physicians, citizens, and clergy —and their task is not an enviable one.

An alternative approach is the transplantation of a living kidney into a patient. Although the surgery itself is expensive, it is far less than the costs of year after year of hemodialysis. There are problems here, too. Giving up a kidney is something that many healthy people are reluctant to do; donors are scarce. More than half the kidneys that are donated come from individuals who have died as a result of accidents. Imagine yourself in the position of a severely afflicted individual, in the ethical dilemma of waiting and half-hoping for someone else's death. Then, too, about a third of all kidney transplants don't take hold anyway, for reasons that will become apparent in the remainder of this chapter.

Is either the treatment or the transplantation really worth the cost? Would you ask the question if you yourself became one of the stricken—and would your answer be the same?

pumps found in the membranes of cells making up the tube walls. Still other examples could be given, but the important point is this:

Urine composition is regulated by hormonal action (which affects the amount of water and solutes excreted in urine) and by a thirst mechanism (which affects water intake).

RESPONDING TO INJURY AND ATTACK

When Blood Vessels Rupture

Suppose you are paring an apple and the knife slips just enough to pare part of your thumb, too. Your blood, being under higher pressure than your surroundings, promptly flows outward. At the sight of such outpourings, have you ever wondered for one spinning moment about what keeps your body fluids from draining completely away? When one of your blood vessels is ruptured or severed, its smooth-muscled walls constrict in a rapid reflex action. In small-diameter vessels at least, the constriction is sometimes enough to shut off blood flow. Many invertebrates and all vertebrates show this muscular response to **hemorrhage** (a bulk flow of blood from damaged vessels). For vertebrates, it is the first of a series of events that help prevent blood loss. For instance, when small blood vessels are damaged, these homeostatic events proceed as follows:

---

1. Constriction of smooth-muscled walls, in an extended spasm that may curtail blood flow.

2. Adherence and clumping of platelets, which temporarily plugs the rupture.

3. Prolongation of the muscle spasm when the clumped-up platelets release vasoconstrictors.

4. Clot formation as blood becomes coagulated (converted to a solid gel).

5. Clot retraction into a compact mass, which draws ruptured walls back together.

Depending on the injury, the initial muscle spasm may last for as long as half an hour. During this time, a platelet plug forms and coagulation begins. Platelets form from cells in bone marrow. Some of the cell cytoplasm protrudes,

budlike, and the "buds" pinch off to become platelets. Most platelets circulate in blood, but about a third are stored in the spleen. When the nervous system detects hemorrhaging, it causes the spleen to contract. Hence stored platelets are propelled into the bloodstream. Once they reach a damaged vessel, the normally disk-shaped platelets take on a spiny appearance (Figure 17.7). These spines are pseudopodia, formed by microtubular growth. Through chemical recognition, the pseudopodia spread out and adhere to exposed collagen fibers in damaged vessel walls. When attached to collagen, platelets release ADP—which attracts still more platelets. They also release calcium ions, which promote clumping into a plug. Such plugs can seal tiny ruptures.

Platelet adhesion is also crucial for repairing larger vessels that are under high hydrostatic pressure. It triggers clumped-up platelets into releasing powerful vasoconstrictors (such as serotonin), which maintain the muscle spasm. It also triggers the release of substances necessary for clot formation.

How does a clot form? First, damaged tissues release a substance called thromboplastin. And clumped-up platelets release a phospholipid. When exposed to calcium ions, these two substances together activate molecules that are precursors for a special enzyme. This enzyme acts on large, rodlike proteins produced in the liver and constantly circulating in blood plasma. These soluble proteins are made to adhere to each other, end to end and side by side. Soon they form insoluble, threadlike polymers. The growing threads become tangled in a net that traps blood cells, platelets, and plasma. This mass is the blood clot.

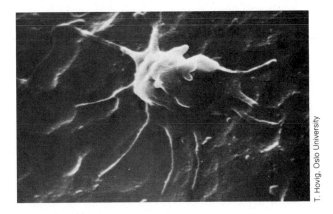

Figure 17.7 Scanning electron micrograph of a spiny blood platelet.

T. Hovig, Oslo University

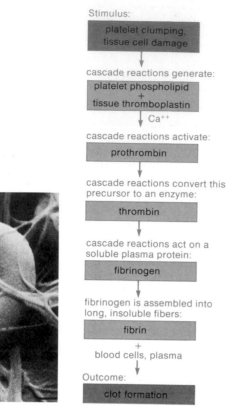

Stimulus:

platelet clumping, tissue cell damage

↓

cascade reactions generate:

platelet phospholipid + tissue thromboplastin

↓ Ca⁺⁺

cascade reactions activate:

prothrombin

↓

cascade reactions convert this precursor to an enzyme:

thrombin

↓

cascade reactions act on a soluble plasma protein:

fibrinogen

↓

fibrinogen is assembled into long, insoluble fibers:

fibrin

+ blood cells, plasma

Outcome:

clot formation

Figure 17.8 Summary of events leading to blood clot formation. The scanning electron micrograph shows a red blood cell trapped in a fibrin net. (From Emil Bernstein, *Science*, vol. 173 cover, 27 August 1971. Copyright 1971 by the American Association for the Advancement of Science)

The picture just given is simplified, in that at least thirteen reactions occur in series before a clot appears. In this cascading reaction series, one substance is switched on. This switched-on form is the activator for another substance, which in turn activates another, and so on down the line. Figure 17.8 summarizes these events.

What could be the advantage of such a cascade of reactions? Perhaps it provides the opportunity for many built-in control points, which would guard against clot formation when none is needed. After all, blood clots circulating in the blood can cause all sorts of circulatory ills (Chapter Sixteen). For some individuals, though, the system's complexity is disadvantageous. For them, the absence or malfunctioning of just one substance in the series can block the entire process. This is what happens in hemophilia. Afflicted people are not genetically equipped to produce one of the substances taking part in the cascade

mentioned above. Hence clots never do form. During emergencies, hemophilics are given a transfusion of normal plasma, which contains the missing substance. It's enough to last for several days. During that time, clotting proceeds and further blood loss is prevented.

Tissue Invasions and the Vertebrate Immune System

Survival, for animals, includes more than response to injury. It also includes response to attacks on normal cell functioning. Sponges aside, all animals have skin and mucous membranes that provide passive barriers against invasions from the outside. Such outer layers are a first line of defense. Other defenses require mobilization of specialized cells such as phagocytes. Among single-celled organisms, phagocytosis is a way of ingesting food. Somewhere during animal evolution, this nutritive function probably proved adaptive in defense. Some cells of the multicellular body were able to ingest and destroy potentially harmful organisms and substances that worked their way into tissues. When circulatory systems developed in some invertebrates, phagocytic cells apparently took to the blood and lymph highways, and began patrolling all tissue regions. Later, among vertebrates, specialized cells appeared that could recognize individual types of invaders. In addition, cell products emerged that could neutralize invaders, or hasten their elimination. Thus there arose lines of phagocytic white blood cells, with appetites ranging from broad to specific—the granulocytes, monocytes, and macrophages ("big-eaters," derived from monocytes), and the pickier lymphocytes. There arose accessory lines, including plasma cells, which assist in battles by way of chemical secretions. All of these cells, and their products, together constitute the **vertebrate immune system**.

What follows is a discussion of the vertebrate immune system—which is not normally part of a brief survey course in biology. It is included here because there just aren't many places where you can get a simple picture of why our bodies are vulnerable to attack and, too often, to great suffering until the time of death. To be sure, death is programmed into the life cycle for our species—*but the pain and suffering of a body under siege are not.* What you will read here concerns research that may someday help give us the option to go through life, and death, with dignity.

Cells of the vertebrate immune system are responsible for *defense* against foreign agents such as bacteria, viruses, and fungi. They are also responsible for *extracellular housekeeping*, in that they help maintain the tissue environment by eliminating worn-out or damaged body cells and structures. Through events called *immune surveillance*, they also remove mutant and cancerous cells—as well as foreign cells of solid tissues that have been surgically transplanted from one individual to another. Their diverse functions may be summarized in this way:

The vertebrate immune system is concerned with recognizing microbial invaders as well as cells or substances that don't belong in a given tissue—and with selectively eliminating or neutralizing them.

Where do the cells of this system reside? They are strategically dispersed through the body. Many patrol blood and lymphatic vessels. Some stand guard over the respiratory, digestive, and urinary membranes exposed to the outside environment. Whole populations are found in tissues of the **lymphoid system**—bone marrow, thymus, lymph nodes, spleen, appendix, tonsils, adenoids, and patches of the small intestine (Figure 17.9). As you will see, lymphoid tissues and organs function as cell production centers and as sites for some immune responses.

Immune responses are of two sorts: specific and nonspecific. *Nonspecific immune responses are vital in first-time contacts as well as later encounters with foreign matter.* It is not that invaders are recognized as being, say, a unique virus or bits of dirt brought in by a splinter. Rather, there is only a general mobilization against the presence of something in the tissue that is "nonself." Phagocytosis and inflammation are nonspecific responses. Phagocytic cells help clean up local tissues that are damaged, cluttered with debris, or under siege by just about anything. These cells may individually patrol a region, or they may become mobilized and travel in force during **inflammation**: a series of homeostatic events that restore damaged tissues and intercellular conditions. From the time of invasion or injury, the inflammatory response proceeds through these events:

---

1. Under local chemical commands, blood vessels dilate, which increases blood flow to the local tissue region.

2. Local capillaries become "leaky" to phagocytic white blood cells and to about fifteen different circulating proteins called **complement**. Complement has an amplifying function. (Some complement proteins enhance blood vessel dilation and permeability. Others can release phagocyte-attracting chemicals. Others coat invaders, thereby making them "tastier" to the phagocytes.)

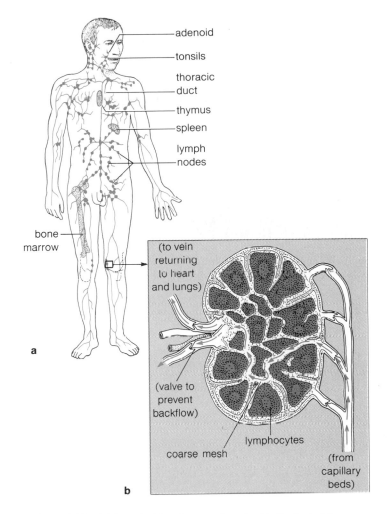

Figure 17.9 (**a**) Lymphoid system, as described in the text. (**b**) Close-up of a lymph node. All lymph passes through at least one lymph node before returning to the bloodstream.

How do we know when an inflammatory response is proceeding? Two outward symptoms are warmth and redness. Another is local swelling. With proteins leaking out, pressure gradients between capillaries and intercellular fluid are altered, and water moves into the tissues, too. The resultant swelling, as well as chemicals released because of the tissue damage, stimulate local receptors that give rise to sensations of pain.

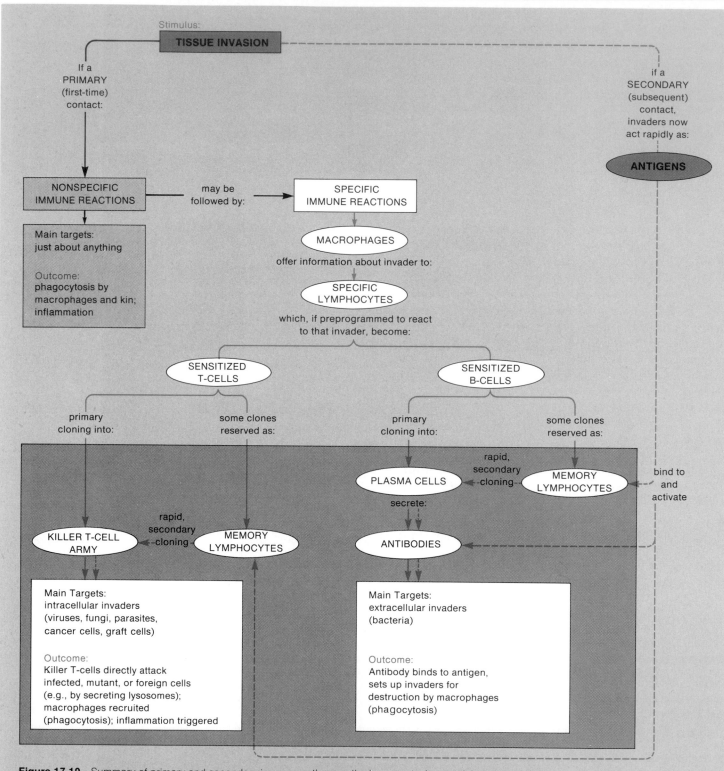

Figure 17.10 Summary of primary and secondary immune pathways: the key events, key participants, and their interactions. Solid brown lines show initial routes; dashed brown lines show secondary routes. All events inside the brown-shaded area may also proceed directly upon subsequent encounters with the same kind of invader; hence secondary immune pathways lead more rapidly to the disposal of foreign matter.

Specific immune responses amplify the nonspecific ones by recognizing and disposing of invaders in a highly discriminatory way. In defending the body against any one of nearly unlimited foreign configurations, they offer a major refinement on the broad distinction between self and nonself. This expansion of the immune arsenal occurs indirectly through chemical warfare agents and directly through lymphocytes sensitized for the kill.

Killer T-Cells and the B-Cell Ways of Chemical Warfare

Two distinct lymphocyte populations carry out the specific immune responses. These lymphocytes are called **T-cells** and **B-cells**. Both kinds arise from stem cells in bone marrow. Some of the stem cell progeny move into the thymus (hence their name, T-cells). In the thymus, they acquire specific factors on their cell surface. These factors will be a complementary "fit" with some surface region of a given type of invader. In other words, T-cells are already precommitted to do highly discriminatory battle. B-cells also become precommitted, but no one knows for sure where B-cells mature and acquire their cell-surface factors.

What triggers T-cells and B-cells into action? Recall that during a first-time invasion, macrophages mount a nonspecific counterattack. Once invaders are ingested by macrophages, they become enclosed within lysosomes (membranous sacs containing hydrolytic enzymes). There, they are degraded into harmless bits. A few bits, though, end up on the macrophage surface. In some way, lymphocytes taste these chemical crumbs and become activated. In addition, foreign matter that invades the bloodstream and lymphatic vessels accumulates in lymph nodes. There, it collects in chambers that house lymphocytes. This contact, too, invites lymphocytes to join the fray.

Only precommitted lymphocytes become sensitized and move into action. They divide and divide again, in a series of clonings that produce a whole army of cells. T-cell clones are mostly killer T-cells; they *directly* attack their targets. B-cell clones *indirectly* attack their targets. First, B-cells differentiate into plasma cells. Then plasma cells secrete protein molecules called **antibodies**, which are distributed by the circulatory system. An antibody can bind to the surface of its target. When it does, it marks the invader for disposal by other forces (Figure 17.10). Any invader that triggers antibody production is an **antigen** (*antibody generator*).

From nonspecific immune responses, to specific T-cell and B-cell clonings, then on through their direct and indirect counterattacks—this is the **primary immune pathway**. It is the route taken during a first-time contact with an antigen. It usually takes about five or six days from the first general battles to successfully mobilize the body's more specific defenses.

What happens if the same kind of antigen turns up again? Then the body typically relies on the **secondary immune pathway**. The secondary pathway only becomes available after an initial battle has run its course. When T-cells and B-cells undergo their initial clonings, some of the clones do not join the battle. Instead they are held in reserve as **memory lymphocytes**. Unlike the rest of the army, memory lymphocytes may last for years—even decades, in some cases. Subsequent contacts with the same antigen can directly trigger memory lymphocytes into large-scale clonings. Thus a plasma cell army may rapidly proliferate, and antibodies may be rapidly churned out in abundance. Similarly, a T-cell army may be raised at once. Because memory lymphocytes are already present and primed for action, antigens are often destroyed before they can take hold and cause disease symptoms. It is not that subsequent antigen attacks are any different, or that response mechanisms of the secondary immune pathways are all that different. *It is that the response mechanisms are amplified.*

Specific immune responses can also be triggered by deliberately introducing a particular antigen into the body. Following such introductions, the body is expected to mount a counterattack and develop an army of memory lymphocytes—which will rapidly destroy like antigens that may later invade the body. Deliberately provoking the production of memory lymphocytes is called **immunization**. The preparation used to provoke their appearance is known as a **vaccine**. Typically, a preparation of some killed or weakened antigen is injected into the body or taken orally. For instance, Salk polio vaccine is prepared from inactivated virus; Sabin polio vaccine is prepared from live but weakened virus. Other vaccines are prepared from the toxic by-products of dangerous antigens. Such preparations offer protection against the bacteria that cause tetanus and other diseases (Chapter Twenty-Two).

Of Mice and Men, and Tissue Transplants

When does the body acquire its capacity to distinguish self from nonself? In some way, its perceptions must become established during embryonic development. Suppose that cells from one mouse species are injected into a mouse embryo of another species while it is developing in the mother's uterus. Now, we know that the adult body perceives as foreign *any* material from a different species, a different individual, even an organ from a different individual, and sets out to destroy it. Yet nonspecies cells injected into a mouse embryo are not perceived as foreign—they become recognized as self! How do we know this? When such mice have been allowed to mature, tissues from the same foreign species have been surgically transplanted into 'them. The transplants have taken hold.

In mice, as in all mammals, a single DNA region influences what the immune response will be to many antigens, how well the body will resist many diseases, and whether the body will reject tissue transplants. This DNA region contains a number of linked genes, called the **major histocompatibility complex** (MHC). The genes control surface recognition factors present on all nucleated cells. Foreign cells are assembled according to different DNA instructions, so they have different surface markers. T-cells recognize the differences and destroy the nonself invaders. But the genetic basis of self is only part of the story, as the mouse transplant experiment makes clear. Another part is when and how the T-cells learn what "self" is supposed to be. Studies suggest it is during embryonic development in the thymus that T-cells learn which surface markers to accept and which to reject. And changes in the thymus microenvironment may cause adjustments in what, exactly, is learned. Unraveling these genetic and molecular puzzles has clinical implications, for the answers may tell us why the body rejects transplants.

The body will normally tolerate tissue and organ transplants between identical twins (who have identical sets of DNA). But not everybody has an identical twin available (and willing) to donate replacements if one's own parts should break down. Given that humans do not inbreed to any great extent, many different allelic combinations are possible in the MHC region that governs tissue rejection. As a result, the odds are about a thousand-to-one that transplants between unrelated individuals will take hold. Simple Mendelian inheritance tells us that the odds are about one-to-four that transplants between close family members will take hold. One goal of current research in this area is to improve the odds. In the meantime, T-cell populations of organ recipients are bombarded with drugs and x-rays, which kills the transplant-fighting T-cells. Unfortunately, other kinds of cells are killed along with them. This stop-gap measure also has the alarming effect of severely compromising the body's ability to mount counterattacks against other kinds of invaders (Commentary).

COMMENTARY

Cancer, Allergy, and Attacks on Self

The vertebrate immune system is an utter marvel of surveillance, a restorer of homeostatic balances that have been upset by all manner of threats. Here is a system that defends the body against myriad sorts of viruses and bacteria, fungi and parasites. Here is a system that sweeps out bits of dirt and old cell structures, the debris of day-to-day living. Here is a system that monitors and disposes of malignant cells that slip from their tissue niche, lodge in improper places, and crowd normal cells to death with their berserk divisions. The amazing thing is how *well* the immune system works, given the diversity and unpredictability of the tissue invasions to which it must respond. Yet the more we learn about the extent of its functioning, the more formidable seems the challenge of setting things right when the system breaks down.

Consider that cancer cells may arise in your body every day, as a result of mutations induced by events such as viral attack, chemical bombardment, or irradiation (Chapter Twelve). As fast as cancer cells do arise, they normally are destroyed before they can multiply enough for anybody to detect them. That, anyway, is the essence of the immunosurveillance theory. The mutant cells are seen to move past circulating lymphocytes, which become sensitized to them. When the activated lymphocytes undergo their rapid cloning in response, they secrete lymphokines—substances that arouse the voracious macrophages. Once macrophages have become activated, they go on the rampage against any and all invaders, including the cancer cells. Sometimes, however, cancer cells slip through the immunosurveillance net. Maybe the mutation doesn't affect cell surface recognition factors; or maybe they are altered so slightly that they escape detection. Perhaps surface markers become chemically disguised. Perhaps they are even released from the cell surface and begin circulating through the bloodstream—and lead the immune fighters down false trails. In addition, a successful immune response depends on how well individuals are genetically

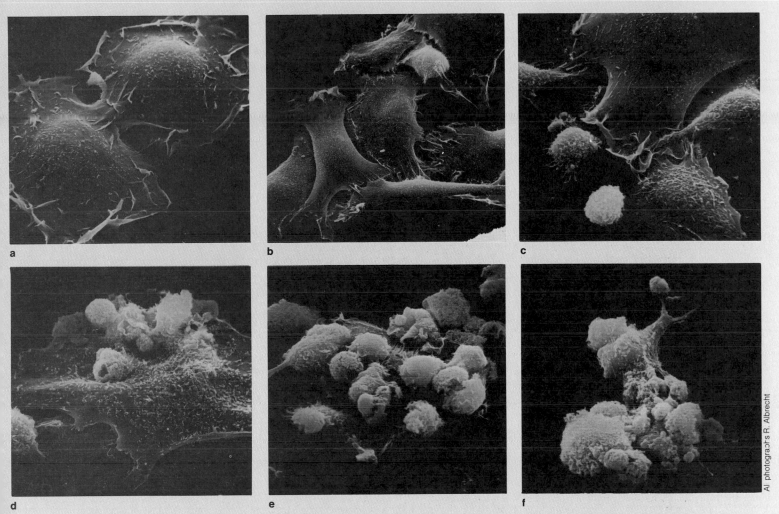

Scanning electron micrographs of the birth and immune-mediated death of cancer cells. (**a**) These cancer cells, dividing in a culture dish, are magnified about 2,000 times. (**b**) After fifty-six hours, their progeny are crowding each other. (**c**) Even with cancer cells all around them, macrophages (smaller spheres) ignore the invaders; they have not yet been turned on. (**d,e**) Activated macrophages cluster around a tumor cell. Recent work by John Hibbs at the University of Utah suggests that macrophages transfer some substance into the invader. (**f**) One or two days after the transfer, the cancer cell dies—and the macrophages ingest the leftovers. (After Krahenbuhl and Remington, 1978)

All photographs R. Albrecht

equipped to respond to specific antigens, how old those individuals are, and their overall health.

When immunosurveillance fails, the only weapons currently available for fighting cancer are surgery, chemotherapy, and radiation treatment. The trouble is, surgery offers little hope if cancer cells have begun their wanderings (Chapter Twelve). And chemotherapy and radiation destroy good cells as well as bad. On the horizon is immunotherapy. The hope is that living microorganisms introduced into cancer patients will set off the general immune alarm. Cancer cells would then be disposed of, along with the deliberately introduced invaders. One obstacle is that, by the time immunotherapy is employed, the cancer may already have outpaced the immune system's ability to respond. Another obstacle is the effect that microorganisms and their metabolic by-products can have on the patients they are indirectly supposed to protect.

Also on the horizon is a naturally occurring agent that is being used to fight infectious diseases such as hepatitis. This agent is **interferon**. Initial trials show that interferon displays some anticancer activity. Research has been somewhat hampered because only minute amounts of the substance, which is being extracted from human white blood cells, are currently available. With gene-splicing methods (Chapter Ten *Commentary*), mass production may soon be possible and research will step up. Cloning experiments should provide information on the number and arrangement of gene coding for given types of interferon. With standard gene-splicing methods, it may be possible to produce interferon molecules with tailored specificity and activity against certain forms of cancer.

Sometimes the immune system can damage the body instead of protecting it. Allergy is an altered secondary response to a normally harmless substance, and may actually cause tissue injury. Some antibody-mediated allergic reactions occur explosively, in minutes; others are delayed. About fifteen percent of the population has a genetic predisposition to become sensitized to dust, pollen, insect venom or secretions, drugs, certain foods, and other seemingly innocuous things. Emotional state, physical conditions (outside temperature and air pressure, for instance), and other kinds of infections are known to trigger or complicate reactions to these antigen-behaving substances. With each recurring exposure to the antigen, antibodies are produced and become attached to mediator cells of the immune system. The mediator cells release histamine, serotonin, prostaglandins, and other chemicals. Histamine causes increased capillary permeability and mucous secretions. Serotonin dilates blood vessels and constricts smooth muscle. Prostaglandins and other factors cause platelet clumping. Together, these chemicals initiate a local inflammatory response. In asthma and hay fever, the resulting symptoms include a drippy nose, sneezing, congestion, and labored breathing. In a few hypersensitive individuals, mediator cells release chemicals in copious amounts, and the inflammatory response spreads. There can be massive constriction of air passages leading to the lungs; circulatory shock may occur when plasma escapes rapidly, from capillaries made too permeable. That is why individuals who are hypersensitive to, say, wasp or bee venom can die within minutes following a secondary immune response to a single sting. Therapy for allergy sufferers includes reliance on "blocking" antibodies. Once the allergy-producing substance has been identified, it can be injected periodically into the body in increasing amounts. The injections provoke antibody production, so that circulating antibodies are available to bind with and inactivate the substance whenever it attacks the body.

One of the most puzzling immune disorders is autoimmune disease, in which the body mobilizes its forces against certain of its own tissues. What causes these self-destructive responses? There are several possibilities. For instance, self-markers on the surface of the body's cells may become altered through mutation or when drugs or pollutants bind to them. Viral infections, too, may alter cellular machinery so that protein components of self-markers are changed. There may also be a genetic predisposition to some autoimmune diseases, an unfortunate promise that may be fulfilled under certain environmental conditions. In juvenile-onset diabetes, individuals produce an antibody that reacts with insulin- and glucagon-secreting cells in the pancreatic islets. Certain histocompatibility alleles are known to occur as much as ten times more frequently in individuals who are afflicted with this disease. At some point, these individuals may contract a viral or bacterial infection. The invader may end up attacking the pancreas along with other tissues—and white blood cells would track them down there. (Recent studies show a correlation between the abrupt onset of this kind of diabetes and the higher incidence of infectious diseases in fall and winter.) White blood cells would become sensitized to the defectively marked islet cells and unleash their weapons against them.

CASE STUDY: THE SILENT, UNSEEN STRUGGLES

The last part of this chapter has been dealing with major topics of widespread interest. Let's now conclude it with a case study of how the immune system helps *you* survive attack. Suppose it's a warm spring day and you are walking barefoot across the lawn to class. Abruptly you stop: you stepped on a staple that had been lying prong-upward. The next morning the punctured area is red, tender, and slightly swollen, even though you had pulled out the staple at once. Later on, the swelling begins to subside. Within a few more days your foot is back to normal, and you have completely forgotten the incident.

All that time, though, your body had been struggling against an unseen foe. You had inadvertently stepped on the place where a diseased bird had fallen the night before, and had lain until a scavenging animal carried it off. Your foot picked up some bacteria that had parted company from their feathered host. As long as the bacteria remained outside your body, they were harmless. But the staple prongs had carried several thousand bacterial cells into the moist, warm flesh beneath your skin. Here they found conditions suitable for growth. And grow they did—they soon doubled in number and were on their way to doubling again. Left alone, they would have threatened your life.

There was nothing vicious in their attack. The bacteria were simply responding to available resources, as all forms of life do. But they were also releasing metabolic by-products that interfered with your own cell functioning. Eventually those substances would begin interfering with your tissue and organ activity.

Yet even with the staple's penetration into your foot, events were set in motion that would mobilize your body's defense systems. Around the wound, some blood had escaped from ruptured capillaries, and it now pooled and clotted. Complement molecules in the blood became activated, and some of the fragments they formed released various chemicals from nearby cells. These chemicals began diffusing through the surrounding tissue. Some of the chemicals acted on precapillary sphincters, causing them to dilate and allow more blood to reach the injured site. And some of the fragments acted as cues for phagocytic white blood cells.

Usually most of your phagocytes remain in the bloodstream. But once complement stimulated them, their behavior changed. They attached themselves to capillary walls and crawled along, amoebalike, until they reached an epithelial junction. Then they slipped out into the tissue. There they crept about between cells and fibers, following the concentration gradient of complement fragments. Like bloodhounds on the trail, they moved in the direction of higher concentration.

Specifically avoiding your own cells, these phagocytes now began engulfing any particle not having a surface marker that meant, "Leave me alone—I'm *self!*" Dirt, rust, bits of broken host cells, bacteria—all were engulfed indiscriminately.

If bacteria had not entered the wound, or if they had been of a type that couldn't multiply rapidly in the tissue, then phagocytosis and inflammation would have cleaned things up. As it was, the bacterial multiplications were outpacing these nonspecific immune responses. It was time for the body's more discriminatory defenders: the lymphocyte lines and their products.

If this had been the first time your body encountered this bacterial species, few lymphocytes would have been around to respond to it. Your body would have had to make a primary immune response—it would have to wait for days while its B-lymphocytes divided enough times to produce enough antibody to control the invasion. But during your childhood, your body did fight off the same type of bacterium, and it still carries vestiges of the struggle: memory lymphocytes. When the bacteria showed up again, they encountered a lymphocyte trap ready to spring: they activated a secondary immune response.

As the inflammation progressed, lymphocytes were among the white blood cells creeping out of the bloodstream. Most were specific for other types of antigens and did not take part in the battle. But each specific memory lymphocyte entering the area encountered its target, bound a few antigens to its surface, and became activated. It moved into lymph vessels with its foreign cargo, tumbling along until it reached a lymph node and was filtered from the fluid. For the next few days, memory cells with bound antigens steadily accumulated in the node. They divided several times a day, so that their numbers and products increased rapidly.

For the first two days the bacteria appeared to be winning, for they were reproducing faster than phagocytic cells, antibody, and complement were destroying them. But by the third day, antibody production reached its peak and the tide of battle turned. For two weeks or more, antibody production will continue in your body until every last bacterium is destroyed. When it finally shuts down, the newly formed memory lymphocytes will go on circulating, prepared for some future struggle.

Readings

Bellanti, J. 1978. *Immunology II.* Second edition. Philadelphia: W. B. Saunders. Outstanding, comprehensive coverage of basic principles, mechanisms of the immune response, and clinical applications.

Guyton, A. 1979. *Physiology of the Human Body.* Fifth edition. Philadelphia: W. B. Saunders. Contains a good discussion of homeostatic mechanisms for preventing blood loss from the body (Chapter 7).

Hickman, C., et al. 1979. *Integrated Principles of Zoology.* Sixth edition. St. Louis: Mosby. Nice look at invertebrate as well as vertebrate defense mechanisms.

Hood, L., I. Weissman, and W. Wood. 1978. *Immunology.* Menlo Park, California: Benjamin/Cummings. Advanced overview of the molecular and cellular bases of the immune response. Paperback.

Krahenbuhl, J., and J. Remington. 1978. "Belligerent Blood Cells: Immunotherapy and Cancer." *Human Nature* 1(1):52–59. Very well written article on immunotherapy as one treatment for cancer.

Notkins, A. 1979. "The Causes of Diabetes." *Scientific American* 241(5):62–73. Excellent summary of current understandings of diabetes and its complications.

Schmidt-Nielsen, K. 1978. *Animal Physiology.* Second edition. New York: Cambridge University Press. Outstanding discussions of invertebrate and vertebrate temperature regulation and osmoregulation.

Vander, A., J. Sherman, and D. Luciano. 1980. *Human Physiology.* Third edition. New York: McGraw-Hill. One of the clearest expositions of the body's defense systems.

Review Questions

1. Define ectotherm and endotherm. In endotherms, what controls help balance the amount of heat lost and heat gained?

2. In your own body, where are thermoreceptors and the main center of temperature control located?

3. All animals have mechanisms for maintaining body fluid concentration and composition. In your own body, which organs cooperate in these tasks? How does a nephron in your kidney resemble an earthworm nephridium?

4. Describe what happens during (a) filtration, (b) reabsorption, and (c) secretion in the kidney's nephron/capillary unit. What do these three processes influence?

5. Would you expect desert animals to have long or short loops of Henle?

6. Most exchanges between blood capillaries and nephrons are obligatory, but the final urine concentration is under complete homeostatic control. What two interrelated mechanisms act as the controls?

7. Define hemorrhage. Describe the events involved in repairing a damaged small blood vessel.

8. Sponges aside, all animals have defenses against attack on normal cell functioning. Define the cells and products that constitute the vertebrate immune system of defense against such attacks.

9. Define inflammation. What happens during an inflammatory response?

10. Define antibody, antigen, and memory lymphocyte. What is an immunization? A vaccine?

11. How do the main targets and outcomes of nonspecific and specific immune responses differ?

12. What is the immunosurveillance theory? An allergy? An autoimmune disease?

With a full-throated croak that only a female of its kind could find seductive, a male frog proclaims the onset of warm spring rains, of ponds, of sex in the night. By August the summer sun will have parched the earth, and his pond dominion will be nearly gone. But tonight is the hour of the frog! Through the dark, a female moves toward the irresistibly vocal male. They meet, they dally; he clamps his forelegs about her swollen abdomen and gives it a squeeze. Out streams a ribbon of thousands of large eggs. At the same time the eggs are released, the male's body expels a milky cloud of swimming sperm. Each egg accepts and joins with only a single sperm. Not long afterward, the sperm nucleus and egg nucleus fuse. With this event a zygote is formed, and **fertilization**—the fusion of the genetic endowment of two gametes—is completed.

Within minutes after fertilization, a drama begins to unfold that has been reenacted each spring, with only minor variations, for countless millions of years. The single-celled frog zygote begins to divide rapidly into two, then four, then eight cells, and many more. These newly formed cells remain attached as a multicellular embryo. The cells themselves do not increase in size during this cleavage. Instead, each becomes smaller and smaller with each successive division. At last there is a hollow ball of tiny cells. And now division begins to slacken. Parts undergo rearrangements that will give rise to a specific embryonic shape. A dimple forms on the ball's surface, where some cells begin sinking into the interior. From these inwardly migrating cells, internal tissue layers will arise.

Soon a groove forms along one side of the embryonic surface. The embryo begins to lose its spherical shape; it grows longer in the direction of the groove. The groove deepens, then the edges fold over and seal to form a hollow tube. From this tube a nervous system—brain, spinal cord, and nerves—will ultimately form. It is the first visible sign of organ formation. Now an eye begins to form on each side of the developing head. Within the trunk, a heart is forming and will soon be beating rhythmically. Fins take shape, a mouth forms and opens to the surface. These developments, appearing as they do at different points in time, speak of a process going on in *all* the cells that were so recently developed from a single zygote: they are all becoming different from one another in appearance and in function!

Soon the embryo becomes transformed into a tadpole (a kind of larva) that can swim on its own. For the first time since fertilization it feeds itself—and it continues to grow. For many months it grows until, in response to environmental cues, the larva embarks on a new course of development. Its body now changes into a form that is

18

REPRODUCTION AND EMBRYONIC DEVELOPMENT

Nicholas Smythe / National Audubon Society / PR

Figure 18.1 Few dramas in life are more compelling than the development of single cells into complex multicellular animals.

in the image of its parents. Legs grow from its body. Its tail becomes shorter and shorter, then disappears. Its mouth, once suitable for feeding on algae, develops jaws with teeth suitable for devouring insects and worms. Its gills, which served so well in trapping oxygen from the pond environment, decrease in size and ultimately disappear; lungs have developed to replace them. Eventually a full-fledged frog leaves the water for life on land. If it is lucky it will avoid hungry predators, bacterial attacks, and other assorted threats through the many months ahead. And come another spring it may find a pond, swollen with the warm waters of the new season's rains, and the hour of the frog will be upon us again.

Watching the promise of the zygote unfold into the reality of the adult, it is difficult not to view development as one of life's greatest mysteries. *How does a simple-looking zygote become transformed into all the specialized cells and structures making up the adult body of a complex animal?* Chapter Twelve surveyed the kinds of explorations that may help us discover the answers to this question. Here we will be turning to the actual transformations in form that occur from the time of reproduction to the emergence of the new individual.

THE BEGINNING: REPRODUCTIVE STRATEGIES

In Chapter Eight, you read that both sexual and asexual reproduction occur among different animal groups. Most commonly, though, animal reproduction is sexual. Hence it typically begins with gametogenesis—the formation of gametes, or sex cells, within the body of parent organisms (Figures 8.18 and 8.19). The very few exceptions to this kind of reproductive beginning include some flatworms. Like single-celled organisms, some flatworms rely on a form of binary fission: their entire body divides transversely, and each half grows and regenerates its missing half. **Regeneration** refers to the capacity of a fragmented or otherwise incomplete parent body to give rise to a complete organism, or even parts of it. Sometimes predators tear away arms or legs from sea stars, crabs, and lizards, and all of these animals can grow replacements. A detached sea star arm can serve as parent to a whole new sea star. Like sponge fragments, bits of sea anemone tissue torn away as the animals tumble across the seafloor can grow into new sea anemones. *Hydra* and some other cnidarians typically rely on **budding**: certain cells begin differentiating

and growing outward from the parent body, then the bud breaks away and develops into a new individual.

As intriguing as these asexual strategies may be, neither budding nor regeneration promotes genetic variability. The offspring are clones (genetically identical copies of the parents). This is advantageous only so long as the parents are highly adapted to the surroundings, and only so long as the surroundings remain stable. But most animals live in changing and unpredictable environments. Not surprisingly, most of them—and this includes cnidarians, sea stars, crabs, and lizards—depend on sex as the basic means of reproduction.

Sex need not be as straightforward as the male-frog-meets-female-frog example opening this chapter. Between clear-cut reliance on asexual or sexual reproduction is **parthenogenesis**: the cleavage and subsequent differentiation of an *unfertilized* egg into an adult. Every so often, between times of sexual reproduction, beetles and aphids may produce fatherless offspring. For these arthropods, a sperm is not the only stimulus that can prod an egg down the developmental pathway. Changes in pH, temperature, salinity, even mechanical stimulation of the egg can trigger parthenogenesis. They can do so not only in insects but in frogs, salamanders and, in some experiments, rabbits. (Weak, sterile rabbits are produced, but they are rabbits nevertheless.)

Or consider the earthworm and the parasitic tapeworm. Each of these animals comes equipped with *both* male and female reproductive organs. Each produces both sperm and eggs. They are **hermaphrodites**. Two earthworms lie head to tail and exchange sperm, thereby cross-fertilizing each other. Parasitic tapeworms can fertilize themselves. One need not look askance at the tapeworm, however, when one realizes that these animals may end up living all by themselves somewhere in the body tissues of a host animal. Given the bountiful supply of resources, hermaphrodism does have its advantages for the lone parasite.

Complete separation into male and female sexes imposes its own demands. After all, there must be some means of assuring that eggs of one individual will get together with sperm from another. First and foremost, the male and female reproductive cycles must be *synchronous*, so that mature gametes are released at the same time. Neurohormones and hormones are at work here, and are themselves released through the animal's perception of such environmental stimuli as lengthening days or some other planetary rhythm. When the appropriate moment arrives, some water-dwelling animals and amphibians release their gametes directly into the surrounding water, or deposit them on some substrate below. Frogs do this; so do sea urchins and salmon. Such *external* fertilization is somewhat chancy. Usually, large numbers of egg and sperm are released, which enhances the probability of the sperm bumping into a receptive egg. In contrast, nearly all land-dwelling animals and even some aquatic forms depend on *internal* fertilization. Obviously, sperm released on dry land do not stand much of a chance of swimming over to an egg. Specializations in the ducts of certain male organs, the **testes**, assure that sperm can be stored and nourished in a hospitable setting, and that some structure will be available to transfer them to the female body. Thus many animals have a **penis**, a copulatory organ at the end of the sperm duct. From this male organ, sperm is ejected into the female **vagina**, a specialized duct leading, along one structural route or another, from the **ovary** (an organ where female gametes mature). Reptiles, birds, mammals —all have reproductive structures of a sort that enhances the probability of succesful internal fertilization.

Animal reproductive strategies range from asexual processes, through development of unfertilized eggs into adults, through self-fertilization, to sexual reproduction between separate male and female forms.

In this chapter, the focus is on the most common strategy—that of sexual reproduction.

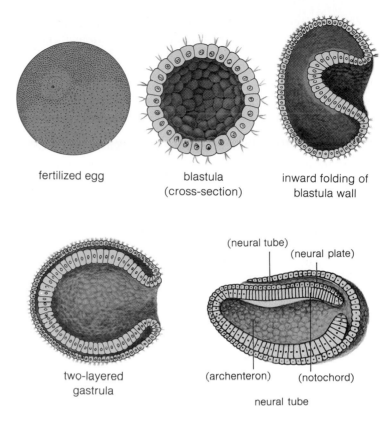

fertilized egg   blastula (cross-section)   inward folding of blastula wall

two-layered gastrula   (neural tube)   (neural plate)   (archenteron)   (notochord)

neural tube

Figure 18.2 Blastulation and then gastrulation in a small, streamlined marine invertebrate, *Amphioxus*, which is thought to resemble the ancestral forms from which the chordates arose. Following gastrulation, neural developments proceed. (After C. P. Hickman, Jr. et al., *Integrated Principles of Zoology*, 1979)

THE EARLY EMBRYO

Let's now take a closer look at the developmental stages following fertilization: cleavage, gastrulation, organ formation, and growth. Recall that tremendous amounts of RNA are transcribed in the egg even before fertilization occurs. Some RNA molecules are translated at once into proteins that will take part in the first cell divisions. Others become attached to ribosomes in distinct cytoplasmic regions. These regionally positioned maternal messages, activated at fertilization, will direct the first cleavages.

Following fertilization, animal development proceeds through stages of cleavage, gastrulation, organ formation, and growth. Initially, development is under the direction of regionally positioned messages in the egg cytoplasm.

Cleavage: Carving Up Cytoplasmic Controls

For most animals, the repeated mitotic divisions of **cleavage** segregate the egg cytoplasm into a cluster of cells. The entire cluster is known as a **blastula**. In frogs and mammals, the cells are rearranged around a fluid-filled cavity (blastocoel). There is no increase in cell size during cleavage. Rather, the one cell divides into smaller, more numerous cells that take up about the same space.

Animal ova contain varied amounts of **yolk**, or food reserves. The amount present influences cleavage patterns. In the eggs of sea stars, mammals, and a marine invertebrate called *Amphioxus*, so little yolk is present that cleavage planes pass completely through the egg, and all the newly divided cells are about the same size (Figure 18.2). Bird and reptile eggs contain so much yolk that cleavage is restricted to a tiny disklike region of cytoplasm near the egg surface (Figure 18.3). Fish and amphibian eggs are about one-half to three-fourths yolk. Here, the yolk is concentrated at one end, called the vegetal pole. The nucleus and most of the cytoplasm reside at the other end, called the animal pole. The animal pole is the metabolically active region; cleavage is more rapid here (Figure 18.4).

Each new cell contains its own DNA. Some of its DNA will be read selectively, according to the set of maternal messages that the cell inherited. The fate of some cytoplasmic regions is known to be established by this time. Consider what happens when localized cytoplasmic regions are destroyed (by surgical removal, by searing with a tiny heated needle, by irradiation with a beam of ultraviolet light). In some animals, the individual that develops will lack one or more structures or cell types. For instance, if a certain region of a squid zygote is irradiated, the left eye of the embryo won't form.

Gastrulation: Layered Hints of Things to Come

During cleavage and blastula formation, cells divide rapidly. But there is little rearrangement of cells to speak of. Now, however, the cell division pace suddenly slackens. In one region of the blastula, cells grow longer and in some cases put out long cytoplasmic threads. Either one by one or while attached to each other, these cells begin a mass migration to the interior. There they create one or more layers of cells. The process of cell regrouping and association into layers is called **gastrulation**. It is a developmental stage in almost all animal groups, and it marks the onset of differentiation of internal and external parts. Figure 18.2 shows how two layers form in *Amphioxus*; Figure 18.4 shows a more complex three-layer formation, typical of amphibians.

The cells remaining on the outside of the embryo are now known as the **ectoderm** ("outer layer"); they will give rise to the nervous system and the animal's skin. The cells forming the innermost layer make up the **endoderm** ("inner layer"). Basically, the endoderm is a hollow tube in the center of the embryo; it will give rise to the gut and to such structures as lungs and glands. Usually, between these two layers will be cells of the **mesoderm** ("middle layer"). The bones, muscles, and other internal organs of complex animals are derived largely from mesoderm cells. It's important to keep in mind, however, that these tissue layers do not function in isolation from one another. Most body parts of the adult animal develop through cooperative interactions among cells of different layers.

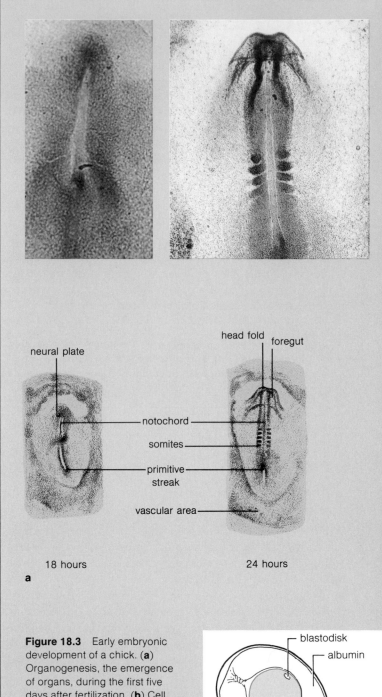

neural plate

head fold
foregut

notochord

somites

primitive streak

vascular area

18 hours

24 hours

a

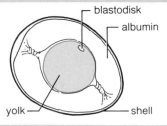

blastodisk

albumin

yolk

shell

b

Figure 18.3 Early embryonic development of a chick. (**a**) Organogenesis, the emergence of organs, during the first five days after fertilization. (**b**) Cell division patterns during blastulation and gastrulation. (a, after Patten, 1971; b, after Hickman et al., *Integrated Principles of Zoology*, 1979)

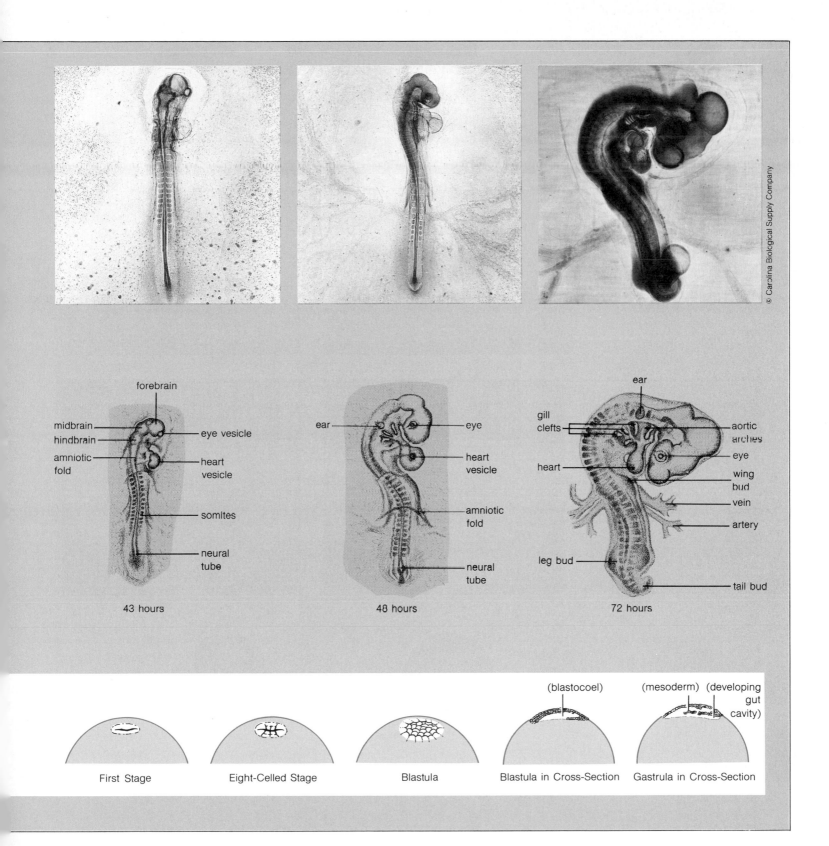

© Carolina Biological Supply Company

forebrain

midbrain
hindbrain
amniotic fold

eye vesicle

heart vesicle

somites

neural tube

43 hours

ear

eye

heart vesicle

amniotic fold

neural tube

48 hours

ear

gill clefts

aortic arches

heart

eye

wing bud

vein

artery

leg bud

tail bud

72 hours

First Stage

Eight-Celled Stage

Blastula

(blastocoel)

Blastula in Cross-Section

(mesoderm) (developing gut cavity)

Gastrula in Cross-Section

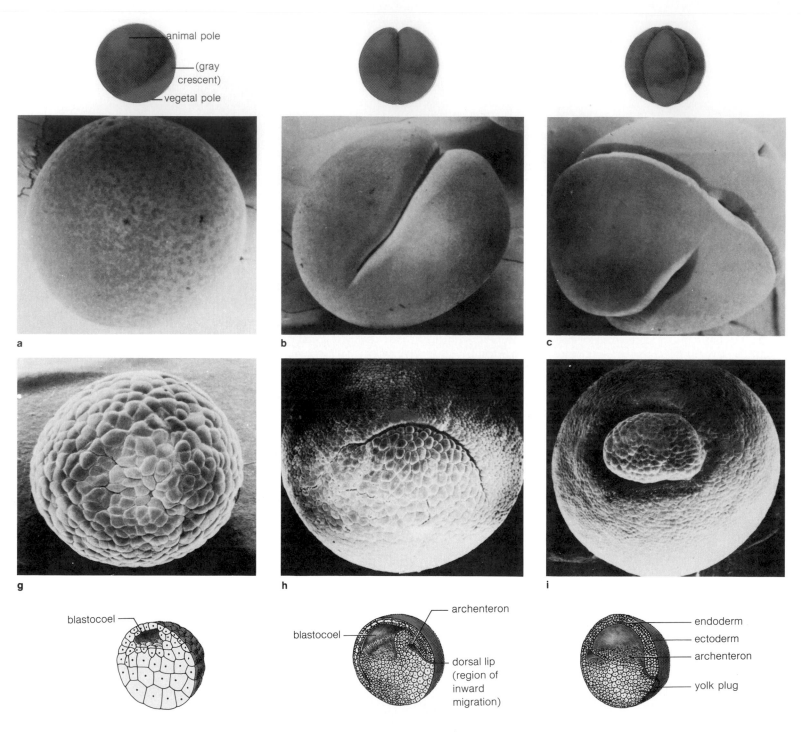

Figure 18.4 Early embryonic development of a frog. For these scanning electron micrographs, the jellylike layer surrounding the egg (see Figure 12.4) has been removed. (**a**) Within about an hour after fertilization, a region of differentiated surface cytoplasm (gray crescent) appears opposite the site where the sperm penetrated the egg. (**b–g**) Cleavage leads to a blastula, a ball of cells in which a cavity (blastocoel) has appeared. (**h,i**) Major cell

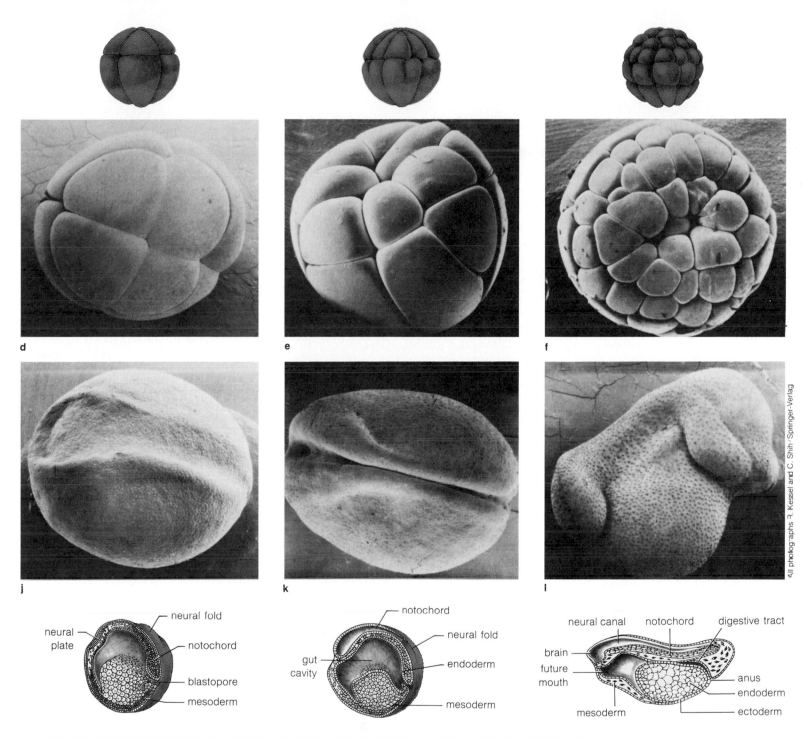

movements and rearrangements occur during gastrulation. Tissue layers form; a primitive gut cavity (archenteron) develops. (**j,k**) Neural developments now take place, and a coelom, the fluid-filled body cavity in which vital organs will be suspended, appears. (**l**) Differentiation proceeds, moving the embryo on its way to becoming a functional larval form.

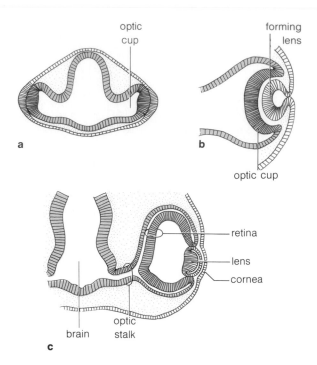

optic cup

forming lens

optic cup

retina

lens

cornea

brain

optic stalk

a

b

c

Figure 18.5 Cross-sectional view of the way the retina of a frog's eye develops as an outgrowth of the brain, and the lens develops as an ingrowth of the skin. (After John W. Saunders, Jr., *Patterns and Principles of Animal Development.* Copyright © 1970 by John W. Saunders, Jr. Macmillan Publishing Co., Inc.)

Embryonic Induction: Fitting Parts Together

In the adult animal, many body parts work only if they fit precisely with adjacent parts. For instance, the retina of an eye must be positioned in a precise way with respect to the lens, which focuses light onto the retina. The remarkable thing about this relationship is that the retina forms in a structure called the optic cup, which develops as an outgrowth of the brain—but the lens develops as an ingrowth of the skin (Figure 18.5).

What coordinates the development of such separate but interdependent parts? The answer came from experiments by the embryologist Hans Spemann at the beginning of this century. Spemann used a salamander embryo in which eye lenses were still unformed, but in which optic cups had started to grow out from the brain. He surgically removed one of the optic cups. Then he placed it just under the skin in the belly region. A lens never did develop where it was supposed to, on the side of the head now deprived of an optic cup. But belly skin cells that had come in contact

with the transplanted optic cup differentiated to form a lens—which fit perfectly into the transplanted part! Spemann concluded that a lens does not develop independently of a retina. Rather, it is caused (or induced) to form wherever the optic cup makes contact with skin. In such **embryonic inductions**, one body part differentiates because of signals it receives from an adjacent body part. These signals are necessary for coordinated development. They are thought to be chemical, but their precise nature has yet to be determined.

To what extent do embryonic inductions dictate the course of development? Half a century ago Hilde Mangold, one of Spemann's students, provided an answer. For her experiments, she used a dark-pigmented species and a light-pigmented species of salamander embryos. From the light-pigmented embryo, she cut out a region known as the **dorsal lip**, a site of inward cell migration during gastrulation (Figure 18.4). She inserted this piece of tissue into the body wall of the dark-pigmented embryo, across from its own dorsal lip. Gastrulation proceeded at both sites—the result being Siamese twin salamander embryos, each with its own set of organs. More than this, almost all the organs of the twin embryo were dark-pigmented. This meant that the extra set of organs had developed from the host, *not* from the transplanted tissue. The transplanted dorsal lip had induced some of the host cells into developing in ways that they normally would never do. Thus, although the gray crescent-dorsal lip region assures some prepatterning, induction must be a key process underlying most of the subsequent organization of individual salamander embryos.

In other animals, such as the squid, far more prepatterns are laid out as localized substances in the egg cytoplasm. For them, embryonic induction is not as important in determining the ultimate body parts. However, the following principle applies to animal development:

In all animals, the coordinated development of body parts in specific regions depends on some combination of (1) the influence of cytoplasmic substances within the unfertilized egg, and (2) inductive interactions between cells at later stages of development.

Embryonic Regulation: Compensating for Slip-Ups

If an embryo is to develop into a normal, integrated adult, its cells must function on schedule, in their proper place.

Sometimes cells become damaged, or they divide too slowly or too fast. Fortunately, embryonic cell behavior may be modified in ways that compensate for missing or extra parts. This capacity is known as **embryonic regulation**.

The control mechanisms underlying embryonic regulation have not yet been discovered. But there is impressive evidence of their existence. For instance, half a century ago, the biologist Ross Harrison identified the region on a salamander embryo where a foreleg was destined to form. He cut out the entire block of cells from this region, and transplanted them to the head. Would a leg form on the head, or where it was supposed to form? Or would some parts form in both places? Surprisingly, a complete, normal leg developed at *both* sites. Cells in the leg region were not committed merely to forming parts of a leg but to forming an entire leg. And they did so despite Harrison's intervention. But apparently *more* cells had become committed than are normally needed in leg formation. Just as apparently, each cell's behavior was somehow modified by its neighbors in such a way that missing parts would be replaced.

To cross-check his findings, Harrison grafted the leg-forming block of cells of one embryo right next to the leg-forming region of another. In some cases, he even grafted two, three, and four of these cell blocks onto the same place. In each case, only *one* leg of normal size and shape developed. Just as the cells modified their behavior and compensated for missing parts, so did they compensate for extra parts.

Dramatic evidence of the capacity for embryonic regulation comes from the splitting of a zygote, either naturally or experimentally, into two equal parts. The result is not two half-embryos but **identical twins**: two complete, normal individuals. (In contrast, nonidentical twins occur when two different eggs are fertilized by two different sperm.)

Differentiation: Emergence of Diverse Cell Types

An adult animal contains dozens or hundreds of differentiated cell types. Few generalizations apply to all of the pathways by which they develop. Some cells, such as those of the vertebrate nervous system, acquire their final form and function before birth. If they are later destroyed, no other cells can replace them or assume their function. Other cells, such as those in skin, blood, and gut, undergo constant renewal. A population of **stem cells** (which are capable of self-replication, differentiation, or both) divide

mitotically and give rise to replacements throughout the life cycle. For instance, about every two days, stem cells replace all cells lining the harsh environment of a mammal's intestine. Still other cells undergo differentiation early in the life cycle but do not mature until much later. Cells that will form eggs are one of the first cell types set aside in a human female embryo. The first completes differentiation a decade or so after birth—and the last may not do so until four decades later!

Cell types undergoing differentiation may be identified by their protein products. For instance, in the final stages of differentiation of your red blood cells, ninety-five percent of the proteins being synthesized is hemoglobin. Although all of your cells contain a gene coding for hemoglobin, the gene is expressed only in red blood cells. The important point is this:

Differentiation requires <u>selective</u> gene expression in certain cells at certain times.

Gene control was discussed in Chapters Eleven and Twelve, and won't be repeated here. It's enough to say that the cytoplasm contains primary control agents. The localized cytoplasmic substances that are parceled out to newly forming cells during cleavage are able to modify gene behavior. When certain genes become active, their products modify the cytoplasm. These modifications presumably cause further changes in the patterns of *which* genes in the nucleus are expressed. Inducer substances from one cell somehow stimulate another cell—hormones do this—and the pattern changes again. Such interactions between the nucleus and cytoplasm are so complex that no one has yet detailed all the steps in the differentiation of even a single cell type!

Morphogenesis: Formation of Functional Body Units

It's one thing to ask how a bone cell becomes different from a muscle cell or red blood cell. It's another thing to ask how one bone becomes different from all other bones. Consider that all the bones in your hand and arm are practically identical in their molecular composition—yet they differ in size, shape, and arrangement. How do such differences arise in different clusters of cells, all of which are committed to making bone?

Morphogenesis is architectural development—the growth, shaping, and spatial coordination of tissues and

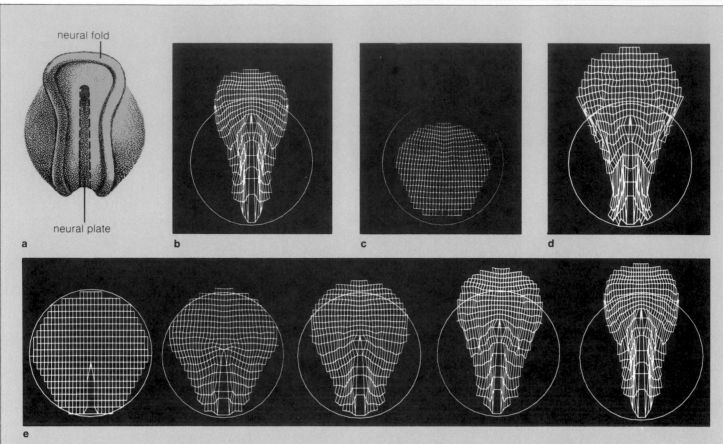

neural fold

neural plate

a

b

c

d

e

Figure 18.6 Computer simulations of forces at work shaping the development of a neural plate in the California newt embryo (**a**). Tissue-shaping is depicted as a distortion of a geometric grid placed over the embryo. (**b**) Computation of two forces—nonuniform shrinkage, and elongation at the neural plate midline—produces a grid that corresponds almost exactly with the actual shape of the neural plate, except for foreshortening. (**c**) Computation of shrinkage alone produces a grid that doesn't have the keyhole shape characteristic of neural plates. (**d**) Computation of elongation alone produces an overlarge keyhole shape. Thus both shrinkage and cell elongation help shape the neural plate. (**e**) A sequence of frames in the computer simulation of the formation of a normal newt embryo plate. The last frame corresponds to the above sketch of the embryo. (a from R. Gordon and A. Jacobson, "The Shaping of Tissues in Embryos," *Scientific American*, June 1978. Copyright © 1978 by Scientific American, Inc. All rights reserved. b–e courtesy Richard Gordon)

organs according to predefined patterns. Once again controls over gene expression are at work. These controls are known to affect the extent, direction, and rate of growth. Some of the variations involve microfilaments and microtubules, which are assembled and disassembled in coordinated programs involving sheets of cells. For example, in the newt embryo, neural plate formation coincides with differential shrinkage in the exposed cell surface area (Figure 18.6). This shrinkage occurs when neural plate cells lengthen. Circling one end of each neural plate cell are contractile microfilaments. When they contract together, like a drawstring, they cause the cell to lengthen.

In addition, interactions with adjacent cells and tissues—chemical communication links, physical attachments, the physical restraints of growing right against or apart from neighboring cells—contribute to shifting directions of growth. In the newt embryo, abrupt boundaries form between the neural plate and the epidermis. These two cellular regions tend to displace each other during their development. The displacement may alter or break chemi-

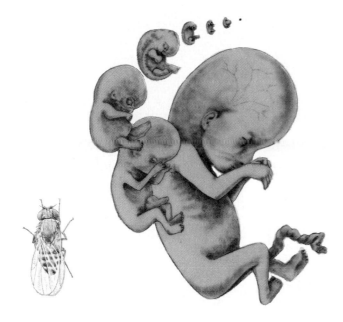

Event	Time for Fruit Flies	Time for Humans
Fertilization of egg	zero	zero
First cleavage division	23 minutes	30 hours
Eighth cleavage division	90 minutes	3 days
Early formation of nervous system	4 hours	3½ weeks
Hatching/birth	24 hours	9 months

Figure 18.7 Timetable of the development of a fruit fly and a human embryo. By the time a fruit fly larva has emerged and is crawling about in search of food, the human zygote has not even divided for the first time into two cells. By the time the infant is born, a fruit fly conceived at the same instant would have been dead many months—but in theory it could have left behind many *billions* of great, great grandchildren!

cal links that previously existed between adjacent cells at this boundary. Perhaps the chemical separation contributes to the subsequent movement of these two domains down different developmental pathways.

Morphogenesis depends not only on gene controls. It depends also on chemical communication links, physical attachments, and physical restraints imposed by neighboring cells in the developing embryo.

DEVELOPMENTAL STRATEGIES

Among many animal groups, it is common for sexual reproduction and embryonic development to follow the cycling of seasons, or other planetary rhythms. In this way, young are born when environmental conditions (such as food availability) converge in the most favorable way for the species. To a large extent, differences among animal groups relate to differences in how developing embryos are nourished. For some animal species, eggs that develop outside the mother's body contain little yolk, but they develop rapidly into free-living larvae. Often a **larva** is an immature form that lives on environmental food sources until it grows and develops into juvenile, then adult forms. Many larvae emerge when certain kinds of food (plankton, new leaves) become abundant. Then they undergo **metamorphosis** (change in form) when those foods dwindle and other kinds become available. Algal-feeding tadpoles quickly metamorphose into insect-eating frogs. Fruit-eating *Drosophila* maggots quickly metamorphose into adult flies (Figure 18.7). Figure 1.4 showed how leaf-eating caterpillars undergo metamorphosis into nectar-feeding moths. A program that proceeds through an adaptive larval detour before metamorphosis into juvenile, then adult form is called **indirect development.**

In contrast, reptile, bird, and mammal eggs undergo **direct development**. There is no larval stage. Either these eggs contain abundant yolk (as bird eggs do), or nearly yolkless eggs attach to and are nourished by the mother's body. For instance, human embryos are nourished within the mother's uterus by the placenta, a vascular structure that is a composite of both maternal and embryonic tissue. Individuals undergoing direct development grow at a somewhat leisurely pace (Figure 18.7) inside or outside the mother's body. In most cases, the individual hatches or is born directly in some juvenile form, which matures into the adult form. Figure 18.8 gives other examples of animals that rely on direct development. The point to keep in mind is this:

Indirect development proceeds through an immature larval stage.

In direct development, the embryo either is nourished (internally or externally) by the mother's body, or is nourished externally by maternally derived food reserves (such as yolk).

The remainder of this chapter will focus on the direct development of human embryos, from the reproductive events that set the stage for successful fertilization to the moment of birth.

Figure 18.8 Some animals that follow the direct developmental pathway (**a**) Alligators show ovoviviparity: fertilization is internal; the fertilized eggs develop inside the mother's body, without additional nourishment; then the young are born live.

(**b**) Birds show oviparity: eggs with large yolk reserves are released from and develop outside the mother's body. The hatchlings depend on parental feeding and protection. (**c**) The duckbilled platypus, a monotreme, is unique among mammals. The females lay eggs, hence are oviparous, yet they secrete milk (as other mammals do) for nourishing the juvenile form. (**d**) Snails are oviparous but all are not doting parents; the eggs are left unprotected and hatch as miniature adults.

The kangaroo (**e**) and opossum (**f,g**), both marsupials, are viviparous: the embryo develops within and is nourished by the mother's body. However, their young emerge in somewhat unfinished form, and undergo further fetal development in a pouch on the ventral surface of the mother's body, where they are nourished from mammary nipples. (**h,i**) Dolphins (cetaceans) are also viviparous, but the young dolphin emerges fully formed, an adult in miniature. Parental care and protection begins at once and continues until the new individual is capable of surviving on its own.

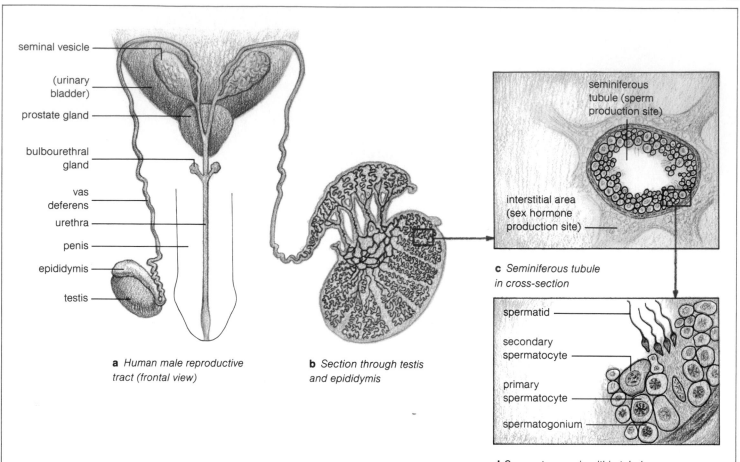

a *Human male reproductive tract (frontal view)*

b *Section through testis and epididymis*

c *Seminiferous tubule in cross-section*

d *Spermatogenesis within tubule*

Labels in figure a (top to bottom): seminal vesicle, (urinary bladder), prostate gland, bulbourethral gland, vas deferens, urethra, penis, epididymis, testis

Labels in figure c: seminiferous tubule (sperm production site), interstitial area (sex hormone production site)

Labels in figure d: spermatid, secondary spermatocyte, primary spermatocyte, spermatogonium

Figure 18.9 (**a**) Human male reproductive tract, and (**b–d**) spermatogenesis. The main male reproductive functions are sperm production (spermatogenesis) and sperm deposition in the female reproductive tract. These functions are carried out by the *testes, epididymis,* ducts such as the *vas deferens,* and *penis,* together with accessory glands.

Each testis contains *seminiferous tubules.* These tubules are lined on the inside with undifferentiated cells (spermatogonia) that give rise to sperm. Surrounding the tubules are *interstitial cells* that secrete the male sex hormone, testosterone.

From about age thirteen onward, spermatogonia are continually replenished by mitotic divisions. Some differentiate through stages that lead to sperm (compare Figure 8.17). Sperm leaving the seminiferous tubule pass into the epididymis. In this coiled duct, sperm mature and develop their motility. Prior to ejaculation, most sperm are stored in the vas deferens. They are expelled from this duct by strong contractions of its smooth-muscled walls.

Before leaving the body, sperm are diluted with *semen.* This fluid is composed largely of secretions from the *seminal vesicles, prostate gland,* and *bulbourethral glands.* Fluid from seminal vesicles contains mucus, along with nutrients for the sperm. Sperm become mixed with seminal fluid, then pass into a duct that leads into the prostate gland. Thin, alkaline fluid from the prostate also empties into the duct. (This fluid probably helps neutralize acidic secretions of the female vagina, which are about pH 3.5–4. When pH increases to about 6, sperm motility and fertility are enhanced.) The fluid now empties into the urethra. In this duct, bulbourethral gland secretions add mucus to the semen. The urethra passes through the penis and empties to the outside.

Sperm can live for several weeks in the vas deferens. Once sperm are ejaculated with semen into the female reproductive tract, they can live for twenty-four to seventy-two hours. On the average, 400 million sperm are present during ejaculation. If only a single sperm is required to fertilize the female gamete (ovum), why are so many produced? One possible explanation follows. The head of each sperm contains protein-degrading enzymes. A single sperm can't do much; but multiplied by millions of sperm, these enzymes apparently may help remove outer layers from the ovum and thereby assure its penetration. (d after A. C. Guyton, *Basic Human Physiology,* second edition, W. B. Saunders Company, 1977)

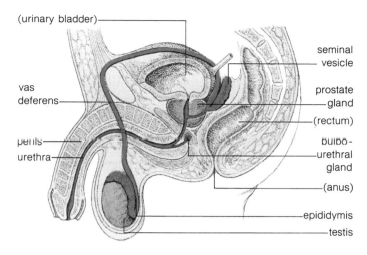

Figure 18.10 Primary and secondary reproductive structures of the human male.

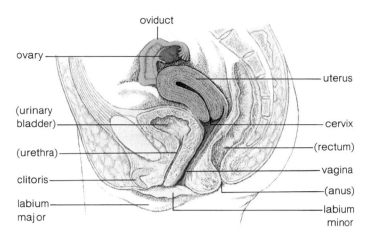

Figure 18.11 Primary and accessory reproductive organs of the human female.

HUMAN REPRODUCTION AND DEVELOPMENT

Structures and Events in Human Reproduction

In human males and females, the primary reproductive organs are called **gonads**; the remainder are **accessory reproductive organs**. The gonads not only produce gametes, they secrete important sex hormones. Accessory reproductive organs include gland-lined channels and structures through which gametes are transported. In both males and females, these organs don't become functional until the onset of puberty (usually between ages ten and fourteen). Figures 18.9, 18.10, and 18.11 show the main components of these reproductive systems.

Male gonads are known as **testes** (singular, testis). Within testes are highly coiled *seminiferous tubules,* in which spermatocytes grow into mature sperm (Figure 8.18). In fact, spermatogenesis in adult males is a continuous process that yields several hundred million sperm each day. In its mature form, a sperm cell consists of a DNA-packed head, a mitochondrial powerhouse, and a tail.

Also in the testes are *interstitial cells*, which produce and secrete sex hormones (Figure 18.9). The most important is the hormone *testosterone*. It is involved in sperm production, in maintaining normal sexuality, and in the growth and maintenance of secondary sex traits in males. These traits include the form of external sex organs, deepening of the voice, skin texture, overall body size, the distribution of hair, fat, and skeletal muscle masses—even certain behavior patterns, which correspond to increasing and decreasing levels of testosterone secretion.

Mature sperm leave the seminiferous tubules and pass through a long, coiled tube (the *epididymis*) and a thick-walled sperm duct (the *vas deferens*), which either store most of the sperm or, during sexual activity, move them along by peristaltic action. As they are transported, sperm become mixed with secretions from glands such as the *seminal vesicles* and *prostate*. These secretions make up most of the *semen*, the sperm-carrying fluid that is ejaculated from the male body during sexual union, or *coitus*.

Sperm are ejaculated from the penis. The human penis contains three cylinders of spongy vascular tissue, arranged around the urethra. At the tip of the penis, the ventral cylinder terminates as a mushroom-shaped structure called the *glans penis*. This structure contains abundant mechanoreceptors which, when stimulated by friction, are responsible for nerve signals that the brain interprets as pleasurable sensations. During normal activity, blood vessels leading into these three cylinders are constricted, and the penis is limp. During early sexual excitation, blood flows into the cylinders faster than it flows out. Blood collects in the spongy tissue, which lengthens and hardens the organ. At this time, blood inflow equals blood outflow. These changes, which can occur within seconds of sexual stimulation, facilitate penetration into the female vagina. During coitus, pelvic thrusts stimulate the penis, as well as the vaginal walls and clitoral region of the female body. The mechanical stimulation triggers rhythmic muscular contractions that force the contents of the seminal vesicle and prostate into the urethra. The contractions, which are involuntary, expel semen into the vagina. (During ejaculation, a sphincter closes off the bladder. As a result, sperm cannot enter the bladder and urine cannot be excreted from it.) The involuntary muscular contractions, ejaculation, and associated sensations of release, warmth, and relaxation constitute the event called *orgasm*.

Female orgasm is characterized by similar physical events, including an intense vaginal awareness, a series of involuntary uterine and vaginal contractions, and sensations of release and warmth. A male or female may or may not reach this state of excitation during sexual union. But if the male ejaculates, the female can become pregnant whether or not she experiences orgasm.

Female gonads are a pair of **ovaries**, where oocytes develop into mature ova prior to fertilization (Figure 8.19). Ovaries produce the important sex hormones estrogen and progesterone. Accessory reproductive organs include the oviducts (or fallopian tubes), uterus, vagina, external genitalia, and mammary glands. The *oviducts* are passageways that channel ova from the ovary into the uterus. The *uterus* houses the embryo during pregnancy. It is also the source of menstrual discharge (the monthly sloughing of blood-enriched uterine lining when pregnancy doesn't occur). Strong muscular contractions of the uterus expel the fetus from the female body at birth.

The *vagina* receives sperm from the male. This organ also forms part of the birth canal, and acts as a channel to the exterior for uterine secretions and menstrual flow. The vaginal orifice is separate from the urethral orifice, and slightly behind it (Figure 18.11). The external genitalia, or *vulva*, include organs for sexual stimulation (such as the clitoris). They also include organs lined with adipose tissue, which cushions these external parts. *Mammary glands*, or breasts, function in lactation (secretion of milk for nourishing the offspring).

Ovarian and Menstrual Cycles

In most mammals, **estrus** is a predictably recurring time when the female becomes sexually receptive to the male. Typically, estrus is limited to a period during spring and early summer, when the female comes into "heat." At first glance, human sexual activity seems to have broken away from the cyclic mammalian plan. Tempered only by social standards (as well as by inclination, opportunity, and endurance), a sexually mature human male is capable of producing and dispersing viable sperm on a more or less continuous basis. A human female, too, has the potential to be physically and emotionally receptive to the male's overtures at any time. But there has been no escaping the cyclic changes that correspond to estrus in other mammals.

During each cycle, the female's reproductive system and the gametes maturing within it are physically prepared for fertilization. Fertilization can occur only within a span of about three days during that cycle. Once it does occur, additional fertilizations will be impossible for another nine months (assuming that embryonic development runs its full nine-month course).

Controlling the structural and physiological changes associated with the female reproductive cycle are hormonal feedback loops between the hypothalamus, anterior pituitary, and ovaries. These hormone-regulated changes unfold in two phases. The **ovarian cycle** consists of (1) the ripening of an oocyte in a fluid-filled cavity inside the ovary, (2) the transformation of the cavity into an endocrine element when it ruptures and the mature ovum escapes from it, and (3) a concurrent production of ovarian hormones. The **menstrual cycle** consists of profound changes in the uterine lining, which are synchronized with events of the ovarian cycle.

Let's look at the ovarian cycle first. Ovaries do not constantly produce new eggs. A human female is born with about 2 million potential ova. Of those, only about 500 will mature and be released during her lifetime; the rest start to degenerate from birth onward. Usually between ages twelve and fourteen, oocytes begin to mature into ova. Mature ova usually are released one at a time, on a monthly basis. (Hereditary defects occur occasionally among children born to women approaching menopause, the end of the period of reproductive potential. These defects are thought to arise from changes in the remaining oocytes, which may be forty to fifty years old by then. Menopause most often begins during the forties and fifties.)

Each oocyte is contained in a spherical chamber called a **follicle** (Figure 18.12). A ripening follicle grows in size, and the cell around the egg secretes a fluid about it. Soon there is so much fluid that a mature follicle balloons outward from the ovary surface. At **ovulation**, the follicle ruptures and the ovum escapes from the ovary. The ovum is then swept by ciliary action into the oviduct, the road to the uterus. It takes about two weeks for a follicle to develop fully. The exact time varies from one female to the next, even from one month to the next in the same female. Following the ovum's expulsion, the ruptured follicle develops into an endocrine element: the **corpus luteum**. Like the follicle from which it forms, the corpus luteum secretes the steroid hormone estrogen. Unlike the follicle, it also goes on to secrete the steroid hormone progesterone.

Changing estrogen and progesterone levels cause profound changes in the uterus. Estrogen causes the uterine

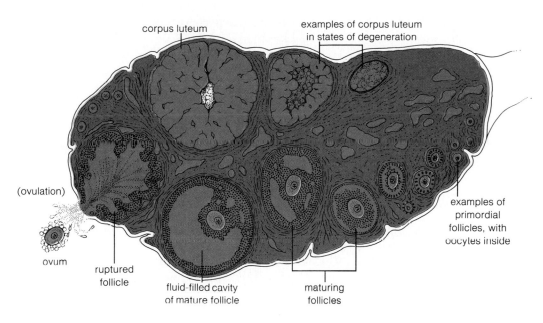

Figure 18.12 A human ovary, drawn as if sliced lengthwise through its midsection. Events in the ovarian cycle proceed from the growth and maturation of primordial follicles, through ovulation (rupturing of a mature follicle, with a concurrent release of an ovum), through the formation and maintenance (or degeneration) of an endocrine element called the corpus luteum. The positions of these structures in the ovary are varied for illustrative purposes only. The maturation of an oocyte occurs at the *same* site, from the beginning of the cycle to ovulation.

muscles and **endometrium** (the epithelial lining of the uterus) to thicken. Under the influence of estrogen, the cervix secretes a clear, thin mucus. The mucus is an ideal medium through which sperm can travel rapidly on their journey to the oviduct. Progesterone goes to work on the thickened tissues, stimulating glands into secreting various substances. In such ways, the uterus is prepared for implantation.

The entire sequence of events begins when the hypothalamus stimulates the anterior pituitary into secreting FSH. As Figure 18.13 shows, FSH stimulates the growth of follicles in the ovary. As a follicle grows and thickens, parts of it secrete estrogen to the interior. Then, about two days before ovulation, the pituitary's LH secretions increase almost tenfold. This may result when the hypothalamus detects the increased estrogen level in blood, which is known to peak about this time. The flood of LH causes the follicle to start secreting progesterone. It is this midcycle peak of LH, and probably a corresponding peak in FSH, that triggers ovulation (Figure 18.14).

Now, continued progesterone secretions are possible only when the corpus luteum is constantly stimulated by

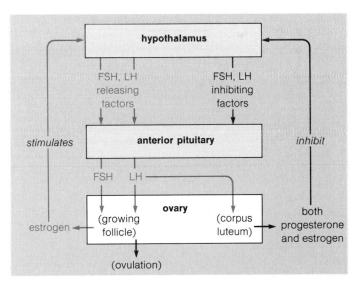

Figure 18.13 Feedback regulation between the hypothalamus, pituitary, and ovary in the female reproductive cycle. High estrogen levels provide positive feedback: they stimulate the hypothalamus into commanding the pituitary to release FSH and LH. High progesterone levels provide negative feedback: they lead to a decrease in FSH and LH secretions.

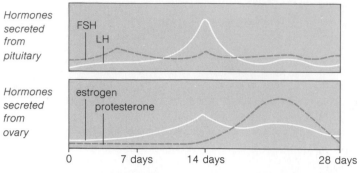

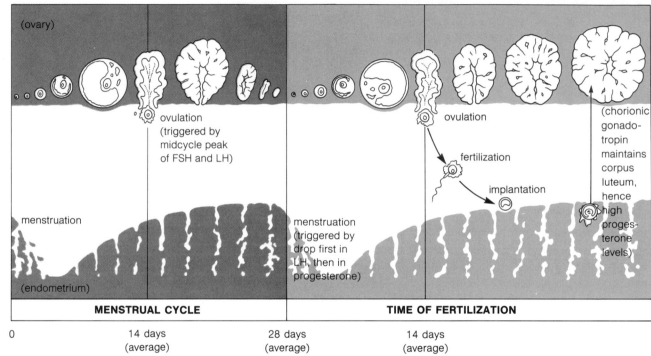

Figure 18.14 Ovarian cycle and menstrual cycle, as described in the text. Changes in the endometrium are correlated with the timing of ovulation and corpus luteum formation in the ovary. Changing hormonal levels that orchestrate these events are depicted in the boxed insets above.

LH. But once ovulation occurs, the hypothalamus detects the rising levels of estrogen and progesterone—and sends out command signals to shut down LH and FSH secretions. This completes the feedback loop. Without the stimulation of FSH and LH, the corpus luteum stops secreting estrogen and progesterone, and it starts to degenerate. Without estrogen, blood vessels in the endometrium become constricted, which means that oxygen and nutrients no longer nourish the thickened tissues of the lining. The endome-

trium begins disintegrating and is gradually cast off, which marks the first day of the menstrual cycle (Figure 18.14). The weakened endometrial capillaries begin to hemorrhage. For about three to six days, blood mixed with epithelial lining is cast off. On the average, 50 to 150 milliliters of blood is lost. During this menstrual period, the level of *all* the sex hormones in the blood is low.

Near the end of the menstrual period, the pituitary begins secreting FSH once more, because it is no longer inhibited by progesterone. Now a new follicle is stimulated into developing and secreting estrogen. The estrogen shuts off the menstrual flow and stimulates the uterus to begin rebuilding for a new cycle. It takes about ten days from the end of the menstrual flow for hormonal activity to prepare the uterus for ovulation.

The interrelated cycles just described may be summarized in the following manner:

The ovarian cycle consists of oocyte maturation in a fluid-filled follicle within the ovary, the transformation of that cavity into a corpus luteum (an endocrine element) when the follicle ruptures and the mature ovum escapes from it, and a concurrent production of ovarian hormones.

The menstrual cycle consists of profound changes in the endometrium (epithelial lining) and muscles of the uterus, as well as glandular secretions, all of which prepare the uterus for implantation. In the absence of implantation, the changes are repeated monthly, as are the events of the ovarian cycle.

From Fertilization to Birth

If sperm ejaculation into the vagina coincides with ovulation—if it occurs at any time between about six days before ovulation and about fifteen hours afterward—pregnancy can result. Within thirty minutes after ejaculation from the penis, sperm are in the oviduct, where fertilization usually takes place. Of the several hundred million sperm entering the vagina, only a few thousand complete the journey, and only one succeeds in penetrating the ovum. Fertilization is completed once the sperm nucleus and egg nucleus fuse to form a human zygote.

For three or four days, the fertilized egg travels down the oviduct. It picks up nutrients from maternal secretions and undergoes some cell division. By the time the fertilized egg reaches the uterus, it has become transformed into a solid cluster of cells. Now the surface cells separate from those inside, which forms a **blastocyst:** a hollow sphere of sticky surface cells with inner cells massed at one end (Figure 18.15). The surface cells (collectively called the *trophoblast)* are important in implantation. They will quickly establish vital links between the embryo and the mother. The *inner cell mass* will give rise to three layers of cells (endoderm, ectoderm, and mesoderm). From these layers, all organs will be derived.

Five or six days after conception, the blastocyst contacts and adheres to the uterine wall. Enzymes secreted by surface cells destroy some of the uterine lining. Their action allows the blastocyst to burrow inside. Another effect is that maternal blood vessels being encountered are ruptured. The entry site then heals over, leaving the blastocyst embedded inside, in a pool of the mother's blood. About this time, two membranes begin to form. The inner **amnion** arises from the inner cell mass; this membrane will hold a fluid in which the embryo will develop freely (Figure 18.15). Later, the outer surface of the amnion and the inner surface of the trophoblast acquire a lining of connective tissue. Blood vessels develop in the connective tissue associated with the surface cells, at which point this structure forms the **chorion,** the second major membrane.

The chorion secretes a hormone, chorionic gonadotrophin. This hormone is much like LH in its effect: it helps maintain the corpus luteum, hence the endometrium. Extensions from the chorion fuse with the endometrial layer of the uterus to form the **placenta,** an organ specialized for mediating the interchanges between mother and fetus. (At childbirth, it will weigh about a pound and will be cast off as the "afterbirth.") In the placenta, maternal blood vessels and embryonic blood vessels lie side by side —separate circulatory systems, but close enough for the transfer of materials through diffusion. Here, the embryo absorbs nutrients needed for growth; here, its metabolic wastes are cast off, to be disposed of through the mother's lungs and kidneys. As the embryo grows, it remains firmly attached to the placenta by the umbilical cord. This cord is a flexible strand of tissue through which blood vessels extend.

By the time the placenta is forming, the inner cell mass has begun to differentiate. Its development proceeds fairly rapidly along the same general course described earlier for all complex animals, except that the human embryo is not spherical. Prior to gastrulation, the embryo itself consists simply of two flat layers of cells, the *embryonic disk* (Figures 18.15 and 18.16h). During the second week, the onset of gastrulation is marked by the appearance of a groove on the upper surface of the embryo. This groove, the *primitive streak,* is the site where cells from the surface move inward to form the mesoderm. When gastrulation is completed, the remaining surface tissue is the ectoderm, which gives rise to the nervous system and skin. Inside, the endoderm forms the respiratory and digestive systems; and the mesoderm develops into internal organs such as the heart, muscles, bone, and blood. By the end of the

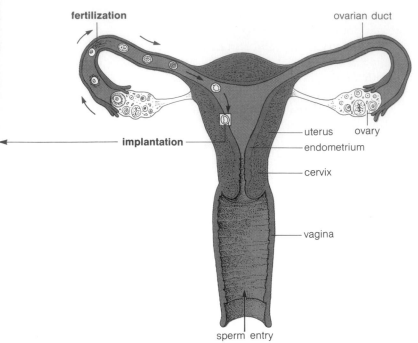

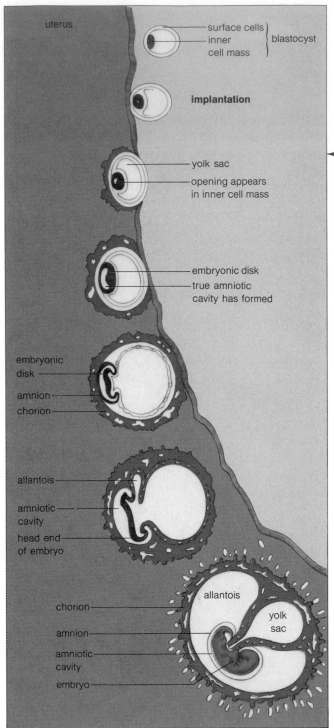

Figure 18.15 A current view of developmental events, from the time of implantation (about eight days after fertilization in the oviduct) to about the end of the first month, as described in the text. The various stages are not drawn to scale relative to one another, the point being simply to show the developmental fate of the inner cell mass.

first month, the embryo has grown 500 times its original size and has begun taking on recognizably human characteristics. This rapid spurt of growth gives way to two months of relatively slow development for the main organs. This three-month period is the **first trimester** (Figure 18.16a-j).

By the beginning of the **second trimester,** all major organs have been formed. The embryo now resembles an adult in miniature—it is about 7.5 centimeters (3 inches) long. Even through the **third trimester** the individual grows considerably, but few new parts form. That is why the term *embryo* usually is reserved for the first trimester of human life, when body parts are being formed. In the remaining time inside the uterus, the developing individual is called a **fetus.**

Not until the middle of the third trimester will the fetus be sufficiently developed to survive on its own if born prematurely or if removed surgically from the uterus. By the seventh month, fetal development appears to be relatively complete, but fewer than ten percent of infants born at this stage survive, even with the best medical care. In most cases they are not yet able to breathe normally, swallow, or maintain a normal body temperature. By the ninth month, survival chances increase to about ninety-five percent.

Case Study: Mother as Protector, Provider, Potential Threat

With the shift to internal fertilization and development, the offspring of mammals have secured a survival advantage. But with this greater protection for offspring has come a greater responsibility for the prospective mother. No longer does her material contribution end with producing the egg. From conception to birth, the developing individual is at the mercy of her diet, her health habits, and her life-style.

Over evolutionary time, many controls and safeguards have been built into the human reproductive system. The placenta, for example, is a highly selective filter. By screening substances present in the maternal bloodstream, it prevents many noxious substances from gaining access to the embryo and, later, to the fetus. More than this, if the mother's diet is in some way deficient, the placenta often can (at her own body's expense) preferentially take in scarce nutrients for the fetus.

But millions of malformed, malnourished infants born each year are testimony to the fact that human development is far from an infallible process. Even in the United States, where women generally have enough food and adequate medical care, between 250,000 and 500,000 infants are born each year with physical and mental handicaps severe enough to warrant special care. Only about half of these handicaps can be corrected by intensive medical care immediately following delivery. Although it is not known how many birth defects could have been prevented in the first place, the number is certainly significant.

Particularly during the first six weeks after fertilization, the embryo is vulnerable to certain damaging influences from the maternal bloodstream, because that is the critical period of organ formation. If, for instance, the mother contracts German measles during this period, there is a fifty percent chance the embryo she is carrying will be severely malformed. The probability of damage diminishes after this stage; the same disease, contracted during the fourth month or thereafter, has no discernible effect on the embryo.

Throughout pregnancy, the developing individual is well protected from all but the most severe bacterial diseases. But certain viral diseases may have their effects if they are contracted during the sensitive first developmental period. During that time, the mother must maintain her body in the best of health and avoid becoming exposed to individuals with virus infections.

During the first trimester, the embryo is also sensitive to drugs—to new, manufactured agents against which

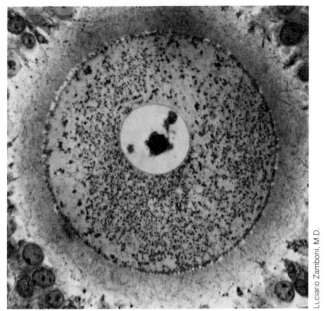

a Human ovum, ripening in follicle. 800x.

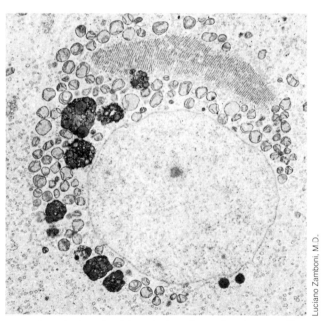

b A closer look at the rich profusion of cytoplasmic organelles forming near the egg nucleus. This metabolic machinery will be called into action once the sperm and egg nuclei fuse.

Figure 18.16 Moments in the continuum of human embryonic development. This illustration continues over the next few pages. (Unless otherwise indicated, photographs are from The Carnegie Institution of Washington and Ronan O'Rahilly, M.D., Director of the Carnegie Embryological Collection)

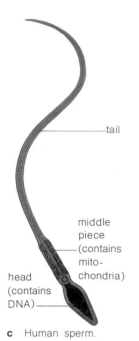

tail

middle
piece
(contains
mito-
chondria)

head
(contains
DNA)

c Human sperm.

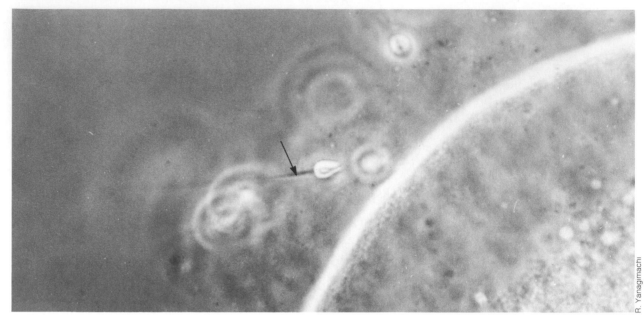

R. Yanagimachi

d Human sperm (arrow) passing through the mucoprotein coat (zona pellucida) of an ovum. 1,950x.

natural selection has not had the opportunity to build defenses. The most shocking example of drug effects came during the first two years after the tranquilizer *thalidomide* was introduced on the market. Women using this drug gave birth to infants with missing or horribly deformed arms and legs. As soon as these deformities were traced to thalidomide, the drug was immediately withdrawn from the market. But there is evidence suggesting that various other tranquilizers as well as sedatives and barbiturates still in use may cause similar, albeit less severe, damage.

Even though certain drugs may cause embryonic malformations only if taken during the first trimester, the embryo does not become impervious to drugs in the maternal bloodstream at any stage. Many drugs pass freely across the placenta and have the same kind of effect on the fetus as on the woman who takes them. For example, infants born to heroin or alcohol addicts are themselves addicted. Without expert detection and prompt medical care, they are likely to die shortly after birth. It is true that certain medicines and drugs may be necessary in certain cases, but the decision to use them at any time during pregnancy must be made by a skilled physician.

As birth approaches, the growing fetus places more and more demands on the mother for essential nutrients. Thus it is during the last phase of pregnancy that her diet profoundly shapes the course of development. Poor nutrition affects all organs of the fetus, but it is most damaging to the brain. It is in the weeks just before and just after delivery that the human brain undergoes its greatest growth. Normal brain growth is assured *only* if there are adequate amounts and kinds of amino acids for building brain proteins.

Obviously, nutrients other than amino acids are needed for normal development. But a balanced diet that provides enough calories and amino acids normally provides all other nutrients in sufficient amounts. One of the most common fallacies of Western culture is that if you pop a vitamin pill into your mouth every day, the rest will take care of itself. Particularly with respect to pregnancy, nothing could be further from the truth. Because the placenta preferentially absorbs vitamins and minerals from the mother's blood, the developing fetus is more resistant to vitamin and mineral deficiencies than she is. And in cases where the diet is marginal, the money spent on vi-

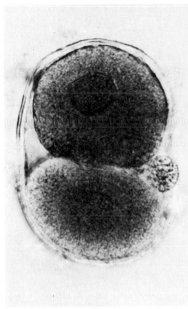

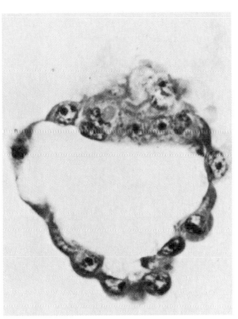

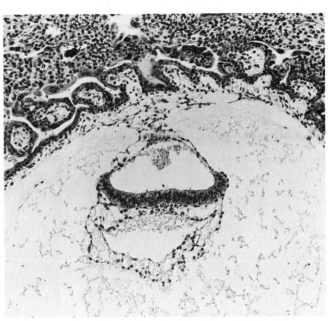

e Between the second and third day following fertilization, cleavage begins, progressing from the two-cell stage to about the sixteen-cell stage.

f Between the fourth and sixth days, a blastocyst forms, with an outer cell layer and an inner cell mass.

g Cross-section of the embryonic disk at about sixteen days.

tamin pills and other food supplements usually does the fetus more good if spent on wholesome, protein-rich food.

There is one other important aspect of food intake during pregnancy that has recently come to light. A few years ago, it was fashionable for a pregnant woman to keep her total weight gain to ten or fifteen pounds. But it is now clear that a woman who limits her weight gain to this level does so at the expense of her child. Particularly during the last trimester, restricted food intake leads to restricted fetal development and to the birth of an underweight infant. Significantly underweight infants face more postdelivery complications than infants of normal weight. They also face a much higher incidence of mental retardation and other handicaps later in life than do infants of normal or somewhat above-level birth weight. In most cases, it is now recommended that a pregnant woman manage her diet to assure a total weight gain of between twenty and twenty-five pounds.

Factors other than diet during the last trimester also affect birth weight, hence the state of maturity of the newborn. Two factors are the mother's height and her long-term nutritional status. Generally, the taller she is and the

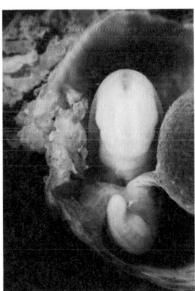

h Eighteen days. Neural folds appear.

i Twenty-eight days. Neural folds have fused; optic vesicles have already formed.

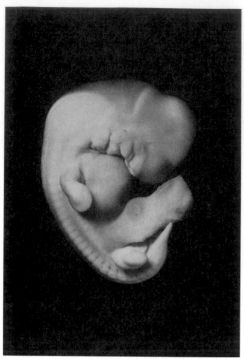

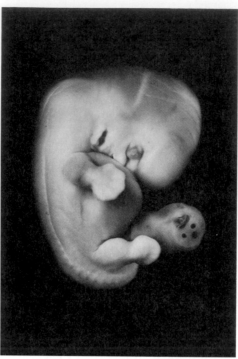

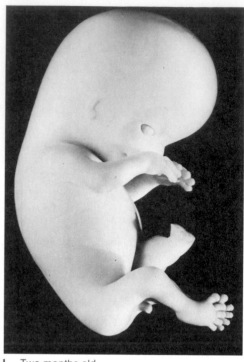

j   Thirty-nine days. Limb buds and foot plates have appeared.

k   Forty-two days. Head begins to enlarge, trunk begins to straighten. Finger rays appear.

l   Two months old.

better her nutrition has been (from before her own birth to the time of conception), the larger and healthier her own infant is likely to be. In developing countries, the effects of marginal or inadequate nutrition persist for more than a generation. Recent studies in Asia show that children of malnourished women are less capable of using food efficiently. And when the female children mature, they in turn tend to be small and to have smaller, less well-developed infants even if their own diet is improved. Studies in Japan suggest that only over the long term will improved diet diminish the effects of malnutrition.

One other factor known to have an adverse effect on fetal growth and development is cigarette smoking. A woman who smokes every day throughout pregnancy has smaller infants even when her weight and nutritional status and all other variables are identical with those of pregnant women who do not smoke. Often it has been stated that such infants are normal in all other ways. But a recent study in the British Isles suggests otherwise. In England,

Scotland, and Wales, records were kept for seven years on all infants born during one particular week. Those infants born to women who smoked were indeed smaller. More than this, they had a thirty percent greater incidence of death shortly after delivery and a fifty percent greater incidence of heart abnormalities. Most startling of all, at age seven such children had an average "reading age" nearly half a year behind that of children born to women who were nonsmokers.

In this last study, the critical period was shown to be the last half of pregnancy. Children born to women who had stopped smoking by the middle of the second trimester were indistinguishable from those born to women who had never smoked. Although the mechanism by which cigarette smoking exerts its effect on the fetus is not known, the fact that it does affect the fetus is further evidence for the inability of the placenta—marvelous structure though it is—to prevent all the assaults on the fetus that the human mind can dream up.

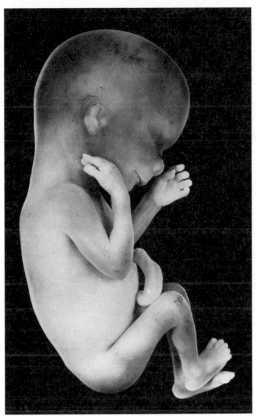

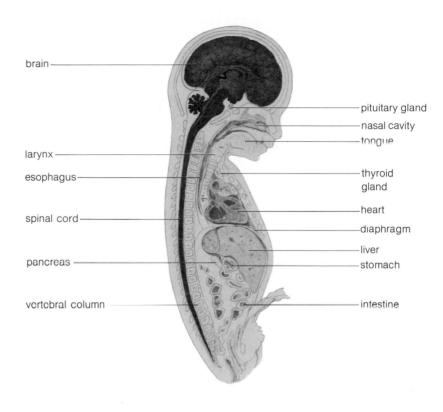

brain

pituitary gland
nasal cavity
tongue

larynx

thyroid gland

esophagus

heart

spinal cord

diaphragm

liver

pancreas

stomach

vertebral column

intestine

m About four months old—the adult in miniature. In this median section, organs such as the lungs and kidneys cannot be seen. (After Hans Elias, 1971.)

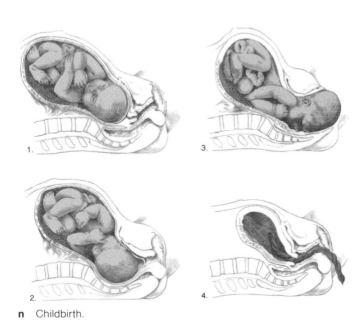

1.

2.

3.

4.

n Childbirth.

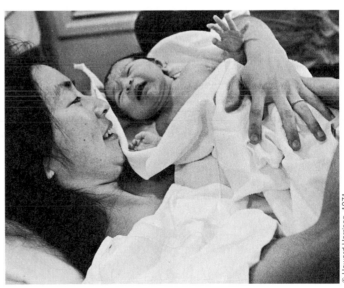

© Howard Harrison, 1971

o The newborn, five minutes old.

COMMENTARY

In Vitro Fertilization: Is It Safe and Repeatable?

(The following is condensed from an article by Gina Kolata in Science, *201:698–699, 25 August 1978.*
Copyright 1978 by the American Association for the Advancement of Science)

Louise Brown, the first baby ever to be conceived in a petri dish, was born in England in July 1978. The technique for external conception (in vitro fertilization) is conceptually straightforward. Needed are ripe ova ready to be fertilized, sperm, a medium in which to mix the two, and a medium in which to support initial embryonic development. To obtain the ripe ova, the woman is given a hormone that causes her ovaries to prepare eggs for release. Then, thirty-three to thirty-four hours later, it is time to try recovering the eggs. If a few extra hours elapse, the eggs will have been released from the ovary and will be unrecoverable. The woman is put under general anesthesia for removal of the preovulatory eggs. A small incision is made in her abdomen and a laparoscope inserted. A laparoscope is a long metal tube containing a light and an optical system, and can be used to view ovaries directly. Preovulatory eggs, which look like bulges on the ovarian surface, are removed by suction.

Before laparoscopy, the woman's husband donates a sperm sample, which is washed and diluted in order to simulate conditions in the oviducts. The sperm are put in salt solution. Within a few hours, they undergo chemical changes that prepare them for fertilization. Sperm-containing droplets are then placed in a petri dish partly filled with inert oil. The oil keeps the droplets intact and confines the ova (which are added to one of these droplets) in a small volume, together with the sperm.

A few hours after sperm and ovum are combined, fertilization occurs. About twelve hours later, the embryo is transferred to a solution that supports embryonic development. Some time between two and four days after fertilization, the developing embryo is inserted into the woman's uterus. The embryo is drawn into a fine tube, the tube is inserted into the uterus, and the embryo is expelled. If all goes well, the embryo may implant.

With all this laboratory involvement in what is usually a natural process of conception and implantation, many have questioned whether the resulting babies will be normal. The geneticist Joseph Schulman says that "There are no data to support the fears that in vitro fertilization will lead to abnormal babies. But more research is clearly desirable." A substantial amount of work has been done with other animals, and there is no evidence that in vitro fertilization leads to genetic or morphological abnormalities in the offspring of any species. According to Schulman, preimplantation animal embryos are surprisingly resistant to manipulation. "You can remove cells from embryos, you can take two embryos and fuse them, or you can freeze embryos. Yet the resulting offspring are reported to be normal."

Even supposing there were an increased risk of abnormalities, Schulman says, the decision to have a child should be left to the prospective parents. This is common medical practice. For example, if a couple has a child with a genetic disease, there is often one chance in four that subsequent children will also have the disease. Yet no one tells such couples that they cannot have children.

Louise Brown no longer remains the only baby conceived outside the human body. Already Western Europeans are working on the problems of implanting human embryos. (In the United States, research involving in vitro fertilization was halted in 1975. At the time of this writing, a clinical laboratory in Virginia has been granted permission to proceed with research into external fertilization.) It is virtually certain that we will soon know whether the technique is readily repeatable and whether the babies conceived in petri dishes are at increased risk of being abnormal.

On Human Birth Control

Few processes in the living world are more inspiring than the gradual transformation of a single zygote into an intricately detailed, coordinated adult of the species. And few processes evoke more profound questions. *When does development begin?* As you read in this chapter, many key aspects of development have already emerged before the union of sperm and egg. But this answer leads to a much more basic question: *When does life begin?* Each human female can produce, during her lifetime, as many as five hundred eggs—and all those eggs are alive. Each human male can produce, during one ejaculation, a quarter of a billion sperm—and all those sperm are alive. Even before they merge by chance to establish the genetic constitution of a new individual, sperm and egg are as much alive as any other life form. In no sense, then, is it tenable to suggest that "life begins" when they fuse. *Life began only once, billions of years ago; and each sperm and egg, each zygote and mature individual, is only a fleeting stage in the continuation of that beginning.* This fact cannot diminish the meaning of conception, for it is no small thing to entrust a new individual with the gift of life, wrapped in the unique threads of our species and handed down through an immense sweep of evolutionary time. Child, man, woman—who among us can witness the birth of a living creature and not know the profound force of the life process? For an instant time stands still; and a sense of past and future descends on us. For an instant our pulse beats in synchrony with the newborn and unseen predecessors.

How can we reconcile this compelling moment of individual birth with our growing awareness of too many births in the biosphere? At the time this book is being written, an average of 2.2 infants are being born each second—132 each minute, 7,920 each hour. By the time you go to bed tonight, there will be 190,080 more humans on this earth than there were last night at that hour. Within a week, the number will reach 1,330,560—about as many people as there are now in the entire state of Massachusetts. *Within one week.* This astounding birth rate has outstripped our resources, and each year millions face the horrors of starvation. Living as we do in one of the most productive lands on earth, few of us can know what it means to give birth to a child, to give it the gift of life, and have no food to keep it alive. Few of us can know what it means to a mother thirty years old, with eight children and with the knowledge that they face poverty and starvation. From a photograph in a magazine her dying children look out at us, and, uncomprehending, we turn the page.

Living as they do with these realities, people over much of the world have practiced abortion, often by the most primitive and dangerous methods imaginable. Long before birth control became a global issue, individuals resorted to infanticide. These are some of the practices followed in some countries to assure that some individuals, at least, have enough food to live.

Such practices mean the deliberate termination of life. Just as surely, indifference to rampant population growth means the deliberate termination of life. Many believe it is wrong to deny life to an unborn embryo; and just as many believe it is no solution to withhold compassion from those—child and adult—who have no alternative but to resign themselves to starving to death. For our species, reproduction and development and maturity flow on, through years, through centuries. But there is no escaping the principles underlying this flow of life, and in some way the birth rate for our kind must be reconciled with available resources. How we decide to control population size is one of the most volatile issues of this decade. Yet the decision must be made if we are to assure our kind of long-term stability in the biosphere. We will return to this issue in Chapter Twenty-Six, in the context of principles governing the growth and stability of all populations.

At present, there simply is no completely effective, completely safe, low-cost way to control births. Instead there are various methods that more or less reliably prevent conception, and more drastic surgical ones that are used to abort the developing embryo. A related problem is the lack of enough educational programs concerning birth control. Each year in the United States we still have about 100,000 shotgun marriages, about 200,000 unwed teenage mothers, and *1,000,000 abortions.* This is the legacy of our confusion over what to do about a sexual revolution that has swept through our society down to the primary school level—but that has not been accompanied by adequate sex education programs. On the one hand, many parents promote boy-girl relationships for younger and younger ages. On the other hand, they close their eyes to the possible outcome in terms of premarital intercourse and unplanned pregnancy. Birth control advice to teenagers is often condensed to a terse, "Don't do it, but if you do, be careful!" In this lack of resolution lies part of the controversy over what birth control represents. For most of the world, birth control is a question of utmost urgency, a matter of sheer survival; for sexually liberated Americans, it is often touted as a matter of convenience. More than 500 million years of sexual evolution have gone into assuring a powerful motivation to engage in sex and thereby reproduce. Overlaid on those 500 million years are a few centuries of moral sanctions that demand suppression of the compelling sex drive—and yet simultaneously demand that we go forth and multiply. We have not managed to suppress sexuality, but we *have* managed to multiply far beyond any previous imaginings. It seems we ought to start thinking about reconciling our biological past with the need for a stabilized cultural present. Given the alternatives, we don't seem to have much choice about it.

Once conception has occurred, the only form of birth control is **abortion**, in which the implanted embryo is dislodged and removed from the uterus. (In miscarriages, the embryo is expelled spontaneously.) Until recently, abortions were generally forbidden by law in the United States, unless the pregnancy endangered the mother's life. Supreme Court rulings in this past decade have held that the State does not have the power to regulate abortions in the first trimester. The outcome has been legalization of abortion in this country and others. Moving the large number of backroom operations to modern medical facilities at least reduces the frequency of dangerous, traumatic, and often fatal attempts to abort embryos, either by pregnant women themselves or by quacks. Newer methods have made it relatively rapid, painless, and free of complications if performed during the first trimester. Abortions in the second and third trimesters will probably remain extremely controversial unless the mother's life is clearly threatened. For both medical and humanitarian reasons, however, it is generally agreed that an acceptable route to birth control is not through abortion but through control of conception in the first place.

The most effective method of preventing conception in the first place is complete **abstention:** no sexual intercourse whatsoever. It is unrealistic to expect many people to follow it. A modified form of abstention is the **rhythm method,** in which intercourse is avoided during the woman's fertile period. The fertile period, recall, begins a few days before and ends a few days after ovulation (the release of an egg from an ovary). It is determined either by keeping records of the length of a woman's menstrual cycle or by taking her temperature every morning when she wakes up. (Just before the fertile period, there is a one-half to one-degree rise in body temperature.) But ovulation can be irregular, and miscalculations are frequent. The method *is* inexpensive (it costs nothing after you buy the thermometer) and it doesn't require fittings and periodic checkups by a doctor. But its practitioners do run a forty percent risk of becoming pregnant.

Withdrawal, the removal of the penis from the vagina prior to ejaculation, is a truly ancient contraceptive method dating at least from biblical times (not because overpopulation was seen to be a problem then, but because our ancestors didn't always live by the rules, either). It requires extraordinary willpower. And even if the mind manages to conquer the body, the method may fail anyway: the fluid released from the penis just before ejaculation may contain viable sperm cells.

Other methods are based on using physical or chemical barriers to prevent sperm cells from entering the uterus and moving up the ovarian ducts. **Condoms** are thin, tight-fitting sheaths of rubber or animal skin that are worn over the penis during intercourse. They are about eighty-five to ninety-three percent reliable (Table 18.1), and they do help prevent venereal disease. But condoms can only be put on over an erect penis, which calls for an interruption of activities at a time when rational behavior somehow seems not as interesting as immediate fulfillment. Also, they have been known to tear and leak, which renders them useless. A **diaphragm** is a flexible, dome-shaped disk used with a spermicidal foam or jelly. It is placed over the opening of the cervix just before intercourse. A diaphragm is relatively effective if it has been fitted by a doctor, and if it is inserted correctly with each use. **Spermicidal foam** or **spermicidal jelly** is packaged in an applicator and is emptied into the vagina just before intercourse. It is toxic to sperm cells, yet it is not always reliable when used without another device such as a diaphragm or a condom. The practice of **douching**, or rinsing out the vagina with a chemical immediately after intercourse, is almost useless. Sperm cells can move past the cervix and out of reach of the douche within ninety seconds after ejaculation. And no panicky flight to the medicine chest and frenzied rinsing is that rapid.

Still other contraceptive methods are based on hormonal control of the reproductive cycle. Most widely used is **the Pill**—an oral contraceptive of synthetic estrogens and progesterones. These synthetic hormones substitute for hormones normally produced by the ovary during the menstrual cycle. They suppress the release of gonadotropins from the pituitary and thereby prevent the cyclic maturation and release of eggs. Birth control pills are prescription drugs. Formulations vary and should be selected to match the individual patient's needs. That's why it isn't wise for a woman to borrow the Pill from someone else.

If the woman doesn't forget to take her daily dosage, the Pill is one of the most reliable contraceptives available. There is no interruption of the sexual act, and the program is easy to follow. Often the Pill corrects erratic menstrual cycles and decreases associated cramping. Even so, the Pill is not without potential side effects. In the first month or so of use, it may cause nausea, weight gain and swelling, and minor headaches. Its continued use may lead to blood clotting in the veins of a small number of women (3 out of 10,000) predisposed to this disorder. There have been some cases of elevated blood pressure and abnormalities in fat metabolism (which may be linked to a growing number of gallbladder disorders among women). Proponents of the Pill argue that for most women, the known risks associated with using it are far lower than the risks associated with pregnancy.

Newer, estrogen-free formulations now eliminate some of the side effects. The potent "morning-after Pill" now becoming available eliminates pregnancy after intercourse. There are certain hazards connected with this drug, however, and it can't be used on a regular basis. An inexpensive form of the Pill—strips of edible paper treated with oral contraceptives—is now being tested on a large scale. It may extend use of the Pill to developing countries for mass birth control programs.

A birth control method that once seemed most promising for use in developing countries is the **intrauterine device**, or **IUD**. A doctor must insert this small plastic or metal device into the uterus. With this foreign object in the uterus, a fertilized egg cannot remain implanted in the uterine wall. In cases where implantation does occur, the IUD stimulates processes that cause the embryo to be dislodged. The IUD is relatively inexpensive, and once it is inserted there is no need to give it much further thought unless it is accidentally expelled from the uterus. Especially with the newer designs, such as the Copper-T, expulsions are rare.

But the IUDs can give rise to other complications. Typically, there is a marked increase in menstrual discharge. Usually the increase is merely inconvenient, but in a few cases there are increased possibilities of anemia or hemorrhaging. Some research also indicates a greater probability of uterine disease. In such cases, supervision by a doctor is necessary; in all cases, a checkup every six months is advised so the woman can verify that the device is in place and functioning properly. Thus, even though the low cost of the IUD makes it seem ideal for mass birth control in developing countries, its potential benefits must be weighed against potential complications—especially where ready access to a doctor, let alone medical facilities, is not readily available to women.

Hormonal control of male fertility is a more difficult matter. The hormonal methods for suppressing female fertility are based on the cyclic nature of the woman's reproductive system. But sperm production in males is not cyclic, and it is under a more diffuse kind of hormonal control.

Table 18.1   Effectiveness of Birth Control Methods Used in the United States

Method	Theoretical Effectiveness (percentage)	Typical Effectiveness (percentage)
Extremely Effective		
Abortion	100	100
Sterilization		
Tubal ligation	99.96	99.5
Vasectomy	99.85	99.0 to 99.5
Highly Effective		
Oral contraceptive	100	98 to 99
IUD with slow-release hormones	100	98 to 99
IUD plus spermicide	99	98
Diaphragm plus spermicide	99	98
IUD		
Copper T	99	98
Older loops	98.3	94 to 97
Condom (good brand) plus spermicide	99.9	95
Effective		
Spermicide (vaginal foam)	97	90
Condom (good brand)	99	85 to 93
Diaphragm alone	98	85 to 87
Moderately Effective		
Spermicide (creams, jellies, suppositories)	90	75 to 80
Rhythm method based on temperature	95	40
Relatively Ineffective		
Condom (cheap brand)	85	70
Withdrawal	85	70
Rhythm method not based on temperature	90	Variable, but normally below 60
Unreliable		
Douche	——	40

Sources: Lane et al. 1976, Ryder 1973, Vaughn et al. 1977.

Various medications have been developed, but they are still experimental. It will probably be several years before they become available, and their effectiveness is in doubt.

Surgical sterilization methods hold the greatest promise for completely reliable birth control. In male **vasectomy,** a tiny incision is made in the scrotum so that the vas deferens (which transports sperm cells) can be severed and tied off. The simple operation can be performed in twenty minutes in a doctor's office, with only a local anesthetic. So far there is no firm evidence that it disrupts the male hormone system, and surveys suggest there is no noticeable difference in sexual activity. The only major objection to vasectomies is that they are generally irreversible. Some recent surgical advances promise to improve the possibility of reversing the operation, if that should be desired. For females, the most common sterilization method is **tubal ligation,** in which the oviducts are cut and tied off. Because tubal ligation is more complex than vasectomy, it usually is performed in a hospital and requires about a two-day stay after surgery.

There is little doubt that surgical sterilization is one of the safest, most effective, and most convenient birth control methods. It is a permanent method right now, however, so it is best considered by couples or individuals who have made a mature commitment to limit their family size. Many communities maintain family planning services that can be contacted for assistance in evaluating the available birth control options.

Readings

Balinsky, B. 1975. *An Introduction to Embryology.* Fourth edition. Philadelphia: W.B. Saunders.

Ebert, J., and I. Sussex. 1970. *Interacting Systems of Development.* Second edition. New York: Holt, Rinehart and Winston.

Gordon, R., and A. Jacobson. 1978. "The Shaping of Tissues in Embryos." *Scientific American* 238(6):106–113. Interesting discussion of computer simulations of forces shaping embryonic morphogenesis.

Hickman, C., et al. 1979. *Integrated Principles of Zoology.* Sixth edition. St. Louis: Mosby. Contains excellent discussions and illustrations of embryology.

Patten, B. M. 1971. *Early Embryology of the Chick.* Fifth edition. New York: McGraw-Hill.

Rugh, R., and L. Shettles. 1971. *From Conception to Birth: The Drama of Life's Beginnings.* New York: Harper and Row. Magnificent illustrations of human embryonic development.

Saunders, J. 1970. *Patterns and Principles of Animal Development.* New York: Macmillan. Lucid book.

Review Questions

1. What are some differences between asexual and sexual reproduction? Describe some forms of asexual reproduction. What are some adaptations associated with external and internal fertilization of sexually reproducing animals?

2. Describe what goes on during each of these developmental stages: fertilization, cleavage, gastrulation, and organ formation. What is embryonic induction? Embryonic regulation?

3. How does the yolk of animal eggs influence cleavage patterns?

4. Which two factors coordinate the development of body parts in all animals?

5. Briefly summarize the controls over the differentiation of similar cells into diverse cell types.

6. Describe morphogenesis and the factors governing it. What two structural elements are known to be involved in some coordinated programs of morphogenesis?

7. Define indirect and direct development. Give some examples of animals that follow each developmental strategy.

8. Study Figure 18.9. Then list the main organs of the human male reproductive tract and identify their functions.

9. List the main organs of the human female reproductive tract and their functions.

10. Define the ovarian cycle and the menstrual cycle. Trace the events by which an oocyte matures into an ova.

11. Study Figure 18.13. Then, on your own, diagram the feedback loops governing the female reproductive cycle.

12. Describe how the development of a placental mammal (such as a human) differs from the development of a frog.

UNIT FIVE

PLANT SYSTEMS AND THEIR CONTROL

19

PLANT CELLS, TISSUES, AND SYSTEMS

On a quiet summer morning in 1883, a cataclysmic explosion literally blew the small South Pacific island of Krakatoa out of the water. In that one small part of the world, all life abruptly came to an end. Only the peak of the volcanic cone remained, smoldering with lava and totally buried in hot ashes and pumice. For about a year, the island was essentially sterile. But winds and water were carrying inadvertent visitors to it from nearby islands. In time, when the island had cooled, some of these visitors settled down and survived. At first, only algae formed mats on the volcanic rocks. Then airborne spores of mosses and ferns took hold in the algal mats. Seeds of flowering plants germinated; grasses and wild primroses thrived in the sunlight and tropical rains streaming down on the volcano's flanks. Over the years, the remains of plants and plant parts such as leaves drifted down, and gradually enriched the soil. By 1896, wild sugarcane and delicate orchids were established. Seeds of coconuts and other trees drifted over on the sea's currents and took root. All the while, birds and other animals were visiting the island—bringing plant seeds from figs and papayas in their digestive tracts. Half a century after the explosion, Krakatoa was cloaked with a thick forest that served as the food base for more than a thousand species of animals.

In the spring of 1980, Mount Saint Helens in southwestern Washington exploded violently. Within minutes, hundreds of thousands of mature trees were blown down like matchsticks near the volcano's eastern flank. Thick ashes and pumice turned the once-thick forest below into a scarred and barren sweep of land. In time, plants will repopulate the land, as they did on Krakatoa. Roots will sink into the soil and absorb water and nutrients; stems will grow upward, carrying leaves that will spread out beneath the sun; and the jarring vision of the earth completely devoid of plant life will be only a memory.

Events of this magnitude dramatize our total dependence on the plants that have come to cloak the land. We could no more survive without them here on earth than we could survive on the rock-strewn surface of the moon. In this unit, you will be reading about these plants—how they can acquire water and nutrients essential for growth, how they can grow upright toward the sun, how food from leaves travels downward and sustains stems and roots, even as water from roots travels upward to the top of the highest trees. Here you will be reading about the tissues and organs necessary in these tasks. In the next chapter, you will look at how these tissues and systems are put to work in response to environmental conditions—the relatively dry soil, winds, and seasons characteristic of land. Plant anatomy and physiology—these are aspects of remarkable organisms living side by side with us, and essential to our survival.

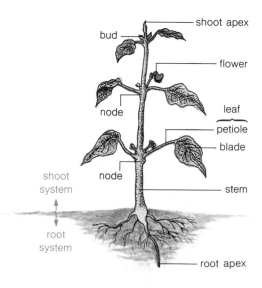

Figure 19.1 Basic body plan for a typical vascular plant.

THE PLANT BODY: AN OVERVIEW

By now, you know about the pitfalls of using a "typical" plant as a textbook example of all plants. A plant may be a tiny string of single cells or a giant redwood. It may live in freshwater, seawater, on land—even high above a forest floor, attached to the woody stem of another plant. Some are **vascular plants.** They have well-developed internal tissues that conduct food and water through the multicelled plant body. Others are **nonvascular plants.** They have no internal transport systems, or what they do have is somewhat rudimentary. As you might imagine, almost all cells of nonvascular plants must remain in direct or nearly direct contact with the environment if nutrients, water, and other substances are to move into and out of individual cells. Most live in water or in moist land settings, and their body plans generally are filamentous or sheetlike. We will be looking at the nonvascular plants in the next unit. Here, examples will be drawn from two kinds of familiar vascular plants that dominate the land: **gymnosperms** (chiefly conifers such as pines, junipers, and redwoods) and **angiosperms** (flowering plants such as roses, cherry trees, corn, and dandelions).

Figure 19.1 shows the body plan of a flowering plant. The plant body differentiates first into three kinds of vegetative organs: the root, stem, and leaf. Later in the life cycle, flowers also appear. A flower is composed of several organs and serves as a reproductive organ system. In the sections to follow, we will be considering the embryonic origins and organization of the cellular building blocks found in these vegetative and reproductive organs.

HOW PLANT TISSUES ARISE

The multicelled plant or animal follows a prescribed journey of growth and differentiation from a single-celled zygote. It grows through mitotic divisions; it differentiates as common predecessor cells give rise to cells and tissues that vary in structure and function. Genetic controls, along with environmental conditions, interact in influencing the journey. However, even though plants and animals are alike in having the same kind of genetic controls over development, they are not exactly alike in *how* they develop.

For animals, all but a very few cells typically become differentiated during embryonic development. Even before birth, most are already committed to traveling down a particular developmental road. In contrast, *plants have the property of open growth.* Even at "maturity," many parts of the vegetative body are composed of nondifferentiated cells. We might say that these are embryonic regions, for new cells continue to be produced here and then become specialized for one task or another. Such embryonic tissue zones are know as **meristems.** Root and shoot tips have dome-shaped *apical meristems.* Here, new cells form rapidly through mitotic divisions. All other plant tissues are derived from apical meristems. First the newly formed cells differentiate into three other kinds of embryonic tissues—protoderm, ground meristem, and procambium. Then cell division, growth, elongation, and differentiation in these tissues produce the **primary plant body.** *Protoderm* gives rise to the plant's surface layers. *Ground meristem* produces most of the plant body's substance. *Procambium* produces primary vascular tissues (Table 19.1).

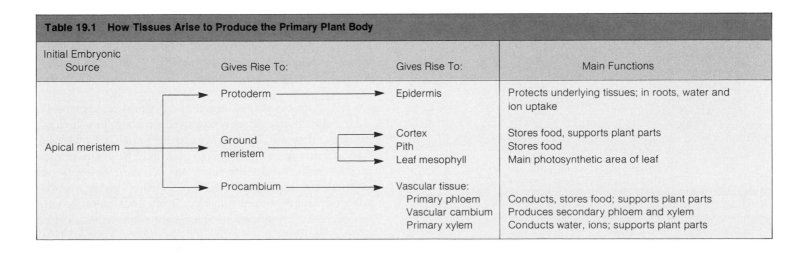

Table 19.1 How Tissues Arise to Produce the Primary Plant Body

Initial Embryonic Source	Gives Rise To:	Gives Rise To:	Main Functions
Apical meristem	Protoderm	Epidermis	Protects underlying tissues; in roots, water and ion uptake
	Ground meristem	Cortex	Stores food, supports plant parts
		Pith	Stores food
		Leaf mesophyll	Main photosynthetic area of leaf
	Procambium	Vascular tissue:	
		Primary phloem	Conducts, stores food; supports plant parts
		Vascular cambium	Produces secondary phloem and xylem
		Primary xylem	Conducts water, ions; supports plant parts

Table 19.1 summarizes the developmental sequence leading to the primary plant body. As you will read later, many plants later increase in thickness as **secondary tissues** are added. Such secondary growth occurs near **cambia**: meristems that run parallel with the sides of roots and stems. For now, the point to remember is this:

The vegetative body of vascular plants shows continued growth through the activity of tissue regions called meristems.

PLANT CELLS AND TISSUES

Identifying animal cells and tissues is fairly straightforward. Nerve, heart, and skeletal muscle cells and tissues are differentiated into distinct types, clustered in distinct regions. But the plant pattern of continued growth from embryonic zones makes it impossible to know exactly where one plant region ends and another begins. Where does a root end and the stem begin? Where does the stem end and a leaf or flower begin? *The same tissue systems are common to all parts of the vascular plant.* All are derived from the same kinds of embryonic tissues, derived from apical meristems. Their structure and function depend on where they are actually located in the mature plant body.

Three main tissue systems extend continuously through the plant. They are called dermal, vascular, and ground (or fundamental) tissue. The **dermal system** consists largely of protective coverings (epidermis or periderm) on the plant's outer surface. The **vascular system** contains food-conducting tissues (phloem) and water-conducting tissues (xylem) that course through the entire plant. The **ground system** makes up most of the plant body. It consists mainly of three tissues (parenchyma, collenchyma, sclerenchyma) that vary in structure and function.

Vascular plants are constructed mainly of three tissue systems—dermal, vascular, and ground—that extend continuously through the plant body.

How are the three basic tissue systems arranged relative to one another? In essence, the dermal system sheathes the ground system, in which vascular tissue is embedded. Vascular tissue is organized in strands. These strands are mostly parallel with the long axis of stems or roots. As you will see, the vascular strands of many plants are distributed in numerous bundles throughout stem ground tissue. Vascular bundles in other plant stems are organized

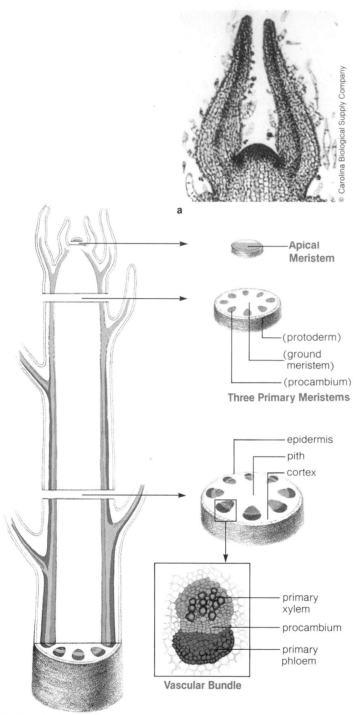

© Carolina Biological Supply Company

a

Apical Meristem

(protoderm)
(ground meristem)
(procambium)

Three Primary Meristems

epidermis
pith
cortex

primary xylem
procambium
primary phloem

Vascular Bundle

b *Sketch of a developing stem, sliced in half lengthwise, and positions of developing vascular tissue*

Figure 19.2 Pattern of stem development for one kind of vascular plant. Apical meristems give rise to the three primary meristem tissues, which in turn give rise to the tissues of the primary plant body.

in a ring (Figure 19.2). Ground tissue inside such rings is called **pith.** Ground tissue between the vascular ring and the dermal surface layer is the **cortex.** Let's now take a look at the component cells of the three tissue systems.

Parenchyma: Photosynthesis, Food Storage

Parenchyma is the main cell type of ground tissue systems. Parenchyma cells form continuous tissues in the stem and root cortex, and in stem pith. They also form the photosynthetic tissue, or *mesophyll,* between a leaf's upper and lower epidermis. Some are even incorporated into vascular tissue systems.

Parenchyma cells vary in size, shape, and wall structure. Some look like rough-cut gems, and press against adjacent cells along their irregularly shaped and numbered facets (Figure 19.3). You'll find these cells in stem pith. All parenchyma cells have relatively thin primary walls. But in some places, such as mature stems, the walls become thicker due to the deposition of secondary cell wall materials. Some parenchyma cells are elongated in several directions. Joined together, they vaguely resemble an immense team of circus acrobats, with the "arms" of one supporting the "legs" of another (Figure 19.3c). This kind of tissue pattern leaves abundant air spaces between cells. As you might imagine, parenchyma cells of this type are important in aerating leaf and stem regions where gas exchange is required in photosynthesis.

Even though parenchyma cells become specialized in such tasks as food storage and photosynthesis, they don't lose the ability to divide. These are the cells responsible for healing wounds, regenerating plant parts (even whole plants), and assuring successful grafts (joining part of one plant to another at an incision made in a stem or root).

Collenchyma: Supporting Young Plant Parts

Just beneath the epidermis of many stems and leaves, you'll find thick-walled yet flexible tissue called **collenchyma.** The overlapping cells of this tissue are longer than parenchyma cells (some are two millimeters long). They also have unevenly thickened walls built up in several patterns (Figure 19.4). When a shoot is first developing, collenchyma cell walls expand in surface area and in thickness. These walls contain cellulose and pectin. Being hydrophilic, pectin attracts water into the walls. Partly because the walls contain considerable water, they remain pliable. Hence stems and leaves can elongate even while gaining structural support from collenchyma cells.

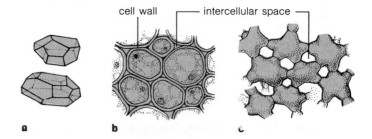

Figure 19.3 Examples of parenchyma cells found in the stem and root cortex (**a,b**) and between a leaf's upper and lower epidermis (**c**).

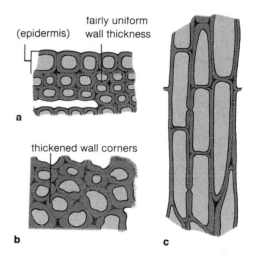

Figure 19.4 Examples of collenchyma cell wall thickening, shown as if the cells were sliced crosswise (**a,b**) and lengthwise (**c**). (From Esau, *Anatomy of Seed Plants,* Wiley, 1977)

Sclerenchyma: Strengthening Mature Plant Parts

Mature stems and leaves, as well as seeds, contain **sclerenchyma.** These cells have somewhat uniformly thick walls. The cells themselves are usually dead at maturity, with only their rigid walls remaining. Sclerenchyma cells occur individually or as small clusters in dermal, ground, and vascular tissue systems.

Sclerenchyma cells are of two sorts: sclereids and fibers. The diverse *sclereids* include cells resembling stones, hourglasses, columns, hairs, and stars. Stone cells (Figure 19.5) give pear flesh its gritty texture. They typically have

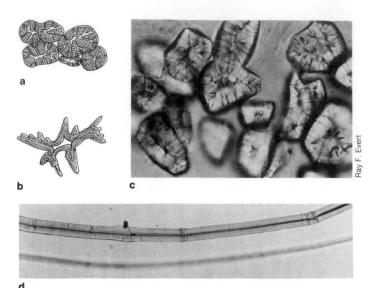

a

b   c

d

Figure 19.5 Examples of sclerenchyma cells. Stone cells, sketched as if sliced through their midsections (**a**) and under the microscope (**c**). Star sclereid, with its arms radiating from a central body (**b**). Stone cells and star sclereid are in the sclereid category. (**d**) Another kind of sclerenchyma cell—a fiber, isolated from a flax stem and sliced lengthwise. (d from C. Carpenter and L. Leney, *Papermaking Fibers*, 1952)

very thick secondary walls impregnated with lignin. *Fibers* are long cells having somewhat flexible walls. These cells support plant parts that have ceased growing. Strands of fibers have great commercial value. They are used in papermaking, in textile and thread manufacture, and in ropemaking. (Cotton "fibers" don't belong in this category; they are really epidermal hairs on cotton seeds.)

Xylem: Water Transport, Storage, and Support

The term **xylem** refers to a highly complex tissue that extends continuously through the plant body. Some of its cells (parenchyma) function in food and water storage. Still others (sclereids and fibers) function in mechanical support. Its most highly specialized cells—**tracheids** and **vessel elements**—passively conduct water and dissolved salts. Both tracheids and vessel elements are more or less elongated cells with relatively thick, lignin-impregnated secondary walls. Both kinds of cells are dead at maturity. Only the cavities that cytoplasm and vacuoles once occupied function as water pipelines. Figure 19.6 shows the structure of xylem's water-conducting cells.

Vessel elements differ from tracheids in having one or more openings called perforations at one end. Vessel elements typically are joined end to end, and water flows freely from one to another through interconnecting perforations. Often the perforations become so large that the end wall disappears altogether. A series of such cells, with the end walls gone, form a continuous tube called a *vessel.* In some flowering plants, vessels are more than a meter long. In conifers, vessel elements are not present. The xylem of these plants is composed mostly of tracheids and fibers.

In contrast to vessel elements, tracheids are not perforated. Water movement between them usually occurs through paired-up **pits:** numerous circular or oval depressions in the cell wall where *secondary* wall material has not been deposited. Apparently, the primary wall in pit regions is not much of a barrier to water and dissolved ions.

During primary growth, xylem is often laid down in strands parallel with the stem or root long axis. In plants showing secondary growth of roots, stems, and larger branches, secondary xylem forms what we call **wood.** In wood, xylem cells are arranged in two patterns. Cells of the *vertical system* run parallel with the long axis of stems and roots. For the most part, its components are the water conductors and supportive cells. Cells of the *ray system* radiate horizontally from the long axis. They store starches and proteins during autumn and winter. During periods of growth, though, these cells transport digestive products of these and other foods through the plant body. More will be said later about these two systems.

Phloem: Food Transport, Storage, and Support

In all vascular plants, another important tissue parallels the xylem highways. Like xylem, it is laid down initially in strands and then, during secondary growth, between these strands. This tissue is called **phloem**. It functions in food conduction, storage, and support. Sclereids and especially fibers in phloem provide mechanical support. Parenchyma cells store food, for the most part. Food-conducting cells found in this tissue are called **sieve elements.** Unlike xylem cells, sieve elements are alive at maturity.

In flowering plants, you'll find sieve elements joined end to end as continuous pipelines. Some wall regions of this pipeline have numerous small pores. Water and dissolved substances flow into and out of neighboring cells through these pores. End walls typically have larger pores than do the side walls. These end regions are called *sieve plates* (Figure 19.7).

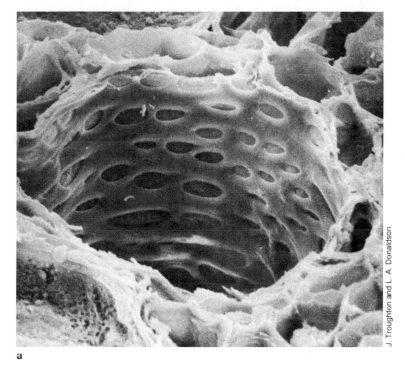

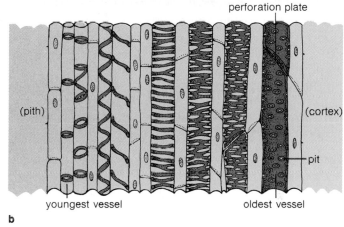

perforation plate

(pith)

(cortex)

pit

youngest vessel

oldest vessel

b

J. Troughton and L. A. Donaldson

a

Figure 19.6 (**a**) Scanning electron micrograph of a vessel element from a cucumber stem. Notice the pits on the side walls. (**b**) Appearance of primary xylem during stem development. Cells are shown as if sliced lengthwise through their midsections.

The oldest (first-matured) cells are the most stretched, having developed during rapid primary growth. The youngest (last-matured) cells developed after elongation ceased; their wall regions will not be stretched further. Vessels in between reflect decreased stretching that corresponds to a decrease in stem elongation. Secondary-wall depositions are shown in dark color. Such depositions prevent collapse as water under tension is pulled through the vessel element. (Sketch from Weier et al., *Botany: An Introduction to Plant Biology*, Wiley, 1974)

Sieve elements of flowering plants are thought to depend on special adjacent parenchyma cells, called **companion cells,** in moving food from photosynthetic sites in leaves to other plant parts. Although the nucleus has been digested away in the mature sieve element, the companion cell nucleus remains functional and might direct activities of *both* cells through the pores between them. Apparently, companion cells help load and unload the phloem pipelines. They move sugars into adjacent sieve elements, *against* concentration gradients. Chapter Twenty explains how such gradients are at the heart of the phloem pipeline operation.

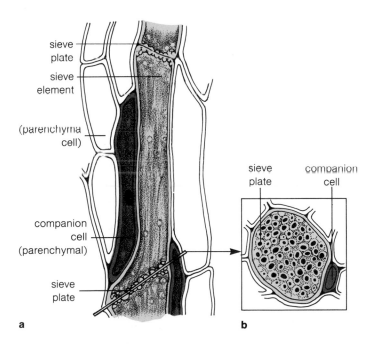

sieve plate

sieve element

(parenchyma cell)

companion cell (parenchymal)

sieve plate

sieve plate

companion cell

a

b

Figure 19.7 (**a**) Mature sieve-tube element and companion cell, sliced lengthwise through their midsections. (**b**) Face view of a sieve plate. Dark regions represent the sieve pores. (After Salisbury and Ross, *Plant Physiology*, 1978)

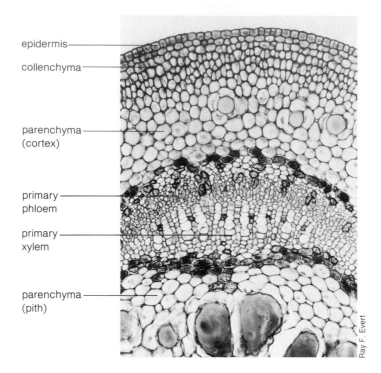

epidermis

collenchyma

parenchyma
(cortex)

primary
phloem

primary
xylem

parenchyma
(pith)

Ray F. Evert

Figure 19.8 Examples of some of the cell and tissue types described so far in the text. This is a section through the stem with mainly primary tissue.

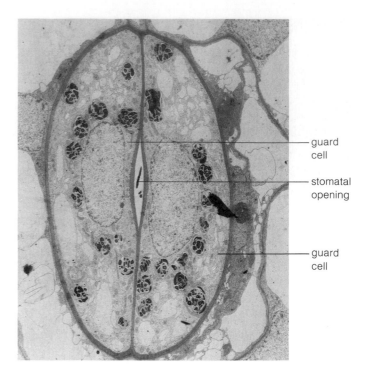

guard
cell

stomatal
opening

guard
cell

Figure 19.9 Paired guard cells of a stoma on the stem of a beavertail cactus (*Opuntia*). Stomata open and close in response to environmental conditions, thereby enhancing the movement of carbon dioxide into the plant, and limiting water loss during drought or at night. (From W. Thomson, "Studies on the Ultrastructure of the Guard Cells of Opuntia," *Amer. J. Bot.* 57(3):309–316, 1970)

Epidermis and Periderm: Interfaces With the Environment

Figure 19.8 shows a section through a stem. The outermost cell layer is the **epidermis.** Epidermis covers the plant, usually as a single, compact, and continuous layer of cells. However, epidermis varies in structure and function, depending on the nature of the external environment. For example, some thin-walled epidermal cells in roots form protuberances that grow in length. Such so-called *root hairs* enhance water absorption from the surrounding soil. On aerial plant parts, the outer walls of epidermal cells are impregnated with waxes and a mixture of substances called cutin. These substances form a **cuticle:** a surface coating that restricts water loss from the plant and may confer some resistance to microbial attack.

Embedded in the epidermal layer are highly specialized cells that represent further adaptations to environmental conditions. In leaves and stems, numerous pairs of guard cells (Figure 19.9) serve as gates across the epidermis. Each of these gates is a **stoma** (plural, stomata). Stomatal opening influences the movement of carbon dioxide into the leaf. The cuticle and stomatal closure influences the movement of water out of the leaf. More will be said about these cells in Chapter Twenty. Leaf epidermis in such plants as mint, lavender, and peppermint also contains oil-secreting structures; hair cells and scales are also common. In the epidermis of insect-eating plants are structures that secrete sticky mucopolysaccharides—as well as enzymes that digest insects trapped by the goo (Figure 19.10). Saltbush and tamarisk plants have epidermal glands that collect excess salt from the plant body and secrete it to the outside. Flower parts, stems, and leaves may contain numerous nectaries (tissues or glands that secrete sugar-rich fluid derived from phloem and xylem). Many more epidermal specializations exist, including such adaptations against predators as hooks on cells that impale insects, and cells that secrete foul-tasting chemicals.

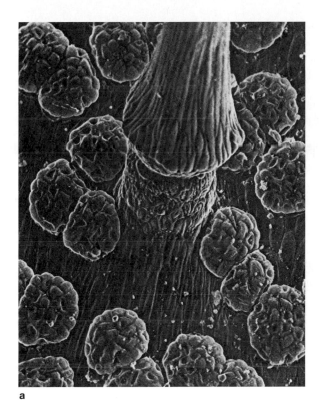

a

b

c

Figure 19.10 Epidermal adaptations to the environment, as seen in the Venus flytrap. These plants grow in nitrogen-poor soil. Their two-lobed leaves open and close like a clam shell; the leaf margin is fringed with spines that intermesh when the lobes close. (**a**) Suppose an insect lands on the leaf and moves against a long epidermal hair (the base of one is shown here). The movement triggers cellular changes at the leaf midrib, and the leaf closes up (**b, c**). Glandlike epidermal cells (the pincushion-like structures in the micrograph) secrete enzymes that digest the proteins of the trapped insect. Nitrogen is released from the proteins so kindly provided. (a from John N. A. Lott, *A Scanning Electron Microscope Study of Green Plants*, C. V. Mosby Company, 1976)

In the thickening stems and roots of gymnosperms or flowering plants undergoing secondary growth, epidermis is replaced by a protective covering. This covering is **periderm**. Its component cells arise from meristem called cork cambium. The outermost cells formed are not alive at maturity; their suberin-impregnated walls remain behind to form a tissue called **cork**. (Cork isn't the same thing as "bark," a nontechnical term that applies to all living and nonliving tissues that have formed outside the vascular cambium.)

THE ROOT

In gymnosperms and flowering plants, the first (primary) root develops from the apical meristem at the root tip of the embryo (Figure 12.2). The primary root increases in length and diameter, and other roots may arise laterally from it. A system based on the primary root and its various lateral branchings is a **taproot system**. In other cases, the primary root may be replaced by *adventitious* roots: numerous branchings that originate not on the primary root

body but, in this case, on the young stem! Adventitious roots also become branched, but all the roots are somewhat alike in length and diameter. They form a **fibrous root system**. Both taproot and fibrous root systems serve in anchoring the plant and in storing photosynthetically derived food. By far their most important function is absorption of water and mineral salts. Roots are long, cylindrical, and highly branched; whole systems of roots penetrate a large volume of soil and absorb a tremendous amount of water from it (Chapter Twenty).

Root Primary Structure

During primary growth, all roots show much the same internal organization. Just in front of the root apical meristem is a thimbleshaped cell mass known as a **root cap** (Figure 19.11). This cap protects the delicate young root from being torn apart during growth through the soil. Behind the root cap, the apical meristem gives rise to cells that differentiate into the three basic tissue systems (dermal, ground, and vascular). In the absorptive regions

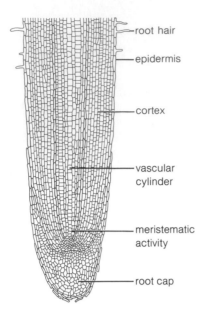

root hair

epidermis

cortex

vascular
cylinder

meristematic
activity

root cap

Figure 19.11 Sketch of a root tip, sliced lengthwise. Compare with Figure 20.19. (From Salisbury and Ross, *Plant Physiology*, 1978)

of a mature root, epidermis with root hairs surrounds the cortex, which surrounds a vascular cylinder (Figure 19.12). The cortex occupies most of the root; it is composed of relatively undifferentiated parenchyma cells. The inner core of the vascular cylinder often consists of pith, but in some cases the pith cells disintegrate and leave a hollow core.

The vascular cylinder is surrounded by a specialized, innermost layer of cortex, called the **endodermis**. This single layer of cells helps control the movement of water and dissolved salts into the xylem pipeline. In the wall of each endodermal cell is a continuous but localized band of fatty suberin deposits. The deposits are like impenetrable concrete between adjacent endodermal bricks that sheathe the vascular cambium. This continuous band, the *casparian strip*, is shown in Figure 19.12. About the only way water gets into the root's xylem pipeline, then, is across the plasma membrane of individual endodermal cells. The plasma membrane, recall, is differentially permeable. Through active and passive transport mechanisms (Chapter Five), water and solute movements can be adjusted in response to environmental conditions.

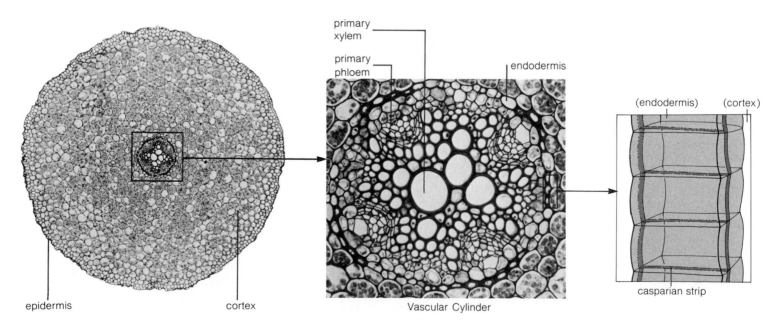

primary xylem

primary phloem

endodermis

(endodermis)   (cortex)

casparian strip

epidermis   cortex

Vascular Cylinder

Figure 19.12 Section through a buttercup (*Ranunculus*) root, sliced crosswise. The insets give closer views of the vascular cylinder and the casparian strip. In this kind of root, pith is not generally formed in the vascular cylinder. (Micrographs from Ray F. Evert. Sketch from Esau, *Anatomy of Seed Plants*, Wiley, 1977)

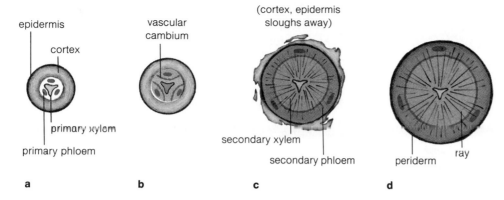

epidermis

cortex

primary xylem

primary phloem

vascular cambium

(cortex, epidermis sloughs away)

secondary xylem

secondary phloem

periderm

ray

a   b   c   d

Figure 19.13 An alfalfa root, sliced crosswise, in different developmental stages. (**a**) Primary growth has given rise to components of the three basic tissue systems (dermal, vascular, and ground). (**b**) Secondary growth begins when the vascular cambium lying between primary xylem and phloem undergoes division. New, secondary xylem and phloem arise from the vascular cambium (see arrows). (**c**) The root grows in diameter as cell divisions proceed parallel and perpendicular to the vascular cambium. The cortex ruptures as diameter increases. (**d**) Epidermis is replaced by periderm. (After Esau, *Anatomy of Seed Plants*, Wiley, 1977)

Secondary Growth in Roots

Gymnosperms and most flowering plants undergo secondary growth, which increases root as well as stem diameter. Figure 19.13 summarizes the processes of secondary growth in roots. Between the primary xylem and phloem in each vascular strand of these plants are narrow strips of meristematic cells (procambium). Eventually, the procambium develops into **vascular cambium**: a one-cell-thick layer of lateral meristem that gives rise to secondary xylem and phloem. Secondary xylem forms on the cambium's inner surface (toward the core of the root). Secondary phloem forms on its outer surface. As the volume of new xylem increases, it pushes the vascular cambium toward the root's periphery. Part of the ground tissue between the phloem and endodermis begins dividing soon afterward. This tissue region is the **pericycle** (Figure 19.12). Cells of the pericycle become aligned with the vascular cambium, and thereby become part of a continuous ring of active cambium one cell thick. Layer after layer of secondary xylem and phloem are now produced by the combined cambial tissue. As you might imagine, the burgeoning mass of new tissue inside the root causes the cortex and outer phloem to rupture. Parts of the cortex are split away and carry epidermis with it. Cork cambium arises on the phloem tissue's outer ridges. Its divisions produce periderm, the corky covering that replaces the epidermis. In some plants, secondary growth of this sort occurs year after year and produces massive, woody roots.

THE STEM AND ITS LEAVES

The stems of vascular plants serve several functions. They are the structural framework for upright growth that gives leaves favorable exposure to light. With their phloem and xylem pipelines, they provide routes for food and water movement between roots and leaves. Some food storage occurs in parenchyma cells of cortex, pith, xylem, and phloem. Both the stem and its leaves represent the shoot system of vascular plants.

Stem Primary Structure

Although stems of the primary plant body are also composed of the three basic tissue systems, they are more complex in structure than roots. At specific points along the stem, leaves are differentiated. These points are called nodes, and segments between them are called internodes. Embryonic shoots called **buds** form in the notches where leaves are attached to the stem (Figure 19.1). These stem buds often grow and differentiate, creating a branched plant body. Complicating the picture further are structural differences associated with different environments. Stems grow underground, in air, or in water. Some stems are thick and free-standing, others creep along the ground; still others climb upward, by elongating in complex ways that allow them to wrap around the stems of other plants.

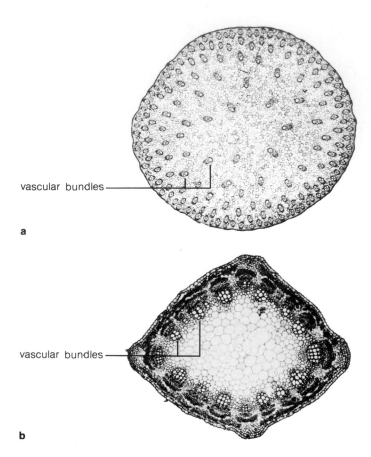

vascular bundles ———

a

vascular bundles ———

b

Figure 19.14 Arrangement of vascular bundles in monocot and dicot stems. (**a**) Section through the stem of a corn plant, a monocot; (**b**) section through the stem of alfalfa, a dicot. (a, © Carolina Biological Supply Company; b, Ray F. Evert)

In the case of flowering plants, stem structure follows not one but two general patterns. Flowering plants are divided into two major classes: Monocotyledonae and Dicotyledonae. It's easier to think of them as **monocots** and **dicots**. (The names refer to the number of cotyledons, or seed leaves, present in the seeds of these plants. Monocots have one such leaf, dicots have two.) Monocot stems usually are uniformly thick along their length; dicot stems are usually tapered. Most monocot stems (and fibrous roots) don't undergo secondary growth; most dicot stems do. As you will see, the primary vascular tissues in monocot and dicot stems are arranged differently. Because of these and other variations, it's possible to give only a generalized picture of stem structure in this brief survey.

Figure 19.14 shows sections through a monocot and a dicot stem. Like roots, both stem types are composed of dermal, ground, and vascular tissue. But in many monocots, the strands of primary vascular tissue are arranged in numerous bundles that are distributed throughout the ground tissue, when seen in cross-section. Once primary xylem and phloem are formed, the procambium does not further develop into vascular cambium; hence secondary xylem and phloem don't occur in these monocots. In most dicots, the vascular bundles are arranged in a cylinder that divides the ground tissue into pith and cortex.

Leaf Primary Structure

During primary growth of a stem, lobelike tissue masses called *leaf primordia* form at genetically prescribed intervals on the sides of an advancing apical meristem. The lobes expand into a stalked blade having a highly branched vascular system (veins). The veins connect with the main primary vascular bundles in the stem itself. Secondary xylem and phloem are not formed in leaves.

Each leaf is an individual organ, an integrated unit of cells and tissues performing a common task. Most are sites for photosynthesis. With its large external surface area—and even larger internal surface areas afforded by individual leaf cells—a leaf provides exceptional contact with the surrounding air. Its structure assures sunlight interception and rapid carbon dioxide uptake even to the innermost photosynthetic cells.

Figure 19.15 shows the tissue layers common to many flowering plant leaves. Uppermost is a protective epidermis, with its cells covered by cuticle. Next comes *palisade parenchyma*, a loosely packed tissue of photosynthetic cells. Below the palisade tissue is *spongy parenchyma*. Between thirty and fifty percent of the leaf consists of air spaces around spongy parenchyma and around most of each palisade cell wall. Below the spongy parenchyma is another cuticle-covered epidermal layer. This layer often contains most of the stomata that permit movement of carbon dioxide into the leaf, and also allow water loss. As you will read in the next chapter, photosynthesis depletes carbon dioxide in the intercellular air spaces. This, in turn, stimulates stomata into opening. When stomata open, carbon dioxide diffuses into the leaf and is rapidly absorbed by palisade and spongy parenchyma cells.

Leaf structure varies considerably among plant species. For example, some succulents (such as *Sedum*) rely on large, water-storing parenchyma cells and leaves with thick cuticles for water conservation. Succulents, all grasses, and

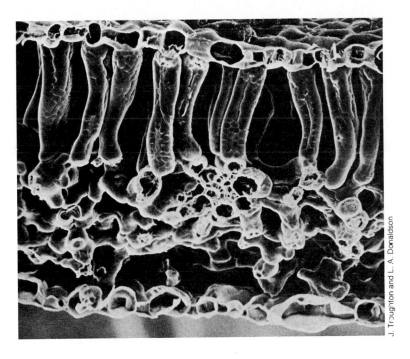

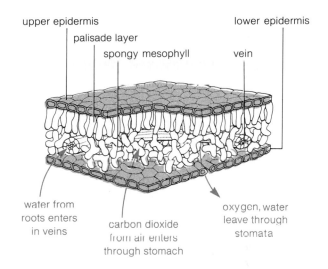

upper epidermis

palisade layer

spongy mesophyll

lower epidermis

vein

water from roots enters in veins

carbon dioxide from air enters through stomach

oxygen, water leave through stomata

J. Troughton and L. A. Donaldson

Figure 19.15 Example of leaf structure. The scanning electron micrograph shows a slice through a mature bean leaf. The sketch identifies the different leaf cells, and shows their arrangement.

conifers have no distinct palisade and spongy parenchyma layers; only cells having appearances intermediate between those shown in Figure 19.15 are present. Such plants normally grow in regions where water is scarce. Some water lilies have leaves that float on water; their stomatal gates are, as you might imagine, exclusively on the upper epidermis. Many plants bear only a solitary leaf at each stem node. Leaves develop from the stem in different directions, the overall arrangement being a spiraling of one leaf after another as they appear at nodes along the stem. Some leaves have a single blade; apple leaves are like this. Others are divided into leaflets; acacia and ash trees have compound leaves of this sort. With all the diversity, though, almost all leaves are alike in being short-lived organs. Even in plant species showing secondary growth, leaves drop away from the stem during certain times of the year.

Secondary Growth in Stems

When older stem regions begin secondary growth, vascular cambium between the xylem and phloem strands in vascular bundles becomes active. The vascular cambium will produce secondary xylem and phloem, much as it does in roots. Two kinds of cells occur in the vascular cambium.

The vertically elongated *fusiform initials* produce xylem and phloem cells that are arranged parallel with the stem-long axis (Figure 19.16). Their products make up the axial system of secondary vascular tissues. The axial system conducts food and water up and down the stem. The *ray initials* are shorter, brick-shaped cells. They produce the ray system of (mostly) parenchyma cells that act as conduits and food storage centers in wood. Through the ray system, water from secondary xylem is fed laterally into the vascular cambium and secondary phloem. Through this system, food from secondary phloem moves into the vascular cambium and the still-living cells of secondary xylem.

In regions having prolonged dry spells or cool winters, vascular cambium becomes inactive during parts of the year. The first xylem cells produced tend to have fairly large diameters and thin walls; they form early wood. As the season progresses and less water is around, the cell diameters become smaller and the walls thicker; these cells form late wood. The last-formed, small-diameter cells of late wood will end up next to the first-formed, large-diameter cells of the next spring's growth. Although we don't see individual cells, there is enough difference in light reflection from a stem cross-section to reveal alternating light bands (early wood) and dark bands (late wood),

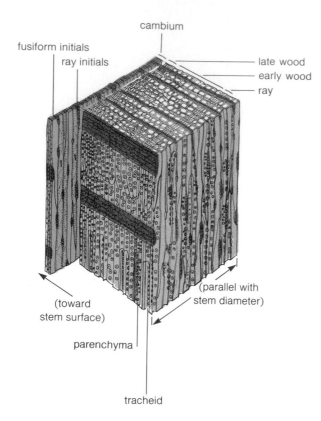

Figure 19.16 Block diagram of a section through white cedar, showing the cells of vascular cambium that give rise to the axial and ray systems of wood. (From Esau, *Anatomy of Seed Plants*, Wiley, 1977)

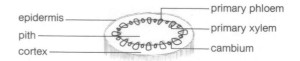

a *Beginnings of secondary growth*

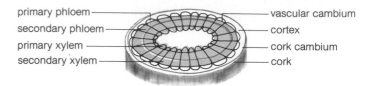

b *Secondary growth under way*

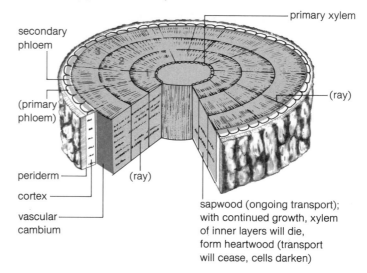

c *Stem after three years of growth.*
Numbers refer to annual growth layers (secondary xylem)

which represent annual growth layers (see Figure 19.17).

In wet tropical regions, there is often a continuous growing season. The annual growth layers of tropical woody plants are either faint or nonexistent. The more pronounced the shifts in seasons, the more pronounced the annual growth layers.

As the mass of xylem increases season after season, it tends to crush the thin-walled phloem cells from the preceding year's growth. That is why new rings of phloem cells must be produced each year, outside the growing inner core of xylem. As you can see from Figure 19.17, phloem in older trees is confined to a thin zone beneath the periderm. If even a narrow band of phloem is stripped all the way around a tree's circumference, the tree will die. The phloem cells will be stripped away and will not be able to transport photosynthetically derived food down to roots, which will starve to death.

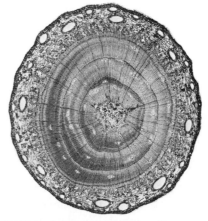

d *Cross-section through a two-year-old pine stem*

Figure 19.17 Stages of secondary growth (**a–c**) and cross-section of a two-year-old stem (© Carolina Biological Supply Company).

REPRODUCTIVE ORGANS OF FLOWERING PLANTS

The roots, stems, and leaves we have been describing are the vegetative organs of the sporophyte, or spore-producing plant (Chapter Eight). In flowering plant sporophytes, the vegetative growth phase gives way (temporarily or permanently) to a reproductive phase that begins with flower production. Figure 19.18 shows the relationship of these two phases in the flowering plant life cycle. Typically, the following events constitute the reproductive portion:

1. Microspore formation	*Microspores develop into pollen grains (immature male gametophytes). Pollen grains later develop into mature, sperm-bearing gametophytes.*
2. Megaspore formation	*Megaspores develop into egg-bearing female gametophytes.*
3. Pollination	*Pollen grains are transferred to female gametophytes.*
4. Fertilization	*Egg and sperm nuclei fuse, and the zygote forms; formation of endosperm (reserve food tissue of the forthcoming seed) is triggered.*
5. Embryo development	*Genetic and cytoplasmic controls (Chapter Twelve) transform the zygote into a plant embryo.*
6. Seed, fruit formation	*The embryo becomes packaged in tissues that aid in its dispersal from the parent plant.*
7. Dispersal, germination	*Seeds and fruits are dispersed (by wind, water, animals). At the proper time and on suitable sites, the seeds germinate (begin growth).*

Germination marks the completion of the reproductive phase. The seedling that now emerges constitutes the immature sporophyte of the new generation. Here we will look briefly at some of the main reproductive organs involved in flowering plant reproduction. The events themselves are discussed in Chapters Twenty and Twenty-Three.

A **flower** is a cluster of reproductive and nonreproductive (sterile) organs. All of these organs are generally regarded as highly modified leaves. Typically, the leaf parts are arranged in whorls on a modified stem tip called a receptacle (Figure 19.19). The outermost whorl consists of *sepals*, often green, sometimes highly colored. The next whorl consists of the *petals*, often conspicuously colored leaves that function in attracting bird and insect pollinators (Chapter Twenty-Three). Petals enclose the **stamens**, male reproductive organs that commonly consist of pollen-bearing structures (anthers), each perched on top of a single stalk (filament). The central whorl of modified leaves consists of one or more **carpels**, or female reproductive organs.

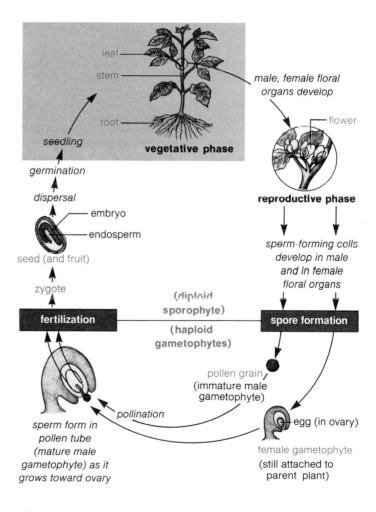

Figure 19.18 States in which vegetative and reproductive structures typically occur in the life cycle of a flowering plant.

A carpel usually has three parts. The expanded base is the *ovary*, which eventually ripens into fruit tissue. Above the ovary is a stalk (called a style). Topping the style is a sticky region, called a *stigma*, which is the landing platform for pollen. Its surface proteins are complementary to those of pollen grains from plants of the same species (Figure 19.20).

The flower just described is a "perfect" flower, having both stamens and one or more carpels. Some plant species have "imperfect" flowers, with only stamens or only carpels. Often they are called male and female flowers. In some species, such as corn, male and female flowers appear on the same plant. In other species, such as poplars and cottonwoods, they occur on separate plants.

In the ovary of each carpel is at least one **ovule**—a structure within which megaspores are produced. Megaspores develop into egg-bearing gametophytes. In the anther of a stamen are one or more structures within which microspores are produced. Microspores develop into sperm-bearing male gametophytes. At some point after pollination, the mature male gametophyte (pollen tube) enters the ovule. A pollen tube has more than one nucleus. One nucleus fuses with the egg, and a zygote forms. Another sperm nucleus fuses with nuclear material in the ovule. The fusion product will give rise to **endosperm**: a mass of tissue that will surround the developing embryo. In dicots, this endosperm remains small or is used up during seed development. The two cotyledons of the embryo usually enlarge and act as food storage centers. In monocots, the endosperm grows considerably; most of the nutrients needed after germination are stored here.

As the endosperm and embryo grow through mitotic divisions, so do tissue layers (integuments) of the ovule and the ovary walls. A fully mature ovule is a **seed**; its integuments form the seed coat. The ripened ovary of one or more carpels constitutes most or all of the structure called a **fruit** (Table 19.2).

When you cut open a mature seed of any flowering plant, you find a similar array of structures. The embryo is already well developed, having followed a limited program of growth and differentiation even while still attached to the parent plant. At one end, the embryo has one or two cotyledons (seed leaves). In many dicots, they are destined to become the first light-trapping structures for the young plant when it begins active growth (Figure 12.2).

Once a seed is completely formed, mitotic divisions cease. The seed dries considerably and enters a state of rest, a period of suspended activity preceding its germination. How long does a plant embryo remain locked within its seed coat and still retain the ability to give rise to a mature plant? The length of time varies from species to species, because plant development is fine-tuned to seasonal fluctuations and to prevailing conditions of different environments. Some delicate seeds will survive only for a few weeks or months; the seeds of some common weeds can endure for decades. Environmental cues shift an embryo from the dormant state to germination, as discussed in Chapter Twenty. These cues set in motion certain internal control mechanisms that govern when embryonic roots will break through the seed coat and begin moving down into the soil, and when the shoot will begin its upward surge. That control is a subject of the next chapter.

Readings

Bold, H., C. Alexopoulos, and T. Delevoryas. 1980. *Morphology of Plants and Fungi.* Fourth edition. New York: Harper and Row. Comprehensive reference book for the serious student.

Cutter, E. 1969. *Plant Anatomy: Experiment and Interpretation.* Reading, Massachusetts: Addison-Wesley. Also available in paperback.

Epstein, E. 1973. "Roots." *Scientific American* 228(5):48–58.

Esau, K. 1977. *Anatomy of Seed Plants.* Second edition. New York: Wiley. Excellent, well-illustrated book, a standard reference in the field.

Foster, A., and E. Gifford. 1974. *Comparative Morphology of Vascular Plants.* San Francisco: Freeman.

Raven, P., R. Evert, and H. Curtis. 1976. *Biology of Plants.* Second edition. New York: Worth. Well-written book, outstanding illustrations.

Ray, P. 1972. *The Living Plant.* Second edition. New York: Holt.

Rost, T., et al. 1979. *Botany: A Brief Introduction to Plant Biology.* New York: Wiley. Excellent abridged version of a classic (T. Weier et al., 1974, *Botany,* fifth edition; same publisher).

Weier, T., C. Stocking, and M. Barbour. 1974. *Botany: An Introduction to Plant Biology.* Fifth edition. New York: Wiley. Time-tested introduction to the world of plants, very well-written and clear illustrations.

Review Questions

1. Sketch the basic body plan of a flowering plant, showing the general location of its vegetative and reproductive organs.

2. How do plant cells and animal cells differ in their development?

3. Define meristem. Which meristem regions produce the primary plant body? Which two kinds of active cambium give rise to layers of secondary xylem and phloem?

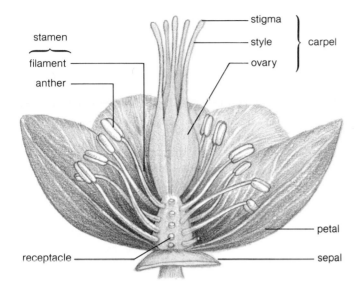

Figure 19.19 Typical arrangement of floral organs. Shown here, a flower of the Christmas rose (*Helleborus*). Five individual carpels are spirally arranged in the center. (From Rost et al., *Botany*, Wiley, 1979)

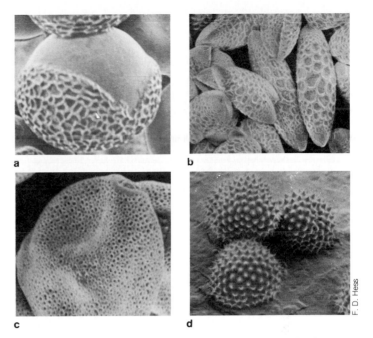

F. D. Hess

Figure 19.20 Scanning electron micrographs of pollen grains from (**a**) iris, (**b**) day lily, (**c**) cucumber, and (**d**) ragweed. Pollen is blown to or transported by birds, insects (even bats) to distant flowers.

4. Which cell types appear in vascular tissue systems? Ground tissue systems? Epidermal tissue systems? With a sheet of paper, cover the column entitled "Main Functions" in Table 19.1. Can you now state the primary function of each plant tissue listed?

5. Is this statement true or false: Unlike a taproot system, a fibrous root system has lateral branchings.

6. Does endodermis occur only in roots? What is the functional relationship between endodermis and the casparian strip?

7. Describe some differences between monocots and dicots. How are primary vascular tissues generally arranged in monocot stems? In dicot stems?

8. In a leaf's interior, are cell layers tightly or loosely packed? How do you suppose this packing arrangement relates to photosynthesis?

9. How are annual growth layers formed in woody stems? Describe what happens when you strip away a narrow band of phloem from a tree's circumference.

10. Draw a flower and label its sterile and reproductive parts. Explain floral function by relating some floral structures to events in the life cycle.

Table 19.2 Kinds of Fruits of Some Flowering Plants

Type	Characteristics	Some Examples
Simple (formed from single carpel, or two or more united carpels)	1. Fruit wall (pericarp) *dry; split* at maturity	pea, magnolia, tulip, mustard
	2. Fruit wall *dry; intact* at maturity	sunflower, corn, wheat, rice
	3. Fruit wall *fleshy*	cherry, peach, olive, lemon
Aggregate (formed from numerous but separate carpels of single flower)	*Aggregate* (cluster) of matured ovaries (fruits), all attached to common receptacle (modified stem end)	strawberry, blackberry, raspberry
Multiple (formed from carpels of several associated flowers)	*Multiple* matured ovaries, grown together into a mass; may include accessory structures (such as receptacle)	pineapple, fig, "ear" of corn, mulberry

It took you eighteen years or so to grow five or six feet tall. A corn plant can accomplish the same thing in three months! Yet how many of us stop to think that corn or any other kind of plant actually does anything at all impressive? Being animals endowed with remarkable mobility, unsurpassed intelligence, and richly varied emotions, we tend to be endlessly fascinated with ourselves—and somewhat indifferent to the immobile and expressionless vegetative plants around us. Besides, from our own experiences and educational biases, we simply have acquired more knowledge about animals than we have about plants. But look outside, beyond the crowded city streets, and you see at once that it is not animal life but *plant* life that dominates the earth's surface. Plants, not animals, sustain whole communities of life by serving as initial food producers in complicated feeding relationships among organisms. Plant function, the subject of this chapter, clearly deserves better than the bias many of us bring to it.

PROBLEMS AND ADAPTATIONS FOR LIFE ON LAND

Plants, like all other organisms, grow and reproduce only as long as they acquire resources from their environment. One of these resources is water. Consider that plants are composed largely of water, and that water absorption is the driving force for plant growth (through cell expansion). Yet the soil from which water must be absorbed is frequently rather dry. Water acquisition, then, is a major problem for most land plants. Or consider that carbon makes up about forty-five percent of a plant's dry weight (what's left over after being heated in an oven to remove most of the water). All of the carbon a plant absorbs must come from carbon dioxide in the air. Atmospheric concentration of this gas averages about 32 parts per 100,000 by volume—which is scarce, to say the least. These examples point to a special problem that land plants have in acquiring nearly all of their resources:

Unlike most animals, land plants generally absorb water and other vital substances from sources in which concentrations of these substances are low.

Plant structure and function speak of the low concentrations of substances in the environment. An obvious example is the thin, broad shape of leaves, which present

20

ADAPTATIONS FOR LIFE ON LAND

a large surface area not only for sunlight interception but for gas absorption. Less obvious is a reliance on large central vacuoles within individual cells as an energetically "cheap" way of increasing cell volume. In a mature plant cell, the vacuole takes up about ninety percent of the space. This organelle is mostly water—which is energetically inexpensive to accumulate, compared with other cytoplasmic substances that might otherwise go into its assembly. With the minimal energy outlay required for assembling a vacuole, cell volume is increased considerably—and so is the cell surface area in contact with the surroundings. In addition, central vacuoles serve as warehouses for essential nutrients. The vacuole membrane controls movement of nutrients to and from cytoplasm. Thus nutrients can be gradually stockpiled when they become available, and they can be metered out when they are scarce.

Plant roots are similarly well adapted to low concentrations of resources. For instance, potassium concentrations in soil water are often fifty times lower than they are in the plant body. What accounts for the difference? As one expert put it, plant roots must mine the soil! Broad, leaflike roots would have a large surface area to do this, but they could scarcely force their way through the soil during growth. (Imagine trying to push a sheet of paper through the soil.) Instead, efficient absorption of dilute substances comes through extensive systems of cylindrically shaped roots—which collectively represent a large surface area, and individually present less area of resistance to the surrounding soil. Root cells, too, rely on large central vacuoles for nutrient storage and increases in surface area.

But now plants have another problem. The large vacuoles in, say, hundreds of thousands of cells that make up the adult plant form greatly increases body size. Thus, as roots grow through the soil and stems grow upward through the air, transporting water and food from one plant organ to another becomes more of a necessity. Plant vascular systems accomplish these tasks. Xylem moves water and salts from soil to stems and leaves; phloem systems move sugars and other compounds from leaves to other plant parts. How do these vascular systems work? What, for example, keeps water running uphill and not downhill in the xylem of plants? And why do some roots grow down, some stems grow up, and still other roots and stems grow at angles? Plants are not so simple after all in their adaptations!

Other problems confronting land plants relate to rapid and sometimes dramatic temperature changes. Plants must adapt to seasonal shifts in temperature, because body temperatures must remain close to those of the environment.

Frank B. Salisbury

Figure 20.1 A dramatic example of interactions between plant hormones and environmental change. The photograph was taken in Germany, well after some severe November frosts. Most of the birch tree leaves had dropped, and most buds were dormant—both responses to the approach of winter. Close to the streetlamp, though, these normal hormone-mediated responses were delayed. Light from the lamp extended the apparent length of daylight for the branches right below it. Change in daylength is an important stimulus in this species' responsiveness to its surroundings. Continued and more severe frosts may well have killed the "confused" buds and leaves near the light, and heavy snows may have broken the leaf-laden branches.

a

b

Figure 20.2 Two plants adapted to different land environments.
(**a**) The snow buttercup (*Ranunculus adoneus*) grows right up through
the edge of a melting snowbank in spring—here, high above timberline
in the Colorado Rocky Mountains. (**b**) The creosote bush (*Larrea
divaricata*) tolerates the extreme conditions of hot, dry deserts in
North America.

Some plants, such as cacti, survive in hot deserts. Here
they are faced with high air temperatures, high heat absorption from incoming sunlight, and low soil moisture.
Many of these plants have a relatively high water content.
Because of water's high specific heat (Chapter Five), these
plants are able to absorb quite a bit of heat without undergoing too much of a rise in body temperature. Similarly,
low surface-to-volume ratios, small leaves, and thick cuticles help minimize water loss.

Still other plants are adapted to cold environments.
Such plants typically outcompete others at high elevations
or in northern latitudes. Most are short, perennial (they

live for several growing seasons), and grow rapidly during
perhaps only one to three months of a cool summer (Figure
20.2). During autumn and winter, these and other plants
become inactive. Their physiological processes slow down,
much like those of a bear hibernating in a cave. However,
there is an even greater difference between these plants
and the bear. The bear body remains warm—but the plant
body temperature may drop to –40°C! How most of a
plant's cells can retain water in the liquid state at such
low temperatures is an interesting (and unanswered) question.

Tolerance to both cold and hot temperatures is not
a suddenly acquired attribute in any individual. Besides
numerous mutations that have conferred tolerance on a
given species, an individual plant even of a cold-tolerant
species may die if it is suddenly transferred from a hot-
summer to a cold-winter environment. Thus, survival
somehow relates to gradual temperature changes during
fall and spring. But temperature changes are somewhat
unreliable guides to what season it is, because weather
can be fickle. Much better guides are the lengths of day
and night—which each year are constant at the same day
of the month in a given region. Plants have mechanisms
that detect and respond to daylength. Plants can anticipate the
coming of winter. Many can form flowers and seeds, develop dormant buds, attain more cold hardiness, and shed
their leaves—all in response to the shorter daylengths of
late summer and early fall. A final part of this chapter
is devoted to how these remarkable responses may be
accomplished.

ACQUISITION AND DISTRIBUTION OF RESOURCES

Vital Nutrients and Their Functions

Apparently, most plants require sixteen essential elements
to grow and reproduce, although most studies have been
carried out with crop plants only. These elements are
oxygen, carbon, hydrogen, nitrogen, potassium, calcium,
phosphorus, magnesium, sulfur, chlorine, iron, boron,
manganese, zinc, copper, and molybdenum. Table 20.1 lists
the form in which a plant absorbs these elements, along
with their approximate tissue concentrations necessary for
adequate growth.

Oxygen is incorporated into organic compounds that
make up the plant's dry weight. It comes from three
sources: water, gaseous oxygen (O_2), and carbon dioxide

Table 20.1 Essential Elements for Most Complex Land Plants

Element	Symbol	Form Available to Plants	Percent Concentration in Dry Tissue	
Carbon	C	CO_2	45	96% of total dry weight
Oxygen	O	O_2, H_2O, CO_2	45	
Hydrogen	H	H_2O	6	
Nitrogen	N	NO_3^-, NH_4^+	1.5	
Potassium	K	K^+	1.0	
Calcium	Ca	Ca^{++}	0.5	
Magnesium	Mg	Mg^{++}	0.2	
Phosphorus	P	$H_2PO_4^-$, HPO_4^{--}	0.2	
Sulfur	S	SO_4^{--}	0.1	
Chlorine	Cl	Cl^-	0.010	
Iron	Fe	Fe^{++}, Fe^{+++}	0.010	
Boron	B	H_3BO_3	0.002	
Manganese	Mn	Mn^{++}	0.0050	
Zinc	Z	Zn^{++}	0.0020	
Copper	Cu	Cu^+, Cu^{++}	0.006	
Molybdenum	Mo	MoO_4^-	0.00001	

(CO_2) in the air. Oxygen, hydrogen, and carbon account for about ninety-six percent of the dry weight—and much more if we look at a normal plant that still contains seventy to eighty percent water. But plant life ceases without the other thirteen elements listed in Table 20.1. (Although the plant requires some 60 million times as many hydrogen atoms, it still can't get along without traces of molybdenum.) These elements are absorbed from the soil, where they are dissolved in water.

Gaseous nitrogen (N_2) represents about seventy-eight percent of all the molecules present in air. Plants can't use it, because they don't have the metabolic machinery for breaking apart the strong bonds holding these molecules together. Unfortunately, insufficient nitrogen frequently limits plant growth. However, some microorganisms that live independently in the soil can break the bonds and use gaseous nitrogen (see *Commentary*, Chapter Twenty-Eight). So can microorganisms that live in nodules of the roots of legumes and certain other plants (Figure 20.3). These nodule residents convert N_2 first to ammonia (NH_3), then use the nitrogen in assembling amino acids, proteins, and nucleic acids required in cell construction and maintenance. They leave some excess ammonia and amino acids for the plants in which they reside, in exchange for food materials transported to them from the leaves.

Potassium ranks behind oxygen, carbon, hydrogen, and nitrogen in terms of abundance. Like most of the remaining elements listed in Table 20.1, potassium serves

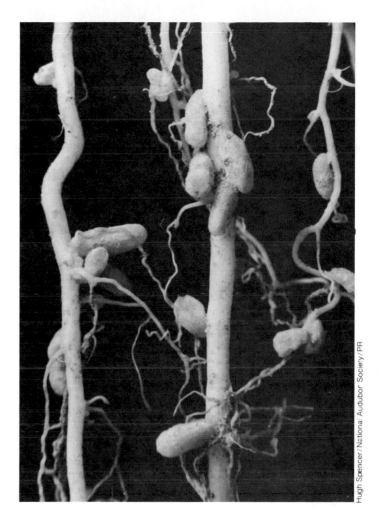

Hugh Spencer/National Audubor Society/PR

Figure 20.3 Clover root nodules, where symbiotic nitrogen-fixing bacteria live. In many plants, groups of nitrogen-fixing bacteria occur within cells of such nodules. Most nodules develop close to the soil surface, where more air (hence more gaseous nitrogen) is present. 15×.

two important functions. First, it activates enzymes involved in protein synthesis, starch synthesis, cellular respiration, and photosynthesis. Second, like other less abundant ions, potassium sets up an osmotic gradient across the plasma membrane. Water moves down such gradients, into the cell (Chapter Five). **Turgor pressure** (the pressure applied to the walls as water is absorbed into the cell) accomplishes one of two things. If the walls are still soft, they expand and the cell grows. If the walls have

Frank B. Salisbury

Figure 20.4 Plant wilt. Turgor pressure allows nonwoody plant parts to retain their normal form. When a concentrated salt solution is poured into a pot containing tomato plants, water moves osmotically from roots into the salty soil. Water from the stem and leaves then follows, and wilting is the result. The severe wilting shown here occurred in less than thirty minutes.

become rigid, the resulting pressure simply allows the plant to sustain a nonwilted form (Figure 20.4).

Salts are important in the formation and activation of plant enzymes. They are also important because their absorption into cells sets up an osmotic gradient—hence water necessary for growth and for maintaining plant shape also moves into cells.

Water Absorption, Transport, and Loss

How do substances actually move into and through the plant body? Let's approach this question by first looking at the movement of water, for this is the substance in which vital elements are dissolved. Liquid water moves into roots, up stems, and into leaves. Some is used to drive metabolism, but most evaporates into the air. **Transpiration** is the name for evaporation from stems and leaves. Transpiration, of course, depends on the presence of water in stems and leaves. How does it get there in the first place? What could possibly make water run uphill, to the top of even the tallest trees? *It appears that water is pulled upward by continuous tensions that extend downward from the leaf to the root.*

First, the (usually) dry air surrounding a plant causes transpiration from leaf cell walls. As some water molecules escape, others move out of the cell cytoplasm and into the walls as replacements. When they do, still other water

molecules move into the mesophyll cells from the xylem pipeline system running through leaf veins. Second, when water moves from xylem's tracheids and vessels, water is pulled out of cells connected to them in the xylem system. Because of the pulling action, water inside all of these dead conducting cells is in a state of tension (negative pressure), compared with the positive pressure in living cells. (The pressure characteristic of each cell type is maintained by strong cell walls.) Third, replacement water molecules move into the xylem from root cells. Water in xylem cells of extensive root systems attract more water from the soil, even when the soil is somewhat dry. Figure 20.5 illustrates the way water is pulled from roots to leaves.

Water moves as a continuous, fluid column along such routes. This may seem puzzling, given all the tension that the molecules are under. Why doesn't the "stretching" cause the molecules to snap away from each other? Some time ago the Irish botanist Henry Dixon came up with an explanation, which has since been named the **cohesion theory of water transport**. According to this theory, hydrogen bonding allows water molecules to cohere tightly enough to keep from breaking apart as they are pulled up through the plant body. Dixon had no way of measuring how much tension exists in xylem (and therefore of convincing skeptics that it really exists). Confirmation of his explanation came much later. In any event, the points to remember are these:

1. The drying power of air causes transpiration (evaporation of water from stems and leaves).

2. Transpiration triggers a state of tension (negative pressure) that is continuous from leaf cells down through the xylem of roots.

3. As long as water molecules vacate transpiration sites, replacements are pulled up along this negative pressure route.

4. Because of the cumulative strength of hydrogen bonds between water molecules (Chapter Three), water can be pulled up as continuous columns to stem and leaf transpiration sites.

Of all the water moving into a leaf, only about two percent gets stored or used in metabolism. The remaining ninety-eight percent is lost through transpiration. As long as roots absorb enough water to replace all that's being lost, the plant functions efficiently. But soil is not always moist, and some days are hotter and drier than others. How do land plants have ways of regulating water loss as environmental conditions change? For one thing, a cuticle

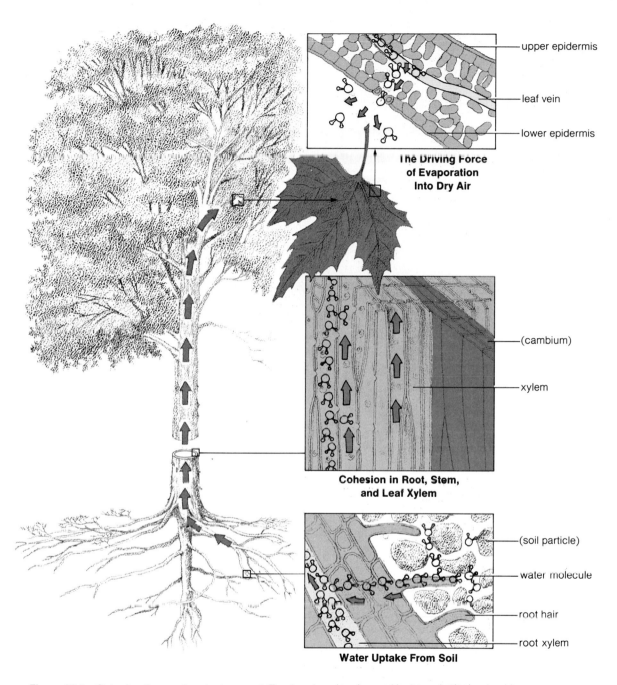

The Driving Force of Evaporation Into Dry Air

upper epidermis

leaf vein

lower epidermis

(cambium)

xylem

Cohesion in Root, Stem, and Leaf Xylem

(soil particle)

water molecule

root hair

root xylem

Water Uptake From Soil

Figure 20.5 Cohesion theory of water transport. Tensions in xylem (caused by transpiration) extend from leaf to root. As a result of these tensions, columns of water molecules hydrogen-bonded to one another are pulled upward (Figure 3.4), much like tiny ropes being pulled up the stem and out of the leaf.

covers the aerial parts of all land plants. Water inside the plant body can barely penetrate the cuticle's waxy layers (see, for example, Figure 3.11). Without a cuticle, plants would rapidly wilt and die during hot, dry spells. Even so, a completely watertight cuticle would be a disaster for plant functioning. *The reason is that carbon dioxide—so essential in photosynthesis—has almost as much trouble getting past a cuticle and into a leaf as water does getting out of it.*

A truly elegant adaptation controls water and carbon dioxide movement into and out of plant parts that are exposed to air. In leaf and stem epidermis are stomata —each with two guard cells (Figure 20.6). When the guard

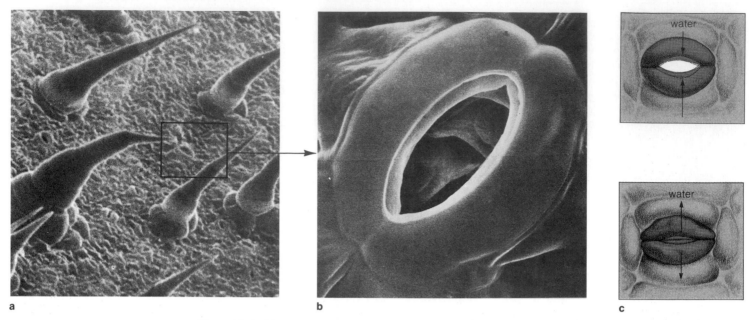

a   b   c

Figure 20.6 Where stomata occur on a typical dicot leaf, and how they function. (**a**) Stomata are found among hairlike structures on a cucumber leaf's lower epidermis. The box identifies one of the stomata. (**b**) Closer look at stomatal structure. Here, we can peer through the gap between the stomatal guard cells and view parts of mesophyll cells inside the leaf. (a, b from John Troughton and L. A. Donaldson)

cells are swollen with water, turgor pressure so distorts their shape that they move apart—hence a gap appears between them. When their water content dwindles, turgor pressure drops and the gap closes. Stomata typically open during daylight hours, when carbon dioxide is being used in the formation of starch and sugars. Indeed, their main function is controlling carbon dioxide levels in the leaf. When stomata are open, of course, the plant loses water even as it gains essential carbon from the air.

Although we don't completely understand the processes involved, this much is clear: *a stoma opens and closes according to how much water and carbon dioxide are present in its two guard cells.* When the sun comes up, photosynthesis begins and carbon dioxide is used up. As carbon dioxide concentrations dwindle in guard cells, potassium ions are actively pumped into them from surrounding epidermal cells (Figure 20.7). As these ions accumulate inside, the osmotic gradient across the guard cell plasma membrane shifts accordingly, and water moves in, too. The resulting increase in turgor pressure causes stomata to open. Transpiration proceeds, and so does carbon dioxide movement into the leaf. Photosynthesis keeps carbon dioxide concentrations low, so the plant continues to lose water and gain carbon dioxide during the day.

When the sun goes down, photosynthesis stops. Carbon dioxide is no longer used, so it accumulates in all cells. Much of the potassium that had accumulated inside guard cells now moves out, and water follows it. Turgor pressure decreases, stomata close, and transpiration is greatly reduced.

As long as soil is moist, stomata can remain open during daylight. But when soil is dry and the air is also dry and hot, land plants close their stomata and absorb little water or nutrients. Although photosynthesis (and growth) slows down as a consequence, these plants still survive drought periods. They can do so repeatedly. Briefly, such stressful conditions trigger the production of a chemical signal—a plant hormone called **abscisic acid**. This hormone is synthesized faster when a leaf is water stressed. When abscisic acid accumulates in a leaf, it somehow causes guard cells to give up potassium ions (hence water)—so stomata close. Thus,

By controlling potassium levels inside the guard cells of stomata, abscisic acid controls stoma activity when leaves are losing more water than roots can absorb. Thus it influences photosynthesis and growth.

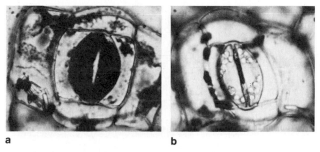

a   b

Figure 20.7 Evidence for potassium accumulation in stomatal guard cells undergoing expansion. Strips from the leaf epidermis of a dayflower (*Commelina communis*) were immersed in solutions containing dark-staining substances that bind preferentially with potassium ions. (**a**) In leaf samples having opened stomata, most of the potassium was concentrated in the guard cells. (**b**) In leaf samples having closed stomata, very little potassium was in guard cells; most was present in normal epidermal cells. (b from T. A. Mansfield)

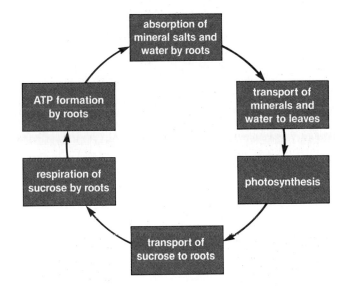

Figure 20.8 Interrelated processes that influence the coordinated growth of roots, stems, and leaves. When one process is rapid, the others also speed up. Any environmental factor limiting one process eventually slows growth of all plant parts.

Solute Absorption and Accumulation

If it is to absorb and retain water, a cell in any part of the plant must maintain a higher overall solute concentration than its surroundings. This means that a cell must expend energy and actively *accumulate* solutes. Without energy outlays, diffusion would equalize solute concentrations on both sides of a plasma membrane. Energy from ATP drives the membrane pumps involved in active transport. These pumps, again, are membrane proteins that move substances into the cell even against a concentration gradient (Chapter Five).

In photosynthetic cells, ATP is formed during both photosynthesis and cellular respiration. But what about nonphotosynthetic cells? How do they get all the ATP necessary for active transport? To give one example, root cells receive sugars (mostly sucrose) from leaves—especially during daylight, when photosynthesis is rapid. These cells absorb oxygen from air in the soil (unless the ground is really waterlogged or flooded). When soil is sufficiently moist, dissolved ions are carried rapidly into roots by way of the transpiration stream. To be sure, more air is present and more oxygen can be absorbed when the soil is quite dry. But insufficient water limits growth and ion absorption no matter how much oxygen is present. At the same time, dry soil causes complete or partial closure of stomata, so leaves absorb less carbon dioxide when water and ions are in short supply. As a result, leaf cells photosynthesize and grow less rapidly—and send less sucrose to roots. Cellular respiration slows down in roots, and so does ion absorption. These feedback relationships among plant organs are summarized in Figure 20.8. They illustrate much of what we know about coordinated growth between different plant regions.

Solute absorption and accumulation are coordinated throughout the plant body in ways that have profound influences on growth.

The roots of many vascular plants have millions, sometimes billions, of root hairs (Figure 12.3). These root hairs enhance contact with the soil and ion absorption from it. In addition, nearly all vascular plants depend on **mycorrhizae**, or "fungus-roots." Figure 20.9 shows a mycorrhiza on a lodgepole pine tree. Mycorrhizae represent a mutually beneficial association between a fungus and a young root. As you will read in Chapter Twenty-Three, the vegetative body of these symbiotic fungi is an extensive mat of very thin filaments. Collectively, these filaments have a tremendous surface area for contact with a large volume of the surrounding soil. Thus, the fungus is most efficient at accumulating water and dissolved ions. The root draws on

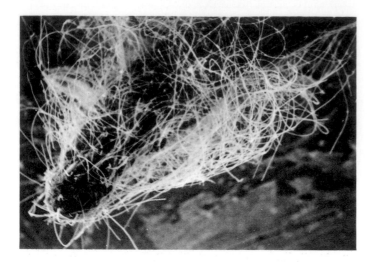

a   b

Figure 20.9 Mycorrhiza (fungus-root) of a lodgepole pine tree. White threads are fungal strands. When you dig up and move a plant to another location, try to include some native soil with the roots. The soil probably contains the proper mycorrhizal fungus. Without it, transplants are frequently retarded in their growth. (From J. G. Mexal, C. P. P. Reid, and E. J. Burke, Botanical Gazette, vol. 140, no. 3. © 1979 by The University of Chicago)

Figure 20.10 Almost all plants benefit immensely from close association with certain fungal species. A five-year-old pine seedling grown without the fungus is shown in (**a**). Six-year-old pines of the same species—grown with the fungus—are shown in (**b**).

some of the water and ions that the fungal mat absorbs. In turn, the fungus absorbs some sugars and nitrogen-containing substances present in the root (Figure 20.10).

Phloem Transport of Carbohydrates and Other Compounds

Leaves are the main organs in which food molecules are assembled. But roots, stems, flowers, and fruits also depend on these molecules for growth. Starch, the dominant food storage product, is deposited as one or more grains in chloroplasts (Chapter Four). Yet starch is a large molecule —much too large to get out of chloroplasts. Besides, starch is too insoluble for water transport elsewhere in the plant body. Fats are another food storage form, especially in many plant seeds. These energy stores are required by the root and stem that develop after a seed sprouts. But fats are also insoluble in water, and they can't be transported out of their plant cell storage sites.

Energy stored in starch and fats is made available through several chemical reactions. In essence, rearrangements of carbon, hydrogen, and oxygen atoms in all such molecules yield a soluble, transportable carbohydrate. In most plants, this carbohydrate is sucrose. A giant redwood

tree trunk, a potato, breakfast cereal, bacon, eggs, cotton shirts—most of the substance of these things was derived, directly or indirectly, from sucrose.

Food transport through the plant body requires that storage starches and fats be converted to soluble, transportable forms— most often, sucrose.

With soluble food molecules available, a plant can nourish its various parts. But how does food travel up to, say, a growing fruit and down to a growing root? The transport of sucrose from one plant organ to another, regardless of the direction, is called **translocation**. Translocation occurs in specialized cells of phloem tissue. Such cells, joined end to end, form long sieve tubes (Figure 19.7). Sieve tubes extend from leaf to root. Unlike xylem cells, these cells are alive at maturity. Numerous small perforations in their end walls allow water and dissolved substances to flow through them at rates up to 100 centimeters an hour. Several sieve tubes lie side by side in a vascular bundle. Flow from a given leaf may be upward in some of these tubes, and downward in others.

Interestingly, the feeding habits of small insects called aphids tell us something about translocation. An aphid

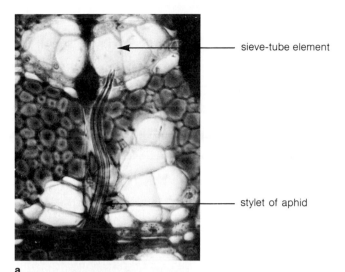

— sieve-tube element

— stylet of aphid

a

b

Figure 20.11 (a) Naturally occurring research tool: the aphid stylet, here penetrating a sieve-tube element. (b) Honeydew droplet at the tail end of a well-fed aphid. (From Martin Zimmerman, *Science*, 133:73–79, 1961. Copyright 1961 by the American Association for the Advancement of Science)

feeds on leaves and stems. It forces a mouthpart (a stylet) into sieve tubes, which contain sugary fluid. Although aphids are often called sucking insects, they are actually passive feeders. Sieve tube contents are under such high pressure—perhaps ten times as much as in an automobile tire—that some "honeydew" seems to be forced right through the aphid gut and clear out the other end (Figure 20.11). Park your car below trees being attacked by aphids and it may get a spattering of sticky honeydew droplets, thanks to sieve-tube pressures. Anyway, in some experiments, feeding aphids were anesthetized. Then their bodies were severed from their stylets, which were left embedded in the sieve tubes the aphids had been attacking. Analysis

of the fluid being forced out of the sieve tubes verified that sucrose is the main carbohydrate being transported through the plant body.

According to the **pressure flow theory**, translocation depends on pressure gradients in the sieve-tube system. By way of analogy, think of what happens when you turn on a faucet connected to a garden hose. Once the faucet valve opens, water flows from the high pressure region (the pipe behind the valve) to a lower pressure region (the empty hose), even when you raise the hose far above the faucet line.

How do large pressure gradients arise in sieve-tube systems? Osmosis is the answer. In a leaf, sucrose is actively transported into sieve tubes. Energy from ATP drives this loading process. Solute concentrations rise accordingly in the tubes. Water that the xylem has delivered to the leaf moves into the tubes, down the osmotic gradient. The resulting increase in local pressure causes water to flow to regions where pressure in the tube system is lower—which may be upward as well as downward from the leaf. Everything dissolved in the water also flows along, if it is small enough to penetrate the end wall perforations in sieve tube cells. And sucrose molecules are small enough.

The key points of the pressure flow theory of translocation may be summarized in the following way:

---

1. Translocation (food transport throughout the plant body) occurs in a pipeline of living cells called the sieve-tube system.

2. Translocation is driven by differences in water pressure from one region of the pipeline to another.

3. Sucrose is actively transported into sieve tube cells present in the leaf, water follows osmotically because of the increased solute concentration, and pressure builds up in the leaf sieve-tube regions.

4. Thus active transport (through ATP expenditure) and osmosis in leaf regions are the basis for food transport to stems, fruits, and roots—all regions of lower pressure.

What maintains the low pressure at the other end of sieve tube pipelines? If pressure were to build up there, the gradient that assures translocation would disappear. Cells maintain low pressures by using up the food molecules delivered to them. For example, some sucrose is converted into cell wall polysaccharides, proteins, and starch; some is used up in cellular respiration. All such metabolic

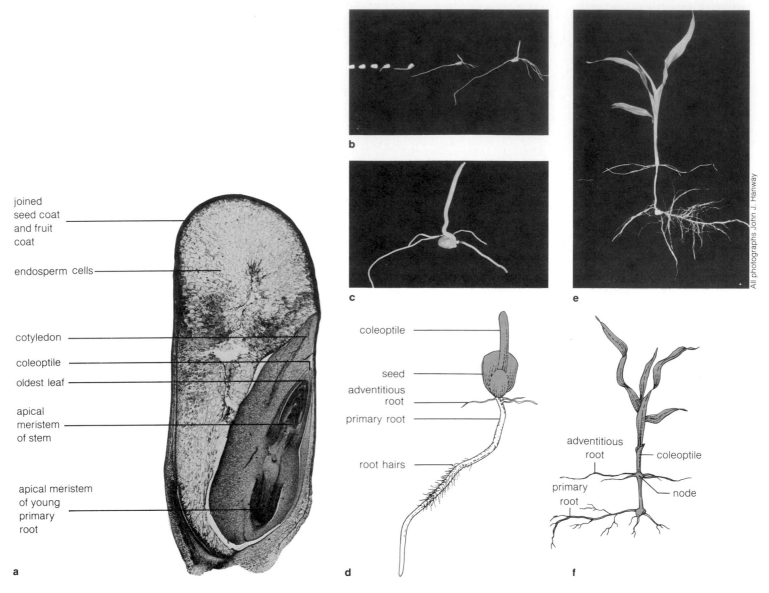

Figure 20.12 Stages in the development of a corn plant (a monocot). (**a**) Corn seed, sliced lengthwise. Most of the seed volume is composed of endosperm cells that store starch and protein. The embryo consists of a cotyledon that, upon germination, secretes digestive enzymes into the endosperm and absorbs digested products in return. These products are transported to the young root and shoot systems. The coleoptile encloses and protects the young leaves as they grow upward through soil. (**b–d**) Seed germination and early seedling development. The radicle grows into the primary root; several adventitious roots develop from the first-formed node at the coleoptile base. (**e,f**) A larger seedling with adventitious roots, and with branch roots formed from the primary corn root. (a Brian Bracegirdle and Patricia Miles, *An Atlas of Plant Structure*, Heinemann Educational Books, 1977)

processes lower the water pressure in these cells, because a low concentration of dissolved solutes means that less water can be held osmotically. Growth also lowers the water pressure. Internal pressure forces cell walls to expand, and this expansion causes the pressure to drop.

The pink cells of a watermelon fruit are a good example of the importance of growth in maintaining translocation. These cells become so large that they are visible to the unaided eye—yet their sweetness verifies that they still contain high sucrose concentrations.

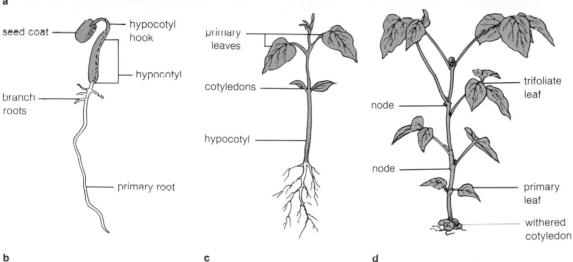

Figure 20.13 Stages in the development of a soybean plant. (**a**) Germination and early seedling development. Branch roots form but, unlike corn plants, adventitious roots do not. In soybean, common bean, and many other dicots, the food-storing cotyledons are carried aboveground by elongation of the hypocotyl below them. The hypocotyl forms a hook (**b**) that makes a channel through the soil. In this channel, cotyledons grow upward without being torn apart. At the soil surface, light causes the hook to straighten. The cotyledons become photosynthetic for several days before they wither and fall from the plant body. In (**c**) and (**d**), the first true (primary) leaves of soybean are positioned opposite each other on the stem, and the leaves themselves are not divided. Afterward, all other leaves are divided structurally into three leaflets (trifoliate leaves). In (**d**), flowers are developing on the axillary buds at the four upper nodes.

PLANT GROWTH: ITS NATURE AND DIRECTION

So far, we have looked at adaptations that keep the land plant functioning at any given moment. Let's now extend the picture in time, by looking at adaptations that allow the plant to follow its prescribed program of growth and development even as environmental conditions change.

Let's consider growth as it occurs in two seed-bearing plants: corn and soybeans. Figures 20.12 and 20.13 show some of the main morphological changes leading from the seed to the mature plant. Inside each organ, many specialized cell types arise through differentiation. From earlier chapters, you know that development involves controlled activation and deactivation of genes. Through these gene controls, specific enzymes required in the formation of specific cell structures are produced, at the appropriate times. The environment, again, is important in gene activation and deactivation. Here, you will learn about some environmental effects on plant growth and form.

If growth begins in the seed, where does it occur and what causes it? Clearly the whole seed doesn't grow much. Rather, it simply swells with incoming water molecules, which hydrogen bonding has attracted to proteins and polysaccharides stored in the seed. In seeds of most species, the first cells to grow are found in the embryonic root, the **radicle**. Most grow mainly in a longitudinal direction. The result is a slender **primary root**, composed of elongated cells. When this root visibly protrudes from the seed remnant, germination is considered to be complete. Later, the primary root branches. These lateral roots also branch, and their products branch again. Depending on its size and species, the mature plant may end up with hundreds or millions of roots.

Plant growth may be defined as a permanent increase in size. Plants grow because their cells grow. Cell division in itself doesn't constitute growth, because the volume of newly formed daughter cells is about the same as that of the mother cell (Figure 20.14). On the average, though, about half of the daughter cells formed enlarge—often by twenty times. Thus, cell division clearly increases the capacity for overall growth. The other half remain meristematic; they grow only as large as the mother cell, then probably divide again. Seeds of many species germinate with no cell division, although division begins soon after in apical meristems of the radicle and of the young stem.

As you have seen, cell growth is driven by water uptake and the ensuing turgor pressure against the cell wall. In some ways, the growth process is like a balloon being blown up with air. Balloons with soft walls are easy to inflate; cells with soft walls grow rapidly under little turgor pressure. An important difference is that the balloon wall gets thinner as it "grows," but the cell wall does not. This means that the chemical reactions leading to formation of the complex polysaccharides in cell walls have to be coordinated with water absorption. Usually, cell growth is accompanied by formation of more cytoplasm, but the increase in water volume is much greater.

Cell walls exhibit two kinds of expansive properties. First, they may be *elastic and reversible*. After stretching, they return to their original shape (much like a balloon when the air is released). Second, cell walls may be *plastic and irreversible*. After stretching, they stay stretched (like bubble gum after the air has been let out of the bubble). Guard cells are highly elastic cells; they swell greatly during stomatal opening, but they return to their original size and shape when the stoma closes. The walls of cells exhibiting true growth are elastic, but they also retain their stretched shape. As you will see, certain plant hormones can make cell walls more plastic, with no reduction of elasticity. The result is that true growth becomes possible.

To a large extent, the arrangement of polysaccharides in cell walls dictates the direction in which the wall yields under turgor pressure. Therefore, how the wall polysaccharides are oriented governs whether a cell becomes long and slender or short and broad. If most cells in a plant

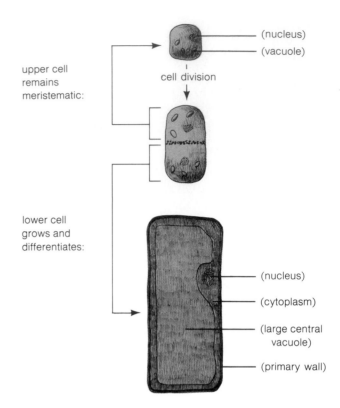

Figure 20.14 The nature and direction of plant cell growth. Through divisions, meristems provide new cells for growth. At a stem tip, small meristematic cells double in size, then divide at right angles to the stem's long axis. The upper cell remains meristematic. The lower cell grows into a mature parenchyma cell of pith or cortex, for example. Tiny vacuoles present in young cells absorb water, fuse, and form the large central vacuole of mature parenchyma cells.

organ are elongated, so is the organ; this is true of roots and stems. If most are more spherical, so is the organ; this is true of many fruits. Thus we have an important principle governing plant growth:

The arrangement of cellulose molecules in a cell wall controls the direction of cell growth. And cell shapes resulting from directional growth help determine plant form.

What controls the direction in which the cellulose molecules themselves become deposited during the formation of cell walls? That we don't know, although microtubular growth (Chapter Four) has been implicated.

PLANT HORMONES

Plant growth, like the growth of other multicelled organisms, is influenced in powerful but poorly understood ways by hormones. **Plant hormones** are information-carrying messenger molecules. Their chemical messages cause certain cell types to change their activities in response to environmental conditions. Plant hormones appear to be less specific than animal hormones in the kinds of physiological responses they trigger. Also, the target tissues themselves may be less specific. Although some plant hormones may cause detectable responses within minutes, none approaches the speed of, say, epinephrine in reaching a target tissue (probably because signals traveling plant transport routes aren't moved as rapidly).

Five hormones (or groups of hormones) are known to exist in vascular plants, and evidence for others is accumulating. These five are auxins, gibberellins, cytokinins, abscisic acid, and ethylene (Table 20.2).

Auxins are best known for their ability to stimulate elongation of stem cells. They also have other effects, which we will be considering. The main auxin is *indoleacetic acid* (IAA). Many investigators believe that IAA is the only auxin in plants, so they use "auxin" and "IAA" interchangeably. However, there is evidence that at least one other compound, *phenylacetic acid*, performs similar roles. It is more abundant than IAA in plants, but its effects are not as pronounced. Several compounds synthesized by chemists behave much like natural auxins. They are used by homeowners, farmers, and horticulturists. One of the most common is *2,4-D* (2,4-dichlorophenoxyacetic acid). When sprayed on the foliage at proper concentrations, this compound is a potent killer of dicot weeds and has little effect on grasses. It is used to kill dandelions in lawns, and numerous dicot weeds that compete for resources in fields of cereal grains. How 2,4-D can kill one plant and not another is poorly understood.

Gibberellins, like auxins, are best known for their promotion of stem elongation. More than fifty kinds have been identified in plants or fungi. Each differs chemically in a small way. The most familiar is a *gibberellic acid*. It is widely used in growth experiments, because of its potency and because a certain fungus serves as an inexpensive factory for its production.

Cytokinins were named for their ability to stimulate cell division (the name refers to "cytokinesis"). But leaf cells also grow larger in their presence. All cytokinins are similar in structure to adenine, a normal constituent of DNA, RNA, ATP, NAD^+, and $NADP^+$. The most common and most abundant cytokinin seems to be *zeatin*, first iso-

Table 20.2	Main Plant Hormones and Some Known (or Suspected) Effects
Auxins	Stimulate cell elongation in coleoptiles and stems; involved in phototropism
Gibberellins	Promote stem elongation (especially in dwarf plants); may help break dormancy of seeds and buds
Cytokinins	Stimulate cell division; promote leaf cell expansion and retard leaf aging
Abscisic acid	Stimulates stomate closure; probably involved in root geotropism; may trigger seed and bud dormancy and abscission of leaves, flowers, and fruits
Ethylene	Stimulates fruit ripening; stimulates abscission of leaves, flowers, and fruits; may play a role in thigmomorphogenesis
Florigen (?)	Arbitrary designation for as-yet unidentified hormone thought to control flowering process

lated from immature seeds of *Zea mays* (corn). Chemists have synthesized an inexpensive cytokinin called *kinetin*.

Abscisic acid (ABA) was once thought to be of primary importance in abscission (the loss of flowers, fruits, and leaves). This role is now questioned, and its importance in stimulating stomate closure, as well as dormancy of buds and seeds, is emphasized.

Ethylene (C_2H_2) is a simple hydrocarbon gas that stimulates fruit ripening. Ancient Chinese knew that a bowl of fruit would ripen faster if incense were burned in the same room, although they didn't know that ethylene present in the smoke was responsible. That plants might produce something that stimulates ripening was first suggested in a 1910 report to the Jamaican Agricultural Department. Oranges, the report stated, should not be stored near bananas on freighters, because some emanation from the oranges caused the bananas to become overripe. Almost surely, this "emanation" was ethylene.

It is not yet possible to separate clearly the effects of hormones from the effects of environmental change on plant development. The problem is that the environment also controls the amount and distribution of hormones. Nevertheless, let's take a look at what *is* known by following the development of those corn and soybean plants shown in Figures 20.12 and 20.13.

HORMONES, ENVIRONMENTAL CHANGE, AND PLANT DEVELOPMENT

Examples of Plant Hormone Action

Hormones are known to govern the upward growth of a grass coleoptile (such as corn) through the soil. A **coleoptile** is a hollow, cylindrical organ that protects tender young leaves growing within it (Figure 20.12d). Without a coleoptile, young corn leaves would be torn apart by soil particles. While underground, both the coleoptile and the tightly curled oldest leaf grow in a coordinated fashion. Once exposed to sunlight, though, the coleoptile stops elongating. Its task is completed. The leaf now breaks through the protective coleoptile cylinder, uncurls, absorbs sunlight, and provides food for the parts below.

In a corn coleoptile and leaf, the youngest cells are at the base, not at the tip. Cells between the base and tip elongate under the stimulation of auxins. Through experiments of the sort described in Figure 20.15, we know that *growth of a coleoptile is controlled by IAA synthesized in its tip.* This hormone moves down the coleoptile and changes metabolic activities of cells below the tip. As a result of these changes, cell wall plasticity is increased—and this plasticity encourages elongation.

What about growth of a soybean stem? Like all dicot stems, it has an apical meristem and young, growing cells near the tip. Cells at the stem base are older and larger. Auxin is synthesized near the stem tip, especially in very young leaves being formed there. Experiments show that a soybean stem stops elongating when the tip is cut off. When auxin is applied to the cut stump, growth resumes. Thus, *auxin synthesized at a dicot stem tip must be essential for growth of cells below.* Auxin applications don't do much to intact stems, probably because small leaves and the apical meristem naturally provide enough auxin for elongation.

What about the effects of other plant hormones? In the stems of many normal plants with tips intact, cells often elongate faster with artificial application of gibberellin. Apparently, *growth of stems that elongate slowly is especially influenced by gibberellin.* Consider dwarf varieties of corn, which have short stems. Often, cells of these dwarf plants have a mutant gene that codes for a defective enzyme. Unlike the normal enzyme, it can't catalyze an important step in gibberellin production. Gibberellin applications to the dwarfs encourage stem elongation. However, they have little effect on a normal corn variety (Figure 20.16).

In contrast to coleoptiles and dicot stems, *root and leaf cells themselves synthesize most or all of the hormones they require.* Normal roots and leaves show little response to hormone

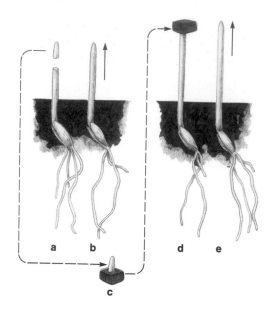

Figure 20.15 Experiment showing that IAA made in the tip of grass coleoptiles promotes elongation of cells below. In (**a**), an oat coleoptile tip is cut off. Compared with a normal coleoptile (**b**), the stump doesn't elongate much. When the excised tip is placed on a tiny block of gelatin for several hours (**c**), IAA moves into the gelatin. When the gelatin block is placed on another de-tipped coleoptile (**d**), elongation proceeds about as fast as it does in an intact coleoptile (**e**).

applications. For roots, this is true whether the growing tip is intact or removed. In fact, root elongation is strongly inhibited by the same concentrations of auxins that promote elongation of auxin-deficient stems! In some plants, gibberellin or cytokinin applications promote leaf expansion, but only slightly.

Effects of Temperature and Sunlight

Both temperature and sunlight influence the growth of coleoptiles, stems, roots, and leaves. Environmental temperature affects cell metabolism and transport of water, mineral salts, and sucrose. Corn and soybean plants grow best in warm regions. Peas may be planted in gardens and fields when temperatures are cooler. But the optimum temperature range for many grasses other than corn is lower than that for many dicots. In general, it's possible to predict the optimum temperature range for a given plant species by observing where it grows best.

How does sunlight influence plant growth? Plants adapted to sunlight but exposed to darkness or shade put

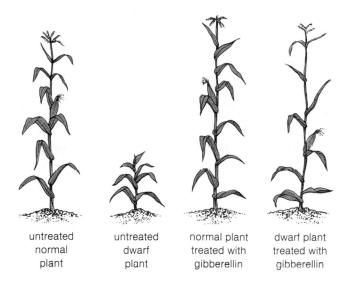

untreated
normal
plant

untreated
dwarf
plant

normal plant
treated with
gibberellin

dwarf plant
treated with
gibberellin

Figure 20.16 Influence of gibberellin on the height of a normal corn plant and a mutant plant that differs in only one gene.

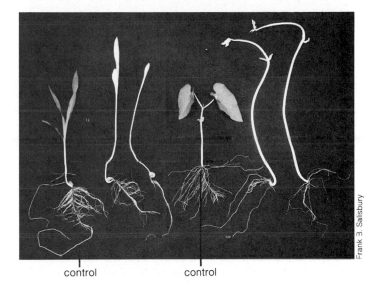

control   control

Figure 20.17 Effects of the absence of light on young corn and bean plants. The plant at the left of each group served as a control; it was grown in a greenhouse. The others were grown in darkness for eight days. Dark-grown plants were yellow; they could form carotenoid pigments but not chlorophyll in darkness. They also had longer stems, smaller leaves, and smaller root systems. (Why do you suppose roots of dark-grown plants grow less, considering that those of light-grown plants aren't exposed to light, either?)

more resources into stem elongation and less into leaf expansion or stem branching (Figure 20.17). That is why we often see tall plants, with few branches, in crowded fields and forests. Their elongating stems are carrying leaves toward the sunlight energy necessary in photosynthesis. Especially for dicots, the following generalization applies:

Exposure to sunlight promotes leaf expansion and stem branching, but inhibits stem elongation.

Phytochrome, a blue-green pigment molecule that is a receptor for light energy, is central to controls over plant growth. *Phytochrome seems to be an on-off switch for hormone activities governing leaf expansion, stem branching, stem length, and, in many plants, seed germination and flowering.* Phytochrome is most active when stimulated by light of red wavelengths. When the sun rises, phytochrome exists mainly in an inactive form (abbreviated Pr) that responds to red light. When Pr absorbs red wavelengths, it is converted to the active form (Pfr). In the shade or at sunset (when far-red light predominates), and at night, Pfr reverts

to Pr. With this reversible reaction to the inactive form of phytochrome, plant responses stimulated by red light are curtailed:

$$Pr \underset{far\text{-}red\ light}{\overset{red\ light}{\rightleftarrows}} Pfr$$

seed germination (+)
stem elongation (−)
leaf expansion   (+)
stem branching   (+)
flowering   (+ or −)

Now, it happens that plants shaded by others are exposed largely to green and far-red wavelengths, because photosynthesizing leaves above them absorb most of the red wavelengths. Thus, shaded plants are Pfr deficient. Often their seeds can't even germinate unless fire or chainsaws remove the taller plants.

How does Pfr promote or inhibit growth of different plant parts? We know that phytochrome resides largely in cell membranes. It may be that conversion of Pr to Pfr controls whether hormones bind to (or are transported across) plasma membranes of certain cells. Even if this turns out to be the case, we have much to learn about which hormones are involved, and how.

The Many and Puzzling Tropisms

Plant **tropisms** are familiar but as yet unexplained growth responses. In these responses, an environmental stimulus acts more intensely on one side of an organ than on the opposite side, and the plant responds with a faster rate of cell elongation on one side or the other.

For example, **geotropism** (or gravitropism) is a response to the earth's gravitational force. It is apparent soon after seed germination, when the plant root curves downward soon after it breaks through the seed coat and when the coleoptile or stem curves upward. Faster cell growth on the upper side of a horizontally growing root leads to its downward curvature. Because root cells are cemented together at primary walls and middle lamellae, the root curves as a unit. Faster cell elongation on the lower side of a horizontally growing stem leads to its upward curvature (Figure 20.18), even in the dark. These responses seem to govern growth of the first (primary) root and main stem of all seedlings. Yet branch roots usually grow somewhat horizontally through more fertile (upper) soil regions. And in shrubs and trees, branches grow more horizontally than the main stem, thereby exposing aerial parts to more light and carbon dioxide.

Which plant hormones take part in geotropism? For a long time, IAA was thought to move from one side of a stem or root to the other in response to gravity. To be sure, IAA is transported from the upper to the lower side of horizontally positioned coleoptiles and dicot stems. But so are gibberellins. And whether transport of either hormone is fast enough and in sufficient amounts to cause the differential elongation is now being questioned. (Epidermal cells on the lower side of a stem may step up their own hormone production and directly speed their elongation.) Besides, it seems clear that auxins have little to do with root geotropism. Although cells of an elongating root region carry out the response, it is the root cap that actually *detects* gravity.

When a root cap is surgically removed, a horizontally oriented root never does curve downward. When the cap is reinstated, curvature is restored. Removing the root cap doesn't prevent cells of the elongating region from growing; if anything, they elongate faster. This suggests that a growth inhibitor is removed along with the cap. Evidence favors the idea that gravity somehow causes abscisic acid to accumulate in cells on the lower side of a horizontally positioned root cap. When this hormone moves out of the cap, toward the lower elongating cells, it inhibits their growth (Figure 20.19).

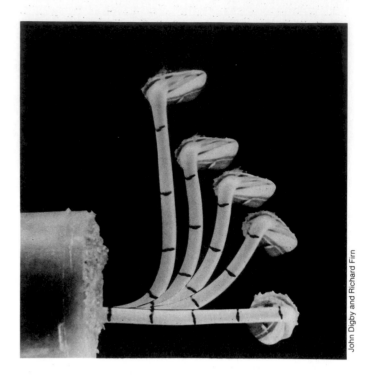

Figure 20.18 Composite time-lapse photograph of geotropism in a dark-grown sunflower seedling. In this plant, the two cotyledons emerge aboveground, because the stem portion just below the cotyledons elongates. Just before this five-day-old plant was positioned horizontally, it was marked at 0.5-centimeter intervals. After thirty minutes, upward curvature was detectable. The most upright position shown was reached within two hours.

Have you ever noticed how the flat leaf surfaces of houseplants turn toward the light from a nearby window? This response is a form of **phototropism**: stem or leaf growth toward light coming in mainly from one side. Stems curve toward the light; the flat surface of a leaf blade becomes perpendicular to the light (Figure 20.20).

In the 1800s, Charles Darwin proved that light must strike a coleoptile tip before curvature will occur. In the 1920s, Frits Went discovered that a growth promoter exists in the coleoptile tip. He called it auxin, after the Greek word meaning "to increase." Went showed that this substance moves down toward shaded elongating cells and causes curvature toward light. Today we know that Went's discovery is IAA. It moves from an illuminated side of a plant organ to the shaded side, then downward, where it especially promotes elongation of epidermal cells. We also know that blue wavelengths are the most important stimulus for phototropism.

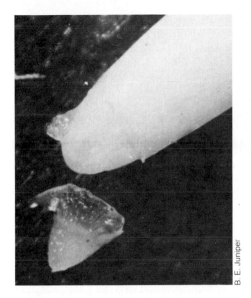

Figure 20.19 Young corn root cap, shown removed from the root tip. When the cap is removed, cells just behind the apical meristem grow faster; when the cap is replaced, their growth slows. These and many other experiments suggest that the root cap synthesizes the growth inhibitor abscisic acid, and that transport of this hormone to lower growing portions of a horizontal root causes the downward curvature responsible for geotropism.

Figure 20.20 Phototropism in bean, pea, and oat seedlings. These plants were grown in darkness, then exposed to light from the right side for a few hours before being photographed. Notice that curvature in the bean and pea seedlings occurs mainly in young (upper) cells that are still growing rapidly. Curvature in oat seedlings occurs in the leaves, where the youngest (still-growing) cells are those near the leaf base.

A large, yellow pigment molecule called **flavoprotein** absorbs blue light. Flavoprotein contains riboflavin (vitamin B_2) attached tightly to a protein molecule. How light absorption by this pigment causes IAA to move horizontally across a coleoptile or a young dicot stem tip is not understood. Complicating the picture is evidence suggesting that other hormones also contribute to phototropism.

Thigmotropism is an unequal growth resulting from physical contact with environmental objects. Stems of peas, beans, and many other climbing vines show this response. These plants are generally so long and slender that they can't grow upright without support. Suppose one side of a pea or bean stem grows against a solid object—a fencepost, a supporting stick in a greenhouse pot, the stem of a nearby woody plant. Cells stop elongating on the contacting side, and the stem begins to curl around the contacted object. It may do so several times before cells on both sides of the stem begin elongating at about the same rate once again. How contact affects elongation remains a mystery, although auxin seems to be involved.

Thigmomorphogenesis is another response to contact, but the result is generally an inhibition of overall plant growth. Contact with rain, grazing animals, farm machinery, even air molecules disturbed during winds causes this plant response. Shaking a plant daily for a brief period can do the same thing, depending on the species. Figure 20.21 shows what happens when you shake tomato plants long enough.

Thigmomorphogenesis helps explain why plants grown outdoors are often shorter, thicker, and more resistant to winds than plants of the same species that are grown indoors. It appears to be an adaptive response to wind stress. Considerable evidence suggests that ethylene is involved in the response. Exposure to very low concentrations of this gas can trigger the same results, and contact or shaking causes a plant to produce more ethylene.

Figure 20.21 Thigmomorphogenesis in tomato plants. The plant to the far left is an untreated control. It was grown in a greenhouse, protected from wind and rain. The center plant was mechanically shaken thirty seconds at 280 rpm for twenty-eight consecutive days. The plant to the far right received two such shaking periods for twenty-eight days.

The Flowering Process

As a flowering plant matures, its physiological processes become directed toward flower, fruit, and seed production. In corn, soybeans, and peas, this activity occurs after only a few months of growth. These plants are **annuals**: they live only one growing season. In the so-called century plant, some ten years pass before seed formation. This plant is a **perennial**: it lives year after year. Still other species produce only roots, stems, and leaves the first growing season, die back to soil level in autumn, then grow a new flower-forming stem from a bud that remained alive in a protected stem region just underground. These plants typically live two growing seasons and are called **biennials**. They include many garden vegetables such as cabbages, carrots, and turnips. Perennial grasses in lawns, rangelands, and pastures form flowering stems each year, but they also die back and regrow in spring from buds near the soil level.

Daylength and low temperature are the strongest stimuli for flowering of land plants. Low-temperature stimulation of flowering is called **vernalization** (after a Latin term meaning "to make springlike"). Unless underground buds of garden biennials are exposed to cold winter temperatures, flowers don't form on the new spring stems. Winter wheats, typically planted in August or September,

produce hardy but nonflowering seedlings. Snow often covers the seedlings until spring. In early summer, new stems bear flowering stalks—but actual flowering is greatly delayed unless the young plants were exposed long enough to low temperatures.

Daylength is a highly reliable cue for flowering and reproduction of many species. The long days of late spring promote flowering of winter wheats and most biennials. Many other species also are adapted to long days and reproduce early in summer. They are known as *long-day* plants. Autumn-flowering species are attuned to shorter daylengths of late summer; they are *short-day* plants. There are also *day-neutral* plants, which flower almost independently of daylength. Some species that grow near the equator—where daylength essentially remains constant—are often day-neutral. So are corn, peas, tomatoes, and other crop plants that have for many generations undergone artificial selection for variants that can flower under a wide range of conditions. Figure 20.22 shows some flowering plants that respond in different ways to long and short days.

Cocklebur is a plant that somehow measures time with uncommon sensitivity. For cockleburs, night length is the key. A single night longer than 8½ hours is enough to cause flowering of most such plants. When a long night is experimentally interrupted with even a minute or two of light, flowering is retarded or prevented. For cockleburs and all other short-day plants, red light is the strongest stimulus, and red light is detected by phytochrome. In short-day plants, Pfr inhibits flowering if it is formed during the dark period. In long-day plants, Pfr is also the active pigment, but extended daylengths provide Pfr for longer periods. The result is faster flowering or more flowers per plant.

We have no idea of how Pfr controls flowering. In theory, at least, synthesis and transport of a hormone (or group of hormones) are involved. In anticipation of its discovery, the elusive hormone thought to control flowering was designated **florigen** more than forty years ago. Much evidence for its existence comes from cocklebur experiments. Only the main stem tip, branch tips, and lateral buds can form flowers. Before they do, they stop producing young leaves. Although these meristematic regions *respond* to daylength by forming flower cells instead of leaf cells, only the leaves *detect* the length of day and night. Presumably, detection involves production of florigen and its transport from leaves to buds; there, young leaves stop forming and flowering begins. If all but one leaf is trimmed from a cocklebur plant, and if this leaf is covered

All photographs Frank B. Salisbury

Figure 20.22 Effect of day length on flowering. Examples include tomato (day neutral), spinach (long-day), and lambsquarter, cocklebur, and Japanese morning glory (short-day). In each photograph, the plant on the left was grown under short-day conditions; the plant on the right was grown under long-day conditions. The tomato plants both flowered.

with black paper for at least 8½ hours, the plant will flower. But if the leaf is cut off immediately after the dark period, the plant will not flower. Apparently, florigen is still in the discarded leaf. Finally, suppose one plant in flower is grafted onto a plant that is not flowering. Suppose the grafted plants are kept under daylength conditions that normally don't promote flowering. What happens? The plant that was not flowering now develops flowers! The same thing happens when a short-day plant and long-day plant are grafted together. Such experiments suggest that "florigen" does indeed exist, but what it is and how the environment controls its formation is not known.

Senescence

Growth of flowers, fruits, and seeds of all plants places strong demands on roots, stems, and leaves for nutrients—nutrients that will sustain seedlings of the next generation. The demands are so strong that reproductive organs actually withdraw nutrients from vegetative organs through connecting sieve tubes. In leaves, newly formed enzymes degrade proteins, chlorophyll, and other large molecules. The breakdown products are transported, along with nutrient ions, to reproductive organs. There they form energy reserves of fats, starches, and proteins in seeds of a new

flowers removed   control group

Figure 20.23 Delay of senescence in soybean plants through the daily removal of flower buds. (From A. C. Leopold et al., *Plant Physiology*, 1958, 34:570)

Figure 20.24 Effect of localized cytokinin applications on senescence in a bean plant. The first two (oldest) bean leaves form on opposite sides of the stem; they are called primary leaves. Normally, the oldest leaves of any plant are the first to senesce. Here, primary leaf senescence was delayed by covering their upper surfaces at four-day intervals with a cytokinin solution. This caused the leaves immediately above to senesce and probably to transport nutrients into both primary leaves, allowing them to stay green and healthy longer. Presumably, cytokinins synthesized by leaves or transported to them from other plant parts normally delay leaf senescence. (From A. C. Leopold and M. Kawase, *American J. of Botany*, 1964, 51:294–298)

generation. In annuals and most perennials, plants are left with tan-colored, dead leaves depleted of most nutrients. In deciduous trees, which shed leaves each autumn, nutrients are also transported to parenchyma cells in twigs, stems, and roots prior to abscission. There they are stored until spring growth begins.

What makes the leaves of deciduous trees fall off, or abscise? Studies suggest that ethylene, formed in cells near the break points, stimulates the abscission not only of leaves but of flowers and fruits. Abscisic acid might also contribute to abscission, perhaps because it causes cells to produce more ethylene.

The sum total of processes leading to the death of a plant or any of its organs is called **senescence**. One stimulus for senescence could be the drain of nutrients during the growth of reproductive organs. Consider that when each newly formed flower is removed from a soybean plant, the leaves and stem remain green and healthy much longer than they otherwise would (Figure 20.23). Gardeners maintain vegetative growth in many plants by removing flower buds. Yet loss of nutrients from vegetative organs isn't the only cause of senescence. If a cocklebur is induced to flower, its leaves yellow regardless of whether the young flowers are pinched off or left on. It is as if some signal formed during short days causes both flowering and senescence of the cocklebur. More evidence for a death signal comes from experiments with soybeans. Such a signal must counteract the effects of cytokinins, which stimulate leaf growth. At least, when the surface of a mature leaf is painted with a kinetin solution (which increases the leaf's cytokinin content), the leaf often remains green longer (Figure 20.24).

Dormancy

As autumn approaches and days grow shorter, stem growth in many evergreen and deciduous trees slows or stops. It stops even if temperatures are still warm, the sun is bright, and enough water is available. This surprising response occurs largely because apical meristem cells in each bud stop forming new stem cells. These buds become tolerant of lower temperatures, and ordinarily they will not grow again until spring. When any plant stops growing under physical conditions that are actually quite suitable for growth, it is said to have entered a period of **dormancy**.

What controls bud dormancy? Short days and long nights are two environmental cues. We can demonstrate that these cues cause bud dormancy in fir plants by interrupting their exposure to a long night with a short period of light—especially of red wavelengths. The plants now

Figure 20.25 Effect of the relative length of day and night on Douglas fir plant growth. The plant at the left was exposed to twelve-hour light and twelve-hour darkness for a year; its buds became dormant because daylength was too short. The plant at the right was exposed to twenty-hour light and eight-hour darkness; buds remained active and growth continued. The middle plant was exposed to twelve-hour light, eleven-hour darkness, and one-hour light in the middle of the dark period. This light interruption of an otherwise long dark period also prevented bud dormancy. Such light causes Pfr formation at an unusually sensitive time in the normal day-night cycle. (From R. J. Downs in T. T. Kozlowski, ed., *Tree Growth*, The Ronald Press, 1962)

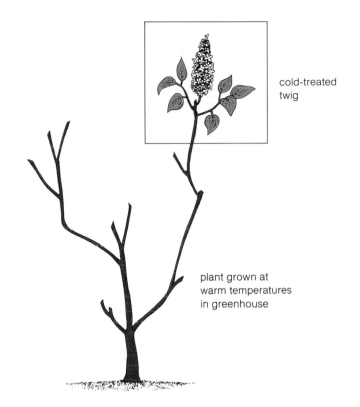

cold-treated twig

plant grown at warm temperatures in greenhouse

Figure 20.26 Effect of cold temperature on breaking bud dormancy in many woody plants. In this experiment, one branch (boxed portion) of a lilac plant was positioned so that it protruded from a greenhouse during winter; the rest of the plant remained inside, at warm temperatures. Only the buds on the branch exposed to low outside temperatures grew again in spring. This experiment suggests that low-temperature effects are localized.

behave as if they were exposed to short nights (long days) and continue to grow taller (Figure 20.25). Phytochrome (Pr) is involved in this response. In this case, conversion of Pr to Pfr by red light prevents dormancy. It may be that buds go dormant because less Pfr can form during the decreasing daylengths of late summer. It may also be that in the absence of Pfr, leaves synthesize a dormancy-triggering hormone (perhaps abscisic acid) that is transported to buds.

Regardless of the hormones involved, understanding how dormancy is controlled has practical implications. For example, the normal daylength cue so essential for survival of many species may be canceled by streetlights (Figure 20.1). Often, trees or tree parts near these lights fail to lose their leaves and become dormant soon enough to avoid the weight of heavy snows and cold winter temperatures.

Branches break, buds are killed, and trees are weakened. High-intensity sodium-vapor streetlights magnify the problem. They illuminate a wider area, and their color (yellow) causes more conversion of Pr to Pfr than older (more blue) streetlights do. Perhaps millions of dollars are wasted annually to replace winter-killed trees in new shopping centers and other areas, simply because many planners don't realize the importance of daylength cues to plant survival.

If short days and long nights increase bud dormancy in trees, what breaks dormancy in spring? After all, spring temperatures and daylengths are similar to those prevailing when buds became dormant the previous fall! Between fall and spring, though, a major *dormancy-breaking mechanism* is at work. This mechanism is low-temperature exposure for hundreds of hours (Figure 20.26). The actual temperature and exposure time varies among species. For instance, Delicious apples grown in Utah require 1,230 hours near 43°F (6°C); apricots grown there require only 720 hours. Generally, tree varieties growing in the southern United States require less cold exposure than those growing in northern states and Canada. So if you live, say, in Colorado and order a young peach tree from a Georgia nursery, the tree might start spring growth too soon and be killed by a late frost or heavy snow.

Both gibberellin and abscisic acid may help control dormancy. When gibberellins are applied to dormant buds, dormancy is often broken. Abscisic acid extends dormancy and partially counteracts the effects of gibberellins. These and other results suggest that the following may be true: *Abscisic acid movement from leaves to buds in late summer may trigger dormancy, and abscisic acid breakdown along with gibberellin accumulation in buds in late autumn and winter may end dormancy.*

Seeds of most native species also exhibit dormancy, which suggests that dormancy may have considerable survival value. But dormancy is rare in highly selected agricultural crops. The mechanisms by which seed dormancy develops and ends are variable. In some species such as honey locust and alfalfa, hard seed coats are formed. These coats prevent absorption of water and oxygen. Dormancy ends when the seed coat is abraded (perhaps as strong winds and rains drive the seed across sand), when it is chemically digested (by bacteria, by fungi, or in the gut of a bird or mammal), perhaps even when fire burns it away.

For species such as lilacs, apples, peaches, plums, and cherries, seeds must be exposed to low temperatures for weeks or months before dormancy ends. Such seeds are shed from the plant in autumn. Built-in controls prevent their germination before spring; without these controls, the seedlings would be killed by frost. In many of these species, gibberellins are known to break dormancy, and abscisic acid is known to prolong it. In fact, abscisic acid powerfully inhibits germination of all seeds that can absorb it, even nondormant ones. These observations suggest that abscisic acid accumulation, together with gibberellin loss, may trigger dormancy in some seed types. It may prove to be that the reverse changes end dormancy, just as in buds.

Finally, many seed types depend on red wavelengths to end dormancy. As you might suspect, phytochrome once again is the essential pigment involved in light detection, and Pfr is the active form. Only in the past few years have we learned about the importance of phytochrome in germination and so many other plant responses, and much more remains to be learned.

CASE STUDY: FROM EMBRYOGENESIS TO THE MATURE OAK

Where the ocean breaks along the central California coast, the land rolls inward as steep and rounded hills. Sixty-five million years ago, these sandstone hills had their genesis on the floor of the Pacific. At that time, violent movements in the earth's crust caused parts of the submerged continental shelf to begin crumpling upward into a jagged new coastal range. Since then, the rain and winds of countless winters have played across the crumpled land, softening the stark contours and sending mineral-laden sediments down into the canyons. Grasses have come to cloak the inland hills, and their organic remains have accumulated and gradually enriched the soil. On these hillsides, in these canyons, the coast live oak—*Quercus agrifolia*—began to evolve more than ten million years ago.

Quercus agrifolia is a long-lived giant. Some trees reach heights of a hundred feet; their evergreen branches may spread wider than that. Some individuals are known to be three hundred years old. In this species, male and female flowers appear on the same tree. In early spring, when the reproductive cycle commences, microspores in the anthers of male flowers develop into pollen grains. The pollen-bearing flowers, clustered together as golden catkins, droop pendulously from the leafy branches. Inconspicuous female flowers also appear among the leaves in small clusters near branch tips. Wind carries pollen from the catkins to receptive parts (stigmas) of female flowers on the same or neighboring trees. From each pollen grain, a sperm-bearing pollen tube emerges and elongates through the style, toward the ovary where the eggs reside. Eventually the pollen tube reaches an ovule in the ovary. There, fusion of a sperm nucleus with an egg nucleus leads to formation of a diploid zygote. As the zygote begins dividing, polarization of its cytoplasmic substances leads to differences in the rate and direction of division. Cells divide again and again, giving rise to root and shoot apices, and to large cotyledons that represent most of the embryo. At the same time, integuments of the ovule form the seed coat, and ovary walls develop into a shell. By early fall, the seed reaches maturity and is shed from the tree as a hard-shelled acorn.

Three centuries ago, long before Gaspar de Portóla sent landing parties ashore to found colonies throughout Upper California, the oaks were shedding the seeds of a new generation. Suppose it was then that a blue-feathered scrubjay, foraging at the foot of a golden hillside, came across an especially shiny, plump, and worm-free acorn. In storing away food for leaner days, that bird used its beak to scrape a small crater in a moist spot, dropped in the acorn, and scraped back just enough soil to cover the prize. Although a scrubjay might remember many such hiding places most of the time, this particular acorn lay forgotten. Within a few days, it germinated.

From the moment of germination, the oak seed embarked on a journey of dynamic, continued growth. At one end of the embryo, the root apical meristem gave rise to cells whose divisions created the primary root. At the other end, the shoot apical meristem gave rise to cells that would create the primary shoot. These meristem-derived cells divided repeatedly. And they enlarged. They grew longer, they increased in diameter. Water pressure drove the enlargement—water taken up osmotically as ions accumulated in the newly forming roots, and as hormones caused a softening of cell walls that otherwise would have been too strong to allow expansion under pressure. Differentiation produced the rootcap that would protect the primary root as growth forced it down through the soil. Differentiation produced cortex and epidermis, as well as a vascular cylinder through which water and ions would flow. From the pericycle (ground tissue just inside the root's endodermis), lateral roots arose under the influence of growth regulators. As these new roots continued to elongate, the plant's absorptive surfaces increased. Even before the shoot developed, the primary root had already pushed down considerable distances through the soil. When the primary shoot did begin its upward surge, separate vascular bundles appeared first near the shoot periphery. Later, activity of the vascular cambium would consolidate these into a continuous cylinder of secondary xylem and phloem.

In parenchyma cells of the developing roots, stems, and leaves, large central vacuoles formed. Vacuole enlargement caused the cytoplasm to be pressed outward as a thin zone against the cell wall. In this way, cytoplasmic contact with the environment was enhanced; essential ions, gases, and water were harvested rapidly in spite of their relatively dilute concentrations in the surrounding air and soil. Myccorhiza—fungal strands symbiotically interlocked with the roots—enhanced the absorption process. In leaves, stomates developed and regulated carbon dioxide movement into each leaf's interior.

As the seedling grew, the xylem pipelines came to form a system that provided functional links among all parts of the plant body. In dead tracheid and vessel elements, thick secondary walls did not collapse under the highly negative water pressure within them. Through xylem, water and minerals moved from roots to stems and leaves. The living phloem cells—sieve tube elements—had no lignified secondary walls. Their primary walls were strong enough to prevent cells from bursting under the strong positive pressures within them. Through the phloem, carbohydrates—sucrose especially—moved upward and downward from leaves to regions where food was being used or stored. Through phloem, various amino acids, amides, and essential elements were redistributed from one part of the oak to another.

At the whim of a scrubjay, the seed actually had sprouted in a well-drained, sandy basin at the foot of several steep hills, in full sunlight. Rainwater accumulated there each fall and winter, and kept the soil moist enough to encourage luxuriant growth during the spring and even through the dry summers. Out in the open, abundant red wavelengths of sunlight activated phytochrome pigments in the seedling. So stimulated, the phytochrome triggered hormonal events that encouraged stem branching and leaf expansion. All the while, delicate hormone-mediated responses were being made to the winds, the sun, the tug of gravity, the changing seasons. Lignin and cellulose strengthened secondary cell walls in secondary xylem and phloem. With this strengthening came resistance to the strong winds racing through the canyon. As the oak matured, phytochrome detected the subtle shifts in daylength throughout the year, which helped the plant respond to changing seasons. The shorter days of late summer promoted bud dormancy, hence resistance to winter cold.

As the oak seedling matured into the adult plant form, more and more roots developed and snaked through a tremendous volume of the moist soil. Branch after branch spread out beneath the sun. Leaves proliferated—leaves where the oak put together its own food from water, carbon dioxide, the few simple inorganic substances it mined from the soil, and the sunlight energy it harnessed to drive the synthesis reactions. Continued activity of vascular cambia increased the girth of roots and branches; with each new spring, terminal and axillary meristems increased the tree's height and diameter. Thus the oak increased in size every season, year after year, century after century. On their way to the gold fields, prospectors of the California Gold Rush rested in the shade cast by its immense canopy. The great earthquake of 1906 scarcely disturbed the giant, anchored as it was by a root system that spread out eighty feet in diameter through the soil. By chance, the brush fires that periodically sweep through California's coastal canyons did not seriously damage this particular tree. Fungi that could have rotted its roots never took hold; the soil was too well-drained and the water table too deep. Leaf-chewing insects were kept in check not only by protective chemicals in the leaves but by the bird predators abounding in the lush growth of the canyon.

During the 1960s, human population underwent a tremendous upward surge in California. The land outside the cities began to show the effects of population overflow as the wild hills gave way to suburban housing. The developer who turned the tractors on the canyon in which the giant oak had grown was impressed enough with its beauty that the tree was not felled. Death came later. How could the new homeowners, newly arrived from the east, know of the ancient, delicate relationships between the giant trees and the land that sustained them? Soil was graded between the trunk and the drip line of the overhanging canopy; flower beds were mounded against the trunk; lawns were planted beneath the branches and sprinklers installed. Overwatering in summer created standing water next to the great trunk—and the oak root fungus (*Armillaria*) that had been so successfully resisted until then became established. With its roots rotting away, the oak began to suffer the effects of massive disruption to the feedback relationships among its roots, stems, and leaves. Eventually it had to be cut down. In their fifth winter, in their red brick fireplace, the homeowners began burning three centuries of firewood.

Readings

Apfel, R. 1972. "The Tensile Strength of Liquids." *Scientific American* 227(6):58–71. A fairly simple treatment of theory and experiments providing evidence that liquids can exist under tension as well as pressure.

Epstein, E. 1973. "Roots." *Scientific American* 228(5):48–58.

Galston, A., P. Davies, R. Satter. 1980. *The Life of a Green Plant.* Englewood Cliffs, New Jersey: Prentice-Hall. A simplified treatment of much of plant physiology.

Hewitt, E., and T. Smith. 1975. *Plant Mineral Nutrition.* New York: Wiley. Techniques and results in studies of plant mineral nutrition.

Leopold, A., and P. Kriedemann. 1975. *Plant Growth and Development.* New York: McGraw-Hill. A thorough treatment of these subjects, with emphasis on experimental results.

Mayer, A., and A. Poljakoff-Mayber. 1975. *The Germination of Seeds.* Second edition. New York: Pergamon Press. A simple treatment of seed biology, and effects of light and other environmental factors on seed germination.

Peel, A. 1974. *Transport of Nutrients in Plants.* New York: Wiley. A short, simple treatment of transport processes occurring in xylem and phloem.

Salisbury, F., and C. Ross. 1978. *Plant Physiology.* Second edition. Belmont, California: Wadsworth. Excellent, comprehensive book covering most plant functions.

Torrey, J., and D. Clarkson (editors). 1975. *The Development and Function of Roots.* New York: Academic Press. One of the few modern treatments of root structure and function.

Vince-Prue, D. 1975. *Photoperiodism in Plants.* New York: McGraw-Hill. Excellent discussion of daylength effects on plant growth and development.

Weaver, R. 1972. *Plant Growth Substances in Agriculture.* San Francisco: Freeman. A much-needed book covering commercial uses of plant growth regulation.

Review Questions

1. Which features enable land plants to absorb water and other vital substances from sources in which concentrations of those substances are low?

2. What specific functions do salts carry out in plants? How does solute absorption and accumulation affect plant growth?

3. Describe transpiration. State how the cohesion theory of water transport helps explain what is going on in this form of water movement.

4. Stomata help regulate water loss from the plant body as environmental conditions change. Can you explain how they work? (For example, what role do potassium ions play? How does the rate of photosynthesis affect stomatal closure and opening?)

5. Why is it important to include some of the native soil around roots when transplanting a plant from one place to another?

6. Look at Figure 20.8. Then, on your own, diagram the feedback relations that influence the coordinated growth of stems, roots, and leaves.

7. Sucrose transport from one plant organ to another is called translocation. Can you explain how it works in terms of the four key points of the pressure flow theory? How did aphids help show that sucrose is indeed the main substance being transported through the phloem pipelines?

8. List the five known plant hormones (or groups of hormones) and the main functions of each.

9. Which of the following plant cells or organs synthesize most or all of the hormones they require in normal growth and development: coleoptiles, dicot stems, root cells, and leaf cells. Describe one experiment that tells us which ones do this, or which ones don't.

10. Explain how sunlight exposure influences leaf expansion, stem elongation, and stem branching during primary growth.

11. What is phytochrome, and what is its role in plant growth?

12. Define plant tropism. How is unequal growth in different parts of the same organ associated with phototropism and thigmotropism? Why are plants grown outdoors often shorter, thicker, and more wind-resistant than plants of the same species grown indoors?

13. Define annual, biennial, and perennial plants. Then describe the difference between long-day, short day, and day-neutral plants.

14. Which factors apparently trigger dormancy, and which factors may help break dormancy?

DIVERSITY: EVOLUTIONARY FORCE, EVOLUTIONARY PRODUCT

21

ORIGINS AND
PATTERNS OF ADAPTATION

By the close of the nineteenth century, "the fixity of species" was crumbling as a scientific concept. By the middle of the twentieth, "the fixity of continents" met a similar fate. Not only was life seen to be an ever changing drama, the stage itself was now seen to be changing in its most fundamental prop: the solid earth beneath our feet. It was not just that the earth's crust has been buckling upward and eroding downward through time. It was not just that the earth has gone through at least four immense glacial epochs and thaws, in which fully a third of the land has been alternately drained and drowned. It was not even that the earth has been steadily spinning slower and slower about its long axis, thereby making something as "basic" as the length of day variable through time. *It was that the whole crust has been divided into vast plates, which have been moving about uneasily on top of a plastic mantle and carrying the continents with them!*

There is no question that this conceptual upheaval is making a shambles of many previously entrenched scientific theories. And yet its implications are exhilarating. Here we are, living at a time when we can anticipate a cohesive theory in which the flow of earth history and life's history are inseparable. In this theory, the chemistry of life, its molecular and cellular architecture, and its myriad multicellular forms and behaviors—all these things have unfolded as parts of the same drama, which has been linked since the very beginning to the stage of a restless chemical earth.

The story you are about to read is sketched in broad outline, simply because there isn't space enough to do much more than this. But these broad strokes should not belie the impressive evidence from diverse fields that has been converging in support of it. This evidence comes from investigators with such far-ranging interests as the nature and evolution of the stars, the solar system, and the earth. It comes from centuries of increasingly precise research into the history of the earth's changing landscapes, oceans, climates, its fossils, its flow of life. We encourage you at the outset to explore and evaluate this evidence for yourself by continuing with the survey books recommended at the chapter's end. At the same time, we encourage you to view this story for what it is—a *tentative* undertaking in our attempts to explain all the observations currently available.

Figure 21.1 The present is often key to the past, if we know where to look. The view northward across Crater Lake in Oregon, a collapsed volcanic cone aligned with Cascade volcanoes reaching into Washington—and all paralleling the Pacific coast.

The excitement of discovery is there, and so must be the legitimate tests of the validity of these newest attempts to find meaning in what is unfolding before us.

ORIGIN OF LIFE: A TRANSITION FROM CHEMICAL TO BIOLOGICAL EVOLUTION

From the remnants of titanic stellar explosions that ripped through our galaxy billions of years in the past, our solar system was born. And through the accretion and gravitational compression of dust and debris swirling turbulently about the primordial sun—a process that continued until about 5 billion years ago—our planet took form. At first it was a cold, homogeneous mass. But through contraction and radioactive heating, it soon had a core that was growing increasingly dense—and hot. By 3.7 billion years ago, it was hurtling through space as a thin-crusted inferno.

How could life have been destined for this forbidding place, with its violent volcanic outpourings and fountains of gases, with its surface quaking from the tumult below and the bombardment of meteorites from above? Yet, in the solidification of the lighter crust over the hot, denser layers below, and in the orbit that the primitive earth set-

tled into around the sun, conditions that could favor the origin of life were assured.

Long before life originated, gases that had been trapped beneath the slowly forming crust or that were released as by-products of reactions in the molten interior were being forced to the outside. Although much of this early atmosphere was lost to space, after millions of years the earth's surface finally retained what is thought to have been a dense, gaseous atmosphere of hydrogen, nitrogen, methane, ammonia, hydrogen sulfide—and water.

Water is so essential for all the reactions and properties of all living things on this planet that it's difficult to imagine life originating here without it. But without enough gravitational mass to retain its atmosphere, the earth would have been as devoid of water as the moon is today. In addition, had the earth come to orbit the sun much closer than it did, its surface would have remained so hot that water vapor never would have been able to condense in liquid form. Had the orbit been much more distant, the earth's surface would have become so cold that any water formed would have become locked up as ice. *Thus the size of the earth, its composition, and its distance from the sun must be considered in identifying conditions that would be favorable for the origin of life.*

Jack Carey

Table 21.1 Main Elements Needed to Build Molecules of Life, and Their Likely Location in the Early Environment		
Element	The Atmosphere	The Oceans
Carbon	Methane (CH_4)	
Hydrogen	Hydrogen gas (H_2)	
Oxygen	Water vapor (H_2O)	Water (H_2O)
Nitrogen	Ammonia (NH_4)	Ammonium hydroxide (NH_3OH)
Sulfur	Hydrogen sulfide (H_2S)	
Phosphorus		Dissolved salts
Potassium		
Iodine		
Calcium		
Iron		
Magnesium		
Sodium		
Chlorine		

At first, however, the promise of liquid water could not be fulfilled. As fast as water vapor condensed in the cool, upper reaches of the atmosphere and began falling earthward, it evaporated in the intense heat blanketing the rumbling crust. Millions upon millions of years passed and the surface cooled. And then it began to rain. For centuries it rained, with interminable torrents pouring over the parched rocks and stripping them of minerals and salts, then streaming through crevices and canyons and thundering downward to the lowest points on the crust to form the hot primeval seas.

Somewhere between 3.7 billion and 3.2 billion years ago, living systems appeared in these mineral-rich waters. Precisely how and where we may never know, for there are few fossil records of the event that are left for us to interpret. Most of the rocks from that period have been melted, solidified into new rocks, and remelted many times over as a result of ongoing movements in the earth's mantle and crust. Some are buried far beneath more recently formed rocks, where they have been so subjected to heat and compression that whatever clue they might have held of this critical period is now altered beyond recognition.

But here is what we do know: with three basic conditions met, life could have originated as a product of chemical evolution. What are these conditions? *First*, the environment would have to contain those elements which make up living things. *Second*, the chemical environment would have to promote reactions between these basic elements so that they could form organic molecules—the building blocks of life. *Third*, there would have to be some physical means for keeping certain molecular building blocks right next to one another. Otherwise, it would be impossible for a self-contained system to arise that would have the chemical and structural complexity characteristic of life, and it would be unlikely that the system ever would be reproducible.

Let's consider the first condition. As Table 21.1 indicates, the ancient atmosphere and oceans could have contained all the elements needed for the synthesis of complex organic compounds. So the question becomes not whether the required elements were available, but whether there was some way for them to combine into the building blocks of life. If the chemical environment of the early earth had been the same as it is today, there would be no way for that to happen. The main deterrent would be the abundance of free oxygen in the existing atmosphere. Organic compounds, recall, are spontaneously broken down into simpler molecules in the presence of free oxygen. It's true that the process appears to occur slowly in terms of human life spans. But over the millions of years it would have taken for the first cells to evolve, even a slow rate of decomposition would have been notable: organic molecules simply could not have accumulated faster than they were being broken down if free oxygen had been present.

Analysis of the chemical composition of the earth's early rocks as well as the atmospheres of other planets does indeed suggest the early atmosphere had little or no free oxygen, but that it contained abundant hydrogen as well as nitrogen, methane, ammonia, and water. Hydrogen, a reducing substance, would have encouraged reduction reactions in which complex organic molecules could be built up. That is why the early atmosphere is thought to have been a "reducing atmosphere" instead of the "oxidizing atmosphere" of today (Chapter Seven). Any organic molecules built up would be less likely to undergo degradation in a reducing atmosphere. In fact, polymerization of small molecules into macromolecules would tend to occur spontaneously. All that would be required would be a source of environmental energy to activate molecules and make them collide with enough force to assure chemical interaction.

What energy sources might have been available? Ultraviolet radiation from the sun undoubtedly was bombarding the earth's surface more strongly at that time than it does today. For in the absence of free oxygen, there would have been none of the ozone (O_3) layer that now envelops the earth and screens out much of the intense ultraviolet wavelengths. Aside from that, levels of light and heat energy reaching the earth would have been higher during that period, when solar reactions were somewhat more intense than they are today. Energy might also have been provided by volcanic activity and by the radiation being released from the abundant radioactive elements making up the rocks in the earth's crust—elements that have largely decayed now to nonradioactive forms. And surely electric storms in the hot, dense atmosphere would have sent lightning crackling incessantly to the earth.

Thus the early earth could well have had the elements, the chemical environment, and a variety of energy sources. With that combination, organic molecules could arise spontaneously. We know this from the pioneering experiments of Stanley Miller and of many other chemists since. In 1953, while a graduate student at the University of Chicago, Miller set up a reaction chamber (Figure 21.2) containing a reducing mixture of hydrogen, methane, ammonia, and water. For a week he kept the mixture recirculating. All the while, he bombarded it with a continuous spark discharge to simulate lightning as an energy source. By the week's end, he found organic molecules had formed—including many kinds of amino acids!

Such experiments have been repeated many times, with variations in elements, in gas mixtures, and in the types of energy sources used. The results invariably show that all the building blocks required for life—including lipids, carbohydrates, proteins, and nucleotides—can form under abiotic conditions. More than this, if inorganic phosphate is present in the starting mixture, even ATP—the energy molecule used by all living systems—will form!

But now we come to the most puzzling question of all. By what mechanism could independently formed organic molecules have combined into a permanent, reproducible living system? There is no universally accepted hypothesis here, although two lines of experiments have given us some interesting things to think about. According to the first hypothesis, abiotic synthesis of increasingly complex molecules led gradually to a DNA molecule. With its double-stranded structure, DNA is intrinsically reproducible. And its formation would be a key step on the road to self-reproducing organisms.

But even though DNA is central to life, it must be kept in mind that *a DNA molecule is incapable of reproducing itself.* To the best of our knowledge, DNA can be replicated only as part of a metabolic system that can provide (1) nucleotide subunits in great enough concentrations, and (2) an enzyme for holding the reactants in place while they combine. It's difficult to envision how a fully self-replicating molecule could have arisen without the support of a

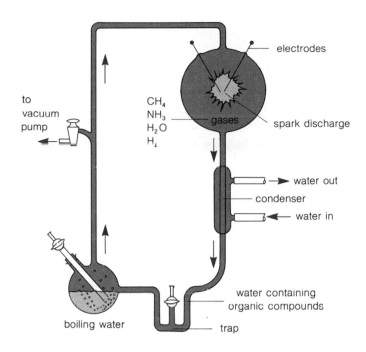

Figure 21.2 Stanley Miller's apparatus used in studying the synthesis of organic compounds under conditions believed to have been present on the early earth.

relatively complex and functionally isolated metabolic system. Even supposing there had been enough nucleotides concentrated in the primordial seas to permit the first DNA molecules to self-reproduce spontaneously, what would it have meant? DNA reproducing in isolation would have been like a book with no words on the page, and no one to read it.

According to the second hypothesis, a closed metabolic system must have appeared first. Metabolism, recall, means chemical control—*and chemical control is possible only with chemical isolation from the random ebb and flow of materials in the environment.* If we assert that chemical evolution alone led to a metabolic system, we must show that molecular boundaries can arise spontaneously, thereby isolating and concentrating the essential molecules from the rest of the environment. Laboratory experiments again suggest how such partitioning might have occurred. When Aleksandr Oparin mixed together proteins, polysaccharides, and nucleic acids, he discovered that these molecules can assemble into small spheres. These **coacervate droplets** are intriguing, for they seem to be differentially permeable. They

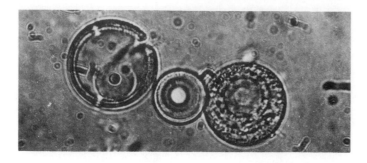

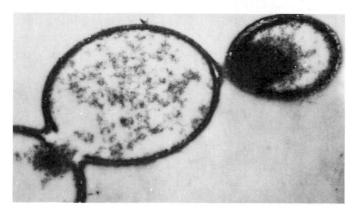

Figure 21.3 Micrographs of microspheres. (Sidney Fox, Institute for Molecular and Cellular Evolution, University of Miami)

can selectively accumulate certain substances in greater concentrations than are found outside. There is a tendency also for droplets to pinch off the main spheres, creating smaller subsystems from them.

In a similar kind of experiment, Sidney Fox and his coworkers began by heating amino acids under dry conditions, which produced a number of long polypeptide chains. Perhaps, Fox suggested, such polymers were formed in hot volcanic regions and then were washed by rains into the seas. Whatever the case, when these chains were added to hot water solutions and then allowed to cool, they assembled into small, stable **microspheres** (Figure 21.3). Like Oparin's droplets, these microspheres tended to accumulate specific substances in their interior. Depending on the substances present in the solution, the spheres could shrink or grow in size. In fact, given the presence of enough polypeptide chains, they could grow to the point of producing budlike growths that could break off into new microspheres. More than this, when lipid molecules were also present in the solution, they tended to combine

with the protein microspheres. *And the outcome was the formation of a lipid–protein film surrounding each droplet that was remarkably like a simple cell membrane.*

It's important to keep in mind that neither coacervate droplets nor microspheres are alive. They are no more than the outcome of relatively simple chemical interactions between water and complex organic molecules. They do demonstrate that some boundary layers, at least, can come to exist spontaneously. But even though a selective boundary layer is one of the prerequisites for life, in itself it does not demonstrate the property of "being alive." Similarly, even though microspheres can grow by simple chemical acquisition of molecules and then fragment after a certain volume is achieved, they still cannot be called alive—for there is a long way to go from passive, random fragmentation to a self-reproducing cell.

At present, we can only speculate that the formation of such structures *could* have marked the beginning of a long series of chemical developments, with individual systems accumulating various environmental resources and carrying out such reactions, probably at comparatively low rates. For if such spheres existed that could grow in size, more molecules could be forced together in their interior. And because of chemical interactions of the sort outlined in Chapters Three and Five, the molecules inside them would tend to interact spontaneously and become spatially *organized* relative to one another. If the spheres grew large enough to break in two, or if the pounding against some ancient shore broke them in two, the fragments could continue to gather materials selectively and thus grow in their turn—only to be broken apart into fragments that again would grow.

Each new fragment would enter a slightly different part of the environment and would accumulate a slightly different array of molecules. Perhaps some sphere even acquired a molecule such as a polypeptide that acted to speed reactions involved in the formation of new membranelike films. The molecule essentially would be a catalyst; it would permit the sphere to grow more rapidly, hence to accumulate materials more rapidly, than neighboring spheres. Such a polypeptide catalyst may have represented the beginnings of enzyme action. Gradually, proto-cells containing enzymes that speeded their own growth would succeed at the expense of less efficiently organized ones. In the nondirected chemical competition for resources in the seas, there would automatically be selection for microspheres having enzymes.

Where does this line of speculation take us? During the many millions of years that lifeless microspheres were accumulating organic molecules, suppose they were also gathering in various nucleic acids and nucleotides from their surroundings. Suppose some of the nucleic acids were no less than short molecules of deoxyribonucleic acid—DNA. If the nucleotides became concentrated near the DNA molecules, there would be a greater likelihood that the DNA could be replicated, given the presence of an enzyme capable of unzipping the double helix.

In itself, this step would not represent a selective advantage. But among all the millions of tiny spheres that contained nucleic acids, perhaps there was one that came to acquire a nucleic acid having a base sequence that was, by chance, no less than a template for a membrane-forming enzyme. It would be a self-enclosed system for promoting the replication of a nucleic acid: the nucleic acid would promote formation of the enzyme, and the enzyme would promote formation of new membrane. Thus the sphere could grow faster and accumulate more resources, which would speed the growth–replication cycle. The spheres into which it fragmented could come to differ slightly in the interdependent cycle of membrane, nucleic acid, and enzyme formation. And at each stage, variants capable of functioning somewhat more efficiently would grow faster and produce more microspheres like themselves. Given the replicative nature of DNA, the perpetuation of new modifications would have been assured. The evolution of a living cell, with its gene-action system and its battery of gene-coded enzymes, would have begun.

It is true that the entire sequence of events just described is indeed speculation. But it is also true that between 5 billion and 3.2 billion years ago, there was time enough and resources enough for chemical evolution of this sort to occur. And by 3.2 billion years ago, we know now that clearly recognizable forms of life had appeared.

A TIME SCALE FOR PATTERNS OF DIVERSIFICATION

In itself, the notion that living systems not much different from prokaryotic bacteria could evolve into forms as complex as ourselves might elicit disbelief. Part of the problem is that we are used to thinking in terms of minutes, hours, days, and years, certainly never much beyond centuries. Besides, until recently there was no way to judge the true age of the earth, hence the span of time in which evolution might have taken place. We could only observe the relative sequences of rock formations and the fossils they contained without knowing how to assign dates to the boundaries between them. Thus the **geologic time scale** (Figure 21.4)

Era	Period	Epoch	Age (millions of years)
Cenozoic	Quaternary	Recent	
		Pleistocene	
	Tertiary	Pliocene	30
		Miocene	
		Oligocene	
		Eocene	
		Paleocene	65
Mesozoic	Cretaceous	Late	100
		Early	130
	Jurassic		185
	Triassic		230
Paleozoic	Permian		265
	Carboniferous		355
	Devonian		413
	Silurian		425
	Ordovician		475
	Cambrian		570
Proterozoic (Precambrian)			1,000
	oldest definite fossils known		3,500
	oldest dated rocks		4,000
	origin of the earth		5,000

Figure 21.4 Geologic time scale, with dates based on radioactive isotopes from rocks of each era. Figure 21.5 describes the naturally occurring process that makes radioactive dating possible.

Main Radioactive Elements Used in Rock Dating			
Radioactive Isotope	Half-life	Stable Product	Age Range (Years)
Rubidium 87	50 billion years	Strontium 87	>100 million
Thorium 232	14.1 billion years	Lead 208	>200 million
Uranium 238	4.5 billion years	Lead 206	>100 million
Uranium 235	0.71 billion years	Lead 207	>100 million
Potassium 40	1.28 billion years	Argon 40	>100,000
Carbon 14	5,730 years	Nitrogen 1	0–60,000

Data from *Handbook of Chemistry and Physics*, 1975.

Figure 21.5 Radioactive dating. For many years, scientists tried to measure the age of rocks by assuming erosion, uplifting of mountains, and so on occurred at a constant rate. All such attempts were frustrated because these processes simply don't occur at a constant, invariable rate. In this century alone, volcanoes have suddenly popped out of the sea, only to crumble and disappear in a matter of months!

More recently, radioactivity has proved to be an accurate timekeeper. A *radioactive element* has an unstable combination of protons and neutrons in its nucleus. It breaks down spontaneously, and it releases radiation (such as x-rays or electrons) when it does. Some combinations are inherently more unstable than others. Thus some radioactive elements break down rapidly, others slowly. *But each radioactive element has its own characteristic decay rate.* And that rate simply can't be modified. Radioactive elements have been subjected to intense heat and pressure, suspended in high vacuum, cooled to nearly absolute zero, rearranged chemically into myriad chemical forms, charged with electricity, enclosed in a vast array of environments, and studied as gases, liquids, and solids. And no one has ever found any evidence that such changes have any effect whatsoever on the rate of radioactive decay. *All radioactive elements go on ticking off the seconds in total disregard for the environment.*

Each radioactive element has a *half-life:* the time required for half the atoms in a sample of the element to decay. How can half-life be used to date ancient rocks? Consider a radioactive isotope of potassium. When it breaks down, it's converted to a lighter element: argon. Argon is stable, but it is a gas at normal earth-surface temperatures. The half-life for the decay of radioactive potassium is 1.28 billion years. In that time, half of a pure, radioactive potassium sample becomes converted to argon. (Since a given rock first solidified, all the argon present in it has been derived from potassium. Any argon present before that time would have escaped to the atmosphere as gas. But once the rock became solid, argon formed by radioactive decay of potassium would have no way to escape.) If the measured ratio of potassium-to-argon is 1:1, then the rock solidified 1.28 billion years ago.

initially was no more than a progression of four broad eras: the *Proterozoic* ("very first life"), the *Paleozoic* ("ancient life"), the *Mesozoic* ("between-ancient-and-modern life"), and the *Cenozoic* ("modern life"). Within the past three decades, however, we have been able to assign fairly firm time boundaries to these geologic intervals by using radioactive dating methods on an enormous number of rock samples taken from all over the earth. All these methods are based on comparing the known, invariant decay rates of such radioactive isotopes as uranium, thorium, potassium, and strontium to the measured amounts of these isotopes and their decay products in different kinds of rocks (Figure 21.5). Some of the most ancient fossilized cells taken from carefully dated rocks are shown in Figure 21.6. These fossils have been found in certain regions of Africa, North America, and Australia, where large land masses have remained fairly stable through episode after episode of geologic unrest.

As we move from these earliest signs of life, the fossil organisms preserved within younger and younger rocks become progressively more abundant, more diverse, and more like modern forms. In many cases, it's possible to trace gradual modifications of form among apparently related organisms. And often there appears to be gradual divergence of related forms into distinct species. Chapter Twenty-Five describes ways in which this progression could have occurred. Here we will simply outline how it may have been largely a result of natural selection.

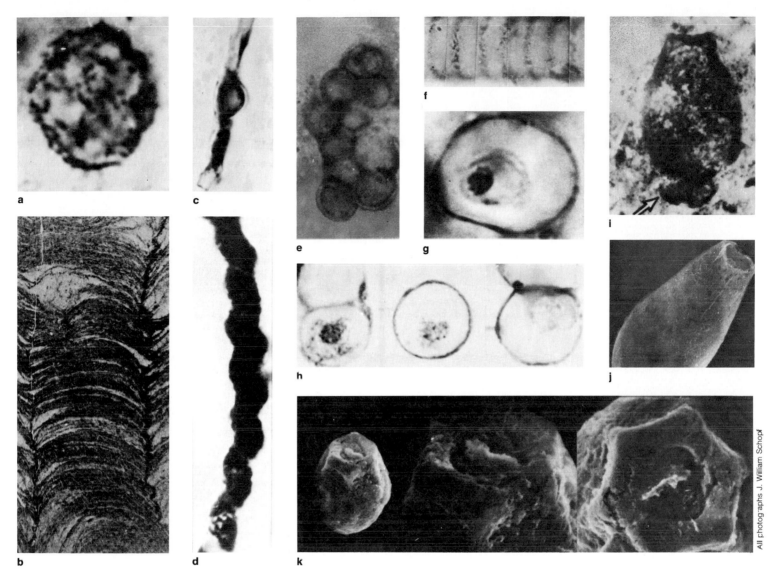

Figure 21.6 A sampling of the oldest known fossils. (**a**) From South Africa, bacteriumlike form about 3.2 billion years old. (**b**) Limestone stromatolite from Rhodesia, about 3.1 billion years old. The stacked organic (dark) and calcium-containing (light) layering is identical with the one laid down by modern communities of photysynthetic microorganisms. (**c**) Cellular form from South Africa, about 2.25 billion years old, and similar to modern blue-green algae. (**d**) From Central Australia, a spiral form 900 million years old. (**e**) Colonial cells found in stromatolites in the USSR, about 650 million years old. (**f**) Filamentary form, about 650 million years old. (**g**) Eukaryotic cell from Central Australia, about 900 million years old; the granules and spots may be remnants of organelles. (**h**) Fossil eukaryotes containing cytoplasmic remnants, about 900 million years old. (**i**) Eukaryotic cell 750 million years old. (**j,k**) Notice the well-developed mouth region in these fossil eukaryotes, also 750 million years old.

Within each population of organisms, there is always a certain amount of genetic variability. This variability slowly but inevitably arises as new mutations occur. In each generation, selective agents favor those best adapted to a given environment, and they will be the ones most likely to reproduce. As environments change, so do selection pressures change—and so does the population change in character. If one population somehow expands into two different settings, different traits will be selected for in each setting. With time, the two populations will accumu-

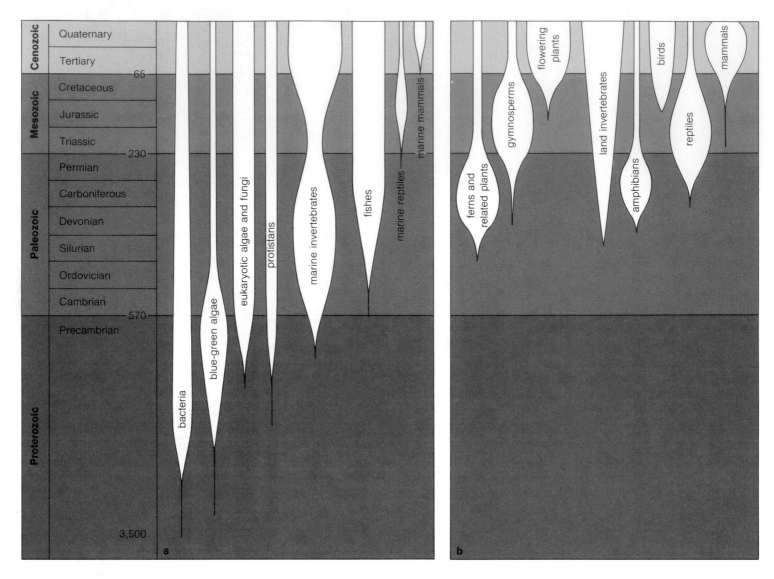

Figure 21.7 Diversification of life in the oceans (**a**) and on land (**b**). The varying width of each evolutionary line represents periods of relative dominance and decline for each group of organisms.

late more and more structural and/or behavioral differences until they reach the point of *speciation:* members of the two populations can no longer interbreed even if they are brought back together.

But not all transitions in the fossil record are gradual, for certain major groups of organisms often appear relatively abruptly—and they are already highly developed and diverse when they do. What could account for such dramatic entrances of whole groups onto the evolutionary stage? First, not all organisms become fossilized when they die.

The fossil record represents only a tiny fraction of all the organisms alive during any span of the remote past. Second, the probability that representatives of any group will be preserved depends partly on how many individuals there are in the group. Thus small numbers of organisms might be successfully adapted to some restricted environmental pocket—even though the chance of finding fossil evidence of their existence would be extremely low. Third, suppose the range of the environment to which an isolated group was adapted suddenly expanded as a result of a comparatively abrupt climatic or geographic change. Then its populations could be expected to expand proportionately. Not only would the change mean an increased probability that fossil evidence would become more abundant, it could well set the stage for rapid diversification. For as the group of organisms radiated into an expanded environment to which it was already adapted, local populations would have new opportunities to exploit existing genetic variability, and they would embark on a rapid course toward speciation. As you will now read, sprinkled through the fossil record of gradual modifications in form over vast spans of time are just such episodes of abrupt environmental change—with a corresponding acceleration in the emergence of new species.

THE AGE OF PROKARYOTES

Until about 3.7 billion years ago, the earth's crust may have been too unstable or too thin—and perhaps the heat flow from the interior was too great—for permanently stable land masses to form. (It would be at least another 1 billion years before there would be stable crustal plates, and even then they would be fringed with mobile, volcanically active zones.) Nevertheless, there probably were myriad volcanic islands rising above the primeval seas. And it may have been in the shallow, nearshore waters of these islands that life appeared. Reeflike formations, perhaps formed by calcium-secreting algae, date from about 3.5 billion years ago. Some of the oldest rocks, formed from lime deposits and black mud (in which oxygen is absent), contain organic compounds such as amino acids and parts of the chlorophyll molecule—and as far as we know, only life forms can organize a chlorophyll molecule. Some organisms apparently were developing the means for photosynthesis (Chapter Seven). Even so, fossils of any sort from the Precambrian are scattered sparsely in limited regions, and they speak only of the most limited diversification. For about 2 billion years, it seems, organisms resembling modern bacteria and blue-green algae had the world much to themselves. If there was little or no oxygen in the atmosphere, anaerobic pathways must have been the style of the day.

Between 2 billion and 1 billion years ago, both heterotrophic and autotrophic prokaryotes grew steadily more abundant, if not more diverse. Apparently there were no pressures prodding life toward greater structural and functional variation. It was only with the convergence of new environmental conditions that new evolutionary roads opened up.

THE RISE OF EUKARYOTES

Between 1.4 billion and 570 million years ago, the first eukaryotes arose. If we are to assign some meaning to the array of fossilized shells, spines, and armored plates they left behind, it would have to be that predatory forms were among them. That being the case, simple coexistence could no longer be possible; change would be the order of the day. Fossils of prey organisms from that time onward are increasingly larger. Fossils of predators are correspondingly larger, and some even sport devices for tearing off pieces of larger prey. In some cases, apparently, the only survivors among certain prey groups were the ones too large to be swallowed. With their selective advantage, they would be the ones more likely to leave progeny. Thus, subsequent generations became larger in size. At the same time, other groups may not have included enough variants capable of withstanding predation, and they became extinct. One of the groups that seem to have been hit hardest by predation were the blue-green algae of the seas. An important group, whose fossilized, matlike remains form stromatolitic rocks (Figure 21.6b), had been expanding steadily in numbers for many millions of years. But with the rise of predatory eukaryotes they began to decline. Perhaps it's no coincidence that living representatives of the group are confined to marine environments having low levels of diversity and low levels of predation.

By 570 million years ago, astonishingly diverse eukaryotes dominated the scene. Why such rapid diversification *then*, and not before? Consider the possibilities. During the period between 1.5 billion and 570 million years ago, oxygen had begun to accumulate to significant levels as a result of the combined activity of billions of photosynthetic microorganisms. (These increased levels have been determined by measuring the extent of oxidation of surface rocks which were solidifying at that time.) The primitive eukaryotes were mobile, and they were heterotrophic; they chased after their prey. The use of cellular respiration (Chapter Seven) would clearly be advantageous in providing an energy source for their movements, and it seems inevitable that the gradually increasing atmospheric levels of oxygen would foster organisms capable of cellular respiration. Whatever the case, by the beginning of the Cambrian, oxygen levels were high enough to allow considerable cellular energy to be devoted to building mineralized shells, spines, and the other hard protective parts we see for the first time in the fossil record. In addition, the rising oxygen levels meant that an effective ozone (O_3) barrier against incoming ultraviolet radiation was forming. Thus

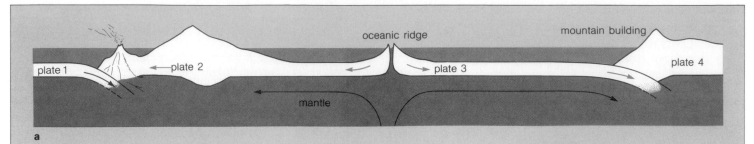

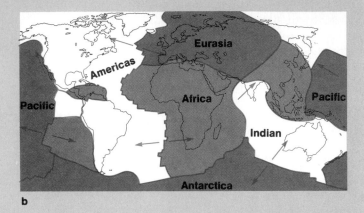

Figure 21.8 Plate tectonic theory. (**a**) The earth's surface is seen to be broken up into rigid plates. Today these plates are drifting toward or away from each other in the direction of the arrows. Boundaries between plates are marked by recurring earthquakes and volcanic activity. (Recent eastward movement of the Caribbean plate, the cross-hatched area in **b**, caused the devastating 1976 earthquake in Guatemala, which, like much of Central America, is part of this plate. Explosive volcanic activity and extensive surface faulting suggest this region of the crust is slowly being torn apart.) (**a**) Seafloor spreading and continental drift. Plate tectonics is based on observations that the seafloor is slowly spreading away from sites called oceanic ridges, and on measured displacements of the continents relative to these ridges. Thermal convection in the mantle is proposed as the mechanism underlying these movements. More heat is seen as being generated deep beneath oceanic ridges than elsewhere. The hotter material slowly wells up, then spreads out laterally beneath the crust (much like hot air rising from a stove, then spreading out beneath the ceiling). Oceanic ridges are places where the material has actually ruptured the crust (as it continues to do in the volcanic Hawaiian Islands). As the cooler material moves away from the ridges, it acts like a conveyor belt, carrying older oceanic crust along with it. Thus plates grow and spread away at oceanic ridges. As the plates push against a continental margin, they are thrust beneath it, which causes the crumpling and upheaval we call mountain building.

shallow offshore waters became more hospitable environments for eukaryotes—which are highly sensitive to these potentially deadly wavelengths.

Correlated with these rising oxygen levels was a geographic change of the first magnitude. By the dawn of the Cambrian, the marine environment expanded dramatically as the climate warmed up in what appears to have been the sudden aftermath of a worldwide glaciation. As the ice melted, shallow seas inundated much of the land. Into these vast seas the photosynthetic algae and eukaryotes radiated.

Before this transitional period in the history of life—a comparatively brief span of about 800 million years—the eukaryotes did not leave behind much evidence of their early adaptations. But with the convergence of conditions favoring the development of hard parts around fragile bodies, the eukaryotes would henceforth commit evidence of their journey to the sediments of time. And so today we pick up the remains of that past age and watch them

fall silently through our fingers; and we begin to suspect at last the meaning of the relatively abrupt appearance of ever more elaborate shells and spines that once harbored vulnerable as well as voracious life in the crashing surf of the Cambrian shores.

UP GO THE MOUNTAINS, OUT GOES THE SEA

If there is one thing that characterized the Paleozoic, it was the number of massive inundations of continents and retreats by shallow seas. By the late Cambrian, there apparently were three major land masses. Two would eventually become North America and Europe. The third, **Gondwanaland,** was a supercontinent destined to become fragmented into Africa, South America, Australia, Antarctica, even parts of Asia and the eastern North American coast (Figure 21.8). But the three land masses were covered in large part by shallow seas. Because uplifted land had been eroded almost to sea level, not far below the water's surface were mud, sand, and more mud. Perhaps that's why this was the golden age of trilobites—of mud-crawling, mud-burrowing scavengers that eventually were 600 genera strong. They, along with algae and myriad other shelled organisms, thrived in this marine setting.

By the Ordovician, even low-lying land masses seem to have become submerged. Abruptly, there was a new burst of adaptive radiation into vacant marine environments. This was what we might call the age of invertebrate experiments. All but one invertebrate phylum had arisen by the close of the Cambrian, but now new varieties were decked out in the most astonishing adaptations—including well-developed eyes for locating predators or prey. Many of the innovations of this period would lead nowhere, but others would survive even into the present (Chapter Eighteen).

For several million years, life flourished in the seas. Then, in the mid-Ordovician, the crustal plates bearing the three land masses began moving toward one another. As they did, they began edging over the crustal plate that formed the basin of the ancestral Atlantic Ocean. At first, the land masses were raised slightly and the shallow seas were drained off. The sudden shrinking of the immense marine havens must have severely tested existing populations, and the fossil record does indeed reflect a corresponding explosion of new forms apparently vying intensely for food and space near the continental margins. By the late Ordovician, the seas inundated the land once more, and a new variety of life forms appeared in vast numbers. The jawless fishes—organisms at the crossroads leading eventually to reptiles, birds, and mammals—seem to have been among them, but they were not yet populous enough to leave behind much more than traces of their appearance. No matter. Toward the end of the Ordovician, storms of mountain building were brewing as the continental plates continued on their collision course. Volcanic outpourings in the mobile zone along the eastern edge of the North American land mass created the immense ancestral Appalachian Mountains as the European plate closed in.

During late Silurian times, some life forms by chance must have developed traits that would eventually permit them to escape from the seas. In the fossil record are transitional forms that appear to have been evolving in ways that would lead to the invasion of land. Regardless of when the first invasions occurred, they could have occurred only when certain challenges were met. Before some organisms could leave the seas, they would have to be equipped with the potential for tapping into new water supplies or carrying water around with them. They would require tough

Figure 21.9 Reconstruction of the lobe-finned fish.

or waxy surface layers to keep from drying out. No more could aerobic organisms absorb oxygen from the surrounding water—they would live on land only with systems for taking in oxygen from the surrounding air. Without the buoyancy of water, different frameworks would be needed to support the body. For animals, new modes of moving about would be required. And how could the balance of salts in body fluids such as blood be maintained once saltwater environments were left behind? There would be selection for new structural and behavioral adaptations. And what about reproduction? Gametes could not be simply dispersed for fertilization in the waters of the sea; there would be selection for new modes of reproduction and offspring dispersal. Adaptations that permitted radiation onto the land are shown in Chapters Twenty-Three and Twenty-Four.

As the Silurian gave way to the Devonian some 400 million years ago, the continents collided. Even as the proto-Atlantic Ocean disappeared, there was a dramatic increase in dry land area. Plants and then animals, equipped by chance with variations that permitted them to survive at the land's edge, began their tentative adaptive forays into this vacant environment. The first land plants were single stalks no taller than your little finger, and they were anchored in wet mud. Crabs and related forms, with their protective shells, were scuttling about in the alternating wet–dry tidal zones. Many groups of fish perished. But certain fish, left behind in drying tide pools or in evaporating freshwater ponds, were able to use their swim bladders (air sacs) as lunglike organs, and their fins as simple limbs (Figure 21.9). With such adaptations, they had the means of crawling from pond to pond. Over prolonged periods of alternating wet and dry conditions, those individuals

a

b

American Museum of Natural History

c

d

best able to breathe air and crawl about in the mud survived even as their water-dwelling relatives perished. Thus there apparently began an evolutionary trend that gave rise to the *amphibians:* organisms adapted to life both in water and on land. Yet, despite the flourishing diversity, the plants and animals of this period had one thing in common: a total dependence on a constant supply of water. The transitional plant forms still needed a wet environment to complete their sexual cycles. Devonian amphibians, like their descendants, had to return to water to lay eggs, for their eggs had no protective shell to keep from drying out.

The Carboniferous was the heyday of the amphibians. And why not? From beginning to end of this period,

land masses were submerged and drained no less than *fifty times!* Besides these major inundations, there were ever-changing sea levels in between. And imagine the conditions along the flat, low coastlines. Immense swamp forests became established; over time the seas moved in and buried them in sediments and debris; they became reestablished as the seas moved out, and then they were submerged again. (The organic mess left behind has since compacted to become the world's great coal deposits.) But some groups survived the inundations, for adaptations had appeared among them that allowed movement onto higher (and drier) land. The gymnosperms (Chapter Twenty-Three) were one such group. These seed-bearing plants

American Museum of Natural History

Peabody Museum, Yale University

Figure 21.10 (**a**) A Jurassic reptile capable of running swiftly on two legs. Such reptiles are thought to be the ancestors of dinosaurs and birds. During the age of dinosaurs, some forms took to the air; (**b**) shown here, pterosaurs had wingspans of 12 meters (40 feet) or more. Other forms became adapted to the water and became impressively large, swimming predators—the long-necked plesiosaurs (**c**) reached 15 meters (50 feet) from head to tail. These forms were doomed to extinction when the vast inland seas dried up. On land, dinosaur diversity reached its peak, with herbivores and carnivores ranging from the size of gophers to behemoths three stories high and half a block long (**d**).

together into one vast supercontinent, called **Pangaea.** The changes in land area and in elevation brought pronounced differentiation in world temperature and climate. To the north, arid lowlands and humid uplands emerged. To the south, glaciers built up and ice sheets spread over the land that had become positioned there. As shallow seas were drained from the massive continent, life forms radiated throughout the land. And everywhere in the shrinking seas: massive extinctions.

After the great Permian marine extinctions, the survivors in the seas had 165 million years of relatively stable environments in which to diversify. On land, the character of the living community changed profoundly. By Jurassic times, the climate was warm and humid; mountains had emerged, along with plains and vast lagoons. In these new settings, gymnosperms and reptiles had the competitive edge. Some groups of reptiles became readapted to life in the water, others came to develop means to take to the air. The ones remaining on land diversified in the most spectacular ways and were unchallenged for the next 125 million years—the golden age of dinosaurs (Figure 21.10).

Their ultimate challengers, though you might never have believed it at the time, would be the little ratlike mammals scurrying through the shrubbery of the Jurassic. These mammals were food for many dinosaurs, so they had to be clinging most precariously to their place in the community. But cling they did, only to explode into the biotic vacuum left by their predators at the end of the Cretaceous. For at that time all dinosaurs disappeared suddenly from the earth. This total, abrupt extinction has yet to be explained. There were no profound shifts in climate. Although there is evidence of extensive mountain building, there apparently were no sudden, massive upheavals in the crust. Other kinds of organisms living in the same places managed to survive. Why not the dinosaurs? The answer to that question is not known. Whatever the reasons, with the land essentially cleared of major preda-

were not restricted to the water's edge; they could complete their reproductive cycles without free-standing water. The reptiles were another: these animals could break away from an aquatic existence because of shelled eggs and internal fertilization. These adaptations meant development could proceed within the eggshell, a moist and (compared to the outside) dependable setting.

Swamp forests and ancestral frogs—as magnificently adapted as many species were to these alternating environmental conditions, many of their adaptations would prove disadvantageous when the environment changed. And change it did, for as the Carboniferous gave way to the Permian, collisions of crustal plates brought all land masses

tors, the mammals eventually diversified, along with flowering plants (Chapter Twenty-Three).

CENOZOIC UPHEAVALS AND THE FURTHER DIVERSIFICATION OF LIFE

It was the best of times, the worst of times; it was the dawn of the Cenozoic and the continents were on the move.

Beginning in the Mesozoic, the supercontinent began to break up (Figure 21.11). Widespread volcanic activity marked the birth of the Atlantic basin as North America, Europe, and Africa began moving their separate ways. But by Cenozoic times, major reorganization was going on among all the crustal plates. Unbelievable amounts of lava began pouring through immense faults and fissures that penetrated to the very basement of the earth's crust. Brittle fragmentation occurred along the coasts; severe volcanism and uplifting gave rise to mountains along the margins of massive rifts and along zones where some plates were thrust under others. The immense volcanoes dominated by Kilimanjaro in Africa, the Sierra Nevada and the Cascade ranges paralleling North America's Pacific coast, the Alps, the Andes, the Himalayas—never before had there been such world-wide mountain-building as crustal plates collided with and jockeyed for position relative to one another.

Correlated with this latest redistribution of the land was the onset of the world's most recent glaciation. Sea levels changed, as did the earth's overall temperature. Change brought about widespread extinctions among some groups even as it opened the door to expansion for others. The separation of continents encouraged diversification of life in divergent ways. In much of the world, the concurrent shifts in climate led to the emergence of extensive, semi-arid, cooler grasslands into which plant-eating mammals and their predators radiated (Figure 21.12). Now the ungulates—hoofed mammals of the sort shown in Chapter One—roamed in vast herds. The ancestors of such modern forms as deer, giraffes, horses, pigs, rhinos, camels, and elephants appeared, as did carnivores such as the saber-toothed tiger. And as the climate continued to change, the vast tropical forests dating from the preceding epoch began to be fragmented into a patchwork of new environments. Many of their inhabitants were forced into new life styles in mountain highlands, in deserts, in the plains. One such evicted form was destined to give rise to the human species—with all our problems, with all our promise. The events surrounding that emergence will be described in Chapter Twenty-Four.

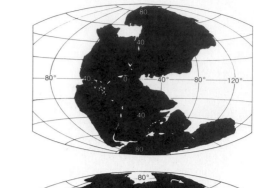

a

b

Figure 21.11 Restoration of the supercontinent Pangaea (**a**) and its breakup, showing the impending collisions of Afro-Arabia and India with Eurasia in Cenozoic times (**b**).

PERSPECTIVE

All living things are composed of molecules made of the same chemical elements—primarily carbon, hydrogen, nitrogen, and oxygen. We can reasonably assume that all these elements were present, in one form or another, in the crust and atmosphere of the primordial earth. It has been demonstrated that these elements can combine spontaneously into carbohydrates, lipids, proteins, nucleic acids, even ATP—into the stuff of life—given the right chemical environment and some source of activation energy. It has been demonstrated that, given a plausible set of environmental conditions, such macromolecules can form differentially permeable structures much like simple cell membranes. It is possible that further experiments into the molecular and biochemical nature of life will throw light on how such structures might have evolved chemically into the first organized, self-reproducing systems—into the first living cells. Whatever the details of such chemical evolution might have been, we suspect from fossil signposts along the way that the progression flowed in this manner to the threshold of biological evolution and the subsequent explosive diversification of life.

Figure 21.12 Reconstruction of North American mammals of the Eocene (**a**) and Pliocene (**b**).

And so today you are sharing the earth with some 10 million different kinds of life forms. They, and you, share allegiance to the same principles of energy flow and chemical interactions. They, and you, share the same molecular and cellular heritage. All these things speak of the underlying unity of life. But more than this, they speak eloquently of its subsequent diversity. For if the environments of the first living things had never changed, if there had never been different horizons waiting the vanguard of inadvertent explorers, perhaps the world today would hold little more than testimony to life's unity. Perhaps there would be little more than single cells of the sort preserved in ancient rocks—cells matted against rocks or suspended in the waters of the sea, quietly soaking up nutrients.

But the record of earth history tells us environments *have* changed, and that organisms within a given population either have been equipped to respond to those changes or have perished. The record also suggests that just as the diversity of life has been a product of evolution, so has it been an evolutionary force of the first magnitude. "Diversity" not only means adaptations to some combination of temperature, chemical balance, available water, light, dark, and living space. "Diversity" means adaptations to different kinds of predators, different prey, different competitors or cooperative groups after the same resource, different behaviors and patterns of fur and feather for attracting mates and assuring reproductive success.

It is possible, then, to view all existing species as the product of interactions with the environment and with one another. These interactions are the focus of paleoecology; they continue to be the focus of modern-day ecology. And this brings us to a concept of profound importance. By the very fact of their continuing existence, all 10 million kinds of organisms now on earth can claim adaptive success. However, as you will read in this unit and the one to follow, *success is assured only as long as there is responsiveness to the environment, and only as long as there is dynamic stability between the requirements and the demands of organisms making their home together.* Both the environment and the community of life change, shifting as imperceptibly as shifting winds over the centuries or abruptly obliterating all trace of forms that have gone before. And therein lies the story of evolution, the story of random chemical competition leading to the first organized, self-reproducing forms of life, of dinosaurs and continents on the move, of simple strategies unchanged since the dawn of life, and of the complex human strategy—as yet unresolved—that can hold a world together or rip it apart. But must we predict gloomily that such unresolved activity on our part will end this magnificent story for all time? We doubt it. For if the record of earth history tells us anything at all, it is that life in one form or another has survived disruptions of the most cataclysmic sort. That life can evolve tenaciously through tests of flood and fire suggests it has every chance of evolving around and past our transgressions, too. *Viva Vida!*

Readings

Dott, R., and R. Batten. 1976. *Evolution of the Earth*. Second edition. New York: McGraw-Hill. Findings from myriad lines of research are distilled into a stunning picture of earth and life history. The authors are masters at initiating the reader into the excitement of discovery. This second edition continues to weather the conceptual storms of a rapidly changing field; you can almost see the authors bracing for the unexpected with wit and good humor. We enthusiastically recommend this book for your personal library.

Barghoorn, E., and J. Schopf. 1966. "Microorganisms Three Billion Years Old from the Precambrian of South Africa." *Science* 152:758–763.

Cox, A. 1973. *Plate Tectonics and Geomagnetic Reversals*. San Francisco: Freeman. Excellent overview of the evolution of plate tectonic theory.

Kay, M., and E. Colbert. 1965. *Stratigraphy and Life History*. New York: Wiley.

Marquand, J. 1971. *Life: Its Nature, Origins, and Distribution*. New York: Norton.

McAlester, A. 1977. *The History of Life*. Second edition. Englewood Cliffs, New Jersey: Prentice-Hall.

Schopf, J. W. 1975. "The Age of Microscopic Life." *Endeavor* 34(122):51–58.

Review Questions

1. What three physical characteristics of the primordial earth could have been favorable for the origin of life?

2. What three conditions must have been met if life originated as a product of chemical evolution?

3. The early earth apparently had a "reducing atmosphere." What effect would such an atmosphere have had on the spontaneous assembly of complex organic molecules? (What effect does our present "oxidizing atmosphere" have?) What kinds of energy sources could have driven the formation of macromolecules of life?

4. How do the Oparin and Fox experiments suggest that cell-like membranous spheres could have evolved? Why does it seem likely that metabolism cannot proceed without boundary layers such as cell membranes?

5. Even if cell-like membranous spheres could have evolved, how could the first *living* cells have originated? For your answer, think about the lifeless molecules called enzymes, nucleic acids, and nucleotides. Assuming these molecules became concentrated in membrane-like spheres, what might have happened?

6. Describe a radioactive element and how it breaks down spontaneously (decays) at its own unique rate. How are these elements used to date rocks?

7. If you assume that evolution into markedly different species proceeds gradually, then what three things could account for the *relatively* abrupt entrance (say, in less than a million years) of many groups of diverse and already well-developed species on the evolutionary stage?

8. What modern-day organisms did the earliest forms of life resemble? Given atmospheric and geologic conditions, what must have been their main energy-acquiring strategies?

9. Between 1.4 billion and 570 million years ago, eukaryotes made their entrance. Given that oxygen was accumulating in the atmosphere as a result of photosynthetic activities, and given the shells, spines, and armored plates these first eukaryotes left behind, what kind of energy sources were they going after? How might the increased atmospheric oxygen have increased their capacity for movement after energy sources?

10. How does plate tectonic theory help explain why there are different assemblages of organisms that characterize the main geologic eras?

11. By the late Cambrian, three major land masses were covered largely by shallow seas. What kinds of organisms evolved in this environment? By the Ordovician, even low-lying land was submerged. What kinds of organisms radiated into the vacant marine environment? What structural and behavioral traits must already have existed in Silurian times, before the geologic upheavals that created vast land environments in which such traits would prove adaptive?

12. What sort of organisms evolved during the Carboniferous, when land masses were submerged and drained no less than fifty times?

13. Describe some of the ancestral forms that probably gave rise to modern-day mammals, beginning with the little ratlike forms scurrying through the dinosaur-dominated shrubbery of the Jurassic.

14. In the evolutionary story, what two kinds of interactions have given rise to all existing species?

We humans, armed as we are with such complex cultural adaptations as thermal underwear, air conditioning, backpacks, submarines, and hot air balloons, can survive in nearly all land environments, remain airborne for a time, even live for awhile beneath the surface of the sea. Can any other kinds of organisms possibly match us in our sheer adaptability to so many settings? Well, yes. Monerans and protistans are merely single cells, not very large and certainly not what you would call complex; yet they are found almost everywhere, including several places where humans would really rather not be. Swamps, sewers, Antarctic snows, near-boiling hot springs, plant and animal tissues, the atmosphere, deserts, the ocean depths, your gut—different species of monerans and protistans teem in all these places, and more. Most are hidden from the unaided eye. A handful of rich, moist soil, for instance, may contain billions of bacteria. For these microscopically small organisms, a handful of soil, a dead leaf, a drop of swamp water are worlds in themselves. Each such world has a unique combination of acidity, oxygen, temperature, moisture, and nutrients. Given the diversity in microenvironments, we might expect to find corresponding diversity in adaptations among these forms of life, and we do. In this chapter, we will first consider some adaptations of the monerans—bacteria and blue-green algae. We will also take a brief look at their stripped-down cousins, the mycoplasmas and viruses. Then we will turn to the puzzling protistans—single-celled eukaryotes that seem to mirror the ancient forms thought to have given rise to the multicellular fungi, plants, and animals.

BACTERIA

Bacterial Adaptations

About 15,000 species of bacteria have been identified. Some bacteria are chemosynthetic autotrophs, extracting energy from inorganic molecules in such seemingly implausible places as oil wells. Some are photosynthetic autotrophs; others are heterotrophs of the sorts called decomposers and parasites. For some bacteria (the obligate anaerobes), oxygen is lethal. For others (the facultative anaerobes), alternatives exist. If oxygen is present, they use it in metabolism; if not, they switch to anaerobic pathways.

All bacteria are built according to the same body plan, with a rigid cell wall surrounding the plasma membrane.

22

MONERANS AND PROTISTANS

Figure 22.1 Scanning electron micrograph of the magnificent glass house of a diatom. 5,800×.

a

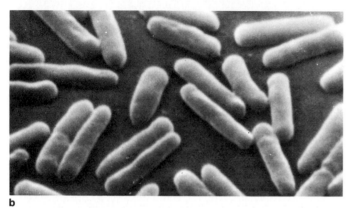

b

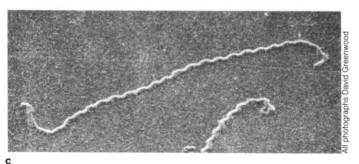

c

All photographs David Greenwood

Figure 22.2 Bacterial body plans. (**a**) The ball-shaped forms are represented here by *Streptococcus,* the causative agent of strep throat which, if left untreated, may lead to rheumatic heart disease and irreversible heart damage. (**b**) Rod-shaped forms are represented by *Pseudomonas aeruginosa,* which is an important nitrogen-fixing bacterium found in nodules of legume roots. (**c**) Spiral shapes are represented here by *Spirochaeta stenostrepa.* A related spiral bacterium is the cause of syphilis, one of the most common infectious diseases in the United States.

Although many bacteria are agents of human disease, the benefits derived from their activity far outweigh the harmful effects some bacteria have on our lives. Because bacteria recycle resources, life without them would be impossible for more complex organisms.

As prokaryotes, bacteria do not have a true nucleus and the profusion of organelles typical of eukaryotes (Figure 4.3). Sometimes the bacterial cell wall is encapsulated in secretions, such as mucopolysaccharides. These capsules afford some protection against bacteriophage attacks as well as against dedicated medical researchers.

Even though all bacteria have the same body plan, they differ somewhat in shape. Some look like rods, others like balls or spiral noodles (Figure 22.2). These three shapes speak of adaptations to the environment. Bacteria must have a moist environment in order to grow and reproduce. Ball-shaped bacteria, or **cocci,** are less apt to shrivel up if their living quarters dry out. Rodlike forms, or **bacilli,** have more exposed surface area per unit volume than cocci. This means they are less resistant to drying out—but it also means they have more plasma membrane for soaking up nutrients when conditions are favorable. Many spiral forms, or **spirilla,** can move more rapidly than either cocci or rodlike forms, much like a corkscrew twisting through a cork. Many spirilla are parasitic, and their twisting is done through the tissues of host organisms.

Although each bacterial cell is a functionally independent unit, bacteria often remain clustered together following division. Many cocci reproduce by dividing in only one direction. Others divide in two directions and adhere, forming a sheet of cells. Still others divide in three directions, and form a cube. At one time, bacteria were classified according to the way that they remained attached to one another following cell division. However, variations in the environment are known to have a profound effect on what the resulting colonies look like. For instance, when some bacterial species are deprived of just one mineral, the chains that they form are far shorter than what you would otherwise observe.

The entire bacterial way of life—growth, reproduction, even becoming dormant—revolves around resource availability. When nutrients are plentiful, all bacteria reproduce rapidly. In fact, their reproductive exploits are mind-boggling: under the right conditions, they may reproduce many millionfold in just one day! Bacteria rely on binary fission, the simplest sort of division process, and one that can be completed every twenty to thirty minutes in some species. Sexual reproduction among bacteria is nonexistent, although genetic exchange occasionally occurs through con-

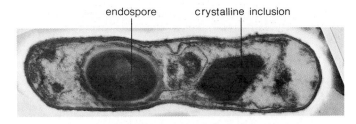

endospore   crystalline inclusion

Figure 22.3 Endospore formation in *Bacillus thuringiensis.* The crystalline inclusion is of a kind that may be "harvested" and used as a natural insecticide for certain insect species. (David Vitale and George B. Chapman, Georgetown University)

jugation, genetic transformation, and transduction (Chapter Eight).

When the environment becomes hostile, many bacteria form an **endospore,** a spore structure within the bacterial cell body (Figure 22.3). The spore coat resists moisture loss, irradiation, disinfectants, even acids. Once an endospore forms, the rest of the cell disintegrates. Such spore formation can cause humans all sorts of problems. Exposure to high temperatures for a few minutes is often enough to rid various foods and equipment of most actively growing bacteria. But if the environment is slightly alkaline, endospores can live through several hours of boiling! That is why hospitals use autoclaves (steam pressure devices that raise the temperature above that of boiling water) for killing all microorganisms. In home canning, the risks are ever present. *Clostridium botulinum,* for instance, is a bacterium that thrives in the anaerobic environment of canned foods. You know that the metabolic activity of any organism leads to certain by-products. In the case of *C. botulinum,* one by-product is extremely toxic to humans. It gives rise to the often-fatal food poisoning, **botulism.** Unless the canning solution is quite acidic, simply boiling the jars of food is not enough to kill these endospore-forming bacteria. If the foods are neutral or slightly alkaline (such as peas, beans, and corn), the bacteria can survive boiling. Either acidic substances must be added or the jars must be processed in a pressure cooker, which is clearly the simplest course. (Every now and then you read about commercially canned foods being recalled because of improper sterilization, but that is a relatively rare event. What is less rare is the accidental denting of cans or jar lids. Because even a small break in a seal may be enough

to let undesirable bacteria inside, dented cans of food that are on sale may prove to be no bargain.)

Bacterial Agents of Venereal Disease

Some bacteria, such as the photosynthesizers, are not adapted for growth in the animal body. But many heterotrophic bacteria can live *only* within animals. Some produce toxic by-products that cause disease in humans. Cholera, diphtheria, bubonic plague, leprosy, tuberculosis—long ago, these and other diseases were traced to such bacteria. Most have been brought under control, at least in industrialized countries. But one type of bacterial disease has reached epidemic proportions even in countries with the highest medical standards. And it has done so in spite of the fact that it can be diagnosed, treated, and quickly cured if it is reported promptly. This is **venereal disease** (VD), of which syphilis and gonorrhea are the most rampant forms. The bacterial disease agents are transmitted to uninfected persons through sexual intercourse, kissing, or intimate body contact with the sexual organs of infected persons. There must be intimate contact for the disease to spread, for the bacteria don't form endospores; they can't live long when away from the moist, warm parts of the human body. They die on exposure to light and air.

Syphilis is caused by a spirochete bacterium, *Treponema pallidum.* These forms possess flexible bodies. The first warning of infection occurs between two and six weeks after exposure. It is a single **chancre,** a painless bump under the skin that soon becomes an open ulcer at the site where bacteria entered the body (typically the genitals, anus, or mouth). Because the chancre is not sore or itchy and is often internal (especially in females), it may not even be noticed. Even without treatment it soon disappears. But this only means the bacteria have gone deeper into the body. Two to six months later, the infected person may experience a round of headaches, enlarged lymph glands, a sore throat and fever, aching joints, and a skin rash. These symptoms come and go over a four-year period. The problem is that too many other common diseases have the same symptoms, and without proper diagnosis the danger may still be ignored. Thus the bacteria have time to become more deeply entrenched. For five years or more, *T. pallidum* may slowly attack every organ in the body. The final stage of syphilis may last for as long as twenty years. By then, the damage is irreparable—nothing can save the victim from a combination of heart damage, blindness, and insanity. In all but the final stage, an infected woman who becomes pregnant can transmit syphilis to the fetus during the second and third trimester. The bacteria move across the placenta and into the unborn child's bloodstream. And the newborn will be either stillborn or syphilitic.

Gonorrhea is caused by the bacterium *Neisseria gonorrhoeae*, which thrives in the mucous membranes of the genital tract (and the eyes, if carried there). With males, there is a greater chance of detecting the disease in its early stages. Within a week, yellow pus is discharged from the uretha. Urination becomes more frequent and painful, because the infection leads to inflammation of the urinary tract. If left untreated, gonorrhea can cause severe bladder infections and sterility. With females, there may or may not be a burning sensation in the genital region; there may or may not be a slight vaginal discharge—and even if there is, it is often mistaken for a simple, common genitourinary infection. As a result, the disease often goes untreated. But the bacteria are there. They spread into the oviducts, eventually leading to violent cramps, fever, vomiting, and often sterility. An infected pregnant female can transmit gonorrhea to her unborn child as it moves through the birth canal during delivery.

The grave consequences of prolonged infection from these diseases *can* be avoided with prompt diagnosis and treatment. But as a preventive measure, those who engage in sexual activity with more than one partner should recognize how important it is for the male to wear a condom (Chapter Eighteen) to prevent the spread of infection. They should undergo medical examinations, complete with blood tests for the diseases, at least once a month. As a long-term measure, school and community programs of sex education should be encouraged. Part of the problem is that the initial stages of the diseases are so uneventful that the true dangers are masked. Another part is an unfortunate tendency to dismiss the problem as being the deserved curse on a few isolated sex freaks. More than this, it is a common misconception that a person cured once of venereal disease is immune for life, which is simply not true. An individual may contract venereal disease on exposure to an infected person no matter how many times he or she has been infected and cured. The result is that every twelve seconds, somebody—young and old, male and female, of any social class—becomes infected. Gonorrhea, in short, is out of control. With 3 million people known to suffer from the disease (public health workers estimate the number may be seven times higher), it has become the leading communicable disease in the United States; syphilis ranks third. Indifference to these appalling statistics can only encourage widespread ignorance and casual attitudes that perpetuate the vicious cycle of venereal disease.

Fifteen Thousand Species of Bacteria Can't Be All Bad

It is tempting, after reading about some of the horrors of bacterial infections, to render all bacteria guilty by association. But the vast majority are not guilty of terrible deeds, and in fact they are an essential underpinning not only of human life but of the biosphere in general. As you will read in Chapter Twenty-Eight, they are a vital link in the cycling of carbon, nitrogen, and other materials in natural communities. But bacteria do more than decompose organic debris in the forest and the sea; they do the same in the digestive tract of most animals, ourselves included. The digestive tract is home to a wide variety and tremendous numbers of bacterial cells. *Escherichia coli* is one of the most prevalent of these bacteria. They secrete enzymes for breaking down food materials, and they build up vitamins, which are absorbed and used by their host. Just how important they are becomes apparent during intensive antibiotic therapy for serious infections. The broad-spectrum antibiotics often wipe out the beneficial bacteria as well as the bad, which leads to digestive and intestinal disorders!

And can you imagine what the world would be like without bacteria to help dispose of human wastes, as well as the pathogenic organisms lurking therein? Modern waste-water treatment centers rely heavily on bacteria for restoring water to some semblance of its former self before it is dumped into the waterways (Chapter Thirty). Waste water from domestic and industrial sources includes solid and liquid substances from sewers, sugar mills, textile and paper mills, food processors, dairies, slaughterhouses, oil refineries, chemical manufacturers, and mining operations. It takes aerobic and anaerobic bacteria to break down the organic residues of this stuff. (That is also why we now have "biodegradable" laundry detergents. At one time, soaps were replaced with synthetic detergents, which brighten clothing. As it turned out, the detergents contained chemical groupings that microorganisms couldn't break down. Streams, rivers, lakes, even parts of the sea began to foam up. Only by redesigning detergents have we started to remove suds from the waterways.)

Finally, consider the way crops are sprayed each year with insecticides and herbicides. For instance, some chlorinated hydrocarbons sprayed on crops may linger in the fields for more than a decade. Thus they accumulate to levels high enough to harm not only insects and weeds but humans (Chapter Twenty-Eight). Certain factors such as leaching of the soil help keep these compounds to tolerable limits—but in some cases, so do bacteria out for a meal of carbon. In their energy extraction processes, they can convert some toxic compounds to harmless ones. As we are pressured into using new kinds of insecticides when old ones become ineffective, we must determine in advance whether they are biodegradable by soil bacteria.

AT THE BOUNDARIES OF LIFE: MYCOPLASMAS AND VIRUSES

If you can imagine what a bacterium would look like without its rigid cell wall, then you have an idea of the appearance of free-flowing **mycoplasmas**, disease agents that are the smallest organisms known. Mycoplasmas have no cell wall, although they typically have a tough plasma membrane which helps keep their contents from dribbling out. They can get along without a cell wall because they live in the osmotically protected sanctuary of an animal's body. When grown in a rich solution in the laboratory, the unwalled mycoplasmas fail to take on any characteristic shape: they form random threads, balls, and blobs (Figure 22.4). When grown on a solid medium such as agar, their options are fewer and they end up looking like tiny fried eggs.

Despite the structural simplicity of bacteria and mycoplasmas, these organisms are capable of metabolism and of transforming energy into usable forms. They also have a complete genetic system for maintaining and reproducing themselves. Hence they unquestionably are alive. Sometimes **viruses** (Figures 10.3 and 22.5) are said to be the simplest of all living things. But viruses are a paradoxical group that many biologists would not consider to be alive. They do have a set of nucleic acids sheathed in a protective protein coat, and their nucleic acids do contain the blueprint for making more virus. But they are not capable of metabolism; neither are they capable of reproducing themselves on their own. It is the infected host cell, *not* the virus itself, that assembles new virus particles. Because viruses can be perpetuated only inside the cells of another living organism, they probably developed at some point long after cellular life originated. One hypothesis is that they are the noncellular remnants of some ancestral bacterial cells that became parasitic. Invariably, when one organism becomes parasitic, it becomes simplified. If the host will do some task for it, there is no longer any selective pressure demanding that the parasite retain the ability to perform the task for itself. Viruses thus may be the ultimate expression of parasitic simplification. They retain only two

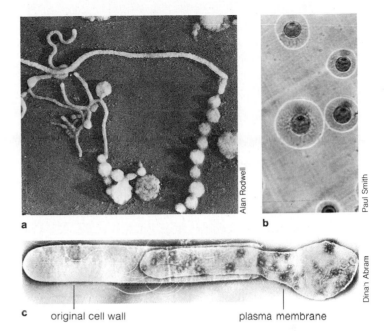

a

b

c original cell wall   plasma membrane

Figure 22.4 (**a,b**) Mycoplasmas, and (**c**) a bacterium undergoing lysis. Three things point to an evolutionary link between mycoplasmas and bacteria. First, under certain conditions, cells resembling mycoplasmas can be made from bacteria. Most bacteria undergo lysis when their cell wall is degraded by an enzyme, which allows water to rush into the cell, expand it, and make it burst. But lysis can be prevented if the wall is removed in a concentrated sugar solution. (The sugar lowers the concentration of free water on the outside and prevents it from diffusing rapidly into the cell.) The organisms so stabilized end up looking rather like mycoplasmas (**c**). Second, although under certain conditions these naked bacteria may revert to their former clothed selves, some never do: they may go on growing like mycoplasmas. Third, the DNA composition of mycoplasmalike organisms closely resembles that of certain bacteria. If past environments somehow counteracted wall-degrading agents, stripped-down bacteria could have given rise to mycoplasmas.

features the host will not provide: a mechanism for recognizing and infecting suitable host cells, and the hereditary blueprints for making new viruses.

Even if viruses are not alive, their influence on the living world is staggering. On the one hand, they are the causative agents of many human diseases such as smallpox, influenza, polio, and possibly cancer; they also cause extensive damage to livestock and crops. On the other hand, they play an important role in keeping certain bacterial populations within bounds in some environments.

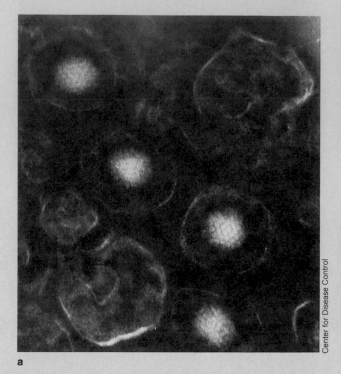

a

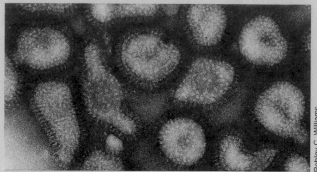

Robley C. Williams

b

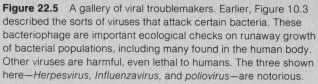

Robley C. Williams

c

Center for Disease Control

Figure 22.5 A gallery of viral troublemakers. Earlier, Figure 10.3 described the sorts of viruses that attack certain bacteria. These bacteriophage are important ecological checks on runaway growth of bacterial populations, including many found in the human body. Other viruses are harmful, even lethal to humans. The three shown here—*Herpesvirus, Influenzavirus,* and *poliovirus*—are notorious.

(**a**) Primary infections by *Herpesvirus simplex* lead to inflammation of gum tissues, the glans penis (in men), and vaginal epithelium (in women). Early on, cold sores or fever blisters erupt at the corners of the mouth, or on the external genitalia. Even though the body may develop antibodies to the infection, the virus is not destroyed; it becomes inactive, only to flare up again during times when the body is stressed by fever, emotional tension, even the common cold. Encephalitis, an inflammation of brain tissue, has been traced to *Herpesvirus;* the attacks may result in extensive brain damage. One form of *Herpesvirus* has been implicated in cancer of the cervix. (124,900 ×)

(**b**) Asian flu, Hong Kong flu, Spanish flu—these diseases occur worldwide, in ten- to forty-year cycles; and localized epidemics occur every year in between. The source of these diseases is one strain or another of *Influenzavirus.* Between 1918 and 1920 more than 20 million humans—half a million in the United States—died under the attacks of a Spanish flu virus. That happened not so long ago; your grandparents were alive then. The disease symptoms begin with sensations of chilling, followed by a sudden rise in body temperature to 100°–104°F. Muscles ache severely; the body feels drained and weakened. The virus attacks the upper respiratory tract. Often, the

weakened respiratory tissues are susceptible to agents of pneumonia. This complication accounts for a major part of the subsequent fatalities. Since the development of sulfa drugs and antibiotics, pandemics seem to be held in check. The viruses undergo major and minor mutation, however, and new strains arise more or less continuously. Not surprisingly, the World Health Organization has monitoring stations throughout the world; as soon as new strains are identified, researchers work rapidly to develop a modified vaccine that will help keep outbreaks under control. (250,000×)

(**c**) *Poliovirus hominis* is a small RNA virus that attacks only human cells in nature (chimpanzees and monkeys can be infected experimentally). Polio was first defined as spinal paralysis in infants and young children. Other forms damage the brain stem, so that muscles throughout the body cannot receive the signals to contract; when respiratory and vascular centers are involved, the consequent paralysis can be fatal. Viral agents of the disease are transmitted mostly in feces, also in saliva. During the 1840s, when the disease was first recognized, raw sewage was not treated. The disease spread largely because of inadequate disposal measures. With increased sanitation, fewer infants were exposed; those who might have developed immunity from the exposure became susceptible teenagers and adults. Between the 1930s and 1950s, these were the groups hit hard by polio epidemics. With the introduction of a killed-virus vaccine (developed by Jonas Salk and his coworkers) and a live-virus vaccine (developed by A. Sabin and others), the disease has been brought under control wherever widespread immunization programs have been implemented and maintained, as they are in the United States. (420,000×)

SUCCESS IN SIMPLICITY: THE BLUE-GREEN ALGAE

Bacteria are one major kind of prokaryote; **blue-green algae** are another. Aside from their photosynthetic activities, the blue-green algae (like some of their bacterial relatives) are nitrogen fixers of no small importance. If deprived of their activities, the biosphere as we know it would probably collapse for lack of enough usable nitrogen compounds for protein synthesis.

Although blue-green algae may exist as single cells under some conditions, most of the 1,500 known species look like filaments because cells remain attached to one another after division (Figure 22.6). In many blue-green algal filaments, we see the first inklings of a division of labor: a process not seen among bacteria, but a feature of more complex organisms. Frequently the filaments consist of two cell types: one specialized for photosynthesis, the other for nitrogen fixation.

Like photosynthetic bacteria, blue-green algae don't have chloroplasts (light-trapping organelles found in some eukaryotic cells). Instead their plasma membrane is folded inward and studded with light-trapping pigments, some of which give each species its typical blue, red-green, or almost black color. Unlike those bacteria, blue-green algae have the *same* kind of chlorophyll pigments found in all true algae and plants. Yet they are so much like their bacterial predecessors in cellular organization, it would probably be less confusing to think of them as "blue-green bacteria." Whatever the name, they appear to be living representatives of a long-since vanished evolutionary bridge between bacteria and the plant kingdom.

Blue-green algae are found throughout freshwater, saltwater, and land environments. Many species thrive in places where other organisms would be hard-pressed even to stay alive—in near-boiling hot springs, in extremely salty lakes, on desert rocks, in permanent snowfields. Two factors underlie their adaptability to such a broad range of settings. First, they have very simple requirements for energy and raw materials. Almost any environment has sunlight, carbon dioxide, some water, and simple mineral salts, and that is all the blue-green algae need in order to grow. Second, they reproduce by the simplest means possible. When conditions are favorable, they divide rapidly by binary fission. When conditions are unfavorable, some species form thick, sporelike walls or a gelatinlike coat that keeps them from drying out. Like bacteria, the blue-green algae grow in good times and wait out the bad. Thus most of the energy-rich nutrients they take in can be used at once for rapid growth and reproduction—which means

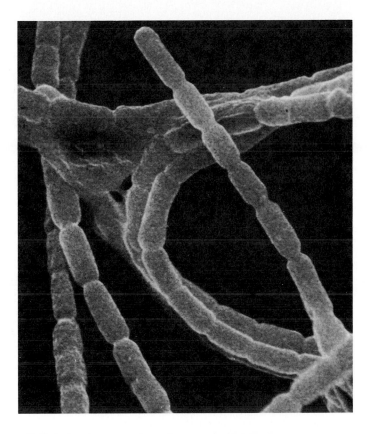

Figure 22.6 Filaments of the blue-green alga *Anabaena*. A mucilaginous sheath, secreted by the individual cells, coats each filament and in some cases causes filaments to stick together. 4,040×. (From John N. A. Lott, *A Scanning Electron Microscope Study of Green Plants*, C. V. Mosby Company, 1976)

they can compete quite well with other organisms. It is a time-tested adaptation. Both bacteria and blue-green algae were among the earliest forms of life on earth, and it's probable they will be among the last.

THE BOUNDARY-STRADDLING PROTISTANS

For the earliest prokaryotic monerans, survival was bestowed on those forms which were adapted for extracting energy-rich molecules from certain environments, for reproducing rapidly, and for being able to go dormant when conditions deteriorated. But by 1 billion years ago, some had evolved into eukaryotic forms adapted for feeding on other organisms. What were those first eukaryotic organisms like? Their membrane-enclosed nucleus allowed for more complex developments (Chapter Four). One such development was the frequent, more elaborate genetic recombination possible with sexual reproduction, which must have accelerated the process of diversification. The fossil record suggests ever more diverse forms evolved rapidly in several different directions, in what must have been intense competition for energy and resources. Thus today we have the animals, which move about actively in search of food energy. We have the plants, which generally stay put, trapping nutrients and sunlight in their immediate surroundings. We have the fungi—nonmotile, nonphotosynthetic, and specializing in absorption of organic nutrients. But more revealing for our question of what the first eukaryotes were like, we also have a diverse collection of single-celled eukaryotes. Some are actively prowling predators without photosynthetic pigments, and in this they resemble animals. In fact, they are known as protozoans (meaning "first animals"). Others are somewhat immobile and have photosynthetic pigments; they resemble multicellular plants. Still others resemble fungi in their life-style. But also in this hodgepodge are many motile, photosynthetic forms such as *Euglena*, which resemble plants, animals, *and* fungi! How do we go about classifying "little green funguslike animals"?

But taxonomic schemes are a little like fences dividing up the expanse of nature. They are artificial boundaries, subdividing a continuum that extends from the distant past. We are left with trying to discern what these eukaryotic microorganisms might represent in the total evolutionary scheme. And that is what the five-kingdom scheme, which we use in this book, has attempted to do. By viewing all the single-celled eukaryotes as a separate kingdom—the Protista—they are recognized not as insignificant cousins left behind in the great flow of life history but as successful adaptive lines in their own right.

PROTISTAN EXPERIMENTS IN PHOTOSYNTHESIS

On the Origins of Chloroplasts

Long ago, certain eukaryotes gained the capacity for photosynthesis. Perhaps, as some biologists have speculated, it was when a predatory animal cell happened to engulf a photosynthetic, bacteriumlike alga. Instead of succumbing to digestion, the organism so engulfed had the means to stay alive—indeed, to thrive in the presence of the carbon dioxide being given off by its engulfer. Why not take up permanent residence, dividing and reproducing even as the diner-turned-host divided and reproduced itself? The animal cell would benefit from the energy-trapping abilities of the smaller cell; the smaller cell would gain protection as well as raw materials for photosynthesis. (Such mutually rewarding dependence is one form of **symbiosis**. As you will read in Chapter Twenty Eight, it occurs today among many organisms.) Over hundreds of millions of years, evolutionary forces worked on this relationship, the outcome being the retention of adaptive structures and the weeding-out of nonadaptive ones. And gradually the photosynthetic organism lost its ability to live outside its host, even as its host came to depend on the energy-trapping guest.

Although this scenario might seem difficult to prove, detailed comparisons of chloroplast DNA and metabolism with those of bacteria and blue-green algae are compatible with such a theory of origin. Chloroplasts have their own complement of DNA and an apparent ability to reproduce parts of themselves with the instructions it contains—in partial independence of the division of nuclear material of the cell in which they reside!

If this was the route by which some eukaryotes acquired the capacity for photosynthesis, it must have been traveled more than once. Probably several different lines of photosynthetic protistans arose long ago, and have persisted into the present. Today, they resemble one another only in the fact that they are photosynthetic. Their chloroplasts are different and resemble different types of blue-green algae. They have different body forms and life-styles. Their accessory pigments are different, so they differ in color. All these lines are called "algae," but they are often distinguished from one another on the basis of color: "golden algae," "red algae," "green algae," and so on. There is a problem in trying to confine these lines to the protistan kingdom, for several have multicellular as well as single-celled members! The green algae are a case in point; they almost certainly gave rise to the multicellular green plants. Thus the green algae will be described in the next chapter as part of the story of plant evolution, even though they include some members that belong with the protistans as much as anywhere else. In this chapter we will look only at some of the algal lines that are largely or completely protistan in their makeup.

Dinoflagellates

The **dinoflagellates** (Figure 22.7) are what you might call primitive photosynthetic eukaryotes. True, they have a membrane-enclosed nucleus. But the composition of their

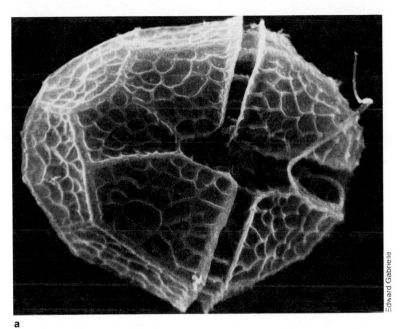
a

b

Figure 22.7 Dinoflagellates. (**a**) The scanning electron micrograph shows the plates making up the cell wall. Assorted accessory pigments give these organisms a yellow-green, to brown, to red color. (**b**) Dinoflagellates such as *Gonyaulax* are a source of episodic red tides. They produce a toxin that affects the central nervous system of fish. If their population levels are high, hundreds of millions of fish may be killed and washed up along coasts. Sometimes dinoflagellate blooms occur in lakes and reservoirs, which makes the water foul-tasting and often toxic. (Many dinoflagellates exhibit bioluminescence, the production of light caused by the metabolic reaction of ATP with specific compounds. This light is usually a pale blue. It often is seen at night.)

chromosomes is much like that of prokaryotes. Their mode of DNA replication seems vaguely prokaryotic. At least, the two DNA molecules formed prior to cell division attach to different parts of the nuclear membrane (just as bacterial DNA molecules attach to different parts of the plasma membrane), then they are separated as the membrane pinches in two.

The dinoflagellates are photosynthetic *and* motile. Most are found in the seas as members of **phytoplankton**: communities of microscopic photosynthesizers that are suspended in the water, where they form the basis for much of the region's productivity. In both fresh water and seawater, populations of dinoflagellates can grow explosively under certain conditions. During these so-called "blooms," there may be many million microorganisms in each liter of water. The sheer numbers usually play havoc with the environment. At night, when photosynthesis is impossible, the cellular respiration of such large populations depletes the dissolved oxygen in the water. The depletion may cause the death of many organisms such as fish, which are sensitive to oxygen levels. Some dinoflagellates of the oceans

produce toxic by-products. Nearly every year, destructive **red tides** occur somewhere in the world. They are the result of blooms of such dinoflagellates as *Gymnodinium* and *Gonyaulax*. As Figure 22.7 indicates, many dinoflagellates exhibit bioluminescence. Other marine organisms show bioluminescence, often because dinoflagellates are living inside them or in close association with them.

Golden Algae and Diatoms

Somewhere around 570 million years ago, oxygen had accumulated to high enough levels in the atmosphere and enough salts had accumulated in the seas to permit the formation of hard body parts (Chapter Twenty-One). There apparently was selection for those forms having the genetic blueprints and the capacity to expend energy for building shells, spines, and armored plates that provided some defense against predation. Two lines of descent from those adaptive forms are the armor-plated **golden algae** and **diatoms**. The golden algae have armor made of cellulose, pectin, and sometimes silicon. Their close relatives, the

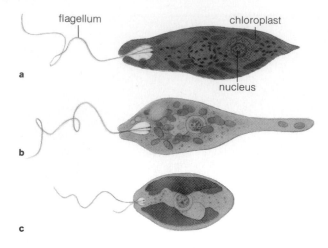

flagellum · chloroplast · nucleus

a

b

c

Figure 22.8 Organisms with features reminiscent of plants, animals, and fungi. (**a**) *Euglena*, a photosynthetic flagellate. (**b**) *Astasia*, a heterotrophic flagellate. (**c**) *Ochromonas*, a photoheterotrophic flagellate.

diatoms, have what is literally a glass shell, made mostly of silicon dioxide. The diatom shell is made of two pieces, much like a glass petri dish, and is often elaborately sculptured (Figure 22.1).

Today, the golden algae exist mostly in fresh water. At one time they numbered in the billions; the Cretaceous fossils of one group make up tremendous chalk deposits in various parts of the world. In terms of sheer numbers, however, the diatoms are the truly long-standing success. Immense diatomaceous deposits date from 70 to 10 million years ago—and these organisms are *still* abundant in almost all aquatic settings. Areas rich in diatoms are rightfully called "the pastures of the sea," for the diatoms, despite their glassy shell, remain the prey on which many marine communities are built.

THE PUZZLE OF LITTLE GREEN FUNGUSLIKE ANIMALS

Several thousand species of **photosynthetic flagellates** exist, and they clearly are related to both plants and animals. *Euglena* belongs to this group. Each *Euglena* contains about ten or so chloroplasts. But its cell body is not surrounded by a cell wall, as plant cells are. Instead, it is enclosed in a semi-rigid layer intimately connected with the plasma membrane. *Euglena* is not completely autotrophic, even with its array of chloroplasts. It needs certain

nutrients such as vitamins—which it absorbs from its environment in a funguslike manner. This photosynthetic flagellate and its relatives are most successful in places where organic molecules are concentrated—in polluted barnyard pools, for instance, which are heavily contaminated with urine and fecal material. They are also abundant in waste-water treatment centers (Chapter Thirty).

Except for their chloroplasts, *Euglena* and its relatives are virtually identical with a number of heterotrophic flagellates (Figure 22.8). In fact, if some species are raised in the dark, they rapidly change into forms having no chloroplasts at all. If these bleached-out forms are later grown in sunlight, they never regain chloroplasts. From then on, they survive only if they can absorb a complete diet of food molecules. Other heterotrophic flagellates are even more puzzling. Consider *Ochromonas* (Figure 22.8). They ingest whole particles through their body wall, much like an amoeba would do. But they also have chloroplasts and are capable of some photosynthesis. They can't grow in the dark, no matter how rich their surroundings are in nutrients. Even so, they are incapable of producing enough food by photosynthesis to support themselves. They must have both light for photosynthesis *and* food molecules to survive. They belong to a rare group of organisms known as **photoheterotrophs**. The resemblances between photosynthetic and heterotrophic flagellates, together with the existence of photoheterotrophs, could signify divergence of many modern flagellates from a common ancestor. It is a strong argument for establishing a protistan category instead of attempting to differentiate in every case between plantlike and animallike forms at this level.

PROTISTAN EXPERIMENTS IN PREDATION

There are four main types of animallike protistans: the heterotrophic flagellates discussed above, the amoebas and their relatives, the sporozoans, and the ciliates. All are unwalled single cells that live by pinocytosis and phagocytosis (Chapter Five). Some species are nearly as small as bacteria; others can be seen with the naked eye.

Amoebas and Their Relatives

As primitive as the **amoebas** seem to be as they move over some rock in a polluted pool, they are quite complex organisms. They move by sending out cellular extensions of themselves, called **pseudopodia** ("false feet"), and surround food and engulf it. Many relatives of the common

Amoeba not only have pseudopodia, they have flagella for swimming about. Amoebas are thought to have evolved from heterotrophic flagellates, which capitalized on the ability to distort the cell surface, extend it, and withdraw it. They developed this capacity to such an extent that they have come to rely on it as a way of life.

The movements of amoebas speak once again of the basic unity of life. Amoebas use the very same proteins you use for movement: the muscle proteins actin and myosin. If actin isolated from an amoeba is mixed with myosin isolated from a mammalian muscle, a jellylike clot forms; if ATP is added to the clot, it contracts forcefully! In amoebas, the actin and myosin proteins are attached in what might seem to be a random fashion to the cell membrane. When contraction occurs in one region, a pseudopodium is slowly extended. Contraction of two pseudopodia causes them to sweep around food and engulf it (Figure 22.9).

Some amoebas provide protection for their fragile body by encasing it in a layer of sand grains. Others secrete a hardened shell that surrounds and protects them. From one or more openings in the shell, pseudopodia poke out. This trend reaches its peak of development in the diverse group known as the **foraminiferans** (bearers of windows). They secrete a hardened case of calcium that is peppered with tiny holes. The organism lives within its beautiful shell. Some can project hundreds or thousands of thread-like pseudopodia through the holes. Most often, though, pseudopodia are extruded through a mouthlike opening in the shell. Because the pseudopodia are sticky, they can trap bits of phytoplankton floating in the seas even while the foraminiferan hides in its house (Figure 22.10).

Sporozoans

Calling an organism a **sporozoan** is not so much a reference to its pedigree as to its life-style. The sporozoans are a diverse group of single-celled organisms that are probably more closely related to several other kinds of protistans than they are to one another. They share one feature: they are all parasitic, living within the bodies of animals. Their host performs all the tasks connected with obtaining food. One group of sporozoans in particular has been prominent in human history: the members of the genus *Plasmodium*, which are the source of the disease malaria (Figure 22.11). Both a human host and a certain species of mosquito are needed if a malaria parasite is to complete its life cycle. Thus a key to eradicating malaria has been to drain and fill in the breeding grounds—swamps and stagnant ponds—of the malaria-carrying mosquitoes.

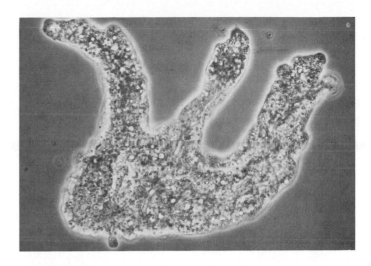

Figure 22.9 *Amoeba proteus* under the light microscope. (From Jensen et al., *Biology*, 1979)

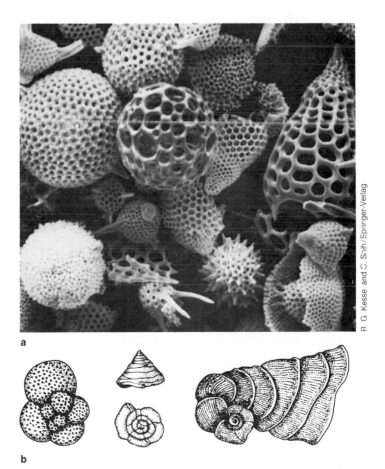

R. G. Kesse and C. Shih/Springer-Verlag

a

b

Figure 22.10 (**a**) Body shells of radiolarians, relatives of the amoeba and foraminiferans (**b**). (b from Jensen et al., *Biology*, 1979)

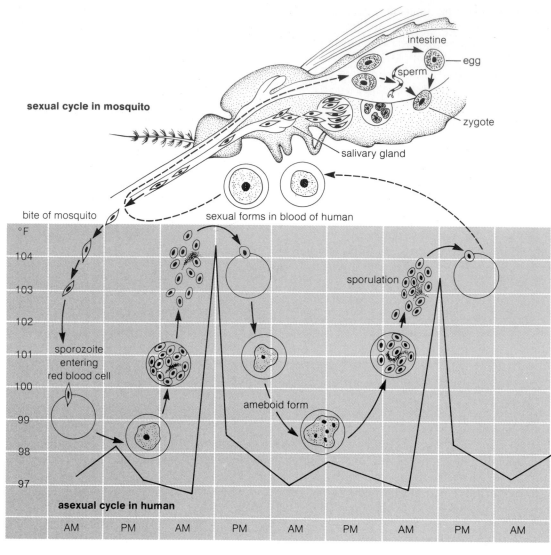

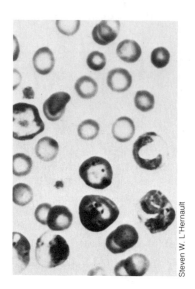

Figure 22.11 Malaria parasites in red blood cells. At this stage, each parasitic form divides repeatedly. At regular intervals, they burst out and reinfect more cells. Recurring, serious fevers accompany this cycle. When a mosquito draws blood from an infected individual, gametes of the parasite undergo sexual fusion to form zygotes, which eventually develop into spindle-shaped forms (sporozoites) within the mosquito. When the mosquito bites again, sporozoites enter the victim and begin the cycle anew. (From Jensen et al. *Biology*, 1979)

Ciliates: Peak of Single-Cell Complexity

Nowhere in the living world are the fantastic potentials of a single cell expressed more fully than they are in the group of protistans known as **ciliates**. Covered with thousands of cilia synchronized for swimming, armed in many cases with hundreds of deadly, poison-charged harpoonlike weapons that can be fired at prey and predator, and possessed of a voracious mouth, the ciliates prowl through or wait in ambush in woodland ponds, the intestinal contents of various organisms, and a variety of other habitats.

Perhaps the most widely occurring ciliate is *Paramecium* (Figure 22.12). Like most ciliates, *Paramecium* has a gullet, a cavity that opens to the external watery world. Rows of specialized cilia beat food particles into the gullet. Once inside, the particles become enclosed in food vacuoles, where digestion takes place. Unusable leftovers are carted off to a region known as the anal pore, which functions to eliminate wastes to the outside. Living as it does in freshwater environments, *Paramecium* depends on contractile vacuoles (Chapter Five) for eliminating the excess water that is constantly flowing into the cell.

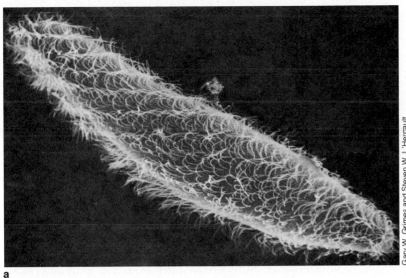

a

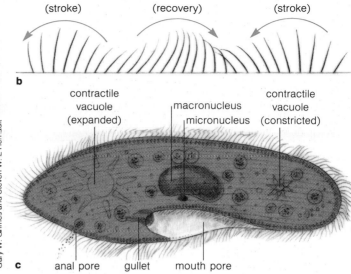

b

contractile
vacuole
(expanded)

macronucleus

micronucleus

contractile
vacuole
(constricted)

c   anal pore   gullet   mouth pore

Gary W. Grimes and Steven W. L'Herrault

Figure 22.12 (**a**) *Paramecium* on the prowl, using coordinating ciliary beating (**b**) to drive its body through its watery environment. (**c**) The sketch hints at the astonishing complexity of this single cell.

Predatory cell that it is, *Paramecium* is built for rapid movement. Between 10,000 and 14,000 individual cilia project like tiny oars from the cell surface. Through the mechanisms for cellular movement described earlier in Chapter Four, each orderly row of cilia beats in a coordinated way with adjacent rows (Figure 22.12). So efficient is this coordination that some species of *Paramecium* can move 1,000 micrometers through their surroundings every second! That rate of movement allows them to far outstrip their prey.

Ah, but that which works for one predatory ciliate works for others. Often *Paramecium* itself is outmaneuvered by another voracious ciliate, *Didinium*, as you saw in Figure 8.4. When a didinium engulfs a paramecium (which is typically larger than the predator), the didinium seems to be all mouth. Less visibly dramatic but equally stunning in the contest between predator and prey are the didinium's **trichocysts**—long contractile filaments anchored within the cell that are used to hold onto and paralyze its dinner.

Both *Paramecium* and *Didinium* are free-swimming, but other ciliates crawl about or simply stay put. *Vorticella* normally rises stalklike from its holdfast on the floors of ponds, lakes, and streams. It is important in removing bacteria from polluted waters. With utmost grace, it turns its highly efficient, cilia-rimmed mouth into the currents flowing past (Figure 22.13). The ciliate *Stylonychia* crawls along in search of prey. Although they are astonishingly

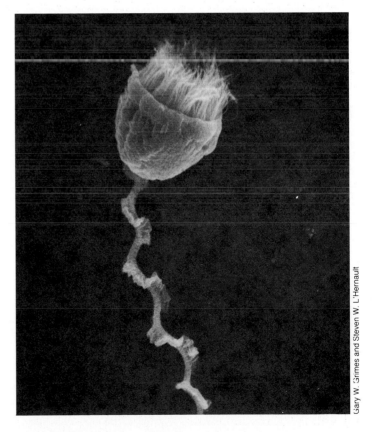

Figure 22.13 *Vorticella*, a sedentary ciliate.

Gary W. Grimes and Steven W. L'Herrault

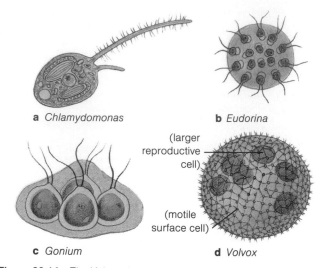

a *Chlamydomonas*   **b** *Eudorina*

(larger reproductive cell)

(motile surface cell)

c *Gonium*   **d** *Volvox*

Figure 22.14   The Volvocales, a group of related organisms that suggest a possible route from unicellularity to a multicellular organism having a division of labor among cells. (**a**) *Chlamydomonas* is a free-living single cell thought to resemble the ancestor of the entire group. In the simplest colonial forms (**b**), all cells—held together in a jellylike sphere—are identical in appearance and function. In somewhat more complex species (**c**), cells at one end of the sphere are larger and retain the capacity for movement (with flagella) and for reproduction. The most complex member of the group, *Volvox* (**d**), has two distinct cell types, each incapable of existence without the other. The smaller cells around the sphere's surface are motile but incapable of reproduction. The larger, interior cells are immotile; their only function is reproduction. Each larger cell is capable of dividing to give rise to a new individual having both cell types.

diverse in form, all such single-celled ciliates are of the same sort: they are the lions of the microscopic world.

ON THE ROAD TO MULTICELLULARITY

We can guess that somewhere among the ancestral lines of single-celled protistans are the long-since vanished gateways to multicellularity. The transitional forms probably arose many times in many evolutionary lines that were radiating into the world's shallow Precambrian seas. Perhaps the earliest were no more than simple colonies, designed by chance when single-celled organisms divided and the newly formed daughter cells failed to separate. Such inadvertent adherence meant increased size. To the

extent that it deterred smaller predators, or improved motility, or offered resistance to strong currents, the factors causing adherence were selected for and perpetuated. Indeed, colonial forms have persisted even to the present. They are structurally diverse, and they clearly are related by way of some such ancestors to protistans of many different types.

In a colony, each cell benefits from the loose association but each acts independently; it is incapable of modifying its behavior according to what is happening to its neighbors. All the cells are sensitive to the same things (light, food, temperature) and all respond to change in the same way. They feed and reproduce in the same way. It is only with the division of labor that there is interdependency, where one type of cell can't exist without the other. As Figure 22.14 indicates, the transition from colonial to multicellular forms could have occurred gradually, with an ever-increasing division of labor. The potential of multicellularity has been realized in three major pathways: the plants, fungi, and animals. These remarkable developments are topics of the next two chapters.

PERSPECTIVE

In the realm of monerans and protistans, we have evidence for one of the most significant evolutionary events. At the molecular level, these two kinds of single-celled organisms are so similar it seems likely that one evolved from the other. With the appearance of protistans, life was destined for stunning departures from the simple prokaryotic body plan. And those departures would be based largely on two things: a more complex genetic apparatus and self-contained energy centers.

Except during replication, prokaryotic DNA is a single, circular strand in the cytoplasm. In contrast, the DNA of eukaryotic protistans is separated physically from the cytoplasm by its own nuclear membrane. Moneran reproduction relies mostly on simple binary fission. Protistan reproduction relies on mitosis and, in sexually reproducing forms, on meiosis to assure the orderly parceling out of far greater stores of DNA into daughter cells. With nuclear isolation, protistans were on their way to a complex cellular division of labor. Soon after, perhaps through engulfment of bacteria and algae that managed to become symbionts instead of dinner, the eukaryotic protistans became equipped with energy-generating mitochondria and with energy-trapping chloroplasts. (Mitochondria are suspiciously like bacteria, just as chloroplasts are suspiciously like blue-green algae.) However nuclear isolation and energy control centers came about, they apparently helped trigger the extensive and rapid evolution into forms called plants, fungi, and animals.

It would be foolish, however, to portray the monerans and protistans as being somehow left behind during this magnificent surge toward complexity. By remaining structurally simple, they have survived astonishing environmental extremes. By retaining their simple strategy of reproducing quickly and in large numbers, they are able to adapt genetically in a short period to new environmental windfalls and pressures. Witness, for instance, how little time it takes for new resistant strains of bacteria to evolve following the introduction of some new antibiotic. In contrast, the body plans of plants, fungi, and animals, with all their potential for meeting many new and different challenges, are too complex for meeting others. Even as new doors were being opened for them, others were being closed behind—and that is why you won't find plants, animals, and fungi flourishing in hot springs. They cannot displace their simpler relatives in such places. In a given range of environments, monerans and protistans became masters of survival long before the appearance of the spectacularly diverse forms of multicellular life, and thus they continue to exist with them even into the present.

Readings

Jurand, A., and G. Selman. 1969. *The Anatomy of Paramecium Aurelia*. New York: St. Martin. A tribute to the astonishing complexity achieved in a single cell.

Margulis, L. 1970. *Origin of Eukaryotic Cells*. New Haven, Connecticut: Yale University Press. Outstanding treatment.

Nester, E., et al. 1978. *Microbiology*. Second edition. New York: Holt, Rinehart and Winston.

Schuhardt, V. 1979. *Pathogenic Microbiology*. New York: Lippincott. Excellent book, easy to read, with good case studies.

Stanier, R., et al. 1970. *The Microbial World*. Third edition. Englewood Cliffs, New Jersey: Prentice-Hall.

Review Questions

1. List four energy-extracting pathways used by different species of bacteria.

2. Although bacterial cells are built according to the same basic plan, what are three structural variations on this plan?

3. Name three kinds of bateria that pose threats to human life. Name three kinds of bacteria that are helpful to humans in particular or to the biosphere in general.

4. How do mycoplasmas differ from viruses? Describe the basic structure of a virus.

5. What is a bacteriophage, and how do some bacteriophages incidentally help humans as a result of their activities?

6. Give three examples of viruses that are harmful, even lethal, to humans.

7. Describe how blue-green algae differ in cellular organization from other algal groups.

8. Can you describe the symbiotic theory of chloroplast origins?

9. Give some examples of the diverse life styles of protistans that make these single-celled eukaryotes so difficult to categorize.

10. What kinds of protistans are photosynthetic? Predatory? Parasitic? Which move by means of pseudopodia?

11. What is one scenario by which single-celled organisms gave rise to multicellular forms?

Redwoods, cacti, orchids, mosses, wheat, kelps, magnolias —in turning to the astonishing variety among plants, we find that the concept of evolution gives insight into how such diversity may have come about. Between 3.5 billion and 400 million years ago, the evolution of life seems to have been confined to shallow waters of the earth. For water-dwelling plants, at least, conditions must have been much the same everywhere, for the fossil record shows comparatively little diversity in plant form. But 400 million

23

PLANTS AND FUNGI: REPRODUCTIVE ROADS TO DIVERSITY

Figure 23.1 Sunlight filtering through a grove of California coast redwoods.

Howard King

years ago the earth's crustal plates began colliding, and vast regions became elevated above sea level (Chapter Twenty-One). As shallow seas drained away, many aquatic forms must have been structurally and functionally equipped in ways that allowed them to survive on dry land.

Today, in almost any water that borders lowlands, you can find aquatic plants. Among these plants are forms that grow as filaments or sheets, only one or two cells thick. For them, simple diffusion and passive transport carry materials and wastes across external cell membranes, which are in direct contact with the surrounding water. On dry land, though, plants rely on special structures for getting and conserving water. Typically, the plant body has undergone expansion and differentiation into roots, stems, and leaves. Many land plants also have stomata and waxy leaf coverings, which control water loss and gas exchange (Chapter Twenty). Such plants survive largely because they can take in soil water and nutrients through roots, transport water through vascular tissue, and control water loss at stems and leaves.

Dependency on water has also influenced the development of plant support tissues. Aquatic plants have much of their body weight supported by the surrounding water. (Even in large, complex species, support tissues are devoted more to keeping the plant body intact during wave action than to holding it upright.) When the first plants had made the transition to land, perhaps support tissues weren't needed. Perhaps those pioneer plants simply sprawled over the ground's surface, soaking up sunlight. But as plants increased in numbers, they must have begun crowding one another. Variant forms that grew a little taller would have captured more sunlight and shaded out their neighbors. We may readily envision that competition for sunlight set in motion a series of developments that led to strong, reinforced stems and vascular tissues.

Even with such adaptations in body form, however, *radiation into vacant land zones would not have been possible without adaptive shifts in reproductive modes.* Consider that the dominant plant body of many existing water-dwelling species is composed only of haploid cells. When sexual reproduction occurs, cells of the haploid body divide mitotically and produce gametes. Every two gametes that fuse form a diploid zygote. In some species, these single-celled zygotes enter a resting stage that carries them through adverse conditions (such as the dwindling water supplies and lower temperatures of winter). When favorable conditions return, the zygotes undergo meiosis. This leads to the formation of haploid **spores**: cells that give rise to a new, conspicuous haploid body (Figure 23.2a).

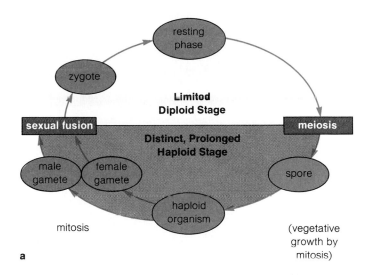

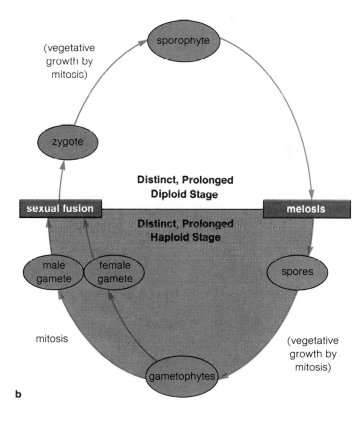

Figure 23.2 Generalized life cycles for (**a**) simple aquatic and (**b**) complex land plants. The haploid stage dominates the more primitive life cycles in both size and duration. In advanced plants, both haploid and diploid plant bodies are distinct and prolonged.

Table 23.1 Overview of Adaptations of Nonvascular Plants (Algae and Bryophytes)

Phylum* (Division)	Some Representatives	Characteristic Environment	Kinds of Tissue Differentiation
Rhodophyta (red algae)	*Porphyra, Nemalion*	Some freshwater; most marine, especially deep tropical waters	Single-celled to branched filaments; some with filaments massed into stemlike and leaflike structures; no vascular (internal transport) tissue
Phaeophyta (brown algae)	*Fucus* (rockweed), kelps, *Sargassum*	Almost all marine, coastal waters especially; some float in open ocean	Filamentous growth; some branched; some with leaflike, stemlike (stipe), and anchoring (holdfast) structures; some with ducts for transporting photosynthetic products to lower plant regions
Chlorophyta (green algae)	*Ulva* (sea lettuce), *Ulothrix, Spirogyra*	Most freshwater; also marine; moist soils	Single-celled to simple sheetlike filamentous and branched forms; no vascular tissue
Bryophyta (bryophytes)	mosses, hornworts, liverworts	Most land; moist, humid sites; some arid sites; a few submerged aquatic	Threadlike anchoring structures (rhizoids); leaflike and stemlike structures, often branched; some with simple water- and food-conducting tissues; pores with guard cells in epidermis

*This classification scheme based on Weier et al., *Botany: An Introduction to Plant Biology*, Wiley, 1974.

In contrast to such primitive life cycles, most plants have pronounced haploid and diploid stages alternating with each other. The zygotes as well as the spores grow through mitotic divisions, which give rise to distinct, vegetative plant bodies in *both* generations. This is the kind of alternation of generations you read about in Chapter Eight. Recall, from that chapter, that the haploid plant body is the **gametophyte**; it produces gametes. The diploid plant body is the **sporophyte**; it produces spores (Figure 23.2b).

Long ago, the emergence of the prolonged diploid stage must have been exploited in some plant groups making the transition to dry land. As you will see in this chapter, the invasion of land has indeed been correlated with increasing reliance on the diploid generation. Tables 23.1 and 23.2 give an overview of the kinds of plants that we will be describing in our evolutionary story. Figure 23.3 introduces some of their representatives.

Plants That Never Left the Water

Algae is a term that originally came into use to define simple aquatic "plants." It no longer has any formal significance in most classification schemes. The organisms lumped together under the term are now generally recognized as belonging in three different kingdoms. Thus we have the following phyla (called **divisions**, with respect to plant classification schemes):

Cyanophyta (blue-green algae)	Monera
Chrysophyta (golden algae, diatoms) Euglenophyta (photosynthetic flagellates) Pyrrhophyta (dinoflagellates)	Protista
Rhodophyta (red algae) Phaeophyta (brown algae) Chlorophyta (green algae)	Plantae

As you can see, the red, brown, and green algae remain in the plant kingdom. These three groups represent the peak of plant diversity in water environments.

Red algae (Rhodophyta) include plants that closely resemble forms that lived at least 600 million years ago, in the shallow waters of the earth. Today, with the exception of a few freshwater forms, the 4,000 or so species of red algae are important members of communities ranging from shallow intertidal zones to the basement of tropi-

Light-Harvesting Structures	Gas Exchange Mechanisms	Water Transport and Conservation Mechanisms	Main Reproductive Strategies
Chloroplasts; phycobilin pigments (trap blue-green light in deep water) plus chlorophylls; some species with radial or bilaterial branches (nonoverlapping exposure to sunlight)	Diffusion of dissolved gases across individual cell membranes	Direct exchange with surrounding water, across individual cell membranes	Alternation of sexual fusion and meiosis; sexual reproduction based on oogamy (gametes different in appearance); also vegetative reproduction from plant body fragments or asexual spores
Chloroplasts; xanthophyll pigments plus chlorophylls; leaflike structures			Alternation of multicelled generations; sexual reproduction based on isogamy (gametes all identical in appearance), or oogamy; vegetative reproduction from plant body fragments or asexual spores
Well-organized arrays of chloroplasts with membrane stacks; chlorophylls dominant			Alternation of generations (some have resting spores); sexual reproduction based on isogamy to oogamy; also vegetative reproduction by fragments or spores
Well-organized arrays of chloroplasts with membrane stacks; leaflike structures show radial and bilateral symmetry	Diffusion of dissolved gases across individual cell membranes; some stomatal control	Direct exchange with moisture-laden air; absorption from moist substrate; simple water-conducting cells; waxy covering (cuticle) retards water loss	Alternation of generations; diploid plant body produces homospores (they produce only one type of spore); gametophyte is dominant generation with dependent sporophyte; vegetative reproduction mostly by asexual reproductive bodies (gemmae)

cal reefs. Some can even be found growing about 175 meters below the surface when water is clear enough for light penetration. Body forms range from single cells, to filaments, to filaments massed tightly into sheets that may fan out a meter or so (Figure 23.3).

In some aspects, red algae resemble the prokaryotic blue-green algae, and may have been derived from them. There are structural similarities in the photosynthetic membranes of these two kinds of organisms. Both contain phycobilins (a class of light-trapping accessory pigments). Depending on the particular pigment, red algae range in color from green, to red, purple, and greenish-black. Their accessory pigments contribute to adaptive success in deeper waters. Chlorophyll functions most efficiently in light containing red wavelengths. But red wavelengths don't penetrate far below the water's surface. There, available wavelengths are mostly blue-green. Accessory pigments in red algae absorb these wavelengths and pass on some of the energy to nearby chlorophylls engaged in photosynthesis.

Reproduction of red algae may occur by vegetative growth of fragments broken away from the parent body. Sexual reproduction is also known to occur in some species. Recall that sexual life cycles require both meiosis and fertilization. Among some red algal species, the zygote itself undergoes meiosis, which leads to the formation of haploid spores. These spores give rise to the mature haploid plant body, which will eventually produce gametes by mitosis. Both spores and gametes must be dispersed by the random flow of currents; red algae generally have no motile cells. In sum, the haploid stage dominates red algal life cycles; the diploid stage is typically limited to the zygote. And spore dispersal as well as sexual fusion of gametes depends on the presence of free-flowing water.

Brown algae (Phaeophyta) include about 1,500 species. An accessory pigment (one of the xanthophylls) gives these plants their olive-green or dark-brown color. Many of the brown algae live along rocky coasts, where they anchor themselves to submerged rocks by structures called holdfasts. Some, including the kelps, grow profusely offshore. The giant kelps sometimes grow 100 meters from their holdfasts. As strangely beautiful as the underwater "forests" of giant kelps may be, more than one scuba diver has been in real trouble by becoming entangled in the dense, waving plants. Some brown algae thrive in the open sea. *Sargassum* (Figure 23.3) floats as immense, tangled masses through the Sargasso Sea, which lies between the Azores and the Bahamas. This brown alga helps support a community of unique marine animals.

Table 23.2 Overview of Adaptations of Vascular Plants

Phylum* (Division)	Some Representatives	Characteristic Environment	Kinds of Tissue Differentiation
Lower Vascular Plants (non-seed-bearing; depend on free water for fertilization)			
Lycophyta (lycopods or club mosses)	*Lycopodium, Selaginella*	Land; wet, shaded sites in tropics and subtropics; some arctic, desert	Sporophyte with true vascular system; roots, stems, leaves; stomata with guard cells in one or both leaf surfaces; vascular system makes possible a higher volume-to-surface ratio than in nonvascular plants
Sphenophyta (sphenopsids)	*Equisetum* (horsetails), only living genus	Land; acid soil; sand dunes, swamps, moist woodlands, lake margins, railroad embankments	Sporophyte with true vascular system; extensive underground stem (rhizome) with roots and aerial shoots at nodes; aerial stems jointed, hollow, branched or unbranched
Pterophyta (ferns)	Sword ferns, lady ferns, *Cyathea* (tree ferns)	Land; some epiphytes (attached to but nonparasitic on other plants); most wet, humid sites	True vascular system; well-developed xylem and phloem; some with creeping rhizomes; columnlike stem in tree ferns
Gymnosperms ("naked-seed-bearing"; not dependent on free water for fertilization)			
Cycadophyta (cycads)	*Zamia*	Limited tropical, subtropical land regions	Well-developed, complex vascular system; stems short and bulbous, or columns; palmlike appearance; fernlike leaves; massive cones (reproductive structures bearing seeds)
Ginkgophyta (ginkgos)	*Ginkgo biloba* (only existing species)	Land; temperate regions	Well-developed, complex vascular system; tall, woody stem (trunk) with long side branches; fan-shaped, deciduous leaves
Coniferophyta (conifers)	pine, spruce, fir, juniper, hemlock, cypress, redwood, larch	Land; widespread through Northern and Southern hemispheres	Well-developed, complex vascular system; some shrubby, others impressively thick, tall-trunked trees; most with whorled or spiral branching; most evergreen
Gnetophyta (gnetophytes)	*Gnetum, Ephedra, Welwitschia*	Land; warm-temperate regions; desert and mountain sand or rocky soil	Well-developed, complex vascular system; some shrubby, branched, with whorled leaves; *Welwitschia* short, bowl-shaped stem, two large, strap-shaped leaves, flowerlike reproductive structures
Angiosperms (flowering, seed-bearing; depend on pollinating agents; not dependent on free water for fertilization)			
Anthophyta			Well-developed, complex vascular system; floral structures
Monocotyledonae (monocots)	Grasses, palms, lilies, orchids, onions, pineapple, bamboo	Almost all land zones; some aquatic	One seed leaf (cotyledon); floral parts generally in threes or multiples of threes; parallel-veined leaves common
Dicotyledonae (dicots)	Most temperate-zone fruit trees; roses, cabbages, melons, beans, potatoes	Almost all land zones	Two seed leaves; floral parts generally in fours, fives, or multiples of these; net-veined leaves common

*This classification scheme based on Weier et al., *Botany: An Introduction to Plant Biology,* Wiley, 1974.

Light-Harvesting Structures	Gas Exchange Mechanisms	Water Transport and Conservation Mechanisms	Main Reproductive Strategies
Well-organized arrays of chloroplasts with membrane stacks; most with spirally arranged leaves; some with palisade layer of photosynthetic cells in leaves	Diffusion of dissolved gases across individual cell membranes; regulation through numerous stomata	Well-developed xylem and accessory cells; cuticle (retards water loss)	Alternation of generations; complex, dominant sporophyte independent of gametophyte; some homosporous (they produce only one type of spore), others heterosporous (produce spores of different types, which give rise to male or female gametophytes)
Well-organized chloroplast arrays; upright growth habit possible through pronounced stem thickening; whorled, scalelike leaves common but most photosynthesis in stems			Alternation of generations; complex, dominant sporophyte free-living (independent of gametophyte); homosporous
Well-organized chloroplast arrays; radially symmetrical leaves and stems common; epiphytes capture light above forest floor			Alternation of generations; sporophyte body initially develops on gametophyte (which later disintegrates); most homosporous
Well-organized chloroplast arrays; complex leaves, arranged in pattern that allows good sunlight exposure	Diffusion of dissolved gases across individual cell membranes; complex stomatal control	Well-developed xylem and accessory cells; cuticle; water storage in roots, stems	Alternation of generations; dominant, woody sporophyte body, heterosporous, well-developed, seed-bearing cones; pollen
			Alternation of generations; dominant, woody sporophyte body, heterosporous; well-developed, seed-bearing "cones"; pollen
			Alternation of generations; dominant, woody sporophyte body (e.g., the pine tree); heterosporous; well-developed seed-bearing cones; pollen
Well-organized chloroplast arrays; in *Ephedra*, most photosynthesis in stems and branches (almost leafless)			Alternation of generations; dominant, woody sporophyte body, heterosporous; "male" and "female" diploid plants; bears small, naked seeds; pollen
Well-organized chloroplast arrays; complex leaves, arranged in patterns that allow good sunlight exposure	Diffusion of dissolved gases across individual cell membranes; complex stomatal control	Well-developed xylem and accessory cells; cuticle; water storage in roots, stems	Alternation of generations; dominant woody or herbaceous sporophyte body; diverse floral structures adapted to pollinating agents (wind, water, insects, birds, bats); all are heterosporous; seeds enclosed within ovary; fruits (ripened ovary) aid in seed dispersal by wind, water, animals

Many kelps show complex organization. Besides having a holdfast, some species have leaflike blades and a stemlike structure (stipe). Sometimes hollow, gas-filled stipe regions (floats) occur at regular intervals and help hold the plant body upright in water. Surrounded as they are by water, brown algae have no requirements for water-conducting tissue. However, one of the giant kelps has tubelike strands similar to phloem. Through these strands, photosynthetically derived food may travel to body regions suspended in dimly lit water below the surface.

Brown algae show alternation of generations. Among simpler, filamentous species, the gametophyte and sporophyte may be much the same in outward appearance. In more developed species, the gametophyte is extremely reduced in size and a large, complex sporophyte dominates the life cycle.

Green algae (Chlorophyta) include about 7,000 species. They range from single motile cells to broad, sheetlike forms such as *Ulva* (sea lettuce), which is shown in Figure 23.3b. The most common body plan is a series of straight or branched filaments, each a single cell thick.

One group of green algae represents an evolutionary line that is thought to have given rise to the first land plants. Existing species are mostly freshwater organisms; some even live on moist land. Their photosynthetic system—which includes chlorophylls *a* and *b*, carotenoids, and xanthophylls—is identical with that of complex land plants. These pigments permit green algae to capture the kinds of light wavelengths available in shallow waters. Thus, ancestral algae could have been candidates for surviving in the shallow bogs and lagoons that formed when continental land masses were rising 400 million years ago.

Some lines of green algae continued to exist in these settings. In other lines, modifications of the reproductive cycle possibly helped set the stage for moves onto land. Like red and brown algae, the green algae show alternation of generations. For most species, the gametophyte is dominant; the diploid body usually is limited to the zygote. In some species, though, the diploid zygote grows through mitotic divisions that lead to a multicellular sporophyte body.

Also, like brown algae, the green algae show variation in the type of gametes produced. Simpler forms rely on **isogamy**: the gametes are identical in appearance. More complex forms rely on **oogamy**: gametes differ in size and motility, much like the motile male gametes (sperm) and larger, nonmotile female gametes (eggs) of animals. Eggs that develop while attached to a gametophyte receive protection from the parent plant. Motile sperm reach the eggs by swimming through the surrounding water. Of course,

a Rhodophyta (red alga).

b Phaeophyta (brown alga *Sargassum*).

g Filicinophyta (true fern, showing frond shaped like fiddlehead)

h Ginkgophyta (*Ginkgo biloba*)

Figure 23.3 Representatives from twelve plant divisions (phyla).

c Chlorophyta (green alga *Cladophora*).

Ray F. Evert

d Bryophyta (liverwort *Marchantia*)

Hill, Popp, and Grove, *Botany*

e Lycophyta *(Selaginella)*

Fay F. Evert

f Sphenophyta *(Equisetum)*

Edward S. Ross

Cycadophyta (cycad)

Jensen and Salisbury, *Botany*, 1972

j Coniferophyta

Earth Scenes/Mark Newman

k Gnetophyta *(Welwitschia)*

Scagel et al., *Plant Diversity*, 1969

l Anthophyta (orchid)

Robert N. Bowman

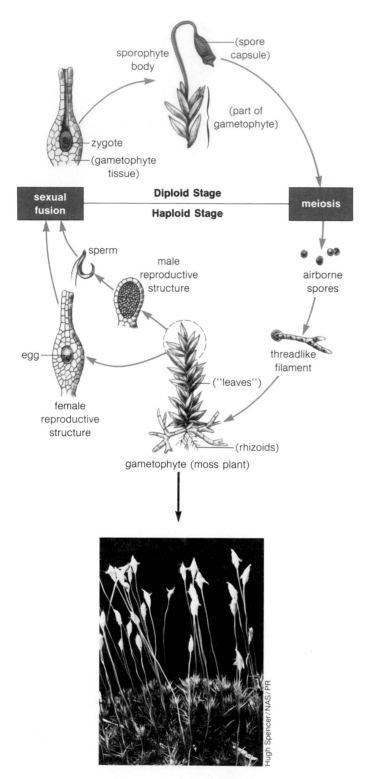

sporophyte body

(spore capsule)

(part of gametophyte)

zygote

(gametophyte tissue)

sexual fusion	**Diploid Stage**	meiosis
	Haploid Stage	

sperm

male reproductive structure

airborne spores

egg

threadlike filament

("leaves")

female reproductive structure

(rhizoids)

gametophyte (moss plant)

Hugh Spencer/NAS/PR

Figure 23.4 Moss life cycle. (As in the other plant life cycles, the various structures are not drawn to scale.) Notice how the sporophyte body remains attached to the gametophyte (moss plant) body.

motile sperm would have a problem in reaching eggs on dry land. However, in certain environments, and through certain "packaging" modifications, their sexual union came to be assured among pioneer land plants. The points here are these:

1. Ancestral green algae were metabolically equipped for radiating into new environments—shallow bogs and lagoons—when the seas began shrinking 400 million years ago. This is the apparent time of origin for land plants.

2. Like some of their modern-day descendants, a few ancestral species probably relied on a relatively prolonged sporophyte generation. A prolonged sporophyte generation is characteristic of most land plants.

3. Like some of their descendants, a few probably relied on oogamy—which, as you will see, would have implications for gamete dispersal in land environments.

Transitional Land Plants

Mosses, liverworts, and hornworts constitute the **Bryophyta**. These nonvascular plants have one of the most ancient lineages. They are about 24,000 species strong. Moss plants are the most common bryophytes. They grow on (and to some extent above) the land's surface. A moss plant body has no roots; it has tiny cellular threads (rhizoids) that serve as anchors. Most mosses have long parenchyma cells into which water moves by slow, cell-to-cell diffusion. They also have a few leaflike structures that often can curl up and dry out when water becomes scarce, only to revive quickly when water becomes available. Mosses can't reproduce sexually unless they live in areas where liquid water is periodically abundant.

When moss spores germinate, they form a green, threadlike filament that resembles filamentous green algae. This gametophyte filament goes on to produce the familiar moss plant (Figure 23.4). A haploid moss plant can have male or female reproductive structures, or both. In most species, they are found on separate plants that are growing near each other. Even this short journey to the eggs is just about impossible for sperm unless a film of free water covers the entire route. When sperm and egg do fuse, the resulting zygote divides mitotically, which produces a diploid sporophyte. In this case, the sporophyte is carried on a stalk growing right out of the tip of the gametophyte body (Figure 23.4); it is *attached* to and dependent upon the haploid plant.

The development of a protected embryo sporophyte—attached to and nourished by gametophyte tissue—must have been an important advance for life on land. It is first seen among the bryophytes, which are structurally the simplest of all land plants. And it is seen among vascular plants.

Vascular Plants

Vascular plants have complex food-conducting and water-conducting tissue systems (Chapter Nineteen). They first appeared about 400 million years ago. Within a relatively short span of 50 million years, the major lines of these plants had become established on land. Some lines later became extinct, but many have modern-day descendants that give hints of ancient adaptations.

Horsetails (Sphenophyta) are relatively simple vascular plants. Their ancient relatives were highly diverse, and included treelike forms about fifteen meters (fifty feet) tall. A single genus, *Equisetum*, survives. Plants of this genus often grow in vacant lots, along railroad tracks, and around ponds or lakes in northern parts of Eurasia and North America. Horsetail sporophytes typically have underground stems (rhizomes) and aerial branches arising at nodes. Scalelike leaves are arranged in whorls about a hollow, photosynthetic stem (Figure 23.5a). Growing out from some shoot tips or from side branches are a kind of cone. A **cone** is a reproductive structure that bears spores. Winds disperse the fragile horsetail spores, which have only a few days to germinate on moist surfaces. Germinating spores give rise to haploid gametophytes. The horsetail gametophyte is a free-living plant about as small as a pinhead. It bears separate male and female reproductive structures, in which sperm and eggs develop. Sperm must travel through free water (dew, rain) to reach the female structures.

Lycopods (Lycophyta) were also highly diverse 350 million years ago. Then, some forms were tree-sized. The genus *Lycopodium* contains modern-day, albeit miniature, descendants. Club mosses growing on forest floors are in this group. A *Lycopodium* sporophyte has roots and small, scalelike leaves that spiral around stems and branches. The sporophyte bears cones (Figure 23.5b). Spores dispersed from the cones germinate to form small, free-living gametophytes.

Ferns (Pterophyta) are another kind of primitive vascular plant with ancient origins. About 12,000 living species have been identified. These plants have extensive root systems and well-developed leaves (Figure 23.6). Except for tropical tree ferns, their stems are horizontal on the ground's surface or just beneath it; only leaves are aerial. The sporophyte is the conspicuous fern plant (Figure 23.6). Its spores are dispersed through air. They germinate and develop into small, heart-shaped gametophytes. Fusion of fern gametes eventually leads to a new sporophyte generation.

Adaptive Importance of Pollen Grains and Seed Dispersal

Horsetail, lycopod, and fern sporophytes have well-developed vascular systems; they are adapted for life on land. But their distribution is largely limited to wet, humid sites, because their short-lived gametophytes usually have no vascular tissue for water transport and conservation. In addition, the male gametes must have free liquid water if they are to reach the eggs that remain in female reproductive structures. Thus the primitive vascular plants have made it only halfway to land. For part of the life cycle they still are, in essence, aquatic.

Yet, among these plants are species that hint at how a new evolutionary road opened up. Consider first that horsetail spores are exactly alike in form. Thus horsetails are **homosporous** (*homo-*, meaning "same"). Most lycopods and ferns are also homosporous. Some species, though, produce two kinds of spores. *Megaspores* give rise to female gametophytes. *Microspores* give rise to male gametophytes.

a   b

Figure 23.5 (**a**) Horsetails (*Equisetum*), showing three cone-bearing stems and one immature vegetative structure between them. (**b**) *Lycopodium* sporophytes with cones.

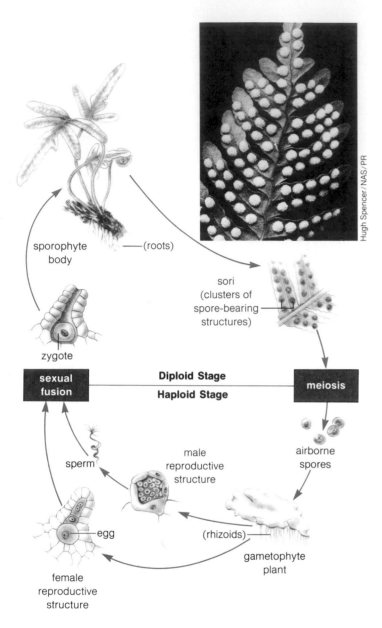

Figure 23.6 Fern life cycle. *Sori*, in which spores are produced, occur on the underside of the fern leaf.

Labels in figure:
sporophyte body — (roots) — sori (clusters of spore-bearing structures) — zygote — sexual fusion — **Diploid Stage** — **Haploid Stage** — meiosis — airborne spores — sperm — male reproductive structure — egg — (rhizoids) — gametophyte plant — female reproductive structure

Hugh Spencer/NAS/PR

Plants that produce spores of two sorts are said to be **heterosporous** (*hetero-*, "different"). Lycopod and fern species that rely on this strategy still require ample water for their gametes to develop and get together. Even so, heterospory seems to have been the route by which the dominant group of vascular plants—the seed plants—arose.

All seed plants are heterosporous. They produce megaspores that develop into female gametophytes, and microspores that develop into male gametophytes. The female gametophytes develop while still attached to the parent plant. In this way, water and food required for their development are provided by the sporophyte—the stage that is well adapted for obtaining these resources on dry land. In contrast, immature male gametophytes, or **pollen grains**, are released from the parent plant and travel to the female gametophyte by winds, insects, and the like. For most species, *free water is not necessary for this transfer.* Fusion of an egg with sperm from the male gametophyte leads to formation of a zygote, which grows into a conspicuous sporophyte plant. More than any other factor, the development of pollen grains, which can reach female gametophytes *without* liquid water, would have allowed vascular plants to radiate from wet lowlands through almost all other land environments.

Seed plants take their name from the manner in which sporophytes are dispersed. A **seed** contains the embryo sporophyte, which is surrounded by a coat that guards against mechanical damage. Generally, a seed has internal food reserves that nourish the sporophyte embryo during germination (Chapter Nineteen). *As you will now read, seeds have taken over the dispersal function of spores.*

Gymnosperms: First of the Seed Plants

The word **gymnosperm** means "naked seed." It refers to the way that seeds of these plants are carried on surfaces of reproductive structures, without being protected by additional tissue layers. Gymnosperms include cycads, ginkgos, conifers, and gnetophytes (Table 23.2). **Cycads** (Cycadophyta) once flourished with the dinosaurs of the Mesozoic. The few surviving cycad species are confined to the tropics or to warm, temperate zones. At first glance you might mistake a cycad for a small palm tree (Figure 23.7a). Despite having similar leaves and stems, palms and cycads are not closely related, for palms have true flowers. Cycads have massive cones. The sharp spines on some cycad cones (along with spiny leaf tips) are probably adaptations that minimize seed predation. Seeds are a sought-after food in areas where water (hence plant growth) is limited. In most cases, it takes years for cycad seeds to mature. Their slow reproductive rate has probably put cycads at a disadvantage, compared with other seed plants of more recent origin.

Ginkgos (Ginkgophyta) are even more restricted in native distribution than cycads. Only a single species (Figure 23.7b) has survived, despite the success of this plant during the Mesozoic. It seems that several thousand

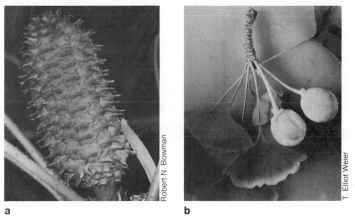

a   b

Figure 23.7 (**a**) Cycad cone. (**b**) Branch of *Ginkgo biloba*, showing fan-shaped leaves and seeds.

years ago, this attractive tree species was extensively planted in cultivated grounds around Asian temples. But the small natural population from which these "domesticated" trees were derived must have become extinct. The near-extinction of this living fossil from the age of dinosaurs is puzzling, for ginkgos seem to be hardier than many trees. Especially in cities, they are planted because they are attractive and because they seem resistant to insects, disease, even air pollution! But cultivated plants are propagated by humans and they grow under protected conditions. In the natural state, ginkgos may be less adaptive.

Conifers (Coniferophyta) are the most diverse and widely distributed gymnosperms. They are cone-bearing woody trees and shrubs, with needlelike or scalelike leaves. Most conifer leaves are retained for several years on the sporophyte body. Some leaves are always present on the sporophytes of most species; that's why conifers are commonly called evergreens. Pine, spruce, fir, hemlock, juniper, cypress, and larch are conifers; so is the redwood (Figure 23.1). Conifer seeds develop on the surface of perhaps a hundred shelflike scales that are arranged into a cone. Each pine cone scale has two sporangia (spore-producing structures) on its upper surface. Sporangia of "male" cones bear microspores that develop into pollen grains—immature male gametophytes. Sporangia of "female" cones bear megaspores that develop into immature female gametophytes.

Conifer life cycles are diverse in their details. Here we will look only at the pine life cycle (Figure 23.8). The pine tree is the sporophyte plant. Each spring, it produces perhaps millions of pollen grains. These immature male gametophytes have winglike projections, which aid in their

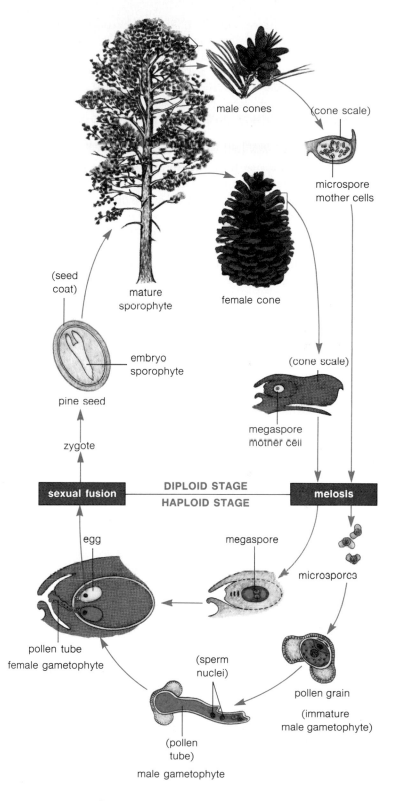

Figure 23.8 Pine life cycle, as described in the text.

dispersal. On these "wings" they float through the air, sometimes as thick yellow clouds. With such extravagant discharges, some pollen grains are bound to land by chance on female cones, and they do. A sticky substance on the female cone scales holds them in place. Here they remain trapped, not too far away from the megasporangia. When the sticky fluid dries, it pulls pollen grains onto the megasporangium surface.

At least a year will now pass between the time of pollination and actual fertilization. During this period, the pollen grain produces a **pollen tube**: a tubelike projection that starts to grow through the megasporangium wall. Eventually, a cell of the pollen grains will divide mitotically, and one of the cells produced will go on to form two sperm cells.

During this same period, events are proceeding in the megasporangia that will lead to a female gametophyte. Within each megasporangium is a diploid "megaspore mother cell." This cell divides by meiosis and produces four haploid megaspores. Three abort; the one remaining divides mitotically into a female gametophyte—an oval mass of cells within the sporangium walls. The megasporangium, tissue-layer covering, and female gametophyte constitute an **ovule**. As Figure 23.8 shows, the gametophyte has reproductive structures, each containing a single egg. As soon as a sperm fertilizes an egg, the stage is set for development of a pine seed. The diploid zygote divides mitotically, initiating the formation of an embryo that becomes the next sporophyte generation. A seed coat forms. The female gametophyte tissue serves as food reserves for the embryo during germination.

Conifers were dominant land plants during the Mesozoic, and their mechanisms of seed production, protection, and dispersal helped assure their success. Since then, their distribution has been gradually reduced. Even so, conifers are still the dominant vegetation in many communities, especially in northern regions and at high altitudes (Chapter Twenty-Nine). They also are major sources of lumber, paper, and numerous other commercial and industrial products.

Flowering Plants

Angiosperm means a seed carried within a "vessel"—namely, within a matured ovary. This descriptive term applies to flowering plants (Anthophyta). Although flowering plants appeared only about 135 million years ago, they quickly rose to dominance. Today, they are grouped into two classes: monocots and dicots (Chapter Nineteen). The

dicots number at least 200,000 species; they are the most diverse group of flowering plants. The 50,000 or so species of monocots include such familiar plants as grasses, palms, lilies, and orchids. They have played a central role in human population growth. The main crop plants supporting the human population are wheat, rice, corn, oats, rye, and barley—all monocots, and all domesticated grasses.

Angiosperms have a unique reproductive system involving flower formation (Figure 19.19). Flowers have microspore- and megaspore-producing structures, as well as accessory parts that have both direct and indirect roles in sexual reproduction. Recall that the ovary of flowering plants contains one or more ovules made of two tissue layers. These surround the megaspore mother cell and are attached to the ovary wall. In some plants, hundreds or thousands of individual ovules may be attached to the inner wall of the ovary or to its inner partitions. Meiosis of the megaspore mother cell occurs *inside* an ovule. Four haploid megaspores form, but three usually abort. This leaves a single functional megaspore that divides and forms a single female gametophyte, which will produce one egg cell (Figure 23.9).

Regardless of the type of flower, sexual reproduction depends on pollen grains being transferred from anthers (male reproductive structures) to stigmas (female reproductive structures). Once a pollen grain has been deposited on the stigma, a pollen tube forms and grows down to the ovarian chamber below. Two sperm nuclei from the pollen grain break through the tube's end and penetrate the female gametophyte. One sperm fuses with the egg nucleus, forming a diploid zygote. In most organisms, fusion of egg and sperm is a singular event in the life cycle. But in flowering plants, **double fertilization** occurs. One sperm nucleus fuses with that of the egg—and the other sperm nucleus usually fuses with two other nuclei in the ovule, forming a single triploid ($3n$) nucleus.

Following double fertilization, the ovule expands and develops into a seed. Outer tissue layers thicken and harden, becoming the seed coat. The fertilized egg develops into a diploid embryo. The triploid nucleus divides repeatedly and forms endosperm (Chapter Nineteen). In some seeds, endosperm serves as food storage tissue, which is used for early growth when the seed germinates. In other plants, the embryo's cotyledons (seed leaves) absorb the endosperm and function in food storage.

In contrast to the "naked" seeds of gymnosperms, the angiosperm seed is contained within the ovary as it develops. The ovarian wall expands and ripens into another structure characteristic of flowering plant reproduction: the **fruit**. We tend to think of fruits as juicy edible structures, as indeed many are. But fruits are any structures that develop from the ovary and contain ovules or developing seeds (Table 19.2). Grains and nuts are dry fruits; tomatoes are fleshy fruits. A raspberry is an aggregate of many fruits from one flower that later separate; in a pineapple, several flowers form a cluster of multiple units that remain together at maturity. A maple fruit is winged, and its wall is dry and intact at maturity. What is the function of such diverse structures? *Fruits, be they edible structures, small capsules, or a variety of other forms, are adaptations for some type of seed protection and dispersal.* Thus they help assure successful reproduction.

Coevolution of Flowering Plants and Pollinators

Flower-bearing plants grow in alpine regions, in forests, in deserts, even in water. More than any other kind of plant, they have spread across the earth's surface and have become extremely diverse. What underlies their adaptive success? For possible answers, let's return to the Cretaceous Period of the Mesozoic, about 130 million years ago. Before then, wind-pollinated gymnosperms were the dominant land plants. But by that time, flowering plants had appeared and had begun to take hold. By the close of the Mesozoic, about 65 million years ago, they had expanded in numbers and kinds through diverse environments.

It happens that this period of rapid expansion was paralleled by expansion among insect groups. Although insects appeared far earlier than flowering plants (during Silurian times, Figure 21.4), they did not undergo rampant diversification at first. But during the early Mesozoic, some beetles apparently were already feeding on pollen or other parts of wind-pollinated plants. They probably visited plants on a haphazard basis, just as beetles do today. Hence they may have been secondary in importance as pollinating agents. Yet, we can imagine that plants being slightly more conspicuous or tasty would attract more insects, and pollen dispersal would have been more effective because of it. There would have been selective advantage for plants having ample pollen, special glands for secreting nectar (a sugary fluid), and flamboyant floral advertisements. However the interactions developed, by the dawn of the Cenozoic, diverse species of insects and flowering plants existed and were locked in interdependence (Figure 23.10).

The arrangement of flower parts, the patterns in which they fuse or remain distinct, their color, odor, nectar, size —all are known to attract beetles, bees, wasps, butterflies, moths, flies, birds, even bats. It is true that many success-

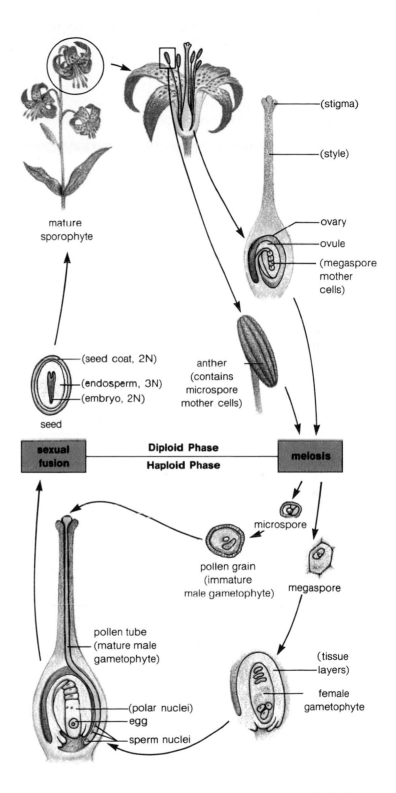

Figure 23.9 Life cycle of a flowering plant, as described in the text. The example here is a lily (*Lilium*).

a   Harlo H. Hadow

b   Harlo H. Hadow

c   Harlo H. Hadow

d   R. Taggart

e   R. Taggart

f   R. Taggart

g   R. Taggart

h   Ted Schwartz

i   Edward S. Ross

j   Edward S. Ross

k   Edward S. Ross

l

m

p

n

o

q

Figure 23.10 Pollination of plants through reliance on winds, deceit, spring traps, and various exotic and luscious inducements.

(**a,b**) Reproductive structures of the saguaro cactus brush against the head of a bird pollinator, the gila woodpecker, and pollen is thus carried from blossom to blossom in return for sips of nectar. (**c,d,e,h**) Plants pollinated by bees tend to be blue or yellow and brightly colored. (**f,g**) Some orchids tend to look like females of the pollinating species (in which case the males attempt to copulate with the flower) or like rival male bees vibrating in the wind (in which case the territorial male insect attacks the flower and pollinates it). (**i,j,k**) Scotch broom is an "explosive" flower. The weight of its pollinator is needed to force open the flower, which releases the pollen-laden stamens that are positioned to strike against the underside of the pollinator. (**l**) Close-range visual guides to nectar. (**m**) Glorybower, with suspended flowers, is pollinated by hummingbirds. Wind-pollinated flowers include cattails (**n**) and grasses (**o**) (**p**) Arctic lupine, an "explosive" flower of the far north.
(**q**) Stephanotis is serviced by night-flying moths; it is light-colored and astonishingly fragrant during the night. (Day-flying hummingbirds are opportunistic when it comes to stephanotis, and are secondary pollinators of it.)

R. Taggart

R. Taggart

Roger K. Burnard

R. Taggart

R. Taggart

Ted Schwartz

Roger K. Burnard

All other photographs Victor Duran

Figure 23.11 A few representative fungi. (**a**) A coral fungus. (**b**) Scarlet cup fungus. (**c**) Bird's nest fungus. (**d**) Shelf fungus growing outward from a tree trunk. (**e**) Common edible mushrooms. (**f**) Edible morels.

ful plants are wind-pollinated; grasses, oaks, birches, and maples are like this. Their flowers typically don't have nectar or perfume and tend to lack colorful petals. But many flowering plants do depend on animals to disperse their pollen (Figure 23.10). Color is an important attractant for many pollinators. For instance, bird-pollinated flowers tend to be red, a wavelength to which bird eyes are more sensitive. Bee-pollinated flowers tend to be blue or yellow, with prominent ultraviolet components. Odors, too, lead the way and bring in pollinators from great distances.

Once a pollinator locates a flower, color patterns and petal shapes guide it to the nectar. They also channel the pollinator's movements in ways that aid pollination. For instance, petals of hummingbird-pollinated flowers often form a long tube, which corresponds to bill length and shape. These flowers often exclude other potential pollinators (such as heavy bumblebees).

Some orchids mimic the female form and coloration of pollinating insects. The male insects attempt to mate with flower after flower. They spread pollen about as they are led ever onward by this deception. Eventually a real female comes along, so the insects do keep reproducing, and the orchids do keep getting pollinated.

In rigorous climates, several different insect species often pollinate the same plant. In tropical environments, many plants and pollinators are linked in highly specialized ways. This says something about the nature of the coevolutionary process between flowering plants and their cohorts. On the one hand, *the more refined the "fit" between plant and pollinator, the more efficient pollination can be.* For instance, less energy has to be channeled into producing abundant pollen grains. On the other hand, *the more specialized the plant–pollinator relationship, the greater the chance that the plant may face extinction if its pollinator should happen to disappear.* In harsh, low-diversity climatic zones, such disappearances may be a real possibility. In attempting to identify the nature of such interactions, we must therefore look to the feedback relationships that may exist not only between coevolved species, but between those species and their environment. More will be said about this in the next unit.

PART II. KINGDOM OF FUNGI

Fungi, recall, are eukaryotes. They share some characteristics with plants. For instance, like many forms of algae, fungi typically have a vegetative body that is branched and filamentous. Because this kind of body allows contact with a large volume of the surroundings, fungi have access to raw materials that may be dilute (as they are for algae) or scarce. Like plants, many fungi have cells with large central vacuoles, and their body cells generally have walls. Like plants, spore formation is an important part of their life cycle. However, fungi lack chloroplasts and differ from plants in other ways. Significantly, *most fungi rely on enzyme secretion that promotes digestion outside the fungal body, followed by nutrient absorption across the plasma membrane of individual cells.* It is largely because of this energy-acquiring strategy that many biologists now place fungi in a kingdom separate from plants. Figure 23.11 shows a few fungi.

Between 80,000 and 200,000 different fungal species are recognized. In this book they are classified as members of a single division, the **Mycota**, which is subdivided in the manner shown in Table 23.3.

Some fungal species are **parasites**: they obtain nutrients directly from *still-living* organisms. Most plant diseases are the result of parasitic fungal attacks. So is athlete's foot, a disease that affects humans. The annual dollar cost of crop loss and food spoilage by parasitic fungi runs into the millions. But most fungal species are **saprophytes**: they feed on remains of *dead* organisms or their by-products (such as leaf litter on a forest floor). Saprophytes first secrete enzymes that break down these organic remains, then they absorb soluble breakdown products. Along with bacteria, saprophytic fungi are decomposers of the first rank. By bringing about the decay of organic material, they help recycle such vital substances as nitrogen, potassium, iron, sulfur, calcium, and magnesium through communities of life. Some saprophytic fungi (yeasts) are a mainstay of baking and brewing industries. Others yield such useful metabolic by-products as penicillin, citric acid, and vitamins (riboflavin, for instance). Some wild saprophytes are hallucinogenic, some are extremely poisonous, and some are good to eat (Figure 23.11). Let's take a look at some of these diverse species.

Table 23.3 Classification of Fungi (Division Mycota)		
Subdivision Myxomycotina	*slime molds*	
Subdivision Eumycotina:	*true fungi*	
Class Oomycetes	*egg fungi*	
Class Zygomycetes	*zygote fungi*	
Class Ascomycetes	*sac fungi*	
Class Basidiomycetes	*club fungi*	
Class Fungi Imperfecti	*imperfect fungi*	

Figure 23.12 One of the slime molds, *Leocarpus fragilis*, showing the vegetative body form (**a**) and the spore-bearing, or fruiting, structures (**b**).

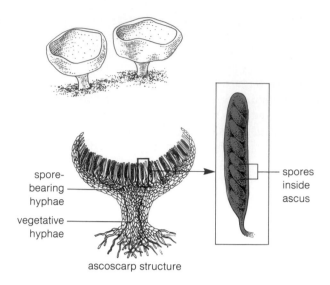

Figure 23.13. Structure of one kind of ascoscarp, a spore-bearing structure that occurs in the life cycle of sac fungi. (From Rost et al., *Botany*, Wiley, 1979)

Slime Molds

Exactly where the slime molds should be placed in classification schemes is controversial. They have funguslike and plantlike features. Some bear strong resemblance to amoebalike protistans. Whatever their affinities, slime molds are studied more by mycologists than by anybody else, and that's why they are described here.

The vegetative body of a slime mold is a mass of cytoplasm, with no cell wall outside the plasma membrane. You can find these masses moving, amoebalike, on dead leaves and on moist rotting logs. There they absorb nonliving organic matter, or engulf bacteria and spore structures of other fungi. When slime molds reproduce, though, their spores have a rigid cell wall—as do the spores of the true fungi. *Dictyostelium discoideum*, described earlier in Chapter Twelve, is a cellular slime mold. At some point in its life cycle, many separate amoebas aggregate into a multicellular mass (pseudoplasmodium). From this mass, a stalked fruiting structure arises and functions in spore dispersal (Figure 12.1). Other slime molds are not composed of clustered cells; they are considered acellular. Each is a multinucleate cytoplasmic mass (plasmodium) that has grown in size from

a single cell, without cellular division. A plasmodium can stream out, veinlike, and back again as it feeds on microorganisms (Figure 23.12). Some can grow into thin sheets, measuring several feet across. Acellular slime molds also produce spore-bearing stalked masses. Spores discharged from these stalks are dispersed by winds.

True Fungi

The vegetative body of most true fungi is a **mycelium**. This is a meshwork of branched, tubelike filaments called **hyphae** (singular, hypha). The hyphae are surrounded by cell walls, which usually contain chitin. In some species, multinucleate cytoplasm runs continuously through the tubes; in others, perforated cross-walls compartmentalize the hyphae into separate cells, yet allow chemical communication and movement of nuclei between body parts.

Most true fungi can reproduce asexually through spore formation, fission and budding (employed by single-celled yeasts), or fragmentation from a parent mycelium. Many also can reproduce sexually. In some species, gametes are all potentially compatible. In others, only gametes of different mating types can fuse.

Egg fungi (Oomycetes) are the only class with motile reproductive cells. They range from single cells to species that develop a highly branched mycelium. In this group are *water molds* (mostly saprophytes) and *downy mildews* (parasites on many plants). One egg fungus, *Phytophthora infestans*, causes a disease called late blight, which will be described later on.

Zygote fungi (Zygomycetes) are structurally the simplest fungi that have no motile cells in their life cycle. They are saprophytic, for the most part. Chitin reinforces the cell walls of their mycelia, which are extensive. The class includes *fly fungi* (which decompose those dead flies on windowpanes and on garage floors) and *bread molds*. One bread mold, *Rhizopus*, forms cottonlike masses on bread and other baked goods.

Sac fungi (Ascomycetes) bear spores in saclike structures called asci (singular, ascus). Most often, these sacs are concentrated in a complex reproductive structure known as an ascoscarp (Figure 23.13). *Yeasts* are in this group. So are the parasitic *powdery mildews* that attack apples, grains, cherries, grapes, and other food crops. *Monilinia* is a sac fungus that infects fruit blossoms and causes severe fruit crop damage. Truffles and morels (Figure 23.11) are edible sac fungi.

Club fungi (Basidiomycetes) are structurally the most complex of all fungi. They include edible mushrooms, extremely toxic ones (such as *Amanita*), shelf fungi, puffballs, many rusts, and smuts. Their complex reproductive structures are called basidiocarps. The common field mushroom may be used as an example of a club fungus reproductive cycle. There is an alternation between haploid and diploid stages. A new generation begins when an airborne haploid spore lands on a moist substrate and germinates. The spore grows into a small, loosely organized mycelium. When two compatible hyphae from these mycelia grow near each other, they fuse—but without actual fusion of nuclei (Figure 23.14). Thus a "dikaryotic" hypha with two nuclei per cell is formed. This hypha develops into an extensive mycelium. The mat is destined to become a long-lived network in soil. At some point in the life cycle, a spore-bearing basidiocarp forms aboveground. Inside this reproductive structure, the pairs of nuclei derived from the two different hyphae fuse. The fused diploid nuclei undergo meiosis at once, and spores are produced. Once spores are dispersed by winds, the mushroom withers. As spores fall on moist ground and germinate, the cycle begins again. It's important to understand that a mushroom is only the reproductive part of the entire organism. Many club fungi are very large.

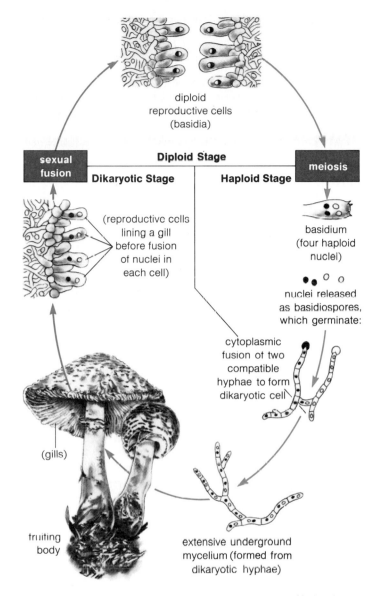

Figure 23.14 Life cycle of the common field mushroom. Notice the dikaryotic stage, in which hyphal cells contain two distinct nuclei.

Fungi as Symbiotic Organisms

Almost all complex land plants depend on intimate association with fungi that help them absorb certain vital nutrients from the soil. Pine trees, for instance, won't grow well unless hyphae of certain fungi form dense mats (mycorrhizae) around and within their roots (Figure 20.8). These fungi absorb carbohydrates from the live trees. At the same time, the thick fungal mats help the trees absorb soil nutrients more rapidly than they otherwise would. Such a mutually beneficial association is one form of symbiosis (Chapter Twenty-Eight).

Or consider the **lichen**, a "composite" organism that is actually a fungus and an alga living together in interdependence. The fungal hyphae form a dense mat above and below the algal cells. In this arrangement, the fungus is assured of photosynthetically derived food. And the alga enjoys improved water conservation (water can be retained longer in the dense mat before drying out), mechanical protection from being blown away, better gas exchange (compare the anatomy of a leaf, Chapter Nineteen), and less overlap between individual algal cells, as there would be in an algal crust. About 15,000 known species of lichens live in such diverse settings as open patches of forests (lichens need some sunlight to survive), deserts, tundras, regions of Antarctica, even on bare rocks of isolated oceanic islands and mountain peaks.

In most communities, however, different species of parasitic and saprophytic fungi compete with other organisms and among themselves for available substrates. In their natural environment, parasitic fungi normally attack organisms that are damaged or weakened, as with age. Young, healthy plants or animals are usually resistant to fungal attack unless environmental conditions shift drastically and somehow make them vulnerable. Their resistance developed over time, as selective agents worked against less resistant strains. Even as saprophytic fungi break down remains of dead organisms, so also do parasitic fungi break down weakened or less resistant organisms. In this way, minerals and other materials are returned to soil or water, where they can be picked up and recycled through the ecosystem.

Late Blight

What happens when we fail to recognize normal interactions among plants and fungi? Consider the potato, a plant native to the cool, dry regions of the Peruvian Andes. This plant was under cultivation almost two thousand years ago. A parasitic fungus, *Phytopthora infestans*, causes the disease **late blight** in potatoes (and tomatoes). The fungus was also native to the Peruvian regions, but it posed no serious threat to crops because environmental conditions kept it in check. The potato was introduced to Europe in the sixteenth century. It became a major food crop. In Ireland it became *the* major food crop. During the growing season, Ireland has cool, moist nights and warm, humid days. Its climate is quite unlike that of the Andes. Evidently, the fungus was not introduced at the same time as the potato in Ireland. Hence potatoes there were not initially selected for resistance to it. When the fungus did gain entry to Ireland, the climate, the practice of planting uninterrupted fields of host plants (monocrop agriculture), and the susceptible strains of potatoes (asexually propagated clones) all contributed to massive crop destruction. The fungus simply went wild. Once a plant was stricken with late blight, its vines rotted within fourteen days. Between 1845 and 1860, a third of Ireland's population starved to death, or died in the outbreak of typhoid fever that followed as a secondary effect, or fled the country.

Wheat Rust

Even though we are beginning to appreciate the complex relationships between organisms and their environment, it is not always a simple thing to come to terms with what we find out. Consider, for instance, that two-thirds of the dry bean crops and almost all pea crops in the United States consist of only two varieties each. The absence of diverse and potentially resistant strains means that these staple crops are extremely vulnerable to epidemic fungal attacks. Consider that **wheat rust** has claimed appalling amounts of annual wheat crops—one of our most basic food sources. Currently as much as ten percent of the United States wheat crop is lost because of this disease. The fungus causing the disease has coevolved with two hosts: the wheat plant and the barberry bush. And it goes through not one but *five* spore stages during the year! It is a never-ending challenge to develop resistant wheat strains. Each time a new wheat strain appears in the fields, the rust may have undergone mutations and developed into forms able to attack the resistant variety. For a time, it was thought that eradicating barberry plants (on which sexual reproduction of the fungus takes place) would interrupt the life cycle. It appears that enough mutations can produce new fungal strains even in the absence of sexual reproduction.

Finding a way out of this precarious situation will not be easy. In economic terms, it is far more efficient to plant, tend, and mechanically harvest monocrops than to have crop diversity. Efficiency *is* crucial for our survival. As long as the human population continues on its course of explosive growth (Chapter Twenty-Six), it is a race against time to produce enough food in the fastest way possible. We are, as a consequence, highly vulnerable in our total dependence on monocrop agriculture. We must continue to meet existing demands with existing technology. *But simultaneously, we must begin to rethink our agricultural strategies so that we will be less in conflict with natural characteristics of and interactions among organisms.*

PERSPECTIVE

How often, if we think of them at all, do we think of plants as little more than greenery in the scenery? How often do we look upon fungi as curious and/or nasty intruders into human affairs? Yet, without these organisms, without their initial invasion of the land and subsequent interactions, we could scarcely have even made it onto the environmental stage. Until they came to cloak the earth, inch by inch, continent by continent, there could be no other forms of life on the vacant land. Their evolution, while not smacking of the drama of, say, the rise and fall of dinosaurs, was remarkable nevertheless when we stop to consider the odds.

The vegetative bodies of plants are immobile. They cannot crawl, leap, run, or fly. Beginning with ancestral forms that could not have been more complex than simple green algae, a long series of mutations apparently led to plants having a few relatively inconspicuous traits. They had a few threadlike anchors to sink into muddy margins of pools and lagoons, a tiny chloroplast-filled stalk, a few scalelike projections that spread out stiffly beneath the sun. And yet, within those anchors, stalks, and scales, hollow tubes eventually appeared. These tubes were forerunners of highly adaptive transport systems that now run through roots, stems, and leaves. The diploid stage of some aquatic plant life cycles proved to have remarkable potential. Resistant as it was to adverse conditions, the diploid sporophyte generation gradually become more and more pronounced in the move to higher and drier land. Among some land plants, gametophytes came to be housed in the parent sporophyte body. The sporophyte transport system could nurture these gametophytes with water and dissolved nutrients. In some lines, seeds took over the dispersal function of spores. Reproductive structures became attuned to seasons, to times of rain and winds. Most astonishing of all, some plants coevolved with insects and other animals. Their life cycles came to depend on animals to carry pollen grains to female reproductive structures, and to disperse seeds. Simultaneously, life cycles of the animal pollinators came to depend on specific plants for food. Over time, the fungi became part of life on land, as they have been in water—decomposers, recyclers of life-giving nutrients for the animals and plants with which they are linked in the web of life. Like plants, most fungi rely on wind, splashing rain, and insects for dispersal. Thus, without means of their own for spectacular motility, plants and fungi have moved over the barren plains, to the mountains, to land everywhere.

Readings

Bold, H., C. Alexopoulos, and T. Delevoryas. 1980. *Morphology of Plants and Fungi*. Fourth edition. New York: Harper and Row. Outstanding source book, as authoritative as any available.

Christensen, C. 1972. *The Molds and Man*. Third edition. Minneapolis: University of Minnesota Press.

Raven, P., R. Evert, and H. Curtis. 1976. *Biology of Plants*. Second edition. New York: Worth. Outstanding text, with beautiful illustrations.

Rost, T., et al. 1979. *Botany: A Brief Introduction to Plant Biology*. New York: Wiley. Excellent abridged version of a classic (T. Weier et al., 1974, *Botany*, fifth edition; same publisher).

Review Questions

1. Adaptive shifts in _____ and in _____ must have been necessary when aquatic plants first began radiating into dry land zones some 400 million years ago.

2. Can you distinguish between the spores and gametes of land plants? As part of your answer, define sporophyte and gametophyte. Which is haploid? Diploid?

3. What three observations suggest that ancestral plants much like modern-day green algae may have given rise to plant lineages that now populate the land?

4. What are the advantages in embryo sporophytes remaining attached to the gametophyte? Name some plant species that rely on this developmental strategy.

5. Distinguish between these terms:
   a. Homospory and heterospory
   b. Megaspore and microspore
   c. Pollen grain and pollen tube
   d. Female gametophyte and ovule
   e. Seed and fruit

6. What is the adaptive significance of pollen grain formation? Of seed formation?

7. If both gymnosperms and flowering plants are seed-bearing, then how do they differ in reproductive modes?

8. What happens in double fertilization? Is this reproductive event characteristic of both gymnosperms and angiosperms?

9. Observe the kinds of flowers growing in the area where you live. On the basis of what you have read about the likely coevolutionary links between flowering plants and their pollinators, can you perceive what kinds of pollinating agents your floral neighbors might depend upon?

10. What is the difference between parasitic and saprophytic fungi? In what ways do we find many saprophytic species beneficial?

11. Describe the reproductive cycle of a club fungus. Is the mushroom the main part of the fungal body?

24

EVOLUTION OF ANIMALS

Figure 24.1 Will the real animals please stand up?

It is the last scene of an epic Western film. The hero has finally tracked down the rustlers, saved the longhorns, fought off a mountain lion with his bare hands, narrowly escaped the lunge of a rattlesnake, and found the rancher's daughter (who had been kidnapped and hastily abandoned in the desert) by spotting the vultures circling in the sky above her. Now, with his faithful dog beside him, he waves goodbye to the girl, mounts his restless stallion, splashes across a shallow stream, and rides off into the golden west.

When you think back over the stories on which we are raised, it's easy to see how we acquire a stereotyped picture of what "animals" are. For instance, how many kinds of animals would you say are mentioned in the pared-down saga opening this chapter? Seven, right? But did you happen to notice that all the animals mentioned are *vertebrates*—animals with backbones? The saga is just a little lopsided: more than ninety-nine percent of all the kinds of animals in the world are backboneless *invertebrates*. Undoubtedly just out of camera range were all manner of worms, grasshoppers, spiders, scorpions, ticks, centipedes, millipedes, water fleas, mosquitoes, flies, beetles, butterflies, moths, aphids, snails, bees, and various other spineless things. Long after we finished counting all the kinds and numbers of vertebrates alive today, we would still be counting up the kinds and numbers of a staggering array of invertebrates. Only by beginning with a few examples of these less conspicuous animals can we gain insight into how all animals evolved—into what they all have in common and how diversity arose among different groups. Only by putting aside our culturally shaped preconceptions can we piece together a picture of the evolutionary progression that led to organisms as complex as horses, dogs, snakes, and longhorns, and ourselves.

A FEW THOUGHTS ON "COLONIAL ANIMALS"

Somewhere between 2 billion and 1 billion years ago, the first animals arose. They left behind nothing of themselves other than a few petrified feces, and a few tantalizing tracks and burrows frozen in ancient rocks. At the dawn of the Cambrian they were leaving behind hard parts, so we know something of what they looked like by then. But by then they were already well-developed multicellular animals, or **metazoans**. Thus we can assume that long before 700 million years ago, there must have been any number of experiments in multicellularity, most likely of the sort described in Chapter Twenty-One.

One such experiment is thought to have led to the **sponges**—diverse, abundant, but relatively simple animals found in shallow seas throughout the world. In one sense, they are like a multicellular vase with holes in the sides. Seawater flows in through the holes (pores), then out through the opening at the top. The inside is lined with flagellated *collar cells*, which trap and ingest microscopic organisms carried in on the water flowing through the various pores.

Collar cells may link sponges to far simpler relatives (Figure 24.2). Certain flagellated protistans also sport a delicate, sticky collar, which collects food particles that the flagellum beats up around it. There are also several colonial forms of collared flagellates that are functionally independent but are held together in a jellylike matrix. Yet sponges clearly are a step above these colonial forms in organization, for they have four distinct cell types that are incapable of independent existence. Besides collar cells, sponges have thin cells lining the outside surface, thick cells lining the pores, and various amoebalike cells creeping about in the body wall. To these amoeboids fall the tasks of food catering (passing on some food from the collar cells to the other cell types) and reproduction.

Sponges do resemble animals—they are multicellular and they ingest other organisms. Yet communication between cells and integration of activities are truly primitive. Water flows into a central cavity, and it flows out; if microorganisms flow in with it, they are captured when they brush against the sticky collar cells. The cells lining the pores show only simple responsiveness to simple changes in internal conditions (they contract and shut off the inward flow of water if it contains certain noxious substances). *Thus sponges differ from all other animals in three important ways: they have no nerve cells, they have no muscles, and they have no gut.*

Does this put sponges outside the animal kingdom? Perhaps not. Sponges possess certain proteins found in all other animals—but not in protistans, plants, or fungi. More than this, cells resembling sponge collar cells have been found throughout the animal kingdom, even in humans. It is possible that sponges are an extremely ancient offshoot of the main line of animal evolution, but animals nevertheless. Whatever their ancestry, diversity among sponges occurs mostly through increased folding in the body wall. Such folds create ever more twisted channels for water flow. For sponges, this body plan has proved adaptive to their environments. For about a billion years, they have survived in their own quiet way in marine and freshwater settings.

RADIAL SYMMETRY IN CNIDARIANS

If not in the sponges, where do we find the simplest organs that might provide clues about early routes in animal evolution? Here we must return to the water-dwelling **cnidarians**—jellyfish, sea anemones, and their kin. Recall, from Chapter Fourteen, that cnidarians show radial symmetry. This arrangement indicates that in open water, food and danger pass underneath the floating cnidarians, and above the attached ones. Thus we find the centrally positioned jellyfish mouth pointing down into the water, and the sea anemone mouth pointing up. Stinging tentacles ring the mouth. The sting comes from the tiny barbs that are discharged from many cellular capsules called **nematocysts**. Each barb caps a coiled thread within a capsule. The touch and chemical taste of predators or prey trip the barb into action. The thread uncoils with explosive force, and the barb shoots out. Sometimes the barbs are drenched in poison. Neurotoxins of the Portuguese man-of-war and some of the large jellyfish are potent enough to harm humans (Figure 24.3).

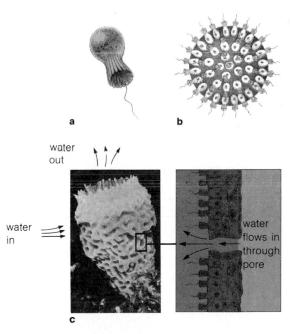

Figure 24.2 Sponges and some simpler relatives. (**a**) Collared flagellates are protistans that resemble collared cells found in sponge body walls. (**b**) Some colonial protistans are composed of collared cells, embedded in a jellylike matrix. (**c**) A sponge has no organs; its tissues are composed of four cell types: collared cells, amoeboid cells, pore cells, and thin outer cells (Douglas Faulkner).

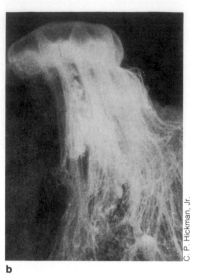

a   b

Figure 24.3 Two cnidarians to avoid. (**a**) *Physalia physalis*, the Portuguese man-of-war, shown here dining on a fish. This ''individual'' is actually a colony of polyps and medusae (see Figure 24.4). Individuals in the colony act in an integrated manner. The name is a reference to a kind of combat ship that once sailed the oceans; the ''sail'' on this cnidarian catches the wind and helps the colony float. (**b**) *Cyanea*, a jellyfish of arctic waters. Its crown sometimes measures two meters across; its tentacles may extend twenty-two meters (about seventy-five feet) from the crown. Both forms secrete toxins that are harmful, and occasionally fatal, to humans. (b from C. P. Hickman, Jr. et al., *Integrated Principles of Zoology*, 1979)

Most cnidarian life cycles proceed through a planula stage. A **planula** is a larval form, composed of an undifferentiated cell mass and an outer layer of ciliated cells. A planula crawls or swims at first. Later it settles on one end. A mouth forms at the other and leads into a gutlike cavity. Tentacles develop around the mouth. The individual has become a sedentary, attached form called a **polyp**. A polyp vaguely resembles an upside-down bell (Figure 24.4). For most cnidarians, a free-swimming, bell-shaped form called a **medusa** emerges during the life cycle.

Like sponges, cnidarians as a group show little deviation from the basic body plan. In nearly 10,000 species, individuals are shaped like an upside-down or right-side-up bell—regardless of whether they are solitary or part of colonies. The body itself is composed of only two epithelial layers, sometimes with mesoglea (a jellylike substance) sandwiched between them. There doesn't seem to be much evolutionary potential in jelly. At least, the interior of these animals never has filled up with complex organs and systems. Cnidarians have undergone adaptive radiation largely through form changes during the life cycle—through reliance on polyps and medusae, each adapted to a different part of the environment.

Even so, *cnidarians have a true gut—along with integration of neural and muscular activity*. These animals have nerve nets, muscle tissue layers, and some sensory organs

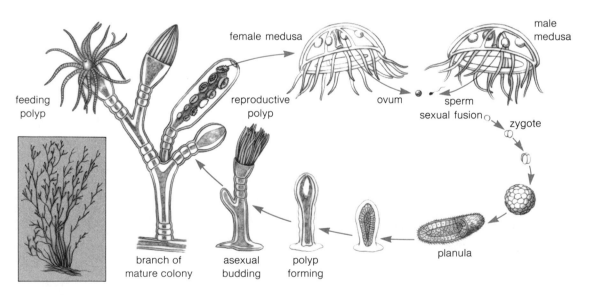

Figure 24.4 Life cycle of a representative cnidarian, *Obelia*. The inset shows a mature colony, actual size. On its branches are feeding polyps and reproductive polyps, both formed by asexual budding. The medusa stage is free-swimming. Sexual fusion of gametes from male and female medusae leads to planula formation. The swimming or crawling planula settles on one end and develops into a polyp, which will give rise to a new colony. (From T. Storer et al., *General Zoology*, sixth edition, 1979, McGraw-Hill)

(Chapters Fourteen and Fifteen). Both medusae and polyps have a gut, even though it is little more than an epithelial cavity. Enzymes secreted from glands in the epithelium break down food that has been stuffed into the sac. Gut wall cells ingest food particles and break them down farther. Soluble food molecules diffuse through the rest of the body; there is no circulatory system for material transport.

BILATERAL SYMMETRY IN FLATWORMS: HINTS OF THINGS TO COME

Anatomically speaking, the **flatworms** are only a step above the cnidarians. Flatworms include free-living turbellarians, parasitic flukes, and parasitic tapeworms. All have flattened bodies, shaped like broad leaves or long ribbons. Many are virtually microscopic. Some are as short as your little toe. Others (such as tapeworms that parasitize dogs, cats, and people) may be eighteen meters long. Instead of mesoglea, flatworms have tissue in which many cell structures are embedded. A middle tissue layer, or **mesoderm**, gives rise to muscular, reproductive, and water- and salt-regulating systems. In many flatworms, these systems show bilateral symmetry (Figure 24.5).

How do we get from radial animals to bilateral ones? Chapter Fourteen gave one hypothesis, which we can recap here. The most primitive flatworms bear striking resemblance to the planula stage of cnidarian life cycles. Long ago, some mutant planulalike animals may have kept on crawling instead of growing up to be a vase. For such mutants, food and danger would most often be bumped into by the leading end. Hence directional movement would have fostered nerve and sensory cell clustering at the head end (the advantage being faster sensing and response). Predators might attack crawlers from above, but not very often from below; hence dorsal–ventral differentiation would have been encouraged. Food and danger would be found as often on one side of a forward-crawler as the other—hence the emergence of bilateral symmetry. *And such a shift from radial to bilateral symmetry could have led to paired organs of the sort seen in many flatworms.*

With these adaptations, flatworms have gone on crawling through just about all shallow-water settings imaginable—through seas, lakes, streams, even moist soil and the moist digestive tracts of complex animals. Adult parasitic flatworms have lost their motile structures (cilia) along the way. These parasites, and the larger, free-living flatworms, rely on muscles for creeping or for undulating motion. Accompanying this muscular activity has been

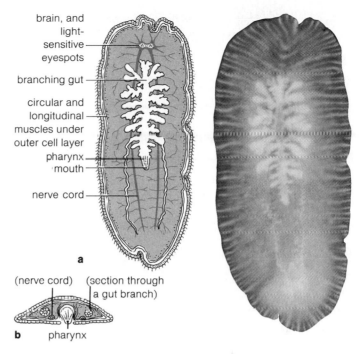

Figure 24.5 Body plan of one forward-crawling, bilaterally symmetrical flatworm, shown here sliced lengthwise (**a**) and crosswise (**b**). (From William H. Amos, Kirchoff/Walberg, Inc.)

greater neural activity, and more nerve cells concentrated in the brain. In larger forms the gut is more branched, so all cells have a way of getting nutrients and disposing of metabolic wastes. In some flatworms, simple tubes pick up dissolved wastes from internal cells and expel them from the body, much as kidneys do for other animals. Flatworms have no respiratory systems. Plenty of oxygen enters and carbon dioxide leaves their cells by diffusion through the thin, leaflike body.

Flatworms have an incomplete digestive system. It has only one opening to the gut cavity, so food goes in and undigested residues go out the same way. However, the development of a second opening to the gut (an anus) would not have been that difficult, given how close the flatworm gut is to the body surface. This development probably emerged many times in ancient flatworm groups. It apparently figured in the evolution of several other animal lines—the ribbon worms, nematodes, nemerteans, and annelids. In fact, the ancestral flatworms are thought to occupy a position of central importance in the evolution of all the more complex animals. *For with one-way traffic through the gut, regional specialization is possible for more efficient food processing.*

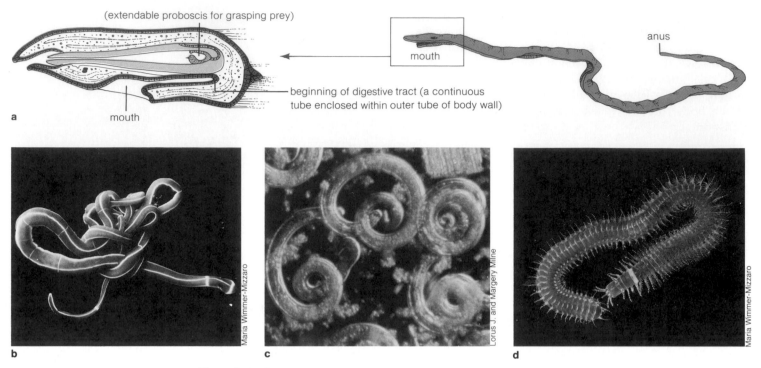

Figure 24.6 Body plans of some animals thought to have evolved from ancient flatworms. (**a**) The tube-within-a-tube arrangement of nemerteans. (**b**) External appearance of one nemertean, a ribbon worm. (**c**) A parasitic nematode, *Trichenella spiralis*, which lives in muscle tissue of rats, cats, dogs, hogs, and humans. This tiny nematode causes the disease trichinosis. Adult worms burrow into the small intestine's lining. There, female worms produce larvae. The larvae travel the bloodstream to muscles, where they coil up and live in cysts for perhaps years. Because undercooked pork can carry live larvae, it's wise to cook pork thoroughly. (**d**) An annelid called a polychaete worm. (a after M. Gardiner, *The Biology of Invertebrates*, McGraw-Hill, 1972)

SUCCESS AMONG THE SAC WORMS

We are, by the nature of our interests, unfolding a story of increased complexity in organ systems that may have figured in our own evolution. But it's important not to lose sight of the many living things that do quite well without complexity. Consider a kind of sac worm, the **nematode** (Figure 24.6c). There may be hundreds of thousands of nematode species. They live in more settings than any other animal group. Countless species live as parasites inside plants and animals. Others are free-living forms found in deserts, snows, hot springs, and ocean depths. And are they abundant! A single rotting apple may house 100,000 nematodes. An acre of rich farm soil may contain 100,000,000,000 nematodes in its top few inches.

It isn't that nematode complexity assures their success. They are little more than a tube-within-a-tube. Although they do have a complete digestive system, the only specialized region is a muscular, horny-toothed, or plated zone behind the mouth. Reproductive cells are crammed in between the gut and body wall, where they give rise to gametes. The nervous system is scarcely more complex than a flatworm's. Nematodes have no circular muscles, only a few longitudinal ones. Thus nematodes thrash about somewhat awkwardly.

Their secret, it seems, is the way they can resist so many environmental insults. Nematodes are covered with a tough cuticle. They can survive high levels of acidity and alkalinity, terrible temperatures, and noxious compounds. Sometimes, when being prepared for microscopic examination, they live for hours in fixatives and preservatives that would instantly kill other animals. Nematodes are remarkably resistant to suffocation. If oxygen is present, they use it. If not, they switch to anaerobic pathways. We should be so blessed, especially during smog alerts. In fact, if we continue to pollute our environment in every way imaginable, we can be sure the nematodes will be among the last to suffer.

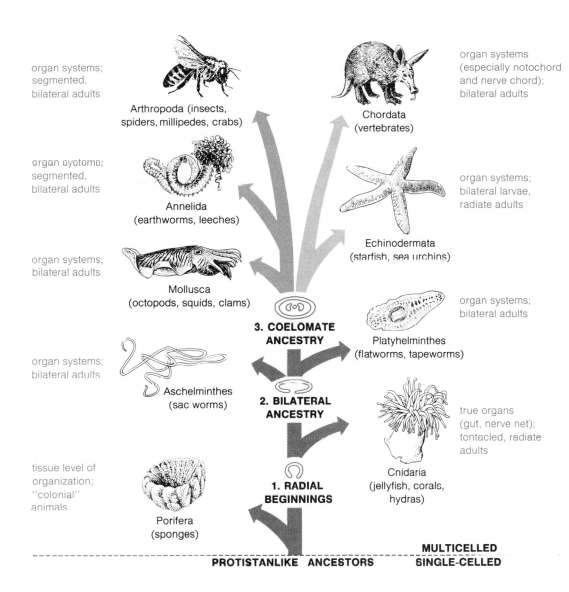

organ systems;
segmented,
bilateral adults

Arthropoda (insects,
spiders, millipedes, crabs)

organ systems
(especially notochord
and nerve chord);
bilateral adults

Chordata
(vertebrates)

organ systems;
segmented,
bilateral adults

Annelida
(earthworms, leeches)

organ systems;
bilateral larvae,
radiate adults

Echinodermata
(starfish, sea urchins)

organ systems;
bilateral adults

Mollusca
(octopods, squids, clams)

3. COELOMATE
ANCESTRY

Platyhelminthes
(flatworms, tapeworms)

organ systems;
bilateral adults

organ systems;
bilateral adults

Aschelminthes
(sac worms)

2. BILATERAL
ANCESTRY

true organs
(gut, nerve net);
tentacled, radiate
adults

tissue level of
organization;
"colonial"
animals

1. RADIAL
BEGINNINGS

Cnidaria
(jellyfish, corals,
hydras)

Porifera
(sponges)

PROTISTANLIKE ANCESTORS

MULTICELLED
SINGLE-CELLED

Figure 24.7 Assumed evolutionary relationships among some major animal phyla. The dark gold arrows show the protostome branchings; the light gold arrows show the deuterostome branchings. Both of these groups are thought to have diverged from a common, flatworm-like ancestor. On the main trunk, the sketch labeled 1 depicts the generalized body plan as if it were sliced lengthwise through the body wall, from the mouth. Sketch 2 depicts a crosswise slice through the body, near the mouth region. Sketch 3 depicts a crosswise slice through the body's midsection.

TWO MAIN LINES OF DIVERGENCE: PROTOSTOMES AND DEUTEROSTOMES

Two groups that may be descended from ancestral flatworms show astonishing variety within their ranks. They differ from each other in the way the second opening to the gut arises during embryonic development. In the simpler **protostomes**, the first opening formed becomes the mouth; the anus forms later. The group includes annelids (such as earthworms), arthropods (such as insects and crabs), and mollusks (such as snails). In the **deuterostomes**, the first opening to the gut becomes the anus and the second, the mouth. Deuterostomes include the echinoderms (sea stars and their relatives) and all vertebrates—fishes, amphibians, reptiles, birds, and mammals.

In itself, the distinction between protostomes and deuterostomes seems trivial. But it helps us identify two major lines of animal evolution that diverged long ago (Figure 24.7). Although many of the structural adaptations in these two lines are strikingly similar, the similarity is not evidence of some recent common ancestry. Rather, it is evidence of parallel solutions to common challenges. Thus, although both beetles and reptiles came to walk the land, they did so with different sorts of legs. Even though butterflies and birds took to the air, they did so with different sorts of wings. Let's take a look now at some likely ancestral sources for these examples of parallel evolution.

PROFOUND ADAPTATIONS IN ANNELIDS

On Coeloms and Circulatory Systems

One apparent line leading away from ancient flatworms came to differ from their forerunners in some important ways. This was the line called **annelids**, the true segmented worms. Living annelids have a complete digestive system—a mouth and anus, with a long, straight gut in between. In addition, they have a fluid-filled space, a **coelom**, between the gut and body wall (Figure 24.8). In contrast, the flatworm gut is joined to the body wall by a solid mass of cells. Hence flatworms are said to be *acoelomate*, or "without a coelom."

Like present-day forms, ancestral flatworms could not have had a vigorously active gut or complex internal organs. Any movement of the gut would have affected the body wall; and any contraction of muscles in the body wall would have shifted the gut and all organs. However, once organs became suspended in a coelom (as they are in annelids), different body parts could move independently. And they could do so without tearing the body apart in the process. *Thus a coelom insulates internal organs from the stresses of body movement—and permits increased size along with more activity.*

A coelom also bathes organs in liquid. However, if coelomic fluid had remained the only exchange medium, annelids might have had limited success as a group. For distribution of materials between the gut wall and the outer body would have to depend on the slow process of diffusion. But early in annelid evolution, a circulatory system appeared. Through various tubes, fluid containing nutrients and wastes traveled from the body's surface to its inner

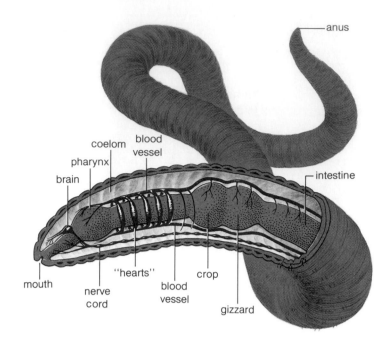

Figure 24.9 Arrangement of organ systems in the segmented body of an earthworm, one of the more common annelids.

regions, then back again. The circulating fluid (blood) provided a means for transporting materials between internal and external environments. *And it provided the potential for larger and more massive bodies.*

In the annelid system, forceful contractions in muscularized blood vessels ("hearts") keep blood circulating in one direction. One-way valves in some of the larger blood vessels assist in the task. As Figure 24.9 suggests, smaller vessels leading away from the main longitudinal channels carry blood to cells of the gut, nerve cord, and body wall. In some annelids, hemoglobin is one of the protein components of blood. Hemoglobin dramatically increases the blood's oxygen-carrying capacity, which increases the capacity for cellular respiration. *And with its high energy yield, cellular respiration means greater potential for muscular activity.*

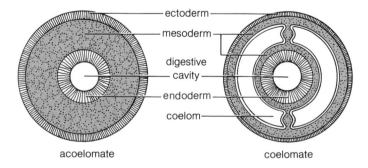

Figure 24.8 Comparison between acoelomate and coelomate body plans, shown in cross-section. In acoelomates such as flatworms, mesoderm forms a solid matrix between the outside layer and the digestive cavity. Coelomates such as earthworms have a central fluid-filled, epithelium-lined cavity in which various internal organs are suspended.

Body Segments and the Fine Art of Walking

In annelids, we see well-developed **segmentation**: a repeating series of body parts. Each annelid segment has a flexible outer wall (a cuticle) surrounding its own coelomic chamber. The repeated chambers form a long, large body. Some segments, especially those near the head, are modified for food-getting and for sensory perception.

Most flexible segments have a set of paired antagonistic muscles and a set of outward-projecting bristles (setae). Waves of contractions and expansions of circular and longitudinal muscles pass through the segmented annelid body. With these waves, bristles on contracted segments grip the ground, which keeps the body from slipping back when other segments expand in length. Thus the earthworm moves forward with far more coordination than the thrashing nematodes. Other annelids have their bristles embedded in fleshy, paddle-shaped lobes called parapodia. These lobes project from both sides of the body wall (Figure 24.10). In some annelids, a third set of muscles, which run obliquely to the others, extends to the lobes. By alternating the contractions of oblique muscles, the lobes can be swung up, forward, down, and backward. *Thus, in some annelids we see paired appendages and a mode of "walking."*

We also see a greater level of complexity in nervous systems. Without precise control over the activation of circular, longitudinal, and oblique muscles, there would be anarchy among body segments. And the worm would get nowhere. But a double nerve cord travels the length of the annelid body. This cord integrates body movements. In each segment, the cords are swollen with cells controlling local activity. The brain integrates sensory input.

HINGED SUITS OF ARMOR FOR ARTHROPODS

The cuticle covering each annelid body segment is thin, strong, and flexible. It doesn't fail, even when such animals bend while feeding and crawling. However, a thin cuticle can be pierced by predators, and some ancestral annelids developed ever more thickened and hardened body coverings. We see this pattern of cuticle thickening in **arthropods**: insects, crustaceans, arachnids, and their relatives on land and sea. Thickening also imposes limits on flexibility. Even though much of an arthropod body segment is covered with a rigid "shell," the cuticle between segments remains pliable and functions as a hinge.

Only the most minor rearrangement of muscles would have been required to accompany the change in segmentation. As in many existing annelids, some long muscles could run continuously between adjacent segments. Contraction of a single longitudinal muscle on one side of the body would bend the body in that direction. Through contractions of some muscles, the body could be bent up, down, or to the right or left at pliable connections between segments. Also, lateral appendages could become segmented into hardened tubes, with flexible connections between them spanned by oblique muscles. Such modifi-

Figure 24.10 An annelid with parapodia: *Nereis*.

cations would be enough to give rise to the jointed leg. (The word "arthropod" means jointed leg.) This motile structure is capable of far more controlled movements than the soft parapodia from which they may have been derived.

In the seas, arthropods known as crustaceans have come to be well represented. But it is on land that the arthropod potential for variation is most pronounced. We may assume that the hardened arthropod exoskeleton was at first an effective means of defense. Later, it may have been capitalized upon for locomotion across the floors of ancient seas. It also turned out to be an adaptation that had considerable survival value when some of these animals radiated into dry land environments. *The arthropod hardened exoskeleton is a superb barrier to evaporative water loss —and it also provides support for an animal body deprived of water's buoyancy.*

The first land arthropods were the arachnids. Their descendants (spiders, scorpions, ticks, and mites) are still with us. Centipedes and millipedes arose later. Insects are thought to have evolved from centipede-like ancestors.

The transition to land required still another adaptation. The larger marine arthropods extract oxygen from water, and they do so with gills: thin tissue flaps richly supplied with blood vessels (Chapter Sixteen). In ancestors of land arachnids, gills gradually came to be enclosed in lungs. Arthropod lungs are chambers below the cuticle, where body fluids keep them moist and facilitate diffusion across epithelial boundary layers. In some arthropod stocks, another means of breathing developed. A series of branching tubes—tracheae—were used as routes for air flow in all directions through the body. These fine tubes became stiffened with chitin, which kept them from collapsing under the pressure of the body's weight. *The success of land arthropods has depended not only on a jointed exoskeleton, but on a series of organ adaptations that led to a system for taking in oxygen.*

a

b

Douglas Faulkner

Douglas Faulkner

c

Lynn M. Stone/NAS/PR

d

Roger K. Burnard

e

Roger K. Burnard

Among insects, tracheal breathing reaches its peak of development. The tracheal tubes branch into ever smaller tubules, which terminate next to interior cells. Hence each cell in the insect body has its own supply of fresh air. This means of oxygenation is the basis for the highest metabolic rates known. These rates are required for the powerful, tiny muscles used in insect flight. When some ancient insects took to the air, the group began an explosive radiation into almost all environments. A distance that might take a spider a lifetime to traverse could now be spanned in minutes.

An insect wing isn't a modified leg, as is the wing of birds. Rather, insects have a modified flap of cuticle that lacks internal muscles. In some insects, muscles are

f

g   h

Figure 24.11 Marine arthropods: (**a**) a banded coral shrimp with delicate sensory structures, (**b**) two spiny lobsters vaguely reminiscent of Las Vegas chorus lines, and (**c**) a male fiddler crab, displaying his oversized claw as a warning to intruders. Usually the claw functions in courtship and competition between males of the same species. Land arthropods: (**d**) a stunning cecropia moth, (**e**) a many-footed millipede, (**f**) a grasshopper with powerful, chevron-striped hind legs, (**g**) two dragonflies mating, and (**h**) a spider on moss.

attached to the wing's base. They are used in making the wing pivot around its attachment. In many other insects, the wing is hinged to body wall plates. When tension distorts the body wall, the wing flaps. In both cases, the wing may move in a figure-eight pattern that lifts the insect and can propel it forward as fast as forty-eight kilometers (thirty miles) per hour! The rate ranges from about four

beats a second for some butterflies to a thousand beats a second for some mosquitoes. Figure 24.11 shows a few representative arthropods.

FLEXIBLE FLAPS FOR MOLLUSKS

It appears that annelids or their predecessors also gave rise to still another successful group: the mollusks. **Mollusks** include about 100,000 species of clams, oysters, abalones, limpets, snails, slugs, squids, octopuses, and an array of other diverse species (Figure 24.12). Recall that arthropod evolution must have been based on hardening of the body wall. *In molluscan evolution, the body wall has stayed flexible and its musculature has become well developed.*

We can speculate that molluscan forerunners became increasingly adapted for movement across the seafloor. The body wall surface in contact with substrates would have become thicker and more muscular, until it became a large foot. On the dorsal surface, a flap formed in the body wall and developed into a **mantle**. In several molluscan lines, mantle cells secrete substances that, together with mineral inclusions, eventually form a rock-hard shell.

In chitons, limpets, and snails, the shell is a protective shield from all but the most persistent, clever, and forceful predators. There is a price to pay for protection, though. The shell slows movement literally to a "snail's pace." Other molluscan lines, notably squids and octopuses, have become adapted for speed rather than for protection. Over time, the shell of these predators has been reduced to a tiny internal structure (cuttlebone) or has disappeared entirely. The mantle has become a conelike envelope surrounding internal organs. It is highly muscularized, and can be contracted with great force. Water moves into the mantle cavity when the cavity expands. Mantle contractions then force water out in a powerful stream that propels the body in the opposite direction. Thus freed from the inertia of a shell, and equipped with a jet-propulsive mantle, the squid has become the most magnificent swimmer of the invertebrate world. Some species are among the largest marine mammals. Individuals may measure eighteen meters (sixty feet) from their dorsal end to the tip of their tentacles. Giant sperm whales prey upon these giant squid, which probably put up quite a fight. Large sucker scars observed on whale bodies suggest there may be squid of monstrous dimensions in the deep oceans.

The nervous systems of squids and octopuses represent the peak of complexity in the invertebrate world. In fact, in terms of sheer size and complexity, the brains of these animals approach those of mammals. As in mammals,

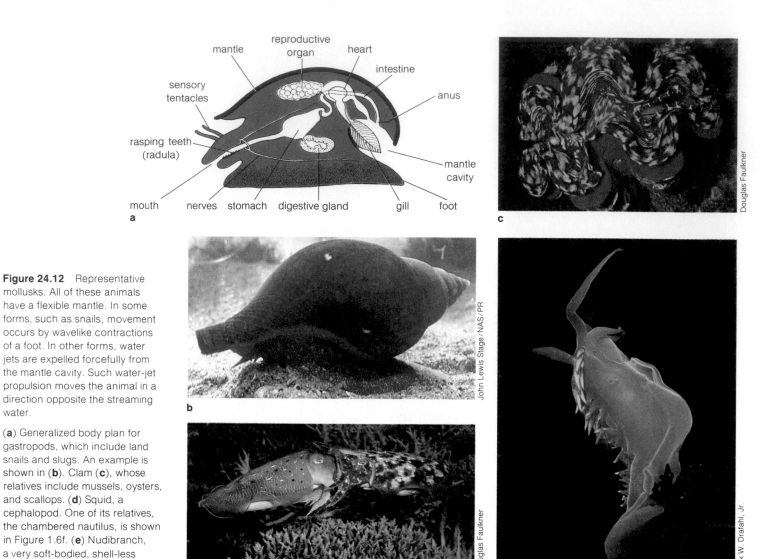

Figure 24.12 Representative mollusks. All of these animals have a flexible mantle. In some forms, such as snails, movement occurs by wavelike contractions of a foot. In other forms, water jets are expelled forcefully from the mantle cavity. Such water-jet propulsion moves the animal in a direction opposite the streaming water.

(**a**) Generalized body plan for gastropods, which include land snails and slugs. An example is shown in (**b**). Clam (**c**), whose relatives include mussels, oysters, and scallops. (**d**) Squid, a cephalopod. One of its relatives, the chambered nautilus, is shown in Figure 1.6f. (**e**) Nudibranch, a very soft-bodied, shell-less gastropod.

the brain includes a cortical region where information is stored. In some species, information is processed and used in modifying behavior. These cephalopods, in other words, have the capacity for learned behavior. When compared with vertebrates, the squid and octopus provide clear examples of parallel evolution. Ancestral to both groups were fast-swimming, predatory forms, which required increasingly refined systems for detecting, pursuing, and capturing prey in dimly lit waters. Hence, like vertebrates, these animals have acute vision and refined motor control. Because of similar evolutionary pressures, the entirely separate vertebrate and cephalopod lines developed many similar neural and sensory structures, having similar functions (Chapter Fifteen).

THE PUZZLING ECHINODERMS

Compared with protostomes, the branch of the animal kingdom in which we reside is far less exuberant in its diversity. Only a small number of deuterostomes have survived to the present. Of those, only two groups are prominent. They are the echinoderms and chordates. **Echinoderm** means spiny-skinned, and refers to the bristling spines on some members of the group (Figure 24.13). Echinoderms are a puzzle. They share many characteristics with bilaterally symmetrical animals—yet they are obviously radial in the body plan. What might explain this combination of complex and simple body organization? Consider the following hypothesis.

Figure 24.13 Representative echinoderms. (**a**) Crinoid, with its flower-like arms. Both its mouth and anus are at the center of the tentacles, on the body's upper surface. (**b**) Brittle star, which moves with rapid, snakelike twists of its arms. (**c**) Sea urchin, with rounded body and bristling spines. (**d**) Cobalt sea star. (**e**) Sea cucumber, the only echinoderm with a long, slender, muscular body that has tube feet running down the sides.

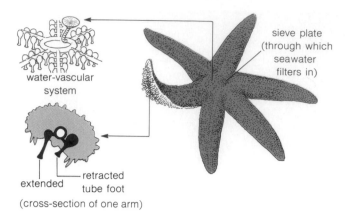

water-vascular system

sieve plate (through which seawater filters in)

extended

retracted tube foot

(cross-section of one arm)

Figure 24.14 Water-vascular system of canals and tube feet, the basis of echinoderm locomotion.

The first deuterostomes may not have been much different from the most primitive flatworms. It seems likely that they used cilia for swimming about. With a one-way gut and a swimming life-style, these animals would have been on an evolutionary road to bilateral symmetry, and to a concentration of nerve and sensory cells at the head end, about the mouth. But at least two deuterostome lines took a different evolutionary route somewhere during the Precambrian. Then, predators were increasing in both size and numbers. These deuterostomes were prey. Over time, some of their descendants came to be equipped with heavy, defensive armor of mineralized plates embedded in the body wall. Heavy armor must have helped them avoid being eaten. However, it also may have forced them to settle to the sea bottom. Whatever the selection pressures might have been, today their descendants attach to the seafloor or slowly lumber about. Radial symmetry (like that of bottom-dwelling cnidarians) has been overlaid on an earlier bilateral heritage.

Most echinoderms still go through a free-swimming, bilaterally symmetrical larval stage. But adults generally show radial symmetry. Some forms, such as sea stars, can move in any direction over a surface. Their unique way of moving about is based on constant circulation of seawater through a **water-vascular system** of canals and tube feet. As Figure 24.14 shows, each tube foot leads from short branches of radial canals. One end is a muscular bulb, the other is suckered. Contraction of the bulb wall forces water into the tube part, making it rigid and extended. Coordinated extension and retraction of many suckered tube feet is combined with directed contraction and relaxation of muscles in the tube foot walls. Together, this activity enables echinoderms to move body parts in specific directions.

DIVERSITY AMONG CHORDATES

Extensive modification associated with a sedentary life-style is not restricted to echinoderms. Similar modifications seem to have occurred among other primitive deuterostomes, and among the most primitive of all living chordates: the tunicates. Let's now take a look at the sorts of evolutionary roads that opened up as a result.

From Notochords to Backbones

When they are tiny larvae, **tunicates** look and swim like tadpoles. They are bilaterally symmetrical, with a front end and a tail end (Figure 24.15). They have a nervous system, with a nerve cord running the length of the body. Also running down through the body is a support structure, a rod of stiffened tissue called a **notochord**. Muscles attached to the notochord contract rhythmically, which causes the body to bend and move through the water. As simple as the notochord appears to be, it represents a profound evolutionary step. The notochord was forerunner to the chordate's internal skeleton, or **endoskeleton**.

After a period of swimming, tunicate larvae settle down and become attached, at the front end, to the seafloor. There they undergo drastic metamorphosis. The tail, notochord, and most of the brain are lost by the time metamorphosis is complete. The adult animal also becomes encased in a tough outer "tunic" (hence the name).

Tunicates practice **filter feeding**. They draw in water through the mouth, then the water is passed out through a series of pores known as **gill slits** (Figure 24.15). Cilia lining the gill slits create a water current, and a sheet of mucus captures food in the water flowing past. The food is then passed into a gut for digestion. Filter feeding is an efficient system, as long as plenty of food floats past. Of course, if a larva becomes attached to a site where food doesn't flow past, it has blown its one chance. Adult tunicates can't detach themselves and move on. Like other kinds of sedentary animals, tunicates do two things: (1) they produce many offspring, which helps assure that at least some will attach to suitable sites and survive; and (2) they produce larval forms with sensory structures that improve chances for dispersing and discovering new sites.

Some zoologists see our ancestral beginnings in primitive tunicates. They picture ancient seas in which great numbers of larval chordates swam about. They picture further that there may have been great variation in how fast the larvae became attached to substrates and metamorphosed into adults. Rare individuals may have failed to make the change altogether, thus retaining the option of

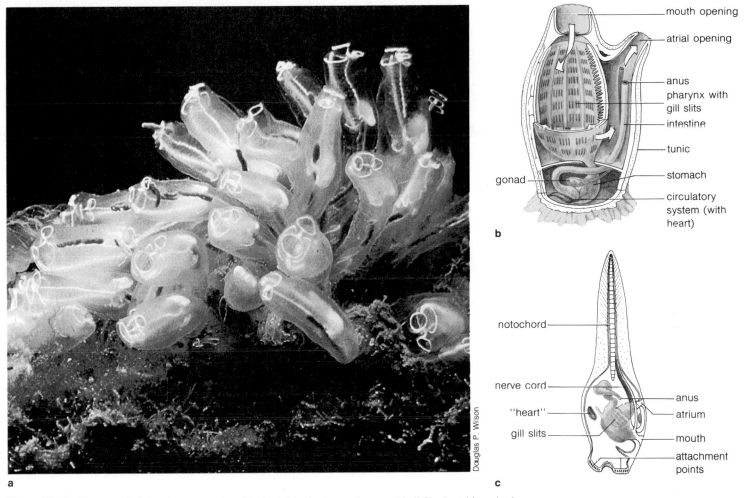

Labels on figure (b): mouth opening, atrial opening, anus, pharynx with gill slits, intestine, tunic, stomach, circulatory system (with heart), gonad

Labels on figure (c): notochord, nerve cord, "heart", gill slits, anus, atrium, mouth, attachment points

Douglas P. Wilson

a

b

c

Figure 24.15 Photograph (**a**) and cutaway view (**b**) of adult tunicates, or "sea squirts." The larval form is shown in (**c**), although it is not drawn to the same scale. (Drawings after W. D. Russell-Hunter, *A Biology of Higher Invertebrates.* Copyright © 1969 W. D. Russell-Hunter. Macmillan Publishing Co., Inc.; photograph Douglas P. Wilson)

escaping from a bad spot. Today, living in seafloor sediments throughout the world are a few animals of this sort. They are the lancelets, or *Amphioxus* (Figure 24.16).

Adult lancelets are fishlike. They retain a bilaterally symmetrical body and the capacity to swim about. They keep the notochord. On locating what appears to be a suitable feeding site, they burrow into the sediments, leaving only the mouth exposed. Then they draw in water and extract food particles from it, much as adult tunicates do. But if food doesn't float past, lancelets can swim away in search of it.

Now consider what would happen if such swimming forms sampled the seafloor as they moved about. With enough food leading them on, perhaps they would end

up keeping a roaming, searching life-style. At that point, a capacity to move effectively through tides and currents would become important. But better swimming abilities would require stronger muscles—and stronger muscles would be useless without stronger body parts to pull against. Suppose, long ago, that free-swimming chordates came to have their notochords strengthened while remaining flexible. *Through such a development, the vertebrates could have emerged.*

The first vertebrates were the **jawless fishes**. The reinforced notochord of their ancestors had been gradually supplemented with a series of hard bones, called **vertebrae**. These bones were arranged in a vertical column that formed a backbone. The vertebrae had three unique fea-

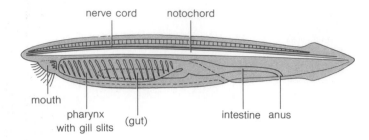

Figure 24.16 Cutaway view of a lancelet (*Amphioxus*), showing the position of its nerve cord and flexible notochord. (After J. Young, *The Life of Vertebrates*, Oxford University Press, 1962)

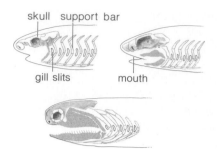

Figure 24.17 Suggested evolution of gill-supported structures, as found in jawless fishes, into the hinged vertebrate jaw. (After J. Young, *The Life of Vertebrates*, Oxford University Press, 1962)

tures. *First*, the shape provided firm attachment sites for muscles running in several directions from the backbone to the body wall. *Second*, the vertebrae fit next to each other with smooth joints, which allowed the backbone to flex and bend. *Third*, because the vertebrae were hollow, they served as a protective shield for the nervous system.

The brain region, too, came to be protected by hard structures that eventually developed into a **skull**. Gradually, modifications appeared in the vertebral column and the skull. Variations arose in the number, size, shape, and position of bones. For example, a piece of bone developed beneath the mouth, and became attached to muscles controlling mouth movements (Figure 24.17). This feature is characteristic of the **true fishes**: vertebrate animals with jaws capable of tearing off chunks of almost anything edible in sight.

The development of formidable jaws was surely an evolutionary force of the first rank. As the number of jawed fishes increased, there must have been selective advantage in being able to discern food in the distance, which could be pursued and consumed. Corresponding to the rise of these predatory, swimming vertebrates were increasingly sensitive eyes, olfactory receptors, and pressure sensors for detecting movements in the water. All such sensory structures were functionally integrated with a brain and spinal cord of impressive complexity. *The trend toward increasingly specialized sensory detectors and neural integration among the true fishes prevailed through descendant forms that moved onto land.* These trends would be the evolutionary legacy of amphibians, reptiles, birds, and mammals.

Lungs and the Vertebrate Heart

Among primitive filter-feeding chordates, gills were devices mainly for feeding. Simple diffusion of oxygen into body tissues, supplemented by some oxygen absorption through gills, was enough for these animals. But with the increasing activity of the jawless fishes came the need for more muscles. And muscle cells had to be supplied with more oxygen. Now gills became more important in respiration. Blood pumped through the thin tissue flaps picked up oxygen from water, then transported oxygen to the rest of the body. With the development of the jaw and the evolutionary lines leading away from filter feeding, the

a

b

Figure 24.18 (**a**) A green tree frog, representing the amphibian line of evolutionary development, and (**b**) a Florida alligator, representing the reptilian line.

gills came to be devoted entirely to the task of oxygenating the blood. They still do this in all true fishes.

But gills must remain moist in order to function (Chapter Sixteen). Because they are located more or less externally, they quickly become ineffective when exposed to air. How, then, did the first vertebrates begin moving onto land during Devonian times? Gradually, reliance on gills gave way to a reliance on the internal **swim bladder**. This already existing structure was being used in maintaining body position and balance in water, through adjustments in gas volume. In the moist, thin-walled air bladder, air could slowly give up oxygen to the surrounding blood and tissues. The vertebrate lung had appeared.

During this period of evolution, though, oxygen was being received simultaneously from gills *and* air sacs. In fishes that apparently gave rise to amphibians, blood traveled from gills to a two-chambered heart. The heart pumped it to the rest of the body, then the blood moved somewhat sluggishly back to the heart. The earliest amphibian hearts were equipped with a second atrium, which received oxygen-rich blood directly from the lungs. The other heart atrium received blood returning from body tissues (Figure 16.6). In this three-chambered amphibian heart, blood from both atria was forced into the same ventricle. Although the ventricle was undivided (as it is in existing amphibians), blood flow from one atrium tended not to mix with blood flow from the other. Thus oxygenated blood was sent to body tissues, and oxygen-depleted blood flowed separately to the lungs.

Later, modifications to this heart structure eventually appeared in two separate lines of reptiles. One of those lines was destined to give rise to birds as well as mammals. For these two vertebrate lines, a partition divided the ventricle into two chambers. Blood flow was now completely separated; systemic and pulmonary circulation now worked separately, in synchrony. *Thus, what began as a single circulatory system among primitive chordates became, in birds and mammals, a pair of separate circulatory systems.*

What was the significance of having separate circulatory highways? For the system to work, blood flow through merely one pair of organs—the lungs—had to be *equal* to blood flow through the rest of the body! The two atria contract as a unit, which means that the volume of oxygen-enriched blood leaving the heart for body tissues had to be the same as the volume of oxygen-depleted blood being sent to the lungs. The division into two separate routes meant that there could be less resistance to blood pressure in the lungs than in all other tissue regions. Thus blood flowed faster through the pulmonary system and kept pace with oxygen demands of the rest of the body.

On the Potential of Fur and Feather

When the world climate began to grow cooler and drier during the Cenozoic, birds and mammals rose to dominance (Chapter Twenty-One). Both groups share one feature that undoubtedly has contributed to their success. They have internal and external adaptations that help them maintain a relatively constant body temperature even when environmental temperatures rise and fall.

Birds and mammals have internal controls over the rate of metabolic activity. They also have mechanisms at the body's surface for increasing or decreasing heat loss (Chapter Seventeen). For example, complex controls governing the blood vessels in such exposed body parts as ears and feet can regulate blood flow, which influences how much body heat is lost to the surrounding air. In most land environments, air temperature is usually below the temperature at which many living cells function best (37°C to 41°C), so heat must be generated from within. Thus birds and mammals must take in food on a fairly constant basis in order to maintain a high rate of metabolism. Beyond this, birds rely on the insulative properties of feathers; and mammals, of fur. With such adaptations, these two major animals groups have radiated into almost all the diverse climatic zones on earth.

THE HUMAN HERITAGE

In these few pages, we have spanned a tremendous range of evolutionary history. We have pieced together a picture of the flow of adaptations leading to nervous systems and brains; to bilateral symmetry and segmentation, foretelling of paddles, fins, and legs; to specializations in head, trunk, and appendages. We have brought other adaptations into the picture: internal circulatory systems for supplying ever larger multicellular bodies with nutrients even while removing metabolic wastes; backbones for strength and flexibility; internal temperature control to match the changing tests of the external environment. In doing so, we have watched the sprouting of the first metazoans and their subsequent growth over 2 billion years into the mammalian stem, one of many stems on the animal phylogenetic tree. It is here that we exist, one branch among a diverse array of other mammalian forms that have in common all these structural and functional antecedents.

With apes, monkeys, and prosimians, we share further adaptations that make us uniquely **primate**. We have a large cerebral cortex, highly integrated systems of muscles and nerves, and flexible shoulder joints. We have forward-directed eyes astonishingly good at discerning color, shape, and movement in a three-dimensional field; we have fingers and toes adapted for grasping rather than running. We have teeth of a varied sort, capable not only of tearing in the manner of jawed fishes but of piercing, crushing, and grinding. We give birth not to large litters but generally to only one offspring at a time.

When and where did these adaptations develop? Here the evolutionary story picks up with a small, perhaps insect-eating mammal that lived in tropical forests at the dawn of the Cenozoic, some 65 million years ago. It surely evolved from the ratlike mammals scurrying over the Mesozoic landscape. But its move to life in the trees called for grasping digits and flattened fingernails rather than pointed, ratlike claws. It called for greater hand-eye coordination and forward-directed eyes for stereoscopic vision, hence better depth perception. Climbing and, much later, reaching for and hanging onto overhead branches were adaptations that led all primate progeny to view the world from an upright position. Long before the striding, two-legged walkers called humans appeared, selective agents were testing out the adaptations of tree-dwelling primates that would make possible the eventual long-range migrations of the human species.

These tree-dwelling primates got around. By the Miocene, apelike forms known as the **dryopithecines** were well represented in parts of Africa, Europe, and Asia. Some seem to qualify as the stock from which modern humans and apes eventually arose. If anything, the dryopithecines were transitional forms. They were capable of climbing about in the trees, yet also capable of walking about on the ground—explorers, perhaps, ever ready to seek the protection of the branches above. Interestingly, tree-dwelling apes alive today are typically plant eaters, gentle in disposition and hiding in the forest canopy rather than venturing into the open. But apes living in open grasslands are of a different sort. They eat plants but they also eat meat—and they are far more aggressive. (For instance, a chimpanzee that is threatened or on the attack is quite capable of throwing stones and making a general ruckus.) In the dryopithecines, we have a possible antecedent for both.

Geologic upheavals of the late Cenozoic may well have been the environmental trigger for the sudden adaptive radiation of many primate lines out of the forests. By that time, Old and New World primates had already diverged and were going their separate ways even as the continents of Africa and South America were going their separate ways (Chapter Twenty-One). But to the east in Africa, repeated crustal fracturing was creating the Great Rift Valley (Chapter One), flanking it with volcanic mountain ranges and carving it up with lakes, swamps, and rivers.

Within this immense valley, a mosaic of geographically isolated environments emerged—cloud forests on wet mountain slopes, tropical forests, savannas, open plains, deserts. And if we know anything at all about environmental diversity, it is that it often is paralleled by diversity in life forms. The microcosms of the Great Rift Valley surely fostered distinct adaptations in the small, scattered bands of early apes.

The fossil record in East Africa does in fact point to a very rapid evolution of apelike forms during this period. Isolation in one region beset by alternating wet and dry climates apparently led to the chimpanzees. Their tree-dwelling, half-terrestrial way of life may well reflect the changing environments to which their ancestors had to adapt and readapt. Isolation in another region of continually warm, humid forests apparently led to the evolution of the gorilla. And in regions of increasing aridity, we have fossil signs of primates in the human family—the first **hominids**. They may have been already adapted to foraging on the ground for seeds, fruits, nuts, and the remains of small animals when they were forced to move into open grasslands. Were they already standing on two legs? Possibly. There would be a strong selective advantage bestowed on those forms which could stride about, with eyes on the horizons not only for potential food but for the carnivores—leopards, lions, tigers, hyenas, jackals—that preyed on them. More than this, an upright stance would free the hands for defense if one were caught out in the open. The first hominids never abandoned their foraging habits. But through use of hand-held objects—protoweapons, prototools—they also developed the capacity to hunt. Hominids had become both predator and prey.

The earliest indisputable hominids were the **australopithecines**, the "southern ape-men" that apparently emerged during the Pliocene (Figure 24.19). These forms at first had an average brain size no larger than a chimpanzee's. Somewhere around 2 million years ago, there were also members of the genus **Homo**—of which we are the only living representatives. Their brain size was intermediate between that of most australopithecines and **Homo erectus**—a later form with jaws and teeth not all that different from modern hominids. During the Pleistocene, at least two species seem to have been coexisting in the same regions. They could not have done so and remained distinct species unless they were not competing for the same things. One, derived from the early australopithecines, was a scavenger and gatherer, perhaps of seeds from wild grasses and trees. The other, *Homo erectus*, was becoming specialized as a tool-using food gatherer—and hunter. Perhaps it was that the smaller sized and smaller brained species came to be recognized as potential food by the hunters. Whatever

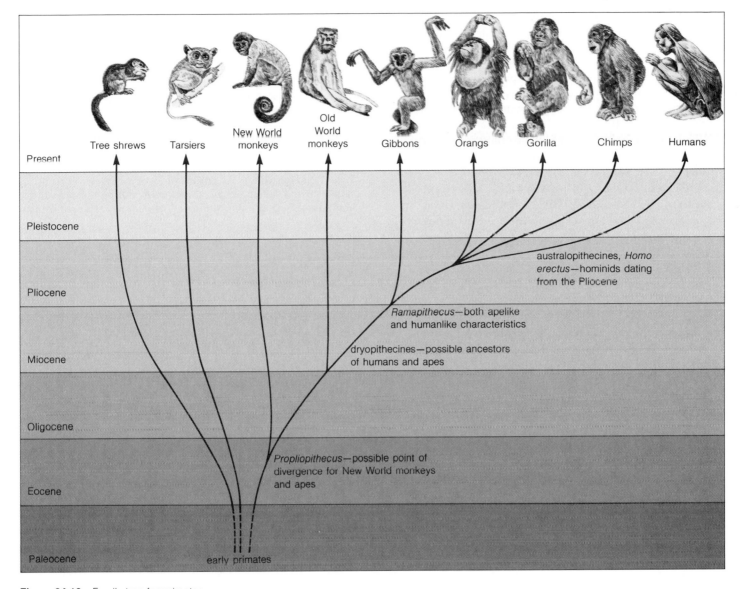

Figure 24.19 Family tree for primates.

the case, by 1.2 million years ago the scavengers had become extinct.

Homo erectus populations evolved past them, and not only in the physical sense. Their repertoire of cultural adaptations included clothing made of animal pelts and the use of fire as well as various types of weapons. With such adaptations they fanned out from Africa to Europe and Asia. All the while, their brain size became increasingly more modern as survival came to depend more on intelligence and the capacity to accumulate and transmit experience. The **Neanderthalers**, a distinct hominid population appearing about 100,000 years ago in Europe and Asia,

were one of our closest relatives. They may have developed in the relative isolation of Europe, cut off from the rest of hominid evolution by the Pleistocene glaciations. They developed a distinctive and considerably complex culture. Their cranial capacity was indistinguishable from our own. Yet, about 30,000 years ago they seem to have disappeared. Did they simply merge with the overall human population when the retreat of the glaciers again permitted migrations, or were they overwhelmed in an early show of our capacity for genocide? We do not know. But by 30,000 years ago, there was one remaining hominid species—**Homo sapiens**, man the reasoner, self-proclaimed man the wise.

PERSPECTIVE

Many thousands of years ago, as human bands radiated across the continents, they began their increasing manipulations to bring both themselves and their environment closer to their visions of a better life. As visions evolved, so did these dispersed human groups evolve in culturally divergent ways. Even while geographic isolation led to different adaptations in superficial physical traits, so also did it lead to the divergence of habits, opinions, adornment, songs, words, foods, and myths. All these evolving forms of distinct behavior might be called the beginnings of cultural speciation. But biological speciation did not follow in its wake. Physical barriers proved permeable, gene flow between populations continued despite cultural barriers, and we have remained one.

Too often we view culture as an inviolate gift separating ourselves and others—with "others" meaning not only members of regionally distinct human groups but all non-human forms of life. What is it about culture that makes us think it comes mysteriously into being on its own? Culture is no more and no less than a culmination of the long evolutionary processes we have been describing throughout this book. It is an expression of a trend toward behavioral diversity and versatility that appears to have begun in the Precambrian seas. Culture did not come suddenly into being with *Homo sapiens*. In the most basic sense, it is simply an ability to accumulate experience, to modify present and future behavior in terms of the past, to learn from the environment and anticipate it. Since the development of the first nervous systems, animals of many sorts have been traveling the route toward increasing reliance on learned behavior. You will be reading about some of these animals in the next unit. For them, each increment in the capacity for learning has proved adaptive and has been selected for. Long before the appearance of humans, other animals developed the capacity for local traditions, so to speak—for behavior learned by direct observation and imitation, then transmitted by repetitive behavior in overlapping generations of social groups. Such traditions are, in a sense, protocultural.

We have evolved far beyond that point, of course, for we have developed the capacity to communicate with symbols. We combine and recombine thoughts in a dream-like interplay of images in our head. Some we translate into artifacts as lasting as the pyramids, others as fleeting as words uttered into the wind. It is a remarkable degree of behavioral evolution we share with no other life form—but it is, in the final analysis, an outgrowth of structures and processes we share with others traversing the long road of animal evolution. It is a magnificent heritage.

Readings

Campbell, B. 1974. *Human Evolution: An Introduction to Man's Adaptations*. Second edition. Chicago: Aldine. Marvelous comparative treatment of hominid form and function; well written and certainly provocative.

Hickman, C., et al. 1979. *Integrated Principles of Zoology*. Sixth edition. St. Louis: Mosby.

Russell-Hunter, W. 1968. *A Biology of Lower Invertebrates*. New York: Macmillan. An excellent little paperback covering biological unity and diversity from coelenterates to mollusks.

———. 1969. *A Biology of Higher Invertebrates*. New York: Macmillan. Same skilled writer, this time exploring the physiology and adaptations of animals ranging from annelids to sea squirts.

Review Questions

1. In what ways are sponges unlike other animals? In what ways are they similar?

2. Can you put together a scenario based on the planula theory of origins for bilaterally symmetrical animals? How do you suppose this theory helps explain why you happen to have two eyes instead of one (like the mythical cyclops), two arms, two legs, two ears? Why one mouth instead of two? Why brain tissue concentrated in your head instead of strung out, noodlelike, through your elongate body?

3. In the anatomical sense, are flatworms more complex than sponges? Than jellyfish? Than earthworms?

4. Nematodes are the most far-flung of all animal groups. What might account for their adaptive success in so many different environmental settings?

5. Flatworms are thought to be living representatives of ancient organisms from which two main lines of animals diverged long ago. Name these two evolutionary lines, and state which of the existing animal phyla each includes. If this divergence did indeed occur, then are insects or echinoderms closer to you on the evolutionary tree?

6. Explain what role the following developments probably played in the evolution of larger and more complex forms of animals: coelom, circulatory system, segmentation, lungs.

7. Arthropods include insects, crustaceans, arachnids, and their kin. Describe the basic body plan that characterizes these diverse organisms. How might some of these features have helped some of their ancient aquatic ancestors radiate onto land?

8. Mollusks include forms ranging from small soft-bodied slugs to hard-shelled clams to giant squid that may be as long as a six-story building is tall. What body features do such diverse animals hold in common?

9. What are some of the characteristics we share with other primates—with apes, monkeys, and prosimians? Using the evolutionary perspective, can you suggest how these shared features may have evolved since the dawn of the Cenozoic? As part of your story, explain how geologic upheavals of the late Cenozoic may have triggered diversification among our closest relatives.

UNIT SEVEN

LIFE IN A CHANGING ENVIRONMENT

25

INDIVIDUALS, POPULATIONS, AND EVOLUTION

Where would the mallard duck be now if its destiny had been placed in the hands of that prominent seventeenth-century cataloger of life, Carl von Linné? There, awaiting classification, was a bird with emerald-green head feathers as well as wings emblazoned with metallic blue patches. There, living in the same ponds and marshes, was a drab little brown-feathered duck bearing no obvious resemblance to the more resplendent waterfowl. Thus did von Linné, on the basis of appearance alone, pronounce the male and female mallard duck as separate species. (It goes without saying that the male and female duck, paying no attention whatsoever to his pronouncement, continued to produce more ducks.)

Now, you may be thinking that it is somewhat unfair to von Linné to dredge up one of his less-than-brilliant insights. But put yourself in his shoes and suppose you had come across the male and female mallard duck for the first time. How would you go about classifying them? The point is, anybody armed with the notion of rigid species slots is going to encounter some problems. Recall, from Chapter Two, that studies of life's diversity were once funneled through the concept that species don't change. The approach was *typological*: an individual was selected as being the perfect type for the species, based on a somewhat arbitrary choice of what "perfect" physical features were. Then individual specimens were compared against this standard to determine if they were the same kind of thing. Small variations among similar individuals were viewed as imperfect renditions of the perfect species plan. Dramatic variations in physical appearance were often viewed as evidence of different species statuses (hence the divorce of the ducks).

Today, you will find that this somewhat rigid approach is being set aside for something called the population concept. It holds that variation is far from unimportant, but rather is the force and the product of evolutionary change. Before exploring this concept, you may find it useful to consider in advance the most general definitions of individuals, populations, and species:

Individual. An individual is an organism having its own set of genes; it is a single genetic unit.

Population. A population is a group of individuals of the same species, occupying a given area at a given time.

Species. In general, a species encompasses all of those actually or potentially interbreeding populations that are reproductively isolated from other such groups.

Exceptions do exist, but these definitions are still a good starting point for the discussions to follow.

VARIATION: WHAT IT IS, HOW IT ARISES

You know, just by thinking about people around you, that individuals differ in their appearance; they show *variation* in such physical traits as leg length. For a given population, it's possible to measure the trait for each individual and determine the average leg length, along with the degree of variance from that value. This gives a *frequency distribution* for the trait. As you will see, evolution means changes in frequency distributions over time. These changes come about through differences that arise in genotype (the genetic basis for an individual's traits). They also arise through differences in environmental conditions which, through interactions with genotype, influence the individual's physical appearance (phenotype). The environmental effects on gene expression were described in Chapter Nine. Here, the initial focus is on some mechanisms that may cause changes in the relative abundance of genotypes.

Hardy, Weinberg, and Pools of Genetic Diversity

The sum total of all the genotypes in a given population has traditionally been called a gene pool. For populations of sexually reproducing individuals, it is more accurate to think of it as a pool of alleles. (An *allele*, recall, is an alternative form of a gene at a given locus.) When you take the whole population into account, you find that for any one locus, some alleles occur more often than others. Thus it's possible to think of variation in terms of **allele frequencies**: the relative abundance of different alleles carried by the individuals in that population.

Early in this century, the mathematician G. Hardy and the physician W. Weinberg independently came up with a simple rule that can be used as a baseline against which changes in allele frequencies may be measured. They proposed the following:

Figure 25.1 Population of snow geese at their wintering ground in New Mexico. In this unit, we leave behind the levels of cells and organisms, and turn to the questions of what populations are in time and space.

Figure 25.2 Hardy–Weinberg equilibrium. Evolution may be defined as change in allele frequencies over time. In itself, the *shuffling* of alleles that occurs among sexually reproducing individuals isn't evolution. Consider that the body cells of a diploid individual generally have two alleles at each gene locus, and that three allelic combinations are possible for the locus. Thus the individual can be homozygous dominant, heterozygous, or homozygous recessive. To determine whether the proportions of alleles in a population change through sexual recombination, let

$$p + q = 1 \qquad \text{(equation one)}$$

where p = the frequency of one kind of allele (A, for instance)
q = the frequency of the alternative allele (a)

This equation simply says that, within the population, all the alleles at this particular gene locus can be accounted for by the dominant A alleles plus the recessive a alleles. However, alleles are not present by themselves in normal diploid individuals. Rather they occur in pairs. During sexual reproduction, the two alleles at a gene locus normally end up in separate haploid gametes. When the random fusion of gametes brings together two dominant alleles ($A \times A$), we have homozygous (AA) dominant individuals. If A alleles are present in the population at a frequency designated p, then the frequency of the homozygous dominants will be

$$p \times p = p^2$$

When the fusion of gametes brings together a dominant and a recessive allele ($A \times a$), we have heterozygous (Aa) individuals; and their frequency within the population will be

$$(p \times q) + (q \times p) = 2pq$$

(Here, the "2" refers to the doubled chance for an encounter: dominant male gametes might fertilize recessive female gametes; and dominant female gametes might be fertilized by recessive male gametes. Refer back to the Chapter Nine discussion of probability.)

Finally, when the fusion of two gametes brings together two recessive alleles ($a \times a$), we have homozygous recessive (aa) individuals; their frequency within the population will be

$$q \times q = q^2$$

Using the three frequencies given, we can expand equation one to show the relative *frequencies* of the three diploid *genotypes* in terms of the frequencies of the alleles:

$$p^2 + 2pq + q^2 = 1 \qquad \text{(equation two)}$$

where all the homozygous dominants, plus heterozygotes, plus homozygous recessives equal the genotypes for the whole population.

Now, suppose that we begin with a population of 1,000 individuals including the following genotypes:

$$
\begin{array}{l}
450\ AA \\
500\ Aa \\
\underline{50\ aa} \\
1,000\ \text{individuals}
\end{array}
$$

Then theoretically, for each 1,000 gametes produced, 700 will be A (450 + 500/2), or $p = 0.7$. And 300 will be a (50 + 500/2), or $q = 0.3$.

Notice that $p + q = 0.7 + 0.3 = 1$.

After one round of random mating, the genotype frequencies in the next generation will be the following:

$$AA = p^2 \quad = 0.7 \times 0.7 = 0.49$$

$$Aa = 2pq = 2 \times 0.7 \times 0.3 = 0.42$$

$$aa = q^2 \quad = 0.3 \times 0.3 = 0.09$$

and

$$p^2 + 2pq + q^2 = 0.49 + 0.42 + 0.09 = 1$$

Notice here that the *genotype* frequencies have changed (remember that the starting frequencies in this hypothetical population were $AA = 0.45$, $Aa = 0.50$, and $aa = 0.05$). But notice also that the *allele* frequencies have not changed:

$$A = \frac{2 \times 490 + 420}{2,000 \text{ alleles}} = \frac{1,400}{2,000} = 0.7 = p$$

$$a = \frac{2 \times 90 + 420}{2,000 \text{ alleles}} = \frac{600}{2,000} = 0.3 = q$$

(Two alleles per diploid individual give us the 2,000 alleles used in the calculations.)

When allele frequencies for a given gene locus remain constant through succeeding generations, the population is said to be at *Hardy–Weinberg equilibrium*. However, Hardy–Weinberg equilibrium is precisely maintained *only* when mating is random (say, an AA male couldn't care less whether the female is AA, Aa, or aa), when mutations don't occur, when there is no gene flow between different populations, when the population is quite large, and when the viability or fertility of one genotype is not different from the others. Rarely are all these conditions met. But many populations are close enough to the ideal that equilibrium equations can be used to good effect.

If all individuals in a population have an equal chance of surviving and reproducing, then the frequency of each allele in the population should remain constant from generation to generation.

This prediction is based on a mathematical formula called the **Hardy–Weinberg rule** (Figure 25.2). You can determine its meaning in a simple way. Look at Figure 25.3, which shows one of Mendel's monohybrid crosses. Notice that as you move from the true-breeding parents to the F_1 generation, then to the F_2 generation, the frequency of the two alleles A and a remains the same. In each generation, half the alleles are A, and half are a. You could go on following the inbreeding of pea plants until you ran out of paper, or patience. As long as you adhered to the one stated condition—that in every generation, *all* individuals in the population have equal probability of reproducing—you would end up with the same result. You would get the same results even if you started out with a population in which the ratio of alleles was not 1:1. For instance, if the population has 90 percent A alleles and 10 percent a alleles, the ratio would remain 9:1 in every generation. Such stability of allele ratios, which would occur if all individuals have equal probability of surviving and reproducing, is called **genetic equilibrium.**

Genetic equilibrium implies a static state. It is a theoretical baseline only. It very rarely exists in nature, where allele frequencies may change all the time. Let's look at some disturbances that may cause changes.

Mutation

A mutation, recall, is a heritable change in the number, kind, or internal structure of chromosomes. In any population, mutations arise very infrequently *but inevitably*, by chance. These spontaneously occurring events, which bring about small changes in allele frequencies, are the original source of all genetic variation. (Sometimes recombination is said to be another original source of variation. But mutations must first give rise to any variant allele that gets shuffled during meiosis. Besides, with each new generation, alleles are shuffled into *new* combinations, so the effects of recombination are mostly short-lived.)

Through mutations, individuals come to vary in structure, function, and behavior. Some mutations have profound effects on survival. For instance, mutation in a key gene, such as the one governing the main protein component of bones, may have drastic repercussions through

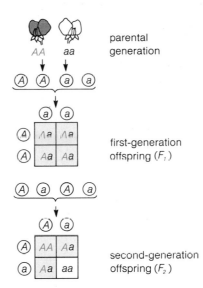

Figure 25.3 The meaning of the Hardy–Weinberg rule, as illustrated by a monohybrid cross between pea plants. If all plants have equal chance of surviving and reproducing, the frequency of the two alleles, A and a, doesn't change through the generations (in this example, half always stay A, and half always stay a).

the developmental program (Chapter Nine). However, other mutations may make their bearers more suitable for operating under existing conditions—for responding effectively to physical aspects of the environment, to predators or prey, to other members of the species. Still other mutations may improve chances for surviving and reproducing if conditions should undergo long-term change. In either case, an advantageous mutation may become fixed in the population. The important point is this:

Over time, allele frequencies tend to change through infrequent but inevitable mutations, which are the wellspring of genetic variation.

Genetic Drift

It also happens that the relative proportions of different alleles may change at random from one generation to the next, simply by chance. Random fluctuation in allele frequencies over time, due to chance occurrence alone, is called **genetic drift**.

When population size is small, the actual frequency of an allele may drift considerably over time from the value that might be expected of a large population. By analogy, there are two outcomes when you toss a coin, each with equal probability of occurring. When the coin is tossed ten times, it could by chance end up tails eight times and heads twice, rather than the ideal 1:1 ratio. However, the laws of probability (Chapter Nine) tell you that when the coin is tossed, say, a thousand times, there is a better chance of approaching the ideal ratio. That is why the effects of genetic drift may be more pronounced in small populations than in large ones. In such small groups, allele frequencies may so change because of accidents that one allele may be completely lost, leaving an alternative allele fixed in the population. The same thing can happen in moderately large populations, though, if an allele is present in only one or a very few individuals.

Even if an allele doesn't contribute to making its bearers suited to prevailing conditions, genetic drift may bring it to preponderance. Such alleles are thought to have a *neutral* effect on survival and reproduction. Genetic drift may even fix a maladaptive allele in the population. For instance, suppose neutral or maladaptive alleles just happen to be closely linked to highly adaptive genes in a chromosome. Then they could still be perpetuated, through meiosis and sexual fusion, and ride on the coattails of the adaptive genes through environmental tests. These possibilities bear importantly on our view of life's diversity, for they suggest that the following may be true:

Some alleles in a population may not be the "best" possible alternatives in contributing to survival and reproduction; through genetic drift, they may have become established just because "better" alternatives were lost by accident or chance.

Gene Flow

Allele frequencies also change through gene flow associated with immigration and emigration. New members may join a population (immigrate), some members may leave (emigrate) for one reason or another. For example, baboons in Africa generally live in troops. Each troop represents a separate allele pool. Commonly, some of the males wander off or are kicked out of one troop (Chapter Twenty-Seven). They may join up with another troop some distance away. Assuming that migrant males encounter receptive females in the newly joined troop, the pool of alleles changes, just as it changes in the troop they left behind. This example reinforces the following point:

Allele frequencies may change when individuals of a species leave or enter a separate population, taking or bringing their particular genotype with them.

Selection Pressure

Finally, and most importantly, natural selection occurs in populations. Individuals vary in genotype and phenotype. This means that they simply are not equipped in quite the same way to deal with environmental challenges. Some will be more suitable, under current conditions, for surviving and reproducing. If their phenotypic differences are genetically based, then the advantageous genotype will tend to occur in more and more individuals with each generation. In this sense, *natural selection is differential survival and reproduction of genotypes within a population.*

Selection pressures may so favor one allele that the allele completely supplants all others for the gene locus. Thus the same form of a trait appears among all individuals. Often, though, two or more alleles may persist at the same time. And they may persist at a frequency that is greater than can be accounted for by newly arising mutations alone. Populations in which two or more forms of a trait persist are said to show **polymorphism** at that gene locus. The section to follow includes examples of polymorphism. In considering the examples, the important points to remember are these:

When individuals of different phenotypes in a population differ in their ability to survive and reproduce, their alleles are subject to natural selection.

When individuals of different phenotypes survive and reproduce at about the same rate, their alleles may not be subject to natural selection; the alleles may be neutral relative to one another.

SELECTION WITHIN POPULATIONS— SOME EXAMPLES

It is important to understand that natural selection is no longer in the realm of pure theory; it has now been observed in many cases. Here we will consider examples of directional, disruptive, and stabilizing selection. We will also look at an example of balanced polymorphism.

a   moth   lichen   moth   b   moth   moth

Figure 25.4 An example of directional selection. (**a**) The light- and dark-colored forms of the peppered moth are resting on a lichen-covered tree trunk. (**b**) This is how they appear on a soot-covered tree trunk, which was darkened by industrial air pollution in certain parts of England.

From the experiments of Dr. H. B. D. Kettlewell

The Peppered Moth

Consider, first, an organism that was mentioned briefly in Chapter One. Populations of *Biston betularia*, the peppered moth, are widely distributed in England. This moth flutters about by night. But during the day, it rests on tree trunks and branches—where it is highly vulnerable to bird predators. Now, within these populations are two polymorphic types, which arise through different allele combinations at one gene locus. One type has speckled light-gray wings. The other, a result of a once-rare mutant allele, has black wings and body.

Before the industrial revolution, the speckled gray moth was the most commonly occurring form. At that time, lichens grew profusely on trees. Speckled moths resting on the lichens were inconspicuous to bird predators, which apparently are the main selective agents working on the moth populations. The dark moth occurred only rarely; it stood out conspicuously on lichens (Figure 25.4). Between 1848 and 1898, though, the dark form increased to *ninety-nine percent in frequency* in certain rural districts—and the speckled form in those districts became exceedingly rare! Industry had moved into these districts, and smokestacks belched soot and other pollutants into the air. Among other

things, the pollutants blackened trees and killed the lichens. With the environment so changed, the previously non-adaptive dark form gained advantage; it blended with the darkened trees! Today, there are controls on air pollution. As a result, tree trunks are less sooty, lichens are becoming reestablished—and the dark forms are selected against.

The transient polymorphism observed among peppered moths appears to be a case of **directional selection**: because of a specific change in the environment, a heritable trait occurs with increasing frequency and the whole population tends to shift in a parallel direction (Figure 25.5).

Pesticide-Resistant Pests

Insect populations that develop resistance to insecticides are another example of directional selection. With the first application of an insecticide, most of the insects it is designed to kill are indeed killed. But variant forms (mutants) in the population may survive because of physiological differences that enable then to resist the insecticide's chemical effects. If the resistance has a genetic basis, it is a trait that is passed on. Then the next generation will contain more of the resistant individuals. Increasing

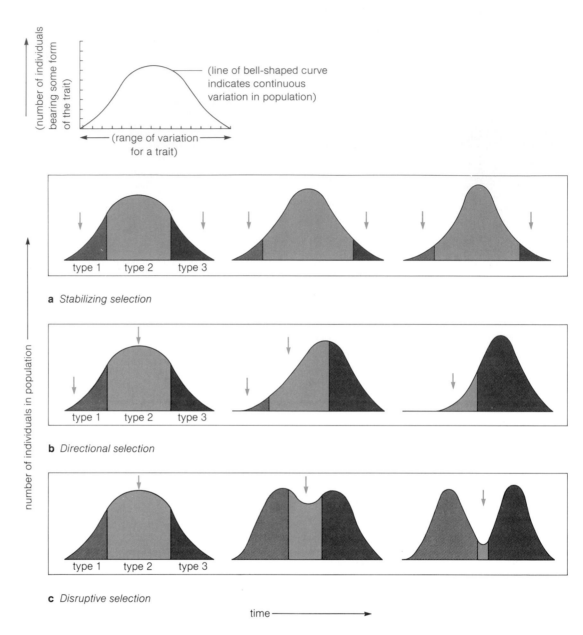

Figure 25.5 Change in populations as a result of selection pressure. Three polymorphic types are used in each example. Types 1 and 3 are extreme variants; type 2 is the common form in the population. (**a**) In stabilizing selection, conditions favor the most common form. The downward-pointing arrows signify that the variant forms are being selected against. (**b**) In directional selection, the character of the whole population moves in a direction corresponding to environmental change. (**c**) In disruptive selection, two or more variant forms are favored and become increasingly represented in the population. (The model graph above these three examples describes the two variables that give rise to their bell-shaped curves.)

numbers of resistant insects often lead to heavier and more frequent applications of the insecticide—which acts as a selective agent that favors the resistant individuals even more. Gradually, the genotypic structure of the population shifts due to differential survival and reproduction.

(Recognition that this kind of directional selection exists has brought about changes in pest control efforts. For instance, many programs now include biological controls. Natural predators may be released at critical points of the pest's life cycle. Sometimes hordes of artificially sterilized males are released during the breeding season, to occupy the egg-laying females that would otherwise interact with normal males and produce offspring.)

Deceptive Female Butterflies

In **disruptive selection**, two or more distinct polymorphic varieties are favored and they become increasingly represented in a population. Unlike directional selection (in which the character of the population shifts as a whole), disruptive selection tends to split up a population. Here we may consider a remarkable case of mimicry among *Papilio dardanus*, one of the swallowtail butterflies found in Africa. (In **mimicry**, one kind of organism bears deceptive resemblance in color, form, or behavior to another kind of organism that has a selective advantage. The first is the mimic, the second is the model. For instance, if a model has warning coloration that identifies it as inedible to potential predators, then a tasty species that mimics it will have a better chance of being left alone, too. The English naturalist Henry Bates was the first to point out that such protection-by-deception would have advantages, hence the phenomenon is known as Batesian mimicry.)

The males of *P. dardanus* populations are not the mimics. Presumably, it's more advantageous for the males to look like themselves and not like males of other species during the breeding season, even if they are conspicuous to predators. The likely selective advantage is the greater recognition by females, hence greater reproductive potential in having unambiguous form and coloration. However, not only one but several female varieties have wing patterns and coloration that mimic those of inedible species found in various regions. Each distinct female variety bears striking resemblance to a different foul-tasting species of butterfly present in its distribution range (Figure 25.6). Around Entebbe, near Lake Victoria, only four percent of the *P. dardanus* females are intermediate forms that don't bear any resemblance to inedible species. And in places where models are absent, so are the mimicking females. Predators are almost certainly the selective agents promoting disruptive selection in these populations.

A Continuum of Crabs?

By now, you may be thinking that it takes a changing environment to trigger selection processes. But even when an environment remains much the same, there may be selection for the *existing* species form—which has proved to be adaptive—and selection against extreme variant forms that do appear. For instance, plowing across the sand in shallow waters off the northeastern Atlantic coast is a

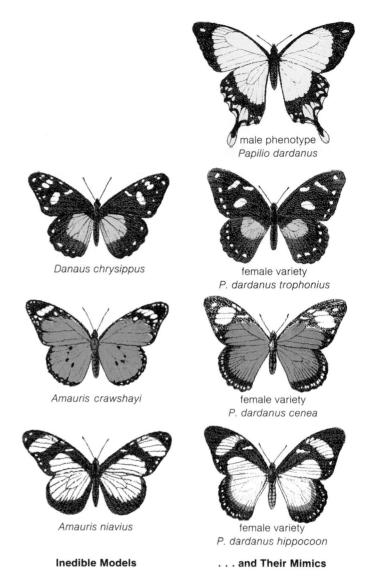

male phenotype
Papilio dardanus

Danaus chrysippus

female variety
P. dardanus trophonius

Amauris crawshayi

female variety
P. dardanus cenea

Amauris niavius

female variety
P. dardanus hippocoon

Inedible Models   **. . . and Their Mimics**

Figure 25.6   Batesian mimicry among females of the African swallowtail butterfly. Disruptive selection is apparently favoring several phenotypic variations, as the text describes. (The males and females of the models look alike. The female, not the male, swallowtail butterflies resemble the models.)

dark-brown horseshoe crab (*Limulus polyphemus*). Its distinctive, somewhat flattened shape (Figure 25.7) is well suited for pushing through mud and sand as the animal scavenges for food, and for affording it protection from predators. Apparently horseshoe crabs are extremely well adapted to such environments. The fossil record indicates that for more than 250 million years, the form of horseshoe crabs has remained essentially the same. This kind of organism probably represents a balanced system of the best working combination of traits in an environment that has remained

a

b

Figure 25.7 Two hundred and fifty million years of stabilizing selection? (**a**) Body casting of a horseshoe crab made that long ago—imprints that could well be made by a modern-day horseshoe crab (**b**), here shown mating and perpetuating the general species form. (a Copyright © 1975 Peabody Museum of Natural History, Yale University)

relatively stable. Presumably, extremely variant individuals were continually eliminated, so that little structural change has occurred through evolutionary time. A process in which a form already well adapted to a given environment is selected for and maintained, even as extreme variants are selected against, is called **stabilizing selection** (Figure 25.5).

The Other Side of Sickle-Cell Anemia

Normally, the red oxygen-carrying pigment in blood of most humans is hemoglobin A, or HbA. A variant form, HbS, results from a single amino acid substitution in a gene coding for hemoglobin. Individuals who are heterozygous for the trait suffer few consequences. But recessive homozygotes may suffer severely from sickle-cell anemia (Chapter Nine, Figure 9.17).

If the HbS mutant allele causes such severe abnormalities, why didn't the action of natural selection remove it from the population long ago? The reason is that variant alleles may not be advantageous or disadvantageous in themselves. *Their survival value must be weighed in context of the environment in which they are being expressed.*

In regions where the death rate from malaria has been high, HbA/HbS individuals held a heterozygous advantage over either kind of homozygote. Malaria is caused by parasites that attack red blood cells—and individuals carrying the HbS allele have increased resistance to the disease. Such individuals are especially resistant to the more dangerous forms, caused by a parasitic sporozoan *(Plasmodium falciparum)*. What is the physiological basis for the resistance? It may be that the parasite has difficulty metabolizing hemoglobin S when sickled red blood cells are clumped in capillaries. (Clumping impairs the ability of these cells to acquire oxygen and get rid of carbon dioxide wastes.) Whatever the basis, heterozygous (HbA/HbS) individuals have an advantage if they do contract malaria, for they have a greater chance of surviving the parasitic attacks than does either kind of homozygote.

In West and Central Africa, there was a clear relationship between the incidence of the more lethal forms of malaria and the HbS allele frequency. The relationship was evident for at least seventy generations. It probably began when agriculture was introduced into the region, more than 3,000 years ago. The increased food production provided a basis for population growth. Before that time, the mutant allele may have been rare in the population living in the region. Afterward, tropical forests were cleared, people became concentrated in villages—and malaria began to spread. As the disease became more prevalent, more HbA/HbA individuals died from it than did heterozygotes before reaching reproductive age. In addition, HbA/HbA women had a slightly lower fertility rate, perhaps because malarial parasites also attack the placenta and can bring about miscarriage. Thus the normal homozygotes contributed fewer genes to succeeding generations—which meant that the HbS allele *frequency* began to rise proportionally in the population. Soon, deaths of HbA homozygotes because of malaria were balanced by deaths of HbS homozygotes brought about by sickle-cell anemia. In this manner, balanced polymorphism at the sickle-cell locus was maintained in populations throughout the region for as long as malaria was prevalent.

Malaria has since been brought under control, more or less. (New strains, alarmingly resistant to antibiotics, are appearing.) In addition, carriers of the mutant allele have emigrated to many regions that are free of this tropical disease. In the absence of malaria as a selective agent, the HbA/HbS heterozygous advantage is diminished. Individuals afflicted by sickle-cell anemia unfortunately must await the development of phenotypic "cures" or, further ahead on the horizon, genotypic cures (Chapter Ten).

DIVERGENCE BETWEEN POPULATIONS AND THE EVOLUTION OF SPECIES

So far, we have considered how allele frequencies within a population can change as a result of various disturbances. In fact,

---

Evolution may be thought of as successive changes in allele frequencies brought about by such occurrences as mutation, genetic drift, gene flow, and selection pressure.

But how do we go about identifying such changes through time in natural settings? To be sure, a population may be sufficiently small and isolated so that we can readily determine its character—its number of individuals, its density (how many individuals are present in a given area), its distribution range, and its general patterns of phenotypic variations. For example, we may be able to find a tiny, isolated population of snails on a remote, small island, or an isolated population of trout in a mountain lake. But often populations span a large geographic range, without interruption. Some grasses of the Great Plains form such populations; so do some grasses of the Argentine plains (Pampas), which stretch from the Atlantic coast to the foothills of the Andes. Still other populations form a long,

more or less continuous narrow band along rivers and seashores.

Trying to track changes in allele frequencies through the Argentine Pampas or a similarly extensive population would be physically impossible. (Can you picture yourself counting and describing the phenotypic character of each new individual grass plant?) Usually the focus of research is on some **local breeding unit**: a population of localized extent within a larger population system. This is the kind of unit we will be looking at here.

Within a single population, shifts in allele frequencies may move the whole population in one direction or another. But sometimes barriers arise between parts of the population, and local breeding units are formed. These barriers may be enough to put an end to gene flow between them. The local breeding units may then diverge genetically from each other. Over time, they may even change enough to give rise to new species. Let's look first at the kinds of barriers that arise, then at the kinds of reproductive isolating mechanisms that can follow in their wake and set the stage for speciation.

Barriers to Gene Flow

The usual barriers to allele exchange between local breeding units of a larger population are those which create **geographic isolation**. Such separation may occur when a few individuals are dispersed far beyond the larger population's normal distribution range. Geographic isolation has also occurred during the millions of years that went into uplifting mountain ranges or moving entire continents apart. It may also occur suddenly. For instance, a population of crawling insects might be split by a change in a river's course. This sometimes happens during major floods or earthquakes. If the individuals so separated can't fly, swim, or float, they are as effectively barred from interbreeding as if mountains stood between them.

Following geographic isolation, it may be that mutation, genetic drift, and selection pressures operate on the different local breeding units. Certainly mutations won't be precisely the same in these isolated populations. It may also be that the physical environment and the array of organisms within it are different in some small or large way, thereby representing different challenges for survival and reproduction. If the populations are very small, genetic drift may have major consequences on allele frequencies. Over time, these forces may lead to **divergence**: a buildup of differences in allele frequencies (or genetic differentiation) between isolated populations of the same species.

Sooner or later, the genotypic differences may directly or indirectly prevent the exchange of genetic information between populations. Perhaps the DNA of parents from different populations can no longer match up, point by point, during meiosis. Perhaps the genetic variation gives rise to reproductive structures or behaviors that don't mesh. Whatever the differences may be, divergence and genetic differentiation have produced new species.

In general, speciation is said to have occurred when some restriction on interbreeding between populations is followed by enough genetic differentiation that, even if individuals from those populations do make contact, they cannot or will not interbreed.

Isolating Mechanisms

The term **isolating mechanisms** refers to some aspect of structure or functioning (as well as living habits) that may prevent interbreeding between populations that are undergoing or have undergone speciation. Partial reproductive isolation indicates that a population may already have embarked on its own genetically unique course. Complete reproductive isolation is evidence that speciation has occurred. Different isolating mechanisms may act singly or in concert with others. Some of the more important ones are outlined here.

Incompatibilities in the structure or functioning of reproductive organs may prevent individuals of different populations from producing fertile offspring. For example, a pollen tube is a reproductive organ that grows from a male gametophyte (pollen grain). A pollen tube of one kind of plant may be so different in size, shape, or length that it can't reach the female gametophyte in the ovary of another kind of plant. As another example, even if fertilization does occur, there may be incompatibilities between the developing embryo and the maternal organism. Because of the incompatibilities, the embryo may die. In some cases, hybrid offspring are produced, but they commonly are weaker in structure, physiology, or behavior than normal offspring of either species. Hence they will not have as much chance of surviving. In other cases, hybrid offspring are vigorous but sterile. (A cross between a female horse and a male donkey produces a mule. A mule is a hybrid that is fully functional *except* in its reproductive capacity.) Finally, even if a first-generation hybrid survives and manages to reproduce, there may be hybrid breakdown in the next generation because the offspring are exceedingly weak. (Crosses between two species of

evening primroses give rise to a partially fertile first generation. But the second-generation plants are slow-growing dwarfs, susceptible to disease and totally sterile.)

Seasonal isolation also occurs between populations or species. The timing of reproduction may act as this kind of isolating mechanism. Related species of plants may have overlapping reproductive seasons that differ only in their peak period; hence they are partially isolated from each other. Or related plant species may come into flower at nonoverlapping times, which means that their reproductive isolation is complete.

Behavioral isolation occurs among many animal species. For example, complex courtship rituals often precede mating. Perhaps squawks, head-bobbing, wing-spreading, and dancing are interwoven into a complex, stereotyped pattern. Such specific behaviors may turn on the females of one bird species—but they may not even be recognized as overtures by females of another bird species.

Finally, ecological isolation may develop and give rise to local breeding units in the same environment. As you will read in later chapters, feeding habits, nest selection, and similar ecological factors may serve to confine incipient species (or closely related species) to different parts of the same area. Such ecological isolation also reduces the likelihood of gene flow between breeding units.

Differences in reproductive structure or physiology, in timing of reproduction in behavior, and in ecological factors may serve as isolating mechanisms between populations that are undergoing (or have undergone) speciation.

Speciation Routes

Local populations that are geographically separated are called allopatric. (The word means "different fatherlands.") **Allopatric speciation** is the most common pattern in nature. Geographic separation of two populations, accompanied by gradual divergent evolution between them, leads to reproductive isolation. Even if their distribution ranges overlap in the future, biological differences between individuals of the two populations serve as isolating mechanisms and maintain their distinct character. True speciation has taken place. Usually, disturbances to the environment are the trigger for the initial separation. Clearing fields for agriculture may have this sort of disruptive effect. Sometimes, too, a few individuals drift away or are blown away to a semi-isolated or isolated spot. Once separated from the main population, they may evolve independently. The few individuals may not be a genetically representative sample of the species. Hence they may undergo rapid genetic differentiation because of the atypical allele frequencies. Such small-scale dispersal and rapid differentiation are known as the *founder effect*.

Distinct local breeding units may coexist in the same geographic range. In such cases, they are called sympatric (having the "same fatherland"). Sometimes they are non-overlapping but adjacent neighbors; sometimes their ranges overlap. You might expect at least some interbreeding to occur between their members, but in rare cases it apparently does not. In **sympatric speciation**, two or more populations occupying the same distribution range are thought to undergo reproductive isolation before genetic differentiation transforms them into separate species.

Especially among plants, **polyploidy** and/or **hybridization** are two other speciation routes. No doubt they have been important in plant evolution: about forty percent of all flowering plant species are polyploid. Recall that during meiotic cell division, improper separation of homologous chromosomes may mean that a gamete will end up with one or more extra chromosome sets. Fusion of the abnormal gamete with another gamete leads to polyploidy. Most often, polyploid offspring cannot inbreed with the parent diploid species; hence reproductive isolation is established in merely one generation's time. Polyploidy may also assure abrupt speciation in crosses that would otherwise yield sterile hybrid offspring. (The reason an interspecific hybrid is usually sterile is that its homologous chromosomes are from different species of parents; hence the "homologues" probably can't pair during meiosis. Because some polyploid hybrids have four or more chromosome sets, the duplicates *can* pair, meiosis can proceed, and viable gametes may form.) Wheat is a successful polyploid hybrid. Kentucky bluegrass hybridizes with a number of related species.

Adaptive Radiation

Apparently the origin of a new major species requires occupation of a new environmental setting, with adaptation and proliferation then giving rise to the major group. When populations of the same species become adapted for exploiting different aspects of the environment in specialized ways, they are said to have undergone **adaptive radiation**. Adaptive radiation may or may not be accompanied by full speciation.

Successful radiation seems to occur in two ways. A small founding population may somehow become dis-

persed into a setting that is devoid of directly competitive groups. The new setting doesn't even have to be all that different from the place they left behind; the advantages lie in the possibilities for isolation of a small group, expansion, and genetic differentiation. The dispersal of Darwin's finches (Chapter Two) is thought to be an example of this manner of adaptive radiation.

Adaptive radiation may also occur when an *existing* trait in some species opens the door to a new adaptive setting even as old doors are closing behind. Fishes ancestral to land vertebrates had *swim bladders*: air sacs that helped maintain balance and position in water, through adjustments in the volume of gases that they contained. When many marine environments gave way to dry land some 400 million years ago, the air sac had the adaptive potential to become modified into air-breathing lungs. Lungs were one of the pivotal characters that permitted some descendants of fishes to invade and radiate through the new land environments. Other "innovative" characters that emerged among the fishes were bony skeletons and jaws—both of which proved to have adaptive potential for fast-swimming predatory lines, and for predators that made the move to land.

Extinction

The millions of different kinds of organisms alive today are a mere fraction of all that have gone before. **Extinction,** the disappearance of groups of organisms, is one of the major patterns of evolution. Often a single group disappears. Periods of mass extinction seem to have punctuated geologic time. In fact, the boundaries said to define geologic epochs were established on the basis of apparent times of mass extinctions in the seas, on land, or both. Following periods of widespread disappearances, other groups expanded into the vacated settings. As you read in Chapter Twenty-One, geologic upheavals and climatic disturbances of a profound sort could well have been triggers for major turnovers.

Regardless of whether conditions change slowly or rapidly, groups may become extinct if they do not have the potential for adapting to the new circumstances, or for migrating elsewhere. The absence of genetic diversity, a narrow distribution range, small population size—any such factor may place a group on the brink of extinction. In recent times, human-induced disturbances are pushing an alarming number of species over the edge, at an accelerated rate that far exceeds anything that has occurred in the past. We will be returning to this consequence of human activities at the conclusion of this book.

COMMENTARY

Gene Flow and the Human Population

In an evolutionary view, the human population must be as subject to agents of selection as all other populations. Human individuals are highly dispersed, from nearly inaccessible pockets of tropical forests, to remote oceanic islands, across the broad and varied regions of the separate continents. There is, to say the least, potential for geographic isolation and genetic differentiation. Overlaid on this widespread distribution is a patchwork of diverse social sanctions and cultural differences that could, theoretically, act as behavioral isolating mechanisms. Is it possible, then, that different segments of this population have been traveling different adaptive routes that have given rise to the large, regional groupings called "human races"?

Human races are said to be distinguishable by variations in such traits as skin color, hair texture, and shape of the nose. Such phenotypic variations do exist. But only a few are definitely correlated with different selective pressures in different environments. Consider the results from studies of the ratio of body weight to body surface. These studies show that humans generally follow the same adaptive trend as other warmblooded animals do. Body weight for human males is lower in hot climates and higher in cold climates, and the difference is related to regulation of body temperature. Studies of the ratio of protruding body parts (such as limbs) to trunk length show the same kind of adaptive trend. The hotter the climate, the longer the limbs (which helps dissipate body heat). The colder the climate, the shorter the limbs (which helps retain body heat).

Other variations, such as skin color, are not that well understood. Skin color is probably one of many traits that must have been tested out in the forerunners of our species—in creatures that lived million of years ago in the sun-drenched tropics. Because cancer can develop in skin that is exposed for prolonged periods to intense sun, we can guess that natural selection long ago favored the abundant production of melanin, a dark pigment that screens out ultraviolet radiation. However,

even though too much sunlight can be harmful, some sunlight can be used to advantage. Its ultraviolet rays are capable of reacting with lipids in the skin to produce vitamin D precursors, which the body needs for normal functioning. We may speculate that when some human groups began moving out of the tropics, they encountered selective agents of other sorts. In northern lands, where the amount of sunlight was relatively diminished, there would have been reduced selection pressure in terms of maintaining high melanin levels in the skin. Perhaps the need for vitamin D synthesis was enough to select against abundant melanin production in these environments. Of course, then we must ask ourselves why dark-skinned populations have existed for many thousands of years in regions of low light intensity. Eskimos, for instance, do not have bleached-out skin. We could counter by saying that the Eskimo diet is based largely on fish—and fish fats are rich in vitamin D. Perhaps Eskimo populations haven't yet encountered selective agents pressuring for change from an ancestral melanic heritage.

There is a lot of this kind of speculation about human variation, but not much substantive evidence one way or the other on what it means. Even so, on the basis of differences in outward appearance, the human population has variously been divided up into racial stereotypes. Sometimes five racial groups are recognized, which are said to correspond to the original populations of Africa, Europe, Asia, Australia, and the American continents. Sometimes as few as three and as many as thirty-two races are recognized, depending on who is setting up the phenotypic criteria.

What about variation at the genotypic level? It turns out that some differences between groups do exist at certain gene loci. For instance, consider the variations in allele frequencies for the ABO blood group. In American Indians, the A and B alleles are almost completely absent; and a gradual variation in the B allele frequency occurs from Europe across Asia. In another (Rh) blood group system, one allele is found among black Africans and is rarely found anywhere else; another allele is found among white Europeans and is almost never found elsewhere. Considerable variation also exists among alleles for more than seventy different enzymes studied, and these variations exist between the groups called caucasoid, negroid, and mongoloid.

So where does this leave us? Polymorphism obviously can be observed in the human population system, and there is some genetic differentiation within that system. The problem is, it is impossible to make much sense out of what the variations mean. For one thing, understanding of the origin and early history of our species is still obscure. The fossil evidence is still being interpreted in different ways. For another thing, ancient migration patterns that led to the distribution of humans throughout the world are by no means identified, even though we do have some reasonably good ideas about them. Finally, and by far most importantly, nobody has even begun to document the *permeability* of barriers to gene flow that surely existed in the past and continue to exist today.

What we have today is a partial isolation of some regional segments of the human population from others, into what is perceived as races. But notice the word "partial." It is true that if you were to be dropped into the middle of China, then plucked away and dropped successively into Norway, New Guinea, Arabia, and Texas, you would certainly notice that people don't all look alike. But if you could walk step by step through every country in the world, you would probably be more aware of the continuum of phenotypic traits as one local population gives way to the next. Even then, in the midst of all the similarities, you are sure to encounter individuals who "look different" yet are simply expressing part of the potential variability of the gene pool for that region. And that potential variation is an open system. There may be mountain barriers or oceans or social sanctions to bar the way between local populations. But the fact is, individuals of our kind are not known to stay put. Communication and encounters and intermarriages between members of different local populations take place all the time, in war and in peace, through permanent moves and transient wanderings. These are the contacts that have bridged the barriers between local populations and have kept the human species what it is today: *one species*, traveling in the same evolutionary direction.

Readings

Ayala, F. 1978. "The Mechanisms of Evolution," *Scientific American* 239(3):56–69. Succinct look at genetic basis of variation within species, and at the processes leading to speciation.

Cavalli-Sforza, L., and W. Bodmer. 1971. *The Genetics of Human Populations*. San Francisco: Freeman. Comprehensive, authoritative look at available data on the genetics of human populations, interpreted through theoretical models. Objective look at the problems of racial differentiation; clear discussion of sickle-cell anemia.

Futuyma, D. 1979. *Evolutionary Biology*. Stamford, Connecticut: Sinauer. Probably the best currently available synthesis of modern evolutionary thought. Futuyma writes beautifully, and cuts through much of the confusion surrounding the development of concepts in this rapidly changing field.

Review Questions

1. What is the typological approach to categorizing species? In what fundamental way does the population concept differ from this approach?

2. Define these terms: individual, population, and species.

3. What is genotypic and phenotypic variation? Describe evolution in terms of frequency distributions for a given trait, and in terms of the underlying allele frequencies.

4. What is the Hardy-Weinberg baseline against which changes in allele frequencies may be measured? What is Hardy-Weinberg equilibrium?

5. Changes in allele frequencies may be brought about by mutation, genetic drift, gene flow, and selection pressure. Define these occurrences, then describe the way each one can send allele frequencies out of equilibrium.

6. What implications might the neutral effect of genetic drift hold for an earlier concept of "survival of the fittest"? As part of your answer, define polymorphism and explain how different phenotypes can persist indefinitely in the *same* population.

7. Natural selection is no longer in the realm of pure theory. Can you recount the ongoing sagas of the peppered moth and the pesticide-resistant pests to explain why it is now considered an operating principle in biology?

8. Define stabilizing, directional, and disruptive forms of selection and give a brief example of each.

9. Before labeling a particular genotype as being advantageous or disadvantageous, what must you first consider? Can you give an example of an allele that is somewhat advantageous under one set of environmental conditions, yet disadvantageous under others?

10. List some barriers that prevent allele exchange between local breeding units.

11. Give two examples of isolating mechanisms, and outline what they accomplish.

12. What is the difference between allopatric and sympatric speciation? Which do you suppose occurs most often in plants?

26

POPULATION GROWTH AND STABILITY

The optimist proclaims that we live in the best of all possible worlds; and the pessimist fears this is true. —James Branch Cabell

Suppose this year the federal government passes legislation limiting the size of your family to three children; no more under any circumstances. Suppose, further, the new law specifies that each father must be sterilized after the birth of the third child—and that if he refuses, he will be jailed for a few years and sterilized anyway, with or without his consent. *It could never happen here,* you might be thinking, and not only because of the furor it would raise over the violation of individual rights. Decades ago, Americans began to suspect that the quantity of life has a pronounced effect on its quality. Since the baby booms following the return of overseas troops after the world wars—an era when a family of twelve was considered a great achievement—our concept of family planning has been slowly changing. As a nation we are now moving in the direction of **zero population growth**, in which the birth rate over the long term equals the death rate.

But you are among the fortunate few in the world who have the luxury of a choice. You belong to a population that presently has enough food and enough money to buy whatever other resources it is lacking. You enjoy superior standards of hygiene and medical care. All these things mean that you don't need a lot of children to help you survive and perpetuate the family line.

Most of the people in the world don't have the luxury of making such "rational" choices. Consider the conflict going on today in India between individual rights to reproduction and the rights of the overall population. Each year 9 million Indians die—but 22 million are born. To a population now standing at well over 600 million, *13 million new members are added annually.* There are more people in this one country than there are in North and South America combined! Yet India does not have enough food for its population now, nor does it have enough money to buy more food and other resources. Most of the population does not have adequate medical care, and living conditions are for the most part appalling. Each *week* 100,000 more people enter the job market, looking for nonexistent jobs.

Because of these circumstances, the Indian government has been supporting population control programs for more than two decades. The programs have not succeeded. For one thing, seventy percent of the people are illiterate and can be educated only by word of mouth. And sustained population control programs are hampered because eighty percent live in remote agricultural villages and are not easy to reach. Finally, Indian villagers don't see the meaning of total population growth; they see only that many children die of disease, of starvation. They believe that only by having more children will they increase their own personal chance for survival. Without large families, they say, who will help a father tend fields? Who will go to the cities in the hope of earning money to send back home? How can a father otherwise know he will be survived by at least one son, who must, by Hindu tradition, conduct the last rites to assure that the soul of his dead father will rest in peace?

Clearly there is serious discrepancy between the way villagers see their world and what their world has actually come to be. Because of this discrepancy, twenty-five years from now the population may reach *1 billion*—which would wipe out all of India's efforts to increase crop production and to break the cycle of poverty, disease, and starvation that plagues it. That is why, in desperation, their federal government passed legislation in 1976 calling for compulsory sterilization. For them, the hypothetical example opening this chapter briefly became reality, until public outrage became so great that the law was rescinded.

Population control, in short, is an issue complicated by the fact that the conditions under which others must make their decision to limit family size are enormously different from the conditions under which we have made ours. The chasm between us is a legacy of the nineteenth century. That was an age of rampant colonization and control by Europeans and, to a lesser extent, Americans over the vast natural resources of undeveloped countries. It was an age when trade agreements were established, generally by force, that benefited powerful nations. For centuries the pattern has remained one of taking raw materials from these countries and giving them mostly advice in return. Even now, as industrialized nations send 2.5 million tons of low-grade protein (such as wheat) to undeveloped nations, they take out 3.5 million tons of high-grade protein (such as fish). *Population control may become more acceptable when there is more of a balance to the flow of resources on which any population depends—a more secure basis for individual life that will make population control seem the rational approach that it is.*

In this chapter, you will be reading about the biological principles underlying the growth and survival of populations. These principles can tell you a great deal about the human condition—where it is now, where it is heading. But this chapter can only hint at the complex reasons why there is resistance in many parts of the world to accepting what these principles mean. For that reason, perhaps you will want to supplement the biological perspective with reading on your own into the histories of East and West. For in those histories, in their convergence, lie clues to what must be undone and what must be built anew before we can hope to achieve some semblance of long-term, dynamic stability for the human species.

POPULATION SIZE

Birth, Death, and Exponential Growth

A population, recall, is a group of individuals of the same species, occupying a given area at a given time. Its character is defined in part by **population size** (its number of individuals) and **population density** (how many individuals are present in a particular area). Population size and density are not static things. They tend to fluctuate because individuals come and go—by birth, death, and inclination—all the time. But how do we pin down the character of a population if it is more or less changing all the time? We can look first to the way it grows in size. **Population growth** is the difference between the birth rate and death rate, plus or minus any inward or outward migration. As a first approximation, we can begin with this principle:

Any population that is not restricted in some way will grow in size at an increasingly accelerated rate.

A convenient way to express the growth rate of a given population is known as its **doubling time** (how long it takes to double in size). Some bacteria can double their numbers every thirty minutes; humans are now doubling their numbers every thirty-five years. Regardless of the unit of time, in the absence of control factors the pattern is always the same. What starts out as a gradual increase in numbers turns into explosively accelerated increases—a pattern of **exponential growth**. All populations, from those of amoebas to humans to whales, grow exponentially when there are no constraints, no matter what the reproductive rate of their species. The reason for exponential growth is simple. As new individuals are added to a population, the number of reproducing members increases. *And the increase enlarges the potential reproductive base.* The next generation enlarges it further, and so on. For a while, many new individuals can be born and nothing seems to be getting out of hand. But then an ominous thing starts to happen. With successive doubling periods, population size actually begins to skyrocket.

We can see how this happens by putting a single bacterium in a culture flask with a rich supply of nutrients. In half an hour the bacterium divides in two; half an hour later, the two divide into four. Assuming no cells die between divisions, every half hour the number doubles. But the larger the population base becomes, the worse things get. After only 9½ hours (nineteen doublings), the population will be over 500,000—and by 10 hours (twenty dou-

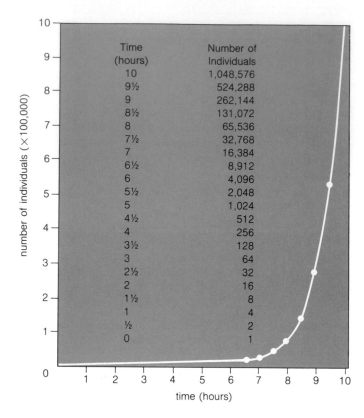

Time (hours)	Number of Individuals
10	1,048,576
9½	524,288
9	262,144
8½	131,072
8	65,536
7½	32,768
7	16,384
6½	8,912
6	4,096
5½	2,048
5	1,024
4½	512
4	256
3½	128
3	64
2½	32
2	16
1½	8
1	4
½	2
0	1

Figure 26.2 Exponential growth of a bacterial population. Division occurs every half hour.

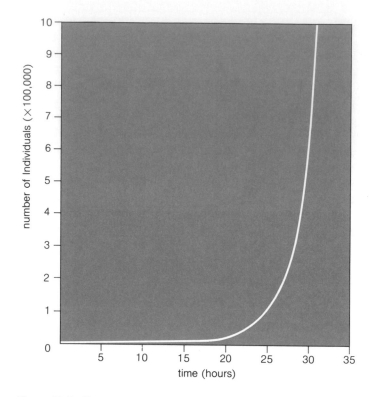

Figure 26.3 Exponential growth of a bacterial population in which division occurs every half hour, but in which twenty-five percent of the individuals die between divisions. Although deaths slow things down a bit, in themselves they are not enough to stop the explosive pattern of exponential growth.

blings), it will simply soar past 1 million! When we plot the course of such exponential growth, we end up with a **J-shaped curve** (Figure 26.2).

For a slowly reproducing species such as a whale, it may take a hundred years for population size to double. Even so, after only twenty doublings (2,000 years), there would be a million whales derived from each original pair. Although the doubling time varies from one species to the next, fantastic population sizes are relatively short-term prospects when the rate of growth is measured on the evolutionary time scale.

Ah, you might say, but a whale doesn't live to be 2,000 years old. Even as some are born, others are dying—which surely must keep population size under control. We can see whether the death rate is a significant control factor by starting over with our bacterium in its nutrient-rich culture flask. But this time let's assume twenty-five percent of the population dies between each doubling time. The death rate does slow things down a bit, in that it takes almost two hours instead of half an hour to double popula-

tion size. But as Figure 26.3 shows, only the time scale has changed—we still have a J-shaped curve! It's just that it now takes thirty hours instead of ten to arrive at a million bacteria. Thus we have another principle:

As long as the birth rate remains even slightly above the death rate, any population grows exponentially.

Carrying Capacity of the Environment

Obviously, you *know* there must be controls of some sort on population size that prevent berserk expressions of exponential growth. For example, chances are you know that when you take a walk through a forest you are not going to be trampled to death by a billion rabbits. Something about a stable natural system such as a forest keeps its populations in check; somehow birth rates are balanced with death rates. But any natural system is bound to have complex interactions going on among its diverse populations. So let's go back to that (by now exhausted) bacterium in its culture flask, where we know we can control the variables. First we will feed it a balanced diet of glucose, minerals, and nitrogen, then we will allow it to reproduce for many generations. Initially the bacterial population goes through an exponential growth phase. Next, population growth tapers off, only to give way to a **plateau phase** in which population size remains relatively stable. Then the population begins to decline—suddenly at first, followed by a more gradual pace until the entire population is dead. In Figure 26.4, curve *a* depicts this characteristic rise and fall in numbers.

What caused the growth curve to be patterned in this way? For these bacteria, glucose meant food and energy— but the culture dish held only so much glucose. We can deduce that, as the population expanded faster and faster, the glucose had to be getting used up faster and faster, too. When glucose supplies began dwindling, so did the basis for exponential growth. Hence we have another principle governing population growth:

When any essential resource is in short supply, it becomes a limiting factor on population growth.

So we try again—only this time we keep adding glucose to the culture dish. And now the growth curve takes on the shape of curve *b* in Figure 26.4. Notice how this curve shoots up higher than curve *a* but then falls more rapidly. Although the extra glucose encouraged further growth, there were no extra supplies of nitrogen and minerals. Different resources had now become the limiting factors. To keep our bacterial population growing explosively, we are obviously going to have to feed it more nitrogen and minerals, too.

But even with all the nutrients it needs, the population not only declines, it crashes! What went wrong? As the bacteria multiplied, so did their metabolic by-products—which drastically changed the nature of the environment. So many waste products were given off that the bacteria actually poisoned themselves to death. Only if the toxic medium were removed every so often from the culture dish and replaced with a fresh medium would things stabilize. Only then would we end up with an **S-shaped curve**: an exponential growth phase that leads into a stable plateau phase (curve *c* in Figure 26.4).

These experiments have helped us identify another principle controlling population growth:

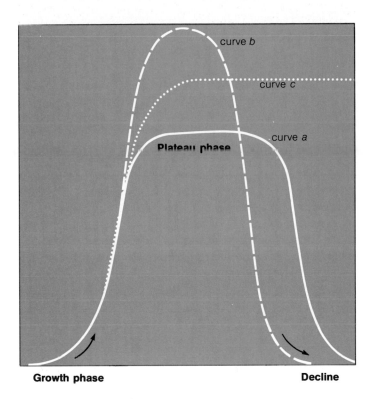

Figure 26.4 Effect of carrying capacity of the environment on population growth curves, as described in the text.

A population of any species can tolerate only a certain range of environmental conditions.

The term **tolerance range** refers to a range of adaptability to environmental conditions. This range controls population distribution. If some environmental factors (such as temperature) exceed or fall below this range, then the population may be effectively excluded from the area.

Resource availability, prevailing conditions, and tolerance ranges interact to dictate where population growth must level off. For a given population, they define the environment's **carrying capacity**: the maximum density of individuals that can be indefinitely sustained in a given area, under a given set of environmental conditions. Food, wastes, living space—such factors influence the maximum size at which a stable population can be maintained. This point is expanded upon in the following sections. In reading through these sections, keep in mind that there is no escaping the limits imposed by an environment's carrying capacity.

For all species, the environment's carrying capacity (the greatest density of individuals that it can indefinitely sustain) is like a brake on runaway growth.

MAINTAINING POPULATION STABILITY

When a natural population is said to be stabilized, that doesn't mean its size is frozen at some level. It means population size typically fluctuates *within a predictable range*, neither dropping to very low levels nor exceeding certain upper limits. It is a dynamic kind of stability, with feedback mechanisms constantly regulating birth and death rates.

There are two ways of looking at these mechanisms. We can see how they relate to population density. We can also see which ones are built into the species and which are dictated by the environment.

Density-Dependent and Density-Independent Factors

Certain control mechanisms come into play whenever populations change in size. Because their impact changes depending on the population density, they are called **density-dependent factors**. When population size increases, various pressures arise that serve to cut it back. When population size decreases, the pressures ease up, which encourages renewed growth. For instance, if population size increases, the amount of food available for each individual declines. The decline in food may lead to impaired health, which may affect the reproductive rate. But if population size falls, the survivors have more food available, their health may improve, and their reproductive rate may rise—so population growth resumes. Food availability, then, may operate in a density-dependent way.

Density-dependent factors work to cut back population size when the density of individuals approaches the environment's carrying capacity. They ease up when population density decreases.

Other control mechanisms are at work regardless of how large or small a population may be; they are **density-independent factors**. You could, for instance, boil a culture flask full of bacteria and the entire population would drop dead no matter how many or how few there were. Such catastrophic events are not confined to culture flasks. They await any natural population whenever environmental

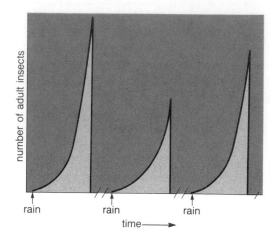

Figure 26.5 Fluctuating population levels of adult insects in a desert environment. Periodic rains trigger the hatching of dormant eggs, and the population level rises rapidly—only to fall rapidly as dry conditions return and bring about the near-simultaneous deaths of all adults.

variables soar above or plummet below tolerance ranges characteristic of the species.

Population size of many desert insects is controlled in this way. For them, brief spring rains break the dormancy of many thousands of insect eggs. In a short time, insects emerge, they feed, they mate, they lay eggs all over the place. They cannot dally, for with the onset of the prolonged dry season, their bodies dry out in the intense heat. The adult populations then collapse. Explode and crash, explode and crash—only by quickly re-creating the thread of dormant, heat-resistant eggs between rainy seasons can these species survive (Figure 26.5).

Density-independent factors work regardless of population size; they are environmental variables that may shift enough to push a population above or below its tolerance range for a given variable.

It is not often, however, that we can clearly isolate density-independent factors from density-dependent ones. They usually interact to such an extent that it's difficult to know where one ends and another begins. For instance, whether a rabbit population tolerates a sudden freeze depends on whether its members have enough food and burrows. Availability of food and desirable burrows are density-dependent factors—yet here they interact with a density-independent change in temperature.

To help shed more light on how interactions such as these control population size, let's take a closer look at both their intrinsic and extrinsic nature. Understanding of these factors has been determined by observations of experimentally controlled and natural populations.

Intrinsic Limiting Factors

Each individual has certain structural, physiological, metabolic, and behavioral traits, and these traits represent **intrinsic limiting factors**:

Intrinsic limiting factors are built-in features of individuals that affect population growth and distribution. They include limitations imposed by the individual's physiology, structure, metabolism, and behavior.

One such intrinsic limiting factor worth mentioning is the **timing of reproduction** in individual life cycles. If reproduction occurs early in life, and if the parents live for some time afterward, they may continue to coexist with several new generations. If reproduction occurs later in life or if the post-reproductive life span is short, there is less overlapping. Reproductive timing often corresponds to the carrying capacity of the environment for different species. Some species are long-lived and start reproducing late in the life cycle. Some go through a long period of embryonic development and give birth to only a few individuals. Perhaps the fullest expression of delayed reproduction is found among species such as Pacific salmon. Their members grow to maturity, then die right after reproducing. Because generation overlap is eliminated, a higher rate of reproduction is possible without adversely affecting the carrying capacity of the environment. More will be said about delayed reproduction in a later section on human populations, for it is sometimes considered to be a possible way to control our own rate of growth.

Perhaps one of the most intriguing demonstrations of how intrinsic factors may operate comes from ecologist John Calhoun's studies of extreme overcrowding among Norway rats. His experiments suggest that endocrinological changes accompany increases in population density; and that these intrinsic changes in turn influence population distribution and reproductive behavior.

For Calhoun's experiments the rats, which are normally territorial, were confined in pens. They were given plenty of food and protection against their normal predators, and disease was kept to a minimum. As population density increased under these artificial conditions, normal

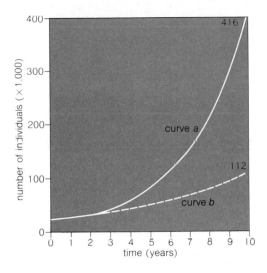

Figure 26.6 Comparison of the growth rates of two hypothetical populations that differ only in the timing of reproduction. In both cases, each individual lives five years, and each breeding pair produces four offspring. In *curve a*, reproduction occurs at age two. *Curve b* shows what happens when reproduction is delayed until age four. Delayed reproduction has some impact on population growth over the short term. But in the long term, *curve b* would also show exponential growth.

territorial behavior began breaking down, only to be replaced by bizarre behaviors. Some rats became cannibals even though there was plenty of other food. Some became frenzied, sexually deviant, or pathologically withdrawn. Underlying these changes was an increase in adrenal gland size. The glandular change probably was related to the steadily increasing aggressive encounters as more and more rats were forced to interact in the restricted space of the pens. Sometimes dominant males would go on a rampage. They would attack not only subordinate males but females and offspring. Individuals low on the social ladder would eat and move about only when the other rats were asleep.

At the same time, there was an overall size reduction in glands other than the adrenals. The reduction was most pronounced in glands concerned with reproductive hormones. Miscarriage was common. So was death of pregnant females during delivery. Among the most stressed groups, only four percent of the offspring survived. Maternal behavior deteriorated as females first became careless in building nests, then began simply to heap up strips of paper as makeshift nests, then finally delivered their offspring directly on the sawdust in the open pen. If females encountered another rat while they were toting their infants about, they would drop them and abandon them. The infants were helpless before the cannibalistic adults.

In studies of experimentally induced overcrowding among mice, the social structure that prevailed was based on varying levels of withdrawal. Mice at the low end of the ladder showed chronic withdrawal symptoms punctuated by short bursts of violence. One would chew on another, which would passively submit to being chewed. Mice at the top of the ladder were healthy and sleek—but absolutely passive. They slept; when presented with food, they ate; when presented with water, they drank. But they never fought, they never went out of their way to find food, and they never copulated. Eventually, survivors of those stressful conditions were placed in a spacious, uncrowded pen. But they only huddled desperately together. They seemed terrified of the less crowded distribution.

(Similar results have been noted in studies of a number of social mammals. It has even been suggested that crumbling social structures and other signs of aberrant behavior in overcrowded human societies may be partly related to the same kind of intrinsic limiting factors. As tempting as the similarities may be, the data available do not indicate that the findings may also apply to overcrowded human populations.)

In natural settings, too, intrinsic limiting factors affect how populations come to be distributed in space. Territorial behavior in natural communities is under intrinsic controls that are tied to the carrying capacity of the environment. So, to a large extent, is migratory behavior. Migration is an intrinsic response to extrinsic limitations of a local region. For instance, elk herds in North America migrate on a daily basis and a seasonal basis. During daylight hours, they bed down in forests or do some light feeding beneath the trees. At night they move into open meadows and do most of their eating. Even with ample food sources, the carrying capacity of a meadow is low because it doesn't provide ample protection against predators. In contrast, the forest provides ample protection but not enough food. The daily movement of the herd creates a composite environment of forest and meadow, with a higher carrying capacity than either area alone can offer. Under the same kind of pressures, the herds have come to migrate between lowland meadows during winter and high mountain meadows from late spring to early fall. Changing weather in the high country gradually drives them back to the lowlands before the onset of winter. The things that keep them from living all year long in the lowlands are summer droughts and limited plant growth. Only because the herds show an intrinsic tendency to migrate is vegetation in both places given time for renewed growth—which supports a larger elk population over the long term.

Figure 26.7 Correlation between the lynx population (dashed line) and the snowshoe hare population (solid line) in Canada over a ninety-year period. These long-term data are derived not from field observations but from counts of the pelts that trappers sold to Hudsons Bay Co. The curves are taken to be a general index of the way predation can control the populations of both predator and prey. (Data from D. MacLulich, University of Toronto Studies, Biology Series 43, 1937)

This figure is a good test of how willing you are to accept conclusions without questioning their scientific basis. (Remember the discussion of scientific methods in Chapter Two?) For example, what other extrinsic and intrinsic control factors could have been influencing the population levels? Were there also fluctuations in climate over this time span? Were some winters more rigorous, thereby imposing a higher death rate on one or both populations? Although this is called a simple predator-prey system, weren't the hares preying on the vegetation, which may have been overbrowsed in some years but not in others? What about owls, martins, and foxes—which also prey on hares? What if some years there were fewer trappers because of such very real variables as Indian uprisings? What about fluctuating demands for furs by the fashion industry? What if some years there was more lynx trapping than hare trapping, or vice versa? And in looking closely at the curves, can you really conclude there is an unvarying correlation between them, or is there evidence here and there of a random drifting apart—which might indicate other factors at work?

Extrinsic Limiting Factors

Also controlling population growth and distribution are **extrinsic limiting factors**, or external conditions—not only temperature, rainfall, nutrients, and such, but the presence or absence of other species. Interactions among species are especially important, for they help define where a species will actually be found in its potential range of distribution.

Extrinsic limiting factors are external variables, both environmental and biotic, that help control population growth and distribution.

Perhaps the most dramatic extrinsic check is the predation of one species on another (Chapter Twenty-Eight). Interactions between the Canadian lynx and the showshoe hare are a classic example of feedback mechanisms at work in a predator-prey system. As Figure 26.7 suggests, the example is somewhat oversimplified, but it still makes its point. Although evidence for this set of interactions is secondhand (Figure 26.7), it has been collected every year for more than a century. And even allowing for unknown variables, certain patterns undeniably are there. The lynx

Ed Cesar/National Audubon Society/PR

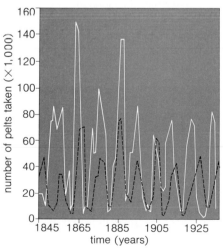

and hare population sizes fluctuate together in cycles, with the lynx population lagging slightly behind that of the hares. When the number of hares is low, there is a period of rapid population growth. The number of lynx, which is also low, rises right after the growing number of hares—then both populations crash. The cycle repeats itself about every decade.

An intrinsic factor (the explosive reproductive cycle of the hares) triggers the cycle. An extrinsic factor (the scarcity of lynx at the start of each cycle) permits rapid fulfillment of that potential. With more hares running around, the lynx are better fed; more lynx survive, and more reproduce. But the intensified predation, along with diminishing food and perhaps with disease, exerts more pressure on the hare population, which soon plummets. The lynx population has only one way to go: down. And once their numbers have also declined, the cycle can start again. In this example, predation helps control the size of *both* populations as they interact with one another.

HUMAN POPULATION GROWTH

So far, we have touched on a few intrinsic and extrinsic controls to show how they can regulate the growth of natural populations. Thus you can see how populations can grow explosively under certain combinations of short-term variables, only to decline in the long term as controls reassert themselves. Let's see how this pattern applies to human population growth.

Doubling Time for the Human Population

In 1975 the human population reached 4 billion. Even if this number represented a stabilized population level, we would still have to contend with monumental problems. In any given year, between 5 and 20 *million* people now die of starvation or malnutrition-related diseases (Chapter Seven).

At present, there is no realistic prospect of advancing the quality of life for all. The future is more grim. By-products of human existence are causing the overall quality of life to deteriorate. As you will read in the last two chapters, our self-inflicted pollutants carry the same danger that the self-inflicted pollutants carried for those bacteria described earlier. In the expansionist mentality of the nine-teenth century, the world was our oyster. Today the world is our culture flask, and no great experimenter in the sky is going to take away our wastes and renew resources for us.

What makes future prospects especially chilling is that our population is by no means static. We are a rapidly growing population of 4 billion! Consider that net population growth for a given period is the difference between the total number of live births and the total number of deaths occurring in that time span. Often, net population growth is calculated on the basis of the number of births and deaths per 1,000 persons in a population at the midpoint of a given year. These values can be used in determining the percent annual growth rate:

$$\text{birth rate} = \frac{\text{births per year}}{\text{total midyear population}} \times 1,000$$

$$\text{death rate} = \frac{\text{deaths per year}}{\text{total midyear population}} \times 1,000$$

$$\text{rate of increase} = \text{birth rate} - \text{death rate}$$

$$\frac{\text{percent annual}}{\text{growth rate}} = \frac{\text{birth rate} - \text{death rate}}{10}$$

This formula can be extended to find how long it will take for the human population to double in size:

$$\frac{\text{approximate}}{\text{doubling time (years)}} = \frac{70}{\text{percent growth}}$$

In 1978, Mexico's growth rate was 3.4 percent. At this rate, its population will double in twenty years. Japan's growth rate was 1.0 percent; its population would double within seventy years. The world average in 1978 was 1.7 percent. If this rate continues, world population will double within forty-one years. *This means that by the year 2019—well within the lifetime of most of us reading this book—world population may soar to 8 billion!* With the most intensive effort, we just *might* be able to double food production over the next thirty-five years to keep pace with growth. But we would succeed in doing little more than maintaining marginal living conditions for most of the world. Under such "ideal" conditions, deaths from starvation might *only* be 10 to 40 million a year! For a while, it would be like the Red Queen's garden in Lewis Carroll's *Through the Looking Glass*, where one is forced to run as fast as one can to remain in the same place. But what happens when resources other than food become limiting factors? What happens in the year 2045, when the population has doubled again to 16 billion? Can you brush this picture aside as being too far in the future to warrant your concern? *It is no farther removed from you than your own sons and daughters; that world is their legacy.*

Where We Began Sidestepping Controls

How did we get into this mess in the first place? For most of our existence as a species, human population growth has been slow. In the past two centuries, there has been an astounding acceleration of growth (Figure 26.8). We have no reason to assume there are special rules governing

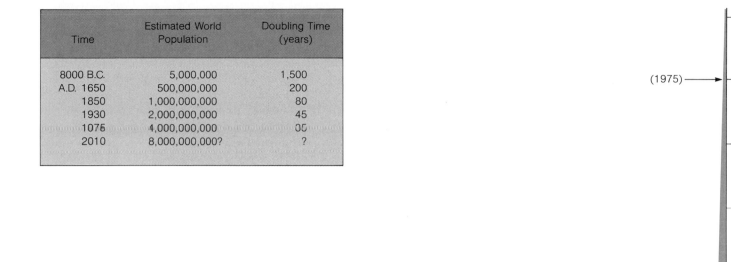

Time	Estimated World Population	Doubling Time (years)
8000 B.C.	5,000,000	1,500
A.D. 1650	500,000,000	200
1850	1,000,000,000	80
1930	2,000,000,000	45
1975	4,000,000,000	05
2010	8,000,000,000?	?

(1975) ⟶

world population (billions)

14,000 13,000 12,000 11,000 10,000 9000 8000 7000 6000 5000 4000 3000 2000 1000   1000 2000

B.C. | A.D.

Figure 26.8 The J-shaped curve of human population growth. The slight dip between 1347 and 1351 shows the time when 75 million people died in Europe as a result of the bubonic plague, or Black Death—a virulent disease spread by fleas that thrived on the massive rat populations in cities. At current growth rates, it would take only thirteen months to replace 75 million individuals.

human population size and growth rate over the long term. Instead, we must assume that the human population is controlled by factors much like those controlling populations of other large animals having relatively long life spans. If that is true, there are only three possible reasons why our long-term growth rate accelerated:

1. We must have developed the capacity to expand steadily into new environments.

2. The carrying capacity of the environment was increased in some way.

3. A series of limiting factors was removed so that more of the available resources could be exploited.

Let's consider the first possibility. We know that by 50,000 years ago, the human species had migrated over much of the world. For most animal species, such extensive radiation into new environments is a much slower process. Not only is there competition with species that may already occupy an area, there must be time to modify tolerance ranges in response to pressures from new environmental conditions. With the human species, environmental con-

straints were bypassed not by mutation but by the application of learning and memory—how to build fires, assemble shelters, create clothing and tools, plan a community hunt. Learned experiences were not confined to individuals but raced like the wind through one human group after another because of language—our ability for cultural communication. It took millions of years for certain animal groups to develop a capacity to fly. Through applied intelligence and through cultural communication, it took less than seven decades from the time we first ventured into the air until we landed on the moon.

What about the second possibility? Since the human species first appeared, there have been profound shifts in climate. Climate has a major influence on the amounts and kinds of vegetation that will grow in a region, hence on the numbers and kinds of animals that vegetation supports. As you read in Chapter Twenty-One, a general trend toward long-term drought may have forced our ancestors from dwindling forest homes and into the more productive grasslands. About 11,000 years ago, an overall warming of the earth apparently triggered a shift from the hunting way of life to agriculture—from risky, demanding moves after the game herds to a settled, more dependable basis for existence in more favorable environments. Even in its simplest form, the agricultural management of food supplies bypassed one of the most basic limits on the carrying capacity of the environment. Populations could expand to new limits. And with each cultural innovation—irrigation, metallurgy, social stratification to provide a labor base, development of fertilizers and pesticides—the limits were expanded and were met again with a resurgence of growth. Thus, with the domestication of plants and animals, and the development of agriculture, the environment's carrying capacity has risen abruptly for human populations.

What about the third possibility—the removal of certain limiting factors from the environment? The potential for growth inherent in the development of agriculture began to be fully realized with the suppression of contagious diseases. Until about 300 years ago, contagious diseases kept the death rate high enough to counteract the high birth rate. Contagious diseases spread like wildfire through crowded settlements and cities; they are density-dependent factors. Without proper hygiene, without

sewage disposal methods, and plagued with such disease carriers as fleas and rats, populations increased only slowly in size over the long term. But plumbing and sewage treatment methods did appear. Bacteria and viruses were recognized as disease agents. Vaccines, antitoxins, and drugs such as antibiotics were developed. Thus one after another major limiting factor on human population growth has been largely pushed aside. Smallpox, plague, diphtheria, cholera, measles, malaria—many diseases have been brought under control in the developed countries. Concurrently, medical technology has been exported in a humanitarian effort to control disease in developing countries as well. With old age and starvation the only remaining checks, population growth has been skyrocketing ever since.

We are faced with two options. Either we make a global effort to limit our numbers, so that our population stabilizes according to the environment's carrying capacity, or we passively wait until the environment does it for us. Our choices and the time we have to make them are far more limited than we would like to think. At our best, we are creatures of compassion, and any move to cause deaths deliberately would mean losing something of what being "human" is about. Even deciding to do nothing at all is to deny part of what we are. At the same time, because we are on a course to alleviate suffering and premature death, medical advances will continue to lower the death rate. Thus the most humane, reasonable option—short of biding time until the crash—is to reduce birth rates as dramatically as we have lowered death rates.

Age Structure and Fertility Rates

Two important factors influence just how much we can expect to slow down the birth rate. The first has to do with **age structure**: how individuals are distributed at each age level for a population (Figure 26.9). Let's explore the meaning of these factors by placing individuals of a population into three categories—those before, during, and after reproductive age. We can use the ages 15 to 44 as the average range of childbearing age. Once we do this, we can create diagrams that readily show the prospects for population growth. The age structure pyramid for a rapidly growing population has a broad base. It is filled not only with reproductive-age men and women but with a large number of children who will move into that category during the next fifteen years. As Figures 26.10 and 26.11 suggest, *more than a third of the world population now falls in the broad pre-reproductive base.* This gives us an idea of the magnitude of the effort that will be needed to control birth rates on

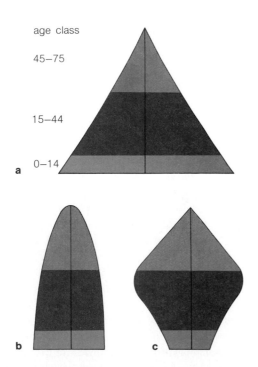

age class

45–75

15–44

0–14

a

b

c

Figure 26.9 Age structure diagrams for (**a**) a rapidly expanding population, (**b**) a slowly expanding population, and (**c**) a declining population.

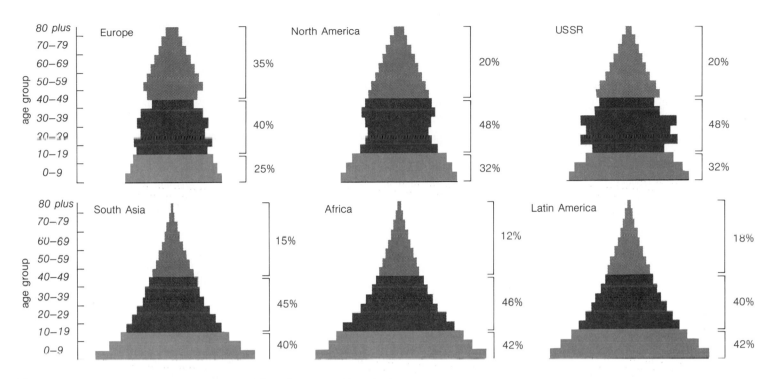

Figure 26.10 Age structure diagrams for the world's major geographic regions.

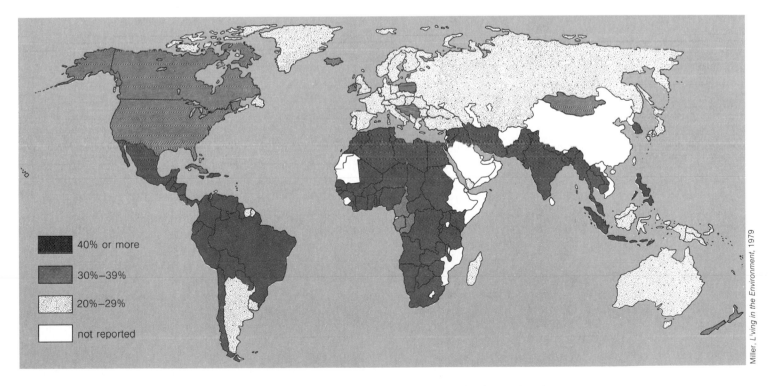

Miller, *Living in the Environment,* 1979

Figure 26.11 Percent of world population under age fifteen, and ready to move into the reproductive age bracket.

a global scale—and of the urgency surrounding birth control implementation.

One other factor influencing the short-term picture of population growth is the **fertility rate**: how many infants are born to each woman during her reproductive age. Frequently, the general fertility rate is calculated on the basis of number of births each year per 1,000 women between ages 15 and 44. Today the average number of children in a family is 2.6 in industrialized countries, and 5.7 in developing countries. A world average of 2.5 children per family is the estimated fertility rate that would bring us to zero population growth. But even assuming we can achieve *and maintain* a world average of 2.5 children per family, it will be 70 to 100 years before our population stops growing. Why? Because an immense number of existing children are yet to move into the reproductive age category!

Even when and if world population does level off, there will be a staggering number of people on earth. Say that a replacement level of 2.5 children per family were achieved by the year 1980. World population would continue to grow until it leveled off, in the year 2070, to about *6.3 billion*. Obviously, it is going to take planning and implementation of birth control programs over the next seven decades to achieve population control. At present, even five- and ten-year plans are virtually nonexistent.

The best we can hope for in the immediate future is to slow the growth rate and buy some time so that we *can* implement more effective measures. One simple way to slow things down would be to encourage **delayed reproduction**—childbearing in the early thirties as opposed to the midteens or early twenties. In Ireland, women customarily marry later in life. In China, the government has raised the age at which marriage is allowed, there are strong economic and social penalties for having more than two children, and rewards for having only one. Such cultural constraints tend to lower the number of children in each family. Figure 26.6 showed how delayed reproduction can slow down population growth rates. But remember that as long as birth rates even slightly exceed death rates, populations experience exponential growth. Short-term strategies ultimately must give way to long-term birth control measures (Chapter Eighteen).

PERSPECTIVE

In this chapter, we began with the premise that all populations have the potential for exponential growth—growth at an ever-accelerating pace to explosively high levels. We proceeded to analyze the nature of limiting factors that keep populations in check and thereby assure stability according to imperatives of the environment. The biological implications of population instability are, as we have seen, enormous. But so are the social implications of achieving stability, of achieving and maintaining zero population growth.

For instance, modern economic systems are based on constant growth, on uninterrupted consumption of goods and demands for services. The growth rate of the gross national product is the standard measure of the nation's economic health. An economic system in which consumption is suddenly lowered would require a totally new approach to government, to employment, to resource distribution, to individual life-styles and expectations. Presently, there are no plans for easing some new system in as old systems are eased out.

Consider, also, that most members of an actively growing population fall in younger age brackets. Under conditions of constant growth, the age distribution means there is a large work force. A large work force is capable of supporting older, nonproductive individuals with various welfare programs, such as social security, low-cost housing, and health care. With zero population growth, far more people will fall in the older age brackets. How, then, can goods and services be provided for nonproductive members if productive ones are asked to carry a greater and greater share of the burden? These are not abstract questions. Put them to yourself. How much are you willing to bear for the sake of your parents, your grandparents? How much will your children be willing to bear for you?

We clearly have arrived at a major turning point, not only in our biological evolution but in our social evolution as a species. The decisions awaiting us are among the most difficult we will ever have to make. Yet we must come to recognize that they *must* be made, and soon.

All species face limits to growth. A host of control mechanisms work to contain the potentially explosive reproduction of all populations. In one sense, we have proved ourselves different from the rest, for our unique ability to undergo cultural evolution has allowed us to postpone the action of most of these control mechanisms. But the key word here is *postpone*. No amount of cultural intervention can hold back the ultimate check of limited resources. We have repealed a number of the smaller laws of nature. In the process, we have become more vulnerable to those laws which cannot be repealed.

Visions of the future are many. All too many foretell of famine, disease, and war, as an overextended population readjusts itself to available resources. Even now, the industrialized world arms itself and talks of war if imports of what it considers to be essential resources are restricted. If you doubt this, think back on the meaning of talk at the local, regional, and national levels of invading the Near East in the wake of oil embargo and oil price hikes. At the same time, developing nations dream of using the weight of sheer numbers to wrest away resources that they, too, so desperately need. All nations of the world are being drawn inexorably into the conflict. There no longer is such a thing as a "regional conflict" over resources in the world we have made for ourselves. With the tangled, interdependent threads of trade and alliance reaching out from the nineteenth century, the specter has become one of another world war.

An individual facing these issues inevitably recoils. After all, what can a single person do to change the momentum of a species? But remember that cultural as well as biological evolution acts through *individuals* in a population. The technology that has made our greatest achievements possible has also originated with individuals—from the vision of the first deliberately shaped tool to a vision of a species at peace with the environment, and with itself. Whether productive human societies continue depends largely on as yet undefined social and economic innovations. These innovations will emerge only when the collection of individuals making up society recognizes the new set of imperatives. If you renounce your responsibility to use resources wisely, if you choose not to consider the urgent need for population control, then we surely must fail. For how can others be expected to sacrifice more than you yourself are willing to sacrifice? *The momentum of our species, be it in the direction of catastrophe or survival, has become the sum total of our commitments as individuals.*

We have come far enough to realize, perhaps for the first time, that what has been considered "good" in the past is not necessarily consistent with survival. It is not merely survival but survival with dignity and purpose that constitutes the greatest long-term "good" for our species. Thus we must now call on an ability we share with no other species—an ability to make a choice regarding limited resources and reproduction of our kind. We are fast approaching real limits to growth, and decisions can no longer be left to the next generation. It is not likely that we will in our time be able to conceive of and implement new social patterns. If we don't make a beginning now, however, the luxury of choice may be lost. Facing our self-inflicted dilemma will mark the end of the childhood of our species. But with enough commitment, it may yet come to represent the finest application of the unique talents with which we have been endowed.

Readings

Ehrlich, P., et al. 1977. *Ecoscience: Population, Resources, and Environment.* Third edition. San Francisco: Freeman. Outstanding source book, somewhat high level.

Frejka, T. 1973. "The Prospects for a Stationary World Population." *Scientific American* 228(3):15–24.

Polgar, S. 1972. "Population History and Population Policies from an Anthropological Perspective." *Current Anthropology* 13(2):203–241. Analyzes often-ignored cultural barriers to programs for population control.

Population Reference Bureau. 1976. *World Population Growth and Response: A Decade of Global Action.* Washington, D.C.: Population Reference Bureau. Superb analysis of the subject.

Scientific American. 1974. *The Human Population.* San Francisco: Freeman. Entire issue devoted to world population problems.

Zero Population Growth. 1977. *The Benefits of Zero Population Growth.* Washington, D.C.: Zero Population Growth. Excellent summary.

Review Questions

1. Why do populations that are not restricted in some way grow exponentially?

2. If the birth rate equals the death rate, what happens to the growth rate of a population? If the birth rate remains slightly higher than the death rate, what happens?

3. Explain the relationship between factors that limit population growth and the carrying capacity of the environment in which that population is found.

4. Contrast density-dependent and density-independent factors that influence population growth. Can you give an example of each?

5. Distinguish between extrinsic and intrinsic limiting factors on population growth.

6. At present growth rates, how many years will elapse before the human population doubles in number?

7. How have human populations developed the means to expand steadily into new environments? How have humans increased the carrying capacity of their environments? How have they avoided some of the limiting factors on population growth—or is the avoidance illusory?

8. If a third of the world population is now below age fifteen, what effect will this age distribution have on the growth rate of the human population? What sorts of humane recommendations would you make that would encourage this age group to limit the number of children they plan to have?

27

THE SOCIAL BASIS FOR POPULATION STRUCTURE

Each year, as biological alarm clocks go off more or less in inevitable synchrony, birds erupt in seasonal songs. Trills, screeches, whistles, crescendos, hoots—in this manner they announce their presence on strategic branches, fenceposts, treetops, chimneys, telephone poles. Through their individual ruckuses they are proclaiming title to food and space. But regardless of variations in number, kind, and volume of notes, all of these songs are meant to carry messages to prospective mates and prospective challengers. In their most fundamental sense, they mean the patterns of distribution for each feathered population are once more being renewed, tested, and redefined.

In turning to the manner in which animals are distributed in space, we must look not only to their interactions with the environment or with predators and prey. We must look also to their interactions with **conspecifics** —individuals of the same species. Underlying all such interactions are two opposing mechanisms. There are **dispersive mechanisms** for spacing individuals far enough apart to provide each one with enough resources for survival. There are also **cohesive mechanisms** for enabling individuals to come close enough together to mate (at the very least) and often for collective food gathering, mutual protection, and the rearing of young. The outcome of these dispersive-cohesive interactions among conspecifics is known as **social behavior**.

In this chapter, we will look briefly at the nature of social groups—how they can be shaped by individual distance and territoriality (dispersive behavior) and by social bonds (cohesive behavior). Along the way, we will look at some of the communicative displays that incorporate chemical, tactile, auditory, and visual stimuli in ways that are significant to the social group.

SELECTIVE ADVANTAGES OF SIMPLE SOCIAL GROUPS

If you have ever watched cars bunch up to a standstill on a freeway during rush-hour, you already know that not all groupings are social. In natural settings, too, simple aggregations can form that are nothing more than a response to some conditions in the local environment. For example, if there is a moist and rotting log in a forest, wood lice tend to collect there—not because they thrive on one another's company but because they thrive in moist and rotting logs. As each individual wanders into the log, it begins to feed and its rate of movement slows to a virtual standstill. Hence it would be there even if there weren't another wood louse in the log.

Consider, by way of contrast, the true social groupings of emperor penguins in Antarctica. During the continuous night of winter, winds there can reach 140 kilometers (90 miles) per hour, and temperatures can plummet to –60°C. The penguins face these icy blasts in huddles of thousands of individuals. Their use of one another's presence to help break up the force of the winds and to keep warm enough to survive is a large-scale social response. But their social behavior includes more than group responses to the environment. It also includes a true division of labor between parents. Parent penguins incubate their eggs during the long winter months, for their offspring must have time enough to grow and develop before entering the surrounding seas during the short summer, when food is abundant. Immediately after laying their eggs, all the females shuffle off to sea, which can be as much as 160 kilometers (100 miles) away. There they feed themselves during the entire incubation period. Meanwhile, back on the ice, each male is left holding an egg on his feet. They make do with what is available: feet. There are no grasses and twigs in Antarctica for building other sorts of nests. The males do more: they incubate the eggs by tucking them up into a warm body fold. But how do the males swim after food in the sea with an egg tucked under their tummy? They don't. They simply stop eating. The males live off their fat reserves until the females return, some two months later, to feed the chicks. Only then do they stagger off to find food.

As unique as the emperor penguin social group seems to be, it shares some basic features with types of social units found throughout the animal kingdom. For one thing, **pair formation** (in which a male and female join together to mate and perhaps share such tasks as food getting, nest building, brooding, and defense) is a common social unit. In some species, pairs last just through copulation or through the reproductive season. In other species they last a lifetime. Among fish such as the seahorse, it is the male alone that raises the young; among hummingbirds, wolf spiders, polar bears, squirrels, and deer, it is the female alone. But in many cases, both parents stay together to raise their young to maturity. Then the social unit may be composed of one or more parents and one or more sets of offspring; they are **family groupings**. Among species such as beaver, there may be as many as three sets of offspring living with the parents at any one time, which creates a rather large and heterogeneous family group. Among species such as wolves and lions, the social units are derived from family groups but the organization is a bit more flexible. Even as some offspring leave, young adults from other family units join up and form a relatively stable hunting unit.

Coherence of a family group bestows certain advantages on the young and thereby promotes their survival. The young are protected from predators and environmental extremes until they are mature enough to fend for themselves. More than this, they have their parents as models in learning where and how to obtain food and to avoid danger. Among certain migratory birds, adults know where to find a local feeding source along the route that a young, first-time traveler with the flock might miss. In effect, prolonged association of adults and offspring extends the amount of information that can be transmitted to offspring. *Thus, in addition to inheriting innate behavior patterns, the young of a family group acquire a set of learned behavior patterns that increase the chances for survival and reproduction.*

EVOLUTION OF MORE COMPLEX SOCIAL GROUPS

We can assume that under certain circumstances, the stable family grouping has offered enough advantages so that larger and more complex social units have evolved. Baboons, for instance, live and travel in large units (troops). All the females born into the troop stay on as part of the family, bearing young of their own when they reach maturity. But none of the young males stays on: eventually they all leave to join other troops. Sometimes the young males succeed in mating one or more times with the females before they depart, but often they do not. Sooner or later, the only adult males remaining to sire offspring are the ones that migrated in from other troops. This kind of structure offers both social stability (by the continued presence of females) and genetic diversity (by the continuing migrations of males).

The larger social units also include schools, herds, and flocks. They vary considerably in size, structure, and permanency. But they are alike in that they all extend selective advantages to the coherent family group. One of the most important of these advantages is a greater degree of protection from predators. For example, a herd of musk oxen (Figure 27.1) will cluster into a defensive formation around offspring at the sight of their natural predator, the wolf. Gazelles of the African savanna will thunder away in an undulating mass as lions launch their attack.

How did such defensive formations evolve in the first place? Usually, larger social units form in open environments—the arctic tundra, savanna, plains, open seas—where food resources are adequate for supporting large populations but where there simply aren't many places to hide. For lack of any other hiding place, it seems that cover is sought among other animals being preyed upon. Faced with such numbers, predators must attempt to separate potential prey from the periphery of the group. Falcons first make mock attacks against a flock of small birds, trying to separate one flying on the fringes of the group before attacking it (and thereby avoiding a possible mid-air collision). In the seas, the schooling behavior of one fish may be a way of taking cover by putting other fish between it and the predator. (Even fish that don't normally travel in schools will clump together in a tank in which there are no hiding places.) By virtue of sheer numbers, the individual apparently attains a measure of protection it would not have on its own. The potential predator usually is restricted in such cases to pursuit and capture of the very young, the weak, the old, or the injured individual that can't keep up with the group.

Both the individual members of a social group and the group as a whole are constantly being tested for their capacity to survive and reproduce in a given environment. As you read in Chapter Twenty-Six, built-in control mechanisms govern the actual size that any given group can maintain. *But within those broader environmental constraints, those individuals having behavior patterns consistent with survival of the group garner some advantage in their own survival and reproduction.* As a result, extremely complex group-oriented behavior patterns have tended to evolve.

In evolutionary theory, the trend toward complexity in social groups shows similarities to the trend that led to multicellular organisms. First there were casual associations of individuals of common descent. Such associations proved significant in competitive and in predatory interactions. The bonds between individuals grew stronger. With a division of labor, the complex animal group called a **society** could have formed. It's true there is some division of labor within many smaller animal groupings. In a baboon troop, for instance, some individuals act as sentinels on the periphery of the main unit. There they must detect and alert the troop to impending danger. Other baboons occupy the center of the troop and concentrate on caring for the young. But it is among the social insects such as bees, ants, and termites that division of labor reaches the level of a complex society (Figure 27.3).

Insect societies show such specializations and interdependence of individuals that they take on the character of "superorganisms." Just as a single cell of a multicellular organism can't survive and reproduce on its own, neither can a single social insect survive and reproduce on its own. In each case, only one insect of the group becomes the queen, the reproductive "organ" for all. But the queen depends totally on the other, nonreproductive members of the society for food, protection, and construction and maintenance of the nest in which young are raised. These nonreproductive members have unique structural and behavioral modifications.

For instance, some members of termite societies have structures and behaviors adapted for constructing and repairing the nest. Others, the "soldiers," have a long proboscis that secretes a sticky substance, which is used in defense of the colony. In some insect societies, the division of labor corresponds not only to structural traits but to the developmental stage. The first ten days of its life, a worker honeybee (Figure 27.3) devotes itself to cleaning honeycomb. Only toward the end of this stage does it begin making short exploratory flights. In the next ten days it not only cleans the hive, it engages in construction activities and transports nectar being brought in by foragers to storage cells. From that point on—from about twenty-one days after birth until the day it dies—it joins the force of foragers collecting the pollen and nectar that sustain the society. Such social groups are noted for their efficiency. When some members become specialized in food gathering, others in defense, and others in reproduction, there is little wasted energy in the group as a whole.

In highly integrated insect societies, selective agents must act upon the cooperative behavior of all the individuals. Suppose a queen bee has genes that enable her to produce large numbers of highly efficient foragers. In itself, the potential advantage for the hive would be wasted unless she also produces enough worker bees that will automatically sink their stingers into invaders of the hive, even though their defensive behavior means certain death (the entrails are ripped out in the process). If the queen does not give rise to a diverse array of cooperative members, she herself probably won't survive and pass on her genes. Only if forms of behavior mesh into coordinated activity will they be perpetuated through the queen's offspring. Again we can draw an analogy with the cells of a multicellular organism. Skin and bone cells, for instance, can't function in the reproduction of a new individual. But the genes determining the adaptive characteristics of skin (toughness) and bone (strength) reside in and are perpetuated by the reproductive cells—which they protect and support.

Insect societies are the most extreme expression of cooperative behavior, for they represent total suppression of the individual. Far more common is some combination of competition and cooperation between individuals in a social group. This dichotomy is expressed in many ways, and helps determine both the spacing of animals within groups and the spacing of groups within the environment.

Figure 27.1 Defensive formation of a family grouping of musk oxen on sighting their natural predator, the wolf.

a

b

Figure 27.2 (**a**) A portrait of male and female wolf, paired in the cooperative venture of raising the young.
(**b**) Greeting ceremony typical of the wolf pack. Guess which one is the dominant male.

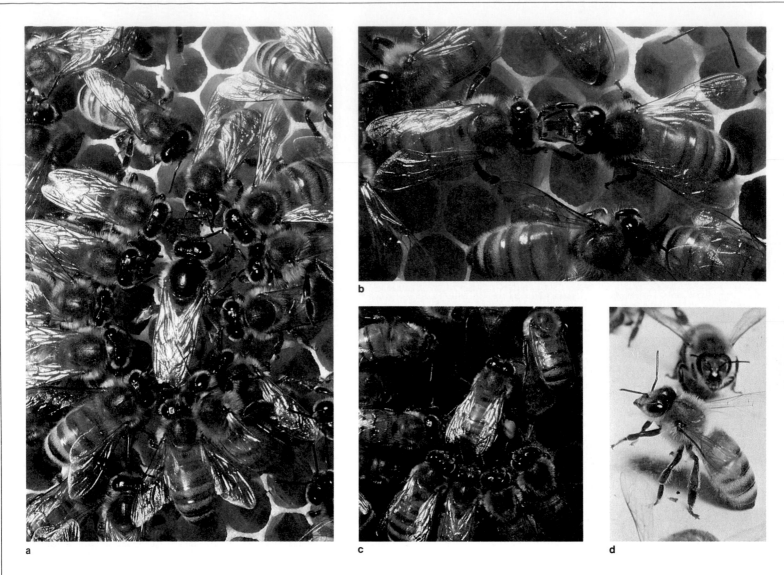

Figure 27.3 Life in an insect society. (**a**) The queen bee and her court. Attracted by pheromones, attendants constantly lick her body. The circle of attendants changes continually, which ensures that, through food transfer, the queen's pheromones (Chapter Fourteen) are passed through the hive. Thus her presence is communicated to all bees in the colony.

(**b**) Transfer of food from bee to bee. Considerable antennae tapping during food transfer helps the bees orient and communicate.

(**c**) Bee dance. Forager bees returning from the field communicate the location of pollen and/or nectar by performing a dance for potential recruits. Together, the angle of the "wag-tail" part of the dance on the comb, its duration, and the sound that the dancer emits communicate the food's location. Floral odors on the dancer's body also help the recruits find the right flowers once they arrive at the location indicated by the dance. Foragers carry pollen in

"baskets" located on their hind legs. One pollen pellet is visible on this forager's leg.

(**d**) Guard bees. Worker guard bees assume a typical stance at the colony entrance. With front legs uplifted and antennae outstretched, they are alert and ready to "examine" all approaching bees.

(**e**) The queen, the single most important member of the colony. She is the only egg layer. During active brooding, she lays about 1,200 eggs each day. There is only one queen for each colony; normally she lives for about two years.

(**f**) The drone. The male member of the colony has no sting. His only function in life is to mate with a virgin queen. The matings take place during flight, with seven to ten drones mating with a single queen. Each mating takes but a few seconds, after which the drones fall dead to the ground. About 300 to 500 drones are found in a colony. Those which do not mate with the queen live about one to two months.

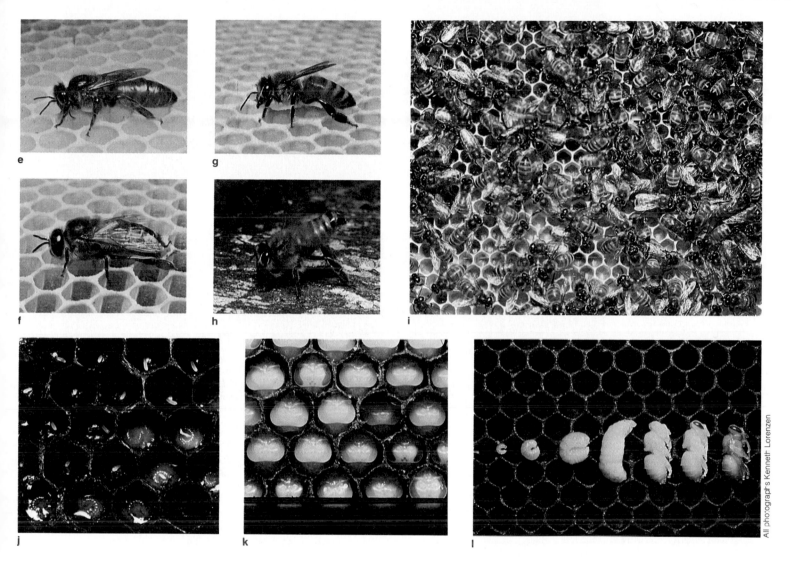

All photographs Kenneth Lorenzen

(**g**) The worker. The workers do virtually all the labor in the colony. They forage, feed larvae, guard the colony, construct honeycomb, and clean and maintain the nest. In the event that the queen is suddenly lost, young worker larvae can be selectively reared into new queens. Between 30,000 and 50,000 worker bees are present in a colony. They live about six weeks in the spring and summer, and can survive about four months in an overwintering colony.

(**h**) Scent-fanning. As air is fanned, it passes over the exposed scent gland of a worker bee. Pheromones released from this gland help other bees orient to the colony entrance or to the queen during swarming.

(**i**) Worker bees on new honeycomb. Worker bees secrete wax, which is used in constructing honeycomb—the site of pollen and honey storage, and brood rearing.

(**j**) Eggs and young larvae. Like other insects, honey bees pass through several life stages prior to becoming adults. These cells have been exposed to show one- and two-day-old eggs and larvae. The larvae are floating on royal jelly, a food secreted by worker nurse bees.

(**k**) Pupae. When cell cappings are removed, the uniformly aligned heads of worker pupae become visible.

(**l**) Developmental stages of the worker honey bee. The small egg is visible on the left. After three days, the larva emerges and grows rapidly for about six days. The larva then stretches out in its cell and becomes a ''pre-pupa'' for two days. The next nine days are spent as a pupa. As the pupa matures, eye pigments are the first to darken, followed by pigments in the rest of the body. After twenty-one days, the adult emerges from its cell, ready to take part in bee society.

a

b

Figure 27.4 (**a**) A stunning show of individual distance among a group of porpoise, fishing through the surf. (**b**) On shore, individual distance among gulls.

Figure 27.5 Ritualized threat behavior of a dominant male cheetah. Notice the exaggerated ruff of fur, the ears pulled back flat against the skull, and oh yes, the fangs.

SIGNS OF DISPERSIVE MECHANISMS: INDIVIDUAL DISTANCE AND TERRITORIES

Typically, individual animals tend to arrange themselves at some generally predictable distance from one another (Figure 27.4). This **individual distance** is defined by the balance between cohesive and dispersive forces acting on each individual. It's not a rigidly defined space. It varies with unexpected events, such as chance encounters with another individual. It also varies with internal rhythms and external rhythms keyed (for example) to the changing seasons. Between late summer and fall, a song sparrow generally couldn't care less about the presence of conspecifics. But come spring and the breeding season, sparrows are most intense about keeping one another at what they consider a suitable distance.

Aggression and Dominance

The most unambiguous way to maintain individual distance is to show **aggressive behavior**, a readiness to do injury to another individual or at least threaten to do so. Interestingly, we call such behavior "aggression" when it's directed at conspecifics, "feeding" when it's directed at prey, and "defense" when it's directed at predators. The same mechanical expressions—pulling, pecking, beating, and so forth—necessarily underlie all three behaviors because an animal has, by virtue of its physical makeup, only certain expressions available to it. Only the adaptive significance of the expression differs in each case.

Aggression to establish individual distance is an expression of competition within a group. Initially there is quite a bit of fighting, which results in *dominance* by one individual and *submission* by others. Such encounters work to separate the stronger animals by a greater distance from the rest of the group. More than this, they work to establish a ranking of its members—a **social hierarchy** of individual distance laden with social meaning. All members of the group with any survival sense at all come to recognize that meaning. As a result, actual fighting dwindles. The mere threat of aggression from top-ranking animals becomes enough to elicit submissive behavior from conspecifics and to maintain social order. *Because less energy is wasted on aggression, more is available for the business of survival. At the same time, selection favors the dominant animal, whose competitive show of strength assures it will get the choicest food and its pick of potential mates.*

A dominant animal advertises its higher status in formal displays, or **ritualized behavior**. Such displays are exaggerations of ordinary functional movements, and they are clearly distinct signals to conspecifics. Thus there is no mistaking a dominant wolf or dog, which carries its head and tail erect and walks with a stiff, formal gait. And there is no mistaking the dominant cheetah (Figure 27.5).

But what becomes of the weaker of the two combatants? Submission finds expression in two ways. In **avoidance behavior**, the submissive individual stays out of the way of the victor. But this maneuver is not necessarily as simple as it sounds. If all must use the same feeding grounds, the same paths, the same watering holes, and sometimes the same females, then the dominant and submissive individuals could be bumping into each other all the time, and the sight of the loser could provoke another attack. So the loser may learn the dominant animal's routine movements—and schedule its own activities at the watering hole, the feeding grounds, or on the paths to avoid confrontation.

Ah, but animals are not automatons, routines are not inflexible, and some confrontations are inevitable despite the best-laid schemes. This is especially true if the group lives in a relatively confined area. In such encounters, the submissive animal may resort to **appeasement behavior** as a further show of deference. No uplifted tails for the losers: instead they use exaggerated displays of submission. For wolves and dogs, appeasement displays include tucking the tail between the hind legs and presenting oneself before the dominant animal in a generally miserable, abject cower. And if those displays don't work, they can always resort to exposing the vulnerable throat area and/or genitals to the dominant animal. Such appeasement behaviors are highly ritualized among baboons (Figure 27.6).

The interesting thing is that, in the natural environment, a dominant animal rarely takes advantage of such total vulnerability. Only under the stresses produced by overcrowding in experimental situations (Chapter Twenty-Six) may the dominant animal carry its **aggressive** behavior through to the kill. *We are left to conclude that in the natural setting, selection pressures must favor the behaviors helping to keep the group together rather than the ones that would promote mutual destruction.* Indeed, there is probably selection against overly aggressive animals. They can frighten away potential mates, perhaps kill their own offspring (hence eliminate their own gene complement), and drive away other members of the population that afford, by their very clustering, mutual protection and assistance. A male baboon, if left alone as a result of its excessive aggression, is much more likely to fall victim to a leopard—which takes its prey from the fringes of the troop or takes those baboons isolated from the troop by their own antisocial behavior.

Territorial Behavior

So far, a dominance hierarchy has been described as being relative to each animal's perception of the individual distance between itself and its conspecifics. But assume environmental conditions are such that the animals are kept constantly on the move. Then individual distance is a portable thing, something each animal carries around with it. Such is the case for the baboon troops of the savanna, where scarcity of food and watering holes sends these animals foraging over a broad area. At the same time, the threat of predators keeps the troop from spreading too far apart, so they move as a unit.

But what happens when animals take up permanent residence somewhere? Then expressions of dominance become attached to a more-or-less geographically definable

a

area—a **territory**—which may be aggressively defended. The nature and size of any territory is determined by many variables—topography, the availability of resources, the number of predators, the number of conspecifics and how they are dispersed, even the time of year. Thus it's difficult to make generalizations about territorial behavior.

For example, in the breeding season a mockingbird becomes highly aggressive in its defense of a certain area, delineated perhaps by treetops, hedgerows, and open fields. From the trees it can scan its entire territory and readily observe any intrusions by other mockingbirds. It sings loudly to announce its presence and its identity, thereby alerting potential invaders that if they approach they will be observed—and will face a fight. How large an area can it carve out? *For any territorial animal, how much space it can defend and still have enough energy left for other essential activities partly defines the upper limits on its domain.*

b

Figure 27.6 Appeasement behavior among baboons. In (**a**), notice the assured position of the dominant animal—and the abject stare and groveling posture of the subordinate one, who is intent on making little conciliatory smacking noises with its lips. In (**b**), a young male presents his genitals to a dominant female; he already is aware of appeasement behaviors that help shape baboon social structure.

In contrast, ground-dwelling animals establish territories and they patrol them—but they can't patrol everywhere at once and they can't see all conspecifics in the territory at once. Often their territories are not even inside a fixed boundary but instead are networks of paths and places where activities occur. Consequently, some territories overlap extensively in space. Thus the scent markings of wolves and dogs are not only olfactory signals of territorial dominance, they also may be traffic signals for maintaining spatial exclusion. Paul Leyhausen suspects that "a fresh mark means 'section closed,' an older mark means 'you may proceed with caution,' and a very old mark means 'go on, but before you use this please put your own mark so that the next one knows exactly what to do.'" In this way, competing animals avoid direct confrontation with one another.

When territories are first established, there usually are aggressive encounters of some sort. Rarely are they fights to the death. Once an animal wins a battle, it accepts submissive behavior of the loser or permits the loser to flee. Such fights are most vigorous near the center of what an animal claims as its territory, as has been shown in the case of the male stickleback fish. As the male moves farther and farther away from the territorial center, the attacks on intruding conspecifics become less and less vigorous. Edward Wilson thus defined territory as "a space in which one animal or group generally dominates others which become dominant elsewhere."

And yet, having a defined living area does not necessarily imply aggression. The whiptail wallabies live in distinct groups. Each group has its own restricted living area. But the areas occupied by such groups are not defended—one group simply avoids neighboring groups. Areas occupied by one group or individual to the exclusion of conspecifics, but which are not aggressively defended, are called **home ranges**. The home ranges of adjacent wallaby groups often overlap. Where such overlapping occurs, members of different groups feed peacefully together. Beyond the normal expression of hierarchical dominance (in this case established to determine access to females), a male invading the home range of another group encounters no aggression whatsoever.

SIGNS OF COHESIVE MECHANISMS: COURTSHIP AND RAISING THE YOUNG

Courtship Usually Conquers All

Many forms of behavior go into dispersing conspecifics and maintaining individual distance, even when that distance reaches the dimensions of territoriality. Thus you might suspect that an aggressive animal would encounter a few problems when it decides to let another animal—most notably a mate—into its private domain. And you would be right. Before mating can take place, the responses of male and female must coincide. The dovetailing of interests occurs not only because of hormonal factors, it occurs also because of specialized behavior patterns called **courtship displays** (Figure 27.7). *The presence of a potential mate may simultaneously give rise to sexual interest and fight-flight reactions; through courtship displays, these opposing behaviors are resolved.*

The male stickleback fish, for instance, can't quite seem to make up his mind about what to do. This highly aggressive fish spends a good part of the time vigorously defending his territory, which centers on the nest he builds. This territory he defends from males whose bellies turn red during the breeding season. The red belly acts as a signal for mutual recognition among males of the species, and it triggers aggression. Ah, but the red belly also tends to attract female sticklebacks. When a female entering the territory is embellished with a certain signal of her own—a silver belly swollen with eggs—then the male proceeds with an astonishing combination of courtship and aggression. It is normal for territorial invaders looking for a fight to turn upward, thereby displaying their belly and their intention. The female, too, turns upward, but her swollen belly deters rather than provokes the male's attack. He (usually) does not attack her and turns away to lead her to the nest, and the female follows. Then the male, who no longer sees the belly stimulus, turns back to attack her; she counters by displaying her belly; and so on until their zigzagging brings them to the nest. When the female peers into it, the male shoves her in and then trembles against her until she discharges her eggs. With that, her swollen belly disappears, aggression once more reigns supreme, and the male drives her away. He then proceeds to fertilize the eggs and, later, to protect the developing young.

At least among sticklebacks, the mate is allowed to escape. Among many species of spiders, males face the titillating possibility of becoming a meal as well as a mate for the larger female. The male's presence in her domain may just as easily evoke her breathtakingly eclectic predatory behavior—which may be one of the reasons why males are a good deal smaller and (hopefully) less conspicuous to her. Understandably, he approaches the dangerous female with elaborate courtship displays and the utmost caution. Sometimes he remains a safe distance away from her and waves conspicuously marked appendages for hours; only if he manages to put her in a trance with these hypnotic movements can he get away with it. One male garden spider announces himself by plucking in a species-specific way at a strand of the female's web. If there is no return signal—or if the web trembles under the pattering of eight rapidly approaching feet—dispersive behavior takes hold, he drops the strand, and runs like the devil. If, however, she responds with a gentle tug on the strand, he proceeds with his mission, tugging seductively on the strand until he is close enough to mate.

Mutual Recognition of Parents and Progeny

With all the aggression surrounding the stickleback's defense of its nest-centered territory, it may come as no surprise that it is an equally attentive parent, up to a point. It is singleminded in the way it protects the eggs from rival sticklebacks (who like to eat them) and in the way it constantly fans fresh water over the eggs and supplies them with oxygen. But come the moment of birth, the male abandons the newly hatched offspring and swims into other territories, with an eye to some other stickleback's yolky treasure.

Delayed aggression is even more pronounced among mouth-breeding fish. One species of catfish holds up to fifty fertilized eggs in its mouth until they hatch (which in itself is no small feat: each egg is about the size of a marble, and the catfish itself is no more than about two feet long). It also carries the newly hatched offspring in its mouth and protects them from predators, spitting them out only to let them forage. Although its parental care is commendable, it is offered only until the young grow to a certain length. That length signals the end of the male's cohesive behavior. Should one then return after being spit out for the last time, it is looked upon as fair game.

Among almost all mammals, bonds between parents and offspring are established by the chemical signals in licking behavior, which takes place during a short period following birth. In fact, in some experiments in which the young were isolated before this behavior could occur, the mother showed no recognition of her own when they were brought back to her. Mother goats will drive such kids away when they try to nurse.

Visual releasers are vital in establishing and maintaining cohesive bonds between parents and offspring of many species, especially birds (Figure 27.8). Among birds that are born blind, weak, and featherless, practically the only motor response of which they are capable is to open their mouth in a wide gape. But it's enough: gaping is a strong visual signal that elicits parental feeding responses. Soon insects or seeds or fruit or meat are shoved into little gullets at an amazing rate. The record may be held by a parent bird that made 800 food-bearing trips to its nest in a single day. Nestlings often can eat their own weight in one day, which can lead to a fiftyfold increase in body size in a matter of weeks.

The young seem capable of trying to hang onto a good thing when they see it. Even when they are first beginning to forage with their parents, they will still try gaping and quivering their wings, demanding the feeding response even when there is food all about them. The parents eventually reach the point of physical exhaustion, so that even gaping stimuli are no longer recognized. Before the coming of the first winter, the young will be dispersed to fend for themselves, either alone or within the hierarchy of a flock, until the cycle of reproduction and the cohesive behavior it implies begins again.

PERSPECTIVE

If we assume natural selection is the basis for evolutionary change, then behavioral as well as structural traits that are the most adaptive, that provide the edge in competition for resources, will be selected for. Behavioral competition among individuals often takes the form of aggressive behavior. On the surface, such behavior would seem to preclude the formation of social groups. Social groups are based, to varying degrees, on mutual cooperation. Aggression is dispersive, cooperation is cohesive. But even though competition and cooperation are opposing behaviors, they are not necessarily mutually exclusive. Both offer advantages that are competitive in terms of survival, and both may be subject to natural selection. Within a given social structure, competition can give dominant individuals first access to resources—food, shelter, space, mates. But within the same structure, cooperation leads to such advantages as the protection afforded by numbers—protection not only for submissive members but for dominant ones as well. It offers the potential for the division of labor, and the possibility for prolonging learning through the developmental period for offspring, which is exceedingly adaptive in variable environments.

Figure 27.7 Courtship behavior among the albatross. (**a**) The male spreads his wings as part of a complex courtship ritual that also includes show of a puffed-up chest (**b**). Bill-touching (**c,d**) represents the breakdown of individual distance barriers and the onset of a year-long pair bonding and the raising of young.

Figure 27.8 Under the attentive eye of their father, and pressing against their mother's body, which helps break the cold offshore wind, these Caspian tern chicks convey the bonds between protector and protected.

In many animal groups, dominance moves past the preservation of individual distance to geographic territories. In other animal groups, dominance maintains individual distances but does not get translated into territorial aggression. More than this, the intensity of aggression and territoriality are not constants. They vary with time, with environmental disruption, with changes in the numbers and distribution of conspecifics. They are only two of many factors—some known, many not yet identified—that shape and balance the behavior of individuals and groups. Finally, sociality as well as aggression appears to have an innate basis in animal behavior.

How do these generalizations apply to human social behavior? We are, after all, products of our biological heritage. Our flesh is like the flesh of all other mammals; our blood flows when we are injured, as their blood flows. We, too, must carve out a share of life's resources to survive. Based largely on a sense of this kind of unity with the nonhuman world, many writers—both trained scientists and popularizers—have attempted to decipher what is the inherent, "natural" behavior of humans. One such writer

has drawn the conclusion that humans are born murderous, aggressive, and warlike in defense of territory. Others have attempted to use highly regimented, socially stratified insect societies as models for human behavior. Both approaches are based on the notion of **biological determinism**: they have in common a tendency to overestimate the role of heredity in determining behavior and status in human societies. At the outset, these approaches often are rooted in observable facts. But as they develop, they bypass the essence of human biology.

No one can deny that the behavior of ants, for example, is encoded in their genes. Given the extremely limited space the brain occupies in the ant body, it could not be otherwise. Either ants will be preprogrammed to behave in a manner that promotes the welfare of the colony into which they are born or they will perish. But to attribute to humanity the same level of biological determinism is a mistake, for it totally ignores the unique contribution of culture. Culture in its broadest sense is the accumulation of the experience of past generations, which can be used to modify present behavior. The *capacity* for culture is determined genetically, for our DNA encodes the blueprint for constructing our storehouse of behavioral potential, the human brain. But that brain happens to be the most complex brain in the living world. It allows us to store the widom and folly of past ages—and learn from them.

In the final analysis, does it truly make much difference whether the first humans were gentle seed eaters or marauding bands preying on one another? For it is here and now that we have both the capacity for competition and the capacity for cooperation. It is here and now that we have the capacity to choose a path of aggression or a path of coexistence. It is true that competition for resources has been a driving force of evolution. But over time, in species after species, the *unit* of competition has shifted as cooperative behavior leads to ever larger social groups.

Among most animals, there are variously defined "in-groups" and "out-groups" that form the basis for their competitive behavior. Among humans, too, there are various definitions of who is "in" and who is "out." But these two terms are relative; they are not constants. For example, there are times when we compete with one another as Northerners and Southerners, as blacks and whites—and then unite as Americans in competition with, say, the Russians. Yet, in rare time of crises—for instance, in the tragic aftermath of a devastating earthquake in a densely populated country—we remember our common humanity. Then, the cooperation we have achieved within the parochial confines of our regional "in-group" flows out to embrace those of us in pain in the distance. Such moments illuminate our capacity for self-recognition.

If there has been any progress at all in human cultural evolution, it has been in our formation of ever more inclusive in-groups and in our attempts to extend our circle of trust. We still have a long way to go, for we all know these circles are as fragile as our shifting priorities. With this in mind, someone joked that we might try populating the planet Mars with imaginary enemies in order to achieve worldwide solidarity, a universal circle of trust against a newly defined out-group. But perhaps in place of this imaginary external threat, we might come to unite before very real ones common to all of us. For we face not only the threat of weapons capable of destroying much of the world, but the threat of self-intoxication with our own technology as it carries the human population further and further away from ecological stability. Overcoming these threats, as you will read in the last chapters, will begin with collective consciousness of the human population as one social group, with our territorial responsibility the entire world.

Readings

Crook, J. 1970. *Social Behavior in Birds and Mammals*. New York: Academic Press. Broad summary of intraspecific social structure as adaptations to the environment.

Esser, A. 1971. *Behavior and Environment*. New York: Plenum. Excellent coverage of social behavior and social space.

Klopfer, P., and J. Hailman. 1974. *An Introduction to Animal Behavior*. Second edition. Englewood Cliffs, New Jersey: Prentice-Hall. Covers historical development of basic concepts in animal behavior.

Lorenz, K. (translator M. Wilson). 1952. *King Solomon's Ring*. New York: Thomas Y. Crowell. Delightful animal lore.

Thorpe, W. 1956. *Learning and Instinct in Animals*. Cambridge, Massachusetts: Harvard University Press. Important survey of natural behavior in all major animal phyla.

Review Questions

1. Define dispersive behavior and cohesive behavior.

2. How do simple aggregations of animals differ from true social groupings?

3. How might the formation of pair bonds and family groupings promote survival and reproduction?

4. What forms of communication promote cohesive behavior in a bee colony?

5. In 1979, sensational newspaper headlines warned of an impending attack of "killer bees"—a highly aggressive species not at all timid about attacking cattle and humans. The killers were reported to be on the wing from South America, through Mexico and, of course, planning to do eventual battle with citizens of the United States. Along the way, the aggressive individuals have been encountering hives of more docile honeybees. Can you speculate on how these encounters would prevent an "attack of the killer bees" from materializing? Incorporate the concepts of social communication and gene flow into your answer.

6. Define individual distance, and explain aggression in terms of it.

7. Distinguish between these terms: ritualized behavior, avoidance behavior, and appeasement behavior. What selective advantage do these forms of behavior apparently offer?

8. Define territory. What helps determine the upper limits of any territory? How does a territory differ from a home range?

9. What opposing forms of behavior are resolved in courtship displays? Give an example of such resolution.

10. What is biological determinism? What factor is not given much weight by proponents of this idea?

28

COMMUNITY INTERACTIONS

Flying through the dense rain-forest canopies of New Guinea is something that is not your typical pigeon. The natives call it *gara*. The *gara* is cobalt blue with plumes on its head, and about as big as a turkey, and flaps so slowly and noisily that its flight has been likened to the sound of an idling truck. It eats fruit, as do eight other sympatric species of New Guinea pigeons. Now, this somewhat enormous pigeon obviously requires more food energy than its smaller relatives do in order to maintain body weight. Could it be that *gara* populations eat so much that the other pigeon populations are severely limited in numbers, or bordering on starvation? Not so. All of these related fruit-eating species have carved out a place for themselves in the same setting. The larger birds sit on heavy branches to dine on large fruit hanging from them —and the smaller birds dine on fruit hanging from thin branches that never would support the weight of a turkey-size pigeon. Hence part of the food supply that is not available to one species supports others in the same forest, even in the same trees. The example reminds us that populations rarely, if ever, exist by themselves. To understand their character, we must look not only within their boundaries, as we have done in preceding chapters, but also at their interfaces with other organisms in the same environment. These interfaces help establish community structure.

A **community** may be defined as all those populations of different species that occupy and are adapted to a given area. In this chapter, we will consider the kinds of interactions that may occur among its constituent species. Then we will turn to some ways in which these interactions shape community structure. We will also be considering the environmental stage on which the drama is played out. Thus, this chapter deals with the whole interlocked system, or **ecosystem**: the community and its physical environment.

You should know at the outset that no one has yet identified all the interlocking relationships in communities or ecosystems. Some guiding principles do exist, however, and they help us understand in broad outline the kinds of events that are taking place. With this understanding has come the dawning of an important idea: that in working with instead of against these principles, we may at last achieve the long-term stability that is potentially ours to share with the world of life.

HOW SPECIES INTERACT

Each environment is a unique physical stage, a particular combination of temperature, soil or water type, and so on. On this stage, different populations interact, and they do so in three basic ways. There may be *competitive interactions* for resources, such as food or living space. Some species may be diner or dinner to one another, and may thus engage in *predator–prey interactions*. Or one species may collaborate with another, readily or reluctantly, in *symbiotic interactions*. It is not yet possible to identify all the complex interactions going on at once in any of these relationships. But a few concepts have been formulated that give us a general idea of what may be occurring. Foremost among these is the concept of the niche.

The Niche Concept

A **niche** is defined by all environmental or ecological requirements and interactions that influence a species in a community. What resources does the species require, and are conditions such that it is able to carry out its life cycle? What energy does it demand, and in what amounts? How much water is needed, and what kind of shelter? What is the range of temperature, wind, shade, and sunlight it can tolerate? How and where are offspring raised? What is the extent of the species' **habitat**—its geographic range, with some characteristic array of organisms? What other species does this particular species depend on, prey upon, avoid, or compete with? How do resources and living conditions vary in time? Such variables define the species' niche.

Given the limited resources and range of conditions in any environment, one species' niche has at least the

John Dominis, Life Magazine. © 1965, Time Inc.

Figure 28.1 Confrontation between predator and prey. As a last resort, the baboon has turned back to face the leopard with threat behavior. The effect has momentarily stopped the predator's advance; under other circumstances it might have meant the difference between capture and escape. Here, in the open, the baboon has run out of alternatives.

potential to overlap the niche of another. For instance, water is a potentially limiting resource for desert communities. For this environmental variable, niche spaces are not constant but are partitioned in time. Some species use a waterhole by day; others use the same waterhole by night. One plant species dominates the habitat during a prolonged dry season; others dominate the same habitat during a brief rainy season. By way of analogy, *think of a niche as an ever-changing cloud*. The cloud shrinks and expands along many axes, in large and small ways, depending on where the cloud happens to be in time and space—and in relation to other niches. What are some of the factors that bring about changes? Let's take a look.

Competitive Interactions

When different species have some requirements or activities in common, their niches overlap. The species may end up in competition with one another, especially when some resource is in short supply. Perhaps competition exists between animal species. (For example, two bird species may compete for nesting places in the same trees, which has direct bearing on reproductive activities.) Perhaps competition exists between plant species. (As you will read later, tall trees that create a dense canopy shade out most other kinds of plants on the forest floor.) Whatever the case, the greater the overlap, the greater the potential for competition between them.

In **competition**, one species or individual exploits the same limited resource as another, or interferes with another enough to keep it from gaining access to the resource.

It may be that when two species are competing for the same thing, one tends to exclude the other from the area of overlap. This prediction is based on a concept called **competitive exclusion**. The exclusion is said to occur because, in theory at least, no two species can be exactly the same in their activities or in their ability to get and use resources. To a greater or lesser extent, one would have the advantage; and the other would be forced to modify its niche. According to this view, no two species in a community can simultaneously occupy the same niche for very long.

Part of the problem with trying to observe competitive exclusion in the natural world is that it may not be an immediate outcome of competition. Whether exclusion

occurs may be influenced by other factors, such as the abundance of resources or population density.

Balancing the tendency to minimize niche overlap is a tendency for each species to expand its niche. It would seem that each species is contantly being tested in terms of its ability to maintain or expand its niche, to fill vacancies that may come to exist in the current community structure (through extinction of one species, for instance), or to compete more effectively than other species when conditions change. More will be said about this concept in the section on community organization.

Predator–Prey Interactions

The term **predator** refers to an organism that actively seeks out and/or captures another living organism, then ingests either the whole thing or parts of it. The sought-after organism is the **prey**. Predators that dine on plants (usually plant parts, such as leaves) are *herbivores*. Those that dine on animals are *carnivores*. The ones that can eat both plants and animals (and other kinds of organisms) are *omnivores*. In Chapter Twenty-Six, you read how predator–prey interactions are extrinsic limiting factors on population size. These interactions help regulate growth and distribution of both predator and prey populations. Here we will look at how these interactions have led to remarkable adaptations that assist in (1) finding and capturing prey, and (2) hiding from, fighting, or escaping predators within the community.

Although exceptions do exist, the following is generally true of predatory activities:

Predators generally focus on prey that are large enough or numerous enough to make the hunt and/or the capture worth the energy outlay.

For instance, tiny shrimplike forms called krill are preyed upon by such behemoths as the blue whale. (One krill population consists of many hundreds of thousands of microscopically small marine animals. If you put twelve elephants together, they collectively would be about as massive as one blue whale.) These whales suck in great quantities of krill-containing seawater, which passes through horny filter plates hanging vertically from the roof of the mouth. Water is forced back out, but the krill get stuck in the plates and drop as a mass to the whale tongue. Apparently it is a large-enough mass to justify the energetic cost of constructing and maintaining the specialized equipment, and of filtering the food. The killer whale, in contrast, gets by with a dozen or two seals, penguins, and dolphins—which certainly are larger than krill but trickier to catch. They represent a larger energy return that more or less equals the larger energy investment required for the hunt and capture.

Prey species respond to predatory attack in four general ways:

Prey species avoid predation by adaptations for flight, hiding, fighting, and/or disguise.

First, some prey species rely on running, swimming, or flying out of reach of predators. Such species typically live in habitats where there aren't many places to hide. For instance, fleet-footed ungulates (hoofed mammals) predominate in the open areas of savannas; fast-swimming fish, seals, and porpoises are found in open seas. Second, where hiding places do exist, there you will find prey species hiding—rabbits under the cover of vegetation, lizards beneath rocks, animals in coral nooks. Third, some prey species will, when cornered, defend themselves with display behavior that may startle or otherwise intimidate a predator (Figures 28.1 and 28.2). Aside from attempts at intimidation, built-in defenses range from spines, to armored plates, to noxious or burning chemical secretions. Many plants have thorns and spines, and some secrete foul-tasting chemicals that discourage herbivores. The beetle affectionately known as the stink bug raises its abdomen and sprays a noxious chemical at attackers. It is an effective adaptation against some predators. But grasshopper mice (Figure 28.2b,c) have learned to get around the defense. These mice pick up the beetles, shove their tail end into the earth, and munch the head end.

Finally, perhaps the most complex of all predator–prey interactions are camouflage and mimicry—"hiding" in the open. Camouflage, or **crypsis**, refers to adaptations in form, patterning, color, or behavior that enable an organism to blend with its background and escape detection. The peppered moth (Figure 25.4) is one example of camouflaging in prey; Figures 28.3 and 28.4 show others. Camouflage also takes the form of countershading: body parts viewed against light backgrounds are light-colored, and those viewed against dark backgrounds are dark. Thus in many surface-feeding fish, dorsal surfaces (which are visible to bird predators looking down on the water) are dark, and ventral surfaces (visible to underwater predators looking up) are light-colored. Crypsis also occurs in plants. In Texas you will find cacti that look like small rocks in shape and color (Figure 28.3). Only during the brief rainy season—when water and vegetation are more plentiful for desert

a

b

c

Figure 28.2 Moment-of-truth defensive behavior. (**a**) A cornered short-eared owl spreads its wings in a startling display that must have worked against some of its predators some of the time; it is part of the behavioral repertoire of the species. (**b**) As a last resort, some beetles spray noxious chemicals at their attackers, which works some of the time but not all of the time: (**c**) Grasshopper mice plunge the chemical-secreting tail end into the ground and feast on the head end.

Figure 28.3 Mimicry among the rocks. Find the plants (*Lithops*) that mimic stones.

Figure 28.4 The fine art of camouflage, as developed in predator-prey interactions. (**a**) To confuse the bird predator, one katydid species has become so adapted to its background that its members look like leaves—even down to blemishes and chewed-up parts! (**b**) Caterpillars of some moth species tend to look like bird droppings because of their coloration and the body positions that they assume. (**c**) By lurking motionless against its like-colored background, the yellow crab spider is essentially invisible to its prey. (**d**) What bird??? With the approach of a potential predator, the American bittern stretches its reed-colored neck and thrusts its beak upward—and even sways gently, like the surrounding reeds in a soft wind.

herbivores—do the "living rocks" put forth bright-colored flowers that attract pollinators. Furthermore, crypsis is not the exclusive domain of prey. Predators that rely on stealth are also extremely well adapted to their backgrounds. Polar bears against snow, tigers against tall-stalked and golden grasses, pastel spiders against pastel flower petals (Figure 28.4c) are a few examples.

In Batesian mimicry, recall, a harmless or tasty species resembles a different and dangerous or foul-tasting species (Figure 25.6). The resemblance probably arose through the selection pressures of predation. In addition, sometimes two or more species of dangerous or noxious organisms resemble each other. Hornets and wasps, for instance, carry a nasty sting; both kinds of organisms sport black and yellow rings on their bodies. Bird predators recognize the warning signals and avoid them all. (This resemblance between *harmful* species has been called Müllerian mimicry. However, no deception is involved, and there is no division into model and mimic.)

Symbiotic Interactions

Today, few species can live for long in isolation from others. But some species happen to be more demanding than others about the kinds of organisms with which they interact. Sometimes one or both members of two species come to rely on the continuing presence of the other for survival. Such a relationship is called **symbiosis** (the word means "living together"). Unlike predatory encounters, symbiotic relations take place between a "guest" species and a "host" species. As you might suspect from your own experiences with visiting relatives, some symbionts are helpful, others are awful, and others are barely noticed.

Symbiotic relationships differ in the degree of collaboration, in how exclusive the attachments are, and in the extent to which one species is helped or harmed by the presence of the other.

The weakest symbiotic attachment between species is **commensalism**. Here, a guest species simply lives better in the presence of its host. The host species does not garner direct benefits from the guest species' presence, but neither does it suffer harm. For instance, robins and fruit flies are both commensal with humans. It's not that they find it impossible to live without us. It's just that, for them, the carrying capacity of the environment is usually higher in places where humans live. There are concentrated resources such as shade trees, earthworm-rich lawns, and berries (which robins enjoy), and overripe fruit (which fruit

flies feed upon). We may find ourselves enjoying the territorial song of a male robin, or we may be mildly disturbed when fruit flies swarm over a fruit bowl. But most of the time we aren't even aware of the presence or absence of either species. Such is the weak involvement in most commensal attachments.

In **mutualism**, bonds between members of two species are stronger than they are in commensalism, for positive benefits flow in both directions. Most of these attachments probably begin in commensalism and gradually turn into mutual dependence. Lichens are like this. A lichen is a composite organism: filaments of a fungal species form a dense mat above and below cells of an algal species. The two parts can reproduce separately, but the lichen can reproduce asexually as a unit, too. The fungal part depends on nutrients from the photosynthetic algal cells; and the algal part depends on the fungal mats for protection from intense sunlight and extreme temperatures found in places where lichens grow.

Examples of mutualism are also found among some flowering plants and their insect pollinators. The pollinating species depend on the plants for food, even as the plants depend on the pollinators for reproductive success (Chapter Twenty-Three). In many cases, their mutual bonds are not exclusive. If honeybees find clover blooming profusely, they concentrate on collecting food from clover. When blossoms of that species fade, they turn to honeysuckle, apple blossoms, and other flowers. Similarly, if honeybees are not around, then bumblebees, butterflies, or moths may pollinate the clover.

As time goes on, mutualism may become more and more exclusive. Two species may coadapt structurally, functionally, and/or behaviorally because each acts as a powerful selective agent on the other; they may undergo **coevolution**. The yucca moth (Figure 28.5) has come to obtain pollen only from the yucca plant; even its larval form dines only on yucca seeds. The yucca plant depends exclusively on this one moth pollinator. Hence the moth's private energy source, available throughout its life cycle, helps assure reproductive success. At the same time, the moth helps assure reproductive success for the plant: the plant's pollen is carried exactly where it must go instead of being randomly spread about by a less picky pollinator that visits different plant species. Whenever selective advantages lead to specialized structures or behaviors that make each partner coadapted to and totally dependent on the other, the two species have entered an **obligate relationship**.

Another form of symbiosis is the one-way relationship called **parasitism**, in which a guest species benefits at the

Figure 28.5 Coevolution in the high desert of Colorado. There are several species of yucca plants (**a**), but each has coevolved exclusively with only one kind of yucca moth species (**b**).

The adult stage of the moth life cycle coincides with the blossoming of yucca flowers. Using mouthparts that have become modified for the task (**c**), the female moth gathers up the somewhat sticky pollen and rolls it into a ball. Then she flies to another flower and, after piercing the ovary wall, lays her eggs among the ovules. She crawls out the style and shoves the ball of pollen into the opening of the stigma.

When the larvae emerge (**d**), they devour about half the seeds of the yucca plant and then gnaw their way out of the ovary to continue the life cycle. The seeds remaining are enough to give rise to a new yucca generation.

So refined is this coevolved dependency that the moth and larva can obtain food from no other plant, and the flower can be pollinated by no other agent.

a

b

c

d

All photographs Harlo H. Hadow

expense of a living host. A parasitic interaction may evolve from commensalism. Perhaps mutations arising in a guest species lead to modifications in form, functioning, or behavior. Perhaps these modifications enable the guest to tap into the host's body as an energy source. Sporozoans such as *Plasmodium*, the culprit in malaria, do this (Chapter Twenty-Two). Perhaps the guest comes to use the host for living quarters or for transportation during some stage of the life cycle. Schistosomes such as flukes do this; they use snails and humans not only as food energy sources but as breeding grounds, as you will see later in this chapter. Whatever the origins, parasites spend less energy

on surviving by allowing their hosts to perform specialized tasks for them.

Compared with their nearest nonparasitic relatives, parasites tend to be structurally simple. Once they have tapped into the energy stores of their living hosts, they are no longer under selective pressure to build and maintain their own energy-procurement systems. For instance, some species of parasitic mistletoe plants sink roots into a species of oak tree and tap the host's phloem tissue. Once they do this, they have easy access to food and water. Some of these parasitic plants do not even bother to form chlorophyll.

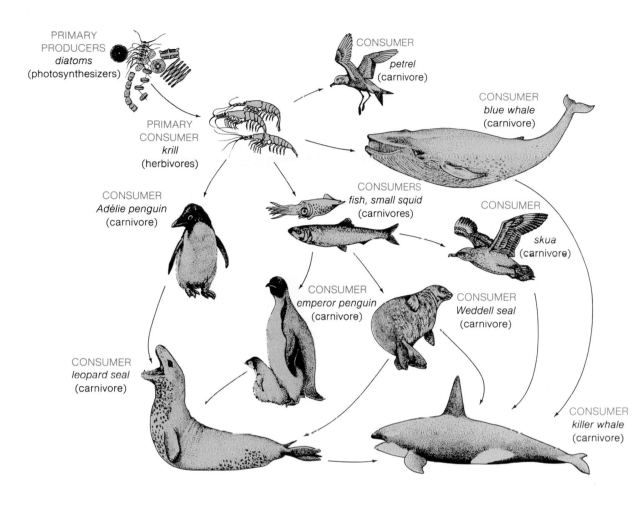

PRIMARY PRODUCERS
diatoms
(photosynthesizers)

PRIMARY CONSUMER
krill
(herbivores)

CONSUMER
petrel
(carnivore)

CONSUMER
blue whale
(carnivore)

CONSUMER
Adélie penguin
(carnivore)

CONSUMERS
fish, small squid
(carnivores)

CONSUMER
skua
(carnivore)

CONSUMER
emperor penguin
(carnivore)

CONSUMER
Weddell seal
(carnivore)

CONSUMER
leopard seal
(carnivore)

CONSUMER
killer whale
(carnivore)

Figure 28.6 Who eats whom in the Antarctic— a few of the participants. (After Ingmanson and Wallace, *Oceanography*, 1979)

HOW SPECIES ARE ORGANIZED IN COMMUNITIES

Food Webs and Trophic Levels

We have been describing one-to-one interactions between different species. How do we go about identifying the ways in which these interactions shape the structure of an entire community? From earlier chapters, you already know something about identifying the community members: the variety of species present, the distribution of their populations through the habitat, their relative numbers. More importantly, you already know that each kind depends on a flow of energy and a supply of material resources. Hence a good way to start would be to poke about and observe who eats whom:

Feeding relationships among species determine the direction and extent of materials use and energy flow through a community.

The feeding relationships among species present in a community are called a **trophic structure**. In its simplest form, a trophic structure is based on three levels of participants: producers, consumers, and decomposers. Figure 28.6 shows the trophic structure in one community.

Producers harness energy (most often, from the sun) and use it in building food molecules from simple inorganic substances. Producers are responsible for the community's **net primary productivity**: most often, the photosynthetically derived food remaining after the producers use what they need for their own growth and development. Other species in the community are **consumers** of one sort or another. Herbivores are *primary* consumers; they feed directly on photosynthetic organisms. Other kinds of consumers are carnivores, which eat the herbivores (as well as other carnivores). Still others are scavengers (which eat organic refuse or carrion), parasites (which eat nutrients supplied by living hosts), and omnivores. Finally, there are **decomposers**: fungi and bacteria that break down organic debris and thereby help cycle nutrients and ions back to producers (Figure 28.7).

Figure 28.7 Simplified version of trophic levels in a natural community. Notice that raw materials may be recycled, but that with each transfer to a new trophic level, some energy is lost (hence the diminishing size of energy flow arrows). Sunlight provides the only energy input for the entire community.

The general sequence of who eats whom is sometimes called a "food chain." But the term implies a simple linear relationship that is seldom seen in natural communities. Imagine a fisherman netting some fish that were feeding on algae near the ocean's surface. Come lunchtime, he fries up some of his catch. Should he later lose his footing on the deck and fall into the water, where other sorts of carnivores are lurking, the "chain" might be portrayed thusly:

$$algae \longrightarrow fish \longrightarrow fisherman \longrightarrow shark$$

Clearly, this chain would be an oversimplification of feeding relationships in the community. It would exclude any number of alternatives. Most likely, krill were also grazing on the algae. Small squids and assorted medium-sized fishes might have been feeding on the krill; some larger fishes may have been feeding on smaller ones. Sharks may have been moving in to feed on the large and the medium-sized fishes. The fisherman might have accompanied his fish fillets with potatoes; he may even have sipped some wine or beer (hence his sloppy footing). Thus he would have been shifting back and forth between herbivore and carnivore—and would even have been more omnivorous in consuming waste products (alcohol) of decomposers. If he had harpooned a shark instead of falling prey to it (and assuming he was not a picky eater), he might have ended up as top carnivore. This example makes an important point:

The pattern of feeding relationships (who eats whom) in a community is better expressed as a complex <u>food web</u> rather than a simple, linear food chain.

Figure 28.6 depicts a simplified food web. With each energy conversion through a food web, there is considerably less energy to be passed on to the next trophic level. Because of built-in conversion losses (Chapter Six), usually less than ten percent of the energy transferred to a new trophic level remains as potential energy. In fact, *a growing animal must take in about 10 kilocalories of photosynthetically derived energy to produce every 1 kilocalorie of stored energy (or potential food energy for predators).* As a first approximation, suppose it took 1,000 kilocalories of energy stored in the

algae to produce 100 new kilocalories stored in the fish, which produced 10 new kilocalories in the fisherman. The small fraction of the fisherman derived from eating the fish would yield only 1 new kilocalorie in the top carnivore, the shark.

Another factor influences the picture of energy conversions through food webs. Energy storage becomes more inefficient over time, because energy must be expended on growth and maintenance. Assume the fisherman was forty years old. Assume that he took in an average of 2,700 kilocalories a day ever since he was a teenager. That amounts to about a million kilocalories a year. Although his food intake was much less during early development, his total consumption probably exceeded 30 million kilocalories. Once he reached adult size and weight, every kilocalorie consumed was used in maintaining his body's cells, and in providing them with energy for work. When he

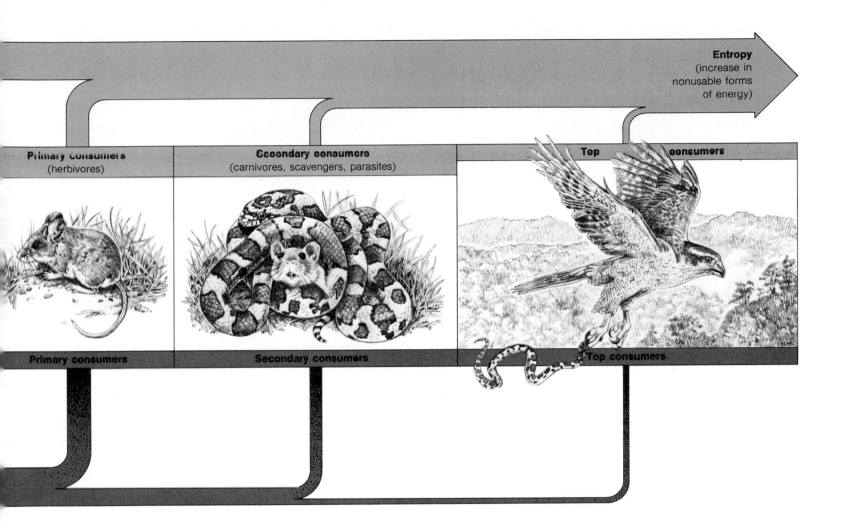

		Entropy (increase in nonusable forms of energy)

Primary consumers (herbivores)

Secondary consumers (carnivores, scavengers, parasites)

Top consumers

Primary consumers

Secondary consumers

Top consumers

fell into the sea, he represented only about 250,000 kilocalories worth of energy. The fisherman, in other words, represented less than one percent of the total energy that had flowed through him during his lifetime! The point is this:

The greater the time interval between energy conversions from one trophic level to the next, the lower the efficiency of the transfer within the community—above and beyond the built-in 10-to-1 conversion loss.

Energy Budgets

Is it possible to determine how much energy flows through an entire community in a given time period? Is it also possible to find out how the energy is divided among different trophic levels? Let's consider a few ways of approaching these questions. In one model, energy flow and the number of individuals in each trophic level are depicted as a pyramid. The pyramid's shape roughly corresponds to actual counts made of all the individuals within a community. For example, we might find the following relationship in a bluegrass field:

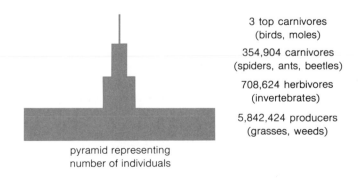

3 top carnivores (birds, moles)

354,904 carnivores (spiders, ants, beetles)

708,624 herbivores (invertebrates)

5,842,424 producers (grasses, weeds)

pyramid representing number of individuals

Figure 28.8 Annual energy flow (measured in kilocalories per square meter per year) for an aquatic ecosystem in Silver Springs, Florida.

The producers are mostly green aquatic plants. The carnivores are insects and small fish, and the top carnivores are larger fish. The energy source (sunlight) is available all year long.

Only 1.2 percent of the incoming solar energy is actually trapped in photosynthesis to generate new plant biomass. And more than 63 percent of the photosynthetic products is metabolized by the plants themselves to meet their own energy needs. Only 16 percent is harvested by herbivores, and the remainder is eventually decomposed by bacteria and fungi. Similarly, most of the herbivore energy is expended in metabolism and goes into the decomposer system; only 11.4 percent is consumed by carnivores. Once again, the carnivores burn up most of the energy they take in and only 5.5 percent is passed on to top carnivores. The decomposers cycle and recycle all the biomass received from all other trophic levels. Eventually all of the 5,060 kilocalories will appear as heat produced during metabolism. (Decomposers, too, are eventually decomposed.)

This diagram has been deliberately oversimplified. No community is completely isolated from all others. Organisms and materials are constantly dropping into the springs. And there is a slow but steady loss of other organisms and materials that flow outward in the stream that leaves the community. Over time, these inflows and outflows balance one another. (After H. T. Odum, "Trophic Structure and Productivity of Silver Springs," *Ecological Monographs*, 27:55–112, 1957. Copyright 1957 by the Ecological Society of America)

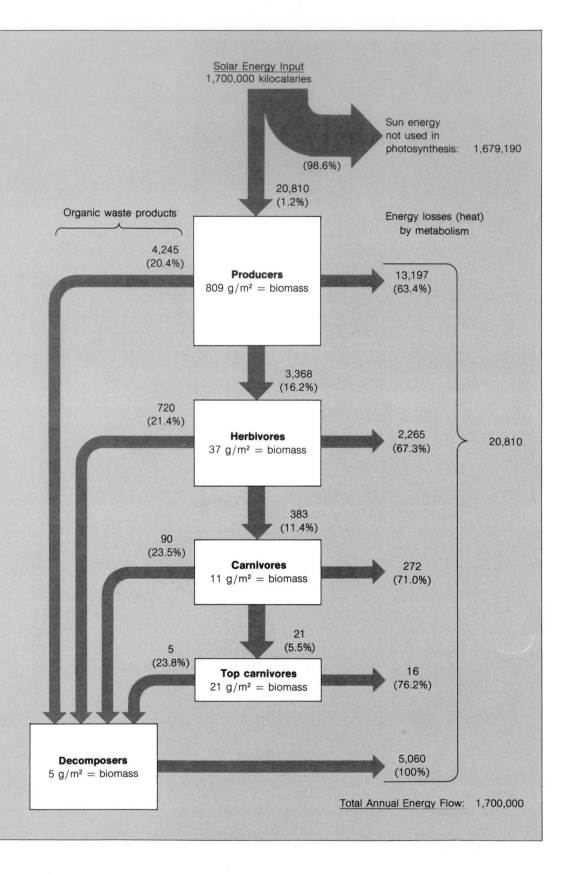

Aside from the monumental patience required to count the organisms in order to prepare the pyramid, the model itself is not very instructive. For one thing, it defines feeding relationships at only one brief moment—*and feeding relationships may change cyclically or permanently over time.* For another thing, a pyramid model doesn't take into account that *the sizes of organisms being counted in each trophic level may vary.* A similar count in a redwood forest would yield a small number of large producers (the trees) which still manage to support a large number of herbivores and carnivores (insects). And one deer would be counted as a single herbivore, as would a single insect—even though a deer eats far more than an insect ever would.

We can overcome a few of these problems if we weigh individuals in each trophic level instead of counting them. This would tell us about the **biomass**: the total weight of all organisms in a given category. Weighing the organisms is better than making a simple head count. But this approach also has shortcomings. For instance, in communities where producers are tiny but grow and reproduce rapidly, the biomass pyramid may be upside-down! Single-celled algae (phytoplankton) in marine communities are producers that grow this way. Here, the consumer biomass at any instant would actually exceed the producer biomass, because algae are consumed about as fast as they reproduce. It is just that the survivors, few as they may be, are reproducing at a phenomenal rate.

Such variations can be accounted for if we calculate the *total biomass for each species over a year's time.* This approach brings us closer to understanding energy flow through a community. But there is one last problem. Not all organisms of the same biomass have the same impact on a community. For instance, a warm-blooded mouse is much more active and consumes far more food than, say, a cold-blooded lizard of the same size and life expectancy. Thus the picture of energy flow will be complete only if we work out the **energy budget** for the entire community: a determination of how much energy the individuals take in, how much they burn up in metabolism, how much remains in their waste products, and how much they store in their bodies. It is a difficult thing to do, but it has been done in a few studies (Figure 28.8). Such studies have revealed that underlying all natural communities—even those of highly diverse sorts—is a general principle:

A community cannot survive indefinitely if its members expend more energy than they take in.

A community cannot spend any more than the energy contained in the annual primary productivity for the eco-system. It may underspend, in that it may not use up all of the available net productivity. For instance, the excess may be stored in the wood of trees or in the leaf litter piled up on a forest floor. But in no way can a community routinely overspend its energy budget.

Functional Roles in a Community

Earlier, you read that each species in a community is constantly being tested in terms of its ability to maintain or expand its niche. To help explain this dynamic aspect of community organization, the ecologist Richard Root proposed that species requiring the same resources play much the same role in maintaining community structure. Such strongly competitive species are said to form functional groups, or **guilds**. Species within some guilds may be interchangeable: if one dwindles in numbers or disappears, another guild member may move in and take its place. For example, in one of Root's experiments, a species of cabbage (kale) was grown in the middle of a natural meadow. Three guilds of herbivores moved in on the cabbage. One guild consisted of nineteen species of insects that more or less grated small pits from the leaves. One consisted of sixteen species of leaf-chewers. The third consisted of fifty-eight species of sap-feeders (including aphids). It turned out that one cabbage aphid proliferated one year, was nowhere to be found the next, then proliferated again in the year following that. In the year of the vacated niche, other species of the guild moved in and rapidly expanded their numbers.

The implications of the guild concept have been summarized by the ecologist Charles Krebs in this manner:

1. Guild species may be functionally equivalent; they may be interchangeable components that function in much the same way in the community.

2. Each community may have a small and perhaps constant number of functional roles (so many producers, consumers, and decomposers of certain types), compared with the number of species actually present.

3. Functional roles may be packed with different numbers of species. However, there may be limits on the number possible for any guild.

Currently, these three hypotheses are being tested experimentally. If they turn out to be valid, they will be a useful tool in studying community interactions. They will mean that the number of component species that have to be considered at a given time during a study can be reduced to conceptually manageable groups.

RESOURCE CYCLING THROUGH COMMUNITIES

Some Major Cycles

So far, our main concern has been with the flow of energy through a community. Let's now consider how material resources become available to its members. Earlier, Figure 7.14 showed one of the great cycles of nature. In this cycle, carbon and oxygen flow through all trophic levels in every ecosystem. The carbon-oxygen cycle is known as a *gaseous cycle*, for the atmosphere is the main storehouse for these two resources. There are also *sedimentary cycles*, whereby resources move from land, to sediments in the seas, and back to the land. In the *hydrologic cycle*, water moves slowly on a vast scale, through the atmosphere, across the land's surface, to the seas, and back again (Chapter Thirty).

The earth's crust is the main storehouse for nutrients that flow through sedimentary cycles. For example, as you might know, the **phosphorus cycle** begins with phosphate rock formations in the earth. Through weathering and other erosive forces, phosphorus is washed into rivers and streams, then moves to the oceans. There, largely on continental shelves, phosphorus builds up as insoluble deposits. Millions of years pass, and geologic events lead to the uplifting of the seafloor in the processes called mountain building (Chapter Twenty-One). Hence these deposits are found in mountain ranges that parallel the coasts. In the United States alone, 2 billion kilograms of phosphorus are mined annually for use in fertilizers. But movement of phosphorus to the land does not keep pace with erosion. Local shortages now exist.

It is one thing to say that some materials are not abundantly available to a community. Added to this, *many materials that are available often are in a chemical form that some species cannot use directly.* For instance, all species depend on proteins, and protein structure depends on nitrogen. Gaseous nitrogen (N_2) makes up about eighty percent of the atmosphere, so it would seem to be abundant just about everywhere. Yet, for most species, nitrogen of the air is totally useless until certain kinds of single-celled organisms convert it to a different chemical form. (Thus, even though we breathe in abundant N_2 molecules, we can't use them for body building; we breathe them right back out again.) In natural communities, nitrogen is harnessed from the air by some bacteria and some blue-green algae, it flows to other microorganisms in the soil, to plants, to animals, then back to the soil. The overall movement is known as the **nitrogen cycle**. This cycle depends on three links: nitrogen fixation, nitrification, and ammonification. These three links are described in the *Commentary*.

COMMENTARY

Nitrogen Availability, Fertilizers, and Running Your Car

Life, recall, depends on proteins. And all proteins contain nitrogen. Where do living things get their nitrogen? Most of the atmosphere is composed of gaseous nitrogen, but the molecules are a form that most organisms can't use. Gaseous nitrogen molecules are held together by stable, triple covalent bonds ($N\equiv N$), and very few organisms have the metabolic equipment for tackling them. Some bacteria and blue-green algae can. In nitrogen fixation, they assimilate nitrogen from the air. They attach electrons (and associated H^+ ions) to the nitrogen through a series of reduction reactions, thereby forming ammonia (NH_3) or ammonium (NH_4^+). These compounds are used in growth, maintenance, and reproduction. When nitrogen-fixing microorganisms die, their nitrogen-containing compounds are released during decay processes. Other bacteria present in soil use the compounds as energy sources. In nitrification, they strip ammonia or ammonium of electrons, and nitrite (NO_2^-) is released as a product of the reaction. Still other nitrifying bacteria in the soil then use nitrite for their energy metabolism, which yields nitrate (NO_3^-) as a product. The diagram shows these relationships.

When ammonia and nitrate dissolve in soil water, they can be taken up by plant roots and incorporated into organic compounds. These nitrogen-containing compounds are the only nitrogen source for animals, which feed directly or indirectly on plants. Another link in the cycle is ammonification. In this process, nitrogenous waste products or organic remains of plants and animals are decomposed by some soil bacteria and fungi. These decomposers use the proteins and amino acids being released for their own growth, and they release the excess as ammonia or ammonium. Some of these by-products are also taken up by plants.

Plants don't have the means of assimilating these nitrogen-containing compounds on their own. They depend on free-living or symbiotic microorganisms. Legumes (peas, beans, and the like) can be used as an example of this dependency. They harbor symbiotic ni-

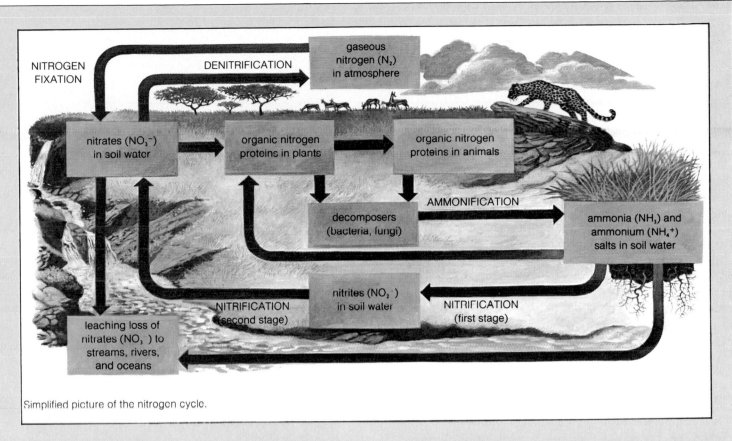

Simplified picture of the nitrogen cycle.

trogen fixers in their roots, supplying their guests with energy-rich sugar molecules even as they derive nitrogen.

Fixed nitrogen may leave the cycle in several ways. During crop harvests, of course, it leaves the field with the plants. Because nitrite is soluble, it may run off in streams with rainwater. Under some conditions, especially when soil is poorly aerated, nitrate is converted to gaseous nitrogen by denitrifying bacteria (they use the oxygen in NO_3^- when confronted with anaerobic conditions). This process is called denitrification.

With any loss of fixed nitrogen, soil fertility—hence plant growth—is reduced. Farms in Europe and the United States have traditionally depended on crop rotation to restore the soil. For example, legumes are planted between plantings of wheat or sugar beet crops. This practice has helped maintain soils in stable, productive condition, sometimes for thousands of years.

Modern agriculture depends on nitrogen-rich fertilizers. With plant breeding, fertilization, and pest control, crop yields per acre have doubled, even quadrupled, over the past forty years. With intelligent management, it appears that soil will maintain such high yields indefinitely—as long as water and fertilizers are available.

The catch, of course, is that we can't get something for nothing. Enormous amounts of energy are needed to produce fertilizer—not energy from the unending stream of sunlight, but energy from oil. As long as the supply of oil was viewed as unending, there was little concern about the energetic cost of fertilizer production. In many cases, we have been pouring more energy into the soil, in the form of fertilizer, than we are getting out of it in the form of food. Unlike natural ecosystems, in which resources are cycled according to principles of energy use, our agricultural systems exist only because of constant, massive infusions of energy.

However, as any hungry person will tell you, food calories are more basic to survival than gasoline calories or perhaps, even, than a car. As long as the human population continues to grow exponentially, farmers will be engaged in a constant race to supply food to as many individuals as possible. Fertilizer enrichment of soils is essential in the race, as it is now being run.

Case Study: Tundra Feedback Loops for Resources

The term **tundra** refers to a region where it is too cold for trees to grow but not cold enough to be perpetually frozen over with snow and ice. The arctic region of northern Alaska is as flat, windswept, and desolate a stretch of tundra as you might ever see (Figure 28.9). Temperatures there struggle up to an average of 5°C (41°F) in midsummer. They have been known to reach a balmy 21°C (70°F), then drop below freezing the next day. Temperatures plummet to –32°C (–26°F) in midwinter. There is nearly continuous sunlight for the three summer months. Plant growth is confined to this brief period. But growth is profuse; and flowering and seed ripening are completed quickly. Then the skies gradually darken with the onset of the long arctic winter, when blizzards close in on the land.

Although the tundra is not completely covered with snow all year long, the brief summer thaw is not enough to warm much more than surface soil. Just beneath the surface is the **permafrost**, a permanently frozen layer 610 meters (2,000 feet) thick in this region! Permafrost forms an impenetrable basement beneath the flat terrain, hence drainage is poor. In fact, the tundra becomes saturated with meltwaters at the onset of summer. By late summer, water evaporating from the boggy soil has created almost perpetual clouds. A cloud cover remains until the first frosts clear the skies, before the next round of blizzards.

What sort of community interactions can you expect to see in this harsh environment? As you might imagine, the arctic tundra ecosystem is far more simple than ecosystems of the savanna or the tropical reef (Chapter One). The main limiting factor is the low level of solar radiation; there simply is not much energy flowing into the community. Another is the effect of permafrost on nutrient cycling. Because of the poor drainage and low temperatures, organic matter cannot completely decompose. It is gradually becoming locked up in soggy masses called peat. In fact, more than ninety-five percent of the carbon in the arctic is now inaccessible to the tundra community. Less than two percent of the total carbon, nitrogen, and phosphorus is found in plants—and most of this is concentrated in underground plant parts.

Thus we have environmental conditions that do not foster much diversity. Most of the plant biomass consists of no more than ten different species. There are only eight main predators, including the snowy owl and arctic fox, which specialize on the main prey—a small, herbivorous rodent called the brown lemming. What this herbivore

a

b

c

d

All photographs this page Roger K. Burnard

e

f

g

Figure 28.9 Participants in the arctic tundra ecosystem. Only ten plant species make up about ninety percent of the biomass. During the brief summer, arctic flowers, mosses, and fungi appear (**a–c**). There is little drainoff in the ecosystem because of the flat terrain (**e**); organic matter accumulates as peat. The lemming (**d**) is the main large prey. Two of its predators: the snowy owl (**f**) and the arctic fox (**f**).

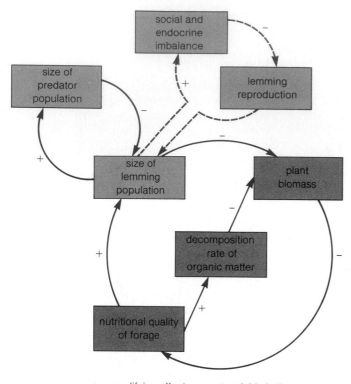

+ amplifying effect on next variable in line
− limiting effect on next variable in line

Figure 28.10 Feedback-loop model for nutrient recovery in the arctic tundra ecosystem, as described in the text. (After A. Schultz, Academic Press, 1969 in Krebs, *Ecology*, second edition. Copyright © 1978 by Charles J. Krebs. Reprinted by permission of Harper & Row)

lacks in size, it makes up for in sheer numbers. Feeding on roots of grasses and sedges, lemmings live throughout the year in underground burrows. Although they consume great amounts of vegetation, they also deposit droppings that fertilize the plants and thereby stimulate plant growth. Every three to five years, the lemming population density reaches high levels, and the population declines sharply.

In the 1960s, the ecologists Frank Pitelka and Arnold Schultz attempted to trace the nutrient cycles through the arctic tundra ecosystem. Their observations formed the basis of a **nutrient recovery hypothesis**. According to this hypothesis, interlocking feedback loops exist between ecosystem variables such as population size for predators and for prey, the total plant biomass, and the decomposition rate for organic matter (Figure 28.10). Each variable is seen to amplify or counteract events and conditions along interconnected pathways.

The following feedback loops between lemmings, plants, and soil are now thought to have major effects on

nutrient cycles in the ecosystem. First, during winter, the lemming population matures, and feeding intensifies with the shift in age distribution. Urine, feces, and litter accumulate in the burrows. Second, at the start of the growing season, melting snow distributes the nutrients throughout surface soil, where they are absorbed by plant roots and underground stems. Part of the nutrients remain tied up in litter and feces, which decompose slowly in the cold climate. Third, as the lemming population increases in size, the plant biomass decreases—by about fifty percent. Fourth, the lemming population density drops, perhaps because of predation or overcrowding with its social and endocrinological disruptions (Chapter Twenty-Seven). Fifth, the decline in lemming numbers is accompanied by a decline in the amounts of urine, feces, and litter deposited over winter months. With the drop in nutrients, the nutritional quality of the plants suffers—which has a feedback effect on lemming reproduction. Finally, over the next few years, the plants slowly recover in nutritional content and in numbers. With their recovery, the lemming population has the basis to rise once more.

SUCCESSION: CHANGE IN COMMUNITIES WITH TIME

So far, we have locked at communities as if their characteristic array of species is always much the same. But how do communities arise in the first place? Once established, do they remain the same over the years? Let's take a look.

From Pioneers to Climax Communities

Glacier Bay in Alaska is remarkable for more than its spectacular scenery (Figure 28.11). Nowhere else are glaciers retreating so rapidly. And nowhere else have changes in newly deglaciated areas been documented so carefully for so long a time. A comparison of maps from 1794 onward shows that the ice has been retreating at rates ranging from

Figure 28.11 Ecological succession in the Glacier Bay region of Alaska. Early visitors include gulls and other birds (**a**). The early pioneer stage includes the flowering plant *Dryas* (**b,c**), horsetails, and fireweed (**d**). The soil (**e**) was covered with ice less than 10 years ago. Within 20 years, young alders, cottonwood, and willows appear in drainage channels (**f**). Within 50 years, (**g**) Sitka spruce has appeared. By 80 years, mature alders are being crowded out by cottonwood spruce (**h**). In an area deglaciated more than 130 years, a dense Sitka spruce, western hemlock forest forms the climax community (**i**).

a

b

c

d

e

f

g

h

i

3 meters annually at the glacier's sides to a phenomenal 600 meters a year at its tip over bays and inlets. At Glacier Bay, biologists have had a prolonged opportunity to observe processes whereby vacant land first becomes populated and goes on to become a stable, complex community.

As a glacier retreats, the constant flow of meltwater tends to deplete the soil of minerals, especially nitrogen. The newly exposed shoreline usually is unable to support plant life. Soon, however, water near the edge is populated with nitrogen-fixing blue-green algae. As algae die, they provide a natural fertilizer for the soil. Horsetails, sedges, cotton grass, and other plants begin to appear along the land's edge. They are accompanied by mosquitoes, diving beetles, and water striders. Sandpipers, gulls, rock ptarmigan, and other birds begin visiting the site. All these organisms are pioneers, adapted to moving quickly into new, barren areas. By their very presence, though, they begin to alter the environment. Organic wastes from birds accumulate and add nutrients to the soil. Meanwhile, the first plants offer some protection for plants of other species—and in this way they set the stage for their own rapid displacement. New species move in, competing more effectively than the pioneers for niche space. Eventually, pioneer species are crowded out. Deciduous trees take hold over the years—alder (which is highly symbiotic with nitrogen-fixing microorganisms), then mature willows and poplars. A few evergreen spruce trees germinate in the shade of the deciduous trees. In eighty years, spruce trees come to dominate the landscape. Decades later, they give way to giant hemlocks. Through all these stages, different species of microorganisms, plants, fungi, and animals make their entrance and then gradually depart. *And all the while, the total biomass slowly increases, offering more possibilities for partitioning of niches.* Eventually the site is dominated by a **climax community**: an array of species locked together in materials and energy use, in ways that allow it to remain relatively stable and self-perpetuating.

What you have just read is a summary of **primary succession**—a gradual, sequential replacement of communities, beginning with pioneer species, until a stable climax stage is reached. The process may take hundreds of years; sometimes it takes thousands.

What we consider to be permanent features of the landscape may include only transitional stages on the way to climax communities. For instance, most North American lakes formed when glaciers of the last ice age began retreating about 11,000 years ago. Succession for these lakes began at that time, and it has been unfolding in the following way. Streams and rivers leading into lakes deposit silt. Aquatic communities produce organic debris. As the lakes become more shallow because of the deposits, different plant species become established in the water or around the lake's edge. This brings in different consumers, and more organic debris. Gradually, deep, clear lakes become so enriched with plant and animal by-products that they develop into marshes. Finally, the lake basins become completely filled in. Once that happens, climax communities characteristic of lake regions become established. The process whereby a body of fresh water becomes enriched in nutrients, increases in productivity, and accumulates organic debris is known as **eutrophication**.

Normally, it takes a few centuries for small lakes to fill in, and more than 10,000 years for large ones. Human activities can bring succession to completion in a lifetime. For example, fertilizer runoff and sewage discharge accelerate the growth of encroaching plants. But lakes rapidly on their way to becoming marshes are not necessarily desirable from aesthetic, recreational, or economic standpoints. Therefore, efforts are being made to slow down eutrophication of the 100,000 or so lakes in the United States.

Secondary succession may also occur when an established community is disrupted in whole or in part. Consider what happens when part of a spruce forest is completely stripped for lumber. Reestablishing a community on the exposed soil is not simply a matter of seeds from nearby spruces drifting over and germinating. Most dominant plant species in a climax community may not grow unless an integrated community structure already exists. Spruce seeds normally germinate only in the litter of a forest floor, and develop into saplings only if an existing forest canopy shades them. So they can't repopulate stripped land at once. Instead, pioneer species tolerant of exposed conditions become established during the first few growing seasons. The first arrivals are annual weeds. They produce abundant seeds, have efficient means of dispersing them, and can germinate in full sun. Perennial weeds and grasses follow them. Year after year, perennials send out more roots and shoots. As they increase in size, they become dominant in the competition for sunlight and resources. As fallen leaves, dead plants, and other debris accumulate, decomposers thrive. Only when a rich layer of decomposing matter has built up do the trees characteristic of the climax stage start to germinate.

Such partial disruptions are one of the reasons why you can see more than one type of community in the same general region. They are passing through different stages of succession. Other variations may exist from one spot to the next depending on soil type, drainage, and local climate conditions.

It is important to note that not all disturbances to natural communities are bad. *Some communities depend on intermittent disruption for maintaining long-term stability.* For instance, the Sierra Nevada mountains of California contain isolated sequoia groves. Sequoias can grow to be giants. Some are more than 4,000 years old. Many sequoia groves are protected as part of national or state park systems. Among other things, protection has traditionally meant minimizing the incidence of fires—not only accidental fires from campsites and discarded cigarettes, but also natural fires touched off by lightning. Fire prevention and suppression programs have been highly successful. The problem is that fires at regular intervals are vital for the community's long-standing stability! Sequoia seeds germinate only on bare mineral soil. If there is extensive forest floor litter, there won't be any new sequoias. Modest fires eliminate the litter. They also eliminate other trees and shrubs that compete with sequoias, yet they do not damage the sequoias themselves. Why? Mature sequoias resist fire because they have bark as thick as your arm is long. The bark burns poorly and insulates the tree against modest heat damage. But when small, periodic fires are prevented, litter builds up. Other species appear that are susceptible to fire. Even though these species don't actually displace the sequoias, the sequoias are no longer reproducing. The litter and undergrowth represent so much potential fuel that fires are hotter than they otherwise would be—hot enough to damage the giants. The point is, without understanding the interactions that maintain a community, efforts to preserve them may have the opposite effect.

Does Diversity Mean Stability?

Through interlocking food webs, materials and energy are cycled through communities. Consumers and decomposers assure that essential nutrients removed from the soil and water by one generation of producers are ultimately returned to nourish a new generation. Hence nutrient cycling helps assure overall community stability.

But this stability is constantly being tested by change —change in the type and number of individuals, change in the environment through normal community activities, through floods, fires, or pollution and the sprawl of human activities. The effect of such disturbances raises two questions. First, will a community be more likely to return to a stable configuration if it has a simple array of species, or a diverse array? Second, will it be more resilient if it has food webs that are rigidly fixed, or that are interchangeable among guild species? These questions are being actively debated at the present time.

The problem is that even the "simple" interlocking feedback loops of the arctic tundra ecosystem are so complex that they are only now beginning to be understood. In ecosystems characterized by high levels of diversity, the loops are monumentally intricate. It may be that the characteristics of individual species are more crucial to stability than some cushion afforded by species diversity. For instance, a patch of disturbed grassland may return quickly to its most stable configuration if the pioneer plant species are able to reseed the area in a short time. It may return quickly if the dominant species of grasses can reproduce fast enough to withstand the onslaught of a fully developed herbivore population that surrounds the patch.

However, if we define stability as being able to persist as conditions change, then some ecosystems are more resilient than others. Think about the limited and highly specialized participants in the nutrient cycles of the arctic tundra. There are no guild species interchangeable with the lemming; nothing is around to take over the lemming's functional role if it should disappear. This ecosystem is vulnerable to disruption.

Or think about a complex temperate forest, which is partitioned in delicate interdependencies. The foundation for its complexity is an array of tree species that capture sunlight energy. By the time sunlight travels from the top canopy of leaves down through the intermediate layers, less than five or ten percent may reach the forest floor. All of these stratified layers of stored energy are the basis of food webs, including insects, parasites, birds, mice, and squirrels. On the forest floor are larger herbivores such as moose and deer, which use the trees not so much for food as for cover against predators. Predators—wolf, lynx, mountain lion—use the forest for shelter against climatic extremes such as blizzards. Dense foliage stabilizes the local microclimate by cutting down winds, by acting as a buffer against temperature changes, and by holding back moisture that otherwise would escape from the community through evaporation. The forest trees also enrich the surface soil with nutrients contained in falling leaves. Extensive tree roots prevent soil erosion. A complex forest ecosystem tends to be tolerant of many minor disruptions. However, it may not be resilient enough to endure, say, the onslaught of lumbering operations that strip all the trees from a mountain, or the introduction of a new leaf-stripping insect that has no natural predators in the area. Whatever the case, major disruptions may destroy mature trees that are the energy foundation for the entire community structure —and that play a key role in recycling nutrients. Depending on the types of species in the community, it might be more than a century before the slow, healing process of succession can return the community to its most stable configurations.

We use and modify the land in many ways, for many purposes. Aside from ethical considerations about what we are doing, it would be to our advantage, at the very least, if we learn to identify the limits of resilience for our "managed" ecosystems. These include the tree farms you will soon read about, and vast commercial monocrops. Monocrops are low-diversity "pioneer" communities. They are high in productivity but low in stability. They give the illusion of stability only because of massive imports of energy, fertilizers, pesticides, and (often) water. Obviously, our dependence on crops is not going to vanish. But we might start thinking about ways of integrating agricultural communities with surrounding natural communities. The ecologist Eugene Odum has advocated intermingling different communities, of different ecological ages. Forests, for instance, are valuable for more than lumber. They are production sites and holding stations for water and minerals, which gradually wash down to replenish surrounding valleys and plains. Strip the forests and we strip the surrounding lands of vital resources, also. Although in the short term there may be lower production levels if forests are preserved, in the long term there would be less vulnerability to insect predators and diminishing nutrients. It may be that a diverse array of crops, complemented by a diverse array of forests (and estuaries, streams, and the like), would be a compromise that offers some stability for *both* agricultural and natural communities.

CASE STUDIES: ECOLOGICAL BACKLASH

It is not only for the sake of our crops that we must come to terms with natural communities. Too often we are finding that our disruptions of natural systems have disastrous consequences not only for our crops but for our economy, even our lives. Four diverse examples will make the point.

DDT and Thee

For much of the human history, we have been at war with insects that destroy crops and transmit diseases. The organic compound DDT, which sends insects into convulsions, paralysis, and death, has been instrumental in bringing many of the worst offenders (such as malaria-transmitting mosquitoes) more or less under control. But DDT is a stable compound that cannot readily be broken down. It can persist in the environment for as long as fifteen

years. Because of its stability, it is a prime candidate for **biological magnification**—the increasing concentration of a nondegradable substance as it moves up through trophic levels. Why does it become so concentrated? Recall there is about a 10-to-1 conversion loss at each energy transfer from one trophic level to the next in a food web. That's the same as saying it takes 10,000 pounds of algae to produce 1,000 pounds of animal plankton that will produce 100 pounds of small fish. These small fish, when consumed by larger fish, will produce 10 pounds of food, which in turn will produce 1 pound of, say, brown pelican. But DDT is nondegradable and insoluble in water—and it dissolves in fat. Thus it tends to accumulate in the fat and oil reserves of the body. It is then passed on to the next species in the food web. Thus the DDT present in all that biomass ends up in the top consumer organism. The consequences have been unfortunate for the brown pelican (Figure 28.12). They have also been unfortunate for us.

Consider what happened back in 1955, when the World Health Organization stepped in with a DDT spraying program to eliminate malaria from the island of Borneo, now a state of Indonesia. That step was not taken lightly. Nine out of ten people there were infected with this terrible disease, which is epidemic proportions by anybody's standards. The program worked, insofar as the mosquitoes transmitting malaria were brought almost entirely under control. But DDT happens to be what is known as a broad-spectrum insecticide; it kills nontarget as well as target species. Sure enough, the mosquitoes had company: flies and cockroaches that made a nuisance of themselves in the thatch-roofed houses on the island fell dead to the floor. At first there was much applause. But then the small lizards that also lived in the houses and preyed on flies and cockroaches found themselves presented with a veritable feast. Feast they did—and they died, too. And so did the house cats that preyed on the lizards. With the house cats dead, the rat population of Borneo was rid of its natural predator, and rats were soon overrunning the island. Now, the fleas on the rats were carriers of still another disease, called the sylvatic plague, which can be transmitted to humans. Fortunately, the threat of this new epidemic was averted in time; someone got the inspired idea to parachute DDT-free cats into the remote parts of the island. But on top of everything else, some of the people of Borneo found themselves sitting under caved-in roofs. The thatch in their roofs was made of a certain kind of leaf that happened to be the food source of a certain kind of caterpillar. The DDT didn't affect the caterpillar but it killed the wasps that were its natural predator. When the predator population collapsed, so did the roofs.

On the Importance of Being a Hippo

Hippopotamuses are short-legged, barrel-shaped mammals with practically hairless skin and a bulging mouth. They live in herds in African rivers and swamps, where they feed on aquatic plants and on plants lining the shore. Sometimes they also leave the water and munch on nearby crops. Sometimes (although rarely) they attack small boats. And for that they have been shot. In fact, in many regions they have been slaughtered in a campaign to clear the waterways.

But now, where hippos have been wiped out, the waterways are burgeoning with plant growth and filling up with silt. It seems the hippos kept the plant populations in check and, in their habit of digging out wallowing holes, they kept the silt from piling up. The now-shallow waterways have become ideal breeding grounds for the snail that is the intermediate host for the blood fluke—whose primary hosts are humans (Figure 28.13). About 200 million people now suffer from **schistosomiasis**, the ravages of the blood flukes. It is one of the most rampant infectious diseases in the world. And its effects are dreadful: stomach cramps, deterioration of vital organs, chronic exhaustion that permits little more than a few hours of activity a day. Sometimes death immediately follows infection. But more typically, the victims can expect only years of pain before their weakened condition makes them susceptible to a killer disease. "The Hippo's Revenge" might well be the title of a most informative documentary film.

The Incredible Spreading Imports

In the 1880s, the water hyacinth from South America was put on display for the New Orleans Cotton Exposition. Flower fanciers from Florida and Louisiana carried home clippings of the blue-flowered plant and set them out for ornamental display in ponds and streams. Unchecked by natural predators and nourished by the nutrient-rich waters, the fast-growing hyacinths rapidly displaced many desirable native plants and choked off the ponds and streams. Then they went to work on rivers and canals. They are still there, and they are still bringing river traffic to a halt.

Species introduction into established communities doesn't always have disastrous effects. Honeybees, mosquitofish, ring-necked pheasant—all have been absorbed into and have become part of community structures. Most of our own food—from apples, cabbage, wheat, oranges, to cattle and chicken—are the progeny of imports from

Figure 28.12 Biological magnification through the food web ending with the brown pelican. The concentration of DDT had the effect of softening the shells of brown pelican eggs, and for a time the populations of these birds dwindled. With the recent ban on DDT, the brown pelicans are making their comeback.

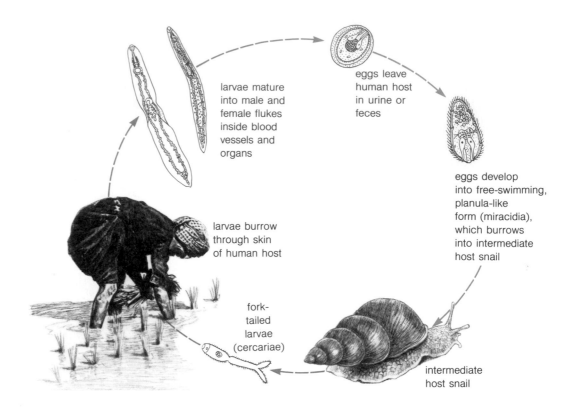

Figure 28.13 Life cycle of a schistosome, a fluke thriving in tropical Africa and Southeast Asia. Adult flukes live in human blood vessels, where they perpetually copulate. The female fluke of some host species prefers to lay eggs in the veins around the host's bladder. Spines on the eggs pierce the vein walls, which creates local hemorrhaging. The blood carries the eggs into the bladder, then out from the host through urine. The eggs hatch into ciliated larvae, which swim about until they find and bore into a water snail. The snail acts as an intermediate host. Feeding on the snail's tissues, the larvae reproduce asexually in enormous numbers. Eventually a larval form (the cercaria) leaves the snail and swims about until it encounters a human (who is working, for example, in a rice paddy). It bores into its new host's skin and from there travels the blood vessels, to begin the cycle anew.

other countries. But as Table 28.1 makes clear, we can't always say that nothing ever goes wrong, or that natural communities always recover.

Aswān and Other Fables

Once there was a country that desperately needed food and energy for its growing population. It happened that one of the most magnificent rivers in the world flowed through this country. Each year the river deposited tons of mineral-rich silt on its fertile floodplain before it reached the sea. Why not dam the river, said the country's leaders, and use the water to irrigate more land, control the annual spring flooding of the river, and provide hydroelectric power all at the same time? The result of this modern-day fairy tale is known as the billion-dollar Aswān High Dam of Egypt, and not all Eqyptians are living happily ever after.

For one thing, as water backed up behind the dam, almost 100,000 Egyptians had to choose between giving up their family homes or being submerged along with ancient and priceless temples that were part of Egypt's cultural heritage. But there have been far more devastating results. Now that the Nile River floodplain is deprived of its annual enrichment with silt, artificial fertilizer has to be trucked in at a cost of 100 million dollars a year—a cost carried by the subsistence farmers who make, on the average, less than a hundred dollars a year each. Furthermore, now there is nothing to wash away the previous year's salt buildup in the soil. And with silt deposits no longer compensating for erosion, the fertile river delta is

Table 28.1 Effects of Introducing a Few Species into the United States

Species Introduced	Origin	Mode of Introduction	Outcome
Water hyacinth	South America	Intentionally introduced (1884)	Clogged waterways; shading out of other vegetation
Dutch elm disease The fungus *Cerastomella ulmi* (the disease agent)	Europe	Accidentally imported on infected elm timber used for veneers (1930)	Destruction of millions of elms; great disruption of forest ecology
Bark beetle (the disease carrier)		Accidentally imported on unbarked elm timber (1909)	
Chestnut blight fungus	Asia	Accidentally imported on nursery plants (1900)	Destruction of nearly all eastern American chestnuts; disruption of forest ecology
Argentine fire ant	Argentina	In coffee shipments from Brazil? (1891)	Crop damage; destruction of native ant communities
Camphor scale insect	Japan	Accidentally imported on nursery stock (1920s)	Damage to nearly 200 species of plants in Louisiana, Texas, and Alabama
Japanese beetle	Japan	Accidentally imported on irises or azaleas (1911)	Defoliation of more than 250 species of trees and other plants, including commercially important species such as citrus
Carp	Germany	Intentionally released (1887)	Displacement of native fish; uprooting of water plants with loss of waterfowl populations
Sea lamprey	North Atlantic Ocean	Through Welland Canal (1829)	Destruction of lake trout, lake whitefish, and suckers in Great Lakes
European starling	Europe	Released intentionally in New York City (1890)	Competition with native songbirds; crop damage; transmission of swine diseases; airport runway interference; noisy and messy in large flocks
House sparrow	England	Released intentionally (1853)	Crop damage; displacement of native songbirds; transmission of some diseases
European wild boar	Russia	Intentionally imported (1912); escaped captivity	Destruction of habitat by rooting; crop damage
Nutria (large rodent)	Argentina	Intentionally imported (1940); escaped captivity	Alteration of marsh ecology; damage to earth dams and levees; crop destruction

From David W. Ehrenfeld, *Biological Conservation*, 1970, Holt, Rinehart and Winston and *Conserving Life on Earth*, 1972, Oxford University Press

shrinking—and an alarming part of what remains has completely dried up. Restoring the delta with pumps, drains, and wells may cost more than the dam itself.

Ironically, evaporation as well as bottom seepage from the new lake filling in behind the dam is so great that the lake basin may never fill up to predicted levels. So nobody can live around the lake because nobody knows for sure where the shoreline will be. More seriously, there is less water to go around than there was before. And even though some 700,000 new acres (about 1.6 million hectares) have been opened up for agriculture, the population outgrew the potential food increase even before the dam was finished. At the same time, with the nutrient-rich flow of the Nile turned off, another major food source—the sardines, shrimp, and mackerel that flourished in the enriched waters off the delta—has declined catastrophically. Worse yet, the lake and the irrigation networks have so accelerated the spread of blood flukes that half the Egyptian populace are now carriers of schistosomiasis. In irrigated areas, where eight out of ten humans live, women can expect to live only to age twenty-seven, men to age twenty-five.

Perhaps in the long run there will be real benefits from this project. Its defenders are quick to point out that even though there isn't enough food to go around now, without the dam and its irrigation system there would be even less. The hydroelectric output should stimulate industrial development. Stocking the immense lake with fish should provide a new source of protein, and perhaps (assuming the blood flukes are controlled) it will even attract tourists. Regardless of the position taken on the dam's ultimate value, regardless of how well-intentioned the planning, development, and international support may have been, there is no defense for one important fact: warnings from ecologists *before* the project began were totally ignored.

The integrated structure and stability of existing communities must be considered in advance in order to determine what the impact will be of any deliberate disturbance—and we must come to recognize how vital it is to seek ways to minimize that disturbance.

Readings

Eibl-Eibesfelt, I. 1975. *Ethology: The Biology of Behavior.* Second edition. New York: Holt. A treasurehouse of detail on animal behavior during predatory and competitive interactions.

Hutchinson, G. 1978. *An Introduction to Population Ecology.* New Haven: Yale University Press. For the stouthearted, the mathematical models and theories of population and community interactions. The section entitled "What Is a Niche?" conveys the frustrations in trying to answer what, exactly, it is.

Krebs, C. 1978. *Ecology: The Experimental Analysis of Distribution and Abundance.* Second edition. New York: Harper and Row. Krebs writes clearly, in a lively manner; he has a knack for distilling complex mathematical theories and models into understandable prose.

MacArthur, R. 1972. *Geographical Ecology.* New York: Harper and Row. Contains excellent material on competitive and predatory interactions.

Odum, E. 1971. *Fundamentals of Ecology.* Philadelphia: W. B. Saunders. A classic in the literature, written by a prominent ecologist.

Ricklefs, R. 1973. *Ecology.* Portland, Orgeon: Chiron Press. Excellent handbook of ecology; good section on community interactions. Hefty paperback.

Wickler, W. 1974. *Mimicry in Plants and Animals.* New York: McGraw-Hill. More on the fine art of mimetic behavior and camouflage. Paperback.

Review Questions

1. Distinguish between these terms: community, ecosystem, niche, habitat.

2. What three kinds of interactions exist between populations? Can you give an example of each?

3. Define competition, and explain how this kind of behavior is incorporated into the concept of competitive exclusion. Why is it difficult to observe competitive exclusion in the natural world?

4. Define these terms: herbivore, carnivore, and omnivore. Give examples of how these different kinds of animals locate, capture, fight, hide, or escape detection from one another.

5. What is the difference between camouflage and mimicry?

6. Commensalism, mutualism, obligatory relationships, and parasitism are all forms of symbiosis. How do these forms differ from one another?

7. How might two species coevolve to the extent that they enter an obligate relationship?

8. What determines the direction and extent of materials use and energy flow through a community?

9. Define these terms: trophic structure, net primary productivity, food web.

10. What happens to energy transfer efficiency in a community if considerable time elapses between energy conversions from one trophic level to the next?

11. Mention some of the inadequacies in attempting to represent energy flow in a community by a pyramid of numbers of individuals in each trophic level. Is there a more informative way of conveying the relationship between energy flow and the members of a community?

12. Can a community routinely overspend its energy budget? What happens if it does? How would a biologist determine the energy budget for an entire community? Would it be possible to determine energy budgets for human-dominated communites?

13. In Figure 28.8, which component of the trophic structure lost the most energy (heat) via metabolism? Which component converted the largest percentage of its prey into its own biomass? Would you say that the percentage of waste products varies greatly or very little from the producer component to the top carnivore component? What percentage of incoming solar energy was actually used in photosynthesis in creating new plant biomass?

14. What are guild species? How might the concept of guild species help us unravel the complex interrelationships going on even in the simplest communities?

15. During succession, what happens to the total biomass, and the size and number of niches?

16. Distinguish secondary succession from primary succession. Name a community whose long-term stability depends on intermittent small fires.

At the northern edge of the Kalahari Desert, in the nation of Botswana, the longest river in southern Africa ends in an immense freshwater delta. The delta supports lush grasses, brush, and groves, which in turn support one of the most magnificent arrays of wildlife imaginable—exotic birds, crocodiles, hippopotamuses, buffaloes, elands, kudus, wildebeests, elephants, lions, and many more. The animals, in their turn, are hosts to bloodsucking mosquitoes and tsetse flies. These insects transmit the dread sleeping sickness to humans and cattle. By their very presence in the delta, the insects have kept humans and cattle out.

Now, it happens that cattle raising is Botswana's economic mainstay. Cattle need water and grazing land, which are becoming increasingly scarce. Pressure is on to open up the delta for the cattle industry by eradicating the tsetse fly and the mosquito. That means clearing the woodlands, burning off the brush, draining the swamps—and destroying the wildlife. Why would people do such a thing? For profit, of course, but also to feed other people. Anyway, how can we demand that they *not* do such a thing? Would Californians so readily turn over the fields and vineyards of their fertile inland valleys to quails and rabbits, coyotes and hawks? Would Texans so readily set aside their open range for the preservation of zebra, kudu, wildebeest, not to mention the occasional lion? Would Nebraskans so readily donate their fields of waving grain to support human populations in Africa, so that the African wildlife can be left alone? The questions we ask are relative to what part of the world we happen to live in. And there simply are no easy answers at the regional level. If there are answers at all, they will probably come with recognition of the whole earth as one system of life—as one **biosphere**—with cooperation a requisite for long-term stability of the whole.

In the preceding chapter, particular communities and ecosystems were used as examples in illustrating some basic concepts. Here, we will look at kinds of prevailing ecosystems throughout the biosphere. At the chapter's end, we will address some basic questions about the effects of human-designed ecosystems on the total world of life.

ECOSYSTEMS IN SPACE: THE WORLD'S BIOMES

Each part of the earth's surface is in some ways unique. It has its own combination of mountains or lowlands, soil and rock, ponds or rivers or ocean currents, high elevations or low, its own nutrients, its own amount and patterns of sunlight and moisture, its own pronounced or slight seasonal changes. All such environmental variables mean

29

STABILITY AND CHANGE IN THE BIOSPHERE

that every ecosystem in the world is in some ways unique. But if we put aside the overlay of differences, we still find that ecosystems correspond to a few general formations, or **biomes**, on the basis of their dominant array of primary producers. These arrays represent stable adaptations to prevailing conditions, such as interactions between climate and topography for a certain land mass. As you read through this chapter, keep this important concept in mind:

When we seek to identify the structure of an ecosystem, we must first look to the primary producers on which it is based.

Primary Productivity in the Seas

The distribution of producer organisms in the vast province of the seas is governed by such variables as light, temperature, salinity, and available nutrients. Most marine ecosystems fall within the shallow **intertidal zone** (between high and low tide marks) and the **abyssal zone** (between 2,000 and 4,000 meters beneath the ocean's surface). Most depend on photosynthesizers as the primary producers. Photosynthetic organisms are found in the upper 100 to 400 meters of water. Only in this **photic zone** (Figure 29.1) is the light intense enough to drive photosynthesis. Temperature and salinity in the deep ocean do not vary much,

mainly because major currents tend to circulate water masses on a global scale. Hence most ecosystems at great depths take on much the same character no matter what their geographic location. But at the surface, temperatures can range from 30°C (86°F) in the tropics to –2°C (28°F) at the poles. Salinity at the surface ranges from high concentrations in such tropical waters as the Red Sea to low concentrations in polar regions (because of melting freshwater ice).

In addition to these environmental variables, *we must consider the way materials are cycled back to the producers in order to understand why marine ecosystems are distributed as they are.* In shallow nearshore waters and coral reefs, decay occurs within the photic zone. In these ecosystems, recycling of resources can proceed in much the same way that it does on land. Over open oceans, wastes and the bodies of dead organisms drift down to the bottom. Once there, this material can't decay quickly because of the low temperatures. Even when the material is finally broken down, it is far removed from the narrow zone of photosynthetic organisms that could thrive on it. This means essential minerals such as nitrates and phosphates are constantly being removed from the surface waters. That's why the primary productivity and species diversity of open-water ecosystems are generally low compared to estuaries, nearshore waters, and tropical reefs.

One notable exception is the geothermal ecosystem of the Galápagos Rift, 2,500 meters beneath the ocean's surface. Seawater filtering through cracks in newly formed ocean crust picks up minerals as it is heated by the hot volcanic rock below. The hot, mineral-laden water then spews, geyserlike, from vents in the seafloor. Around these vents, sulfur-oxidizing bacteria thrive. They are the primary producers for a remarkable community, including tube worms, crabs, giant clams, mussels, and limpets.

Sometimes strong vertical currents, or **upwellings**, transport mineral-rich water from deeper zones to the surface. Upwellings commonly are created by offshore winds. The winds drive warm surface water away from a coast, which allows colder water from below to move up and replace it. Productive ecosystems off the coasts of California, Peru, and Mauritania exist because of such upwellings. Similarly, the Antarctic Seas, cold as they are, have strong circulating currents moving up nutrients. They, too, are highly productive regions.

Estuaries, coastal waters, upwellings—all these regions are being commercially fished at or above levels that are dangerous to their stability. More than this, as we turn to the seas to find more oil, gas, and minerals, the underpinnings of this stability may be weakened even more.

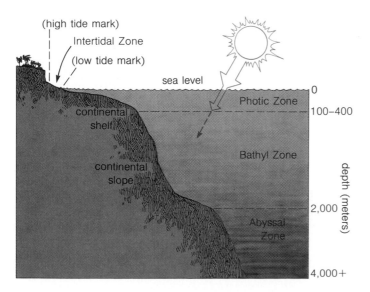

Figure 29.1 Zones of biological activity in the ocean. (The horizontal scale is enormously contracted here.) The lower limit of the photic zone varies, depending on biological, chemical, and physical factors.

Spartina (cord grass)

high tide

low tide

peat

sand

Peabody, Museum, Yale University

Jensen and Salisbury, *Botany*, 1972

Jensen and Salisbury, *Botany*, 1972

Chuck Nicklin

a

b

c

d

e

Figure 29.2 (**a, b**) The estuary—where waters draining from the land mix with seawater carried in on tides. Here, a New England salt marsh, partly inundated by high tides, with *Spartina* rooted in clay and silt deposits from a river. This marsh grass is the dominant primary producer. (**c, d**) Examples of life in the intertidal zone of rocky seashores, where red, green, and brown algae predominate. The sea palm (*Postelsia palmaeformis*) is one brown alga (**c**); the large-bladed *Laminaria* is another (**d**). Just offshore, a California sea lion hovers above kelp, algae, and eelgrass (**e**).

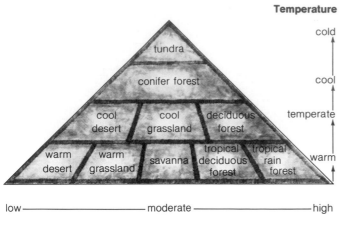

Temperature

cold

cool

temperate

warm

tundra

conifer forest

cool desert | cool grassland | deciduous forest

warm desert | warm grassland | savanna | tropical deciduous forest | tropical rain forest

low ——————— moderate ——————— high

Rainfall

Figure 29.3 Relation between temperature and rainfall, just two of many factors that help dictate the distribution of biomes on the earth's surface. Several major kinds of biomes are shown here. They are arranged as you might find them in traveling from the poles (top of the pyramid) to the tropics (its base), or in descending from the top of a high mountain. The colder the temperature, the less effect rainfall has on the type of primary producers at the site. But as temperatures warm up, the amount of rainfall becomes more important in determining what can grow in a given region.

Primary Productivity on Land

On land, the distribution of biomes is mainly influenced not only by the amount of sunlight but also by temperature and rainfall (Figure 29.3). Moderating these basic limiting factors are other variables. For instance, the amount of incoming sunlight varies with latitude (distance from the equator), with slope exposure, with the seasons, with recurring cloud covers or clear skies. Cloud covers and clear skies depend on such factors as mountain barriers, land masses, and wind and ocean currents. These factors in turn influence air temperature. In coastal zones, moist air and fog tend to hold in heat energy, so there is less temperature variation from day to night and from one season to the next. Inland desert zones, with their clear skies and sparse rainfall, show extreme temperature variations from day to night. The land itself—its elevation, its mineral content, whether its soil holds moisture or encourages runoff, whether it is rolling or flat or mountainous—helps dictate what kinds of life will be found in what places.

Given such variables, ecosystems vary considerably even in the same general region. Yet it's still possible to

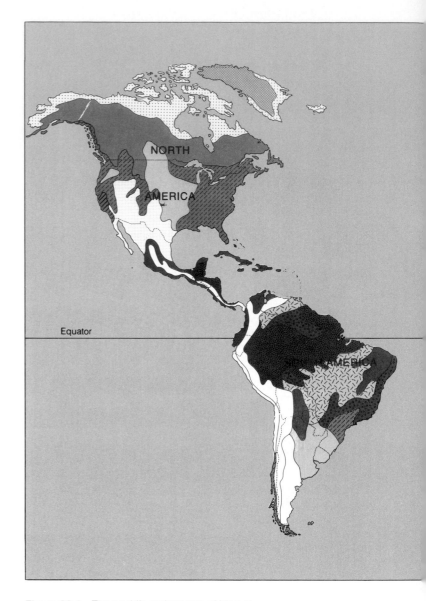

Figure 29.4 The world's major types of biomes.

characterize vast tracts of land as being generally of one type of biome or another (Figure 29.4). You are already acquainted with two major types—the **tundra**, with its primary producers limited to grasses, mosses, lichens, and a few dwarfed woody plants such as willows; and the **savanna**, with its primary producers of grasses and scattered trees adapted to prolonged dry spells. Here we'll focus on five more types of biomes (grassland, desert, tropical rain forest, deciduous forest, and coniferous forest) to illustrate the primary producers of different regions.

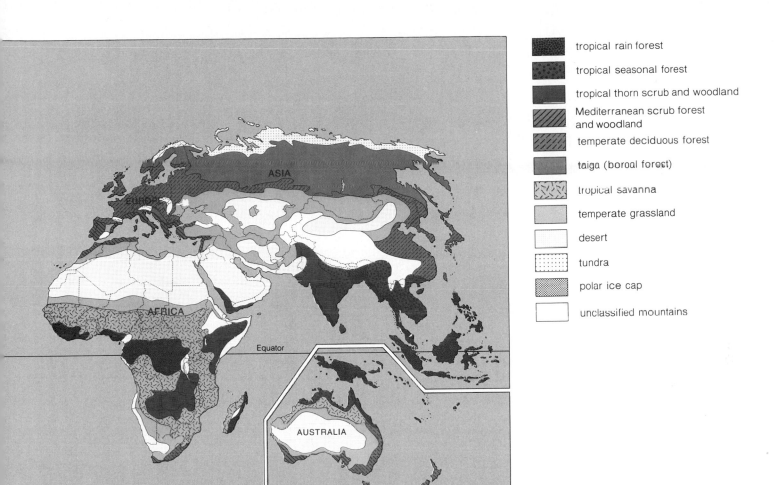

tropical rain forest	
tropical seasonal forest	
tropical thorn scrub and woodland	
Mediterranean scrub forest and woodland	
temperate deciduous forest	
taiga (boreal forest)	
tropical savanna	
temperate grassland	
desert	
tundra	
polar ice cap	
unclassified mountains	

Grassland

In parts of the United States, Canada, Russia, and Argentina, the land is flat, temperatures are moderate, and rainfall is limited by mountains that bar most of the storms moving in from the sea. Even though there is not enough rain to support forests, there is enough to support rolling seas of grass (Figure 29.5). At one time, **grasslands** extended westward from the Mississippi River region to the Rocky Mountains, from Canada through Texas. Where rain was sparse, short grasses were the dominant primary producers. Where there was a little more rainfall, grasses taller than the buffalo living there stretched out to the horizons in all directions.

Today, undisturbed grasslands are rare. For century after century, decaying leaves and litter had enriched the soil of tallgrass regions, making them inevitable targets for conversion into vast fields of corn and wheat. The shortgrass regions did not have enough rain to support these crops, but they seemed ideal for imported cattle.

a

b

Figure 29.5 The grassland: (**a**) rolling shortgrass prairie with mountain barriers to rain in the distance, and (**b**) tallgrass prairie, as it often appears in the vast flat interior of continents.

Enter beef cattle, exit the buffalo (the natural dominant herbivore of the system), and exit the Indian—one of the few modern human groups ever to become an integral part of a stable ecosystem. (The Indian's way of life has long since vanished, but the buffalo may yet make a comeback. With the cost of supplementary cattle feed skyrocketing, the buffalo is looking more and more attractive because of its ability to subsist entirely on the grasses with which it coevolved.) But early cattle operations led to overgrazing, and attempts to farm shortgrass regions led to erosion, on an immense scale, that crippled the primary productivity of much of this land. John Steinbeck's *Grapes of Wrath* and James Michener's *Centennial* are two novels that speak eloquently of this disruption and its consequences.

Figure 29.6 Warm, dry desert near Tucson, Arizona.

Desert

On almost every continent, you can find **desert** biomes: arid regions that support little primary production. About five percent of the world's deserts are extremely torrid during the day. But because there is little moisture in the air, heat loss is tremendous during the night, which can be extremely cold. Few kinds of organisms can survive these extremes. The Sahara Desert of Africa and Death Valley in the United States are such deserts. About 100 meters below some desert regions is a zone of sediments and rocks saturated with water (groundwater). In the few spots where these water-bearing rocks break through the surface at low elevations, springs and oases form.

Where cold winters are followed by prolonged drought, you find *cool deserts*. Here, primary production is limited to the springtime growth of grasses, herblike plants, and shrubs able to withstand frost followed by arid conditions. Cool deserts in the western United States support sagebrush communities.

In *warm deserts* (Figure 29.6), the temperature typically ranges from mild to hot, and the primary producers vary widely from one region to the next. Even so, there is a remarkable similarity in the kinds of plant and animal adaptations to these temperatures. Plants such as the cactus, for instance, remain metabolically active during the long, dry spells between rains. They can do so because they have no leaves, which means they don't lose much water through

a

transpiration (their stems carry on photosynthesis). When the brief spring rains arrive, these plants depend on extensive root systems just below soil surface.

During the day, a warm desert appears devoid of animal life. But where there are producer organisms, so are there consumers. During the summer day, the heat exceeds their range of tolerance, so consumers are found in burrows, under rocks, wherever they can escape the sun. It is during the cool hours of dusk and dawn, and to some extent the middle of the night, that the desert comes to life.

Ironically, it is precisely because of the relentless sun that the desert may be vulnerable to ecological disruption. Hard on the heels of water lines are human populations, out for retirement or fun in the sun. On weekends and holidays, the deserts of Southern California face human hordes on that symbol of American freedom, the motorbike, tearing up erosion-resistant grasses and surface soil, chasing down and running over rabbits or coyotes in a hip Western version of fox-and-hounds. And heaven help the tarantula that wanders into Sun City.

Rain Forest

When warm temperatures combine with abundant and fairly uniform rainfall, the **tropical rain forest** emerges. It is a dense forest of stratified communities, dominated by trees spreading their canopy far above the forest floor. Figure 29.7 shows a tropical rain forest.

At different levels beneath the tall canopy in such forests are plant species adapted to ever diminishing amounts of sun. On the forest floor, few plants are effective at photosynthesis, although any break in the canopy encourages exuberant growth of tree saplings and other plants. Competition for light is intense, but the abundant rains make possible adaptations that never would be seen under drier conditions. Vines of all sorts clamber up toward the sun. Mosses, orchids, lichens, and bromeliads (plants related to the pineapple) grow on tree branches. They obtain minerals from the falling leaves and debris, and from the wastes of animals that live high up in the canopy. The food webs sustained by all these producer plants are as-

Figure 29.7 Tropical rain forest showing aerial plants and ever-present vines in the canopy (**a**), and the dense, lush growth on the forest floor (**b**) which occurs only where light breaks through.

Doria Hutchins

b

tonishing. Entire communities of insects, spiders, and amphibians can live, breed, and die in the small pools of water that collect in the leaves of plants growing high above the forest floor. Many insects, birds, and monkeys spend most of their lives at a single level in the canopy. (For instance, from the time of birth until two months later, spider monkeys hang on for dear life to mother, who keeps her arms free for swinging through the trees and who rarely drops to the ground.) In such places, the kinds of organisms living in or on a single tree often exceed the kinds of organisms living in an entire forest to the north!

With all this diversity, however, a tropical rain forest is one of the worst places to grow crops. Despite all the producers, there is practically no organic debris. Decomposers quickly break down organic debris and return useful nutrients to the soil. There, layers of roots take up the nutrients for rapid conversion into new biomass. Thus, when the forest is cleared of plants and animals for agriculture, most of the nutrients are permanently cleared off with them.

Even with **slash-and-burn agriculture** (in which the forest biomass is reduced to ashes, then tilled into the soil), most of the nutrients are soon washed away because of the heavy rains and poor soils. After a few years, cleared fields become infertile and usually are abandoned—and the forest is left to heal itself through the extremely slow process of succession.

Deciduous Forest

The natural ecosystems of Southeast Asia resemble those of Central America. In these **tropical deciduous forests**, most plants lose their leaves during part of the year as an adaptation to the dry seasons that alternate with the wet monsoons. It happens that small shifts in the jet-stream patterns of air circulating far above the earth's surface occur naturally, and they determine whether the monsoons will fall, say, in India or in the oceans to the south. The natural ecosystems of such areas as India have evolved over thousands of years, and organisms have come to live there that

Spring

Autumn

can survive years of drought. They are little affected by these short-term variations. But the human population is particularly dense in Southeast Asia, and the agricultural systems providing them with food depend utterly on the timely arrival of the monsoons—a dependence that often leads to shortages and famine when vulnerable crops die.

Throughout much of eastern North America are **temperate deciduous forests**, where the dominant trees drop their leaves not in response to dry seasons but as protection against winter cold (Figure 29.8). Few undisturbed ecosystems of this type still exist. Depending on patterns of human settlement, they were cleared or cut over, often

Summer

Winter

All photographs Thomas E. Hemmerly

Figure 29.8 The changing character of a temperate deciduous forest in spring, summer, autumn, and winter.

several times. Today the forests of the Southern Appalachian Mountains are the best example of this disappearing biome. In precolonial times, herbivores included squirrel, rabbit, and deer, which were prey for bobcat, wolf, bear, and mountain lion. Today, because of the human population in the area, the natural predators have been largely exterminated on the grounds that they are dangerous and don't belong there. Populations of large herbivores such as deer are kept in check through controlled hunting seasons—although unlike natural population controls, hunters often seem more interested in taking the healthiest animals instead of the old and weak.

a

Figure 29.9 Two distinct coniferous forests: (**a**) the moist, temperate coastal forest of the Pacific Northwest in summer, and (**b**) the subalpine forest of Yosemite beneath the first snows of winter.

Timothy Ransom

Coniferous Forest

Below the tundra line in northern regions is an extensive zone known as the **northern coniferous forest**, or **taiga** (Figure 29.9). Although winters are prolonged and extremely cold, with a fairly constant snow cover, the summer season is warm enough to promote the dense growth of trees. Evergreen conifers (spruce, fir, pine, and hemlock) have the competitive edge, for they are able to carry out photosynthesis just as soon as temperatures rise above freezing. They also have thick bark and heavily waxed, needlelike leaves, which enable them to withstand extreme temperatures and to conserve water during both winter cold and drought.

Food webs built around the conifers include insects that feed on the bark, buds, cones, leaves, and tender shoots of the trees. Decomposers are not as prolific here as they are in deciduous forests, and the resin-rich needles of conifers decompose much more slowly. The carnivores feeding on insects include birds of all sorts, whose niches are carved out in zones ranging from ground level to the uppermost canopy. Climbing about on the branches and trunks are seed-eating squirrels and mice. Larger herbivores, such as hare, beaver, moose, and deer, do not use the conifers as their primary food source, even though they are sometimes mistakenly included as part of conifer food webs. Instead they forage on low-growing vegetation and deciduous trees that spring up around the region's many lakes, streams,

Ansel Adams

and marshes ("taiga" is Russian for "swamp forest"). They use the forest primarily as protection against predators.

Is it this same recognition of sanctuary that draws us to a forest? It's interesting that, at a time when our grasslands were first being tilled under, a conservation movement began to save the forests. In fact, the first bureau of conservation was the United States Forest Service. Partly as a result of its initial efforts, a third of our country is still heavily forested; of that, twenty-seven percent is preserved in national, state, and local parks. The Forest Service manages all wildlife, watersheds, recreation, and lumbering in the national forest system. Part of its program is based on **tree farming**, the replacement of climax communities with monolithic stands of fast-growing softwood trees.

There is increasing concern that simplified tree farms are replacing too much of the stable, 30- to 200-year-old climax communities either because of or in spite of Forest Service management. As simplification is permitted to occur, the system becomes more vulnerable and the trees must be protected like other crops, through widespread use of insecticides and fertilizers. There is also reliance on clear-cutting to keep up with demands for wood products (the average American directly or indirectly uses approximately 560 pounds of wood products such as paper each year). This practice leaves areas scarred and vulnerable for decades. Yet ninety percent of our national forests are potentially open to clear-cutting. This is another area where our short-term demands eventually must be tempered.

Table 29.1 A Few Characteristics of Urban Ecosystems

Benefits

Employment
Mass manufacturing and production
Excitement and stimulation of ideas
Entertainment

Internal Instabilities

Reliance on imported energy
Reliance on imported food
Reliance on imported water
Reliance on imported materials
Air polluted with heat and chemicals
Water polluted with sewage and industrial wastes
Accumulation of solid wastes
Noise pollution
Congested transportation pathways
Poverty
Potential for disease epidemics
Crowded habitats and concurrent individual stress
Crime, drug abuse

External Urban-Derived Instabilities

Simplification of natural ecosystems
Pesticides and agricultural wastes
Disturbances to all stable ecosystems
Surface mining

After G. T. Miller, Jr., *Living in the Environment*, Wadsworth, 1979.

IS THE CITY AN ECOSYSTEM?

Ocean, savanna, tundra, grassland, desert, forests of different character—*as diverse as all the world's biomes are, as simple or complex as their dominant communities and ecosystems may be, they have in common a self-contained, self-perpetuating stability.* They are essentially self-contained arrays of consumers and decomposers that have achieved dynamic balance with the producers of the site by recycling nutrients even as they themselves are nourished.

For the past 11,000 years, a new kind of ecosystem has been evolving that knows few geographic or climatic boundaries. It is based on human communities that are increasingly overlapping and merging together, first into cities and then into sweeping urban areas called **megalopolises** (Figure 29.10). An urban region contains not only human consumers and their pets, but also such scavengers as rats, cockroaches, pigeons, starlings, and sparrows. An urban center interacts with the land, the forests, the rivers, even the local climate. It *is* an ecosystem. But it is not self-contained, and it is not stable (Table 29.1).

Figure 29.10 The city as ecosystem.

The preceding photographs of a few land biomes were presented not merely to make you gasp in wonder over the glories of natural ecosystems. And it was not out of indifference to the arrays of consumers and decomposers of each ecosystem that we excluded photographs of a squirrel or wolf here, a rabbit or mushroom there. Rather, the photographs are meant to convey that *the foundation of any biome is its characteristic array of producer organisms.* This is where all webs of life begin; without them, life ends.

© Bill Grimes

Urban centers such as cities have no producers. They have only consumers and, compared to the human biomass, they have a negligible number of decomposers. The environmental conditions are much the same for a city as they are for the surrounding biome. But soil is paved over, plants are bulldozed under, and water as well as other resources are diverted for direct and indirect human consumption.

A city endures as an ecosystem only as long as it imports energy and materials from someplace else.

Repercussions of an Unbalanced Energy Budget

The need to import materials and energy into an urban center puts amplified pressures on ecosystems far outside its boundaries. A typical United States urban center having a population of 1 million requires basic daily imports of 625,000 tons of water, 9,500 tons of fossil fuels, and 2,000 tons of food, not to mention material resources for the construction and industry needed to sustain human habitats. Immense tracts of land outside the urban sphere must be converted to agriculture. Vast transportation networks must be created to ship in agricultural products. Entire regions are mined over for metals of all sorts. Water and oil are brought in from sources that span the globe.

What does the city offer in return? Employment and manufacturing (essentially to sustain the artificial environment), entertainment, a stimulating environment are some of the answers given. Table 29.1 lists a few more. The diversion of available world energy resources into cities is mostly one-way, an energy flow that leads not to stability but to amplified demands of the urban population.

Everything Has to Go Somewhere—Pollution Through Nonrecycled Resources

In natural ecosystems, the by-products of life may not accumulate because they are recycled. What we call **pollutants** are by-products of our existence that are not recycled through the webs of life. When they accumulate to high levels, they usually have harmful effects, or at least perplexing ones. How, for example, can one urban center containing 1 million people absorb its 2,000 tons of daily trash—bottles, cans, newspapers, toilet paper, plastic spoons and forks, paper containers, "disposable" diapers, "disposable" razors, plastic bags and wrappers, cellophane and aluminum wrappers, cartons, heavy-duty lawn litter bags, letters, envelopes, ad infinitum? How can 500,000 tons of daily human sewage and industrial wastes—lead, chromium, mercury, sulfur—be recycled through the community life? Which community members really want to recycle 950 tons of daily air pollutants in their lungs? Where to put 500 tons of dog feces and 700,000 gallons of dog urine that daily find their way to open storm drains, then lakes, rivers, and oceans?

The answer is that we cannot recycle all these things. We can reduce, collect, concentrate, bury, and burn wastes, or spread them out through other ecosystems, but we can never be completely rid of pollution. For the past four decades, for example, 350 million gallons of sewage from metropolitan New York and New Jersey have been dumped each day off the shore of Long Island where, being out of sight, it was easy to put out of mind. Then, in the summer of 1970, a dead sea of sewage unaccountably began moving back toward the land, bringing with it the specter of hepatitis, encephalitis, and other terrible diseases to the megalopolis that created it.

But the point of this chapter is not to list examples of abuse after abuse created by the artificial urban ecosystem. They do exist, they are being studied in an attempt to bring them to controllable levels, and it's important for you to know about them, for it is your life as well as all life that is being affected by them. However, our focus here has not been on the symptoms but on the fundamental source of the problem. We have presented biological models of energy flow and materials reuse that work; we have raised the idea that *existing* human-designed models inherently cannot work, because they have strayed too far from the biological models. Where do we go from here? The last chapter of this unit presents some alternatives.

PERSPECTIVE

Molecules, cells, tissues, organs, organ systems, whole organisms, populations, communities, ecosystems, the biosphere. This is the architectural stuff of life—a sequence that reflects a progressive complexity in the adaptations of many living systems that have appeared over the past 3.5 billion years. We are relative latecomers to this immense biological building program. Yet, in the comparatively short span of 50,000 years, we have been attempting to force our own blueprints for living upon it. We are now restructuring the stuff of life at all levels—from recombining DNA molecules of different species to changing the nature of the thin, life-giving atmosphere that envelops the entire earth.

Of course, it would be presumptuous of us to think we are the only organisms that have ever changed the nature of living systems. As you read earlier, evolutionary theory suggests that the activities of photosynthetic organisms irrevocably changed the course of biological evolution, and that the appearance of predatory organisms meant simple chemical evolution would come under the dominion of biological competition. Thus biological systems have been evolving for 3.5 billion years through natural selection and through the correlated prodding of geologic unrest. Change is nothing new to this program. What *is* new is the accelerated, potentially devastating change being brought about by our cultural evolution.

For most of evolutionary history, natural ecosystems have not been characterized by sudden, cataclysmic change. Barring outside disruptions, they have been and continue to be governed by feedback control mechanisms that keep these systems in balance even while allowing room for change. For instance, potentially runaway growth of the population in an ecosystem is kept in check by the carrying capacity of the environment. But feedback control mechanisms can come into play only when deviations already exist in the system itself. The system has already changed before the controls work to return it to dynamic stability.

Feedback control is not enough for human-designed ecosystems. Our designs are far too vulnerable because of their utter dependency on a one-way flow of materials and energy, which gives the illusion of a stable basis for explosive population growth and resource consumption. A sudden shortage of food, a shortage of raw materials—such deviations in the system can come too fast for us to correct. And they can have too great an impact for us to know whether the imbalance will be irreversible.

What about feedforward control mechanisms? Many organisms, as you read earlier, have early warning systems. Skin receptors tell of the outside air growing colder. Each sends messages through the nervous system to the metabolic apparatus for raising internal body temperature before the body itself becomes dangerously chilled. With feedforward control, corrective measures can begin before a change in the external environment has significantly altered the system itself.

But even feedforward controls are not enough for human ecosystems, for they go into operation only when change is already under way. Consider, by way of analogy, the DEW line—the Distant Early Warning System, our nation's sensory receptor for intercontinental ballistic missiles that may be launched against us. By the time it detects what it is designed to detect, it may be far too late, not only for humans but for the rest of the biosphere.

We have, as a species, moved beyond the stability afforded by feedback and feedforward controls over natural systems. It would be naïve to assume we can ever reverse who we are at this point in evolutionary time, to de-evolve ourselves culturally and biologically into becoming less complex in the hope of achieving more stability. *But there is no reason to assume we cannot achieve stability by moving forward.* For we have available to us a third kind of control system, and it is uniquely our own. We have the capacity to anticipate events *before* they happen. We are not locked into responding only after change has begun. We have the capacity for anticipating our future—it is the essence of our visions of utopia or of nightmarish hell. Thus we all have the capacity for adapting to a future which we can partly shape. We can, for example, learn to live with less; far from being a return to primitive simplicity, it would be one of the most complex and intelligent behaviors of which we are capable.

But having that capacity and exercising it are not the same thing. Our ecosystems are already on dangerous grounds because we have not yet mobilized ourselves as a species to work toward stability. Our survival depends on predicting possible futures, on designing and constructing ecosystems that are in harmony not only with what we define as basic human values but with the biological models available to us. Human values *can* change; our expectations can and must be adapted to biological reality.

For the principles of materials and energy flow—which govern the growth and survival of all systems of life—do not change. It is our biological and cultural imperative that we come to terms at last with these principles, and with what will be the long-term contribution of the human species to the unity and diversity of life.

Readings

Flanagan, D. (editor). 1970. *The Biosphere.* San Francisco: Freeman. Collection of articles from *Scientific American.*

Krebs, C. 1978. *Ecology: The Experimental Analysis of Distribution and Abundance.* Second edition. New York: Harper and Row. Excellent book, very well written and with clear illustrations.

Miller, G., Jr. 1979. *Living in the Environment: Concepts, Problems, and Alternatives.* Second edition. Belmont, California: Wadsworth. This is probably your best bet as an introduction to environmental science. Miller writes clearly, in a lively manner, with excellent analogies. The illustrations are superb.

Ricklefs, R. 1976. *The Economy of Nature.* Portland, Oregon: Chiron Press. Ricklefs is one of the poets of ecology.

West, S. 1980. "Smokers, Red Worms, and Deep Sea Plumbing." *Science News* 117(2):28–30. An account of deep-sea exploration and discoveries in the Galápagos Rift.

Review Questions

1. What is a biome? How are primary producers related to biomes?

2. List the primary producers of an intertidal zone. Are there any primary producers in the abyssal zone of the world ocean? What is an estuary?

3. What are the main factors influencing the distribution of biomes? Characterize the primary producers of each of the following:
   a. Tundra
   b. Savanna
   c. Grassland
   d. Desert
   e. Tropical rain forest
   f. Deciduous forest
   g. Coniferous forest

4. List a few of the consumers that are found in each of the seven biomes listed above.

5. In which biome are there many different species adapted to ever diminishing amounts of sunlight?

6. Why are warm deserts considered to be vulnerable environments?

7. Can a city be considered a self-supporting, stable ecosystem?

30

HUMAN ECOLOGY

Once there was a bullfrog who found a lovely shallow pond, and he staked it out for his own. There he sat, ribbeting away, trying to attract a female frog to this really wonderful place. Well, as it turned out the bullfrog had laid claim to a frying pan full of water. Beneath the frying pan was a campstove turned on low. Slowly the water began to heat up—so slowly that the bullfrog didn't even notice. Gradually the water got warmer, but it was such a gradual thing and the frog was so busy ribbeting his little heart out that he still didn't notice. Soon there was no way *not* to notice. But by then it was so hot his muscles no longer responded to signals from his nervous system. And so it got hotter, his croaks got weaker, until he dropped dead. The moral of this story is not (as you might suspect) that you can't build a marriage on the promise of sex alone. The moral is that it is possible to live in the midst of a gradual trend toward unlivable conditions and not even be aware of what is going on.

Like the bullfrog, we generally are so busy concentrating on so many other pressing matters that we are indifferent to ominous changes going on in the world around us. Where is indifference leading us? More than this, can ecological awareness really take us in another direction? Now that we have looked at some basic ecological principles, and have some idea of energy flow and materials recycling through the natural ecosystems, let's see what we have as some of our alternatives.

ON THE NATURE OF POLLUTION

Solid Wastes

Lisa is entering college in the fall. On the outskirts of Del Mar, the town where she lives, a tiny plot of land has finally been set aside as a recycling center for newspapers, glass, and aluminum cans. Last week Lisa decided to take part in the recycling program. There was a brief period of adjustment: all the members of her family had to get used to stacking up the newspapers in the garage, to putting glass jars and bottles in one trashcan and aluminum cans in another. By the week's end, she was mildly surprised at how much had accumulated; somehow she had assumed it would take months to stack up. It took only a few minutes to run everything down to the recycling center. But along the way she became aware, for the first time, of just how many trashcans were lined up along the streets, waiting to be carted out of sight, even though the recycling center had been open for several months. Two or three trashcans sat in front of each house. There are about 5,000 homes

Figure 30.1 Who are these organisms building their own ecosystems, and why are they doing it?

in town, she mused. That meant about 12,000 trashcans a week had to be emptied somewhere. In the surrounding metropolitan area, there are about 200,000 homes. Could it be—400,000 to 600,000 trashcans *each week?*

After Lisa deposited everything in the collecting bins, she decided to take a run out to the county "sanitary landfill station" where all the nonrecycled solid waste ends up. She was stunned by what she saw. One trash-filled truck after another sped through the gate, antlike in their line to the dumping ground. Bulldozers were hastily shoving mounds of all manner of refuse down the sides of what had once been chaparral-covered canyons. What, thought Lisa, will happen when these canyons are filled? To the east are the rolling hills of Rancho Santa Fe; there are too many moneyed people there to let the bulldozers in. To the south and north, hundreds of tract homes are being built—even on top of the filled-in canyons as fast as the bulldozers level off the site. To the west, the ocean. *Surely not the ocean!* This same thing must be happening all over the nation, she thought. What, then, could her isolated commitment possibly mean? How could one small action help turn such a tide of solid waste pollution? Somebody ought to start a local campaign for public awareness, she mused. They should call it "Take Someone You Love to the County Dump." But with college starting soon, how could she get involved? Somebody else would have to do it.

Which "somebody" is going to tackle the 4.5 billion metric tons of solid wastes we dump, burn, or bury each year in the United States? Who is going to decide where the landfills go next? Conversely, is recycling itself a workable alternative? Recycling is part of the answer, but it can't be a hit-or-miss thing. For instance, in her solitary drive to the recycling center, Lisa used up some gasoline—a nonrenewable energy source which ultimately must be figured as part of the energy cost of the program. Although the energy cost of recycling is lower than it would be to extract and use new raw materials, it still can't be ignored—especially when you multiply the energy cost by all the individuals driving separately to the center.

What it is going to take, in the long run, is a change in our basic living habits. We have been brought up to follow the patterns of a "throwaway" culture: use it once, discard it, buy another. For instance, between fifty and sixty-five percent of urban wastes are paper products—of which only nineteen percent is now being recycled. If we were to recycle merely half the paper being thrown away each year, we would do more than conserve trees. For the *energy* it takes to produce an equivalent amount of new paper could be diverted to provide electricity to about 10 million homes! Or consider that about 60 billion beverage

Table 30.1 Energy Needed to Produce or Recycle a Few Raw Materials

Resource	Pounds of Coal Needed To Produce One Pound of Metal	
	From virgin ore	From recycled material
Aluminum	6.09	0.17 to 0.26
Steel	1.11	0.22
Copper	1.98	0.11

Data from Oak Ridge National Laboratory.

Table 30.2 Comparison of Three Programs for Handling Solid Wastes

Item	Throwaway System	Recycling System	Ecologically Based System
Glass bottles	Dump or bury	Grind, remelt; remanufacture; convert to building materials	Ban all nonreturnable bottles and reuse (not remelt and recycle) bottles
Bimetallic "tin" cans	Dump or bury	Sort with magnets; remelt	Limit or ban production; use returnable bottles
Metal objects	Dump or bury	Sort; remelt	Sort, remelt; but tax items lasting less than ten years
Aluminum cans	Dump or bury	Sort; remelt	Limit or ban production; use returnable bottles
Paper	Dump, burn, or bury	Incinerate to generate heat	Compost or recycle; tax all throwaway items; establish national standards to eliminate overpackaging
Plastics	Dump, burn, or bury	Incinerate to generate heat or electricity	Limit production; use returnable glass bottles; tax frivolous throwaway items and packaging
Garden and food wastes	Dump, burn, or bury	Incinerate to generate heat or electricity	Compost; return to soil as fertilizer or use as animal feed

From G. T. Miller, Jr., *Chemistry: A Contemporary Approach*, Wadsworth, 1976.

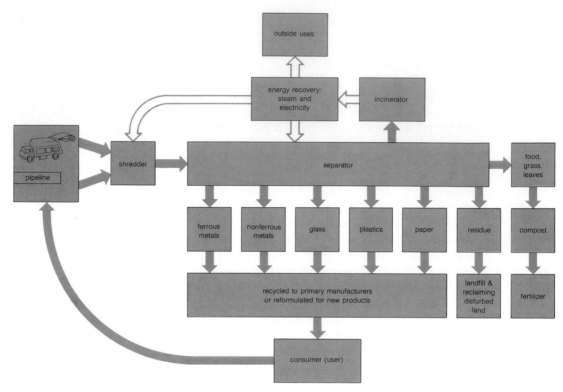

Figure 30.2 Generalized resource recovery system. Solid wastes are treated as urban "ore." Bulk items are sorted out, and the remainder is shredded and separated. Electromagnets could be used to extract steel and iron; air blowers could send plastic and paper to different recovery chambers. Mechanical screening, flotation, and centrifugal hurling could sort out metals, glass, and garbage.

Once sorted, metals could be returned to mills, smelters, and foundries; glass to various glass processing plants. Organic matter could go to compost centers, later to be used as fertilizers or soil conditioners, perhaps as fuel or animal feed. Wood could be used as fuel. Paper could be recycled or used as fuel. Even residues from incinerated materials (including particles removed from smoke to limit air pollution) could be processed into road or building material (such as bricks).

Miller, *Living in the Environment*, 1975

containers are sold each year in the United States. About 50 billion are nonreturnable cans and bottles, too many of which are discarded in public places. These containers represent three-fourths of the litter that must be picked up along highways—a time-consuming, energy-draining activity that costs thousands of barrels of petroleum (not to mention hundreds of millions of tax dollars) each year.

In the early decades of this century, before it became "easier" and "cheaper" to throw things away, beverage containers were used over and over again, sometimes as often as fifty times. The same system could be made to work again. Even now, some states are passing laws requiring deposits on all beverage cans and bottles. The refundable deposit has helped check the waves of litter that typically washed over Yosemite National Park each year. Oregon, Vermont, Maine, and Michigan have had similar success with their programs.

Adherence to ecological principles of materials reuse is long overdue. A transition from a "throwaway" life-style to one based on conservation and reuse is already economically feasible, and we have most of the technology needed to do it. The question becomes whether we are willing to commit ourselves to crossing the threshold. We can, for instance, bring consumer pressure to bear on manufac-

turers by refusing to buy goods that are lavishly wrapped, excessively boxed, and designed for one-time use. We can ask our local post office how to go about turning off the daily flow of junk mail, a flow that represents an astounding waste of paper, time, and energy—and higher mail delivery rates for everyone. We can work to see that our local city and county governments plan to develop large-scale resource recovery centers of the sort described in broad outline in Figure 30.2. In such systems, existing dumps and landfills would be viewed as urban "mines." Nearly one-quarter of our past and present solid wastes might be potentially recoverable from these mines. When will such conservation and recovery programs start?

Water Pollution

Adán is not happy about the housing development scheduled to be built right across the freeway from the small Arizona town in which he lives. Forty thousand homes constructed over the next five years. He saw sketches of the development in the local newspaper; the area is destined to become a high-density suburb. What's worse, there is no way existing public facilities—schools, roads, power plants, waste disposal systems—can absorb the demands

Direct Personal Use
*160 gallons, or 8 percent
of your daily use. Examples:*

Bath *30–40 gallons*
Shower *5 gallons/minute*
Shaving *3 gallons, water running*
Washing clothes *20–30 gallons*
Preparing and cooking food *8 gallons*
Washing dishes *10 gallons*
Housecleaning *8 gallons*
Flushing toilet *3 gallons/flush*
Leaking toilet *35 gallons/day*
Watering lawn *80 gallons/8,000 square feet*

Appalling Little Fact
Total daily faucet and toilet
leaks in New York City alone
200,000,000 gallons

Indirect Agricultural Use
*600 gallons, or 33 percent of
your daily use. Examples:*

One egg *40 gallons*
One ear corn *80 gallons*
One loaf bread *150 gallons*
One gallon whiskey *230 gallons*
Five pounds flour *375 gallons*
One pound beef *2,500 gallons*

Indirect Industrial Use
*1,040 gallons, or 59 percent of
your daily use. Examples:*

Cooling water for electric power plant *720 gallons*
Producing consumer goods *320 gallons*
Sunday paper *280 gallons*
One pound synthetic rubber *300 gallons*
One pound aluminum *1,000 gallons*
One pound steel *35 gallons*
One gallon gasoline *7–25 gallons*
One automobile *100,000 gallons*

Miller, *Living in the Environment*, 1975

Figure 30.3 Average daily use of water by an average American: 1,800 gallons, compared with 12 gallons for an average person in developing nations.

of the increased population. Even putting aside the question of who will pay for expansion of facilities, where will the by-products of the increased population go? Solid waste disposal is only one aspect of the problem. What about something as basic as waste-water treatment? The present facilities are operating to capacity, and there simply isn't enough available water in this part of the country to process more raw sewage before releasing it into the few existing waterways. At the same time, more and more people are being attracted to Arizona because of its magnificent climate. Adán remembered the pointed remark of a real estate agent when he questioned the wisdom of permitting such massive development before these issues were resolved: "Did you want to keep people out *before* or *after* you moved yourself?" Yet, with real limits on natural resources as basic as water, isn't it insane to pretend otherwise? Can more water be brought in? Even assuming someone will pay for it, where will the water come from?

For the first time in his life, Adán was facing tangible evidence that there *are* limits to growth. True, there is a tremendous amount of water in the world, locked into one vast closed system called the **hydrologic cycle**. For as short a time as it takes for water vapor to form over the oceans and fall on the continents as rain, for as long a time as

it takes for water from lakes and rivers and glaciers to flow again to the oceans, there is the promise of movement through the biosphere. Still, the promise is greater for some than for others: three out of every four humans alive today do not have enough water or, if they do, their supplies are in some way contaminated. Aside from the unequal distribution of water throughout the world, most of it (ninety-five percent) is in the oceans and can't be used for human consumption or agriculture; it's too salty. Of the remainder, nearly ninety percent is locked in glaciers or in polar ice caps. Most of the rest is deep in the ground or suspended in the atmosphere. In fact, for every 1 million gallons of water in the world, only about 6 gallons are readily available in a usable form!

Pure water is one of our most precious and most rapidly disappearing resources. It is not that water is leaving the hydrologic cycle. Rather, it is that more and more is becoming polluted. **Water pollution** refers to a condition in which a body of water has been altered chemically or physically in such a way that renders it useless for a given purpose. We have, in effect, tapped into the hydrologic cycle and are using it directly or indirectly as a dumping ground for by-products of human existence. Thus water becomes unfit to drink (even to swim in) because it is polluted with human sewage and animal wastes, which in turn can encourage the growth of large populations of disease-causing bacteria and viruses. Through agricultural runoff, water becomes polluted with sediments from land erosion, with substances such as pesticides and herbicides, and plant nutrients (such as nitrate and phosphate from fertilizers). Through industrial activity and power-generating plants, water becomes polluted with chemicals, radioactive material, and excess heat (thermal pollution).

Table 30.3 highlights some of the ways to start reversing the trend toward unmanageable water pollution. But to give you an idea of how difficult the task will be, let's take a closer look at the problems inherent in **waste-water treatment** methods used today. To begin with, about half the waste water in the United States is not being treated at all. The other half is more or less purified and restored while it is channeled through primary, secondary, and/or tertiary treatment centers before it is discharged into waterways and seas. In **primary treatment**, mechanical screens and sedimentation tanks are used to force coarse suspended solids out of the water to become a sludge. In a process called flocculation, chemicals such as aluminum sulfate are added to accelerate the sedimentation process. About 30 percent of the waste water in the United States goes only through primary treatment before the liquid (the *effluent*) is discharged. Although it is treated with chlorine to kill disease-causing microorganisms in the effluent, chlorine alone is not really enough to get rid of them completely. **Secondary treatment** depends on microbial action to degrade the sludge. Either the sludge is sprayed and trickled through a large bed of exposed gravel in which assorted monerans and protistans live, or the sludge is aerated with pure oxygen to promote microbial activity. As it happens, the microorganisms on which these treatment processes depend are vulnerable to pollution themselves: toxic substances that enter the treatment center can destroy them. Then treatment activities must be shut down until new microbial populations are established. Because levels of pollutants are continuing to rise, the search is on to find chemicals that can degrade sludge. Such chemicals will permit us to be less dependent on microorganisms for handling waste water.

Although secondary treatment is an improvement over primary treatment alone, a host of substances still remains in the effluent: oxygen-demanding wastes, suspended solids, nitrates, phosphates, and dissolved salts such as heavy metals, pesticides, and radioactive isotopes. The water discharged from secondary treatment plants can be used in irrigation and industry, and (if it has been held for a period in settling lagoons) it is often dumped into whatever local body of water may be available.

Only with **tertiary treatment** is there ecologically adequate reduction of the pollutants remaining in the effluent. The process involves advanced methods of precipitation of suspended solids and phosphate compounds, adsorption of dissolved organic compounds, reverse osmosis, stripping of ammonia to remove nitrogen from it, and disinfecting the water through chlorination or ultrasonic energy vibrations. It is largely in the experimental stage and it is expensive; hence tertiary treatment is rarely used.

What all this means is that most treatment centers are not properly treating existing levels of waste pollution. What, then, is going on? A typical pattern, repeated thousands of times along our waterways, is this: water for drinking is removed *upstream* from a town, and wastes from industry and sewage treatment are discharged *downstream*. It takes no great leap of the imagination to see that pollution intensifies as rivers flow toward the ocean. In Louisiana, where the waters drained from the Central United States flow toward the Gulf of Mexico, pollution levels are considered high enough to be a real threat to public health. Water destined for drinking does get treated to remove disease-causing bacteria—but the treatment doesn't remove poisonous heavy metals, such as mercury, which are dumped into the waterways by thousands of factories upstream. More than this, it is at each river's end—the estuary bridge to the sea—that most of the animal life of the sea is born. It is here, in the region we depend most upon for the continued productivity of the world's oceans, that water pollutants are becoming most concentrated.

We must move beyond the mentality of treating an ever-growing mountain of sludge and then quickly dumping it into the environment. We must come to see the folly of dumping industrial wastes into the waters on which we and all life depend. For instance, instead of sending barges of sludge to sea (Chapter Twenty-Nine), we could search for ways to convert it safely to fertilizers or compost. The methane gas that arises from anaerobic activities of microorganisms during secondary treatment could become a fuel source. A few progressive industries and cities have initiated tertiary treatment, and the water they discharge into waterways is actually more pure than the water they originally take out. Tertiary treatment, again, is not without cost. But can we afford *not* to pay the price?

Air Pollution

Denise was about to throw up. The flight from Dallas to Chicago was so crowded that she had to sit in the "Smoking Allowed" section of the plane. When the man sitting next to her lit up a cigar, when the stenchy blue smoke began

Table 30.3 Major Water Pollutants: Sources, Effects, and Possible Controls

Pollutant	Main Sources	Effects	Possible Controls
Organic oxygen-demanding wastes	Human sewage, animal wastes, decaying plant life, industrial wastes (from oil refineries, paper mills, food processing, etc.)	*Overload depletes dissolved oxygen in water; animal life destroyed or migrates away; plant life destroyed*	Provide secondary and tertiary waste-water treatment; minimize agricultural runoff
Plant nutrients	Agricultural runoff, detergents, industrial wastes, inadequate waste-water treatment	*Algal blooms and excessive aquatic plant growth upset ecological balances; eutrophication*	Agricultural runoff too widespread, diffuse for adequate control
Pathogenic bacteria and viruses	Presence of sewage and animal wastes in water	*Outbreaks of such diseases as typhoid, infectious hepatitis*	Provide secondary and tertiary waste-water treatment; minimize agricultural runoff
Inorganic chemicals and minerals	Mining, manufacturing, irrigation, oil fields	*Alters acidity, basicity, or salinity; also renders water toxic*	Remove through waste-water treatment; stop pollutants from entering water supply at source
Synthetic organic chemicals (plastics, pesticides, detergents, etc.)	At least 10,000 agricultural, manufacturing, and consumer uses	*Many are not biodegradable; chemical interactions in environment are poorly understood. Some create noxious odors and tastes; others clearly poisonous*	Prospects are poor at present. Push for biodegradable materials; prevent entry into water supply at source
Fossil fuels (oil particularly)	Two-thirds from machinery, automobile wastes; pipeline breaks; offshore blowouts and seepage; supertanker accidents, spills, and wrecks; heating; transportation; industry; agriculture	*Varies with location, duration, and type of fossil fuel; potential disruption of ecosystems; economic, recreational, and aesthetic damage to coasts*	Strictly regulate oil drilling, transportation, and storage; collect and re-process oil and grease from service stations and industry; develop means to contain spills
Sediments	Natural erosion, poor soil conservation practices in agriculture, mining, construction	*Major source of pollution (700 times more tonnage than solid sewage discharge), fills in waterways, reduces shellfish and fish populations, reduces ability of water to assimilate oxygen-demanding wastes*	Put already existing soil conservation practices to use

to cling to her nostrils and even her clothing, she suddenly compared herself to a condemned woman about to be gassed to death. The flight attendant asked the man to put out the cigar (only cigarette smoking was allowed), but the stench remained. Denise barely made it to Chicago. Taking deep breaths of the cold night air, she walked to the parking lot where she had left her old car. She quickly started up the engine and drove away, cursing all cigarette and cigar smokers for their outrageous insensitivity. Clouds of exhaust billowed from the tailpipe all the way home, although Denise wasn't aware of it. When she got home she turned on all the lights and turned up the furnace, but she didn't stop to think about the fuel burning at the region's power plant, which was belching all manner of

pollutant-filled smoke into the air as it worked to provide business, industry, and homes everywhere with energy.

In a way, the finite space inside the airplane cabin is something like the finite space of our atmosphere. Denise was acutely aware of the tobacco smoke because it had reached high concentrations in a relatively short time. But the same kind of thing is happening to the air around us. In fact, if we were to compare the earth with an apple from the supermarket, the atmosphere would be no thicker than the layer of shiny wax applied to it. *Yet into this thin, finite layer of air we dump more than 700,000 metric tons of pollutants each day in the United States alone!*

If we define **air pollution** as air that contains substances in concentrations high enough to damage living and nonliving things, then we are already in trouble. It isn't just that the pollutants can make the air smell, or cut down visibility, or discolor buildings. Air pollutants can *corrode* buildings. They can ruin oranges, wilt lettuce, stunt the growth of peaches and corn, and damage leaves on conifers hundreds of miles away from their source. They also can cause humans to suffer everything from headaches and burning eyes to lung cancer, bronchitis, and emphysema (Chapter Sixteen).

What are the major sources of air pollution? Tobacco smoking is a localized source, affecting the smokers themselves and, to some extent, nonsmokers who are stuck in confined quarters with them. The pollutants that affect life everywhere are by-products of fossil-fuel burning in transportation vehicles, home heating units, power plants, and the countless furnaces of industry. Among the most obvious are the **particulates** (solid particles or liquid droplets suspended in the air), which include dust, smoke, ashes, soot, asbestos, oil, and bits of poisonous heavy metals such as lead. Among the most dangerous air pollutants are the **oxides of sulfur**. They attack marble, metals, mortar, rubber, and plastic; they cause extensive crop loss near large cities, especially in Southern California, New York, New Jersey, Pennsylvania, Connecticut, and Delaware. At times they become so concentrated in the moist air of downtown St. Louis, they form droplets of sulfuric acid that dissolve holes in stockings of women who step outside office buildings at lunchtime! At best, sulfur oxides can make you cough and wheeze; at their worst they can figure in asthma, bronchitis, and emphysema. When rains wash them into rivers, lakes, and streams, they can destroy entire aquatic communities.

In terms of sheer tonnage, **carbon monoxide** from cars and trucks is the most common air pollutant. Because it can combine firmly with hemoglobin and displace oxygen, it can cut down the supply of oxygen being transported throughout the body. The results? The heart and lungs are forced to work harder to accomplish gas exchange, which does not bode well for people with heart or respiratory disorders. In the prolonged, concentrated fumes of rush-hour traffic, commuters may be subject to headaches, nausea, stomach cramps, and impaired coordination and vision. **Oxides of nitrogen**, other by-products of transportation, also reduce the blood's oxygen-carrying capacity and contribute to similar disorders.

It is in air pollution that we have an example of how environmental interactions can amplify problems. Short-term air pollution depends largely on the weather and topography. Los Angeles, for instance, is in a basin ringed with mountains that is open to the moist marine air from the west. Often a layer of dense, cool air gets trapped beneath a layer of warm air (something called a **thermal inversion**). When that happens, pollutants can't diffuse into the higher atmosphere, and they accumulate to dangerous levels right above the city.

Thermal inversions are most deadly in large urban areas. Here, car exhaust contains nitric oxide that reacts with oxygen in the air to produce nitrogen dioxide. When hydrocarbons and nitrogen dioxide are exposed to sunlight, they become converted to noxious substances collectively called **photochemical smog**. It is featured in Los Angeles, Salt Lake City, and Denver; it is brown, it smells, and it is hazardous to living things. There is another kind of urban air pollution, called **industrial smog**, in which fossil fuel burning gives off particulates and sulfur oxides (produced by oxidation of sulfur contaminants in coal and oil). It appears over Chicago, New York, Baltimore, Birmingham, and Pittsburgh; it is gray, sooty, and potentially lethal. Industrial smog was the source of London's 1952 air pollution disaster, in which 4,000 people died.

Denise can complain to the airlines in an attempt to ban smoking on airplanes. But where can she direct her complaints when it comes to air pollution from transportation and industry? Believing that we suddenly are going to stop relying on fossil fuel burning is about as silly as expecting us to return to a reliance on the horse (which, with all the millions of horses that would be needed, undoubtedly would create a pollution problem of another sort). Denise could keep the thermostat on her furnace turned down five degrees. She could also push for improved mass transit instead of depending on her private automobile. But is she really ready to moderate her personal comfort in such ways? Is she really willing to be a little inconvenienced by waiting around for a bus late at night at the air terminal, after a long flight home from Dallas? What about you?

SOME ENERGY OPTIONS

Paralleling the J-shaped curve of human population growth is a steep rise in energy consumption. It is not only that more population means more demands on available resources. An individual can survive on about 2,000 kilocalories of energy a day. But in the highly industrialized United States, the average individual directly or indirectly uses more than 200,000 kilocalories a day. In the United States, energy means not only survival, but survival draped in a bizarre array of nonessential creature comforts. For instance, in one of the most temperate of all coastal climates, a major university constructed seven- and eight-story buildings with narrow, sealed windows. With this design, a comfortable temperature is maintained not by relying on prevailing ocean breezes but on massive cooling systems—and by relying on massive heating systems instead of aligning the buildings to take advantage of the almost year-round warm sunlight! In support of this kind of life-style, more than half the energy (consumed in the form of nonrenewable fossil fuels) is given off in the form of entropy (wasted heat) and polluting by-products. *Thus the energy problem is not only a result of* overpopulation; *it is also a result of* overconsumption *and waste.*

What are our energy options? In exploring our alternatives, we can't escape the realities of the laws of energy; we can't get something for nothing, and we can't ever break even. The **net energy** available from any source is really what is left over after we have subtracted the energy it takes to locate, extract, ship, store, and deliver energy to the consumer. The cost of conventional nuclear energy must also include supplemental input of fossil fuels, the cost to mine and process uranium fuels, the cost of transporting and storing nuclear wastes. Harnessing the seemingly free energy from the sun must be measured also in terms of the energy it will take to build and maintain solar collectors as well as energy storage and transmission systems. When you hear talk of how much energy is available from different sources, keep in mind that there is often an enormous difference between the *total* amount and the *net* amount available.

Fossil Fuels

Three hundred million years ago, the trees of immense coastal forests spread their leaves, netlike, into the energy streaming from the sun. They grew, they decayed; they were buried. For hundreds of millions of years, photosynthetic algae thrived in the oceans and served as food for billions of protistans, the remains of which accumulated in layers on the ocean floor. Thus the sun helped give form and substance to living treasures that would come to be buried and transformed, under the sediments and ooze and compression of time, into **fossil fuels**: coal, oil, and natural gas. For many generations, fossil fuels have been a basis for human population growth. But in merely the past three decades, we have used up more fossil fuel than in all our preceding history combined! If we continue to expand our use of oil and natural gas at the rate we have been using them for the past fifty years, these resources may be almost completely exhausted in the United

Table 30.4 U.S. Energy Reserves and Expected Depletion Time

Resource	Quads*	Year
Petroleum	930	About 1990
Natural gas	790	About 2010
Uranium**		
Known reserves	610	About 2000
Most optimistic prediction	2,600	About 2040
Coal	14,000	?

*One quad equals one quadrillion BTUs (standard energy units).
**Uranium reserves are expressed on the basis of the amount of energy produced in existing reactors. If used in breeder reactors (which have not yet been shown to be feasible), the same uranium reserves would supply about sixty times as much energy as shown.

States by the year 2000 and in the world by 2010 (Table 30.4). Even with stringent conservation efforts, we will be running out of these fuels early in the next century. In addition to the petroleum reserves listed in Table 30.4, there are in the states of Colorado, Utah, and Wyoming vast **oil shale** deposits that probably contain more oil than the entire Middle East. Oil shale is buried rock that contains kerogen (a hydrocarbon compound). The trouble is, by the time the kerogen is collected, concentrated, heated, and converted to shale oil, it may produce only a slight net energy yield (possibly even an energy deficit). It would take as much energy to refine the shale as could be recovered in the refined oil. And it would come at a tremendous environmental cost. The land would be disfigured, water and air pollution would rise, and there would be less surface water to go around in regions already facing water shortages. One of the main by-products of the extraction process is a known cancer-causing agent (benzopyrene). And it would be produced by the ton, even though there is no known way to use it or dispose of it safely. More than this, oil shale processing produces 12 percent more solid waste than the space the original rock formation occupied. Where do the leftovers go? Some have suggested that controlled atomic blasts deep in the rock formations might distill the kerogen in place. But nobody has yet figured out what six underground atomic blasts a day (the number needed to supply merely 10 percent of our current demands for energy) would mean in terms of not only ecological but geological stability of the surrounding lands.

Perhaps the most impressive figure in Table 30.4 is the one given for the coal reserves in the United States. These reserves represent fully one-fourth of the world's known coal supply. In principle, they are enough to meet all the world's energy needs for at least several centuries. The problem here is that coal burning has been the largest single source of air pollution in industrialized nations (which is one of the reasons why worldwide use of petroleum has been escalating). Much of the world's coal reserves is in the form of low-quality, high-sulfur material—which means high levels of sulfur oxides. A recent study at Stanford University suggests that for every new coal-burning power plant of modest size, built anywhere near a city, we can expect an average of seventy more deaths a year from the effects of sulfur oxides. If expensive (and energy-consuming) devices are installed to lower the pollutant levels at the source of emission (smokestacks), deaths and damage from pollution might eventually be reduced by ninety percent. Locating the power plants far from the cities will further decrease human deaths due to coal burning. But it could be at the expense of crop damage.

To keep pace with our current energy demands, we almost certainly must burn more coal over the next two to four decades. But the consequences will extend far beyond air pollution. Even now, more than *1 billion* dollars a year is being paid in benefits to coal miners (and their families) afflicted with black-lung disease. Modern air-quality standards in mines minimize but do not eliminate this dangerous consequence of breathing coal dust. And mine tragedies—explosions, collapsed mine shafts, poisonous gases—are still frequent. For such reasons, pressure is on to permit widespread strip mining. Some of our coal reserves are close enough to the surface to be gouged out of the earth. But how many millions of acres are we willing to have scarred and, until healed, rendered useless for agriculture in order to maintain our present life-style? Who will pay for restoral, and will restoral be complete? As Table 30.5 indicates, there are other options for use of fossil fuels, but they are at best limited options, with moderate-to-serious consequences for the environment.

Nuclear Energy

As Hiroshima burned in 1945, the world recoiled in horror from the destructive potential of nuclear energy. But optimism soon replaced the horror as nuclear energy became publicized as an instrument of progress. That was the beginning of Operation Plowshare—a massive effort to harness the atom for peacetime use. Now, three decades later, more than sixty nuclear-powered, electricity-generating stations dot the American landscape. They produce 8 percent of the nation's electrical energy. Yet today, plans to extend our reliance on nuclear energy have been canceled or delayed. The reason? Serious questions have been raised about the cost, efficiency, environmental impact, and safety of nuclear energy.

One part of the problem with costs has to do with construction. Nuclear plants must safely contain radioactive fuel that gives off extreme amounts of heat energy. At the outset, it was clear that nuclear power plants would be more expensive to build than coal-burning plants, but the thinking was that its less expensive fuel costs would more than offset high construction costs. In the past decade, however, construction costs have soared faster than inflation. This means as time goes on, there must be greater and greater fuel savings to make nuclear power economically feasible. A second part of the cost problem relates to the fuel itself. How expensive is nuclear fuel, how efficiently is it used, and how large are the reserves? The total cost of uranium fuel used to power existing reactors is impossible to determine because of hidden costs of government subsidies to the nuclear fuel enrichment program. On the basis of costs now being paid by the power companies, nuclear-generated electricity should be considerably less expensive than coal if the nuclear plants are operating (as conventional coal-burning plants do) at eighty percent capacity. Through 1976, however, nuclear plants were operating at an average of fifty-nine percent capacity; forty-one percent of the time, they were shut down because of technical problems. Because of their high construction cost, the less electricity a nuclear generator produces in a year, the more expensive that electricity must be. In short, nuclear-generated electricity is still somewhat cheaper than coal-generated electricity, but it is by no means the bargain it was predicted to be.

What about efficiency? How much *net* energy does the nuclear power program produce? Once again the answer is difficult to find, for the government does not release information about the nuclear fuel enrichment program (because a portion of nuclear fuel is diverted to military purposes). Current estimates are that nuclear enrichment (the processing needed to convert uranium to a form that can be used as fuel) takes nearly three percent of the electricity used in the United States—even though nuclear power plants themselves produce only eight percent of the electricity. Aside from the costs of enrichment of nuclear fuel, there are mining, refining, transportation, and generator construction costs. All these things require a large portion of the energy that nuclear fuel will produce in the form of electricity—which means the net energy yield will probably be low.

But the major concern is for safety. What about release of radioactive by-products during normal operation? Or release of radioactive materials as a result of malfunctioning? Or escape of radioactive material from stored wastes? The amount of radioactivity escaping from a nuclear plant

Table 30.5 Evaluation of Energy Options for the United States

Option	Estimated Availability* Short Term (Present to 1985)	Intermediate Term (1985 to 2000)	Long Term (2000 to 2020)	Estimated Net Energy	Potential Environmental Impact**
Conservation	Fair	Good	Good	Very high	Decreases impact of other sources
Natural gas	Good (with imports)	Fair (with imports)	Poor	High but decreasing***	Low
Oil					
Conventional	Good (with imports)	Fair (with imports)	Poor	High but decreasing***	Moderate
Shale	Poor	Moderate to good?	Moderate to good?	Probably very low	Serious
Tar sands	Poor	Moderate? (imports only)	Good? (imports only)	Probably very low	Moderate
Coal					
Conventional	Good	Good	Good	High but decreasing***	Very serious
Gasification (conversion to synthetic natural gas)	Poor	Good?	Good?	Moderate to low	Very serious
Liquification (conversion to synthetic oil)	Very poor	Poor to moderate?	Good?	Moderate to low	Serious
Wastes					
Direct burning	Poor to fair	Fair to poor	Fair	Moderate (space heating) to low (electricity)	Fairly low
Conversion to oil	Poor	Fair to poor	Fair	Moderate to low	Low to moderate
Hydroelectric	Poor	Poor	Very poor	High	Low to moderate
Tidal	Very poor	Very poor	Very poor	Unknown (moderate?)	Low
Nuclear					
Conventional fission	Poor	Good	Good to poor	Probably very low	Very serious
Breeder fission	None	None to low	Good?	Probably low	Extremely serious
Fusion	Poor	Moderate to low?	Moderate to low	Unknown (could be low)	Unknown (probably moderate to low)
Geothermal	Poor	Moderate to low?	Moderate to low	Unknown	Moderate to low
Solar	Poor (except for space and water heating)	Low to moderate?	Moderate to high?	Unknown	Very low
Wind	Poor	Poor to moderate?	Moderate to high?	Unknown	Very low
Hydrogen	Negligible	Poor	Unknown****	Unknown (probably moderate to low)	Unknown****
Fuel cells	Negligible	Poor	Unknown****	Unknown (probably moderate to low)	Unknown****

*Based on estimated supply as a fraction of total energy use and on technological and economic feasibility.

**If stringent safety and environmental controls are not required and enforced.

***As high-grade deposits decrease, more and more energy must be used to mine and process lower-grade deposits, thus decreasing net energy.

****Depends on whether an essentially infinite source of electricity (such as solar, fusion, wind, or breeder) is available to convert water to hydrogen and oxygen gas by electrolysis or direct heating. Impact will vary depending on the source of electricity.

during normal operation almost certainly is far less of a threat to someone living right across the fence than the pollutants escaping from a coal-burning plant pose for *its* neighbors. But when it comes to the question of accidents, the picture is far from clear. It isn't that a nuclear plant is a potential bomb about to go off. The fuel of existing nuclear plants can't explode like a bomb under *any* circumstances. But there is potential danger of a **meltdown**. As nuclear fuel breaks down, it releases enormous amounts

of heat. In the most common type of nuclear energy plant, the heat normally is removed by water that circulates under very high pressure over the nuclear fuel. The water so heated is used to produce steam, which is used to drive the turbines that generate electricity. Should a leak develop in the water-circulating system, the water level around the fuel might drop rapidly, and the temperature of the nuclear fuel would rise rapidly until it exceeded its melting point. The fuel would then melt and pour to the floor of the generator. There, it would contact the remaining water and instantly convert it to steam.

Depending on the amount of molten fuel and the amount of water, the steam so suddenly produced might be enough to blow the system apart, spewing radioactive fuel to the exterior. If the force of the blast were great enough, it conceivably might even damage the concrete building in which the nuclear reactor is located and permit radioactive material to escape. If that were to happen, it could cause many deaths and extensive environmental damage for miles around. How likely is such an event to occur? The question is basically unanswerable. It's asking how likely is something to happen that has never happened before. All nuclear reactors have secondary cooling systems that are supposed to flood the reactor with water at once if the initial cooling water is lost. A one-minute loss of water may be enough to start a meltdown. But such systems have never been put to the test. The first facility to test an emergency cooling system is now being built, but it will be many more years before it is operational. By then, about a hundred reactors will already be in use, relying on the untested system.

One comprehensive study suggests there is only one chance in a million reactor-years of an accident severe enough to kill 70 people, cause acute radiation sickness in 170 more, and do nearly 3 billion dollars worth of property damage. There is only one chance in a billion reactor-years of an accident severe enough to kill more than 2,000 people at once, cause radiation poisoning in another 5,000 or 6,000, and cause 6 billion dollars worth of property damage. These estimates are lower than the better-known probabilities (for example) that a major dam will break, killing many more people and inflicting much more damage. But are the assumptions valid, and are the risks that low? Many would suggest the risks have been underestimated by a hundredfold. In 1979, at the Three-Mile Island nuclear power plant, some things did go wrong. Some radiation leaked into the atmosphere and thousands of residents were evacuated from the area. The malfunction was corrected, no lives were lost—but the very fact that it occurred at all brought the issue home to individuals throughout the nation.

What of nuclear wastes? Are they a threat? Nuclear fuel can't be burned to harmless ashes, like coal. After about four years, the fuel elements of a reactor are "spent." At that point, it still contains about a third of the useful uranium fuel, but it also contains hundreds of new isotopes of various elements that have been produced during the operation of the reactor. Many of these isotopes could make the reactor nonfunctional. One of the products is an isotope of plutonium which, like the leftover uranium, could be used as fuel for other reactors if it were removed and purified. (At the moment, the United States does not have a plant capable of recovering usable materials from radioactive wastes.)

Altogether, the wastes are an enormously radioactive, extremely dangerous collection of nasty materials. As they undergo radioactive decay, they produce tremendous heat. Thus they are immediately plunged into specially constructed water-filled pools at the power plant, and there they are stored for several months. The water cools the wastes and keeps radioactive materials from escaping. During this period, the level of radioactivity and the amount of heat being generated decrease by as much as 99.9999 percent as the short-lived radioactive isotopes break down. But even at the end of this time, the remaining levels of radioactivity due to long-lived isotopes are extremely lethal. The wastes must then be transported in special equipment to centralized facilities. There they must be held in temporary storage for another five years, at least. But the decay rates of some of the remaining isotopes mean they must be kept out of the environment for *a quarter of a million years!*

Since the beginning of the nuclear age, no radioactive wastes have been put into permanent underground storage in the United States. But such storage will soon be initiated. The nuclear wastes will be mixed with molten glass. The glass will be cast into rods a foot in diameter and ten feet long. (One year's wastes from one nuclear power plant may produce ten such rods.) Then the rods will be encased in sealed steel cylinders. Ultimately these cylinders will be inserted in holes of the same size that have been drilled in salt deposits, deep underground. There are several reasons given for using salt deposits. First, the very existence of a salt deposit indicates long-term absence of underground water that could slowly dissolve the wastes and move them about. Second, salt deposits are typically found in areas that have been geologically undisturbed (for example, by earthquakes) for millions of years. The assumption is they will continue to be undisturbed. Third, salt tends to flow slowly under pressure to reseal any cracks introduced as the drilling and storage take place. It is widely—but by no means uniformly—believed that all such wastes can thus be kept out of the way until they are no longer dangerous.

Although the dangers of nuclear power will be a source of genuine concern for some time, those associated with present systems are said by some to be manageable. But a move to develop a new kind of power source poses dangers of a wholly different magnitude. As Table 30.4

indicates, if nuclear power plants are built and operated at the rate of current projections, the known reserves of high-grade uranium could run out around the year 2000. To forestall nuclear fuel depletion just as the world comes to depend on it, intensive research and development efforts are now directed to the development of **breeder reactors**. As these reactors consume a rare isotope of uranium, they convert a much greater amount of the common, unusable form of uranium to an isotope of plutonium—which is a usable nuclear fuel. Indeed, the plutonium could then be isolated and used for conventional reactors as well as for another breeder reactor. Such reactors would produce more nuclear fuel than they consume. If they come into use, the world's uranium reserves could be used to generate nearly a hundred times as much energy and would last a hundred times as long as conventional reactors. But if breeder reactors were to expand the operating period for our uranium reserves, they would also expand the dangers.

Breeder reactors could not be cooled with water. In the design being most actively pursued, liquid metallic sodium would be used as the coolant. It is a highly dangerous, corrosive substance. The problems of containing the molten sodium and of maintaining the reactors in a functional state are horrendous. More than this, although conventional reactors cannot explode like atomic bombs, breeder reactors potentially could do so. Certain mishaps could cause the entire reactor to explode, spreading radioactive wastes equivalent to those produced by 1,000 Hiroshima-sized bombs over many square miles. In any event, the problems associated with the development of breeder reactors are so complex that some former advocates of breeder technology are now gloomy about the prospects of developing functional breeders before the uranium is gone.

Plutonium, in the form produced and used in reactors, is the most toxic substance known. A speck the size of a pollen grain almost certainly will cause fatal lung cancer. It is also the stuff that makes atomic bombs. Making a bomb from uranium requires a major scientific and industrial effort that can be mounted only by a country with substantial resources. But fashioning a bomb from plutonium is, by comparison, relatively simple. A small amount of plutonium in the hands of terrorists could be used for extortion; a somewhat larger amount could be used for mindless devastation. If we enter an era of breeder reactors, we will surely enter an era of plutonium economy in which all aspects of the nuclear power industry—processing plants, transportation systems, reactors, disposal sites—will require utmost security precautions.

A third type of nuclear power source that may become available some day is **fusion power**, in which hydrogen atoms are fused to form helium atoms, with a considerable release of energy. It is a process somewhat analogous to the reactions that cause the heat energy of the sun. But the problems associated with developing fusion power are so great that, without a major breakthrough, we cannot expect fusion power before more than four decades.

In sum, nuclear reactors of the type now in use are probably safer than many of their critics proclaim, but far less safe than their advocates claim. The most negative aspects are that they have a low net energy yield, achieved at a much higher cost than expected. Besides, the supply of naturally occurring fuel will be quickly exhausted. Breeder reactors could get around the problem of fuel depletion, but only if we are willing to gamble on the possibility of not having major accidents, of not encountering sabotage or terrorist activities. Fusion, while ultimately a possibility, is so far into the future as to hold no hope for the energy problems that will be with us this century.

Wind Energy

In certain geographic regions, winds are predictable enough and strong enough to be a source of energy. Such high prevailing winds sweep through the vast plains of the American Midwest, from the Dakotas to Texas. **Wind energy** is thus seen to be one of the most attractive energy options; not only would it be an unlimited energy source, it would not generate pollution (and it certainly would be harmless in the hands of terrorists). In one proposal, perhaps 300,000 turbine towers could harness the winds of the plains and thereby provide fully half of our present energy needs. It is true that rows of giant turbines erupting like warts from one end of the nation's heartland to the other could be something of an aesthetic insult. On the other hand, they would be nonpolluting; they would not even generate an appreciable amount of waste heat. Besides, many windmills presently being introduced in Europe are of a simple design that will disfigure the landscape far less than several other energy alternatives. Turbine towers would take up less land than, say, solar-energy generating plants, and they most assuredly would be far less disruptive of the environment than, say, strip mining for coal. It is also possible that windmills constructed on top of existing electrical transmission towers could feed their electrical output directly into regional utility lines as a supplemental energy source. The possibility of harnessing the winds is economically and technologically feasible right now. But interest is essentially zero, for reasons that will become apparent in the following section.

Solar Energy

Without question, a virtually limitless, free energy source is potentially available to us, as it has always been to all life on earth. **Solar energy** ranks with wind energy as the most abundant, the cleanest, and the safest of all potential energy sources now being considered as alternatives to fossil fuels. Solar power plants would not release heat as entropy to any significant extent, as coal-burning and nuclear power plants do. They would pollute neither the earth nor the atmosphere. We have yet to tap directly into the sunlight energy streaming all around us. It has been estimated that if 39,000 square kilometers (15,000 square miles) of desert were set aside in Arizona and California for solar energy collectors that convert sunlight energy directly into electricity, we could develop a system for providing half our nation's annual energy needs even decades from today.

Until recently, the seeming abundance of fossil fuels lulled us into thinking there was no need to carry our search for energy to other sources, other places. But more than this, in an economic system based on profit motivation, there has not been much of a commitment to finance research that is unlikely to yield a sustained profit. Who is going to design and develop more efficient solar energy collectors and storage devices for home use? Utilities corporations? Oil companies? What would they "gain" if every home, every building had a self-sustaining or at least a supplemental energy source of its own, and therefore required less fossil fuel and electric power? To date, research funds have come mainly from environmentally concerned groups and citizens. Today, owning solar heaters and collectors is a financial luxury. No one is going to pay homeowners to install their own; it will be a cost out of their own pocket. But times are slowly changing. In 1977 the Energy Research and Development Agency elevated solar energy research from obscurity to the status of poor relative of nuclear energy research. Some commercial operators are developing and marketing home solar-collector units; here, industry is ahead of government.

ON HUMANS AND OTHER ENDANGERED SPECIES

You may have noticed, now that you have just about completed this book, that we did not include a token section on vanishing species, along with a list of most endangered ones in ascending order of vulnerability. It is not that the problem is overrated; far from it. The slow process of natural selection is not keeping pace with the accelerating pressures of human activities on the biosphere, and species are disappearing fifty times faster than even a century ago. Many outstanding organizations are working to buy time for the species we threaten most with extinction. But this is not an isolated problem, with an isolated solution. Throughout earth history, the nature of the community of life has changed constantly, with new species arising to take the place of those which have been left behind in time. But always, since the beginning, it has been a *community* of life forms, whatever the kaleidoscopic shifts in its character. *Like all life forms in the past, we are all potentially endangered species, kin to one another in vulnerability because of intricate, often invisible threads of ecological interdependence stretched through the biosphere.* Timber wolf and puffin, humpback whale and golden eagle, ourselves and our children—this is our shared hour on earth. It is not for the sake of one but for the sake of all that we must make our commitment. We must face not just the isolated symptoms but the major sources of stress that have brought about the heightened state of vulnerability that now marks the entire biosphere:

Human overpopulation—by far the most serious—with its consequent stresses on available energy, food, space, and other resources, with its consequent pollution that the environment cannot absorb.

Overconsumption, founded on throwaway goods, planned obsolescence, minimal recycling, and acquisition of goods not essential for a life of dignity and quality.

Ecosystem oversimplification, with the low-diversity crops of human agriculture, with its disruption of functional and self-sustaining webs of life.

Massive industrial technology, wielded without regard for biological and environmental consequences.

Lack of leadership, with moves toward the future mired in misplaced priorities, pork-barrel politics, and allegiance to totally unchecked economic growth.

Me-first behavior, with regard only for the individual's present, with nonregard for future generations and for the biosphere.

All these sources of stress on the biosphere can be traced, ultimately, to indifference to the principles of energy flow and materials reuse.

We have, most of us, been so steeped in the philosophy of growth and consumption that it might seem almost silly to say we should make attempts at a transition to a new way of life. But we can and must make these attempts, you and I, as individuals:

We can become conscious of our expectations, and learn to ask where they are leading us. We have, in general, been lulled into thinking that no matter how bad things look today, "technology" will do something to save us tomorrow. In all ways, large and small, even technology can't get us something for nothing. Medical advances applied on a global scale can wipe out many horrible infectious diseases, and alleviate much human suffering. But in doing so, they contribute also to explosive population growth—and which is the "better" choice: death by disease or starvation? And so technology moves on to agriculture to find better ways of feeding the multitudes—and ecosystems are stripped for monocrop agriculture, and left vulnerable to weeds and insect predators. Technology moves to chemical control—DDT, malathion—and the effects of biological magnification extend back to those who don't stop to ask what it is we really are expecting to get. Technology can and must be used wisely to help us out of this twentieth-century dilemma. But we must get into the habit of looking for the hidden costs; we must learn to follow through by asking what applied technology may mean to long-term stability, not short-term magic.

We can make an ecological impact statement for our <u>own</u> way of life. In recent years, one of the best steps our government has taken is to insist that no major new construction or development projects proceed until some attention is given to their ecological impact. Think of how much could be accomplished if we *all* did the same thing in our daily lives. What if we got into the habit of asking ourselves: What will be the ecological impact if I drive a car instead of walk? Buy milk in plastic bottles instead of returnable bottles? Buy a home in a new housing development built on what had been the rich soil of a productive farm? Become a parent once, twice, three times? Take this job, instead of that one? Vote for one candidate for public office instead of another? Speak out (at the risk of possible embarrassment) or keep my mouth shut?

We can avoid tokenism. It is so easy for each of us to take a few small steps toward achieving the goal of a better world of life, and then to reward ourselves for our virtue by lapsing back into the life-style that has brought us to our present predicament. If you feel you really want and deserve some of the luxuries of the industrialized society you live in, try to give up enough other luxuries so that you come out, on balance, contributing to the solution instead of the problem.

We can get involved. We can't assume that just because we develop an ecological perspective for our personal life-style, we have done all we can for the future. We can join public interest groups, demanding that public and private institutions alike serve the public interest in a long-term, ecologically sound way. As you can gather by looking over the list of readings at this chapter's end, there are many individuals and groups already working effectively to bring about change.

We can expect an ecological perspective from our elected representatives. Many of the crucial decisions affecting both the short-range and long-range future of the biosphere are going to be made by our elected representatives. At the turbulent close of the 1960s, many students in this country lost faith in the capacity of our system of government to respond to the people. They developed the cynical attitude that a system so corrupt would not—could not—be responsive. Since that time, there has been enough evidence that the system *can* be made responsive *if enough people demand it.* If we wish for other goals, we must work to make our presence felt. *Our system of government has become far more responsive and open than any other system in the recent history of human experience.* If it fails to move in the directions we think best, then more than ever it will be the fault of the people. In the near future, more than the recent past, we will have the kind of government we ask for and deserve. If we feel strongly about issues of public policy, it is up to us to keep our local, state, and national representatives informed of our positions.

In the words of the human ecologist G. Tyler Miller, *the secret of sustained action is to think and work on two levels at once.* We can constantly whittle away at making major changes in our political and economic systems, and in our world view. At the same time, we can do a number of little things each day to help return to dynamic compatibility with the total world of life. Each small act can be used to expand awareness of the need for basic changes in our political, economic, and social systems over the next few decades. They also can help us avoid psychic numbness every time we think about the magnitude of the job to be done. We can begin at the individual level and work in ever widening circles. We can join with others and amplify our commitment. This is the way of change.

Readings

Audubon (bimonthly). National Audubon Society. 1130 Fifth Avenue, New York, New York 10028. Conservationist viewpoint; more than bird-watching. Exceptionally well written, outstanding graphics. Gives a current picture of environmental issues the world over.

BioScience (monthly). American Institute of Biological Sciences. 3900 Wisconsin Avenue NW, Washington, D.C. 20016. Official publication of AIBS; major coverage of biological concerns.

Catalyst for Environmental Quality (quarterly). 274 Madison Avenue, New York, New York 10016. Popular treatment of all aspects of environment, including population control. Reviews books and films for environmental education.

Family Planning Perspectives (bimonthly). Planned Parenthood—World Population. Editorial Offices, 666 Fifth Avenue, New York, New York 10019. Free on request. Detailed, wide-ranging, liberal attitude. Useful for persons concerned with the problem of overpopulation.

Mother Earth News (6 issues yearly). Mother Earth News, Inc. P.O. Box 70, Hendersonville, North Carolina 28739. Good articles on organic farming, alternative energy systems, and alternative life-styles.

National Wildlife (bimonthly). National Wildlife Federation. 1412 Sixteenth Street NW, Washington, D.C. 20036. Good summaries of wildlife issues.

New Scientist (monthly). 128 Long Acre, London, W.C. 2, England. Excellent journal on general science.

Population Bulletin (bimonthly). Population Reference Bureau. 1755 Massachusetts Avenue NW, Washington, D.C. 20036. Non-technical articles on age structure, fertility rates, migration, and mortality.

Science (weekly). The American Association for the Advancement of Science. 1515 Massachusetts Avenue NW, Washington, D.C. 20036. Outstanding forum for American science. Probably the single best source for keeping up with research—and with what researchers are thinking in terms of applications and consequences of research.

Scientific American (monthly). 415 Madison Avenue, New York, New York 10017. Excellent journal; accessible writing style, excellent graphic presentation of scientific research and discoveries.

The Sierra Club Bulletin (monthly). The Sierra Club. 1050 Mills Tower, San Francisco, California 94104. Good coverage of a range of environmental issues and citizen action.

Review Questions

1. Under which conditions would an ecologically based system for handling solid wastes benefit the culture more than a recycling system?

2. What is the basic plan of a generalized resource recovery system?

3. What is meant by primary, secondary, and tertiary waste-water treatment?

4. Where do organic oxygen-demanding wastes come from, how do they affect organisms, and how can this type of pollutant be controlled?

5. Categorize the principal air pollutants, their sources, and the methods of controlling each.

6. How can energy overconsumption and waste be reduced by the average U.S. citizen?

7. When are the U.S. petroleum reserves expected to run out if consumption continues at present rates? Natural gas? Uranium?

8. Which is the only energy option for the United States to adopt that has *good* short-term, intermediate and long-term (until 2020) availability?

9. What is a meltdown? Can a nuclear power plant explode like an atom bomb?

10. Why do oil and gas companies buy up most of the patents for more efficient solar energy collectors as soon as they are awarded?

11. List some of the sources of stress that have endangered the populations of so many earthly species, including ourselves.

12. How can you personally become involved in determining that public and private institutions serve the public interest in a long term, ecologically sound way?

APPENDIX I
BRIEF SYSTEM OF CLASSIFICATION

This appendix, based on Robert Whittaker's scheme (*Science*, 1969, 163: 150–160), reflects current understandings of phylogenetic relationships. It is by no means all-encompassing; it simply includes the kinds of organisms you have read about in the text. The numbers refer to some pages on which representative organisms for the category are discussed.

Kingdom Monera Single-celled prokaryotes; autotrophs and heterotrophs lacking true nucleus and other internal, membranous organelles

Phylum Schizomycetes. Bacteria	411–415
Phylum Cyanophyta. Blue-green algae	417

Kingdom Protista Single-celled eukaryotes; heterotrophs and some autotrophs with true nucleus and other internal, membranous organelles

Phylum Pyrrhophyta. Dinoflagellates	418–419
Phylum Chrysophyta. Golden algae, diatoms	419–420
Phylum Euglenophyta. Photosynthetic flagellates	420
Phylum Sarcodina. Amoebas, foraminiferans	420–421
Phylum Sporozoa. Parasitic sporozoans	421–422
Phylum Ciliata. Ciliates	422–423

Kingdom Plantae* Mostly multicelled eukaryotes; autotrophic forms able to build own food molecules through photosynthesis

Phylum Rhodophyta. Red algae	428–429
Phylum Phaeophyta. Brown algae	428, 429, 432
Phylum Chlorophyta. Green algae	428, 432, 434
Phylum Sphenophyta. Horsetails	430, 435
Phylum Lycophyta. Lycopods	430, 435
Phylum Pteropsida. Ferns	430, 435–436
Phylum Spermopsida. Seed plants	430, 436
Class Cycadae. Cycads	430, 436
Class Ginkgoae. Ginkgoes	430, 436–437
Class Coniferae. Conifers	430, 437–438
Class Angiospermae. Flowering plants	430, 438
Subclass Monocotyledonae. Monocots; grasses, palms, lilies, orchids	430, 438
Subclass Dicotyledonae. Dicots; most common flowering plants not mentioned above	430, 438, 441

Kingdom Fungi Multicelled eukaryotes; most are heterotrophs relying on extracellular digestion and absorption

Division Mycota	
Subdivision Myxomycotina. Slime molds	443
Subdivision Eumycotina. True fungi	443, 444
Class Oomycetes. Egg fungi	443, 445, 446
Class Zygomycetes. Zygote fungi	443, 445
Class Ascomycetes. Sac fungi	442, 443, 445
Class Basidiomycetes. Club fungi	442, 443, 445, 446
Class Fungi Imperfecti. Imperfect fungi	443

*The phyla in this kingdom are also known as divisions in plant classification schemes.

Kingdom Animalia Multicelled eukaryotes; heterotrophs that ingest other organisms for food

Subkingdom Parazoa	
Phylum Porifera. Sponges	10, 449
Subkingdom Radiata. Radially symmetrical animals	232–233
Phylum Cnidaria. Coelenterates, cnidarians	449–451
Subkingdom Protostomia. Protostomes	
Phylum Aschelminthes. Sac worms	452, 453
Class Nematoda. Nematodes	452
Class Rotifera. Rotifers	
Phylum Platyhelminthes. Flatworms, tapeworms	451, 453
Phylum Annelida. Annelids; segmented worms	454–455
Phylum Mollusca. Mollusks	457–458
Phylum Arthropoda. Arthropods	7, 455–457, 500
Subkingdom Deuterostomia. Deuterostomes	453
Phylum Echinodermata. Echinoderms	458–460
Phylum Chordata. Chordates	453, 460
Subphylum Urochordata. Tunicates	460–461
Subphylum Cephalochordata. Lancelets	461
Subphylum Vertebrata. Vertebrates	461
Class Agnatha. Jawless fishes	461–462
Class Chondricthyes. Cartilaginous fishes; sharks, rays	
Class Osteoichthyes. Bony fishes	11, 462
Class Amphibia. Amphibians; frogs, toads, salamanders	4, 462, 463
Class Reptilia. Reptiles; lizards, snakes, turtles	20, 24, 462
Class Aves. Birds	17, 20, 25, 462, 463, 507, 508
Class Mammalia. Mammals	
Subclass Prototheria. Egg-laying mammals, spiny anteater, duck-billed platypus	332
Subclass Metatheria. Pouched mammals; opossum, kangaroo, koala	333
Subclass Eutheria. Placental mammals	339
Order Insectivora. Insect-eating mammals; tree shrews, moles, hedgehogs	
Order Edentata. Toothless mammals; sloths, armadillos	
Order Rodentia. Rodents; mice, rats, squirrels	487, 513
Order Perissodactyla. Odd-toed ungulates; horses, rhinoceros	14
Order Artiodactyla. Even-toed ungulates; deer, cattle, camels, hippopotamus	14, 499, 535
Order Proboscidea. Elephants	17, 18
Order Lagomorpha. Rabbits, hares	489
Order Sirenia. Manatees, dugongs, sea cows	
Order Carnivora. Wolves, dogs, cats, bears, seals, weasels	277, 499, 502, 511
Order Cetacea. Whales, porpoises	125, 502
Order Chiroptera. Bats	252, 253
Order Primates. Lemurs, monkeys, apes, humans	463–465, 504, 511

APPENDIX II
UNITS OF MEASUREMENT

A Few Metric Equivalents		
Length:	kilometer	0.62 mile
	meter	39.37 inches
	centimeter	0.39 inch
	millimeter	0.039 inch
Mass:	gram	0.035 ounce
	kilogram	2.205 pounds
Volume:	cubic centimeter	0.061 cubic inch
	liter	1.057 quarts

Units of Measure Used in Microscopy				
Unit	Equivalence in Millimeters	Equivalence in Micrometers	Equivalence in Nanometers	Equivalence in Angstroms
Millimeter (mm)	1	1,000	1,000,000	10,000,000
Micrometer (μm)	0.001	1	1,000	10,000
Nanometer (nm)	0.000001	0.001	1	10
Angstrom (Å)	0.0000001	0.0001	0.1	1

The micrometer is used in describing whole cells or large cell structures, such as the nucleus.

The nanometer is used in describing cell ultrastructures (those ranging from, say, mitochondria downward), even of macromolecules. Lipid molecules, for example, may be about 2 nanometers long; cell membranes may be 7 to 10 nanometers thick; ribosomes are about 25 nanometers across.

GLOSSARY

abortion Induced expulsion of the embryo from the uterus before the third trimester of pregnancy.

abscission (ab-SIH-zhun) In some plants, leaf (or fruit or flower) drop after hormonal action causes a corky cell layer to form where a leaf stalk joins the stem, which shuts off nutrient and water flow.

acid A substance that releases hydrogen ions (H^+) in a water solution, where it has a pH of less than 7.

action potential Nerve impulse; an all-or-none, brief reversal in the resting potential of a neural membrane. Once triggered, its strength does not change even if the stimulus strength changes.

active site Three-dimensional groove or pocket in an enzyme to which a specific set of substrates becomes temporarily bound and thereby is made more reactive.

active transport Movement of individual ions and molecules across a cell membrane, against a concentration gradient, by ATP expenditure. The ion or molecule is moved in a direction other than the one in which simple diffusion would take it.

adaptation An existing structural, physiological, or behavioral trait of an individual that promotes likelihood of survival and reproduction under prevailing environmental conditions.

adaptiveness In individuals, the capacity for adjustment to shifting environmental conditions. Among populations and species, adjustments to change through successive generations.

adaptive radiation Specialization in exploiting different aspects of the environment by different populations of the same species. Adaptive radiation may or may not be accompanied by speciation.

adenine (AH-de-neen) A purine; a nitrogen-containing base found in nucleotides (adenosine phosphates, nucleotide coenzymes, and nucleic acids).

adenosine diphosphate (ah-DEN-uh-seen die-FOSS-fate) ADP, a molecule involved in cellular energy transfers; formed by hydrolysis of ATP.

adenosine triphosphate ATP, a molecule that serves as a major energy carrier in all living cells.

aerobic pathway (air-OH-bik) In a cell, a pathway of energy metabolism that depends on oxygen as an electron acceptor.

allele (uh-LEEL) An alternative form of a gene at a given gene locus.

allele frequency The relative abundance of different alleles carried by the individuals of a population. Also called gene frequency, which is something of a non sequitur.

allopatric populations Local populations of the same species that are geographically separated from one another.

alternation of generations In many plant life cycles, the alternation of a diploid, spore-producing generation (sporophyte) with a haploid, gamete-producing generation (gametophyte).

alveolus, plural **alveoli** (ahl-VEE-uh-luss) One of many small, thin-walled pouches in the lungs; sites of gas exchange between air in the lungs and the bloodstream.

amino acid (uh-MEE-no) A molecule containing an amino group (NH_2) and an acid group (—COOH); a subunit for protein synthesis. Twenty amino acids commonly occur in living things.

amnion (AM-nee-on) In reptiles, birds, and mammals, a membrane that arises from the inner cell mass of a blastocyst; becomes a fluid-filled sac in which the embryo develops freely.

anaerobic pathway An energy-extracting pathway that does not depend on oxygen as an electron acceptor; both fermentation and glycolysis are anaerobic pathways (although glycolysis *can* proceed in the presence of oxygen).

anaphase In mitosis, the stage at which the two sister chromatids of each chromosome are separated from each other and moved to opposite poles of the cell.

angiosperm (AN-gee-oh-sperm) Flowering plant.

animal Multicelled organism that ingests other organisms for food.

annual plant Vascular plant that completes its life cycle in one growing season.

anther In flowering plants, the pollen-bearing part of the male reproductive organ (stamen).

antibody A type of protein molecule, carried by the bloodstream, that binds to specific agents invading the body (antigens) and thereby marks them for disposal by other defense forces.

antigen Foreign cell or substance that has penetrated the body or some tissue and that triggers antibody production.

anus In some invertebrates and all vertebrates, the terminal opening of the gut through which solid residues of digestion are eliminated.

aorta (ay-OR-tah) Main artery of systemic circulation; carries oxygenated blood away from the heart to all regions except the lungs.

apical meristem In vascular plants, embryonic tissue zones where cells are produced by mitotic divisions; these cells then enlarge and differentiate into all other plant tissues.

asexual reproduction Production of new individuals by any process that does not involve gametes.

atom Smallest unit of an element that still retains the characteristics of that element.

atomic number A relative number assigned to each kind of element based on the number of protons in one of its atoms.

atomic weight The weight of an atom of any element relative to the weight of the most abundant isotope of carbon (which is set at 12).

autonomic nervous system (auto-NOM-ik) Those aspects of the vertebrate central and peripheral nervous system that regulate cardiac cells, muscle cells, smooth muscle cells (such as those of the stomach), and glands; generally not under conscious control.

autosome One of those chromosomes that are of the same number and kind in both males and females of the species.

autotroph "Self-feeder"; an organism able to build all the complex organic molecules it requires as its own food source, using only simple inorganic compounds. Compare *heterotroph*.

axon Nerve cell process serving as a through-conducting pathway for messages that must travel rapidly, without alteration, from one body region to another.

bacillus, plural **bacilli** (BAH-sill-us, BAH-sill-eye) Rodlike form of bacterium.

bacteriophage (bak-TEER-ee-oh-fahj) Category of viruses that infect and destroy certain bacterial cells.

basal body Centriole; gives rise to microtubule systems of cilia and flagella. Self-reproducing structure that apparently occurs in eukaryotic species having motile cells during the life cycle.

base Any substance that combines with hydrogen ions in solution, where it has a pH above 7.

behavior Any coordinated neuromotor response to changes in the external and internal states; a product of the integration of sensory, genetic, neural, and endocrine factors, and may be modified by learning.

disaccharide A carbohydrate; two simple sugar molecules bonded covalently.

disruptive selection The increasing representation of two or more distinct polymorphic varieties in a population; tends to split up a population. Compare *directional selection*.

divergence A buildup of differences in allele frequencies (in other words, genetic differentiation) between isolated populations of the same species.

dominant allele In a diploid cell, the expression of one allele to the extent that it masks expression of its partner.

dormancy In plants, cessation of growth under physical conditions that are actually quite suitable for growth.

ecology Study of the interactions of organisms with one another and with their physical environment.

ecosystem A community of life and its physical environment.

ectoderm In an animal embryo, an outermost cell layer destined to give rise to skin and nerve tissue.

effector A muscle (or gland) that responds to nerve signals by producing movement (or chemical changes) in response to changes in internal and/or external conditions.

electric charge A property of matter that enables ions, atoms, and molecules to attract or repel one another.

electron Negatively charged particle that orbits the nucleus of an atom.

electron transport chain In cell membranes, an organized sequence of molecules that transfer relatively high-energy electrons from one molecule to the next in line. Energy released during the transfers may be used to drive such reactions as phosphorylation of ADP.

element Substance composed of a single kind of atom.

embryonic development Those processes by which cells of an embryo become different in position, developmental potential, appearance, composition, and function.

embryonic induction Processes by which one group of embryonic cells signals an adjacent group, thereby inducing it to differentiate.

embryonic regulation Modification of embryonic cell behavior in ways that compensate for abnormal developments that result from cell damage, or from local cell divisions that proceed too rapidly or too slowly.

endergonic reaction (en-dur-GONE-ik) Reaction to which energy from an outside source must be added before it proceeds.

endocrine element (EN-doh-krin) Cell or gland that produces and/or secretes hormones into the bloodstream.

endocrine system System of cells, tissues, and organs functionally linked to the nervous system and whose chemical secretions help control body functioning.

endocytosis Type of bulk transport in which vesicles form around particles at a cell surface, then discharge their contents in the cell interior. Phagocytosis is an example.

endoderm In an animal embryo, the innermost cell layer, which differentiates into internal organs such as liver and stomach.

endometrium (EN-doh-MEET-ree-um) Mucous membrane of the uterus, consisting of epithelium and uterine glands.

endoplasmic reticulum, rough Ribosome-studded membrane system on which proteins are assembled and within which they may be temporarily stored.

endoplasmic reticulum, smooth Membrane system concerned with isolating and transporting materials, and with assembly of fats.

endoskeleton In chordates, the internal framework of bone and/or cartilage.

endosperm Mass of tissue that surrounds embryo in a seed; in monocots, storage site for nutrients needed after seed germination.

energy Capacity to do work.

entropy A measure of how much energy in a system has become so dispersed (usually as evenly distributed heat) that it is no longer available to do work.

enzyme A kind of protein that speeds up the rate of a metabolic reaction by lowering the activation energy required for that reaction.

epidermis Outermost tissue layer of a multicelled animal or plant.

epistasis Interaction whereby one gene pair masks the effect of one or more other pairs.

epithelium Sheet of cells, one or more layers thick, lining internal and external surfaces of multicelled animal body.

equilibrium A stable condition. The point at which a chemical reaction runs forward as fast as it runs in reverse, so that there is no further net change in the concentrations of products or reactants.

erythrocyte (eh-RITH-row-site) Red blood cell.

estrus (ESS-truss) Among mammals, the cyclic period of a female's sexual receptivity to the male.

estuary (ESS-chew-airy) A region where fresh water from a river or stream mixes with salt water from the sea.

eukaryote (yoo-CARRY-oht) A cell that has membranous organelles, most notably the nucleus.

eutrophication (yoo-trofe-ih-KAY-shun) Process whereby a body of fresh water becomes enriched in nutrients, increases in productivity, and accumulates organic debris.

evagination In membranes, an outfolding such as a fish gill.

evaporation Behavior of water molecules at the surface of a body of water heated beyond its boiling point; hydrogen bonds between molecules break, energy is released, and some molecules escape into the surrounding air.

evolution Successive changes in allele frequencies in a population, as brought about by such occurrences as mutation, genetic drift, gene flow, and selection pressure.

exergonic reaction (EX-ur-GONE-ik) Chemical reaction in which energy is released from the reactants, so that the products contain less chemical potential energy than the reactants.

exocytosis Type of bulk transport in which vesicles formed within cytoplasm fuse with the plasma membrane and discharge their contents to the outside.

exoskeleton An external skeleton, as in arthropods.

exponential growth (EX-poe-NEN-shul) Increasingly accelerated rate of population growth due to an increasing number of individuals being added to the reproductive base. In the absence of control factors, all populations show exponential growth.

extrinsic limiting factor External variables, both environmental and biotic, that help control population growth and distribution.

facilitated diffusion Molecular transport process driven by concentration gradients but dependent on proteins that act as channels across the plasma membrane.

fat A molecule composed of glycerol and three fatty acid molecules.

fatty acid Component of fats and waxes; long-chain hydrocarbon with an acid group (—COOH) attached.

feedback inhibition Control mechanism whereby an increase in some substance or activity inhibits the very process leading to (or allowing) the increase.

fermentation pathways Those pathways in which pyruvate from glycolysis is converted to alcohol, lactic acid, or similar end product.

fertilization Fusion of sperm nucleus with egg nucleus. In flowering plants, an additional sperm nucleus fuses with two other nuclei present in the ovule, forming a single triploid nucleus that will divide and give rise to endosperm; such fusions are called *double fertilization*.

first law of thermodynamics There is some total amount of energy in the universe, and that amount never changes.

fixed action pattern Stereotyped, innate motor response linked to relatively simple environmental stimuli.

flagellum, plural **flagella** (fluh-JELL-um) Structure involved in rapid cell movement through the environment. Longer and less numerous than cilia; contains system of microtubules.

fluid mosaic membrane structure Current model of membrane structure, in which diverse proteins and other molecules are suspended in a fluid lipid bilayer. Lipids with short or kinky tails disrupt the straight hydrocarbon-chain packing typical of much of the bilayer, making it somewhat fluid; lipids and proteins together constitute the "mosaic."

follicle In a mammalian ovary, one of the spherical chambers containing an oocyte on the way to becoming a mature egg.

food web Complex patterns of who eats whom among producers, consumers, and decomposers in a community.

fruit In flowering plants, the ripened ovary of one or more carpels, sometimes with accessory structures incorporated.

fungus Multicelled organism of a type that generally secretes enzymes that break down organic material in the outside environment; the resulting small molecules are then absorbed by individual cells.

gamete (GAM-eet) Mature haploid cell that functions as a sexual reproductive cell.

gametogenesis (gam-EET-oh-JEN-ih-sis) Formation of gametes.

gametophyte (gam-EET-oh-fight) Haploid, gamete-producing phase in a plant life cycle.

ganglion Organized knot of neurons forming an integrative center. Ranges from tiny swellings in multiple body segments of earthworms to the mass called the human brain.

gastrula Among some animals, the embryonic stage following blastulation; a two- or three-layered structure enclosing a central cavity that has an opening to the outside.

gene A unit of inheritance; a specific DNA region coding either for an RNA molecule or for the translation product of an RNA molecule (a polypeptide). Actual expression of a gene may be influenced by interactions with other genes and by conditions in the internal and external environments.

gene flow Change in allele frequency as different individuals move into or out of a population (in other words, through emigration and immigration).

gene frequency More precisely, allele frequency: the relative abundance of different alleles carried by the individuals of a population.

gene pool Sum total of all genotypes in a given population. More accurately, allele pool.

genetic code Basic language of protein synthesis, by which nucleotide triplets in the DNA molecule call for specific amino acids used in protein synthesis.

genetic counseling Providing prospective parents with information on the likelihood that their children may have specific genetic diseases; based on family histories and tests.

genetic drift Random fluctuation in allele frequencies over time, due to chance occurrence alone. Thought to be basis of neutral selection, even of perpetuation of maladaptive traits.

genetic equilibrium Theoretical baseline for measuring change in allele frequency in a population; the assumption is that stable allele ratios occur if mating is random, and if all individuals have equal probability of surviving and reproducing.

genetic transduction In some bacteria, the transfer of genes from bacteriophage-destroyed cells to new cells, with the viruses acting as the gene-carrying agents.

genetic transformation In some bacteria, the incorporation of parts of DNA molecules from dead cells into the DNA of living cells.

genotype (JEEN-oh-type) Genetic constitution of an individual. Compare *phenotype*.

genus, plural **genera** (JEEN-us, JEN-er-uh) A broad category into which similar yet distinct species may be grouped, based on implied descent from a fairly recent common ancestor.

geotropism (GEE-oh-TROPE-izm) Plant response to the earth's gravitational force; also called gravitropism.

gill In fish and other aquatic animals, a thin, moist surface membrane endowed with a capillary bed and outfolded into a bump, flap, or filament; each represents a large surface area for gas exchange.

gland Cluster of secretory cells into a ductless organ.

glomerulus (glow-MARE-yoo-luss) Cluster of capillaries in Bowman's capsule of the kidney.

glucagon Animal hormone secreted by pancreatic cells and essential in breakdown of glycogen (a polysaccharide) to glucose subunits.

glycerol (GLISS-er-ohl) Three-carbon molecule with three hydroxyl groups attached; combines with fatty acids to form fat or oil.

glycogen In animals, a starch that is a main food reserve; can be readily broken down into glucose subunits.

glycolysis (gly-CALL-ih-sis) The initial breaking apart of sugar molecules such as glucose, with the release of energy. Glycolysis may proceed under aerobic as well as anaerobic conditions; but it does not require oxygen to do so.

Golgi complex (GOAL-jee) Membrane system in which membranes and substances from endoplasmic reticulum are modified, then released for distribution; products include lysosomes and secretory sacs. Also site of polysaccharide assembly.

gonad (GO-nad) Reproductive organ in which gametes are produced.

graded potential Short-range nerve signals that show variations in size. Those arriving at a given neuron may together be strong enough to generate further message conduction, or too weak to have their message sent on. A major factor in diverse responses shown by animals having complex nervous systems.

green revolution Term for attempts to improve crop production in developing countries by creating high-yield crop varieties, and by encouraging use of modern agricultural practices, fertilizers, and equipment.

ground meristem Embryonic tissue zone, formed from apical meristem, that produces most of the plant body (excluding surface tissue layers and vascular tissues).

guild species Strongly competitive species that form a functional group in a community; they function interchangeably in the scheme of who eats whom, for example, with the species most adaptive at any particular time being the one most dominant.

gymnosperm Plant in which seeds are carried on reproductive structures, without protective tissue layers. Conifers such as pines are examples.

habitat The physical environment, along with its characteristic array of organisms, in which a species lives and reproduces.

haploid (HAP-loyd) State in which a cell contains only one of each type of chromosome characteristic of its species. Compare *diploid*.

Hardy-Weinberg rule If all individuals in a population have an equal chance of surviving and reproducing, then the frequency of each allele in the population should remain constant from generation to generation.

heart Muscular organ that acts as a pump for circulating blood through the animal body.

hemoglobin (HEEM-oh-glow-bin) Iron-containing protein that gives red blood cells their color; functions in oxygen transport.

hemorrhage Bulk flow of blood from damaged vessels.

herbivore Plant-eating animal.

heterogenous nuclear RNA Large pool of RNA strands of different lengths present in the nucleus during all stages of development.

heterospory In plants, production of spores of different types, which give rise to male or female gametophytes.

heterotroph (HET-er-oh-trofe) Organism that depends on substances produced by other organisms for its food source.

heterozygous (HET-er-oh-zie-guss) In sexually reproducing organisms, having nonidentical alleles for a given trait.

homeostasis (HOE-me-oh-STAY-sis) For cells and multicelled organisms, maintaining internal conditions within some tolerable range even when environmental conditions change.

home range Area occupied over long periods by one group or individual animal but not aggressively defended against others of the species. Compare *territory*.

hominid Any primate in the human family. *Homo sapiens* is the only living representative.

homologous chromosome In diploid cells, one chromosome of a pair that have equivalent gene sequences. For any gene locus in that linear sequence, the pair may have identical or nonidentical alleles. (Typically the two chromosomes of a homologous pair are derived from two different parents, but exceptions do occur, as in the case of self-fertilizing plants.)

homologous structure Body part constructed of the same kinds of materials according to the same basic, functional plan, yet occurring in entirely separate species.

homospory In some plants, production of only one type of spore rather than differentiated types. Compare *heterospory*.

hormone Endocrine cell product, generally transported by bloodstream, that triggers a specific cellular reaction in target tissues and organs some distance away. Compare *neurohormone, neurotransmitter*.

hydrogen bond Weak attraction between a hydrogen atom already bound covalently into a molecule, and an atom (oxygen, nitrogen, or fluorine) already bound covalently in the same or another molecule.

hydrolysis (high-DRAWL-ih-sis) Breaking of certain molecular bonds and the attachment of H^+ to one reaction product and $-OH$ to another. (The H^+ and ^-OH are derived from a water molecule.)

hydrophilic Having an attraction for (able to hydrogen-bond with) water molecules; refers to a polar substance that readily dissolves in water.

hydrophobic Repelled by water molecules; refers to a nonpolar substance that does not readily dissolve in water.

hypha, plural **hyphae** Branched, tubelike filament; structural component of the meshlike, vegetative body of true fungi.

hypothalamus In vertebrates, forebrain region that, together with the pituitary gland, serves as the neuroendocrine control center.

immune system, vertebrate Phagocytic white blood cells, lymphocytes, and plasma cells and their products that take part in the body's defense against foreign configurations (for example, bacteria, viruses, or body cells that are mutant, damaged, or cancerous).

independent assortment Mendelian principle (later modified) that segregation of alleles for a given trait into gametes is independent of segregation of alleles for other traits. Independent assortment can occur from random alignment of homologous pairs of chromosomes at the spindle equator during meiosis. But linkage groups on a given chromosome, and crossing over, may also influence allele segregation.

indirect development For some animal species, progression through an immature larval stage before metamorphosis into the adult form.

inflammation Nonspecific immune response involving mobilization of phagocytic cells and complement; a series of homeostatic events that restore damaged tissues and intercellular conditions.

integration, neural Moment-by-moment summation of all excitatory and inhibitory synapses acting on a neuron; occurs at each level of synapsing in a nervous system.

intermediate neuron Type of neuron concentrated in brain and spinal cord; functional link between sensory neurons leading into and motor neurons leading from the central nervous system.

interphase Time interval (variable among species) in which a cell grows and maintains itself but does not divide. DNA is replicated at some point during interphase, prior to nuclear division.

interstitial fluid In vertebrates, all body fluids excluding that portion (about three-fourths) inside individual cells. In other words, intercellular fluid and the plasma portion of blood.

ion, negatively charged Unit of matter formed when an atom (or group of atoms) gains one or more electrons and thereby takes on a negative charge.

ion, positively charged Unit of matter formed when an atom (or group of atoms) loses one or more electrons and thereby takes on a positive charge.

ionic bond Attraction between ions of opposite charge.

isogamy (EYE-suh-gam-ee) Form of sexual reproduction in which all gametes are identical in appearance.

isolating mechanism Some aspect of structure, functioning, or behavior that may prevent interbreeding between populations that are undergoing or have undergone speciation.

isotope Individual atom that contains the same number of protons as other atoms of a given element, but that has a different number of neutrons; hence isotopes of that element have different mass numbers.

kidney Intricately structured organ of salt and water regulation; its nephron/capillary units are concerned with filtration of water and other noncellular components of blood, selective reabsorption of solutes and most of the water, and tubular secretion (through active transport) of certain substances from the capillaries.

kinetic energy Energy of motion.

Krebs cycle Stage of cellular respiration in which pyruvate fragments are completely broken down into carbon dioxide; molecules reduced in the process can be used in forming energy carriers.

larva, plural **larvae** Immature form of an animal that undergoes metamorphosis to the adult form.

larynx Organ of voice production that lies between the pharynx and trachea.

learning Capacity for modification to innate forms of behavior, to the extent that future responses to stimuli can be altered on the basis of cumulative past experience.

leucoplast In some plant cells, colorless plastid in which starch grains and other substances may be stored.

life cycle For any species, the genetically programmed sequence of changes in form and function, from the moment an individual emerges as a distinct entity to the moment of its reproduction.

light reactions First stage of photosynthesis, concerned with harnessing sunlight and using it as an energy source for ATP and/or $NADPH_2$ synthesis.

linkage mapping Determining the positions of genes relative to one another on a given chromosome.

lipid Hydrophobic molecule to which at least some hydrophilic groups are attached; a lipid thus has both water-insoluble and water-soluble parts.

lymphocyte Phagocytic white blood cell that carries out specific immune responses by recognizing and disposing of foreign agents or substances in a highly discriminatory way.

lymphoid system Tissues and organs that function as cell production centers and as sites for some immune responses; bone marrow, thymus, lymph nodes, spleen, appendix, tonsils, adenoids, and patches of small intestine.

lymph vascular system Network of vessels that supplements the blood circulation system; reclaims water that has entered interstitial regions from the bloodstream; also transports fats from small intestine to bloodstream. Fluid in its vessels is called *lymph*.

lysosome Membrane-bound organelle containing hydrolytic enzymes that can dispose of malfunctioning or worn-out cell parts and foreign particles.

mantle In mollusks, a body wall surrounding internal parts; secretes substances that form the molluscan shell.

mass number Total number of protons and neutrons in the nucleus of atoms of a given element.

maternal messages Messenger RNA that becomes regionally positioned in the cytoplasm of maturing eggs and that becomes activated upon fertilization; their messages direct the primary events of embryonic development.

mechanoreceptor Sensory cell or cell part that detects mechanical energy associated with changes in pressure, position, or acceleration.

medusa (meh-DOO-sah) Free-swimming, bell-shaped stage in cnidarian life cycles.

meiosis (my-OH-sis) Two-stage nuclear division process in which a diploid number of chromosomes is halved; at the end of meiosis, the chromosome number in each daughter nucleus is haploid. Basis of gamete formation. Compare *mitosis*.

menopause End of the period of a human female's reproductive potential.

menstrual cycle Recurring physiological changes in the uterine lining, synchronized with the events of the ovarian cycle.

menstruation Periodic sloughing off of the blood-enriched lining of the uterus when pregnancy does not occur.

meristem In land plants, embryonic tissue zones giving rise to cells that go on to divide, enlarge, and differentiate into all other plant tissues.

mesoderm In some animal embryos, a tissue layer between ectoderm and endoderm; it gives rise to bones, muscles, and other internal organs.

metabolic pathways In all living organisms, sequences of coupled reactions in which products from one reaction serve as starting substances for others, and among which energy is transferred.

metabolism All activities by which organisms extract and transform energy from their environment, and use it in manipulating materials in ways that assure maintenance, growth, and reproduction.

metamorphosis (met-uh-more-FOE-sis) In eukaryotic development, genetically programmed change in form of an organ or structure; also, drastic changes of a larval stage of an animal into the adult form.

metaphase In mitosis, stage when microtubules increase in number and become organized into mitotic spindle, which is responsible for separating sister chromatids from each other.

metazoan Multicelled animal.

microbody Organelle in which fats and some amino acids may be converted to other substances.

microfilament Extremely fine cell structure composed of actin and myosin; involved in cell shape, motion, and growth.

microtubule Hollow cylinder of (mostly) tubulin subunits; involved in cell shape, motion, and growth; functional unit of cilia and flagella.

miscarriage Spontaneous expulsion of an embryo from the uterus before the third trimester of pregnancy.

mitochondrion, plural **mitochondria** (MY-toe-KON-dree-on) In eukaryotic cells, membranous organelle in which the main energy-extracting pathways of cellular respiration occur.

mitosis (my-TOE-sis) Nuclear division process in which a diploid number of chromosomes is maintained; at the end of mitosis, the chromosome number in each daughter nucleus is diploid. Basis of reproduction of single-celled eukaryotes; basis of physical growth of multicelled eukaryotes.

mitotic spindle During nuclear division, a microtubular structure that separates sister chromatids of a chromosome from each other.

molecule Two or more atoms linked by chemical bonds.

monocot Short for monocotyledon; a flowering plant in which seeds have only one cotyledon, whose floral parts generally occur in threes (or multiples of threes), and whose leaves typically are parallel-veined. Compare *dicot.*

monosaccharide A simple sugar molecule that typically has a skeleton of five, six, or seven carbon atoms.

morphogenesis (MORE-foe-GEN-ih-sis) Process by which groups of similar cells become spatially coordinated, producing structures of genetically programmed shapes and patterns.

motor neuron Type of neuron that carries action potentials to branched axonal endings that synapse with cells of muscles or glands; leads out from the central nervous system.

multiple allele system In a given population, all the possible alternatives for a given gene locus.

mutation Any permanent change in the kind, structure, sequence, or number of the component parts of a DNA molecule.

mutualism Type of symbiotic relationship from which both species benefit.

mycelium, plural **mycelia** (my-SEE-lee-um) In true fungi, a matlike, often underground structure formed from hyphae.

mycoplasma (MY-coe-PLAZ-mah) One of the smallest organisms known; typically a disease agent lacking a cell wall and living in the moist tissues of animals.

myofibril (MY-oh-FIE-brill) Threadlike structure composed of myosin and actin; structural component of muscle cells.

NAD⁺ Nicotinamide adenine dinucleotide, oxidized form.

NADH Nicotinamide adenine dinucleotide, reduced form.

NADP⁺ Nicotinamide adenine dinucleotide phosphate, oxidized form.

NADPH₂ Nicotinamide adenine dinucleotide phosphate, reduced form.

natural selection Result of selective agents in environment acting on phenotypes in ways that lead to differential survival and reproduction of genotypes. Traits that are most adaptive in a given environment become increasingly represented among individuals in a population, for their bearers (which in some way have a better chance of surviving and reproducing) contribute proportionally more offspring to the next generation.

negative feedback mechanism Homeostatic control whereby an increase in some substance or activity sooner or later inhibits the very processes leading to (or allowing) the increase.

nematocyst Among cnidarians, cellular capsule containing a tiny barb that can be discharged against predator or prey.

nephridium, plural **nephridia** (neh-FRID-ee-um) In invertebrates such as earthworms, organ for excreting fluids and metabolic wastes.

nephron In the kidney, one of numerous functional units involved in filtration of water and other noncellular components of blood; in selective reabsorption; and in tubular secretion.

nerve In peripheral nervous system, a bundle of axons held together by connective tissue. Such nerves and their branchings transmit signals to and from distinct body regions.

nerve impulse Action potential.

nerve net In some invertebrates, neurons dispersed through epithelium yet functionally linked to sensory cells, each other, and muscle tissue. Permits diffuse response to stimuli; unlike nervous system, has little orientation to information flow (the neurons can carry signals in either direction).

nervous system Constellations of neurons oriented relative to one another in precise message-conducting and information-processing pathways; depending on degree of organization, may provide highly integrated response to stimuli.

net primary productivity In a community, the potential energy available from autotrophically produced food, less the amount the autotrophs themselves require for growth, maintenance, and reproduction.

neuroglia (NUR-oh-GLEE-uh) Cells intimately associated with neurons; in vertebrates they represent at least half of the volume of the nervous system, yet the function of most is not well understood.

neurohormone Secretory product of certain nerve cells whose messages travel, by way of the bloodstream, to nonadjacent cells. Compare *hormone.*

neuron In most animals, a cell whose orientation relative to others like itself and whose excitable plasma membrane allow it to receive and send messages about changing environmental conditions.

neurotransmitter Secretory product of nerve cells whose messages travel only a short distance, across a synaptic cleft, to an adjacent nerve or muscle cell.

neutral selection Result of genetic drift; even though a given allele in a population may not be the "best" possible alternative in contributing to survival and reproduction, it may become established when "better" alternatives are lost by accident or chance.

neutron Subatomic particle of about the same size and mass as a proton but having no electric charge.

niche (nitch) All environmental or ecological requirements and interactions that influence a species in a community.

nicotinamide adenine dinucleotide Nucleotide that functions as an electron acceptor and transporter in oxidation–reduction reactions.

nicotinamide adenine dinucleotide phosphate Nucleotide that functions as an electron acceptor and transporter in several metabolic pathways, most notably photosynthesis. The energy-rich molecule NADPH$_2$ apparently is produced when NADP$^+$ combines with the two electrons and two hydrogen ions (H$^+$) released from a water molecule at the start of noncyclic photophosphorylation.

nitrification Process by which certain soil bacteria strip ammonia or ammonium of electrons, and nitrite (NO$_2^-$) is released as a reaction product.

nitrogen fixation Among some bacteria and blue-green algae, assimilation of gaseous nitrogen (N$_2$) from the air; through reduction reactions, electrons (and associated H$^+$) become attached to the nitrogen, thereby forming ammonia (NH$_3$) or ammonium (NH$_4^+$).

noncyclic photophosphorylation Photosynthetic pathway in which new electrons derived from water molecules flow through two photosystems and two transport chains, the result being formation of ATP and NADPH$_2$.

nonidentical twins Individuals resulting from fertilization of two different eggs by two different sperm.

notochord Somewhat flexible rod of cartilage; probable forerunner to the chordate endoskeleton. During embryonic development of complex vertebrates, it is replaced by a vertebral column.

nuclear envelope Double membrane at the surface of a eukaryotic nucleus; in some cases, outer membrane seems to be continuous with endoplasmic reticulum.

nucleic acid (new-CLAY-ik) Long-chain, single- or double-stranded nucleotide; DNA and RNA are examples.

nucleolus Within the nucleus, a mass of proteins, RNA, and other material used in ribosome synthesis.

nucleoplasm Semifluid substance enclosed by the nuclear envelope.

nucleoprotein DNA and certain proteins complexed as a single, intact fiber; each chromosome is thought to be a single nucleoprotein.

nucleotide (NEW-klee-oh-tide) Molecule containing at least a five-carbon sugar (ribose), a nitrogen-containing base (either purine or pyrimidine), and a phosphate group.

nucleus In atoms, the central core of one or more positively charged protons and (in all but hydrogen) electrically neutral neutrons. In eukaryotic cells, the membranous organelle that houses the DNA.

obligate relationship Symbiotic interaction in which one species depends on another for survival.

oogamy (oo-AH-gam-ee) Form of reproduction in which gametes differ in size and motility. One gamete typically is small and motile (a sperm, for example); the other is larger and nonmotile (the egg).

oogenesis (oo-oh-JEN-uh-sis) Formation of female gamete, from a diploid reproductive cell (oogonium) to a mature haploid ovum.

operon A set of control elements and structural genes operating as a unit.

organ Body unit composed of different tissues having a common function.

organelle Specialized functional unit, usually with its own surface membrane, embedded in cytoplasm.

organic compound Carbon-containing substance.

osmosis (oss-MOE-sis) Movement of water molecules across a differentially permeable membrane in response to a concentration and/or pressure gradient.

ovarian cycle Recurring events synchronized with menstrual cycle; the formation of a mature ovum in a fluid-filled cavity inside the ovary, the transformation of the cavity (which ruptures when the ovum is expelled) into an endocrine gland, and the concurrent production of ovarian hormones.

ovary The primary female reproductive organ in which oogenesis occurs.

oviduct Passageway through which ova travel from the ovary to the uterus.

ovule In seed-bearing plants, the structure destined to become the seed; includes the female gametophyte (with egg cell) and tissue-layer covering.

oxidation The loss of one or more electrons from an atom or molecule.

oxidation-reduction reaction Process by which an atom or molecule (an electron acceptor) picks up an extra electron from another atom or molecule; the electron acceptor thus becomes reduced and the donor becomes oxidized.

oxidative phosphorylation Use of electron energy being released during oxidation reactions to phosphorylate (tack a phosphate group onto) a molecule such as ADP (which yields energy-rich ATP).

parallel evolution In two or more entirely separate lines of descent, the development of similar structures having similar functions as a result of the same kinds of selection pressures.

parasitism Extreme form of symbiosis in which one species benefits at the expense of another.

parasympathetic nerves Nerves that operate antagonistically with sympathetic nerves; generally dominate internal events when environmental conditions permit normal body functioning.

passive transport Movement of a substance across a cell membrane without any direct energy outlay by the cell. Diffusion and bulk flow are passive transport mechanisms.

pathogen (PATH-oh-jen) Disease-causing organism.

penis Male accessory reproductive organ from which sperm are expelled into the female reproductive tract during sexual intercourse.

perennial A plant that lives year after year.

pericycle In roots of gymnosperms and flowering plants, ground tissue (between phloem and endodermis) that shows meristematic activity during secondary growth. Its cells become aligned with the vascular cambium and help form a continuous ring of active cambium.

periderm In stems and roots of gymnosperms and flowering plants, a protective covering that replaces epidermis during secondary growth.

peripheral nervous system (per-IF-ur-uhl) Cell bodies of sensory neurons and all nerves (bundles of axons leading to and from distinct body regions).

peristalsis (pare-ih-STALL-sis) An alternating progression of contracting and relaxing muscle movements along the length of a tubelike organ (such as the esophagus).

pH Whole number referring to the number of hydrogen ions present in a liter of a given fluid.

phagocytosis "Cell-eating"; a form of endocytosis whereby vesicles form around particles or organisms at or near a cell surface, and then discharge their contents in the cell interior. Amoebas and certain white blood cells are phagocytic.

pharynx Muscular tube that is the gateway to the digestive tract and to the windpipe (trachea).

phenotype (FEE-no-type) The physical form and functional behavior that are expressions of an individual's genotype; arises from interactions between genes, and between genes and the environment.

pheromone (FARE-oh-moan) A chemical, secreted by exocrine glands, whose target receptor cells reside in other organisms of the same species. Pheromones may trigger behavioral changes; some are sex attractants, trail markers, or alarm signals.

phloem (FLOW-um) In vascular plants, a tissue that transports food through the plant body; some of its components also function in storage and structural support.

phospholipid Lipid molecule composed of glycerol, fatty acids, a phosphorus-containing compound, and (usually) a nitrogen-containing base. Phospholipids are the foundation for cell membranes.

phosphorylation (FOSS-for-ih-LAY-shun) Addition of one or more phosphate groups to a molecule.

photolysis First step in noncyclic photophosphorylation, when water is split into oxygen, hydrogen ions, and their associated electrons; photon energy indirectly drives the reaction.

photoreceptor Light-sensitive sensory cell.

photosynthesis In most autotrophs, the trapping of solar energy and its conversion to chemical energy, which is used in manufacturing food molecules from carbon dioxide and water.

photosystem Functional light-trapping unit in photosynthetic membranes; contains pigment molecules and enzymes.

phototropism Movement or growth curvature toward light.

phytochrome Light-sensitive pigment molecule whose activation and inactivation trigger hormone activities governing leaf expansion, stem branching, stem length, and, in many plants, seed germination and flowering.

phytoplankton Community of photosynthetic microorganisms in freshwater or saltwater environments.

pinocytosis (PIN-oh-sigh-TOE-sis) "Cell-drinking"; form of endocytosis in which the plasma membrane of a cell dimples inward around particles adhering to it and forms a vesicle around them, which then moves into the cytoplasm.

placenta (play-SEN-tuh) In the uterus, an organ made of extensions of the chorion and the endometrium. Through this composite of embryonic and maternal tissues and vessels, nutrients reach the embryo and wastes are carried away.

plant Generally, multicelled organism able to build its own food molecules through photosynthesis.

plasma Liquid component of blood.

plasma membrane Outermost membrane of a cell; the differentially permeable boundary layer between the cytoplasm and the extracellular environment.

plasmid In some bacteria, a small circle of DNA in addition to the main DNA molecule.

plastid In some plant cells, a storage organelle; some plastids also function in photosynthesis.

platelet Component of blood that functions in clotting.

pleiotropism Multiple phenotypic effect of a single gene; the action of the gene affects many developmental or maintenance activities.

point mutation Addition, deletion, or substitution of individual nucleotide subunits in the DNA molecule.

pollen grain In gymnosperms and flowering plants, the immature male gametophyte (gamete-producing body).

pollination The transfer of pollen grains to the female gametophyte of a plant.

pollutant A naturally occurring or synthetic substance that accumulates to levels that are harmful to living things.

polymorphism In a population, the persistence of two or more forms of a trait, at a frequency that is greater than can be accounted for by newly arising mutations alone.

polyp (POH-lip) Vase-shaped, sedentary stage of cnidarian life cycles.

polypeptide Chain of amino acids linked by dehydration synthesis.

polyploidy Presence of more than two sets of chromosomes in a cell.

polyribosome During protein synthesis, a clustering of ribosomes engaged in translation on a messenger RNA molecule.

polysaccharide Three or more simple sugar molecules bonded together covalently.

population Group of individuals of the same species occupying a given area at a given time.

population growth The difference between the birth rate and death rate, plus or minus any inward or outward migration.

potential energy Energy in a potentially usable form that is not, for the moment, being used.

predator A free-living organism that captures and ingests other living organisms as a means of obtaining nutrients.

primary growth Following seed germination, the cell divisions, elongation, and differentiation that produce the primary plant body.

primary structure For proteins, the sequence of amino acids that forms the protein backbone.

procambium In vascular plants, embryonic tissue (meristem) that gives rise to vascular tissue.

producer An autotrophic organism; able to build its own food molecules from simple inorganic molecules present in the environment.

prokaryote (pro-CARRY-oht) Single celled organism that has no membranous nucleus; a bacterium or blue-green alga.

prophase In mitosis, the stage when chromatin coils up into compact chromosome bodies.

protein Molecule made of one or more chains of amino acids.

protistan Single-celled eukaryote that not only resembles the simpler monerans but that may show similarities to plants, animals, and/or fungi.

protoderm In vascular plants, embryonic tissue (meristem) that gives rise to epidermis.

pseudopod (soo-doe-pod) "False foot"; a nonpermanent cytoplasmic extension of the cell body.

pulmonary circulation Pathways of blood flow leading to and from the lungs.

pulse pressure Measured difference between systolic and diastolic readings; generally measured at a main artery. Refers to the difference between maximum pressure exerted on arteries when blood is forced from heart and the lowest pressure, which occurs just before blood is pumped out. The difference is usually expressed in terms of how much it would raise a column of mercury (Hg).

purine Nucleotide base having a double ring structure. Examples are adenine and guanine.

pyrimidine (pih-RIM-ih-deen) Nucleotide base having a single ring structure. Cytosine and thymine are examples.

pyruvate (PIE-roo-vate) A three-carbon compound produced during glycolysis (the initial breakdown of sugar molecules).

radial symmetry General arrangement of body parts in a regular pattern about a central axis, much like spokes of a bike wheel.

radicle In plant seeds, the embryonic root; typically, longitudinal growth of its cells gives rise to a slender primary root.

ray system Primarily parenchyma cells that act as conduits and food storage centers in wood.

receptor Sensory cell or cell part that may be activated by a specific kind of stimulus in the internal or external environment.

recessive allele In the heterozygous state, an allele whose expression is fully or partially masked by expression of its (dominant) partner; recessive alleles can be fully expressed in homozygous states.

recombination, genetic Production of individuals having new combinations of genes that they acquired, one way or another, from more than one parent cell. Occurs through sexual reproduction, conjugation, genetic transformation, and genetic transduction.

reduction The gaining of one or more electrons by an atom or molecule.

reflex A sequence of coordinated events, elicited by a stimulus, along some neural pathway.

releaser A relatively simple environmental stimulus that triggers a fixed action pattern.

reproduction, asexual Production of new individuals by any process that does not involve gametes. Binary fission, budding, and mitotic cell division are examples.

reproduction, sexual Production of new individuals that requires the formation of haploid sex cells (gametes) in diploid parent organisms.

respiration In cells, aerobic pathway by which energy (released from fragments of glycolysis) is used in forming energy carrier molecules (such as ATP and $NADPH_2$). In most animals, the overall exchange of gases between cells, the blood, and the environment.

resting membrane potential Charge difference that exists across a neuron at rest (one that is not being stimulated).

ribonucleic acid (RYE-bow-new-CLAY-ik) (RNA); a category of nucleotides used in translating the genetic message of DNA into actual protein structure.

ribosome In both prokaryotic and eukaryotic cells, an organelle made of RNA and proteins; the site of protein synthesis.

ritualized behavior Formal behavioral display that is an exaggeration of ordinary functional movements, but that is a clear signal laden with social meaning for members of the same species.

salt Ionic substance containing neither H$^+$ nor OH$^-$.

saprophyte (SAP-row-fight) Heterotroph that feeds on nonliving organic matter.

sarcomere Fundamental unit of contraction in skeletal muscle; repeating bands of actin and myosin that appear between two Z lines.

secondary growth In vascular plants, an increase in stem and root diameter, made possible by cambial activity that gives rise to secondary xylem and phloem.

secondary structure In protein chains, a helical or extended sheetlike pattern created by repeated hydrogen bonding.

seed In gymnosperms and flowering plants, a fully mature ovule (contains the plant embryo), with its tissue layers (integuments) forming the seed coat.

segregation, allelic Mendelian principle that two units of heredity (alleles) exist for a trait, and that during gamete formation, the two units of each pair are separated from each other and end up in different gametes.

semen Sperm-bearing fluid expelled from the penis during male orgasm.

semiconservative replication Manner in which a DNA molecule is reproduced; formation of a complementary strand on each of the unzipped strands of a DNA double helix, the outcome being two "half-old, half-new" molecules.

senescence Sum total of processes leading to death of a plant or any of its organs; cause appears to be built into the life cycle of the species.

sensory neuron Type of neuron that carries signals about changing environmental conditions to the central nervous system.

sex chromosomes In most animal and some plant species, chromosomes that differ in number or in kind between males and females. All other chromosomes are called autosomes.

sieve element Food-conducting cell in phloem.

sodium–potassium pump Enzymes that serve as an active transport mechanism for maintaining the resting membrane potential of a neuron; helps maintain the sodium and potassium concentration gradients across the neural membrane.

solute (SOL-yoot) A substance dissolved in some solution; for example, a sugar that dissolves in (forms hydrogen bonds with) water.

solvent Fluid in which one or more substances is dissolved.

somatic nervous system All sensory pathways and all motor-to-skeletal muscle pathways; usually under conscious control.

speciation Outcome when some restriction on interbreeding between populations is followed by enough genetic differentiation that, even if individuals from those populations do make contact, they cannot or will not interbreed.

species All of those actually or potentially interbreeding populations that are reproductively isolated from other such groups.

sperm Mature male gamete.

spermatogenesis (sperm-AT-oh-JEN-uh-sis) Formation of a mature sperm from a diploid reproductive cell (spermatogonium).

sphincter Ring of muscle that serves as a gate in some tubelike system (such as the one between stomach and small intestine).

spindle Structure assembled from microtubules; physical basis for movement of chromosomes during mitosis and meiosis.

spirillum, plural **spirilla** (spih-RILL-um) Spiral form of bacterium.

sporangium, plural **sporangia** (spore-AN-gee-um) In plants, hollow single-celled or multicelled structure in which spores are produced.

spore In plant life cycles, a cell that develops into a haploid body, which grows and differentiates into the gametophyte (gamete-producing body).

sporophyte Diploid, spore-producing stage of plant life cycles.

stabilizing selection Result of selection pressures that favor an existing species form and work against extreme variant forms that appear; occurs when environment remains much the same.

stamen In flowering plants, the male reproductive organ; commonly consists of pollen-bearing structures (anthers) positioned on single stalks (filaments).

steroid Lipid consisting of multiple carbon ring structure to which different atoms may be attached.

stimulus Any detected change in an organism's external environment or within its body. Every stimulus is some form of energy change (for example, a change in heat, sound wave, chemical, or light energy).

stoma, plural **stomata** Paired guard cell that serves as a gate across leaf epidermis; controls movement of carbon dioxide into the leaf.

structural gene Gene that codes for the structure of a protein.

substrate Molecule or molecules of a reactant on which an enzyme acts.

succession, primary Gradual, sequential replacement of communities, beginning with pioneer species, until a stable climax stage is reached (then, the array of species is locked in materials and energy use in ways that allow it to remain relatively stable and self-perpetuating).

succession, secondary Reestablishment of a climax community that has been disrupted in whole or in part.

surface-to-volume ratio In cells, a physical constraint on increased size: as the cell's linear dimensions grow, its surface area does not increase at the same rate as its volume (hence each unit of plasma membrane would be called upon to serve increasing amounts of cytoplasm).

symbiosis A form of attachment between individuals of two or more different species, in which at least one party benefits.

sympathetic nerves Nerves that usually operate antagonistically with parasympathetic nerves; in times of stress, danger, excitement, and heightened awareness, they dominate internal events and mobilize the whole body for rapid response to change.

sympatric speciation Outcome when two or more populations occupying the same distribution range have undergone reproductive isolation before genetic differentiation has transformed them into separate species.

synapse, excitatory Functional connection between a neuron and another neuron or muscle cell, whereby neurotransmitters increase the receiving cell's permeability to potassium and/or chloride, which drives its membrane potential away from threshold. Inhibitory synapses decrease the likelihood that a receiving cell will fire an action potential.

synapse, inhibitory Functional connection between a neuron and another neuron or muscle cell, whereby neurotransmitters increase the receiving cell's membrane permeability to sodium and potassium, which drives its membrane potential toward threshold. Excitatory synapses increase the likelihood that a receiving cell will fire an action potential.

synapsis In meiosis, the point-by-point alignment of the two sister chromatids of one chromosome with the two sister chromatids of its homologue.

systemic circulation Pathways of blood flow leading to and from all body parts except the lungs.

telophase Final stage of mitosis, during which the separated chromosomes uncoil and become enclosed within newly forming daughter nuclei.

territory For some animal groups, a more-or-less geographically definable area to which behavioral expressions of dominance may become attached.

tertiary structure For proteins, the asymmetrical bending and folding of the protein backbone in space as it assumes its final shape.

testis, plural **testes** Male gonad; reproductive organ in which male gametes and sex hormones are produced.

thigmomorphogenesis Plant response to physical contact, the result being an overall inhibition of growth.

thigmotropism In plants, unequal growth in some structure (such as a stem) brought about by inhibition of cell elongation in the part that makes physical contact with an environmental object. Unequal growth causing a vine to curl around a tree trunk is an example.

threshold value In neurons, the minimum change in membrane potential necessary to produce an action potential.

thymine Nitrogen-containing base found in some nucleotides.

tissue Group or layers of cells united in structure and in functioning.

trachea, plural **tracheae** A tube for breathing; in land vertebrates, the windpipe that carries air between the larynx and bronchi (which leads to lungs).

tracheid Typically elongated cell, dead at maturity, that passively conducts water and solutes in xylem.

transcription Synthesis of a messenger RNA molecule having a nucleotide sequence that is complementary to the DNA region on which it is assembled.

translation Assembly of amino acid subunits into a specific protein, directed by a specific sequence of information contained in a messenger RNA molecule.

translocation In vascular plants, the transport of soluble food molecules (mostly sucrose) from one plant organ to another; occurs in sieve tubes of phloem tissue.

transpiration Evaporative water loss from stems and leaves.

trophic levels For a community, an interlocked array of producer, consumer, and decomposer species through which materials and energy flow.

turgor pressure In general, osmotically induced internal pressure. In plant cells, the pressure applied to cell walls as water is absorbed.

urinary system In mammals, a tubular network of organs concerned with regulating water and solute levels in the body; consists of paired kidneys, paired ureters, a bladder, and a urethra.

uterus A chamber in which the developing embryo is contained and nurtured during pregnancy.

vacuole, central In plant cells, a membrane-bound, fluid-filled sac that may take up most of the cell interior; main function is to increase cell size and surface area, thereby enhancing absorption of nutrients from a relatively dilute external environment. In animals, vacuoles serve as storage organelles.

vagina Female accessory reproductive organ that receives sperm from the male penis; forms part of the birth canal, and acts as a channel to the exterior for menstrual flow.

variables Different but related conditions that might affect the outcome of a scientific test.

vascular cambium In vascular plants, a lateral meristem that increases stem or root diameter.

vascular tissue In multicelled organisms of many species, internal conducting tissue for fluids and nutrients.

vein Vessel that carries blood back to the heart.

vernalization Cold-temperature stimulation of the flowering process.

vertebra, plural **vertebrae** One of a series of hard bones that form the backbone in most chordates.

vertebrate Animal having a backbone made of bony segments called vertebrae.

vesicle In cytoplasm, a small, membrane-bound sac in which various substances may be transported or stored.

vessel element Typically elongated cell, dead at maturity, that passively conducts water and solutes in xylem.

vestigial structure Body part that no longer has any apparent role in the functioning of an organism.

virus An infectious particle consisting of nucleic acid encased in protein; incapable of metabolism or reproduction without a host cell, hence is often not considered alive.

vision Precise light focusing onto a layer of photoreceptive cells that is dense enough to sample details concerning a given light stimulus, followed by image formation in the brain.

white matter Myelin sheaths that surround axons of neurons in the spinal cord.

xylem (ZEYE-lem) In vascular plants, a tissue that transports water and solutes through the plant body.

zygote (ZEYE-goat) In most multicelled eukaryotes, the first diploid cell formed after fertilization (fusion of nuclei from a male and female gamete).

INDEX